P9-APJ-061

THE EARTH IN PROFILE

CANFIELD PRESS **SAN FRANCISCO**
A DEPARTMENT OF HARPER & ROW, PUBLISHERS, INC.
NEW YORK HAGERSTOWN LONDON

THE EARTH IN PROFILE

A PHYSICAL GEOGRAPHY

DAVID GREENLAND UNIVERSITY OF COLORADO

HARM J. DE BLIJ UNIVERSITY OF MIAMI

The editors of this book were Howard Boyer, Malvina Wasserman, and Patricia Brewer; the designer was John Sullivan; the graphic illustrations were drawn by Ayxa Art; the maps were drawn by Larry Jansen; the copyeditor was Rebecca Smith; and the photo researcher was Kay Y. James. Nick Keefe supervised production.

Library of Congress Cataloging in Publication Data

Greenland, David, 1940–
The Earth in profile.

Bibliography: p.
Includes index.
1. Physical geography—Text-books—
I. De Blij, Harm J., joint author. II. Title.
GB55.G63 551 76-54965
ISBN 0-06-383615-7

THE EARTH IN PROFILE: A PHYSICAL GEOGRAPHY

77 78 79 10 9 8 7 6 5 4 3 2 1

Credits begin on page 451.

OUR PLANET'S SPHERES

CONTENTS

THE ATMOSPHERE

THE
BIOSPHERE

THE LITHOSPHERE

THE EARTH IN PROFILE is a physical geography text with some important differences. We hope that the book will strike you as being more than a compilation of concepts and discourses. It is designed to make some of the truly exciting and significant developments in physical geography accessible to the beginning student. Every aspect of the book's planning and execution is aimed at this admittedly ambitious goal. Physical geography, as a growing, changing field, deserves our attention, not only because of what it tells us about our physical environment, but because of what it tells us about our place within the environment—the kinds of interaction we have with it. We would like to give you some ideas and suggestions about reviewing this book that might assist you in determining whether we have met our goal.

HOW IS *THE EARTH IN PROFILE* ORGANIZED?

The organization reflects our view of the earth environment and its component parts. The first of the four parts provides an introduction to the subject of physical geography and some of its more important techniques and concepts. The book's emphasis on a systems framework begins right here, tying diverse realities together and dramatizing the dynamic characteristics of our environment.

The next three parts focus in turn on the atmosphere, the biosphere, and the lithosphere. The hydrosphere, rather than occupying a separate part, is integrated through each chapter. We feel that the role of water—in all its forms—is so thoroughly intertwined with the other major components of the earth environment that an integrated approach is the most realistic one.

An important common denominator interconnects the parts: You will notice that the parts are organized to emphasize the processes of that sphere, followed by a chapter highlighting variation within that sphere—in terms of space or time. It is particularly important that the student appreciate the role of change in the physical environment.

CLIMATE AND THE HUMAN COMPONENT

We have given substantial attention to climatology. Climates are described not only with respect to traditional patterns of temperature and precipitation but also in the context of water and heat energy balances. We also introduce the human component, a theme that carries across

into two unique chapters. Chapter 8 on the urban climate surveys the comforts and discomforts of life in the modified environments of cities. Chapter 9 on our changing climate reviews the evidence of past climatic changes and indulges in some speculation about the likelihood of another ice age.

IMPORTANT DEVELOPMENTS

What are the major schools of thought, breakthrough theories of physical geography today? By giving a carefully structured background, we are able to present the student with some of the revolutionary changes in physical geography. For example, meteorologists have for some time regarded surface cyclones and fronts as effects of upper air dynamics, and this orientation can be introduced to beginning students. Plate tectonics is the theoretical basis of a veritable revolution in geology and geomorphology. Our entire view of landscapes and many other aspects of crustal formation must be reevaluated in this context. After introductory sections on topography and lithology, we give major emphasis to continental drift and crustal deformation. The chapters on the role of water in processes of weathering and erosion culminate in a discussion of the problems of erosion surfaces and their identification. Our experience proves that it is productive to introduce students to the various (if simplified) models of planation, because they stimulate interest in observation and personal interpretation. This theme is further developed in the summary chapter on physiographic regions, which concentrates on North American landscapes.

The discussion of soils and vegetation at this introductory level always presents a difficult challenge, if only because of terminology and classification. We emphasize the ecological aspects of flora and fauna and focus on the flow of energy through ecological systems. Soils are classified according to the USDA Comprehensive Classification System, an outgrowth of the Seventh Approximation that has the merits of clarity and simplicity. This new system—itself representing advances in soil science—is gaining acceptance and is quite manageable at the beginning level.

COMMUNICATING PHYSICAL GEOGRAPHY

How do we communicate physical geography to the beginning student? Surely, the majority of students enrolled in introductory physical geography programs are nonmajors, with no long-term interest in the subject. For these students—perhaps the most important students—*how* you present the material is as critical as *what* you present. We attempt to pursue a "discovery" approach, giving the student an appreciation of the magnitude of some of the questions confronted by physical geographers, and their approaches in the search for answers.

Physical geography is not simply a collection of data; it is a perspective and approach. We look "inside" the geographer's methodology in the four elements called "From the Geographer's Notebook." These conclude each part with a hands-on introduction to the ways in which geographers organize their problem-solving and interpret data. For the majority of courses that do not have a laboratory program to accompany them, "From the Geographer's Notebook" conveys some of the excitement and challenge of studying our natural environment in a scientific manner.

Harmonizing with this approach are the brief boxed items throughout the book. By dramatizing the human component in the earth's systems and showing the impact of natural events on all of us, these items give the students both a "rest" from the more demanding aspects of the text and a valid context into which they can place what they have learned.

PRACTICE AND REVIEW

We have planned the end-of-chapter materials—summaries and review questions—to structure the student's understanding of the material. By carefully using the summaries, the student is reminded of the salient points of the chapter, both their sequence and comparative emphasis. The questions are intended to be thoughtful ones; the student neither answers them automatically nor will he or she find them empty exercises. A glossary at the end of the text has been included for additional review and clarification of key terms.

Why have we written THE EARTH IN PROFILE with these goals and criteria in mind? We would like to make a contribution to the teaching of introductory physical geography by enlivening it and transforming it into an adventure in awareness.

David Greenland
Harm J. de Blij

January 1977

David Greenland majored in geography and geology at the University of Birmingham before completing a master's degree in meteorology and climatology at the same institution. His principal professors were D. L. Linton, F. Shotton, E. T. Stringer, and F. K. Hare. He completed his doctorate at the University of Canterbury with a dissertation entitled *Heat Balance Studies in the Chilton Valley, Cass in the New Zealand Southern Alps.* Dr. Greenland has taught physical geography and climatology courses at the University of Georgia, Canterbury, California (Berkeley), and Colorado. His publications include "*Geography of the British Isles—Climate*" (1968, 1973), "Observations on the Growth of Needle Ice" (1970), "The Application of Climatonomy to an Alpine Valley" (1973) and "A Study of Radiation in the New Zealand Southern Alps" (1975). He has also worked for the Atmospheric Environment Service of Canada where he prepared a monograph on "Aspects of the Mesoclimatology of Toronto and Surrounding Region." At present he is affiliated with the Institute of Arctic and Alpine Research and Department of Geography at the University of Colorado.

Harm J. de Blij majored in geology and geography at the University of the Witwatersrand, where he studied climatology with S. P. Jackson and physiography with J. H. Wellington. At Northwestern University his doctoral dissertation, entitled *Physiographic Provinces and Cyclic Erosion Surfaces in Swaziland*, was written under the direction of William E. Powers; his field work was guided by L. C. King of the University of Natal. Dr. de Blij has taught physical geography and geomorphology courses at the University of Natal, Northwestern University, and Michigan State University. Among his publications in this area are "The Concept of the Physiographic Province Applied to Swaziland" (1960); "The Southern Extent of the African Rift Valley System" (1966); "Cyclic Erosion Surfaces in Swaziland" (1970); and "Continental Drift and Present Landscapes" (1974). At present he is Professor of Geography, University of Miami.

The authors and publisher wish to acknowledge with thanks the review work of Gary I. Anderson, Santa Rosa Junior College; Val Eichenlaub, Western Michigan University; John E. Costa, University of Denver; Richard Dastyck, Fullerton College; John Donahue, University of Montana; Jeff Dozier, University of California, Santa Barbara; Peter K. Gunderson, Cañada College; Richard N. Hammer (Ashland, Oregon); Lawrence S. Kalkstein, University of Delaware; David S. McArthur, California State University at San Diego; Geoffrey McBoyle, University of Waterloo; John E. Oliver, Indiana State University; Laurence J. Onesti, SUNY, Buffalo; Merrill Ridd, University of Utah; Steven Rowland, University of California, Santa Cruz; Jerry D. Ryan, Muskegon Community College; Joseph B. Schiel, Jr., University of Oklahoma; Curtis J. Sorenson, University of Kansas; Rodney Steiner, California State University at Long Beach; W. J. Switzer (Bonita, California); Thomas P. Templeton, Mesa Community College; Keith H. Topps, California State University at San Francisco; Terrence Toy, University of Denver; Stanley W. Trimble, University of California, Los Angeles; Howard Vogel, Shoreline Community College.

The authors are grateful for their constructive advice but remain solely responsible for any errors. David Greenland would like to gratefully acknowledge the support and advice of Risa Palm while the book was in preparation and the cooperation of the staff of Canfield Press, together with the many other people who assisted in various ways.

ACKNOWLEDGMENTS

PART ONE

OUR PLANET'S SPHERES

Unless you are an astronaut, you live every second of your life on or near the earth's surface. From birth to death you are affected by the environments found on our planet. It is therefore not surprising that people have studied the earth's surface and its environments for thousands of years. Such inquiry has led to the development of many specialized subjects such as meteorology, biology, and geology. But only one discipline has kept its focus on the surface of the earth and in particular on the interrelationships and spatial distributions of phenomena found on and near the surface. That field of inquiry is physical geography—the subject of this book.

Part One explains why we study physical geography and describes its nature. We examine some of the many techniques used in investigating the earth's surface. We introduce a method of analysis called general systems theory. We also follow the stories of how humans have mapped their planet, how the spatial arrangement between the earth and the sun gives us our seasons, and how we may regard time.

All this will provide a good starting point. Later we divide our study into three parts. First, we look at the air above the surface, called the atmosphere. Second, we look at the soil and life layers. We call this part the biosphere, although the soil more strictly falls into a third part—the lithosphere, the name given to the rocks of the earth's crust. A fourth part is intermingled throughout the text, that is, the waters of the planet. These are found in the atmosphere, biosphere, lithosphere, and in the oceans. Together they are called the hydrosphere. These four spheres, then, make up an introductory profile of a planet.

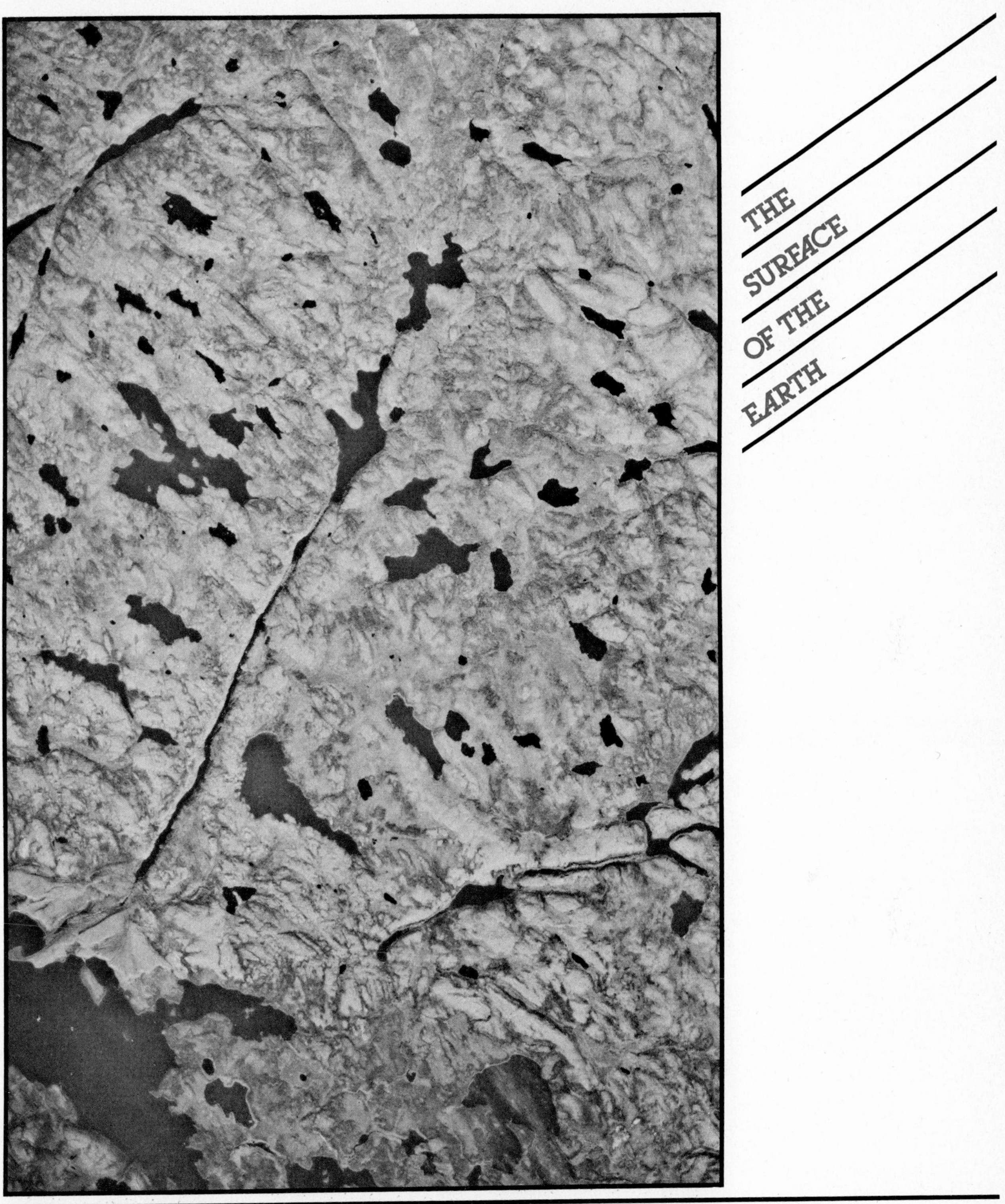

THE SURFACE OF THE EARTH

CHAPTER 1

There is something uneasy in the Los Angeles air this afternoon, some unnatural stillness, some tension. What it means is that tonight a Santa Ana will begin to blow, a hot wind from the northeast whining down through the Cajon and San Gorgonio Passes, blowing up sandstorms out along Route 66, drying the hills and the nerves to the flash point. For a few days now we will see smoke back in the canyons, and hear sirens in the night. I have neither heard nor read that the Santa Ana is due, but I know it, and almost everyone I have seen today knows it too. We know it because we feel it. The baby frets. The maid sulks. I rekindle a waning argument with the telephone company, then cut my losses and lie down, given over to whatever is in the air. To live with the Santa Ana is to accept, consciously or unconsciously, a deeply mechanistic view of human behavior.*

It is perhaps unusual to start a book on physical geography by describing a feature of life in one of the world's most famous urban areas. Urban life seems so divorced from the physical environment. But the environment is a backdrop for every human activity, and often its role is far from passive. Wherever we live, we live on part of the earth's surface, and there is a two-way relationship between us and our surroundings. We are immediately confronted, therefore, with a theme that for many centuries has been central to the study of physical geography—the relationship between humans and their environment.

*Joan Didion, *Slouching Towards Bethlehem* (New York: Farrar, Straus and Giroux, 1968), p. 217.

HUMANS AND THEIR ENVIRONMENT

To illustrate the major themes of physical geography, we could start almost any place. But perhaps the best place to start is where you live. Wherever that is, you are affected by the rocks and soil beneath you, by the vegetation, landforms, and other structures around you, and by the atmosphere around and above you.

From time to time, scientists have suggested that the environment totally determines how humans live. This view, "a deeply mechanistic view of human behavior," is called environmental determinism. The core of this idea has some supporters even in the twentieth century. In contrast, many others have suggested that human nature determines the particular ways we live in any area and what we do with our environment.

THE ENVIRONMENT: BACKDROP OR SCULPTOR?

Is our physical environment a static backdrop against which human history is played out? Or is it a creative shaping force which leaves its stamp on the people who operate within it?

The pioneering American geographer Ellen Churchill Semple put forward some extremely influential theories about the role of the environment in affecting human behavior. Writing at the turn of the century, she declared:

> Man is a product of the earth's surface. This means not merely that he is child of the earth, dust of her dust; but that the earth has mothered him, fed him, set him tasks, directed his thoughts, confronted him with difficulties that have strengthened his body and sharpened his wits, given him his problems of navigation or irrigation, and at the same time whispered hints for their solution. . . . On the mountains she has given him leg muscles of iron to climb the slope; along the coast she has left these weak and flabby, but given him instead vigorous development of chest and arm to handle his paddle or oar. . . . Up on the wind-swept plateaus, boundless stretches of grasslands and the waterless tracts of the desert, where he roams with his flocks from pasture to pasture and oasis to oasis, . . . his ideas take on a certain gigantic simplicity.

Much of Semple's theory did not survive critical analysis in succeeding decades. After all, there are hundreds of kinds of environmental factors and different life-style possibilities within the environment. Surely, the impact of the physical world cannot be interpreted this simplistically.

Yet anthropologists have found that people who live in specific kinds of terrains and climates have distinct kinds of artwork, religions, and other cultural features which can be directly explained in terms of their physical environments.

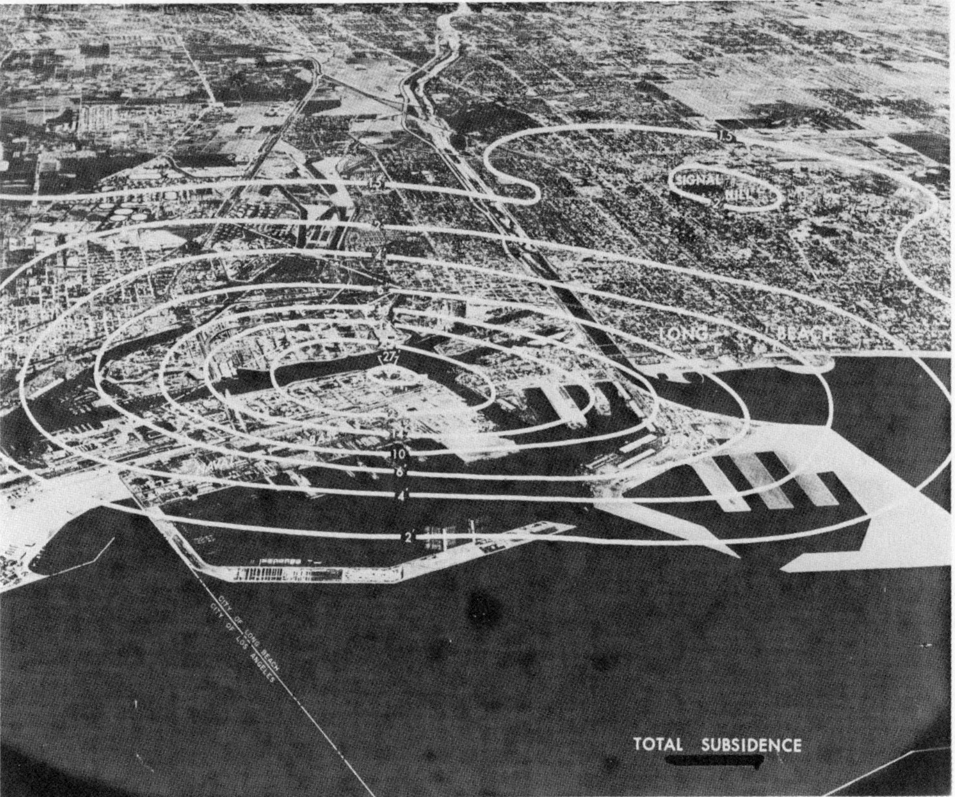

FIGURE 1-1

Sinking of land around Long Beach, California. When oil, gas, and water were taken from the ground, the area became susceptible to flooding.

The answer to the debate lies somewhere between the two viewpoints. In the last 20 years or so, as we have come to realize the seriousness of our population growth we have had to acknowledge the fact that people have a one-to-one relationship with their surroundings, that neither completely determines the other, and that we must learn to live in harmony with our environment. In addition, the vast growth of technology has provided us with an awesome power that has exposed the delicate balance and fragility of our planet. We can now literally move mountains, resculpture large areas of the earth, change climate, and annihilate whole species, including our own. It is vital, therefore, that we be sensitive to and learn about our relationship with our environment.

The lack of space, water, energy, and clean air in parts of urban southern California illustrates the need for respect for the environment. Figure 1-1 shows another result of our carelessness. By 1962, the withdrawal of oil, gas, and water from the ground beneath the city of Long Beach, California had led the land to sink as much as 8.1 m (27 ft). The subsequent threat of flooding in this area reminds us that the environment can respond to human activity and demonstrates the two-way relationship between humans and their environment. As you can see from this case, we have an urgent need to recognize the great value of our earth. The study of physical geography can help us.

THE NATURE OF PHYSICAL GEOGRAPHY

Physical geographers study the distribution and interrelationships of the elements in the physical environment. The elements they most commonly study are the earth's climates, soils, vegetation, and landforms.

For example, in the high-altitude photograph of southern California shown in Figure 1-2, a physical geographer would immediately begin to divide it into areas such as sea, land, highland, lowland, and so on. (A simple division is diagramed in Figure 1-3.) The geographer is interested in the arrangement in space of

FIGURE 1-2

The Los Angeles area, photographed by the Apollo spacecraft in the Apollo-Soyuz Test Project, taken at an altitude of 193 kilometers (120 mi).

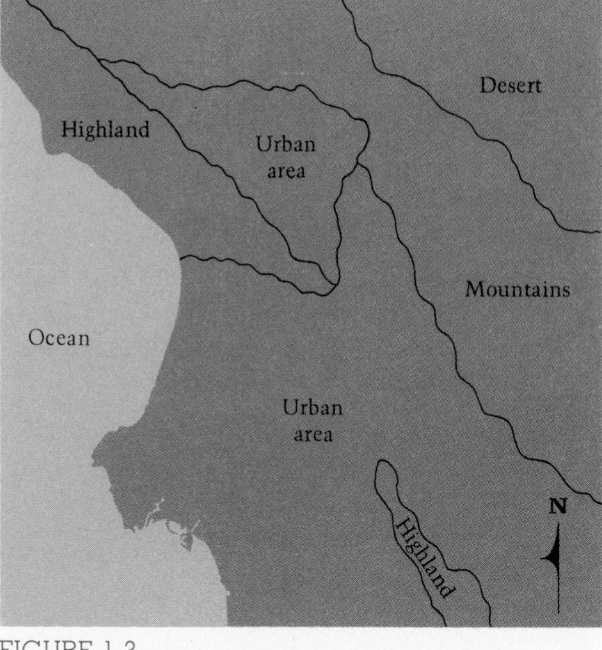

FIGURE 1-3

Simple sketch map derived from the photograph, showing the major areas a physical geographer might pick out.

some of these elements. He or she might then note that desert and vegetated land are unusually close together. A geographer with some training knows that the mountains surrounding Los Angeles are responsible. In this part of California, most of the moisture-bearing winds come from the west. The moving air cools and condenses as it hits the coastal and inland mountain ranges. Clouds develop and rain falls, which helps to maintain the vegetation on the western side of the mountains. But on the eastern side, the air, which has lost its moisture, warms and descends. There is little chance of rain coming from this hot, dry air, and so the area east of the mountains is desert. The spatial arrangement of the vegetated area and the desert is thus caused by an interrelationship between the atmosphere and the mountains of the land surface.

When geographers study the distribution of vegetated land and desert in southern California, they have more than a general interest in the relationships among physical elements. If they can understand why two elements of the physical world are related in one place, they can be reasonably sure that the same relationship exists in similar geographical areas. For example, if they know how the hot, dusty Santa Ana wind operates in southern California, they can apply their knowledge to the chinook wind of the western United States and Canada, to the föhn wind of the European Alps, to the hamsin of Israel, to the nor'wester of South Island, New Zealand, and to many other similar winds in different parts of the world.

Another reason for studying physical geography is that it gives a broad overview of the environment. Professional physical geographers often have a specialized knowledge of one part of the physical world, such as vegetation, but still have a wide knowledge of the other parts. They therefore find themselves well equipped to tackle broad environmental problems and are valuable members of environmental study groups at any governmental level.

Where should we start in a field that covers almost the whole world? The answer is that we first divide the physical world into four major components.

COMPONENTS OF THE PHYSICAL WORLD

Physical geographers divide the physical world into four realms—the atmosphere, lithosphere, biosphere, and hydrosphere. The first two of these are real, in the sense that we can see them or parts of them. Although we can also see parts of the second two, they are more abstract. The *atmosphere* is the layer of air that begins a few meters within the soil or at the water surface and extends to about 60,000 km (37,000 mi) above the earth. *Lithosphere* comes from a Greek word meaning "stone," and in physical geography, it refers to the rock layers at and beneath the earth's surface. The *biosphere* is the part of the earth containing all life, including plants, animals, and all other organisms in water and air. The water of the oceans covers 71% of the earth's surface, and water also exists on and within the surface and in the atmosphere above. All this water together is called the *hydrosphere.*

These, then, are the main components of the physical world. They are linked together in any one area and over the earth as a whole. As Figure 1-4 demonstrates, the physical geographer studies the phenomena within these realms and the relationships among them.

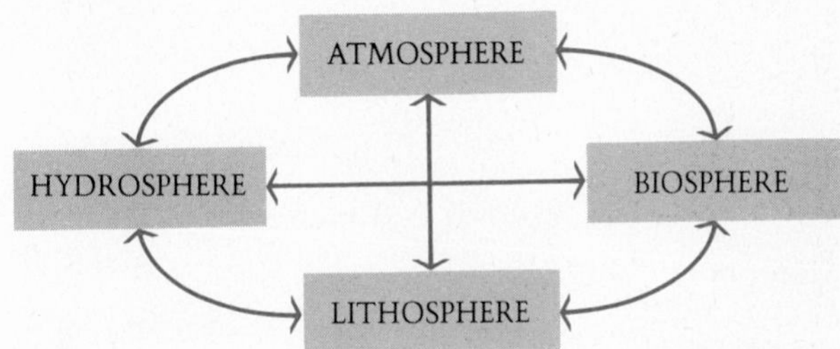

FIGURE 1-4
Relationships among the four spheres of the earth system.

THE EARTH AND TIME

Another factor that adds excitement to the study of physical geography is that the earth is always changing. The patterns we see on the surface today have evolved from patterns of the past. For example, 15 million years ago the area of present-day Los Angeles lay under about 1500 m (5000 ft) of water. Slowly the land rose, stopping occasionally to form beaches, like those in Figure 1-5, that are now called marine terraces. Parts of the area were under water as few as 100,000 years ago and possibly more recently. If speculations about the melting of the world's icecaps are correct, then Los Angeles may once more be under water in about 100 years. (Pause for a moment to consider how you might make these numbers—15 million, 100,000, and 100 years—meaningful to yourself. Try relating them to distance.)

Sometimes the earth changes more quickly. To most who inhabit temperature-regulated homes and drive comfortable cars to college or work, it often seems that the world does not change. But this opinion would not be shared by people who happened to be traveling on the freeways in the San Fernando Valley at 6:01 AM on February 9, 1971. The earthquake that rocked their

WHERE DOES THE MUSIC COME FROM?

John Denver sings "Rocky Mountain High," Bob Dylan delivers "Blowin' in the Wind," and Elton John records "Indian Sunset." Before them the Beach Boys immortalized the Southern California surf. At performances of *Hair,* the audience joined the cast in "Let the Sun Shine." The power of the Columbia River, the devastation of "The Great Dust Storm," the redwood forests, and the Gulf Stream waters were shaped into lyrics by Woodie Guthrie. The environment underlies all kinds of music. Composers of rock hits, folk songs, and Broadway show tunes have all found themes in earth, wind, and water. Music, as well as the rest of human culture, bears the imprint of our physical surroundings.

Composers of classical music captured images from nature in abstract ways. The Russian composer Borodin created a vast plain in sound by sustaining a single violin note throughout "In the Steppes of Central Asia." Debussy composed both "*Clair de lune*" (Moonlight) and "*Reflets dans l'eau*" (Reflections in the Water).

In the 1870s the Czech composer Smetana used a river in his homeland as the motif for a symphonic poem called "The Moldau." The program notes attached to the original score recreate in words what Smetana suggested in music: "Two springs pour forth in the shade of the Bohemian forest, one warm and gushing, the other cold and peaceful." They merge to become the river Moldau. "Coursing through Bohemia's valleys, it grows into a mighty stream. Through thick woods it flows as the gay sounds of the hunt are heard. It flows through grass-grown pastures and lowlands where a wedding feast is being celebrated. At night, wood and water nymphs revel in its sparkling waves." The river surges on, "finally flowing on in majestic peace toward Prague. Then it vanishes far beyond the poet's gaze."

FIGURE 1-5

An air view of marine terraces in southern California.

world killed two men driving on the highway and crumbled several bridges and overpasses, as shown in Figure 1-6.

Whether the earth changes slowly, as with the varying sea level, or quickly, as with earthquakes, it does change. Therefore, besides studying the patterns and spatial arrangements of the elements, the geographer also studies how patterns developed. This interest has a practical application, because if the geographer can say how distributions occurred, he or she may be able to say something about what will happen in the future.

CONCEPTS IN PHYSICAL GEOGRAPHY

As we are beginning to realize, the physical world is rather complex. Physical geographers use many specialized techniques and concepts to deal with its complexity.

The construction of maps is one technique that geographers have developed over the centuries. Chapter 2 describes the development of this skill and how geographers mapped the earth. Many other concepts used by geographers have been borrowed from elsewhere. One of these is a way of thinking called general systems theory.

GENERAL SYSTEMS THEORY

A *system* has two characteristics. First, it has boundaries, real or abstract, that are defined by the investigator. Second, a set of interrelated events take place within these boundaries. *General systems theory* is a body of laws and a method of approach relating to systems.

A reservoir with water entering it and leaving it might be regarded as a system. In this case, the boundaries of the system are the boundaries of the reservoir. The most important event to take place within this system is the rising and dropping of the water level.

FIGURE 1-6

Freeway damage by the San Fernando Valley earthquake.

The city of Chicago could be described as a large and elaborate system. Every day the system receives an inflow of energy, food, water, and vast quantities of consumer materials. Most of this energy and matter is used and changed in form. At the same time, huge amounts of energy, manufactured material, sewage, and other waste is exported from the city.

Open and closed systems There are two major types of system. The simplest is called a *closed system* because there is no exchange of energy or matter through the system's boundaries. If you have ever tried to grow plants in a completely sealed glass jar, you know that the water and gases inside circulate continually and that only light energy comes from the outside. It is difficult to find a true closed system on the earth's surface, although some problems are most easily handled by assuming that the system is closed. Engineers might assume that there are a fixed number of cars on the West German superhighway system and then solve the problem of moving them around most efficiently. Such an approach may give clues to the solution of the real problem.

One concept associated with systems, particularly closed systems, is the idea of entropy. *Entropy* may be regarded as a measure of the inability of energy in a system to do work. Energy can exist in many different forms, and it continually changes to its least useful form. When energy has reached its least useful form, we say that the largest amount of entropy has been reached. For example, at some time in the distant future, our sun will have transformed all its useful nuclear energy into random heat energy, which will disperse throughout outer space. No work can be obtained from dispersed energy. When the nuclear energy has completely dispersed, our solar system, which is virtually a closed system, will run down. It will have reached a state of maximum entropy. It will no longer work. In a closed system, maximum entropy is achieved sooner or later.

WRIGHT AND ORGANIC ARCHITECTURE

Frank Lloyd Wright rejected classical traditions in architecture and urged his students to find ideas not in textbooks and museums but in nature. Wright drew many of his ideas on shape, structure, and materials from the physical environment. His style came to be known as organic architecture. Wright recognized that every site contained a unique combination of physical elements, and he tried to blend the lines and materials of his buildings into their setting.

The flat, open prairies of the Midwest provided the backdrop as well as the stimulus for many of the houses he designed. His consciousness of the "earth-line" led him to use a broad, horizontal plane as the basis of many house plans. Today the long, low ranch style appears all over America. Inside, Wright developed the theme by eliminating divisions between rooms and creating a large, open living space. His "open plan" revolutionized domestic architecture.

Taliesin, his family home and studio, lies in the rolling countryside of Wisconsin. Wright visualized the buildings as the brow of a hill. The lower walls were built from stratified rock. Wright wanted the walls above to be an abstract pattern of native stone and wood that resembled the patterns in the landscape. Plastered stretches of wall were painted tan to mirror patches of sand along the river below. Wright eliminated gutters so icicles would hang from the eaves in winter. His neighbors built steep roofs so snow would slide off, but Wright used a gently sloping roof so snow would cover the house just as it blanketed the surrounding hills.

A second kind of system is called an *open system*. Energy and matter transfer across the boundaries of an open system. The reservoir and the city of Chicago, mentioned earlier, are representative of open systems. There are many other examples of open systems throughout this book, such as a river drainage basin or a beach system. Indeed, the earth itself is an open system.

Any system can have one or more subsystems. A *subsystem* is a part of a larger system. A subsystem can act entirely independently, but it acts within and is related to the larger system. The fresh water supply and the sewage system are subsystems of the total Chicago system.

Systems and subsystems have boundaries that are called *interfaces*. The transfer or exchange of energy and matter takes place at these interfaces. Sometimes the interfaces are visible. You can see where sunlight strikes the roof of an apartment building. But often they are not visible. You cannot see the movement of groundwater, a part of the hydrologic system, through the minerals of the geologic system. Many geographers focus attention on these interfaces, visible or invisible, especially when they coincide with the earth's surface. It is here that we find the greatest activity of our dynamic world.

Dynamic equilibrium and feedback Two other ideas commonly used in general systems theory are those of dynamic equilibrium and feedback. A system is in *dynamic equilibrium* when it is neither growing nor getting smaller but continues to be in complete operation. *Feedback* occurs when a change in one part of the system causes a change in another part of the system. Let us look at some examples.

If you were to look at the sand in a specific area of Santa Monica Beach, near Los Angeles, you would barely perceive that the currents traveling along the shore are taking away some sand and bringing in more. There is continual movement, yet over a period of weeks, the beach apparently stays the same. The beach is thus said to be in a state of dynamic equilibrium.

An example of feedback, or a feedback mechanism, would be the case of solar radiation being reflected from a downtown Los Angeles sidewalk. The sidewalk's surface receives energy from the sun, but it also reflects and radiates it back into the atmosphere, as well as losing it in other ways. The more energy the sidewalk receives, the more it reflects and reradiates. Because of the reflection and reradiation, the sidewalk does not become increasingly hotter. Without the feedback mechanisms of reflection and reradiation, it would certainly be impossible to walk on the sidewalk at midday in the summer.

A feedback mechanism that operates to keep a system in its original condition, like the reflection of radiation from the sidewalk, is called a *negative feedback mechanism*. The opposite case, where a feedback mechanism makes an increasingly greater change from the original condition of a system is called a *positive feedback mechanism*. Chapter 8 explains why the urban growth of Los Angeles, or any city, leads to higher average air temperatures. A change of this kind is an example of positive feedback.

MAKING MODELS OF THE WORLD

Another way that physical geographers study the earth's phenomena is to make models of them. The satellite photograph of southern California is a smaller version of what is almost the real thing, and it is very detailed. It can be called an *iconic model*, a smaller representation of reality. *Icon* comes from a Greek word meaning "image." The simplified version of the photograph, which shows highland and lowland areas (see Figure 1-3), is a second stage of abstraction. A model that involves a second stage of abstraction is called an *analog model*. The diagram in Figure 1-4, showing the four main spheres of the earth, is quite abstract in nature. No one has ever seen a biosphere, but the word is used symbolically to represent something. A *symbolic model* is a model where real-world phenomena are represented by abstract, verbal, or mathematical expressions.

FIGURE 1-7
Symbolic model of the types of models used by physical geographers.

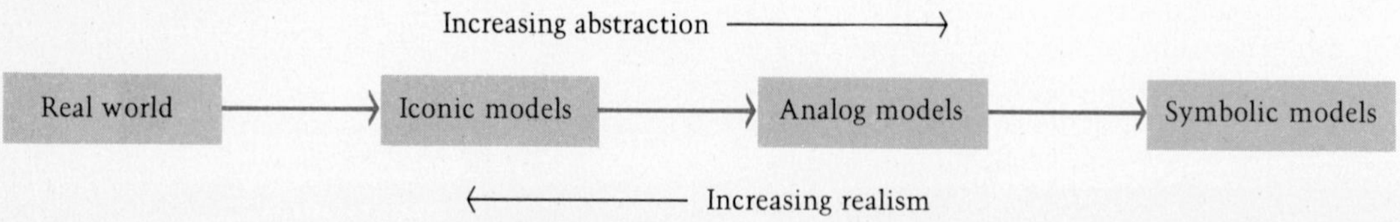

The type of model used depends on its intended purpose. Each model is a convenient way to organize thinking about a given phenomenon or set of phenomena. The types of models, and consequently the ways we think, are related to one another. Moving from the real world through iconic, analog, and symbolic models, we increase abstraction and decrease realism. Figure 1-7 is a symbolic model of these relationships. At each stage, the amount of information to be included must be traded off against the degree of generalization. One skill that the geographer must acquire is knowing just what level of trading off will produce the most efficient explanation of the phenomena he or she is studying.

You can see by now that geographers have developed or applied a large number of concepts to help them study the earth's surface. We introduce other ideas as we progress. Examples have been taken principally from southern California but as you read this book you will discover that the concepts are applicable to many individual areas, and to the world in general. This attention given to southern California is not because the area has any special merit but because it is usually easier to understand somewhat abstract principles if they are attached to a specific place. It is probably more appropriate for you to start your study in the place where you live. Consider for yourself how these notions may be applied to your own area.

FIGURE 1-8

The earth system and its subsystems. As an open system, the earth receives and passes off energy and matter. A regional subsystem encompasses parts of the other four subsystems.

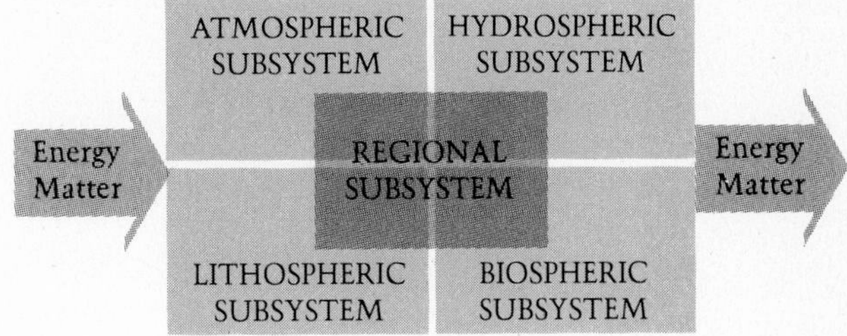

THE PHYSICAL WORLD AS A SYSTEM

Earlier we divided the physical world into components: atmosphere, lithosphere, biosphere, and hydrosphere. We are now in a position to call these spheres subsystems of the earth system and to refine the ideas further. Figure 1-8 illustrates how we can regard the earth surface system in this way. It is an open system, with energy and matter entering and leaving.

Within the total system, there are five subsystems. Four of these subsystems are already familiar. They all interconnect at any given point in the earth's system to form a fifth subsystem, which may be called a regional subsystem. We are interested in the special ways in which the four major subsystems work together to form a regional subsystem. In one sense, each regional subsystem is unique. Let's take an example. The Santa Ana wind travels over rock types that are somewhat different from those crossed by the föhn wind. But there are enough similarities between these winds and other physical elements for us to group some types of regional subsystems together. Regional subsystems are grouped in some of the other chapters, such as that on regional climates.

SLIDING SCALE

Imagine a man sunbathing on Malibu beach. We can photograph him occupying a square of sand about one meter on a side. If we move the camera higher, a square of 10 (10^1) m reveals his friends. When we sharpen the focus on a 100 (10^2)-m area, we can see a crowd of people on the beach. A picture of 1000 (10^3) m includes the beach, some sea, and some land. One with an edge of 10,000 (10^4) m captures part of the city of Los Angeles.

Moving the camera still farther, we shoot a picture of a 100,000 (10^5) m square. It encompasses the southern California region. The next step is 1,000,000 (10^6) m. This photo takes in the entire state, some neighboring states, and a part of the Pacific. A photo showing a square of 10,000,000 (10^7) m covers most of the earth.

And if the camera is far enough away to focus on a square of 100,000,000 (10^8) m, we see the earth as a small globe. Somewhere on it is a man lying on a beach in Malibu.

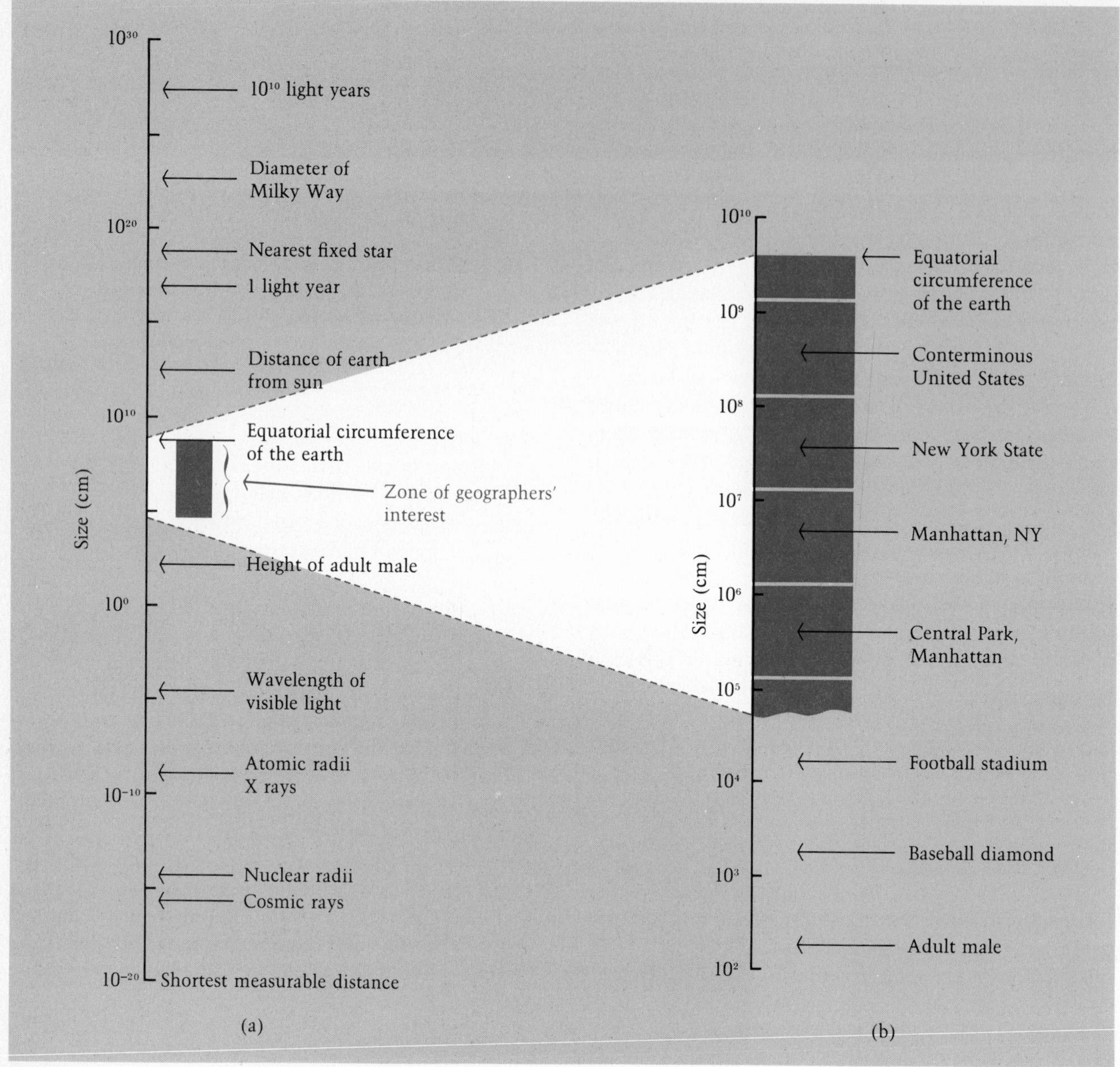

FIGURE 1-9

Orders of geographic magnitude. Geographers are usually interested in the things shown on the right side of the figure, but sometimes they must think in much smaller or much larger terms, as shown on the left side.

The other thing that we must consider in going into the real world is the size of the objects or phenomena that interest us. Even speeding at 900 km per hour (560 mi per hour), one may become uncomfortable on a ten-hour flight from Los Angeles to London because, in purely human terms, the world is such a big place. But in studying the world, we must not think of distance and magnitude in purely human terms.

Let us consider different sizes, or *orders of magnitude.* Figure 1-9 shows the different orders of magnitude with which we must become familiar. The scale in this figure is written in *exponential notation.* This means that 100 is written as 10^2 (10×10), 1000 is 10^3 ($10 \times 10 \times 10$), 0.01 is 10^{-2} (1/100), and so on. This notation saves us from writing numerous zeros. The scales geographers use most often go from 10^5 or 10^6 cm (10,000 yd), the size of Central Park in Manhattan, up to about 10^9 cm (62,000 mi), the order of magnitude of the circumference of the earth.

Physical geographers sometimes have to expand their minds even further. Cosmic rays with wavelengths of 10^{-16} cm may affect our climate. The nearest fixed star is approximately 10^{18} cm (25 trillion mi) away. Brighter stars, even further away, sometimes help when navigating a path across the earth's surface. Occasionally we have to perform mental gymnastics to conceive of such distances, but one of the beauties of physical geography is that it helps us see the world in a different way.

At this point we must pause for a moment before embarking on our detailed study of the earth. Central to our effort is an attempt to understand the environment at the earth's surface. We must live with it in a one-to-one relationship. It is all too easy for us to forget this. Perhaps we can leave the last words to Joan Didion. She tells how, on nights when the Santa Ana wind blows, "every booze party ends in a fight. Meek little wives feel the edge of the carving knife and study their husbands' necks. Anything can happen. . . . The wind shows us how close to the edge we are."

SUMMARY

There is a close relationship between humans and their environment. Elements of the physical world affect us even in urban areas. A study of physical geography helps us to learn about this relationship.

Physical geography is the study of the distribution of the elements of the physical environment and their relationships. The physical elements of the earth are divided into the atmosphere, lithosphere, biosphere, and hydrosphere. Physical geographers also study the development of patterns through time.

Many techniques and concepts are used in studying physical geography. General systems theory is one set of ideas, and the use of models is another.

In studying the physical geography of the real world, we sometimes have to be able to think of very large and very small distances. But in all our studies, we have to remember our one-to-one relationship with our environnent.

QUESTIONS

1. Environmental determinism, in its extreme form, held that the physical environment largely determines the type of culture that exists in it. Ellen Churchill Semple wrote, in *The Influences of Geographic Environment:*

> Man is a product of the earth's surface. This means not merely that he is a child of the earth, dust of her dust; but that the earth has mothered him, fed him, set him tasks, directed his thoughts, confronted him with difficulties that have strengthened his body and sharpened his wits, . . . On the mountains she has given him leg muscles of iron to climb the slope; along the coast she has left these weak and flabby, but given him instead vigorous development of chest and arm to handle his paddle or oar. In the river valley she attaches him to the fertile soil, circumscribes his ideas and ambitions by a dull round of calm, exacting duties, narrows his outlook to the cramped horizon of his farm. . . . Man can no more be scientifically studied apart from the ground which he tills, or the lands over which he travels, or the seas over which he trades, than polar bear or desert cactus can be understood apart from its habitat.

How would you design an experiment to prove or disprove her ideas?

2. We tend to become more conscious of our environment when we have reached a crisis due to our misuse of it. Modern technology places immense disruptive power in our hands, but we have caused great harm in the past by the injudicious use of simpler technologies. Explain the American agricultural disaster of the 1930s, the Great Dust Bowl, in terms of our ability to change our environment and the technology we use to effect that change.

3. Concern for the environment is often seen as opposition to economic growth. Few areas of the country are without such conflicts, and many are of national importance. Which have attracted your attention recently, and how would you resolve the conflicts involved?

4. In studying the air photo of southern California (Figure 1-2), how would the approach of a physical geographer differ from that of a geologist?

5. What type of phenomena might we expect to find geographers studying in each of the four main components of the physical world?

6. What topics might we examine if we were studying the interrelations between the hydrosphere and the biosphere?

7. The time scale we use to examine phenomena is related to the rates of change involved in the processes creating them. If Los Angeles was under 1500 m (5000 ft) of water 15 million years ago, what is the minimum average rate of change in the process that uplifted Los Angeles to its present elevation near sea level?

8. What would be the appropriate time scale for an examination of phenomena associated with an Ice Age? For the formation of a mountain range? For the rate of cooling of the earth? For reporting changes in

climate to farmers planting their spring crops?

9. Systems nest within one another. What may appear to be a closed system at one scale often is also an open subsystem of one or more systems at a higher scale. What larger systems might the subsystem of a stand of pine trees on a mountain slope belong to?

10. Many, if not most, of the systems found in nature attempt to achieve a state of dynamic equilibrium. What general conditions would have to be fulfilled for a local ecosystem to be in dynamic equilibrium? How could this balance be upset, and how might the system respond?

11. What kinds of models are the following, and why: A relief map of the United States? A block diagram showing the types of faults, such as Figure 16-10? A globe with the parallels and meridians marked on it?

12. What types of models would you use to represent the following: The interrelationships between dynamic equilibrium and positive and negative feedback? Glaciation in an alpine valley? The shortest route between Los Angeles and London? The relationship between elevation and vegetation types?

13. If we regard the earth as an open system, what are two larger systems of which it is a part?

14. How would we write the following in exponential notation: The mass of the earth, approximately 6,000,000,000,000,000,000,000,000,000 g? The diameter of the earth, about 12,740,000,000 cm? The volume of the Atlantic Ocean, about 31,800,000,000,000 cm^3? The fraction of H_2 in the atmosphere, .0000005?

15. Different phenomena and processes occur at different scales. If we know that denudation of the earth's surface in an area occurs at an average rate of 1 meter per thousand years, while uplift of the same area occurs at an average rate of 10 meters per thousand years, over a period of one million years what is the net effect on the area's structure? Which process is most important in the formation of the landscape? What scale will we use to describe the change?

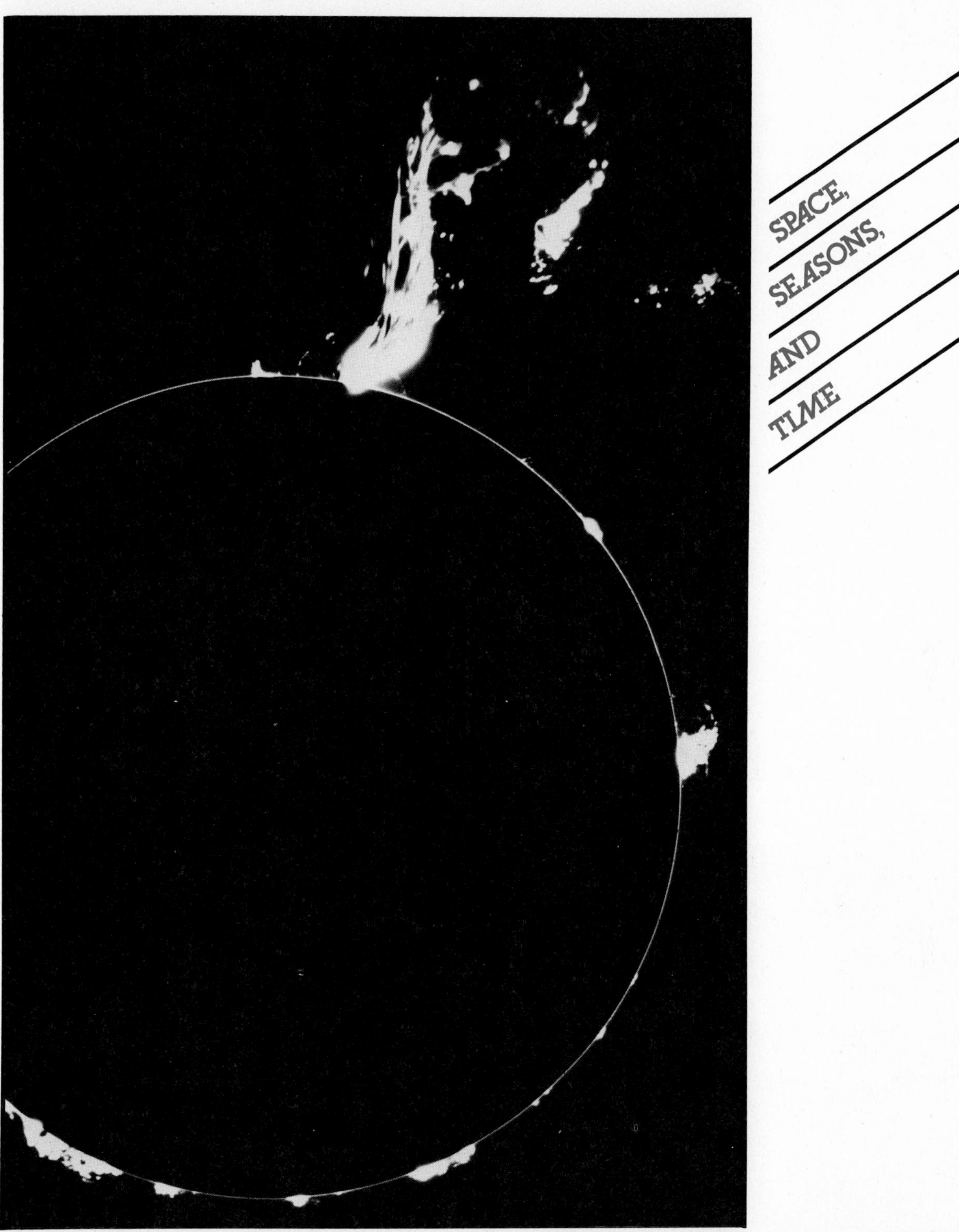

SPACE, SEASONS, AND TIME

CHAPTER 2

We have all thought at one time or another about the concepts of space, seasons, and time. While standing on a hill or a tall building, our thoughts turn to how far we can see and we wonder what is beyond. Seasons have particular meanings for each of us. We talk of a football season, a theater season, a dull season. We use the seasons to mark the passage of time. We also use watches and atomic clocks. Yet it still seems that some times pass more quickly or more slowly than others. With all these variations, it is sometimes hard to tell what we really mean when we talk about space, seasons, and time.

We can partially clear up matters by talking about space, seasons, and time in terms of individual disciplines, such as geography. For example, space can be thought of as the arrangement of such features as land and sea on the surface of our planet, as we learn in this chapter. Geographers might ask how much surface such features occupy or how far apart they are. They are particularly interested in knowing how we can represent space so another person will understand the relationship of one position to another.

In the same way, the most interesting aspects of seasons—at least to geographers—are the changing positions of the sun and the earth and the varying amounts of energy arriving at different points on the earth's surface at different times. A study of the seasons shows space to be closely connected with time.

The last part of this chapter examines the relationship of time and space on the earth's surface. For example, the time in New York is three hours ahead of the time in San Francisco. Time and space are inextricably mixed.

FIGURE 2-1

Stick charts from the Marshall Islands. Shells are used to represent islands and the palm sticks represent the direction of predominant ocean waves or swells. a) General chart. b) Sectional chart.

(a)

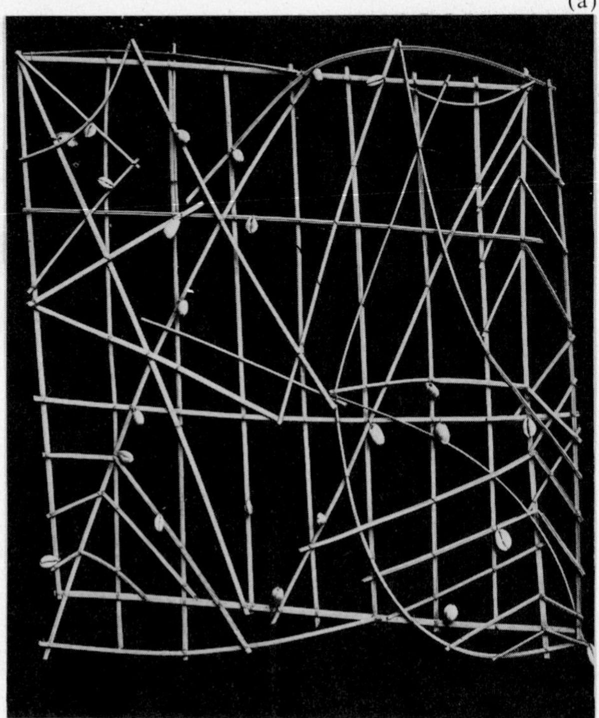

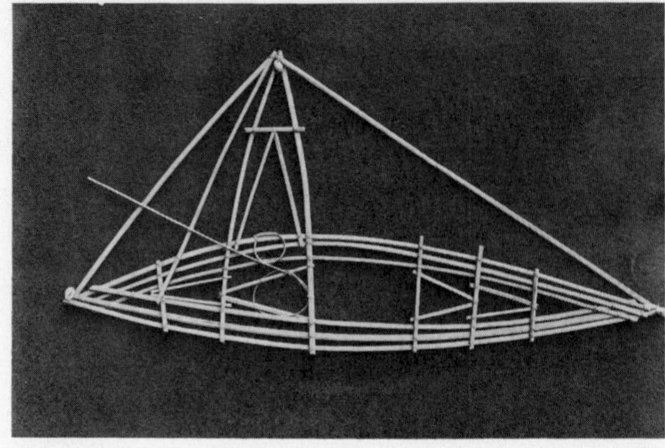

(b)

MAPPING THE EARTH

When you go to a new city, your process of learning about it begins at the hotel you stay in or in your new home. You then locate the nearest important service areas, such as bus stops and supermarkets. Gradually your knowledge of the city and your immediate area grows. This slow pace in learning about the space you live in repeats the history of every human society as it learned about the earth.

Very soon in the learning process, directions concerning the location of a particular place have to be taken from or given to someone. One of the most common forms of conveying such information is the map. We have all seen sketched maps directing us to a place for a party. The earliest maps were of a similar simple nature, ranging from maps scratched in the dust or sand to the stick charts made by the Marshall Islanders, which are illustrated in Figure 2-1.

Many of these simple maps have not lasted. In fact, you probably threw away the last map someone drew for you after you went to the party. Thus, although humans have been drawing maps throughout most of their history, the oldest surviving maps date only from about 2500 BC. They were drawn on clay tablets in Mesopotamia (the area that is now Iraq) and as Figure 2-2 shows, represent individual cities, the entire coun-

FIGURE 2-2

Early Mesopotamian plan of the city of Nippur on a clay tablet.

try of Mesopotamia, and the early Mesopotamian view of the world (Figure 2-3a).

The Mesopotamian world map is especially interesting. The principal features are diagramed in Figure 2-3b. Quite clearly it is not the first map ever produced, because it shows a number of already conceived ideas. For example, Babylon is assumed to be at the center of a flat earth. The Tigris and Euphrates rivers flow into the Persian Gulf, which in turn drains into a sea encircling the globe. The map demonstrates that humans had already begun to explore and chart their surroundings.

The religious and astrological text above the map in the figure indicates that the Mesopotamian idea of space was tied up with ideas of humankind's place in the universe. This is a common theme in cartography (the science of maps). Even today, maps of newly discovered space, such as star charts, raise questions in our minds of where we, as humans, fit into the overall scheme of things.

THE EARTH AS A SPHERE

If you look out your window, there is no immediate reason for you to suppose that the earth is anything but flat. It takes a considerable amount of traveling and observation to reach any other conclusion.

The first philosopher to suggest that the earth might not be flat was a Greek called Anaximander. In about 550 BC, he suggested an earth shaped like a column. Fifty years later, it was thought that the upper surface of the earth column was curved. After another 200 years, the travels of such Greeks as Alexander the Great had gradually expanded knowledge about the location of land and sea.

(a)

(b)

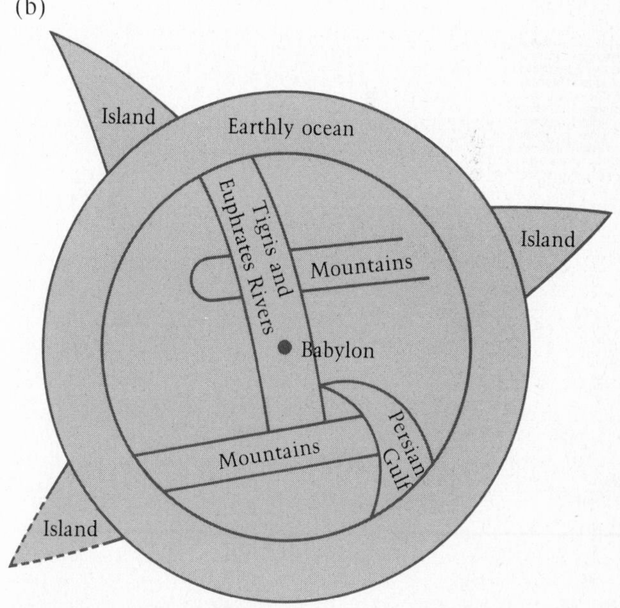

FIGURE 2-3

The Mesopotamian world map. a) The original was made about 2500 BC. b) A modern diagram of the original more clearly indicates its principal features.

About the same time (350 BC), philosophers pondering these travels came to the idea that the earth was in fact spherical. Three of them calculated the size of its circumference. Aristarchus of Samos actually thought that the earth travels around the sun—an idea that was not accepted for another 1800 years because it clashed with religious concepts.

The idea that the earth had a curved surface, probably spherical, was solidified by the work of the head librarian at Alexandria, who died in 196 BC. Eratosthenes was not thought of as a great scholar by his contemporaries, but his measurement of the circumference of the earth is now regarded as one of the great achievements of Greek science. Eratosthenes observed that the rays of the sun, at midday on the day of the year when the sun was highest in the sky, fell directly over Syene (Aswan) and that no shadow was produced by a rod placed vertically in the ground. He also noted that at Alexandria, at the same time of day and year, a similar vertical rod produced a shadow. Figure 2-4 diagrams his observations.

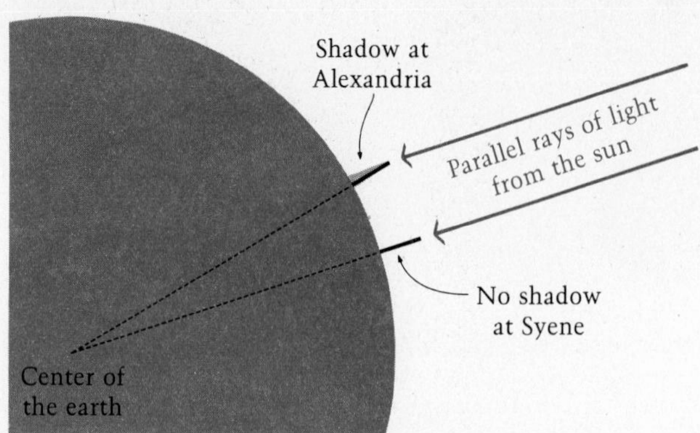

FIGURE 2-4

Eratosthenes' basic observations about the shadows thrown by the sun's rays. His calculations of the earth's circumference were amazingly accurate.

Assuming that the sun's rays were parallel as they struck the earth's surface and using the geometry developed by Euclid, Eratosthenes then calculated the circumference of the earth from the length of the shadow cast at Alexandria. His calculation is now thought to be accurate to within 320 km (200 mi) of the actual circumference, an accuracy of 0.8%.

The geometry of a sphere When a spherical surface, such as an orange, is cut into two parts, the edges of the cut form circles. Once the earth was assumed to be spherical, it was natural to divide it by means of circles. Sometime between the development of the wheel and the measurement of the earth's circumference, mathematicians had decided that the circle should be divided into 360 parts by means of 360 lines radiating from the center of the circle. The angle between each of these lines was called a *degree.* For such a large circle as the earth's circumference, a further subdivision became necessary. Each degree was divided into 60 *minutes,* and each minute was further subdivided into 60 *seconds.*

With a system for dividing the curved surface of the earth, the problem became where the earth, or any circle on it, actually starts. Two sets of information could be used to attack this problem. First, a sense of

INDIANA IS PINK

A conversation between two of the most famous characters in American literature, Tom Sawyer and Huck Finn. The locale: in a balloon, somewhere over Illinois.

"I know by the color. We're right over Illinois yet. And you can see for yourself that Indiana ain't in sight."

"I wonder what's the matter with you, Huck. You know by the *color?*"

"Yes, of course I do."

"What's the color got to do with it?"

"It's got everything to do with it. Illinois is green, Indiana is pink. You show me any pink down here, if you can. No, sir; it's green."

"Indiana *pink?* Why, what a lie!"

"It ain't no lie; I've seen it on the map, and it's pink."

"Huck Finn, did you reckon the States was the same color out of doors as they are on the map?"

"Tom Sawyer, what's a map for? Ain't it to learn you facts?"

"Of course."

"Well then, how's it going to do that if it tells lies? That's what I want to know."

Abridged from Mark Twain [Samuel Clemens], *Tom Sawyer Abroad* (New York: Charles L. Webster & Co., 1894), pp. 42–43.

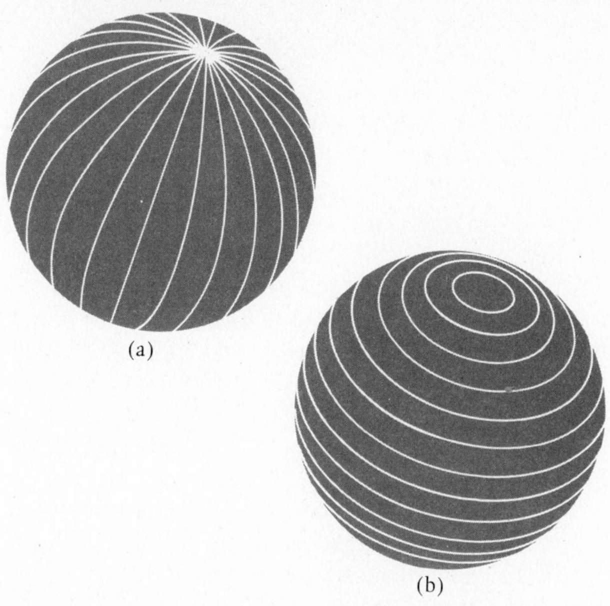

FIGURE 2-5

Meridians and parallels on a globe. Meridians run north and south, and parallels run east and west.

direction had been gained by studying the movements of the sun, moon, and stars. In particular, the sun at midday was always located in the same direction, which was called south. Knowing this, it was easy to arrive at the concepts of north, east, and west. The division of the circle could refine these directional concepts. The second piece of information was that some geographical locations in the Mediterranean area, fixed by star measurements, could be used as reference points for the division of the earth.

Using this information, it was possible to imagine a series of lines on the earth's surface, some running north–south and some running east–west. The lines running east–west are still called *parallels,* the name the Greeks gave them. Those running north–south are called *meridians.* As we can see in Figure 2-5, the two sets of lines differ: Parallels never intersect one another, whereas meridians meet at the top and bottom of the sphere.

Latitude and longitude The Greeks had the beginnings of an earth grid when Eratosthenes divided the known earth with a parallel passing through the Strait of Gibraltar and a meridian passing through Rhodes. The two lines intersected, thus dividing the earth into four parts. Hipparchus, working in the second century BC, divided the globe into many more divisions. He used an imaginary grid of equally spaced parallels and meridians crossing one another at right angles (90 degrees).

Our present-day division of the globe stems directly from the work of Hipparchus. The parallels are called lines of *latitude.* The parallel around the middle of the globe is called the *equator* and is defined as 0° latitude. As Figure 2-6 explains, the other lines of latitude are

FIGURE 2-6

Latitude and longitude. a) Viewed from the side, lines of latitude (including the equator) are horizontal parallels. b) Lines of longitude, the meridians, appear to radiate from a center point when viewed from above the North Pole. They converge again at the South Pole.

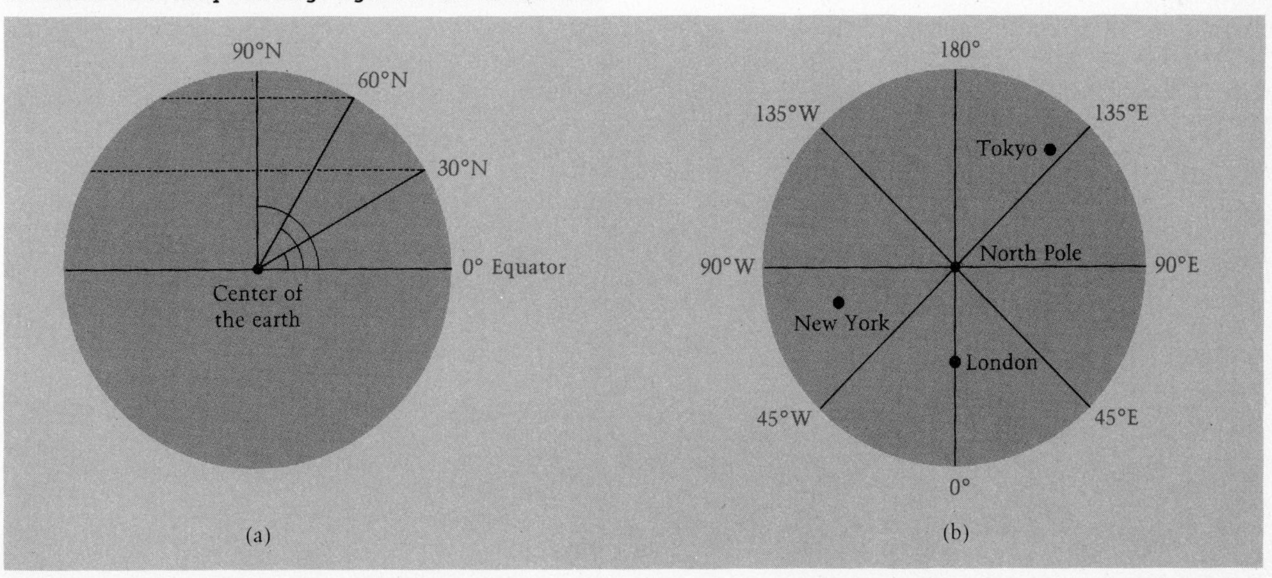

defined by estimating the angle between a line joining the earth's surface, at any particular latitude, with the center of the earth and the line from the center to the earth's surface at the equator. (In Figure 2-6a, two other lines of latitude—the 30° line and the 60° line—are shown in relation to the equator.) Lines of latitude going both north and south of the equator are defined this way. The "top" of the earth, the North Pole, is at latitude 90°N, and the "bottom," the South Pole, is at 90°S.

The meridians are called lines of *longitude.* In 1884 the meridian that passes through the Royal Observatory at Greenwich near London was established as the starting point for measuring longitude. This meridian is called the first, or *prime,* meridian and is defined as having a longitude of 0°. The others are divided as if we were looking down on the earth from above the North Pole, as Figure 2-6b shows. We measure both east and west from the prime meridian. Thus New York City has a longitude of 73 degrees 58 minutes (73° 58′) west of Greenwich, and Tokyo (Japan) has a longitude of 139° 40′ east of Greenwich.

Because meridians converge at the North and South poles, the actual distance in one degree of longitude varies from 111 km (69 mi) at the equator to zero at the poles. In contrast, the length of a degree of latitude is about 111 km (69 mi) at all locations. We say "about" because the earth is not a perfect sphere. It bulges slightly at the equator and is flattened at the poles. This particular shape is called an *oblate ellipsoid.* However, Hipparchus and the Greek scholars knew nothing of this. They had a more immediate and difficult problem: How could they represent the spherical surface of the earth on a flat chart?

THE DILEMMA OF MAPPING

If you have ever tried to cut the skin of an orange, or any spherical surface, and then lay it flat on a table, you realize that it is not an easy task. At least some part of the skin must be stretched to make it flat. Cylinders or cones can easily be cut to be laid out flat, but not a sphere.

Once the Greeks had accepted the idea that the earth was a sphere, they had to determine how best to represent the earth's surface on a flat surface. There is no completely satisfactory solution to this problem. However, by the time of Hipparchus, many of the partial solutions in common use today had already been invented.

The Greeks had noted that a light placed at the center of a globe casts shadows along the meridians and parallels. These shadows, which form lines, can be projected outward onto some surface that can later be made flat. The resulting series of lines on the new surface is called a map projection. More strictly, a *map projection* may be defined as an orderly arrangement of meridians and parallels produced by any systematic method that can be used for drawing a map of the spherical earth on a flat surface.

Any map projection has three variable properties: scale, area, and shape. *Scale* is the ratio of the size of an object on the map to the size of the object it represents in the real world. For example, consider a model globe. Suppose that this globe has a diameter of 25 cm (10 in). It represents the real earth, which has a diameter of 12,900 km (8000) mi). Therefore, 1 cm of the globe represents 12,900/25 km, or 516 km, or 51,600,000 cm, on the real earth. So we say the scale of the model globe

FIGURE 2-7

Construction of the gnomonic, stereographic, and orthographic projections. The light is placed in a different position relative to the globe for each of the three projections, creating three different shadows.

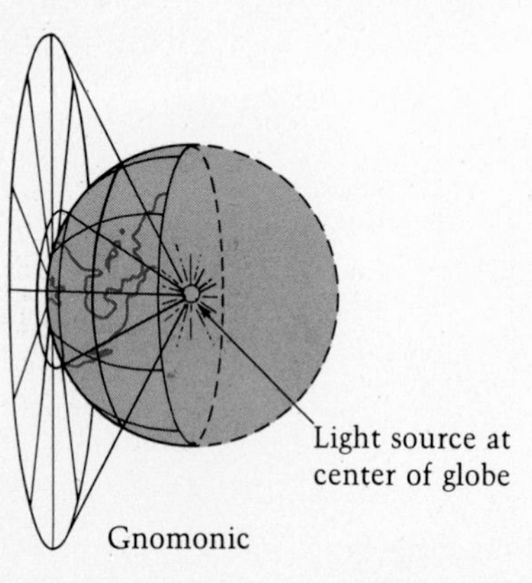

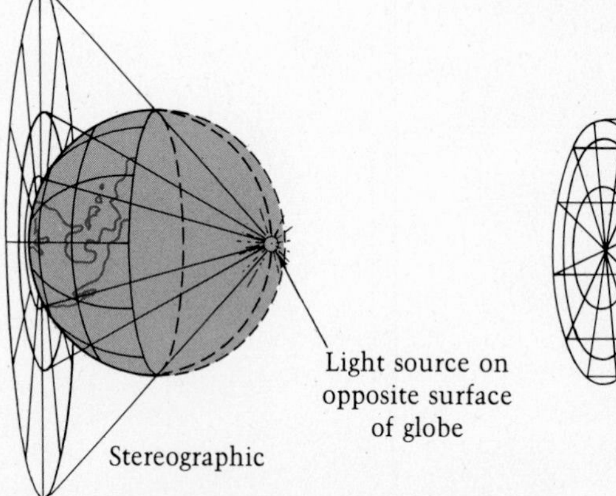

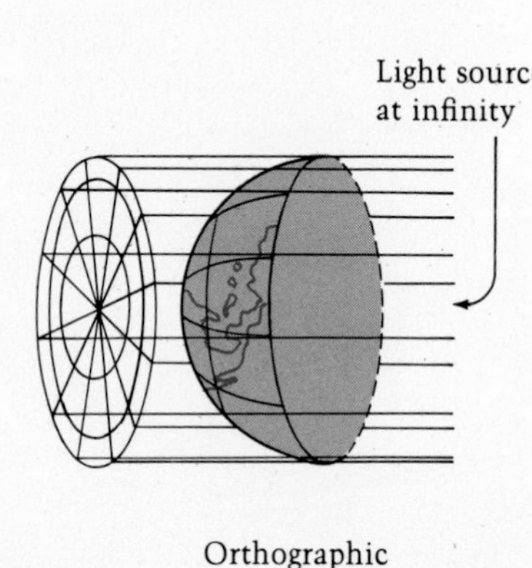

is 1 to 51,600,000 or 1 : 51,600,000.

The *area* of a piece of the earth's surface is found by multiplying its east–west distance by its north–south distance. Of course, the calculation is simple for rectangular pieces of land. In many map projections, real area can be well represented simply by scaling down the distances. Colorado has the same area relative to other states on the model globe as in the real world. The only difference is that the scale has changed.

Now consider *shape*. Shape can often be preserved in map projections—but not always. In the real world, the shape of Wyoming is almost a rectangle. On a map with a scale of 1 : 50,000 for north–south distances and a scale of 1 : 200,000 for east–west distances, Wyoming is squeezed in the east–west direction and stretched in the north–south direction. When a map projection preserves shape, it is said to be *conformal*.

In moving from a globe to a map projection, we can preserve one, or sometimes two, of the properties of scale, area, and shape. But it is not possible to preserve all three at once over all parts of the map. Try it for yourself by drawing a country onto a plastic ball and then cutting the ball to make a flat map. You will have to compromise. To solve this problem, the Greeks followed the rule we still use: Produce a map projection most suited to a particular purpose.

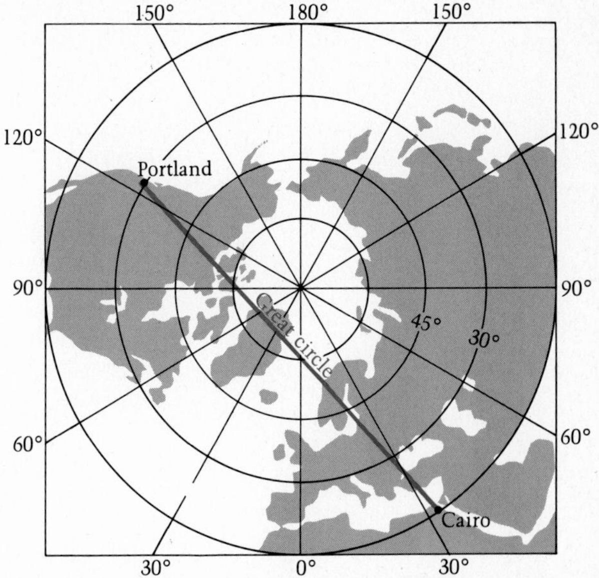

FIGURE 2-8

The great-circle route from Portland to Cairo. On a polar gnomonic projection, the great circle becomes a straight line.

The map projections developed in ancient Greece

The first map projection developed by the Greeks was the *gnomonic* projection, which appeared about 500 BC. The gnomonic projection is constructed by having an imaginary light at the center of the globe projecting outward onto a flat surface that touches the globe at one point. Hipparchus is believed to have constructed the *stereographic* and *orthographic* projections three centuries later. In the stereographic projection, the imaginary light is on the surface of the globe opposite the flat surface. In the orthographic projection, the imaginary light is regarded as being at infinity. Figure 2-7 compares these three projections.

Gnomonic, stereographic, and orthographic projections, which are centered around a point where the imaginary surface touches the globe and which have a wheellike symmetry, are called *zenithal* or, alternatively, *azimuthal* projections. They usually have a curved border and are best used to show half a globe, although they can be extended mathematically to show the whole globe. True compass directions are represented at the center point. Either of the poles, any point on the equator, or any particular point of latitude and longitude may be selected as the center point. Thus the zenithal projections may be called polar, equatorial, or tilted depending on where the flat surface touches the globe.

Although developed more than 2000 years ago, zenithal projections are popular today because of one special property they possess. A straight line on the gnomonic projection is the shortest route between two points on the earth, which has notable implications in this age of air travel. Intercontinental air traffic attempts to follow the shortest routes, and the shortest route between two points on the spherical earth is found by imagining the earth to be cut exactly in half along a line that runs through the two points. When a sphere is divided exactly in half, the circles formed by the two cut edges are called *great circles.* (If the sphere is cut into two unequal portions, then the circles formed on the edges of the cut are called *small circles.)* Air traffic follows great circles on what are called great-circle routes. Figure 2-8 shows the great-circle route from Portland, Oregon to Cairo, United Arab Republic. Although it passes over Greenland, it is depicted as a straight line on the polar gnomonic projection.

About 300 years after Hipparchus, a cartographer called Marinos of Tyre invented another new map projection. Instead of projecting meridians and parallels onto a flat surface, he projected them onto the inside of a cylinder that was later made flat, as demonstrated in Figure 2-9. When the cylinder was flattened, parallels and meridians appeared as straight lines intersecting at right angles.

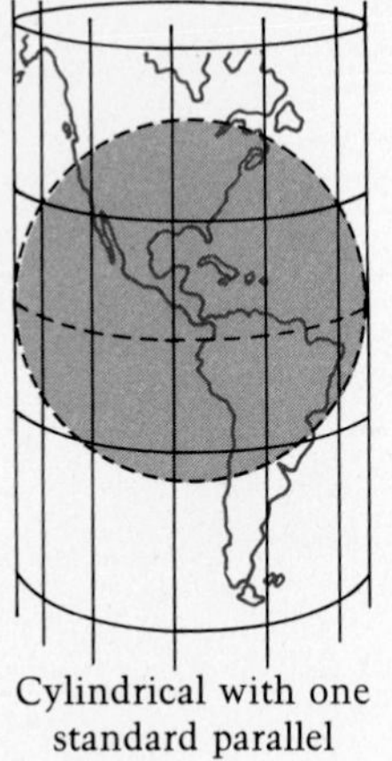

Cylindrical with one standard parallel

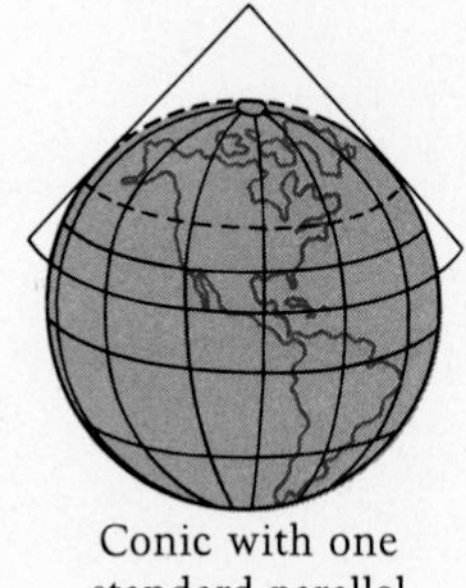

Conic with one standard parallel

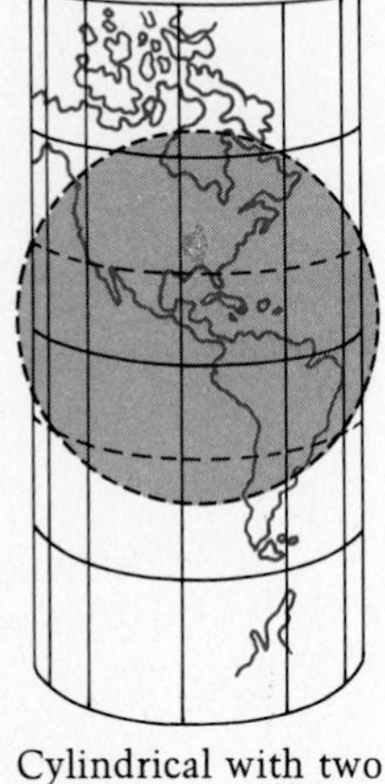

Cylindrical with two standard parallels

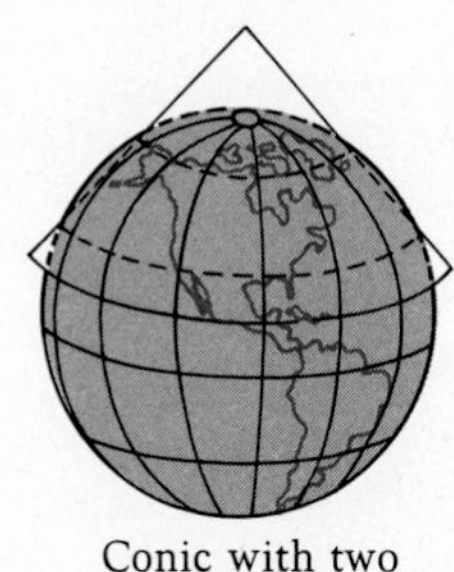

Conic with two standard parallels

FIGURE 2-9
The construction of cylindrical and conical projections with one and two standard parallels.

Marinos' chart was based on the latitude of Rhodes, in the Mediterranean Sea, where the cylinder was imagined to touch the earth. Today we call his projection *cylindrical* with one standard parallel. Along this parallel, the latitude of Rhodes, scale is maintained. But north and south of this parallel, the scale is progressively distorted by lengthening. Cartographers who followed Marinos tried to overcome this deficiency by imagining that the cylinder passes through the globe, thus creating true scale on two lines of latitude. This produces a cylindrical projection with two standard parallels, a kind of map that is useful for representing low-latitude areas.

The contribution of Ptolemy About 400 years after Eratosthenes, the head librarian at Alexandria was Claudius Ptolemaeus, a scholar better known as Ptolemy. Ptolemy lived during the height of Greek geographical knowledge. So many travels had taken place that Ptolemy could list the latitude and longitude of over 8000 places from the Canary Islands to China. He also suggested dividing the world map into large-scale sectional maps.

In addition, Ptolemy invented the *conical* map projection. For this (Figure 2-9), the meridians and parallels are projected onto a cone instead of a cylinder. Figure 2-10 shows that the projected surface, when flattened out, gives straight meridians and that the parallels are arcs of circles with the same center point. The scale is true where the cone touches the surface, but as with the cylindrical projection, it is distorted north and south of the standard parallel (or parallels, if two are used).

Some of Ptolemy's world maps had much more detail than is shown in Figure 2-10, but the figure still gives us a clear idea of the extent to which humans had mapped the earth by about AD 200. The ancient Greeks made an astounding number of discoveries. They conceived the idea of a spherical earth, made measurements of its circumference, divided it into latitude and longitude, produced map projections and maps of different scales, and established a world map that included large parts of Europe, Africa, and Asia. Significant new advances in cartography by Western countries were not made until another 1300 years had passed.

Mercator's map Around AD 1500 explorers based their knowledge on the maps drawn by Ptolemy. However, monarchs such as Prince Henry of Portugal soon were sponsoring exploratory voyages. Christopher Columbus, John Cabot, Pedro Cabral, and Amerigo Vespucci charted eastern ports of the Americas around the turn of the fifteenth century, and Ferdinand Magellan sailed into the Pacific Ocean in 1520.

In 1569 Gerhardus Mercator, a Flemish geographer, drew a map of the earth that looks much more as we know it today. Mercator's map was noteworthy not only for the emerging geography of the new world but also for the new projection he devised. The *Mercator* projection, shown in Figure 2-11, cannot be classified as zenithal, cylindrical, or conical and thus falls into a miscellaneous class that today contains many mathematically derived projections. Although Mercator evolved his projection empirically—by trial and error—it was soon explained mathematically.

Examination of the parallels and meridians in Figure 2-11 shows them to be straight and to intersect at right angles. The spacing of the parallels increases by specified amounts from equator to poles to give the projection two important properties. The first is that shapes around a point are correct, thus making the projection conformal.

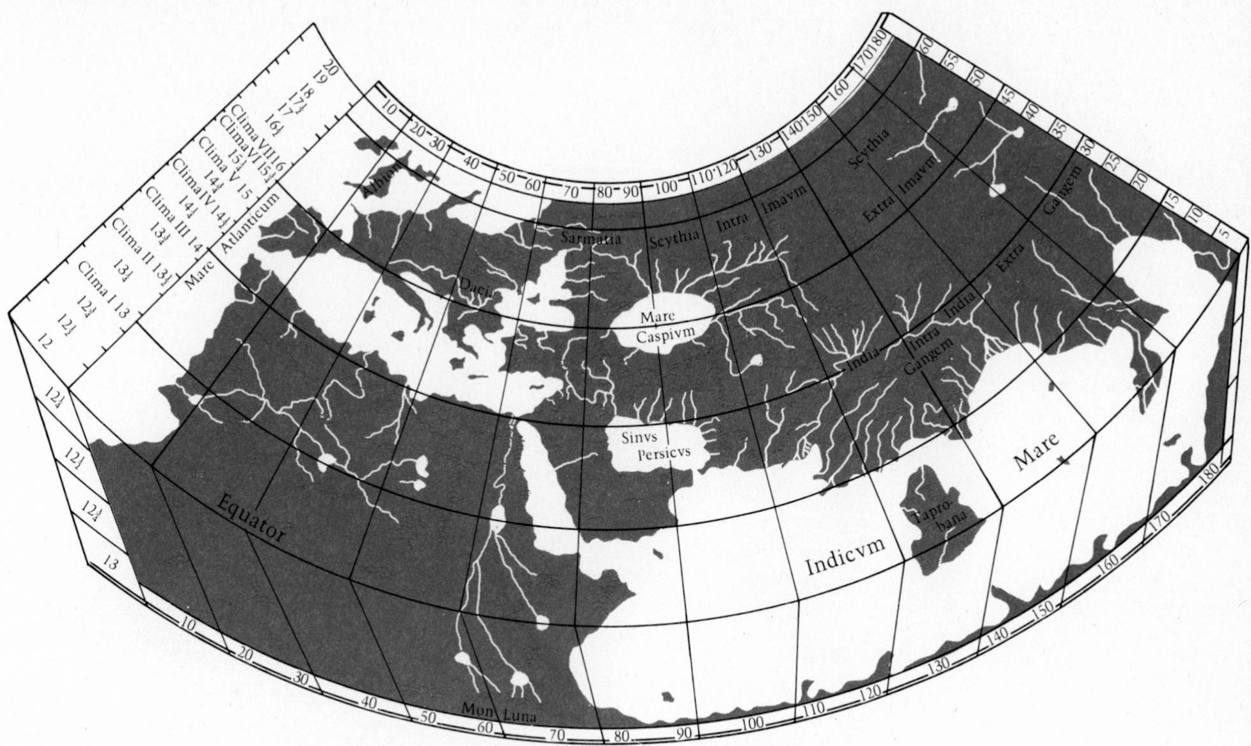

FIGURE 2-10

A redrawing of Ptolemy's world map on his simple conic projection.

The second property, unique to this projection, is that straight lines on the map are lines of true compass bearing. Such lines are called *rhumb lines* ("rhumb" meaning any of the points on a mariner's compass). These rhumb lines are of great importance for the navigator. Once the navigator has determined from the Mercator projection the compass direction he wants to travel, the ship can be kept constantly on this course.

MODERN MAPPING

By the end of the eighteenth century, some important technical and geographic discoveries had been made. The process of fixing the location of places by using compass bearings from two ends of a base line was refined. This system is called *triangulation*. By working out from one base line, whole countries and even continents can be accurately mapped.

Many new navigational instruments had been invented by 1800. The most important may have been a clock accurate enough to lose or gain less than three seconds in 42 days. This clock was fundamental in determining the longitudinal position of a traveler.

Numerous outstanding voyages of discovery and charting were undertaken by explorers. Each of these voyages was an adventure in itself. Captain James Cook was more responsible than anyone else for mapping the coastlines of the Pacific Ocean.

In the time before 1800, ground also was laid for the thematic, hydrographic, and topographic traditions of cartography. *Thematic* maps serve a specialized purpose or show a particular subject. A map of the wind patterns of the earth is a thematic map. *Hydrographic* maps are those of special use to sea travelers. *Topographic* maps show the relationship of places, usually on land, to one another.

These traditions still appear in modern mapping, and they have developed to a high degree of sophistication. In the past two centuries, mapping the earth has been characterized by an increasing degree of accuracy in the maps drawn, a great amount of organization in cartographic institutions (usually on the national and international level), and a vast increase in the techniques and technology available for producing maps.

Organizations for mapping Today almost every country has its national mapping organization. In the United States, mapping and surveying was officially started in 1785 as a result of a report by Thomas Jefferson to Congress. The Public Land Survey was started in eastern Ohio, where unsettled land was

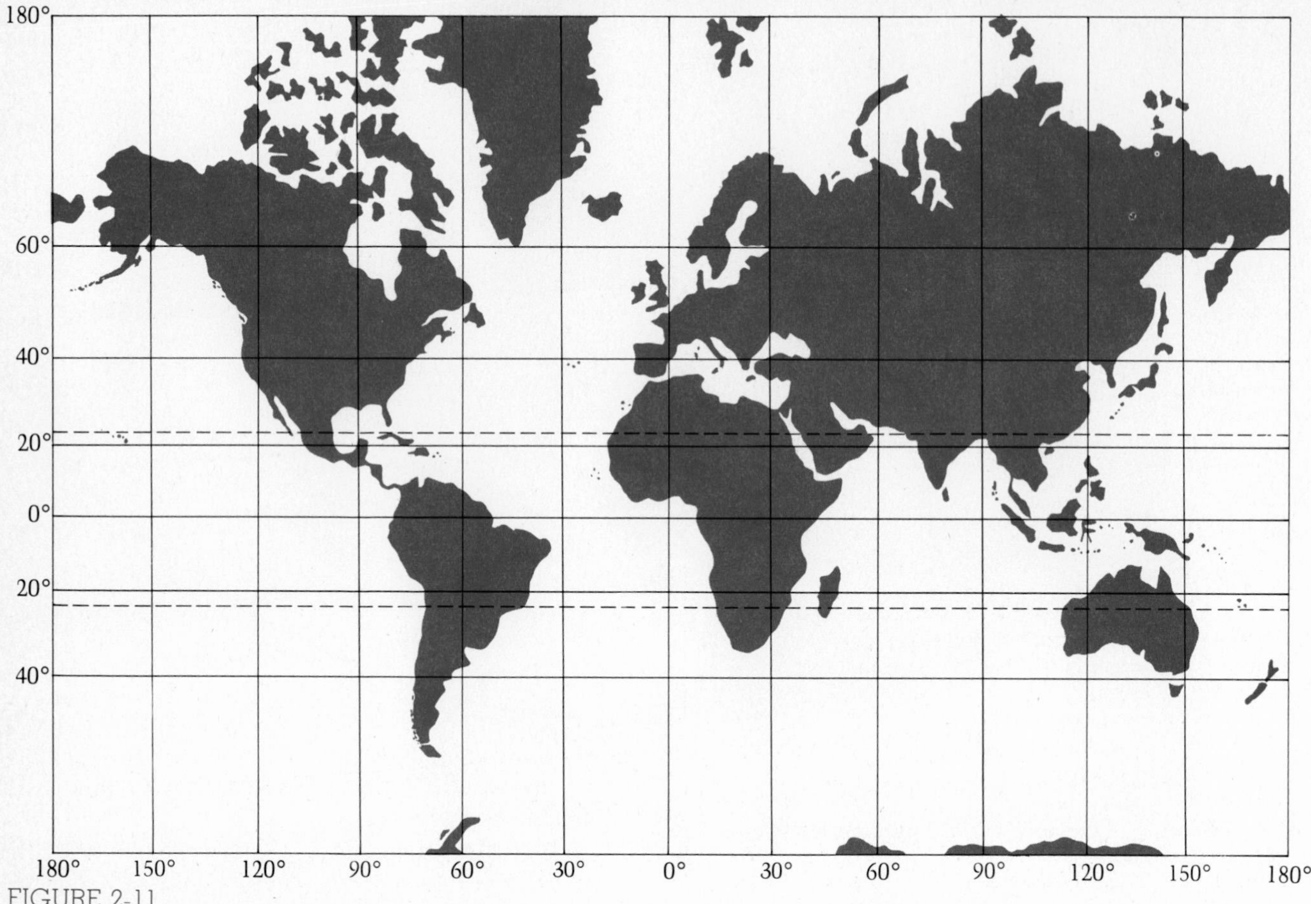

FIGURE 2-11

The Mercator projection. The projection produces straight parallels and meridians.

subdivided into mile-square sections. Townships, made up of 36 sections, were oriented along meridians and parallels. The cultural geography of most of the United States, apart from the east, is today heavily influenced by this north–south, east–west alignment. Many private and public maps based on the Public Land Survey were made for special legal and administrative purposes.

In 1879 the United States Geological Survey was founded. It became the principal national agency for producing topographic maps. The United States is so large that, after a hundred years of endeavor, only a quarter of it has been mapped with maps of standard quality.

On a world scale, even less progress has been achieved. A proposal for an International Map of the World (IMW) was made in 1891. It was to have a scale of 1 : 1,000,000. This has been supplemented by a World Aeronautical Chart (WAC) on the same scale. The development of both maps is now under the control of the United Nations. The WAC is complete, but the IMW charts are still lacking for some areas. The task of mapping the world is by no means finished.

The perspective from above The modern procedure of mapping the world is eased by powerful new tools. The Greek legend of Icarus, who flew too close to the sun and melted his wax-and-feather wings, shows how much the ancient philosophers would have liked to see the earth from the sky. Today it is possible.

One of our tools is *remote sensing* technology, the ability to scan the earth from remote manned or unmanned observation stations. Aerial photography falls into this category. Several methods have been developed that permit maps to be produced directly from series of photographs taken from aircraft. An even more significant extension of this technology is the photography possible from orbiting satellites or space stations. (You can get some idea of the beauty and importance of such photographs from the southern California photograph in Chapter 1.) Satellite photography can give information about the tempera-

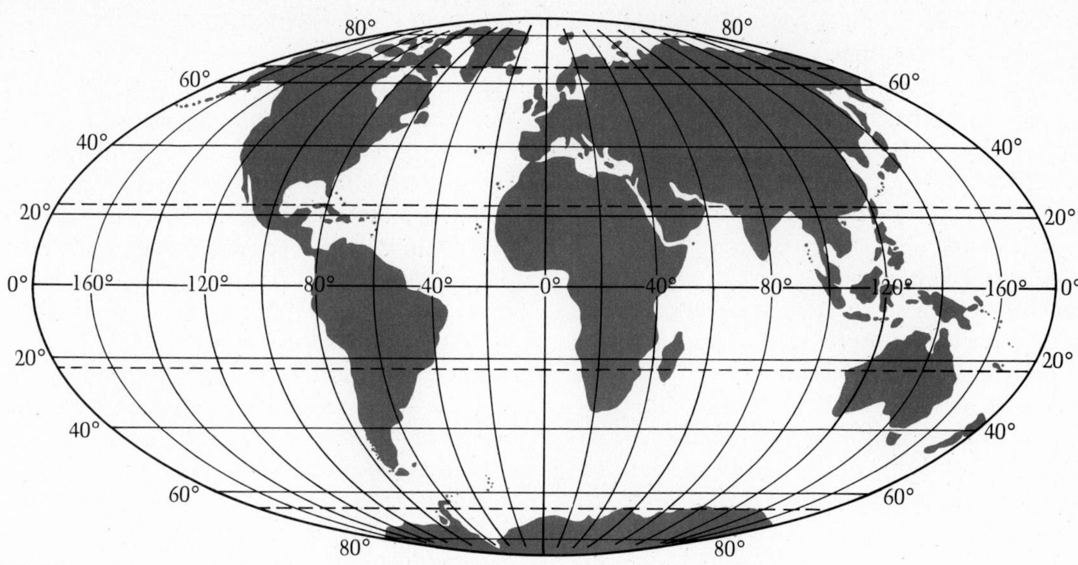

FIGURE 2-12

Projections based on mathematical principles. a) The Mollweide homolographic projection is distorted only in the polar areas.

tures of the earth and its atmosphere, snowfall and rainfall, geologic activity, and land use and vegetation cover. Radar also is used, for detecting rainfall and for "seeing" through clouds to the earth below. Even snow and ice depths can be measured with airborne devices.

Mathematical projections The increased technology for gaining information has accompanied a great development in the techniques used to produce maps. Computers can now be programmed to draw maps directly from raw data. In more traditional areas of cartography, there are many new methods for representing analyzed data and for constructing new map projections.

Two of the most widely used projections fall into the miscellaneous class, which is based on mathematical principles. The *homolographic* projection in Figure 2-12a was developed by K. B. Mollweide in the nineteenth century. It is an equal-area projection, in which parallels are straight and all the meridians except the central one are halves of ellipses. It suffers, however, from distortion in the polar areas.

b) Goode's interrupted homolosine projection, which combines sine curves for meridians and the homolographic projection, reduces the distortion.

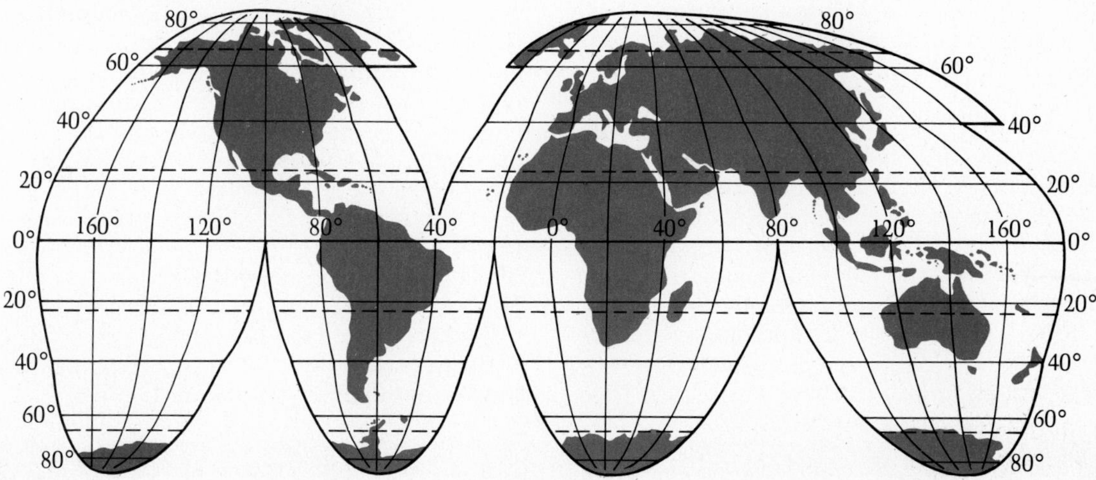

The *homolosine* projection in Figure 2-12b, which is also an equal-area projection, was invented by Paul Goode in 1923. It was one of the first projections to combine two other projections. Between the fortieth parallels sine curves instead of ellipses are used for the meridians. If we compare Figures 2-12a and 2-12b, we can see that, whereas an ellipse is curved throughout its length, a sine curve almost becomes a straight line in part of its length. Poleward of the fortieth parallel, the homolographic system is adopted. Polar distortion is overcome to some extent by centering important land areas to their own central meridian and fitting the parts together. When a projection is broken into parts, as in Figure 2-12b, it is said to be *interrupted.*

The story of map making from Greek times to the modern day runs alongside another one, the discovery of the earth's relationship to other objects in the solar system. This is no less exciting, but here we will only review what is known at the present time.

THE EARTH IN THE UNIVERSE

Naturally, we know much more about our earth than any other heavenly body. For example, we know that the surface of the earth consists of 71% oceans and 29% land. We also know that the earth bulges slightly at the equator because of centrifugal force. The spinning of the earth forces equatorial matter outward. The equatorial diameter of 12,757 km (7927 mi) is only 43 km (27 mi) greater than the polar diameter of 12,714 km (7900 mi). Surface irregularities, such as mountains and ocean trenches, are quite small compared with the diameter of the planet, ranging only 19.5 km (12.1 mi) between the top of the highest mountain and the bottom of the deepest ocean.

The earth has one satellite, the moon. The moon and the earth are attracted to each other by a force known as *gravity.* Sir Isaac Newton discovered that this invisible attractive force can be determined by dividing the product of the size of the mass of the two bodies by the square of the distance between them. Gravity keeps the moon in orbit around the earth rather than allowing it to shoot off into space. The same force acts between the moon and the oceans on the earth to create ocean tides.

Gravity also keeps the earth revolving around its star, the sun. Of nine planets revolving around the sun, ours is the third closest, as Figure 2-13 shows. The earth

THE DYMAXION AIR-OCEAN WORLD PROJECTION

Each new map projection offers us a new view of the world. Most projections are developed to achieve a specific kind of representation of the earth's surface; each is a compromise between specific degrees of accuracy and distortion. A new map projection refocuses our view of the earth.

Most new map projections are developed to meet a specific "practical" purpose. Yet some also combine a philosophical side with the utilitarian. They reflect not only a purpose but also the cartographer's concept of our planet as the human environment.

R. Buckminster Fuller, the renowned American designer/planner/architect, is famous primarily for his geodesic domes. These distinctive structures are based on the triangle, a geometric form with exceptional strength. The Fuller domes have appeared at dozens of world fairs and expositions and are used throughout the world as homes, observatories, convention halls, public buildings, and storage facilities.

In 1954, Fuller copyrighted a projection that clearly reflected his own philosophy. It was a decade of developing jet travel, and Fuller wanted a view of the earth that emphasized not just the earth's surface, but the totality of the earth-air-ocean system in which humankind was traveling at increasing speeds.

The projection is based on the icosahedron, a 20-sided figure, each face of which is a triangle. Within each triangle, the projection is completely distortion-free. Spread out flat, the map shows most of the major landmasses along one side, and most of the bodies of water along the other side. Because of the lack of distortion, it is possible to attach consistent time/distance values to each edge of each triangle. Each edge is equivalent to:

(1) 3,806 nautical miles;
(2) 8½ hours by jet (in 1954);
(3) 14 hours by conventional plane;
(4) 7 days by ship.

If you're curious to have one of these maps, you can order from R.B. Fuller Dymaxion Maps, 3500 Market Street, Philadelphia, PA, 19104.

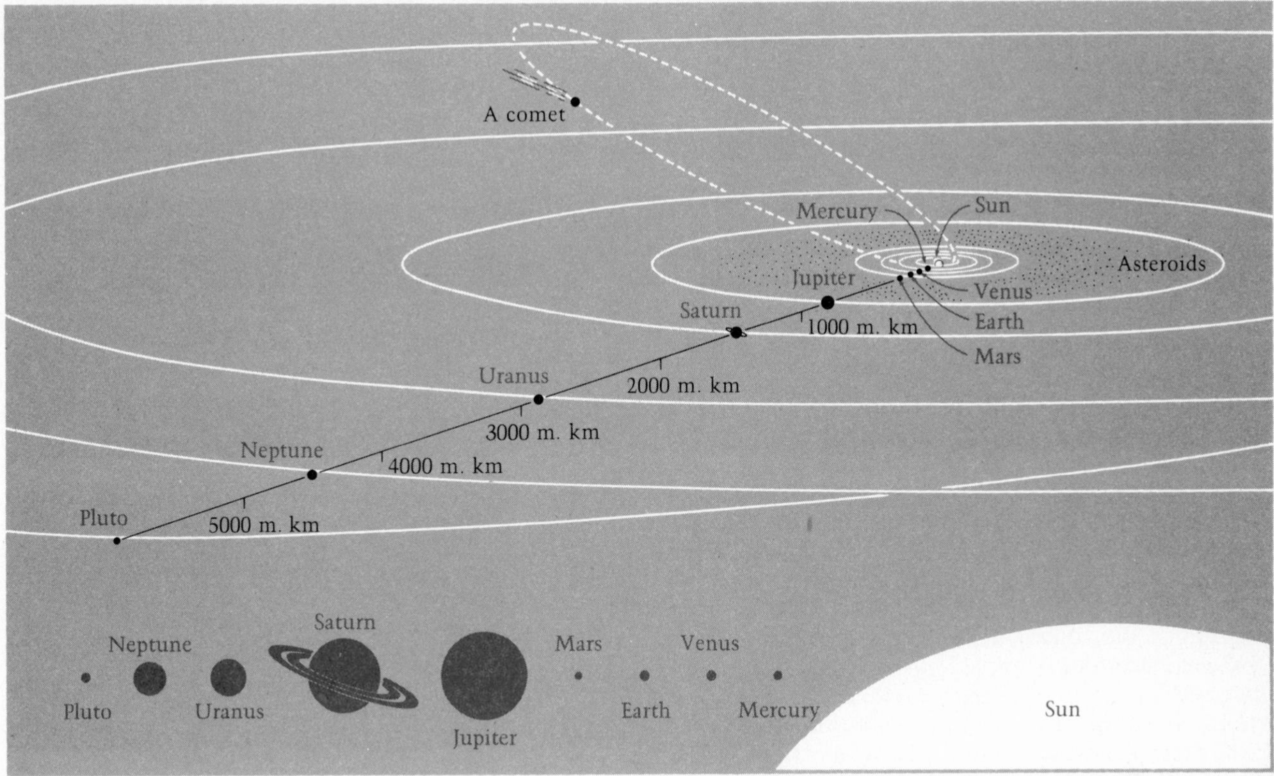

FIGURE 2-13

The relative sizes of the planets and their orbits around the sun.

is the fifth largest of these planets and revolves at an average distance of 150 million km (93 million mi) from the sun. Many of the other planets have satellites like the moon. In addition, relatively small planetlike bodies that usually orbit the sun between Mars and Jupiter are known as *asteroids. Comets,* whose orbits do not often cross the paths of the planets, are bodies of frozen gases and other materials. These are the basic elements of our *solar system:* one star (the sun), nine planets and their 32 satellites, thousands of asteroids, and billions of comets.

The sun is so large that it contains nearly 99.9% of all the matter in the solar system. Yet the sun is a rather small star, just one of 100 billion others in a star system known as a *galaxy.* Our galaxy, called the Milky Way, is one of more than a billion others within the limited range of our telescopes.

We can get some idea of the enormity of the universe by considering time and space together. The speed of a ray of light is 300,000 km per second (186,000 mi per second). Therefore, a ray of light could travel a distance equal to 7½ times around the earth in one second. Even at this speed, it takes a ray eight minutes to travel from the sun to the earth. Light from the nearest star takes 4 years to reach us. If the nearest galaxy exploded, we would not know it for another 2 million years, because it takes light that long to reach the earth. Our telescopes show us history.

These facts can leave no doubt that we humans are comparatively small and unimportant. Early civilizations coped with the enormity of the universe by associating parts of its events with gods who were assumed to act in human ways. The most important gods were the most obvious—the sun and the moon. Activities of the gods also were thought to govern the seasons, periodic phenomena that vitally affect all human activities.

THE SEASONS AND THE EARTH

The word "season" originally meant a time of sowing. Yet even before humans began cultivating crops, they were certainly aware of the periodic times of heat, cold, rain, and drought and of the way plants and animals adapted to these times. Some of the earliest attempts to understand the seasons may still be seen in such structures as Stonehenge in England, which is

THE UNIVERSE OF THE YUROK

Is the earth flat, round, or square? Humans have always wondered about the shape of the earth and its position in the universe. Each group bases its description on what they can see, just as we do.

The Yurok Indians of northern California believe that the world is a flat disk surrounded by water. The Klamath River bisects their earth. The center of the world lies at a point along the banks of the river. Here, the sky was created. The sky is a solid dome. A ladder reaches from the earth to the sky country above the dome. Below the earth lies the land of the dead. Many cultures share the idea of a lower region for the dead. The Yurok Indians imagine that the path to this lower realm is through a lake.

Because the Yurok are fishermen who rely on the Klamath River for food and transportation, it seems appropriate that the river is prominent in their portrait of the universe. The river cuts their world into two symmetrical parts. The Indians do not orient themselves by the directions north, south, east, and west as we do. Instead they refer to places as upstream or downstream. Because the river winds crookedly through the territory, upstream could be almost any point on the compass, but the tribe understands the predominant movement of the water.

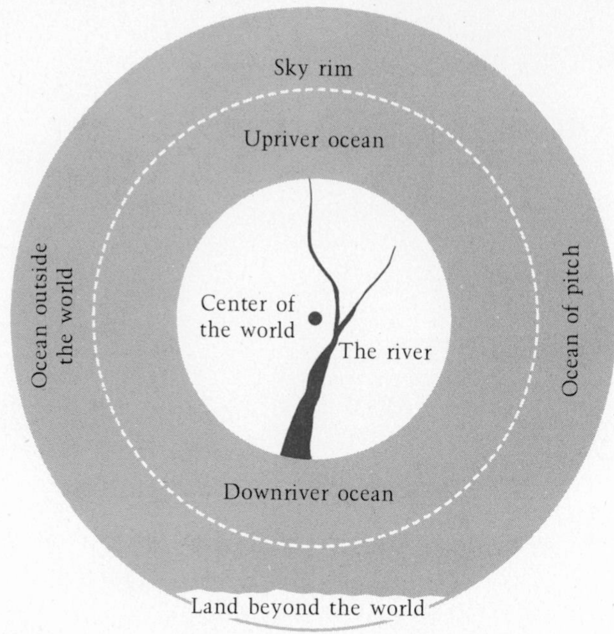

Adapted from Yi-Fu Tuan. *Topophilia*. Englewood Cliffs: Prentice-Hall, 1974, p. 36.

pictured in Figure 2-14. Many temples and other structures had both religious and astronomical significance. But real understanding of seasons had to wait until after the sixteenth century, when it was finally accepted that the earth revolves around the sun.

ORBIT AND AXIS

The earth revolves around the sun in an elliptical orbit, a path where the earth is closer to the sun at some points in the year than at others. About January 3 of every year, the earth is closest to the sun, a position

FIGURE 2-14

Stonehenge in England. The original structure consisted of two circles of stones encircling two other series of stones in a horseshoe shape. The positions of the stones predicted the days of high and low sun and the positions of moonrise and moonset.

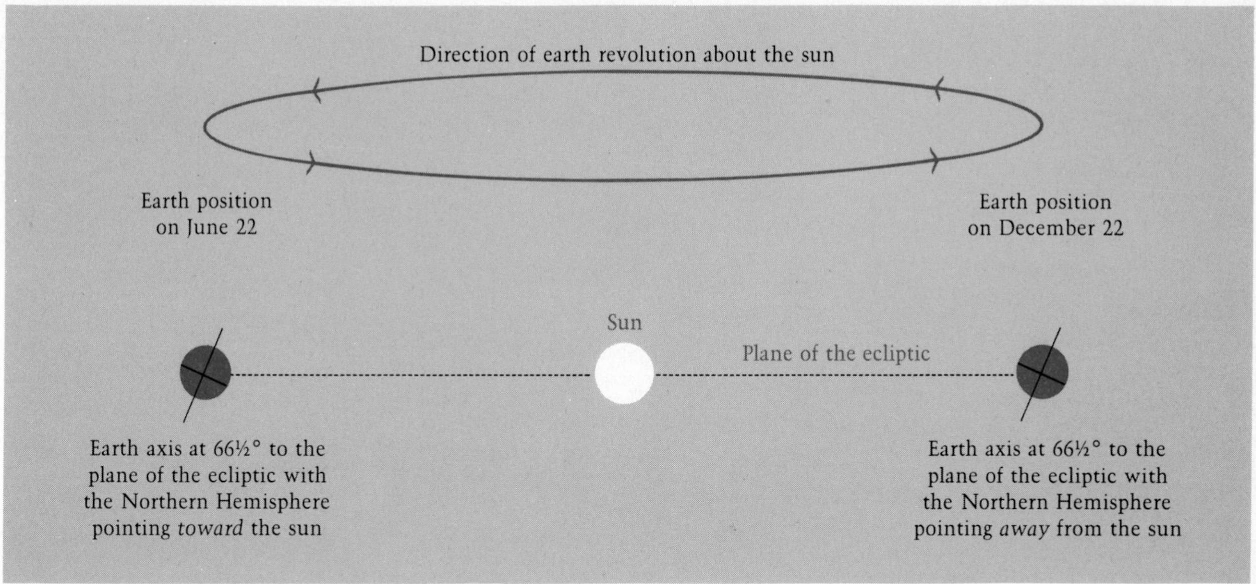

FIGURE 2-15

The extreme summer and winter positions of the earth with regard to the sun. The earth's axis is tilted at the same angle throughout the year.

called *perihelion*. The distance between the sun and the earth at perihelion is 147 million km (91.5 million mi). The farthest the earth gets from the sun, the *aphelion*, is 152 million km (94.5 million mi). This occurs about July 4.

The earth takes 365¼ days to travel around its orbit, and this traveling time defines the length of a year. In our calendars we call 365 days a year and add an extra day every fourth year (a leap year) to allow for the extra quarter day every year.

If you draw onto a flat piece of paper the orbital path of the earth around the sun, the paper can be described as a plane. The actual plane in space of the earth's movement around the sun is called the *plane of the ecliptic*. The seasons occur because the earth is tilted with respect to the plane of the ecliptic.

An imaginary line drawn through the earth from the North Pole to the South Pole is called the earth's *axis*. The earth rotates around this axis, giving us night and day. The axis of the earth is always tilted at an angle of 66½° to the plane of the ecliptic, no matter where the earth is in its orbit. Thus at one point in its orbit, on June 21 or 22, the northern half of the earth, the Northern Hemisphere, is tilted toward rays of light coming from the sun. At this time, the Northern Hemisphere receives a much larger amount of heat and light than the Southern Hemisphere does. When the earth has moved to the opposite part of its orbit six months later, on December 22 or 23, the Northern Hemisphere points away from the sun and receives the least heat and light. This accounts for the seasons of heat and cold, summer and winter. Figure 2-15 shows how these seasons occur at opposite times of the year in the Northern and Southern hemispheres.

It is worthwhile for us to examine these extreme positions in more detail. Figure 2-16 shows that on June 21 or 22 parallel rays from the sun fall vertically on the earth at latitude 23½° N. This latitude is given the name *Tropic of Cancer*. The fact that Eratosthenes' vertical pole at Syene was near this latitude accounts for the lack of shadow thrown at midday. We can also see that areas north of latitude 66½°N, the *Arctic Circle*, remain in sunlight during the earth's 24-hour rotation. If a vertical pole were placed at the equator at this time of the year, the sun would appear to be northward of the pole, making an angle of 23½° with the pole and an angle of 66½° with the ground.

Six months later, the position of the earth relative to the sun causes the sun's rays to fall vertically at noon at 23½°S, the latitude called the *Tropic of Capricorn*. The other relationships between earth and sun are exactly reversed, so that areas south of the *Antarctic Circle*, at 66½°S, receive 24 hours of sunlight. Regions north of the Arctic Circle are in perpetual darkness.

SOLSTICES AND EQUINOXES

To us on earth, it appears that the midday position of the sun gets lower in the sky as the seasons progress from summer to winter. If you were to plot the position of the sun throughout the year at a point

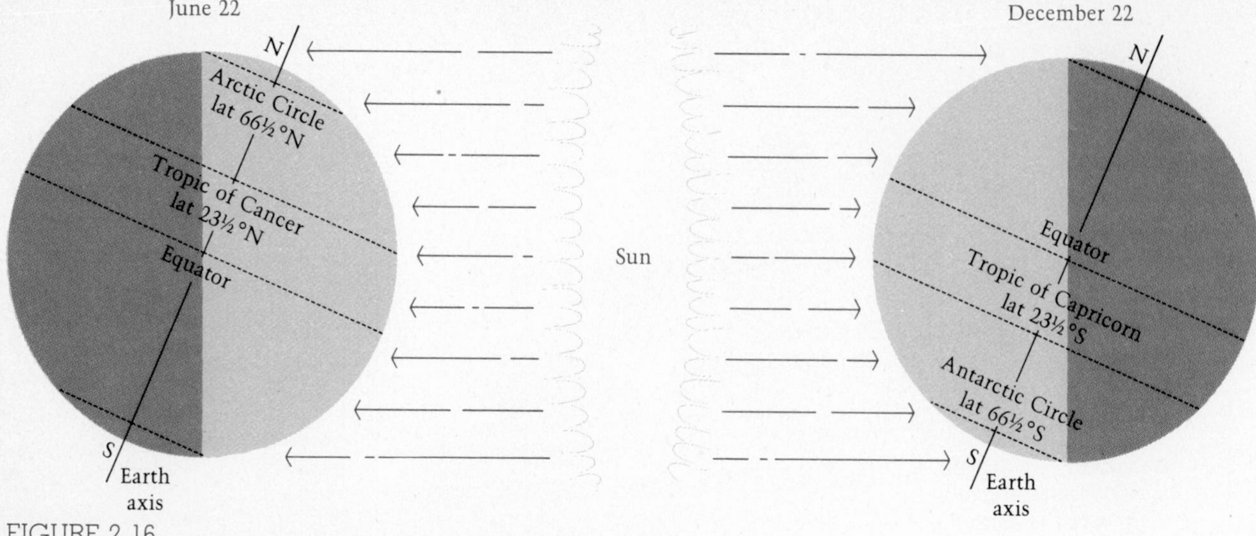

FIGURE 2-16

The relative positions of the earth and the sun on June 22 and December 22. The sun's rays strike the earth at different angles throughout the year.

north of the equator, it would seem to climb higher and higher until June 22, when it would appear to stop. Then it would move lower and lower, until it stopped again at December 22, before once more beginning to climb. South of the equator the dates are reversed, but the phenomenon is the same. The Greeks plotted the apparent movement of the sun and called the points at which the stops occurred solstices ("sun standstill"). Today we use their word, calling the earth–sun position of June 21 or 22 the *summer solstice* and that of December 22 or 23 the *winter solstice*.

You can simulate the movement of the earth around the sun in a darkened room with a light at the center. Carry a globe around the light, with the globe's axis always pointed toward some imaginary position a long way above the ceiling of the room. You can simulate night and day by spinning the globe as you carry it around. Under these conditions, you can clearly see the positions of the summer and winter solstices. The Arctic Circle has 24 hours of sunlight during the summer solstice and 24 hours of darkness during the winter solstice. This happens at opposite points on the earth's path.

Halfway between these two points there are two positions where the spinning globe receives 12 hours of sunlight and 12 of darkness at all latitudes. These positions occur on March 21 and September 23. Because of the equal lengths of night at all latitudes, the positions are called *equinoxes*, which is a Latin word meaning "equal night." At these times the sun's rays fall vertically over the ground at the equator, and the sun sets and rises due west and east. The equinox of March 21 is called the *spring* or *vernal equinox*, and that of September 23, the *autumnal equinox*.

You can achieve a clear idea of the causes of the seasons if you imagine you are looking down on the earth's orbit around the sun from a point high above the solar system, a position depicted in Figure 2-17. The North Pole always points to your right. At the summer solstice, the Arctic Circle receives sunlight during the entire daily rotation of the earth, and all parts of the Northern Hemisphere have more than 12 hours of sunlight. These areas collect a large amount of solar energy in the summer season. At the winter solstice, the area inside the Arctic Circle receives no sunlight at all, and all parts of the Northern Hemisphere receive less than 12 hours of sunlight. The winter is thus a time of cooling. However, at both the spring and autumn equinoxes, the Arctic Circle and the equator are equally divided into night and day. Both hemispheres receive an equal amount of sunlight and darkness, and energy from the sun is equally distributed.

Thus the annual revolution of the earth around the sun and the constant tilt of its axis give our planet its different seasons of relative warmth and coldness. (Chapter 7 shows that the seasonal periods of rain and drought are also associated with this annual rhythm.) Both the rotation of the earth itself and its revolution around the sun have constituted one of our most secure reference points for time. Let us look more closely at some aspects of time that are important in the study of our planet.

KEEPING TRACK OF TIME

One of the most obvious ways to start dealing with time is to use the periods of darkness and light resulting from the daily rotation of the earth. One rotation of the earth, one set of daylight and nighttime hours, constitutes one day. The idea of dividing the day into 24 equal hours dates from the fourteenth century.

The most common procedure for defining how long it takes the earth to make one rotation is to use the period of time between the sun's highest points in the sky on two successive days. Unfortunately, this period is variable throughout the year, because the earth's orbit is not circular and because the sun appears to move along the plane of the ecliptic rather than a plane defined by the earth's equator. As a result, the sun appears to be 14 minutes fast in mid-February and 17 minutes slow at the beginning of November, with other variations between these two extremes.

This variation can be corrected mathematically by applying an equation to tell us how fast or slow the sun is on any particular day. The equation is called the *equation of time.* To apply this equation, we define "mean sun" as a sun that moves at a rate equal to the average rate of the real sun. The rotation of this mean sun is called 24 hours, or a *mean solar day.* Most of our activities, watches, and clocks are attuned to the mean solar day.

Some scientific purposes require measuring a day by the passage of the stars, rather than the sun, across any given meridian. Because the earth is moving around the sun but is fixed in relation to the stars, measuring a day by the stars is different from measuring it by the mean sun. The length of a day determined by the stars is 23 hours 56 minutes of the mean solar day. We call this period a *sidereal day,* and time based on it is *sidereal time.* Sidereal is a Latin word meaning "star"

FIGURE 2-17
The march of the seasons as viewed from a position above the solar system.

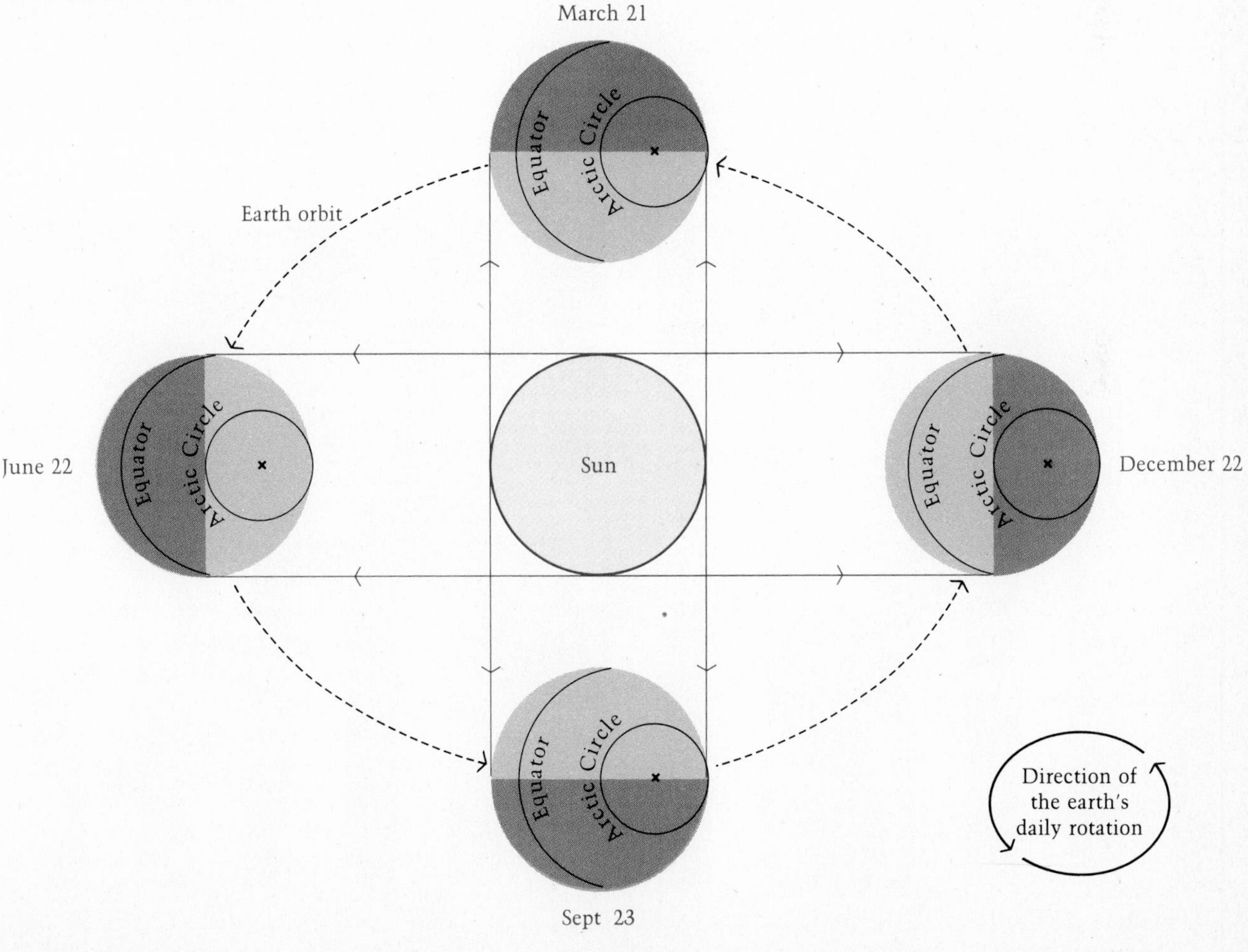

or "constellation," and in practice the first point of the constellation Aries is the star used for sidereal time calculations.

Another problem in measuring time arises because the sun is rising in one part of the earth as it sets in another. Mean solar time varies by four minutes at each degree of longitude. This was first seen to be a problem when sailing ships began to make trans-oceanic voyages, and by the time railroads crossed the United States, it was apparent that some system of time organization among different regions was needed. Before describing the solution used today, we will consider the problem of the sailing ships, because it relates closely to the longitudinal aspects of space on earth.

In a voyage such as that of Columbus in the *Santa Maria*, it was always relatively simple to find out the latitude of the ship. Columbus had only to find the angle of the sun at its highest point during the day. Then by knowing what day of the year it was, he could calculate his latitude from a set of previously prepared tables giving the angle of the sun at any latitude on any particular day. But, it was impossible for him to estimate longitude. To estimate longitude, he would have to know accurately the time difference between the time at some agreed meridian, such as that passing through Madrid, and the time at the meridian where the *Santa Maria* was located. Until about 1770, there was no portable mechanical clock that was accurate enough to keep track of the time difference. The accuracy needed to find longitude to within half a degree, which is equal to 55 km (34 mi) at the equator, is plus or minus three seconds per day on a six-week voyage. Following the invention of a sufficiently accurate timepiece, navigation became much less hazardous.

TIME ZONES

The problem of having different times at different longitudes was finally resolved at an international conference on October 22, 1884 at Washington, DC. Here it was decided that all earth time should be standardized against the time at a prime meridian of 0° longitude, which passed through the Royal Observa-

FIGURE 2-18
Time zones of the earth.

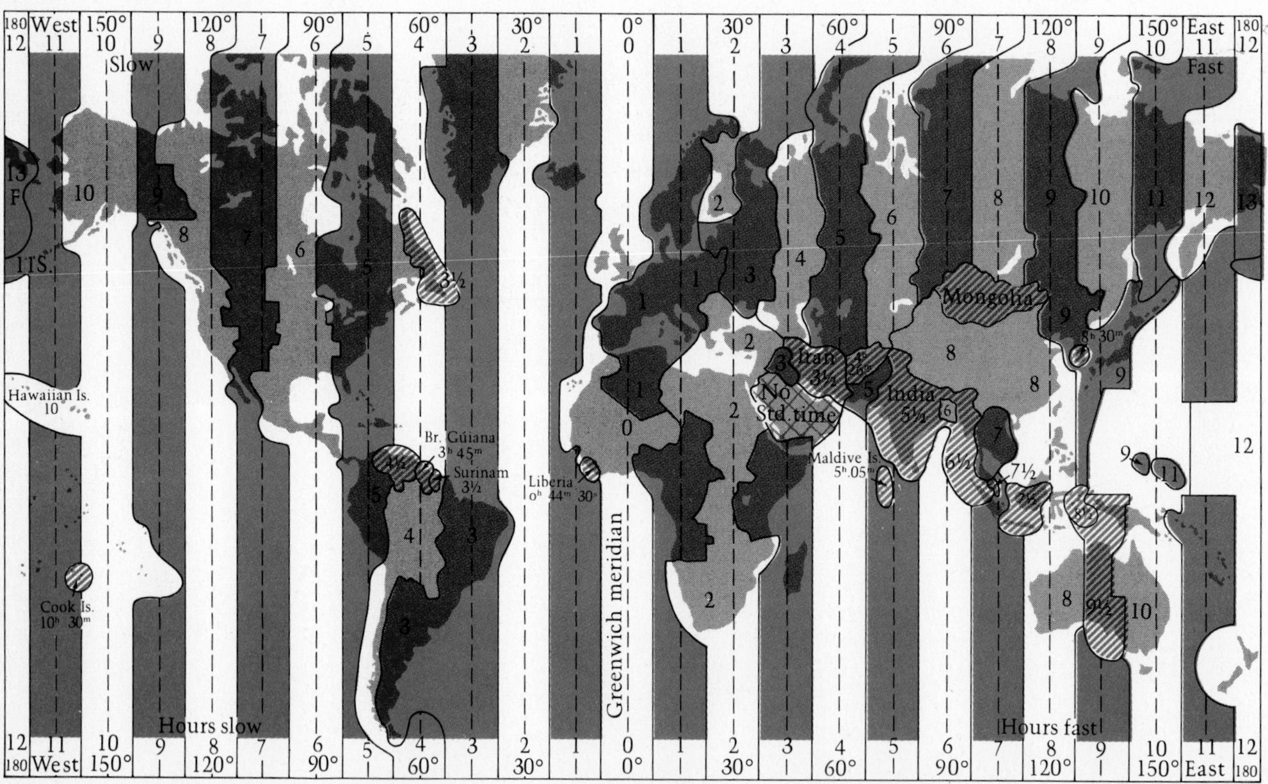

JET LAG

Does an airline passenger who lives in New York feel more tired after a flight to San Francisco or to Lima? Although the distances are similar, on the flight to Lima he or she stays in the same time zone; on the flight to San Francisco he or she crosses three time lines. In a rapid transition across time zones such as a plane flight from New York to San Francisco our body rhythms fall out of step with the time at our destination. This causes tiredness and irritation for a few days until our body can adjust to the new time. The medical name for this condition is flight dysrhythmia. Travelers call it jet lag.

The study of jet lag became an important medical interest after World War II, when high-speed airline travel became commonplace. Indeed, one of the first systematic studies was performed in the early 1950s and studied the effects of jet lag on the natural hormonal cycles of airline personnel.

When we suffer from jet lag we become uncomfortably conscious of how closely our bodies are tuned to the movements of the earth. Our daily and monthly cycles are biorhythms. Humans and other animals have adapted so completely to the daily pattern of light and dark that even in a cave or sealed environment their bodies continue to function on a fixed cyclical schedule.

We may not be aware of how many of our biological processes follow a daily pattern. Our temperature, for instance, rises and falls in a regular cycle. Our liver and kidneys adhere to predictable schedules of activity and rest. This means that medicine may affect us quite differently at 8 AM than at 8 PM. Even our senses of smell, taste, and hearing become sharper at the same time every evening. No wonder a flight from New York to San Francisco is so exhausting. Doctors believe that our body needs about one day to adjust to each hour of time change.

---

tory at Greenwich near London. The earth was divided into the 24 time zones shown in Figure 2-18, each using the time at standard meridians located at intervals of 15° of longitude. Each time zone differs by an hour from the next, and the time within each zone can be related in one-hour units to time at Greenwich, which was designated *Greenwich Mean Time.* When the sun rises at Greenwich, it has already risen in places east of the observatory. Thus the time zones to the east are called *fast.* Time zones to the west of Greenwich are said to be *slow.*

This solution leads to a peculiar problem. For example, if it is noon at Greenwich on January 2, 1977, it will be midnight on January 2 at 180°E and midnight on January 1 at 180°W. However, 180°E and 180°W happen to be the same line, and this was assigned the name of the *international date line* by the Washington conference. It was agreed that travelers crossing it in an eastward direction, toward North America, should repeat a calendar day and that those traveling westward across it, toward Asia, should skip a day. The international date line did not pass through many land areas, thereby avoiding severe date problems for people living near it. Where it did cross land, the line was arbitrarily altered to pass only over ocean areas. Similarly, some flexibility is allowed in the boundaries of other time zones to incorporate such factors as state boundaries. Some countries and states choose to have standard times differing by a half or quarter of an hour from the major time zones. These variations are shown in Figure 2-18.

A further arbitrary alteration of time zones is the adoption in some areas of *daylight-saving time.* In this, all clocks in a time zone are set forward by one hour from standard time. The reason for this adoption is that many human activities start well after sunrise and continue long after sunset, thus usually using energy for lighting and heating. Energy can be saved by setting the clocks ahead of the standard time. When a daylight-saving time of two hours was first introduced in Britain during World War I, there was much criticism for interfering with "God's own time" by people who did not realize that God's own time was established in Washington, DC in 1884.

TIME AND THE EARTH

Throughout history, there have been two important differences in the way humans have conceived of time. Some groups, such as the Mayas of Central America, who had a calendar more accurate than our own, believed that time was *cyclic.* History was expected to repeat itself in 260-year cycles. Other groups, including modern Western societies, believe that time is *linear.* There is a constant progression from past, through present, to future. Time never repeats itself. Actually, both of these concepts are useful to geographers.

This book contains many examples of the cyclic

nature of physical events. A *cycle* is one performance of an action, or series of actions, of or on an object so that the state of the object is the same at the beginning and end. One swing of a pendulum backward and forward constitutes a cycle. The daily rotations of the earth and revolutions of the earth around the sun lead to countless cyclic events. On a different scale, as other chapters show, the process of sculpturing the earth's landforms appears to repeat itself constantly. The land surface is continually being upthrust and worn away. The founder of the modern study of landforms, James Hutton, first realized this in 1785 and could see in earth processes "no vestige of a beginning—no prospect of an end."

The earth is also the scene of unrepeatable events which suggests a linear nature to time. Large-scale weather patterns have never exactly repeated themselves, nor is it possible to predict the exact speed of the gust of wind that may buffet you on a street corner tomorrow. Even the length of one day does not stay the same. Studies of fossil corals show that 600 million years ago the length of the earth "day" was less than 21 hours. In the future it is likely to be longer as the friction of the tides gradually slows the earth's rotation.

The nature of the earth is such that we have to learn to accept whatever concept of time is most useful. It is best for the student of the earth to regard time as cyclic in some cases and linear in others.

TIME AND SPACE

Apart from the notions of cyclic and linear time, many other curious events have happened to the concept of time. A strongly linear view of time was once put forward by the English archbishop James Ussher, who calculated in 1650 that God created the world at 9 AM on Sunday, October 23, 4004 BC. Albert Einstein proved that time would stop if it were measured by a clock that traveled at the speed of light. Many modern physicists now believe that there is some similar form of linear cosmic time in the universe. However, in studying the earth, the most useful concept to have been refined in recent years is that of the relationship between time and space.

At a scientific gathering in Cologne in 1908, Hermann Minkowski pointed out that "nobody has ever noticed a place except at a time, or a time except at a place." He advocated an absolute "world" that was later called *space-time*. The scene of real events is not space, which can be located by measuring distances along three directions or dimensions, but a four-dimensional world in which space and time are linked together inseparably. We have seen many examples of this link in this chapter. It has taken at least a 4000-year time period to explore and map the space of the earth. The seasons that mark time depend on the spatial relationship of the earth and the sun. The correct estimate of longitude needs an accurate assessment of time. Thus the elements in our world seem pervasively related. The remainder of this book describes in greater detail the individual spheres of the earth and the interrelated events that operate within them.

SUMMARY

We have explored the concepts of space, seasons, and time as they relate to the study of physical geography. We have followed the story of how the earth is mapped, from ancient times to the present. A large part of this history concerns the problem of how to represent the spherical surface of the earth on a flat surface. During the last 2500 years, map projections of various kinds have been developed to address this problem. The physical geographer is mainly concerned with the spatial relationships on the surface of the earth.

Seasons, also important to the physical geographer, are caused by the spatial relationships between the earth and the sun and by the tilt of the earth's axis. Seasons are of interest to the student of the earth not only because they give rise to times of maximum and minimum energy input, but also because they mark the passage of time. We have examined the organization of time zones on the earth's surface and have also noted the linear and cyclic concepts of time. Throughout the chapter, there have been many examples of the links among space, seasons, and time.

QUESTIONS

1. The earliest maps did not portray information as our modern maps do. Yet these early maps were useful to their makers and contained elements perceived as important. How could maps that portray the earth as flat or that show other information considered incorrect by present standards suffice at an earlier period? What important application does this have for modern cartographers?

2. What types of evidence do you think the ancient Greeks used to conclude that the earth was spherical? What further evidence do we have today?

3. Where is the place defined by latitude 0°, longitude 0°? Where is the place defined by latitude 0°, longitude 180°W? What relationship is there between the two places?

4. Knowing that a point on the earth is a certain distance in kilometers north of the equator and a certain distance in kilometers east of the prime meridian, would it be easier to find the latitude or the longitude of the point? What property of the system we have developed for describing location and direction on the spherical earth causes this?

5. Which of the following bases for measurement refer to actual phenomena on the earth, and which are purely arbitrary: 0° latitude? 0° longitude? 25°N latitude? the Arctic Circle? the South Pole? standard time?

6. A map projection should be chosen that is most suited to the particular purpose of the map. Which of the three properties of map projections—scale, area, and shape—would you preserve in mapping the following: A world map of national population density (population as a function of a country's size)? Distance between major cities in the United States? The route taken by Magellan in circumnavigating the earth?

7. Which of the projections we have seen in this chapter (gnomonic, cylindrical, conical, Mercator, homolographic, and homolosine) would be most useful for the following purposes, and why: Mapping equatorial Africa? Mapping the direction of cities from Chicago? Mapping the world to show the comparative sizes of its countries?

8. Why was Marinos' chart useful for navigation for such a long time?

9. Ptolemy suggested dividing the world up into large-scale sectional maps. Besides the larger scale for identifying features, what other advantages might such maps have? Give some modern examples of this type of map.

10. What widespread beliefs in Europe contributed to the decline in cartography during the Dark and Middle Ages?

11. If a globe is an accurate representation of the actual shape of the earth, why don't we use more spherical maps instead of flat maps?

12. Mercator's projection was an exceptionally important aid to early navigation. Why is it not used more extensively now?

13. We encounter thematic maps almost daily. What are some examples of thematic maps, besides those in this book?

14. Explain why the discovery that the earth revolved around the sun, instead of the sun around the earth, had such importance in the scientific understanding of the seasons. What further information was necessary before the seasons could be adequately understood?

15. Explain the "midnight sun" of the arctic region in terms of the tilt of the earth, the earth's revolution around the sun, and the solstices.

16. Why do scientists often use sidereal rather than solar time as the basis for their calculations? How are the two different?

17. In *Around the World in 80 Days* Jules Verne wrote of Phileas Fogg, who bet that he could travel around the world using modern (late nineteenth century) technology, in 80 days. Fogg left London, traveling eastward, and arrived back in London having traveled 81 days, according to his reckoning. Yet Fogg won the bet. How was this possible?

18. Why do some countries or states choose to have standard times differing by a half or quarter of an hour from the major time zones? (Think about the solar definition of noon—when the sun is highest in the sky—and the definition imposed by the Washington conference—12:00 PM according to the time zone's position relative to Greenwich, England. Are the two definitions likely to be equal throughout the entire width of a time zone?)

19. Which concept of time, linear or cyclic, is most important for studying earth-sun relationships? For studying the seasons? For examining the gradual slowing of the earth's rotational speed?

SUPPLEMENTARY READING FOR PART ONE

Part One is an introduction to the subjects of mapping, seasons, and time, as well as to physical geography itself. One of the best books dealing with geography as a subject, and the place of physical geography within it is *Geography: A Modern Synthesis* by P. Haggett, published by Harper & Row. The second edition was published in 1975. For a more comprehensive description of systems you can look at *Physical Geography: Environment and Man* by J.F. Kolars and J.D. Nystuen published by McGraw-Hill in 1975. A delightful exposition of the role that maps have played in human history is given by N.J.W. Thrower in *Maps and Man*, published by Prentice-Hall in 1972. A more detailed description of map projections may be found in A.N. Strahler's *Physical Geography*, published by John Wiley & Sons. The same source also provides a good description of seasons and time. Highly recommended also is the book by G.J. Whitrow called *What Is Time?* This was published by Thames and Hudson in London in 1972 and gives some fascinating insights into the nature of time.

FROM THE GEOGRAPHER'S NOTEBOOK

AN EXAMPLE OF INTERACTING SPHERES

When explorer Alexander Von Humboldt and his contemporaries traveled through previously unfamiliar areas in South America, Africa, and elsewhere, they gathered and recorded information to report to their colleagues in Europe. Those nineteenth century travelers were trained in several scholarly fields, and their powers of observation were extraordinary. Von Humboldt's descriptions, and those of Darwin, Wallace, Livingstone, and others, are as fascinating today as they were a century ago. In addition to their vivid writing (and in the absence of cameras), the explorers of the earlier times made beautiful and exact color drawings and sketches of the landscapes, peoples, animals, and plants they saw.

OBSERVING THE EARTH

The modern age—just a century later—is the age of the Mars probe and moon landings. The art of observation and description has largely been replaced by the science of sampling, testing, and analysis.

Physical geographers, however, still find reconnaissance (even in its nineteenth century form) a productive enterprise. It is a challenge that sharpens our powers of observation and tests our ability to select and categorize important and relevant data.

Imagine yourself to be in an area never seen by others of your own society. What will you observe, record, and report? What will be the focus of your narrative? In the absence of the convenience of cameras and tape recorders, how will you transmit your information to your readers and listeners?

Nowadays, we often pay little heed to the nature of the environment. At this very moment, do you know the prevailing wind direction? The outdoor temperature? How humid it is, or how dry? In another age (and still today in some other cultures) these environmental conditions play a crucial role in people's everyday existence. But to most of us—except in an environmental crisis—these circumstances are incidental.

This rather selective attention is one of the causes of our present disregard for our planet's health. Only recently has a renewed concern for the earth's many environments begun to arise. Such a concern can only lead to productive action when we have substantial knowledge of the elements of our physical world.

So try your own reconnaissance. Take a short walk—say, two kilometers (a little over a mile) near your home or apartment—and record what you can of the environment around you. Write about the air and the clouds, plants and soil, hills and slopes. You can do this even in a city. What you cannot interpret is as important as what you can. The pages in the rest of the book will help you do the job better, because they will introduce you to some terminology, but more importantly because they will give you some understanding of the processes at work around you.

Let us take an example of what you might do. Besides observing and noting phenomena on your walk, choose a point that gives you the best possible view of your surroundings and describe what you see as best you can. In doing so, things that you have never seen before will become apparent. Say you have the scene shown in the photo on the next page. Let us try to describe it in terms of the layers of the physical world, as simply as possible.

The atmospheric layer consists of a clear sky with one high, flat cloud. We automatically ask ourselves if this is normal weather for this location. If it is, or even if it is not, we cannot help but wonder how it can be explained. On the day when this picture was taken, the wind was coming from the mountains and the cloud was also moving out toward the flatter land. If this is normal weather for the location, perhaps air has moved over the mountains and left its rainfall on the side the wind first struck. Certainly the air felt dry when it touched the skin of the photographer. If we lived in this area, we might already know that it was common for the wind to come from the mountains.

SIGNS OF THE BIOSPHERE

We can focus next on the biosphere. What can we observe of the biosphere from this photograph? First, there are large areas of bare rock. It is unlikely that there is much life in these areas. Second, there are two major kinds of vegetation. In the middle distance are dark trees. They are probably conifers—or at least they are trees with needle leaves and trees that do not shed their leaves in winter. It looks as though the same trees are also located above the areas of bare rock, but we would have to move closer to tell for sure, and also to tell if the vegetation at the foot of the mountains was the same as on the plains. The dominant vegetation is grassland. The grass is not the lush kind of grass familiar to people living in the eastern United States. The grass suggests that the dryness implied by the atmosphere is a normal event. Another obvious biospheric phenomenon in the scene is the dwellings. Clearly humans use the land, and the sparsity of the houses suggests in this context that livestock ranching is the normal occupation. There is no sign of bare areas in the grassland, so we might assume that this practice fits in well with the natural environment.

What can we say about the lithosphere? Unfortunately we can say little about the soils from this particular photograph. You will be at an advantage on your walk because you may be able to see areas of exposed soil that you could describe. We can say much more about the other parts of the lithosphere. For example, there are two major divisions. First, there are the mountains with steep slopes, bare rock faces, and steep, narrow valleys. Second, there are areas of plains. But these are not completely flat. We can distinguish an intermediate slope at the foot of the mountains. In addition there are some sharp changes of slope. Some of these mark the edges of river valleys such as in the left foreground. Others, such as those in the middle distance on the right (see sketch) do not appear to be near flowing rivers at the present time. We cannot really explain these breaks of slope without studying the geologic history of the area in more detail. Without waiting for further study, could you suggest how these breaks of slope might have formed? Obviously, there is a great deal of informative detail in the landforms of this area.

A USEFUL RECORD

Perhaps, by now, you are beginning to get the idea of the kinds of things you might observe. In real life you can see much more than on a photograph like this.

FROM THE GEOGRAPHER'S NOTEBOOK

Make a sketch of the view you choose and mark it to bring attention to the points of interest that catch your eye. Notice that, in doing so, you are making decisions about what you think is and is not important in the scene. You are analyzing and organizing data. If you have a friend looking at the same scene, you will certainly disagree about the amount of detail to put in, and possibly about the important features of the scene. The sketching exercise will again help you to see things you have never noticed before, and it will bring to your mind a host of questions.

Last but not least, read your description again and take another look at your sketch a few days later. Look for the interrelationships between the spheres. In our photograph, for example, the area of trees seems to coincide with the intermediate slope at the foot of the mountains. Why? Usually coincidences like this are not accidental in the physical world. If your report were read by someone who has never seen this particular place, would he or she get a clear idea of what you were describing?

Nothing you see stands alone or exists by itself. The color and light of the vegetation relate to the depth and fertility of the soil and to the character of the season's weather. The local movement of the air (an element of the weather) relates to larger-scale patterns of circulation in the atmosphere. Humidity and rain sustain the plants and replenish the soil. In turn the leaves of the plants replenish the atmosphere's supply of oxygen. A web of systems produces what you see, and what you observe is only a visible fragment of the great network of interaction that is our physical environment.

PART TWO

THE ATMOSPHERE

During a typical year in the United States, there are a dozen tropical storms and hurricanes; two dozen additional severe storms along the coasts or over the Great Lakes; three dozen major blizzards, heavy snowstorms, and ice storms; six dozen major floods of disastrous proportions; hundreds of tornadoes; and thousands of thunderstorms.

These extreme events of the atmosphere are estimated to cost over 1000 lives and $10 billion in damage every year. Yet this same atmosphere can boast a sapphire hue on a cold winter's morning, a rainbow arching through the sky, a halo around the moon, the mysterious northern lights, or a sunset of spectacular color. Despite all this drama, we usually take the atmosphere for granted. We seldom consider its importance in our daily lives.

The life-sustaining layer of the atmosphere is our subject in this part of the book. We will examine its content and the flow of energy and water between it and the earth. We will see how the grand global currents of air and water form the general circulations of the atmosphere and oceans and how the atmospheric circulation carries the weather systems that give us our daily weather. The long-term operation of weather systems and the general circulation, together with energy flows to and from the earth's surface, cause the climates of our planet. These climates vary across the face of the earth and throughout time. Humans have always been affected by these continual changes in the atmosphere, but recently we have begun to affect the atmosphere as well. Remember that, even as you read about the atmosphere, it continues to move and change, painting a pageant of the skies.

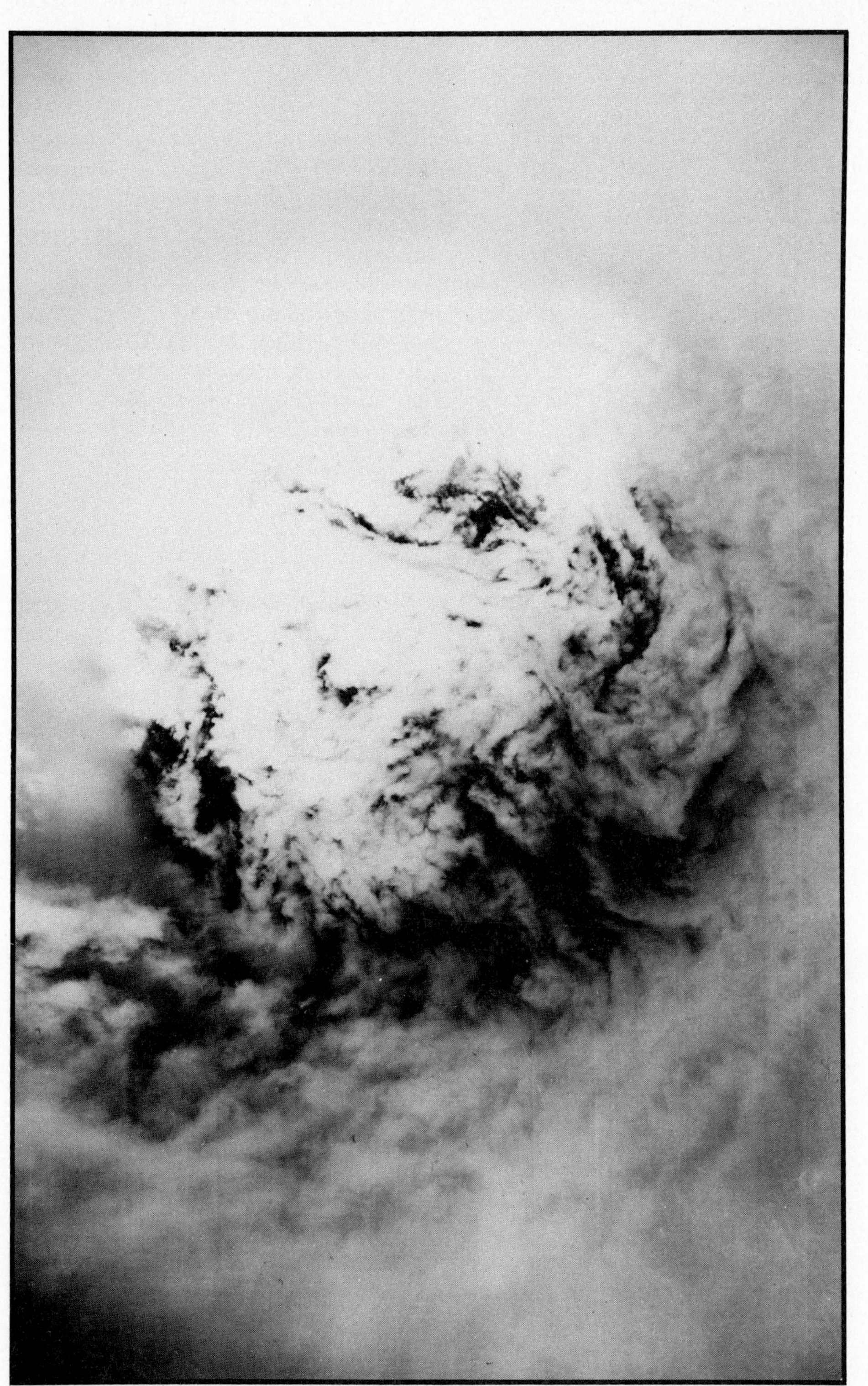

THE ATMOSPHERE: OUR OCEAN OF AIR

CHAPTER 3

The photographs taken by the astronauts in their Gemini spacecraft show us a thin blue and white shell around the earth. This shell is the atmosphere, the spherical, life-sustaining layer of air surrounding the planet. The atmosphere extends from a few meters within the soil in land areas or at the water surface in ocean areas to a height of about 60,000 km (37,000 mi), where some particles can still be considered part of the earth.

To the physical geographer, the most important parts of the atmosphere are the lower parts—those below 50 km (31 mi) and especially below 10 km (6 mi). Important flows of energy and matter occur within these lower layers. Here too great currents of air redistribute heat across the earth. These currents are a part of the systems producing our weather. The weather and the flows of heat and water across the earth are eventually translated into our surface patterns of climate. We are affected by both the climate and atmosphere and have the power to alter them to some degree. In the past, climatic changes have occurred without human intervention, but in the future, and even today, humankind may take a more active role.

These topics are all examined in the chapters in this part, but here we are most interested in the blue-and-white shell itself. We will begin by examining the composition of the atmosphere. Then we will look at the atmosphere as a system with various inputs and outputs. Within this system, there are individual subsystems, or cycles of matter and energy, that have particular value for life on earth. We will learn that most of the mass of the atmosphere is concentrated near the earth's surface and that the temperature of the air changes as we move away from the earth's surface. The lowest atmospheric layer, characterized by its density and ability to mix air vertically and horizontally, is the most important for our daily lives. This is where we find water. But we will also see that the higher layers are becoming increasingly interesting to us because of our recent ability to travel through them and possibly to alter them with our developing technology.

COMPOSITION OF THE ATMOSPHERE

The layers of the atmosphere can be sorted according to different criteria. The most common criterion is change of temperature as a factor of height. (We will examine this later in the chapter.) In terms of composition, another criterion, the atmosphere may be separated into two main layers. A layer extending from the earth's surface to 80 or 100 km (50 or 62 mi) beyond the earth has a uniform composition. It is called the *homosphere*. Beyond this layer, the chemical composi-

AIRCRAFT AND OUR DELICATE ATMOSPHERE

Controversy over the SST has raised a grim question among environmentalists: Are aircraft causing permanent damage to the atmosphere?

Today water vapor is the biggest villain. At airbases in Alaska, Greenland, and other frigid regions water vapor continually disrupts visibility. The take-off of one plane can spew enough water vapor into the air to blanket the field in a cold fog. For many years the Army has experimented with methods to disperse the fog.

At ground level the water vapor problem hampers visibility, but it can be solved. What about the water vapor emitted by thousands of aircraft in the troposphere? In the next few years we may see more high clouds and the thickening of existing clouds. Evidence from several areas of the United States confirms that high clouds have increased since 1965. We do not know how these clouds might alter the temperature at the earth's surface, but even a small change could be disruptive of the existing delicate balance.

Supersonic planes, or SSTs, may create even more serious problems because they fly in the low stratosphere at about 20 km (65,000 ft). The backers of the Concorde, the European-built SST, have won trial landing rights in Washington, DC, but they face a storm of criticism. Environmentalists fear that water vapor and nitric oxide in the exhaust may deplete the ozone. Even a small reduction in the ozone would let in much larger amounts of ultraviolet light. The Federal Aviation Administration estimates that an additional 200 cases of skin cancer each year in the United States could result from six flights a day. Although scientists find it difficult to prove the risks, the damage from SSTs could be disastrous.

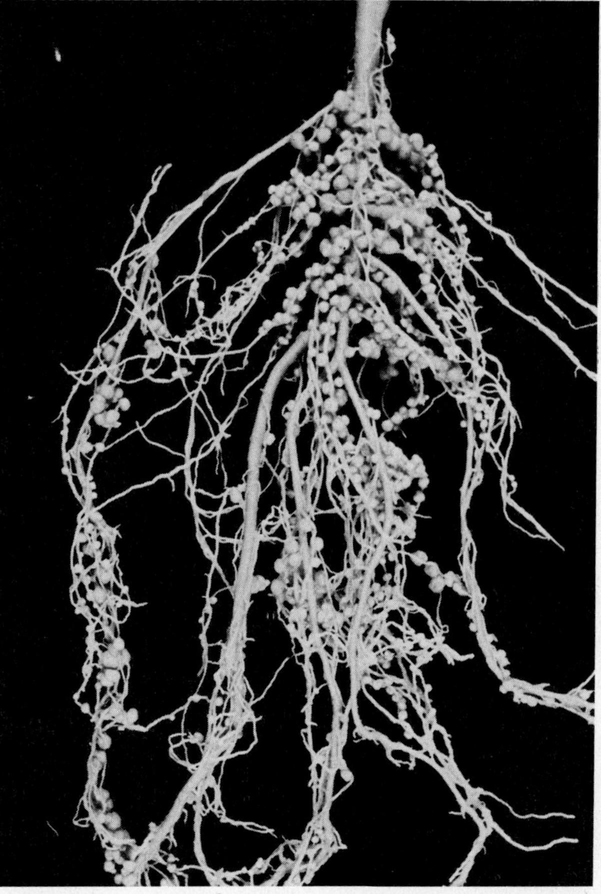

FIGURE 3-1

The vital role of nonvariant gases in the atmosphere. a) At high altitudes, where the air is thin, mountaineers cannot perform their tasks without supplementary oxygen.

b) The roots of plants contain bacteria able to extract nitrogen from the air for later use in forming food compounds.

tion of the atmosphere changes in a layer called the *heterosphere*. In addition to these major divisions, there is a concentration of ozone between 15 and 50 km (9 and 31 mi) that is called the *ozonosphere*.

The homosphere is the most interesting layer to humans, because we live in it. If you experimented by taking many samples of the homosphere, you would find that it contains three major groups of components—nonvariant gases, variant gases, and impurities. The nonvariant gases are always found in the same percentages, but the variant gases and impurities are present in different quantities at different times.

NONVARIANT GASES

As we can see in Table 3-1, the major nonvariant gases (Figure 3-1), making up 99.03% of the air, are nitrogen and oxygen. Both are extremely important for us. The bulk of the atmosphere that we breathe consists of *nitrogen*. As far as we are directly concerned, nitrogen is important only because it is not poisonous. Indirectly it is vital: Atmospheric nitrogen is converted by bacteria into other nitrogen compounds essential for plant growth. Without the food we derive from plants, we could not survive.

TABLE 3-1

The composition of dry air below 25 kilometers in terms of the percentage of total volume

CONSTITUENT	CHEMICAL SYMBOL	CONTENT
Nitrogen	N_2	78.08
Oxygen	O_2	20.95
Argon	A	0.93
Carbon dioxide	CO_2	0.03
TOTAL		99.99+

FIGURE 3-1

c) Argon is often used inside "neon" lights.

More directly necessary to our survival is *oxygen.* We absorb oxygen into our bodies through our lungs and into our blood. One of its vital functions there is to "burn" our food so that its energy can be released. Such "burning" is actually the chemical combination of oxygen and other materials to create new products. The biological name for this process is *respiration,* and the chemical name is *oxidation.* One example of rapid oxidation is the burning of such fossil fuels as coal and oil. Without oxygen, this convenient way of releasing the energy stored in fuels would be lost to us. Slow oxidation can also occur, as in the rusting of an iron bar. Therefore oxygen is essential not only for our breathing, but also for its part in numerous other chemical processes.

In 1894 scientists removed oxygen and nitrogen from the air and noted that a gas remained that seemed chemically inactive. It would not combine with other gases or compounds. A gas that behaves in this manner is called *inert.* The discoverer named the gas *argon* and found that it makes up almost 1% of dry air. Nowadays it is used in neon signs.

The first gas to be discovered in air was *carbon dioxide,* which forms only 0.03% of a sample of dry air. Despite the small amount present, carbon dioxide fulfills two vital functions for the earth. The first is in the process of photosynthesis, in which plants use carbon dioxide and other substances to form carbohydrates. Carbohydrates are a significant part of the food and tissue of both plants and animals. The second function of carbon dioxide is to absorb some of the heat energy rising to the atmosphere from the earth's surface, which reflects heat from the sun. The other nonvariant gases are poor absorbers of this heat; therefore it is primarily carbon dioxide that keeps the atmosphere at temperatures that permit life. Carbon dioxide also plays several less vitally important roles. It helps dissolve limestone, which leads to the intriguing features of such limestone-based landscapes as the Carlsbad Caverns in New Mexico. Furthermore, some scientists believe that carbon dioxide plays a role in minor and major climatic changes.

This last fact raises the question of whether or not carbon dioxide actually is a nonvariant gas in the atmosphere. It has been estimated that, between 1900 and 1935, the total quantity of the gas in the atmosphere rose by 9%. The rise since 1860, which is shown in Figure 3-2, is believed to be a result of increased industrialization and the burning of fossil fuels. Because the presence of carbon dioxide can warm the atmosphere, its continued production might have a significant effect on the climate of the earth. We

FIGURE 3-2

Carbon dioxide content of the atmosphere from 1860 to 1960.

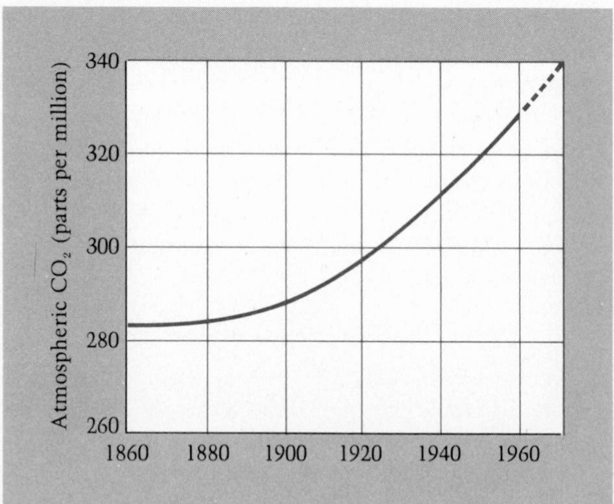

encounter the question again in Chapter 9, but now we will turn to the variant gases in the atmosphere.

VARIANT GASES

The ability to absorb heat energy from the sun is also found in the most important variant gas of the atmosphere—*water vapor,* the invisible form of water. Water vapor is more successful than carbon dioxide at absorbing heat energy because it not only can absorb heat but also can store it. When water vapor is moved around by currents of air, stored heat is transported along with it. This is part of an essential process by which the surface temperatures of most parts of the earth are prevented from reaching extremes intolerable for human habitation.

In deserts or cold regions, water vapor makes up only a small fraction of 1% of the air. But over warm oceans or low-latitude jungles, it might make up as much as 3% or 4%. In general, the warmer the air, the more water vapor it can hold. Because the parts of the atmosphere near the earth's surface have relatively high temperatures, this is where we find most of the water vapor. Without water vapor, there would be no clouds or rainfall. Thus most parts of the earth would be too dry to permit agriculture. Without the great cycle in which water moves from the earth's surface into the atmosphere and back again (which is discussed in Chapter 4), little life of any kind would be found on our planet.

The other variant gases in the lower parts of the atmosphere are found in much smaller quantities than water vapor. The most important is *ozone,* the type of oxygen molecule formed of three atoms rather than two. The greatest concentrations of ozone are found between about 20 and 25 km (12 and 15 mi), although it is usually formed at higher levels and transported downward. Even where most concentrated, ozone often constitutes less than six parts per 100,000 of the atmosphere. But like carbon dioxide, it is very important. It too has the ability to absorb solar energy, particularly the form called *ultraviolet* radiation. Ultraviolet radiation can give us a suntan, but large doses cause blindness, sunburn, and skin cancers. The ozonosphere shields us from excessive quantities of these deadly rays.

Small quantities of many other variant gases are present in the atmosphere. The most noteworthy gases are hydrogen and helium, sulfur dioxide, oxides of nitrogen, ammonia, methane, and carbon monoxide. Some of these are described in Chapter 8 as air pollutants derived from cities and industry. They can have harmful effects even when the concentrations are one part in a million or less. Other pollutants are found in the form of solid bodies, which can be classified as impurities of the atmosphere.

IMPURITIES

If you were to take air samples in or near a city, the number of impurities you found might be very large. One set of measurements showed typical country air to contain about four particles of dust per cubic millimeter, whereas city parks often had four times that amount. A business area in the city could have 200 particles in a cubic millimeter, and an industrial area, 4000 particles.

Both smoke and dust particles are common in urban air, but dust particles are the most prevalent type in the air over rural regions. Bacteria and plant spores are found in all parts of the lower atmosphere. An important additional impurity are salt crystals. Quantities of these usually are formed by evaporation of breaking ocean waves. They are important in the formation of rainfall because they attract water vapor.

Together the impurities play an active role in the atmosphere. Many of them help in the development of raindrops. Furthermore, the small particles can affect the color of the sky. Air and the smallest impurity particles scatter more blue light from the sun than any other color. This is why the sky looks blue. But when sunlight travels a long way through the atmosphere, as at sunrise or sunset, most of the blue light has been scattered. We see only the remaining yellow and red light. Consequently we observe the beautiful reds of sunsets. Occasionally, when there is an abnormally large amount of impurity particles in the atmosphere, such as after the eruption of a large volcano, the process is carried to spectacular extremes.

Such beauty is not always a positive sign. Large amounts of impurities can significantly reduce the quantity of sunlight as well as lead to problems of pollution. However, when left unaltered, the atmosphere has many ways of maintaining an efficient balance in the constituents of the atmosphere. It is this balance that we consider next.

THE ATMOSPHERIC SYSTEM

During the formation of the atmosphere, its major constituents reached a state of dynamic equilibrium. The composition we have just learned about is even today the result of constant inputs and losses of the major—and some of the minor—components. Part of

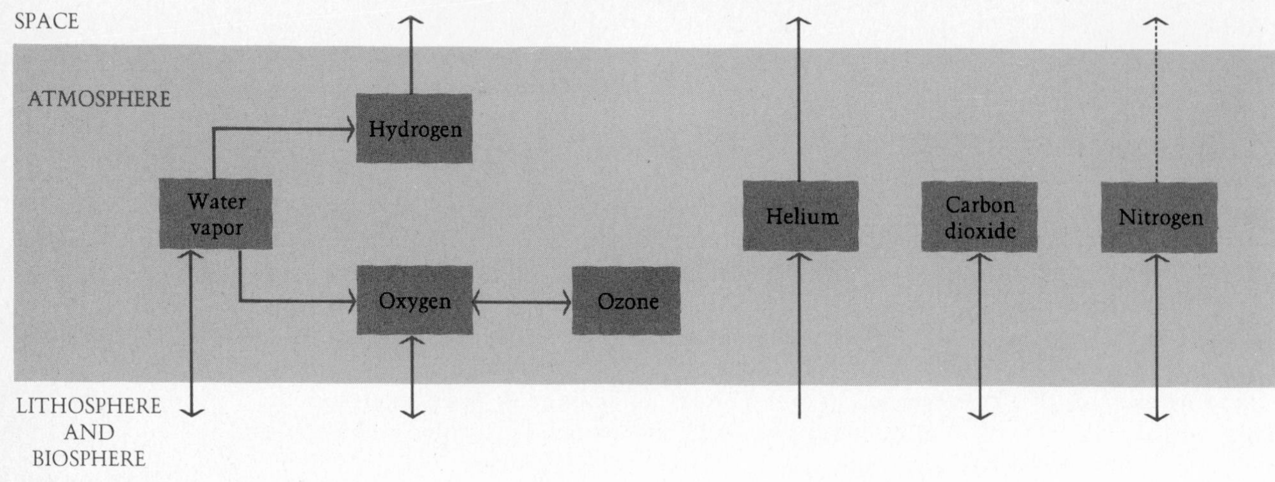

FIGURE 3-3
The atmosphere as an open system.

this flow of components takes place because the lower boundary of the atmosphere adjoins the lithosphere and the biosphere. At its upper boundary, the atmosphere dissipates into space. Thus the atmospheric system shown in Figure 3-3 is an open one. Some of the interesting smaller points in this system are the breakdown of water vapor and oxygen together with the losses and gains of hydrogen and helium. In addition, there are four vital subsystems.

Within the upper atmosphere, rays of sunlight split water molecules into free atoms of hydrogen and oxygen. Some of the hydrogen escapes into space. Rays of sunlight can also break up the two atoms of regular oxygen (O_2), which leads to the formation of ozone (O_3). But ozone is unstable and is often destroyed by the re-creation of oxygen or by additional light rays. These processes help to maintain the small proportions of hydrogen in the atmosphere and the larger amounts of oxygen and ozone. At the same time, helium produced by radioactive decay of uranium and thorium in the lithosphere is added to the atmosphere. It eventually escapes the earth's gravitational field and is lost to space.

But these processes are overshadowed by the atmospheric subsystems of the water, oxygen, nitrogen, and carbon dioxide cycles. The water cycle is so important that a large part of Chapter 4 is devoted to it. In the oxygen cycle, oxygen is put into the atmosphere as a by-product of photosynthesis and is lost when it is inhaled by animals and is chemically combined with other materials in oxidation. This process accounts for the major part of the oxygen cycle. In comparison, the two-way exchange between oxygen and ozone within the atmosphere is small.

THE NITROGEN CYCLE

Erupting volcanoes give some nitrogen to the atmosphere, and a little is lost to outer space. But the greatest part of nitrogen by far is cycled between the lithosphere, biosphere, and atmosphere over and over again.

The roots of plants contain bacteria that can extract nitrogen from the atmosphere. Farmers use these *nitrogen-fixing* bacteria—found in such plants as peas, clover, alfalfa, and soybeans—for maintaining soil fertility. The bacteria convert the atmospheric nitrogen into the organic nitrogen compounds of the plants, especially organic protein. Some of the organic material may be transferred to animals when the plants are eaten. When the plants and animals die, the nitrogen compounds are transformed by other bacteria and microorganisms first into ammonia, urea, and nitrates and eventually back into the gaseous form of nitrogen. In this form nitrogen escapes back into the atmosphere, and the cycle continues.

THE CARBON DIOXIDE CYCLE

The carbon dioxide cycle, a closed system, is illustrated in Figure 3-4. The greatest exchange of carbon dioxide is between the atmosphere and the oceans. Carbon dioxide enters the sea by absorption from the atmosphere, by animal and plant respiration, and by oxidation of organic matter. Carbon dioxide is released from the ocean during the photosynthesis of countless millions of small organisms known as *plankton*.

The second-largest exchange of carbon dioxide takes place between the atmosphere and land plants of the biosphere. Once more, carbon dioxide is taken from the atmosphere by plants during photosynthesis and is released by them in the course of respiration and decay.

In addition, some carbon dioxide passes from the soil to the atmosphere, especially when fields have recently been plowed. More is released when we burn fossil fuels, and some is naturally passed into the atmosphere from the earth's interior by means of volcanoes and hot springs. Certain weathering processes, which break rock into smaller particles or other materials, offset the atmospheric gain of carbon dioxide.

We can see that the total flow making up the carbon dioxide cycle is itself composed of many subcycles. The magnitude of the subcycles is specified in the diagram. The frequency of the subcycles varies. It may take only five to ten days for carbon dioxide to be cycled between the ocean and the atmosphere. Yet when some of it is incorporated into the rocks of the lithosphere, it may not circulate around the atmosphere again for many millions of years.

The circulation of oxygen, nitrogen, carbon dioxide, and water throughout the atmosphere to maintain the dynamic equilibrium of its composition is a beautifully balanced system. Its complexity is seldom appreciated when we breathe, fly kites, or check a weather vane. Nor are we usually aware of another sort of equilibrium in the atmosphere—the equilibrium of forces on individual air molecules.

ATMOSPHERIC PRESSURE

In a tub full of water, any water molecule near the top cannot move to a lower level because of all the molecules beneath it. It has a force acting upward on it from the lower molecules. At the same time, the earth's gravity is pulling the water molecule downward. These forces act until an equilibrium position is reached, where the molecule cannot move because the upward and downward forces are equal. In a similar way, any individual air molecule in our ocean of air is balanced between the pressure of the lower air molecules pushing upward and the earth's gravity pulling

FIGURE 3-4

The carbon dioxide cycle, a closed system. The numbers indicate the amount of carbon dioxide in metric tons either stored in the spheres or involved in the exchanges between the spheres. (English tons may be calculated by multiplying by 1.1.)

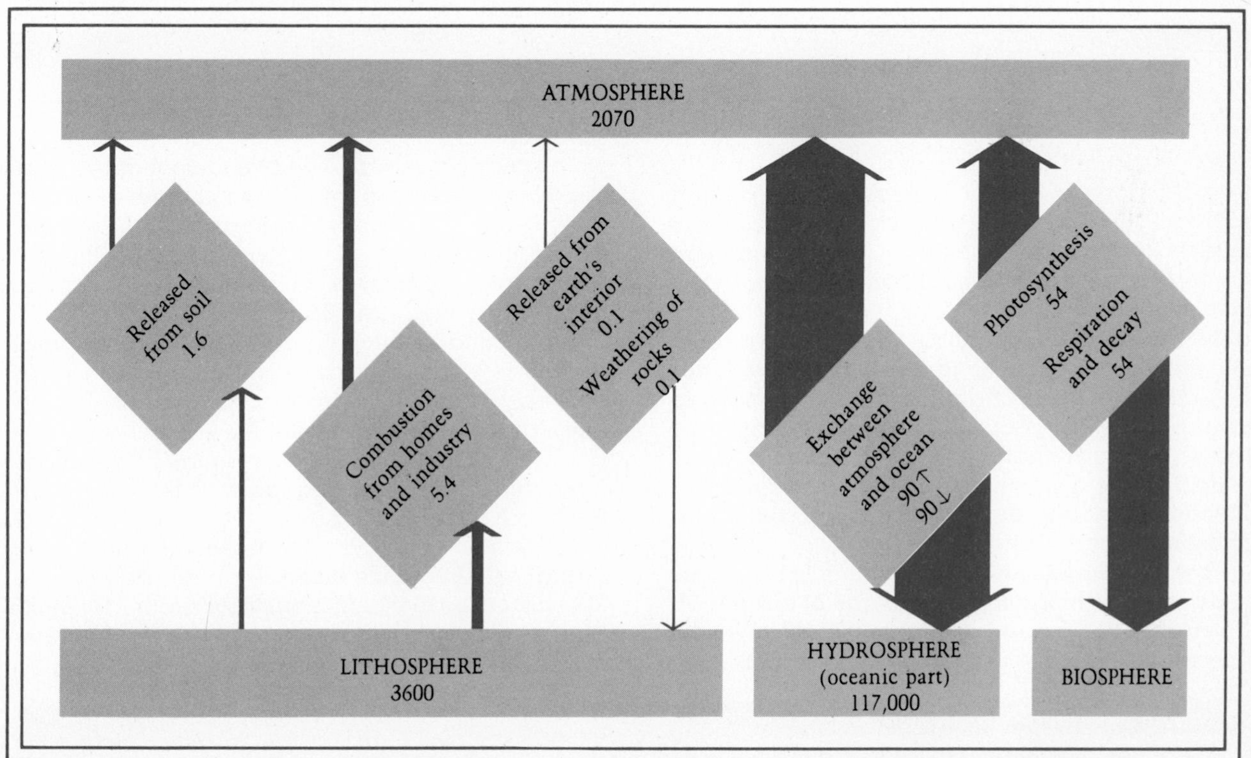

downward. One might say that the sky has already fallen as far as it will go.

THE CONCEPT OF PRESSURE

Imagine an enclosed box containing only molecules of air. They are likely to be moving constantly in all directions, in what is called *random motion*. Some of the molecules hit against the side of the box and create a force to move it outward. The strength or size of this force divided by the area of the side of the box is called *pressure*.

Because forces often imply the movement of an object, we may define a force by the movement it might produce. Sir Isaac Newton, the seventeenth-century British scientist, discovered that the size of any force equals the rate of change it makes in an object's speed multiplied by the mass of the object. Using this law, it is possible to define a basic unit of force, the *dyne*, as the force that produces an acceleration—rate of change of speed—of 1 cm per second on an object 1 g in weight.

The air molecules in our atmosphere create a force on the earth's surface. On a given area of the surface, say a square with sides of 1 cm (0.39 in), the force produces a pressure. The unit of pressure used in atmospheric studies is the *millibar* (mb), which equals 1000 dynes per square centimeter. The average weight of the atmospheric air pressing down on the earth's surface (not to mention your body) is 1013.25 mb (14.7 lb per square inch).

It is sometimes useful to imagine an infinitely long tube placed perpendicular to the earth's surface. The weight of the air pressing down on the surface is the atmospheric pressure. As we have seen, air is composed of several different gases. Given an average atmospheric pressure of 1013 mb, about 760 mb comes from nitrogen molecules, 240 mb from molecules of oxygen, and about 10 mb from water vapor. The remaining 3 mb may be ascribed to the other constituents of the air.

The pressure of the atmosphere may be measured by examining the length of a column of liquid it will support. In 1643 the Italian scientist Evangelista Torricelli performed an experiment in which he filled a glass tube with mercury, put his finger over the end, and then placed the tube upside down in a dish of mercury. Figure 3-5 depicts his experiment. Instead of the mercury in the tube rushing out, the atmospheric pressure pushing down on the mercury in the dish supported the liquid still in the tube. The height of the column in the tube was directly proportional to the atmospheric pressure: The greater the pressure, the higher the column in the tube. Torricelli had invented the first pressure measuring device, known as a *barometer*.

WE ARE ALL UNDER PRESSURE

Even a 98-pound weakling can support 10 tons! It's not an advertisement for a super body-building program. We all bear the weight of the atmosphere pressing down on us. At sea level, the pressure can be 9 to 18 metric tons (10 to 20 English tons) depending on our size.

Like the deep-sea creatures who live their entire lives with the weight of hundreds of meters of water above them, we have adapted to functioning and moving efficiently in our particular environmental pressure. Only a sharp change—a Miami, Florida resident taking a vacation trip to Aspen, Colorado—reminds us of our dependence on a particular atmospheric environment.

There are two factors that explain this sensitivity to altitude. The first is the percentage of oxygen. At higher altitudes the "air is thinner," that is, there is more space between oxygen molecules and fewer molecules themselves. We have to do more work to get the minimal amount of oxygen needed to sustain life. If you recall the Mexico City Olympic games in 1968, you might remember that the low density of oxygen was a factor in the low times recorded by the participating athletes.

The second factor is the response of our internal organs to changes in atmospheric pressure. Our ears may react first as we climb to a higher altitude. The "pop" we hear is actually the clearing of a small tube that allows the pressure between the inner and middle ear to equalize. This prevents our eardrums from bursting.

At higher elevations, we travel in pressurized airplanes. Astronauts also use pressurized cabins. When they leave their capsules for strolls in regions that have no measurable atmosphere, they require suits to maintain a safe pressurized environment.

ATMOSPHERIC PRESSURE AND HEIGHT

Once scientists had found they could measure atmospheric pressure, they set about testing its properties. They soon found that it does not vary greatly horizontally but that it decreases very rapidly with height. Measurements of atmospheric pressure from both land and balloons showed dramatic results. Thus the pressure at sea level (1013 mb) decreases to about 840 mb at Denver, Colorado, where the altitude is 1584 m (5280 ft). On top of Mount Kennedy in the Canadian Yukon, at 4170 m (13,900 ft), the atmospheric pressure is approximately 600 mb, and on top of Mount Everest in the Himalayas, 8742 m (29,141 ft) above sea level, the pressure is about 320 mb.

Because the air pressure depends on the number of molecules in motion and it is highest in the lower layers, we may deduce that most of the molecules are near the earth's surface. This is verified in Figure 3-6, which shows the percentage of the total mass of the atmosphere below certain elevations. For example, 50% of the air of the atmosphere is found below 5 km (3.1 mi), and 90% lies within 16 km (10 mi) of the surface.

Once again, imagine a box containing air molecules. The number of molecules in a box of a specified volume is the *density* of the gas in the box. The molecules move around in a random way, so the more molecules there are in the box, the more likely it is that they will collide with one another. The average distance they move before collisions occur is called the *mean free path* of the molecule.

If we extend this idea to the atmosphere, we begin to see how thin it becomes with increasing altitude. At sea level, an air molecule moves only about 0.0000008 cm before a collision. About 100 km (62 mi) above the surface, the mean free path is about 2.5 cm (1 in), but at 320 km (200 mi), collisions occur approximately every 24 m (80 ft). This example shows how density and atmospheric pressure rapidly decrease with altitude. There are also dramatic changes of temperature with height.

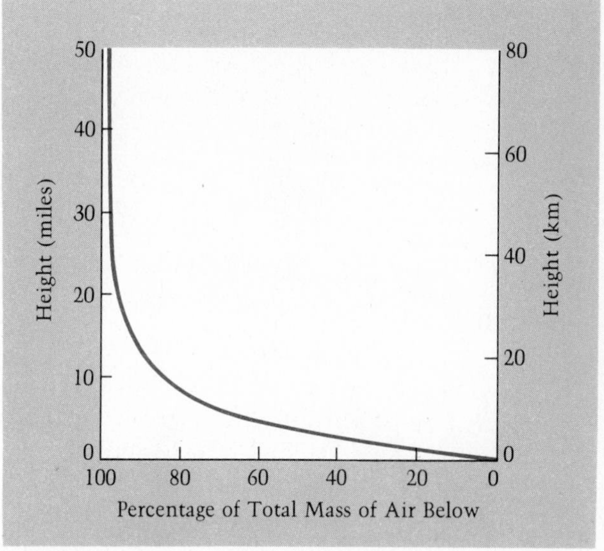

FIGURE 3-6

The mass of the atmosphere as a function of height. A greater proportion of the atmosphere is concentrated near the earth's surface. Atmospheric pressure depends on mass, so it also decreases with altitude.

FIGURE 3-5

The mercury barometer invented by Torricelli. The greater the atmospheric pressure, the higher the column of mercury.

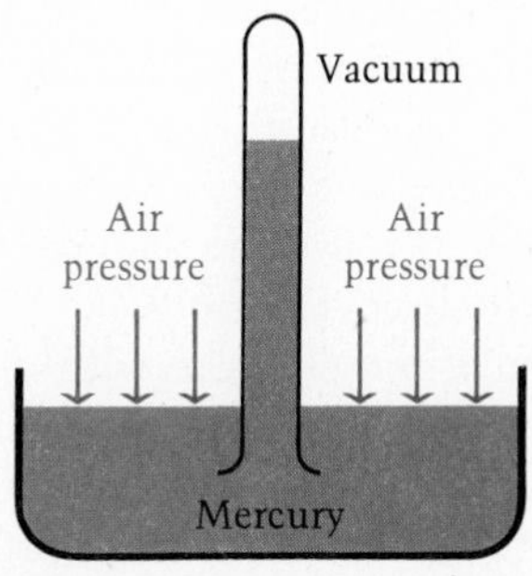

TEMPERATURE IN THE ATMOSPHERE

On the rather cloudy day of June 26, 1863, British scientist James Glaisher and his assistant climbed into the basket of a balloon in Wolverton, England. This flight, one of their 28 flights between 1862 and 1866, lasted an hour and a half. They ascended to 7050 m (23,500 ft) and traveled 80 km (50 mi) before descending at Ely. On the way they encountered rain, snow, and fog (Figure 3-7).

One of the main purposes of this flight was to note the temperatures along the way. In Glaisher's own words, these varied from the "extreme heat of summer" to the "cold of winter." In fact, the temperatures varied from 19°C (66°F) at the ground to −8.3°C (17°F) at 7050 m. The pair had good reason to beware of the hazards of high altitudes: On a flight the previous year, Glaisher had fainted at 8700 m (29,000 ft) from a lack of oxygen. His assistant, arms paralyzed with cold, climbed the rigging of the balloon to release the gas control with his teeth. These flights and heroic actions

FIGURE 3-7
James Glaisher and his assistant Coxwell on a balloon flight on September 5, 1862.

established the fact that both pressure and temperature decrease with height, at least up to 8700 m. Before we examine further the temperatures of the atmosphere we must briefly digress to inquire more exactly what we mean by temperature.

WHAT IS TEMPERATURE?

In our imaginary box of air, the molecules moved because they had the energy of movement, called *kinetic* energy. The more kinetic energy molecules have, the faster they move. The index we use to measure their kinetic energy is called *temperature.* Thus temperature is an abstract term describing the energy, and therefore the speed of movement, of molecules. In a gas such as air, the molecules actually change their location when they move. But in a solid, like ice, they only vibrate. Nevertheless, the speed of vibration is also described by their temperature.

It is almost impossible to examine individual molecules, so we usually use an indirect method of measuring temperature. We know that changes of temperature make gases, liquids, and solids expand and contract. Therefore, the most common measure ment of temperature uses the expansion and contraction of mercury in a glass tube. Such an instrument is called a *thermometer;* you are probably familiar with the medical and weather varieties. The mercury thermometer is placed in a mouth or the air or some other place where it can come into thermal equilibrium with the medium whose temperature it is measuring.

A thermometer is calibrated according to one of several internationally accepted scales. The most commonly used in the world is the *Celsius,* or *Centigrade,* scale. On this scale, the boiling point of water is set at 100°C, and its freezing point is 0°C. On the *Fahrenheit* scale, presently used in the United States, water boils at 212°F and freezes at 32°F. Scientists also use an *absolute,* or *Kelvin,* scale, which is based on a temperature of absolute zero. Scientists theorize that a gas at absolute zero would have no volume, no molecular motion, and no pressure. The Kelvin degree is the same size as the Celsius degree, but water freezes at 273°K and boils at 373°K.

It is important to distinguish temperature from *heat.* Temperature merely measures the kinetic energy of molecules. It does not measure the number of molecules in a substance or its density. But the heat of a substance depends on its volume, its temperature, and its capacity to hold heat. Thus a bowl of soup, with a high *heat capacity,* would burn your tongue at the same temperature at which you could easily drink hot water. A large lake with a water temperature of 10°C (50°F) contains much more heat than a cup of hot coffee at 70°C (158°F). Now that we know something about temperature, we may reexamine the temperatures of the atmosphere.

TEMPERATURE AND HEIGHT

Glaisher's observations for the lower parts of the atmosphere were correct: Temperature does decrease with height. However, later unmanned balloon observations showed that, after about 12 km (40,000 ft), temperature stops decreasing with height and begins to increase. Nobody believed this at first, and only after several hundred balloon ascents was it accepted. Later ascents to even higher altitudes showed even stranger events, as we will discover shortly.

The variation of temperature with altitude, as we now know it, is shown in Figure 3-8. The bottom layer of the atmosphere, where temperature decreases with height, is called the *troposphere.* The rate of a drop in temperature is called a *lapse rate.* In the troposphere the average lapse rate is 6.5°C per kilometer (3.6°F per 1000 feet). The point in the troposphere where temperatures stop decreasing with height is called the *tropopause.* Beyond this, in a layer called the *stratosphere,* temperatures either stay the same or start increasing with height. If temperatures increase with height, we may still talk of a lapse rate, but in this case the lapse rate is positive. Positive lapse rates are frequently called *inversions,* because they invert what we, on the surface, believe to be the "normal" state of temperature change with elevations, a decrease with height. Above the stratosphere, after about 52 km (32 mi) above the earth, temperatures remain constant with altitude. This area is called the *stratopause,* and it is followed by a layer known as the *mesosphere.* In the mesosphere, temperatures fall with height, as they did in the troposphere. Eventually the decrease in temperature stops, at a point you might rightly guess to be called the *mesopause.* This occurs about 80 km (50 mi) above the earth's surface. Above the mesopause, temperatures once more increase with height in a layer called the *thermosphere.*

Soon we will investigate these upper layers in more detail. But as physical geographers, we share James Glaisher's interest in the lowest layer. Let us look more closely at the troposphere.

FIGURE 3-8

The variation of temperature with height. Temperature is given in absolute degrees. On this scale water boils at 373°K and freezes at 273°K.

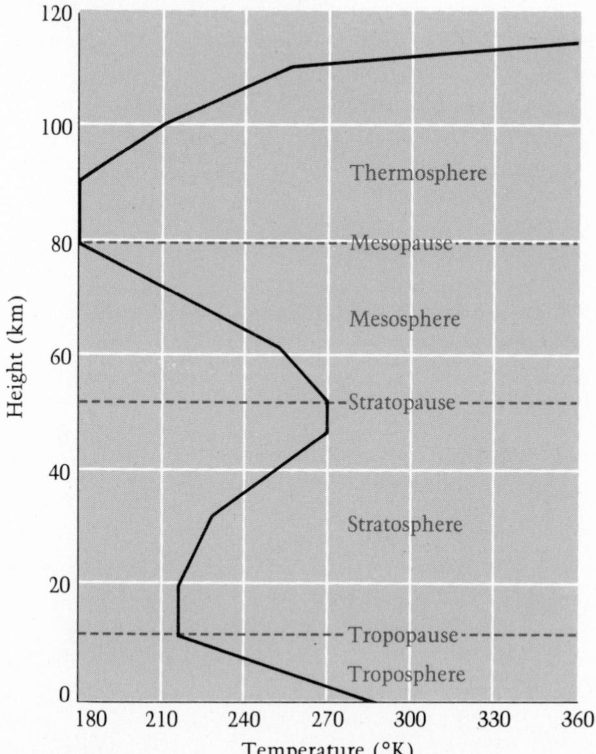

TROPOSPHERIC TEMPERATURE AND STABILITY

The troposphere is the atmospheric layer we live in. It is in the troposphere that the weather events and climates affecting humans occur. As we have seen, its temperature decreases with height, and it contains water vapor. Another distinctive feature of the troposphere is the possibility and frequency of vertical, as well as horizontal, movement of air.

Other chapters give many reasons for the vertical movement of air in the troposphere. But whatever the cause of the initial vertical movement, any long continuation of it depends on the rate of change of temperature with height, the lapse rate. The lapse rate determines the *stability* of the air.

The concept of stability may be illustrated by the wedge of wood in Figure 3-9. When it is resting on its side, a small push at the top may move it horizontally, but its vertical position remains the same. It is therefore stable. When it rests on its curved base, a similar push might rock it, but it will still return to its original position. It is still stable. But if we balance the wedge of wood on its pointed edge, a small push at the top topples it. It does not return to its original position; hence it is unstable.

We use the same terminology to refer to the vertical movement of a small parcel of air. If it returns to its original position after it receives some upward force, we say it is stable. But if it keeps moving upward after receiving the force, then we say it is unstable.

In order to understand air stability, we must first consider that air is a poor conductor of heat. Without any air movement, it takes a long time for heat to pass from one air molecule to the next. Therefore, a parcel of air of one temperature that is surrounded by air at

FIGURE 3-9

The concept of stability. The block of wood, like parcels of air, is considered stable as long as it returns to its original position after a small push.

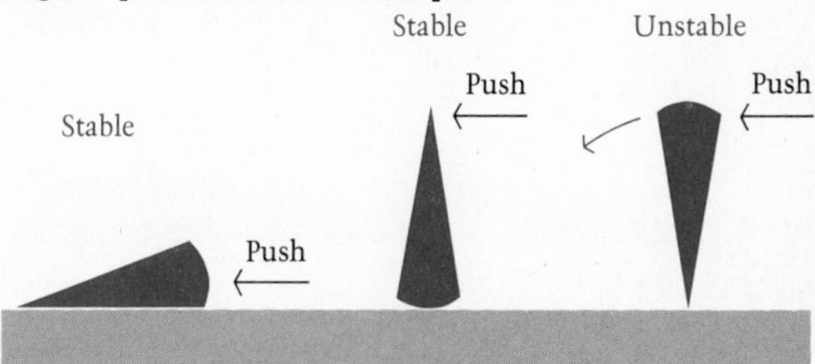

another temperature will neither gain nor lose heat in a small period of time. When heat is neither gained from nor lost to the surrounding air, we call the process *adiabatic*.

Lapse rates If you have ever used a bicycle pump, you remember that, when the air contracts at the bottom of the pump, the air's temperature rises. The bottom of the pump gets hot. The opposite is also true if the volume of a given mass of air is expanded: The temperature of the air decreases.

A similar thing happens in the atmosphere. If a parcel of air rises to a higher altitude, it can expand because of the lower pressure at the greater height. When expanding, it cools. This is an adiabatic process, because the air neither gains nor loses heat from its surroundings. Hence we call the lapse rates in the troposphere *adiabatic lapse rates*. When air is completely dry, it cools with height at a constant rate of 1°C per 100 m (5.4°F per 1000 feet). This is the *dry adiabatic lapse rate* (DALR).

But because air is not always dry, and for other reasons as well, any particular atmospheric lapse rate may not be the same as the DALR. The lapse rate at any particular time or place is the *environmental lapse rate* (ELR).

Now let us go back to our parcel of air and see what happens to it in two different environments with two different ELRs. In Figure 3-10a, a parcel of air rises off the ground. Because it neither gains heat from nor loses it to the surroundings, it cools at the DALR, decreasing its temperature 1°C for each 100 m of ascent. But in this case, the ELR is 0.5°C/100 m, so after rising 100 m, the air parcel has a temperature 0.5°C lower than its surroundings. The colder the air, the more dense it is. Therefore, the air parcel becomes denser and heavier than the surrounding air and tends to fall back to earth. This would happen even if the parcel rose to 300 m, where it would be 1.5°C colder than its surroundings. The cooled parcel therefore returns to its original position. We would say the whole of the air in that environment is stable.

In contrast, Figure 3-10b shows the ELR to be 1.5°C/100 m. In this case, an air parcel rising and cooling at the DALR would be warmer than its surroundings. The warmer the air, the less dense it is, so the air parcel that is lighter and less dense than the surrounding air continues to rise. We would call the air in this environment unstable, because it does not return to its original position.

We can often tell if a portion of the atmosphere is stable or not by looking at it. A stable atmosphere is characterized by clear skies or by flat, layer-like clouds. An unstable atmosphere is typified by puffy clouds, which sometimes develop to great heights. A Gemini spacecraft photograph of Florida, shown in Figure 3-11, illustrates the two states. Over the sea, the ELR is lower than the DALR, so parcels of air stay near the sea surface. But the higher temperatures of the land surface make the ELR higher than the DALR. The resulting instability allows parcels of hot air to rise in the atmosphere, forming puffy clouds as they cool.

Moisture and lapse rates The situation is slightly different when the air contains water vapor that is changing to water droplets. Heat is given out when the state of water changes this way, and the resultant lapse rate is lower than the DALR. This new lapse rate is called *moist* or *saturated adiabatic lapse rate* (SALR). A typical value for the SALR at 20°C (68°F) is 0.44°C/

FIGURE 3-10

Environmental lapse rate conditions. a) When the ELR is lower than the DALR, an air parcel is stable.

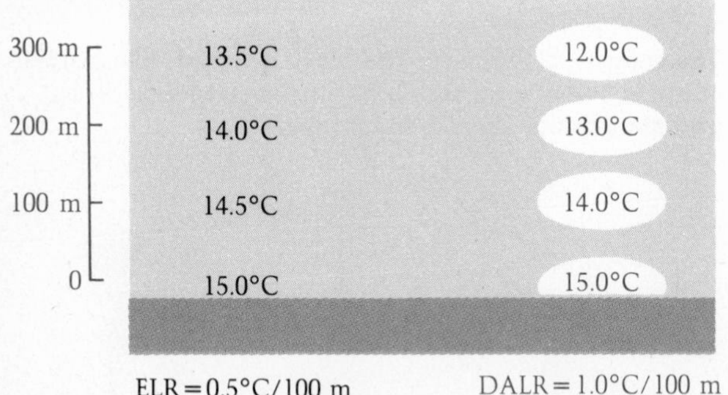

(a) Stable air

b) When the ELR is greater than the DALR, an air parcel is unstable.

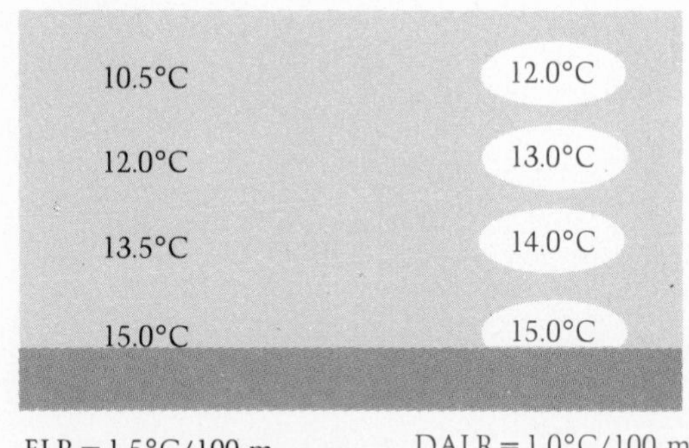

(b) Unstable air

WHERE THERE'S SMOKE

Scientists are not the only ones who make careful observations of the atmosphere. Artists also notice details of weather and landscape. The delicate colors and changing shapes of smoke fascinated the French painter Degas. He wondered how smoke could be painted on canvas. In his notebook for 1860 and 1861 he wrote a note to himself to reflect "On smoke. Smoke of people smoking pipes, cigarettes, cigars, the smoke of trains, of high chimneys, of factories, of steamboats, etc. The crushing of smoke under bridges." He must have studied the density and movement of smoke for months before he was able to capture it on canvas.

Degas may have known what geographers know about smoke. Besides looking at the shape of clouds, watching the plumes of smoke from smokestacks gives us a quick assessment of the stability in the lower atmosphere. When the smoke billows up and down in a snakelike pattern, it signals instability. Smoke that spreads out horizontally at one level in the air indicates stability and often lapse rate inversions. At times several combinations of these two patterns can be seen at once at different heights.

100 m (2.4°F/1000 ft). As a rule, we may assume that the atmosphere will be stable if the ELR is lower than the SALR and unstable if the ELR is greater than the DALR. If the ELR lies between the SALR and DALR, the atmosphere is said to be *conditionally unstable.* The conditions depend on whether the water vapor in the air changes into liquid water and thereby adds heat to the air.

FIGURE 3-11

Gemini V photograph of Florida, looking north. The clouds over the land surface indicate an unstable atmospheric environment compared to that over the sea surface.

In dealing with the stability of the troposphere, we are really considering the possibility of vertical mixing of air. This has many practical implications, such as how well pollutants or insecticides will disperse once put into the air. The possibility, and frequency, of vertical mixing is one of the major characteristics of the troposphere. We can now examine the higher layers of the atmosphere.

HIGHER LAYERS OF THE ATMOSPHERE

We have seen that the atmosphere may be divided into the troposphere, stratosphere, mesosphere, and thermosphere on the basis of varying lapse rates. The particular lapse rates within these layers lead to some of the special properties they possess, such as the tendency for vertical mixing of air in the troposphere. Further characteristics of these layers are explained by their chemical nature and by the electrical changes of their atomic structure. Thus temperature change and other considerations must be taken into account in describing these layers, as Figure 3-12 indicates.

We have learned that the troposphere is divided from the stratosphere by the tropopause. Although the tropopause is at an average height of 12 km (40,000 ft), this height varies with latitude. Over the poles, the tropopause is usually found at about 8 km (5 mi) above the earth's surface. But over the equator, it reaches an altitude of about 16 km (10 mi).

There are usually two distinct breaks in the tropopause, which are characterized by regions of variable lapse rates. These breaks are generally found at latitudes of about 25° and 50° north and south. The breaks, associated with fast-flowing winds in the upper

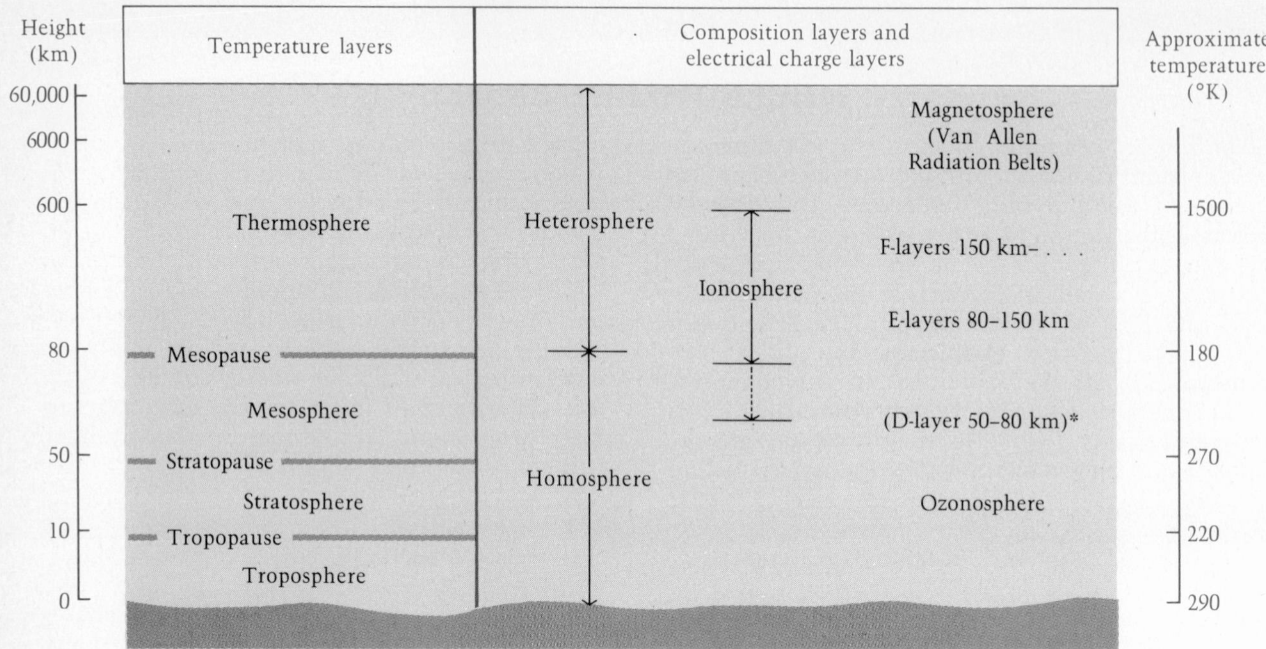

FIGURE 3-12
The layers of the atmosphere. Temperature, composition, and electrical charge vary according to layer.

atmosphere, are important because through them the troposphere and the stratosphere exchange materials. Small amounts of water vapor may find their way into the stratosphere at these points, and ozone-rich air and potentially dangerous radioactive particles from nuclear explosions may penetrate the troposphere through them. Above the tropopause we arrive at the calm, clear air of the stratosphere.

THE STRATOSPHERE

Many modern jet aircraft fly in the stratosphere because it provides the easiest flying conditions. The nearly total absence of water vapor prevents the formation of clouds, thus providing pilots with good visibility. And temperature inversion prohibits vertical winds, so the winds in the stratosphere are almost always horizontal to the earth's surface. Thus flights are less bumpy.

As Figure 3-13 indicates, stratospheric winds in low latitudes always blow from the east and are therefore called easterlies. But at high latitudes, the direction reverses from summer to winter. In the summer, stratospheric winds come from the east and are known as the *polar easterlies.* In winter they blow from the west and are called the *polar night westerlies.* The change between winter and summer in this rarified atmosphere is rather dramatic. Sometimes sudden warmings over a two-day period can push the temperature at about 25 km (15.5 mi) from –80°C (–112°F) to –40°C (–40°F).

Another strange event discovered by recent probings of the stratosphere is a wind reversal that occurs principally at low latitudes. This *stratospheric biannual wind regime* takes place in the layer between 18 km (11 mi) and 30 km (18.5 mi) above the surface. Here the wind blows from the east for 12 to 13 months and then for the next 12 to 13 months comes from the west. Scientists cannot yet explain this peculiar activity.

The region where ozone is concentrated lies within the stratosphere. Here ozone is naturally produced by the action of ultraviolet sunlight on oxygen and naturally destroyed when it is turned back into oxygen. Ozone is also transported by natural processes from one part of the stratosphere to another. These chemical and transportation processes create a constant balance of ozone in the stratosphere.

The importance of the ozonosphere in shielding us from ultraviolet radiation has already been noted. At the same time, the absorption of ultraviolet radiation heats the stratosphere. Thus it is important to maintain the proper ozone balance. Water vapor and nitric oxide emitted by aircraft engines may produce changes in this balance, but the data that would provide a definite conclusion are lacking. More recently it has been suggested that the fluorocarbon gases used in aerosol cans accumulate in the stratosphere and also affect the ozone balance. Once more, further data are required to prove this claim.

THE MESOSPHERE

Above the stratosphere and between about 50 to 80 km (31 to 50 mi) lies the layer of decreasing temperatures called the mesosphere. Over high latitudes in summer, the mesosphere sometimes displays *noctilucent clouds,* very high, wispy clouds that appear at night. They are thought to be sunlight reflected from meteoric dust particles that become coated with ice crystals. The unstable lapse rates in the mesosphere would permit small quantities of water vapor to move upward and solidify.

Another common event in the mesosphere occurs when sunlight reduces molecules to individual electrically charged particles. This process is called *ionization,* and the resulting particles are called *ions.* The ions in the mesosphere concentrate in a zone called the D layer, which reflects radio waves sent from the earth's surface.

THE THERMOSPHERE

The thermosphere is found above 80 km (50 mi) and continues to the edge of space. The temperature rises spectacularly in this layer and probably reaches 900°C (1,650°F) at 350 km (220 mi). However, because the air molecules are so far apart at this altitude, these temperatures really apply only to individual molecules and do not have the same kind of environmental significance as they would on the earth's surface.

Ionization also takes place in the thermosphere, producing two more belts that reflect radio waves. These are the E layer, between 80 and 150 km (50 and 83 mi), and the F layer, above 150 km (83 mi). Occasionally the ionized particles penetrate the thermosphere, creating vivid shows of light called the *aurora borealis* in the Northern Hemisphere (see Figure 3-14) and the *aurora australis* in the Southern Hemisphere.

Even higher in the thermosphere, at about 4000 km (2500 mi) and 20,000 km (12,500 mi), there are further concentrations of ions. These have been called the Van Allen radiation belts. This outer layer is sometimes referred to as the *magnetosphere* because here the earth's magnetic field is often more influential than its gravitational field.

IMPORTANCE OF THE OUTER LAYERS

Scientists know so little of the layers above the troposphere that these outer fringes have sometimes been humorously called the "ignorosphere." These regions of concentrated ozone, electrically charged particles, bitter cold and violent heat, meteoric dust, and weirdly illuminated clouds lie at the frontiers of our knowledge. But as the space research program continues, we will certainly need new data on the conditions of the earth's upper atmosphere, which may affect space capsules. We also are ignorant of the process by which a large burst of heat on the sun blacks out all shortwave radio communication here on earth. Do such solar storms also affect the weather in the troposphere? We do not know. It may be even more important for us to know how much such terrestrial activites as exploding bombs and releasing unnatural particles into the higher layers will affect these possibly delicate outer fringes of our planet.

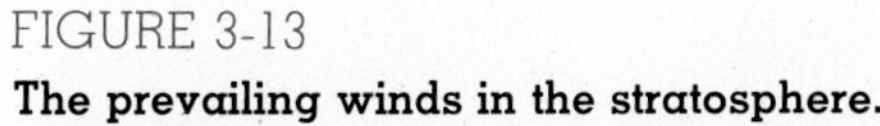
FIGURE 3-13
The prevailing winds in the stratosphere.

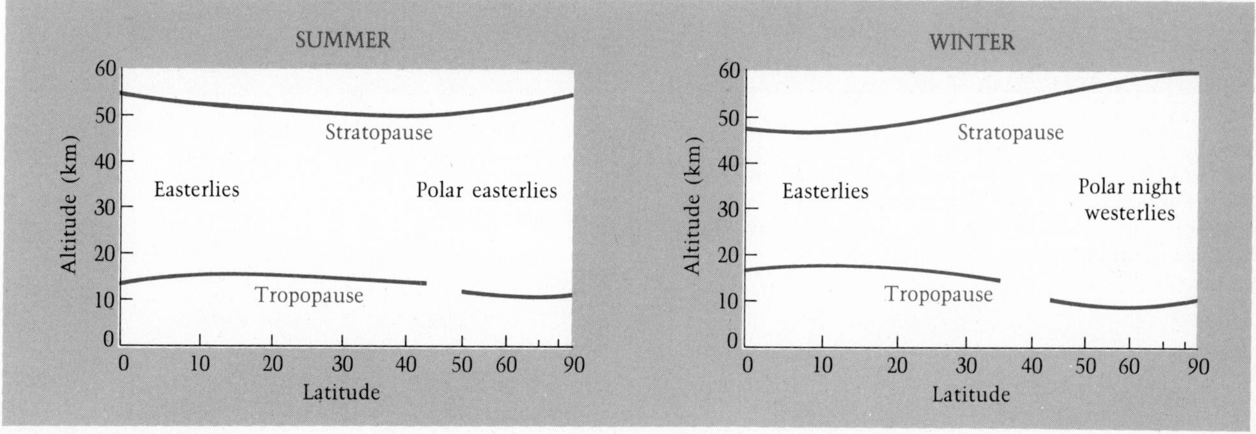

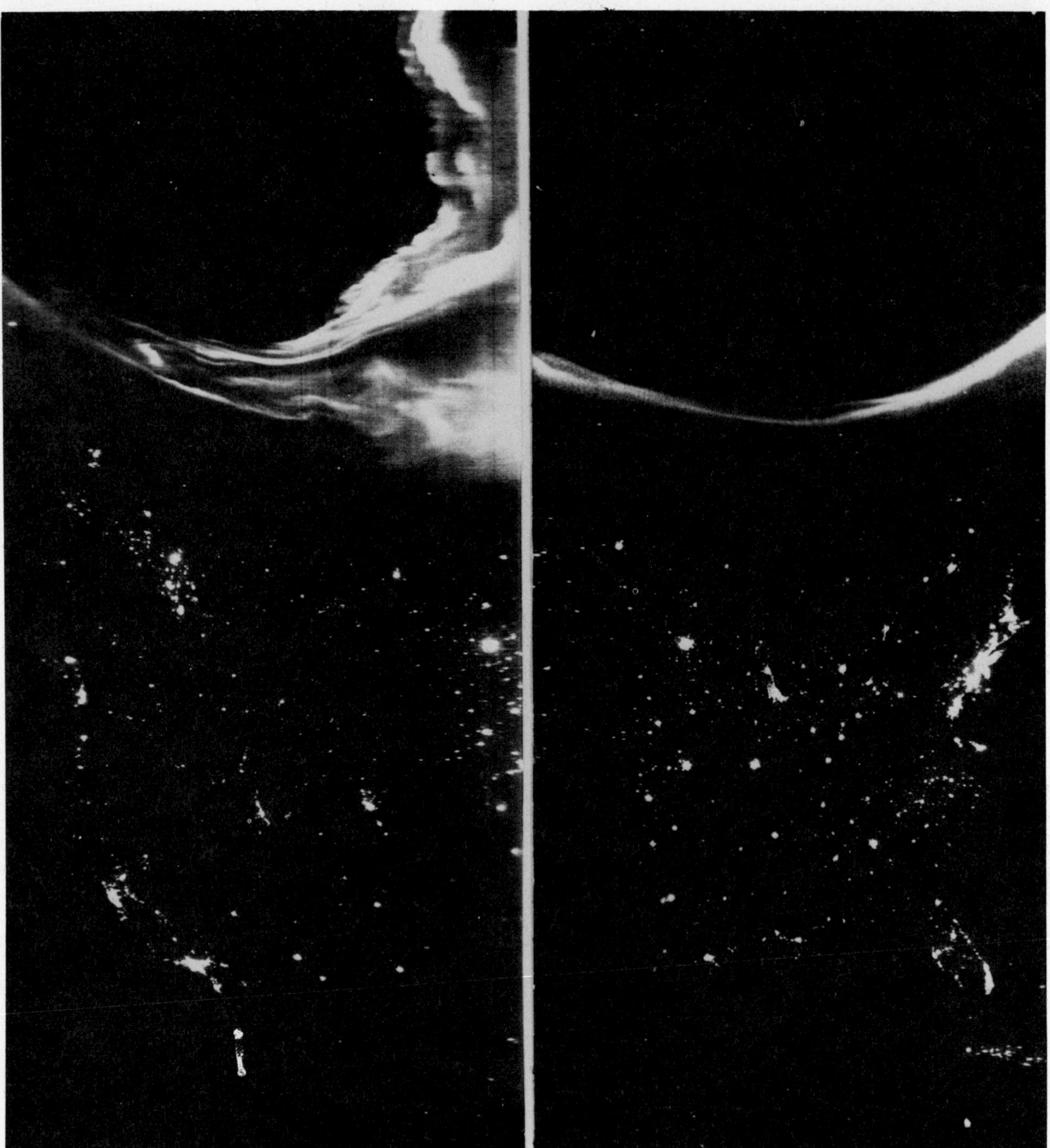

FIGURE 3-14

Satellite pictures of the aurora borealis. a) From a vantage point far above North America, we can clearly see the magnitude of the phenomenon. Lights in the western and eastern areas of the continent show the location of major cities.

b) The aurora as seen from the ground.

SUMMARY

We have examined the thin blue shell that is the earth's atmosphere. We have seen that it is composed of nonvariant and variant gases and a number of impurities. Each of the gases and impurities play important roles in the atmospheric processes that affect life on earth.

The atmosphere can also be considered as a system, containing inputs and outputs of gases and other materials. Within this system, there are important subsystems concerned with cycles of nitrogen and carbon dioxide.

The mass of molecules in the atmosphere exists in an equilibrium of upward and downward forces that results in atmospheric pressure. Pressure decreases rapidly with height, and in the lower atmospheric layer known as the troposphere, decreasing pressure leads to decreasing temperatures. We have seen how this variation of temperature with height controls the stability of the atmosphere and its propensity for vertical mixing.

Above the troposphere, there are other atmospheric layers. Their temperatures do not always decrease with height. Each of these layers has properties that affect our activities on earth in one way or another.

QUESTIONS

1. What are the major nonvariant gases of the homosphere, and what role do they play in terms of human survival on the planet earth?

2. The nonvariant gases are used and produced in many terrestrial systems, yet they are always found in the same percentages in the homosphere. What does this tell us about the homosphere in system terms?

3. Where on the earth's surface would we expect to find the highest concentrations of water vapor? What important role does water vapor play in these and other areas?

4. Recent controversies over the Supersonic Transport (SST) and the use of fluorocarbons as aerosol propellants have centered on the effect these two technologies might have on the ozonosphere. What surface effects might we expect from degradation of the earth's ozone layer?

5. Impurities in the air play a large role in determining the color of the sky, due to the differential scattering of light. The smallest particles scatter the blue wavelengths of light, while larger impurities tend to scatter the longer yellow and red wavelengths of light. Knowing this, where would you expect to find the most spectacular sunsets on a regular basis, and why?

6. The carbon dioxide cycle is one of the four major subsystems that contribute to the make-up of the earth's atmosphere. It is considered a closed system, despite some evidence that concentrations of CO_2 in the homosphere are increasing. Explain how this is possible in system terms, using the major subsystems of the carbon dioxide cycle and their respective time scales.

7. Explain how the nitrogen cycle, one of the four major subsystems of the earth's atmosphere, is a closed system. How long would we expect the complete nitrogen cycle to take? How does this compare with the length of time for completion of the various subcycles of the carbon dioxide cycle?

8. We cannot see pressure, yet it is one of the most important concepts in studying climate. What type of model do we use to explain pressure? What is the unit of pressure used in atmospheric studies, and how do we measure it?

9. At the average atmospheric pressure of 1013 mb, 760 mb is from nitrogen, 240 mb is from oxygen, 10 mb is from water vapor, and 3 mb is from other constituents of the air. How are the millibar figures for the pressure attributable to the various gases derived?

10. What is the relationship of air pressure and height? What connection does this have with the relationship between temperature and height?

11. What are the different parts or layers of the atmosphere in terms of temperature? Why are atmospheric temperature characteristics sufficiently important to form the basis for one type of classification system?

12. We have two different sets of names for parts of the atmosphere. On the one hand we have the homosphere, ozonosphere, and heterosphere; on the other we have the troposphere, stratosphere, mesosphere, and thermosphere. What is the reason for the two different sets of names?

13. What characteristics of the troposphere make it especially important to human life?

14. In the example concerning the parcel of air in Figure 3-10b, if the ELR is 1.25°C/100 m instead of 1.5°C/100 m for the first 600 m above the earth's surface, and at 600 m the ELR changes to .75°C/100 m, how high would the parcel of air travel before reaching a stable condition?

15. What other information would we need to determine the probable stability of the air over a tropical jungle in the rainy season besides the environmental lapse rate and the dry adiabatic lapse rate? Would we need the same information if we were examining the air over the Sahara Desert in July? Why or why not?

16. What factors might account for the varying distance of the tropopause, having a minimum altitude of about 8 km (5 mi) over the poles, and a maximum altitude of about 16 km (10 mi) over the equator? (Remember, the troposphere is principally defined by its lapse rates, which vary but allow for vertical mixing of air.)

17. What characteristics of the stratosphere make it a particularly good level for long-distance flights by jet aircraft? What negative characteristics might a pilot have to take into consideration depending on the direction of flight and the time of year?

18. Which layers of the atmosphere are most important to ham radio operators? What process in these layers makes them important, and what is their relevance to long-distance communications, especially in the days before orbiting communications satellites?

19. As we learn more about the upper atmospheric layers, we are gaining a new appreciation for the role they play in maintaining conditions closer to the earth surface. We are beginning to realize that the atmosphere contains more than the air we breathe. What do you feel are the most important questions still to be answered concerning the atmosphere?

FLOWS OF ENERGY AND WATER

CHAPTER 4

The physical world—earth and universe—is characterized by ebb and flow. Materials and energy are never destroyed; they continually pass from one place to another. The cyclic movement of carbon dioxide and nitrogen, described in Chapter 3, is one example of such flow. But perhaps the best examples are the flows of water and energy on earth.

In this chapter, we consider the continual movement of water among various earth spheres that leads to a balance of water on the earth's surface. This circulation of water is powered by radiant energy from the sun, a form of heat. The inflows and outflows of radiant energy and heat also balance at the earth's surface. Water, like air, can transport heat, as you may notice when a lake "steams" on a frosty morning. In addition, heat flows into and out of the ground. The constant flow of heat leads to the daily, annual, and spatial variations of temperature that we experience on earth. As we will see in the following pages, these flows and balances are interdependent, and such interdependence is a fundamental part of the grand scheme of nature.

THE WATER BALANCE

Our bodies are 70% water. We each need 1.4 liters (l) (1.5 qt) of water a day to survive, and our food could not grow without it. Water is everywhere. We breathe it, drink it, bathe in it, travel on it, and see beauty in it. We use it as a raw material, a source of power, and a medium for waste disposal.

The single greatest factor in water's widespread importance is its ability to exist in three forms (shown in Figure 4-1) within the temperature ranges normally encountered near the earth's surface. The solid form of water, *ice*, is composed of molecules linked together in a regular fashion. The bonds that link molecules of ice can be broken by heat energy. When enough heat is applied, ice changes its state and becomes the liquid form we know as *water*. The molecules in the liquid are no longer arranged in a regular fashion but exist together in a random form. In the liquid state, the individual molecules are freer to move around. More heat completely frees individual molecules from the liquid, and they move into the air. The molecules in the air are now a gas called *water vapor*.

Water is so common and consistent in its physical behavior that we use it to measure heat. We say that 1 *calorie* of heat is the energy needed to raise the temperature of 1 g (0.04 oz) of water by 1°C (1.8°F). It takes about 80 cal per gram of water to change the state of water from solid to liquid, a process we call *melting*. When it was first discovered that 80 cal of heat were needed to break the molecular bonds in solid water, the required heat appeared to be hidden, or latent. Thus the heat involved in melting is called the *latent heat of fusion*. In a similar way, it takes 597 cal of heat to change the state of 1 g of water at 0°C (32°F) from liquid to gas. This change is called *evaporation* or vaporization, and the heat associated with it is known as the *latent heat of vaporization*. Sometimes ice can change directly to water vapor. In this process, called *sublimation*, the heat needed is the sum of the latent heats of fusion and vaporization.

One of the beauties of these processes is that they are completely reversible. Water vapor can change back to water in the *condensation* process, water can change to ice in freezing, and water vapor can change directly to ice (also called sublimation). In reversing these processes the same quantities of latent heat are given out. Chapter 3 reveals an important consequence of this in the smaller value of the saturated adiabatic lapse rate compared with the dry adiabatic lapse rate.

FIGURE 4-1

A schematic view of the molecular structure of water in its three states and the heat energy relationships between the states.

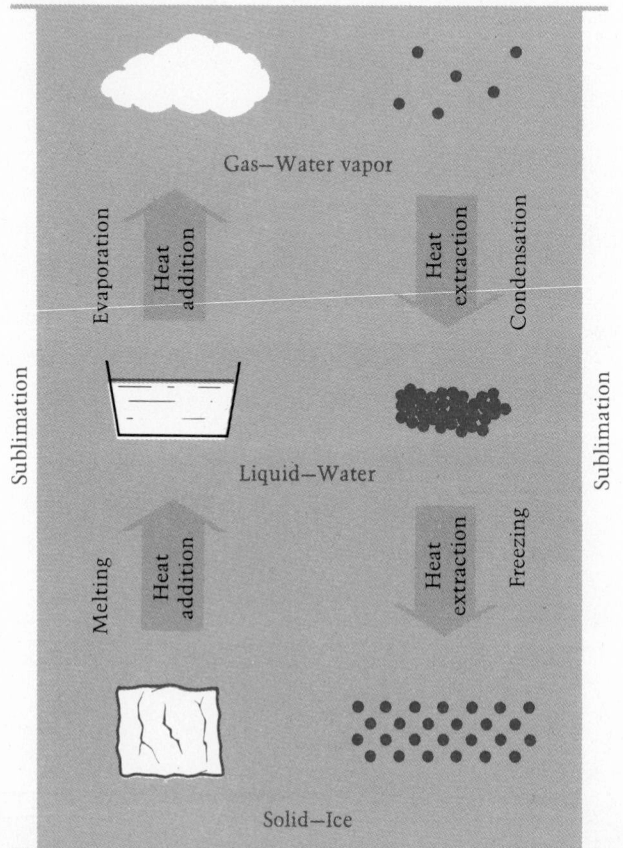

MEASURING WATER VAPOR

Lord Kelvin, the inventor of the absolute temperature scale, once said that we do not know anything about anything until we can measure it. How do we measure these three states of water? The measurement of solid ice and liquid water is quite straightforward; we simply weigh them. We also use this method indirectly for water vapor when we measure the *vapor pressure* of water vapor in a column of air. However, other measurements of water vapor are of value. The most useful of these are relative humidity, specific humidity, and mixing ratio.

One way of making water vapor condense to liquid water is to cool it and the surrounding air. Condensation takes place when the air contains more vapor molecules than it can hold. In this state we say the air is *saturated* or has reached its saturation point. This happens around a cold soft-drink bottle, where we can see that the vapor in the air condenses to liquid water droplets on the bottle. The same process occurs when the earth cools at night, subsequently cooling the air near it and condensing water vapor to the droplets we call *dew.* The temperature at which the air becomes saturated and condensation occurs is therefore called the *dew-point temperature.*

Relative humidity tells us how close a given sample of air is to the saturation or the dew point. Thus we may define *relative humidity* as the proportion of water vapor present in a sample of air relative to the maximum the air could hold at the same temperature. We express relative humidity as a percentage, so air of 100% relative humidity is saturated and air of 0% relative humidity is completely dry. Relative humidity often varies in opposition to air temperature. It is usually lower in the early afternoon when the temperature is high. At night, when the temperature falls, the colder air can hold less water vapor, and so the air approaches its dew point, and relative humidity is consequently higher.

The other terms used to assess the amount of water vapor in the air need less explanation. *Specific humidity* is the ratio of the weight of water vapor in the air to the combined weight of the water vapor plus the air itself. The *mixing ratio* is the ratio of the mass of water vapor to the total mass of dry air.

Both the mixing ratio and the specific humidity, as well as relative humidity, may be found by using a *psychrometer,* a device with two thermometers. The bulb of one thermometer is surrounded by a wet cloth; the air takes as much water vapor from the cloth as it can. The thermometer reaches a temperature called the *wet-bulb temperature.* The other thermometer, not swaddled in a cloth, indicates the *dry-bulb temperature.* The difference in temperature between the wet-bulb and the dry-bulb thermometers is computed. Then the relative humidity or the mixing ratio is found in an appropriate set of humidity tables.

Now that we are armed with some basic terminology and an understanding of the ways water changes from one state to another, we can proceed to find out how it circulates through the earth system. The cycle describing this circulation is called the *hydrologic cycle.*

THE HYDROLOGIC CYCLE

Early scientists believed that the wind blew water from the sea through underground channels and caves and into the atmosphere, purifying the water in the process. Today scientists talk about the hydrologic cycle, in which water comes to the earth from the atmosphere through *precipitation* and eventually returns to the atmosphere by *evaporation,* a process some early scientists were not aware of.

The hydrologic cycle consists of a number of stages, and Figure 4-2 shows the relative amounts of water in each. The largest amounts of water in the whole cycle are those involved in direct evaporation from the sea to the atmosphere and in precipitation back to the sea. Evaporation from plants and land surfaces, called *evapotranspiration,* combines with precipitation of water onto the land surfaces to play a numerically smaller, but possibly more important, part in the hydrologic cycle. If precipitation at the land surface does not evaporate, it runs off the surface in the form of streams and rivers, a phenomenon called *runoff.* In Figure 4-2, the runoff value includes some water that moves underneath the surface and eventually finds its way to a river or ocean.

The hydrologic cycle can easily be viewed as a closed system, in which water is continuously transported among the spheres of the earth system. For example, the values in Figure 4-2 indicate that the amount of water transported in the atmosphere over the continents equals the amount transported by surface runoff back to the ocean. Water flows among the atmosphere, the lithosphere, and the oceans of the hydrosphere. The system can also be split into two subsystems, one dealing with precipitation and evaporation over the oceans and the other concerned with evapotranspiration and precipitation over land areas. The two subsystems are connected by horizontal movement in the atmosphere, known as *advection,* and by runoff flows.

The time required for a complete cycle is usually quite brief. A molecule of water can pass from the ocean to the atmosphere and back again in a matter of days. Over land, the cycle is less rapid. Groundwater goes into the soil or subsurface and can remain there

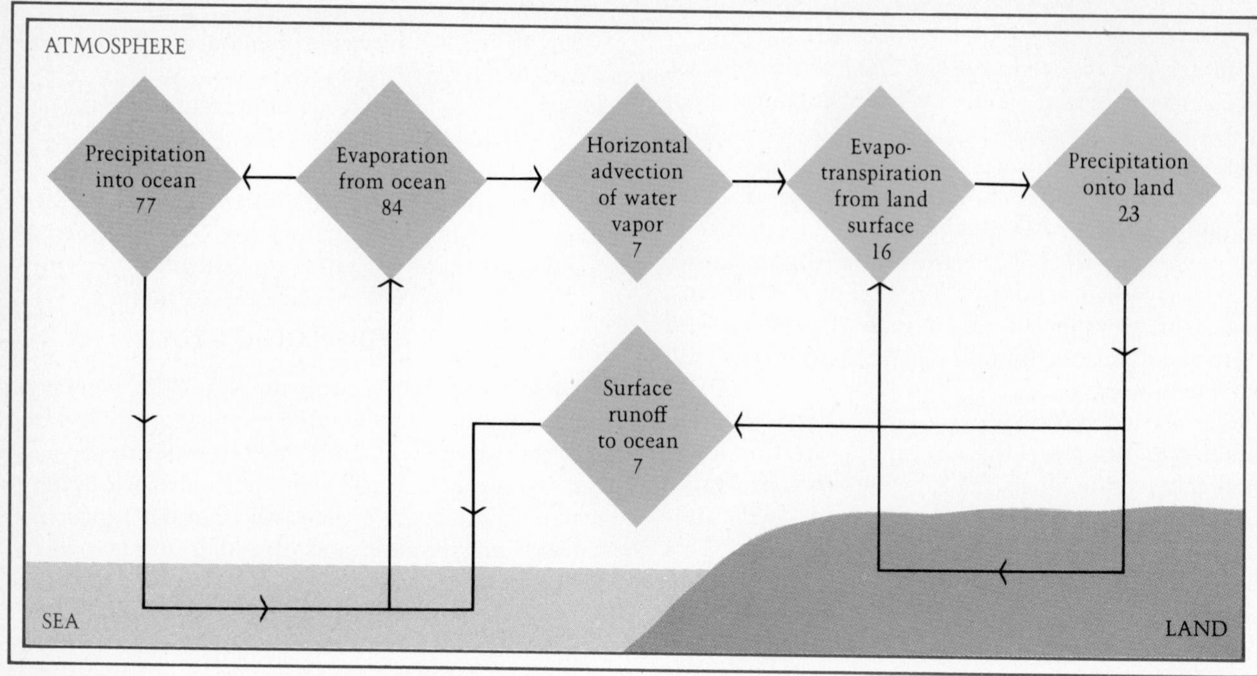

FIGURE 4-2
The hydrologic cycle. The numbers attached to each stage of the cycle show the percentage of the total water annually circulating in the system that is involved in any one stage.

for weeks or months. The circulation is even slower when water in the form of ice is concerned. Some water has been locked up in the major ice sheets of the world for about a hundred thousand years.

Water is very unequally distributed within the hydrosphere, as Figure 4-3 shows. The oceans contain 97% of all water, but the high salt content makes it of little direct use to humans. Of the 3% of fresh water, 75% is locked up in glaciers and ice sheets. The next largest proportion, 14%, is accessible only with difficulty, because it is groundwater found below 750 m (2500 ft). The fresh water needed most urgently for domestic, industrial, and agricultural uses must be taken from the relatively small supplies found at or near the surface—in the soil, rivers, lakes, and atmosphere. These storage areas contain less than 11.5% of the 3% of fresh water in the hydrosphere. This small supply of fresh water must keep circulating within the hydrologic cycle through evaporation and precipitation.

EVAPORATION

We cannot see evaporation occurring, but its results are sometimes visible. When you see steam rising from a hot tub or from a lake that is warm compared to the cold air above it (see Figure 4-4), you are seeing liquid water droplets that have already condensed. It is not too hard to imagine molecules of invisible water vapor rising upward in the same way. Evaporation occurs when two conditions are met. First, heat energy must be available at the water surface to change the liquid water to a vapor. This heat is sometimes provided by the moving water molecules, but most often the radiant heat of the sun, or both sources together, provides the necessary heat energy.

The second condition is that the air must not be saturated; it must be able to absorb the evaporated water molecules. Normally air near a water surface has a large number of vapor molecules. Thus the pressure caused by the movement of the vapor molecules, the *vapor pressure,* is high. But air at some distance from the water surface has fewer molecules of vapor and, subsequently, a lower vapor pressure. In this case, we say there is a *vapor-pressure gradient,* or difference, between the two locations. A vapor-pressure gradient exists above a water surface as long as the air has not reached its saturation point. Just as people tend to move from a very crowded room to a less crowded one, molecules of water vapor tend to move down the vapor-pressure gradient, to the area of less pressure, which usually means moving higher in the atmosphere.

Knowing the requirements for evaporation—a heat source and a vapor-pressure gradient—we can guess the kinds of situations that would yield maximum evap-

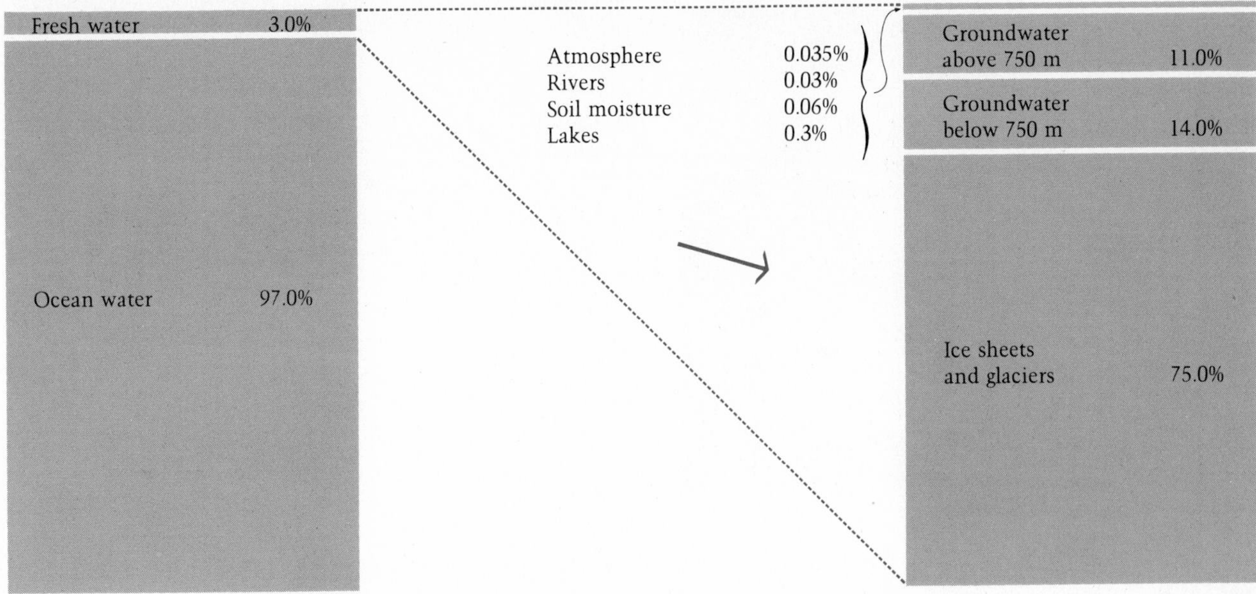

FIGURE 4-3

The distribution of water in the hydrosphere. The column on the right shows the distribution of the 3% of total hydrospheric water that is fresh.

oration. Because radiation from the sun is a major heat source, large amounts of evaporation can be expected where there is a great deal of sunlight. This is particularly true of the tropical oceans. Water molecules also move most easily into dry air, because of the large vapor-pressure gradient. So the drier the air, the more evaporation will occur. Evaporation is even faster in windy conditions, when the air containing new vapor molecules can be continually replaced. Thus higher wind speeds also promote evaporation. All these principles are demonstrated by the action of a blowing hair dryer.

FIGURE 4-4

In Sun Valley, Idaho, steam and invisible water vapor molecules rise from a swimming pool generally kept above 40°C (100°F).

Evapotranspiration We have seen that most of the evaporation into the atmosphere occurs over the sea. Over the land, water evaporates from lakes, rivers, damp soil, and other moist surfaces. But the water plants lose to the air during photosynthesis is another major source. The passage of water through leaf pores is called *transpiration*, and the term "evapotranspiration" covers all the processes by which water evaporates from the land surface.

Physical geographers draw a distinction between potential evapotranspiration and actual evapotranspiration. *Potential evapotranspiration* (PE) is the maximum amount of water that can be lost to the atmosphere from a land surface with no lack of available water. *Actual evapotranspiration* (AE) is the amount of water that can be lost to the atmosphere from a land surface with any particular soil moisture conditions. AE can equal PE when the land surface is saturated, but when the soil moisture is less than its maximum value, AE is usually less than PE.

Evapotranspiration puts the molecule of water vapor into the air. There it can be transformed into a liquid droplet to start the chain of events that lead to precipitation.

A few of the possible snowflake configurations.

PRECIPITATION

Two preliminary factors are necessary for precipitation: cooling of air, so water vapor will condense, and the small particles called *condensation nuclei*, around which liquid droplets can form. Parcels of air may cool enough to produce condensation when they rise to the higher, cooler parts of the atmosphere or when they come into contact with colder air or a colder surface. The condensation nuclei are almost always present in the atmosphere in the form of dust or salt particles.

When water droplets first form, they are so small that the lightest upward air current keeps them airborne for long periods. The basic problem is how they can grow large enough to fall out of a cloud. There are two main processes.

Probably the most common process, called the *ice-crystal process*, was first identified in the 1930s by the scientists Tor Bergeron and Von W. Findeisen. This process requires both liquid droplets and ice particles in a cloud. Ice particles are normally present if the temperature is below 0°C (32°F) and if there are small particles, called *freezing nuclei*. Freezing nuclei perform the same function for ice particles as condensation nuclei perform for water droplets. When both ice particles and water droplets are available, the water droplets tend to evaporate and then sublimate directly onto the ice crystals. This is because the pressure of saturated water vapor over the liquid water droplet is greater than that between the water vapor and the ice crystal. Thus the ice crystal grows at the expense of the liquid droplet. The ice crystals become larger and often join together to form a snowflake. When the snowflake is large enough, it drops out of the cloud. On its way down, it usually encounters higher temperatures and melts, eventually reaching the earth as a liquid raindrop.

Most rainfall and snowfall in midlatitudes is formed by the ice-crystal process, but in tropical areas the temperature of clouds does not necessarily reach freezing point. Thus a second process is thought to make raindrops large enough to fall from clouds. The *coalescence process* requires some liquid droplets to be larger than others, which happens when there are giant condensation nuclei. As they fall, the larger droplets overtake and capture the smaller ones. When the larger droplets incorporate the smaller ones, *direct capture* is said to take place. But narrowly missed smaller droplets may still be caught up in the wake of the larger ones and drawn to them. This is called *wake capture*. In either case, the larger droplets grow at the expense of the smaller ones and eventually become heavy enough to fall to the ground.

Precipitation reaches the ground in many forms, as Figure 4-5 indicates. *Dew* is the fine moisture that nightly bejewels grass and trees. Fine water droplets suspended in air near the ground are called *fog*. Large liquid water droplets form the familiar *raindrops*. If the ice crystals in the ice-crystal process do not have time to melt before reaching the earth's surface, the result is *snow*. A mixture of snow and rain is called *sleet* in British meteorology, and in the United States it refers

FIGURE 4-5

Some forms of precipitation—snow, rain, sleet, and hail.

to pellets of ice produced by the freezing of rain. Soft *hail pellets* can form in a cloud that has more ice crystals than water droplets and eventually fall to the surface. True *hailstones* result when ice crystals make several journeys between the lower, warmer part of a cloud, where they gain a water surface, and the higher, freezing part of the cloud, where the outer water turns to ice. This often occurs in the vertical air circulation of thunderstorms.

As far as we humans are concerned, the most crucial portion of the hydrologic cycle is at the surface of the earth. Here, at the interface between earth and atmosphere, evaporation helps plants grow and precipitation provides the water for evaporation. And at this surface we may measure the water balance.

AT THE EARTH'S SURFACE

An accountant keeps a record of financial income and expenditure and ends up with a balance. The balance is positive when savings have been made and negative when excessive debts have been incurred. We can describe the balance of water at the earth's surface in a similar way.

Water may be gained at the surface by precipitation or, more rarely, by transport across the surface in rivers or in soil or groundwater. Water may be lost by evapotranspiration or by running off across the surface or just beneath the surface. The balance is achieved by adding the gains from precipitation and subtracting the losses through runoff and evapotranspiration.

When actual evapotranspiration is used for the computation, the balance is always zero, because no more water can run off or evaporate than is gained by precipitation. However, when potential evapotranspiration is taken into account, the balance may range from a constant surplus of water at the earth's surface to a continual deficit. Figure 4-6 illustrates this range.

Bellary, in the center of southern India, exemplifies the water balance at a deficit. Throughout the year, the potential evapotranspiration exceeds the water gained in precipitation. On the average, even in the times of the autumn rains, the soil contains less water than it can hold. Because plants depend on water, the vegetation in this region is sparse, except where irrigation is possible. At Bogor, on the Indonesian island of Java, the situation is reversed. During every month of the year, rainfall, sometimes as much as 450 mm (18 in) in one month, exceeds the amounts of water that can be lost in evapotranspiration. The surplus water provides all the water needed for luxuriant vegetation and still leaves copious quantities to run off the land surface. An intermediate state of affairs exists in Berkeley,

MAKING SNOW IN SKI COUNTRY

The development of snowmaking equipment triggered a ski explosion in North America. More than 200 ski areas in the United States and 70 in Canada now hold snowmaking licenses. Relying on natural snow gives resort operators ulcers. Although the season may have an average snowfall, heavy snow might not arrive until after the Christmas holidays, or it might thaw, or be covered with freezing rain. Even in heavy snow regions artificial snow is insurance. Without a good snow cover, resort owners lose thousands of dollars, and skiers despair.

The United States granted the first patent for snowmaking in 1950. In the early days mixing the right combination of compressed air and water in the nozzle was a trial and error process. Instead of snow, crews often got exploding pipes, frozen equipment, and drenched operators. Now the adjustments can be handled fairly precisely at a master control panel. About 104,500 l (27,500 gal) of water will cover one acre with 10 cm (4 inches) of wet snow. The amount of compressed air needed depends on the temperature—less air, at lower temperatures. Another method involves feeding jets of water into a fan to create a fine mist.

In both processes the snow falls like a fine hail. On the ground it becomes granular, and it is difficult to distinguish from natural snow. In theory snow can be created any time the temperature falls below 0°C (32°F), but slope crews usually wait until it is slightly colder. Humidity, air pressure, and wind also affect the quality of the snow and how quickly it can be produced.

FIGURE 4-6

The water balance in deficit, surplus, and intermediate states. Bellary, India shows a constant deficit. A constant surplus is found at Bogor, Java. Berkeley, California experiences a combination of deficit and surplus at different times of the year.

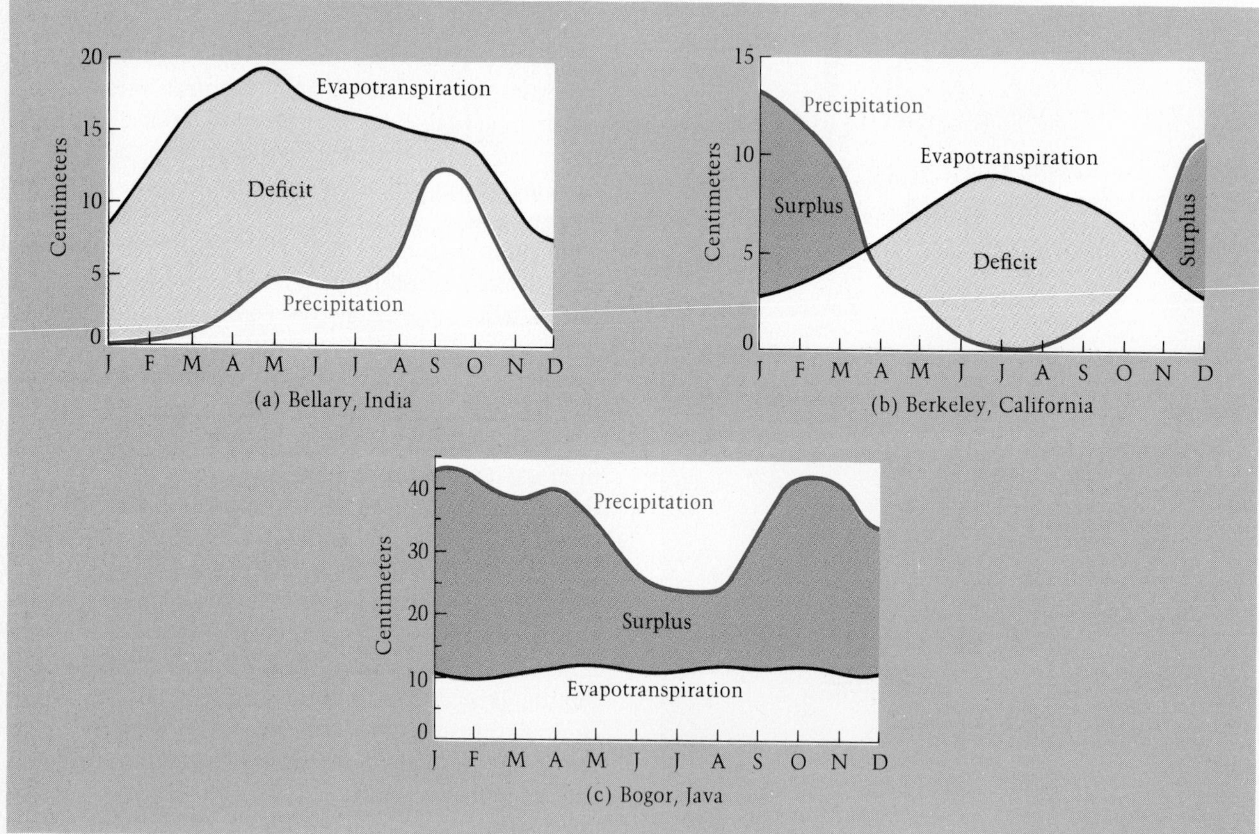

California. From November to May, the precipitation exceeds the potential evapotranspiration, but from April through October, there is a water deficit.

Starting in April, when potential evapotranspiration exceeds precipitation, soil moisture from below the earth's surface is used in evaporation. Most of it is drawn up through the roots of plants and evaporates from their leaves. This process continues until the end of October, when rainfall once more exceeds potential evapotranspiration and the stock of available soil moisture is recharged. At this time runoff is more plentiful from the winter storms.

The amount of runoff in any location cannot exceed the amount of precipitation, and usually there is much less runoff than precipitation. This is because some water almost always evaporates. In the United States, runoff approaches the amount of precipitation only in parts of Washington and Oregon. In the southwestern states, runoff is quite small compared to precipitation.

Over the whole globe, the values of precipitation, evaporation, and runoff vary greatly with latitude. As Figure 4-7 shows, annual precipitation and runoff are highest near the equator. Evaporation also is high at this location, but it is greatest in subtropical latitudes (20° to 35°). High midlatitudes (45° to 60°) feature water surpluses, and precipitation, runoff, and evaporation are lowest in high latitudes.

FIGURE 4-7

The average annual latitudinal distribution of precipitation, evaporation and runoff for the whole globe.

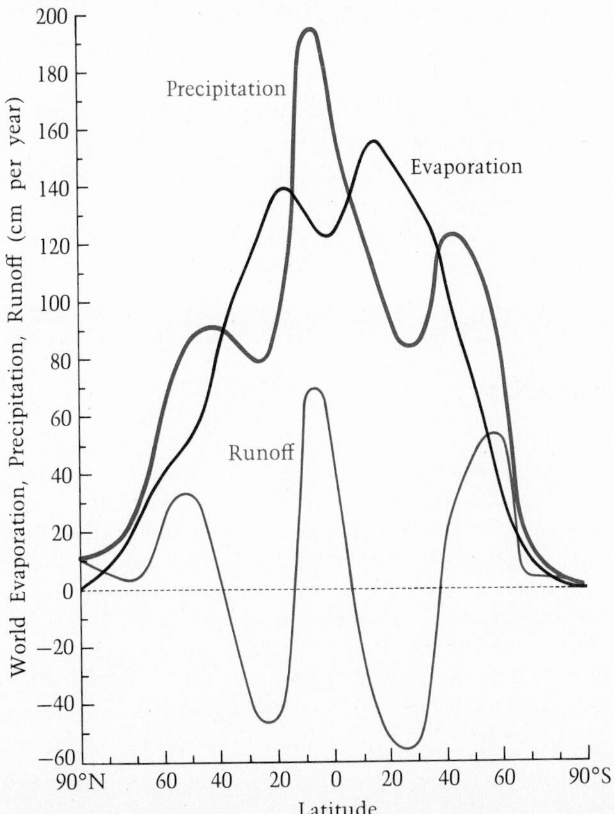

The reasons for this distribution lie mainly in the differences in radiant heat that reach an area and in the large flows of air around the globe collectively known as the general circulation. Near the equator, radiant heat from the sun leads to high evaporation rates. It also causes air to rise, cool, and thereby yield large quantities of precipitation. In subtropical areas, descending and warming air gives clear weather. There are high evaporation rates over the oceans but little evaporation over the land surfaces, because of the lack of moisture to be evaporated. In the higher midlatitudes, eastward-moving storms provide moderate amounts of precipitation in most areas, and moderate amounts of radiant heat evaporate much of the water that falls. In high latitudes, the cold air can hold little water vapor and thus there is little precipitation. Low amounts of radiant heat lead to minimal rates of evaporation. Figure 4-8 diagrams the global pattern of evapotranspiration, and Figure 4-9 shows the distribution of precipitation.

If we consider the distribution of the earth's population, we find that most people live in areas where there are neither great surpluses nor deficits in the water balance. Water is undoubtedly a matter of life and death. Equally, if not more, important is the radiant heat we receive from the sun.

THE RADIATION BALANCE

The hydrologic cycle, which continually replenishes our water resources, could never work without the driving energy it receives from the heat of the sun. Without radiant heat, or radiation, there would be no evaporation and the land surfaces would be barren. Our life-giving sun provides 99.97% of the energy required for all the physical processes that take place on the earth and in the atmosphere, and most of the rest of the energy comes from within the earth. Different types of radiation flow throughout the earth system, and inputs and outputs of radiation strike a balance at the earth's surface.

We may regard radiation as a transmission of energy in the form of waves. The wavelength of the radiation is the distance between two successive wave crests. This wavelength varies in different types of radiation and is inversely proportional to the temperature of the body that sent it out. The higher the temperature at

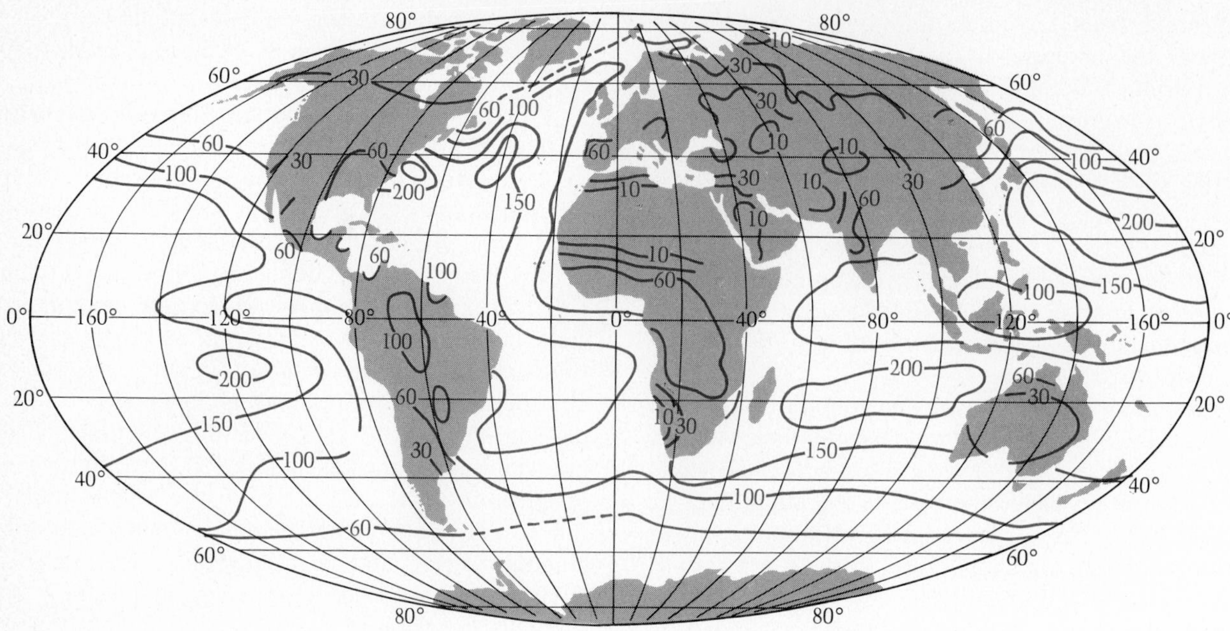

FIGURE 4-8
The global distribution of annual evaporation and evapotranspiration. The broken lines represent uncertainty of data. Isolines are not continued over mountainous areas.

which the radiation is emitted, the shorter the wavelength of the radiation. The sun has a temperature of about 6000°K at the surface, but the earth's average surface temperature is only about 300°K. Thus radiation coming from the sun is called *shortwave radiation*, and that emitted from the earth is called *longwave radiation*. There is in fact a large variety, or *spectrum*, of radiation of different wavelengths, which is depicted in Figure 4-10. This spectrum varies from very short waves, such as x-rays and gamma rays, to very long waves, such as radio waves.

RADIATION FROM THE SUN

The most recent measurements indicate that, on the average, 1.95 cal of energy per square centimeter (0.16 square inch) are received every minute at the earth's outer atmosphere. This value, called the *solar constant*, would equal in a day all the world's industrial and domestic energy requirements for 100 years.

When radiation travels through the atmosphere, several things may happen to it. Of all the incoming rays, 31% travel directly to the earth's surface as *direct*

THE MONSOONS

Cherrapunji, India averages 11,467 mm (450 in) of rain each year. In the record-breaking year of 1861, 26,467 mm (1042 in) fell. During the month of July the villagers were drenched by nearly one-third meter (one foot) every day.

The most spectacular rainfalls in the world come with the monsoons. For six months of the year dry winds blow from the interior of the continent and parch the earth. In Asia these winter winds come from the Gobi desert. Then in June the wind abruptly reverses and sweeps in from the ocean bringing cool air and heavy rain.

Nearly half the world's population relies on the monsoon rains to nourish their crops. From West Africa to China, farmers wait anxiously for the wind to change direction and signal the beginning of the rainy season. In 1972 the monsoon arrived three weeks late, and India lost nearly a third of its food crop. In 1974 the monsoon was again erratic. In some parts of India it came late and caused famines, but in Bangladesh heavy rains caused disastrous floods.

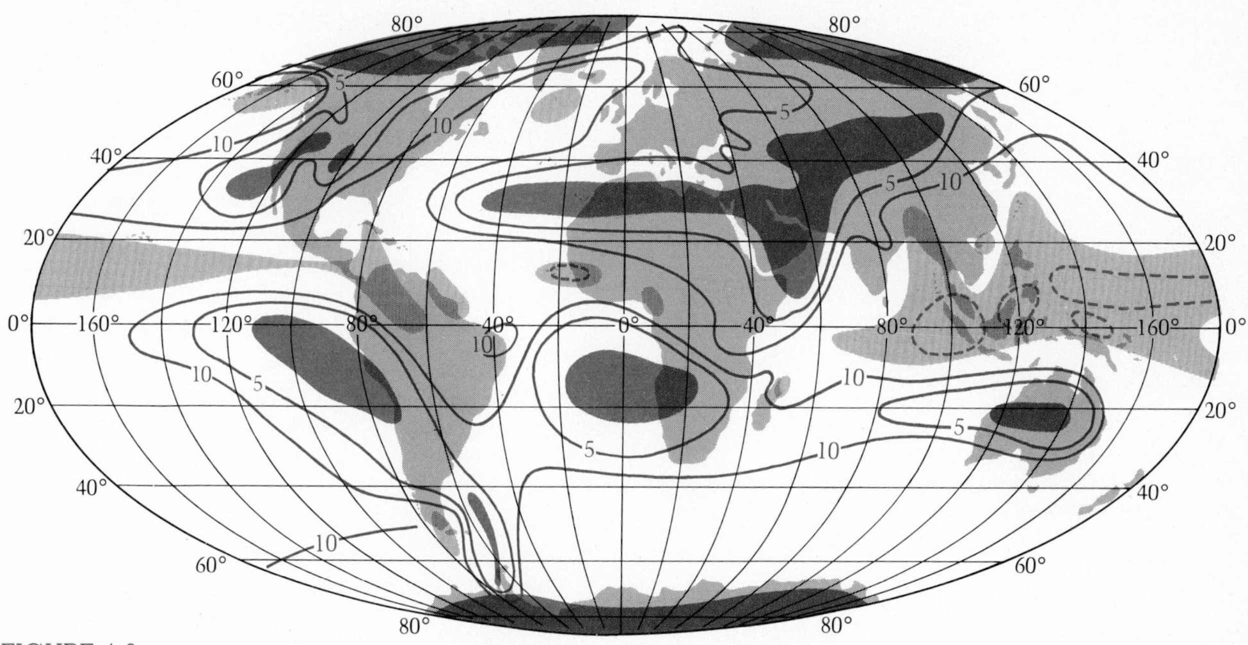

FIGURE 4-9
The global distribution of precipitation, in decimeters (10 cm). Light shading shows areas with precipitation above 20 decimeters, dark shading shows precipitation below 2.5 decimeters. The dashed lines are 30-decimeter lines.

radiation. An almost equal amount, 30%, is reflected and scattered back to space by clouds and particles in the atmosphere. A ray may also be one of the 17% that are absorbed by clouds or other components of the atmosphere. Some of the scattered rays, 22% of them, eventually find their way to the earth's surface and are collectively known as *diffuse radiation.* Altogether, just over half (53%) of the rays coming to the outer atmosphere reach the earth's surface as either direct or diffuse radiation. The rest are either absorbed by the atmosphere or scattered back to space.

The amount of radiation that falls on a surface depends on how much the surface tilts toward the incoming radiation. For example, in Figure 4-11 there are three boxes, each sending three rays of radiation from the sun to the earth at the time of an equinox (when day and night are equally long everywhere on earth). The boxes of radiation are directed toward three thin sheets of cardboard of equal area—one placed at the equator, one at latitude 45°N, and one at the North Pole. Each piece is flat on the earth's surface. At the equator, all three rays in the box strike the

FIGURE 4-10
The complete radiation spectrum.

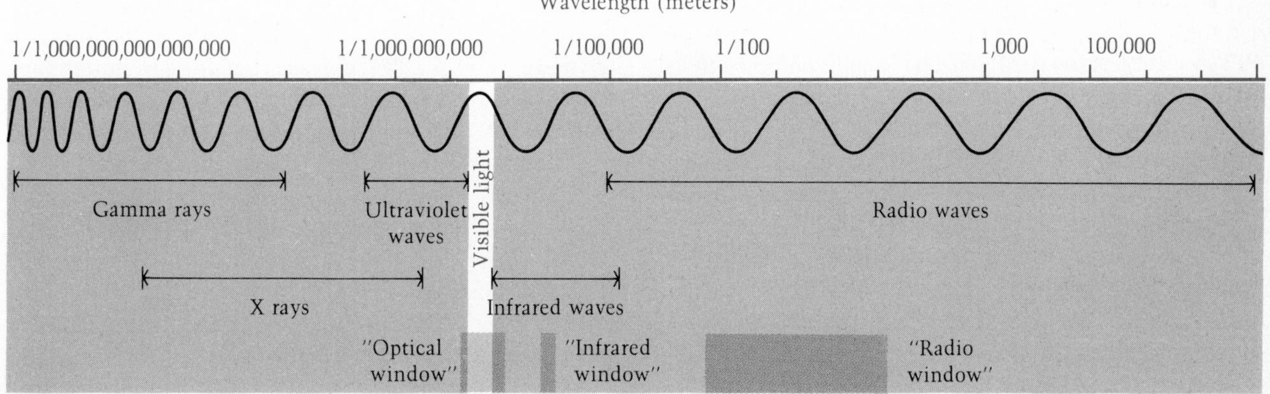

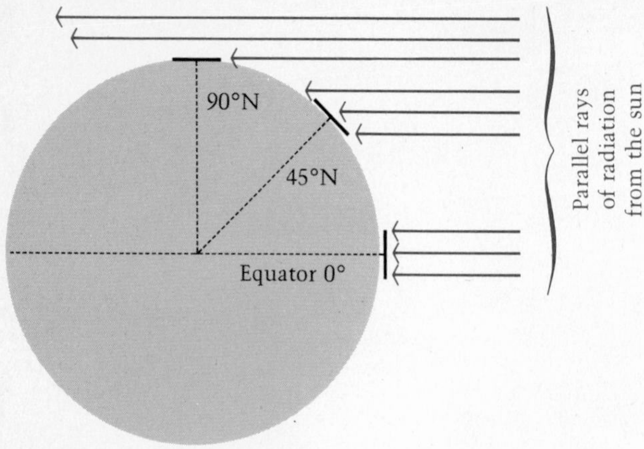

FIGURE 4-11

The reception of radiation. A surface at high latitudes receives less radiation than a surface of equal area at low latitudes.

cardboard. But at 45° N, two of the three rays miss the cardboard, and at the North Pole, none of the rays fall directly onto the cardboard, although one ray runs parallel to the surface. More radiation strikes an equal area of the earth at the equator than at a higher latitude; radiation is more intense at low latitudes than at high latitudes. This effect, which occurs to a greater or lesser extent throughout the year, is fundamental in explaining the amount of solar radiation at various latitudes as well as the general circulation of the atmosphere and the climates of the earth.

No matter where a ray of radiation falls on the earth, one of two things can happen to it: It can either be absorbed by, and thereby heat, the earth's surface, or it can be reflected by the surface, in which case there is no heating effect. The amount of radiation reflected by the surface depends mainly on the color, composition, and slope of the surface. A ray falling on the cardboard at the equator, because it strikes perpendicularly, is less likely to be reflected than one falling on the cardboard at 45°N. And if the cardboard at the equator is a dark color, say black, the ray is less likely to be reflected than if the cardboard has a light color, such as white. The proportion of incoming radiation that is reflected by a surface is called the *albedo*, a word derived from the Latin word *albus*, meaning "white." The albedo of a snowy surface, which reflects most of the incoming radiation, might be 80%, whereas the albedo of a dark-colored forest, which reflects very little radiation, might be as low as 10%. Of all the solar radiation entering the atmosphere, only 47% is absorbed by the earth's surface. We now turn to the other side of the coin—the radiation from the earth itself.

RADIATION FROM THE EARTH

The earth does more than absorb or reflect radiation from the sun; it constantly gives off longwave radiation or *infrared radiation* on its own. Two fates may befall radiation leaving the earth's surface: Either it can be absorbed by the atmosphere, or it can escape to space.

The major atmospheric constituents that absorb the earth's radiation are ozone, carbon dioxide, and water vapor. All these gases absorb radiation of certain wavelengths but permit other wavelengths to escape, through an atmospheric "window." Up to 9% of all terrestrial radiation is thereby lost to space, except when the window is shut by clouds. Clouds absorb or reflect back to earth almost all the outgoing earth radiation. Therefore a cloudy winter evening is likely to be warmer than a clear one.

The atmosphere is heated by the longwave radiation it absorbs. Most radiation is absorbed at the lower, denser levels of the atmosphere, a fact that helps account for the higher temperatures near the earth's surface. The atmosphere itself, being warm, can also emit longwave radiation. Some goes off into space, but some, known as *counter radiation*, is reradiated back to the earth. Without this counter radiation from the atmosphere, the earth would be 30°C to 40°C (54°F to 72°F) colder. The atmosphere therefore acts as a blanket.

The blanket effect of the atmosphere is similar to the action of radiation in a garden greenhouse. Shortwave radiation from the sun, entering the greenhouse through the glass, heats the inside. The inside surfaces then emit longwave radiation. But the longwave radiation cannot pass through the glass and is trapped within the greenhouse, which raises the temperature of the air inside. The same thing happens on the earth, with the atmosphere replacing the glass of the greenhouse. It is called the *greenhouse effect*.

NET RADIATION

The radiation accounts for the earth as a whole are given in Table 4-1. We can see that similar amounts of shortwave and longwave radiation come to the earth's surface but that the outgoing radiation is dominated by the longwave radiation emitted by the earth. The amount left over when all the incoming and outgoing flows have been tallied is the *net radiation*, about one-fourth of the shortwave radiation originally arriving at the atmosphere's outer layers.

The albedo of the earth's surface and its temperature play particularly important roles in determining the final value of the net radiation. For example, there is

TABLE 4-1

The annual radiation balance (net radiation) of the earth's surface in thousands of calories per square centimeter

INCOME	
Shortwave radiation reaching the top of the atmosphere	263
Longwave counter radiation from the atmosphere absorbed at the earth's surface	206
	469
EXPENDITURE	
Longwave radiation emitted by the earth	258
Shortwave radiation reflected into space by the atmosphere and the earth's surface	94
Shortwave radiation absorbed by the atmosphere	45
	397
BALANCE (income minus expenditure)	72

usually a difference in albedo and surface temperature between an area of land and sea at the same latitude. As is shown in Figure 4-12, this difference results in a difference in net radiation values over the land and sea. The difference in net radiation over land and sea is greatest in low latitudes and lessens in higher latitudes. Overall, net radiation is highest in low latitudes and smallest, or even negative, in high latitudes.

Net radiation may well be the most important single factor affecting the earth's climates. It is certainly basic to the majority of physical processes that take place on the earth, because it provides their initial driving energy.

For example, net radiation is by far the most important factor determining the evaporation of water. The amount of water evaporated and the quantity of available net radiation together can largely explain the distribution of vegetation across the land surfaces of the earth, from the dense forests of the equator to the sparse mosses and lichens of the subarctic areas. Furthermore, net radiation is the input into the third and final vital balance of the earth's surface that we are concerned with in this chapter—the heat balance of the earth.

THE HEAT BALANCE

Climate is often considered to be something derived from the atmosphere, and it is true that the climate of a place is essentially the result of the redistribution of heat energy across the face of the earth. However, the events of the atmosphere are greatly affected by the

FIGURE 4-12

The annual distribution of net radiation at the surface of the earth. Values are in thousands of calories per square centimeter.

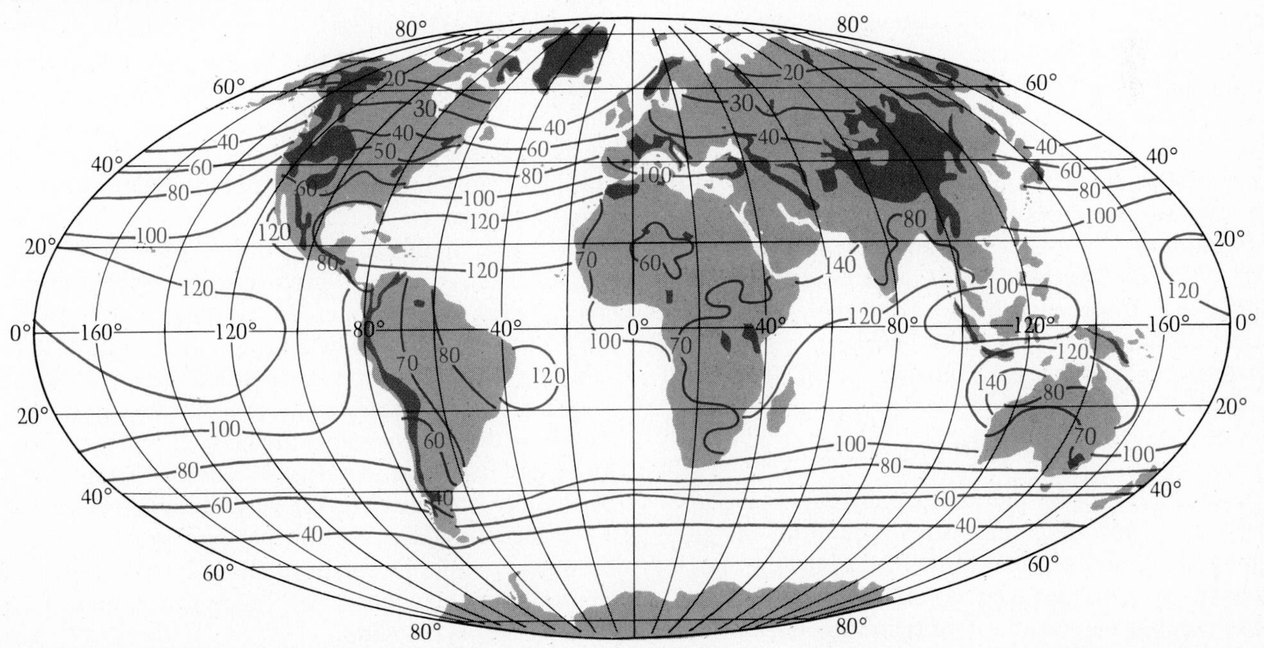

earth's surface itself. Flows of heat energy to and from the surface are as much a part of the climate of an area as the winter snow or summer thunderstorm—more so, in fact, because they exist continually.

The heat energy balance of the earth's surface is composed, in its most simple form, of four different kinds of flows. Two of these, the flow of evaporative heat and the composite flows of radiant heat making up net radiation, are already familiar to us. The other two, sensible heat flow and soil heat flow, are new. Chapter 3 describes how parcels of air can move vertically. The air molecules contain heat energy, the heat that we feel on our skins, and this heat is given the name *sensible heat.* Usually, during the day, the ground warms the air above it, and parcels of air move upward, thereby causing a sensible heat flow. We see the results of this process in the shimmering air above a parking lot on a very hot day. Sometimes, when the ground surface is colder than the air above it, sensible heat flows downward. This often happens at night or in winter in cold climates.

Whereas sensible heat flow depends on the vertical passage of parcels of air, the heat that flows into and out of the ground depends on *conduction*, the transport of heat energy from one molecule to the next. The heat that is conducted into and out of the earth's surface is collectively called *soil heat flow* or *ground heat flow.* These terms are used for convenience, even though the heat sometimes travels into buildings or plants or into the ocean. (In some cases the medium it flows through is specified by name.) Soil heat flow is the smallest of the four heat balance components. Generally the heat that passes into the soil during the day is approximately equal to that flowing out at night. Thus, over a 24-hour period, the balance of soil heat flow is often small enough to be neglected.

Except for the usually small amount of energy used by plants in photosynthesis, the total heat balance of any part of the earth, say that part outside your window, is made up of the flows of radiant heat (composing net radiation), latent heat from evapotranspiration, sensible heat, and heat flow into and out of the soil. We could examine the heat balance of a single leaf or a corn crop or a continent, but we are more concerned here with explaining climates through the heat balance approach.

CLIMATES IN BALANCE

At any location, the temperature of the atmosphere depends on how much heat is involved in the radiant, evaporative, and sensible heat flows and how much heat flows into and out of the soil. Usually net radiation is a source of heat for the earth, and the heat gained this way is used mainly in the latent heat required for evaporation or in a sensible heat flow into the air. But there are significant variations on this theme across the earth's surface, and these lead to significant variations in climate.

Deep in the equatorial rainforest, 1100 km (680 mi) from the mouth of the Amazon River, lies the town of Manáos, Brazil. Its hot, humid climate is explained by the high amounts of net radiation it receives, which evaporates much of the large quantities of rainfall. If we examine the heat balance for Manáos, shown in Figure 4-13a, we can see that most of the heat received in net radiation is lost through the evaporative heat flow. A rather small amount is left over for the passage of sensible heat into the air. These conditions are almost unchanging throughout the year.

In contrast, in the higher latitudes of Aswan, where Eratosthenes placed a pole to measure the earth's circumference, the value of net radiation varies throughout the year, being largest in summer (see Figure 4-13b). There is little surface water to be evaporated, so the loss by latent heat is virtually absent. But Aswan's scorching temperatures would be even higher if most of the heat gained by net radiation did not pass, by sensible heat flow, higher into the atmosphere.

Paris, the capital of France, lies within the middle latitudes and has another type of heat balance, as Figure 4-13c indicates. The seasonal variation of net radiation is again a factor, but in Paris the loss of latent heat is only somewhat greater than the loss of sensible heat. However, a rather curious event occurs in Paris in the winter months. The net radiation becomes negative; more radiant heat is lost than is gained. Net radiation is no longer a heat source. Fortunately, this loss is offset. Air that has been warmed in its journey across the Atlantic Ocean can now provide heat to warm the earth. Therefore, during the winter months, the sensible heat flow is directed toward the earth's surface, as shown by the negative values in Figure 4-13c. The sensible heat flow is responsible for keeping winter air temperatures mild.

In central Siberia this does not happen. Air coming to Turukhausk in the winter has not traveled over a warm ocean but over a cold continent. Although the air passes some sensible heat toward the ground, it does not pass enough to offset the large net radiation deficit experienced in winter near the Arctic Circle (see Figure 4-13d). The result is the body-chilling temperatures described in the writings of Boris Pasternak and Aleksandr Solzhenitsyn. Yet here the seasonal change of climate is extreme. The balance of heat in the summer months is, paradoxically, rather like that in central Brazil. There are many such variations of heat balance across the earth.

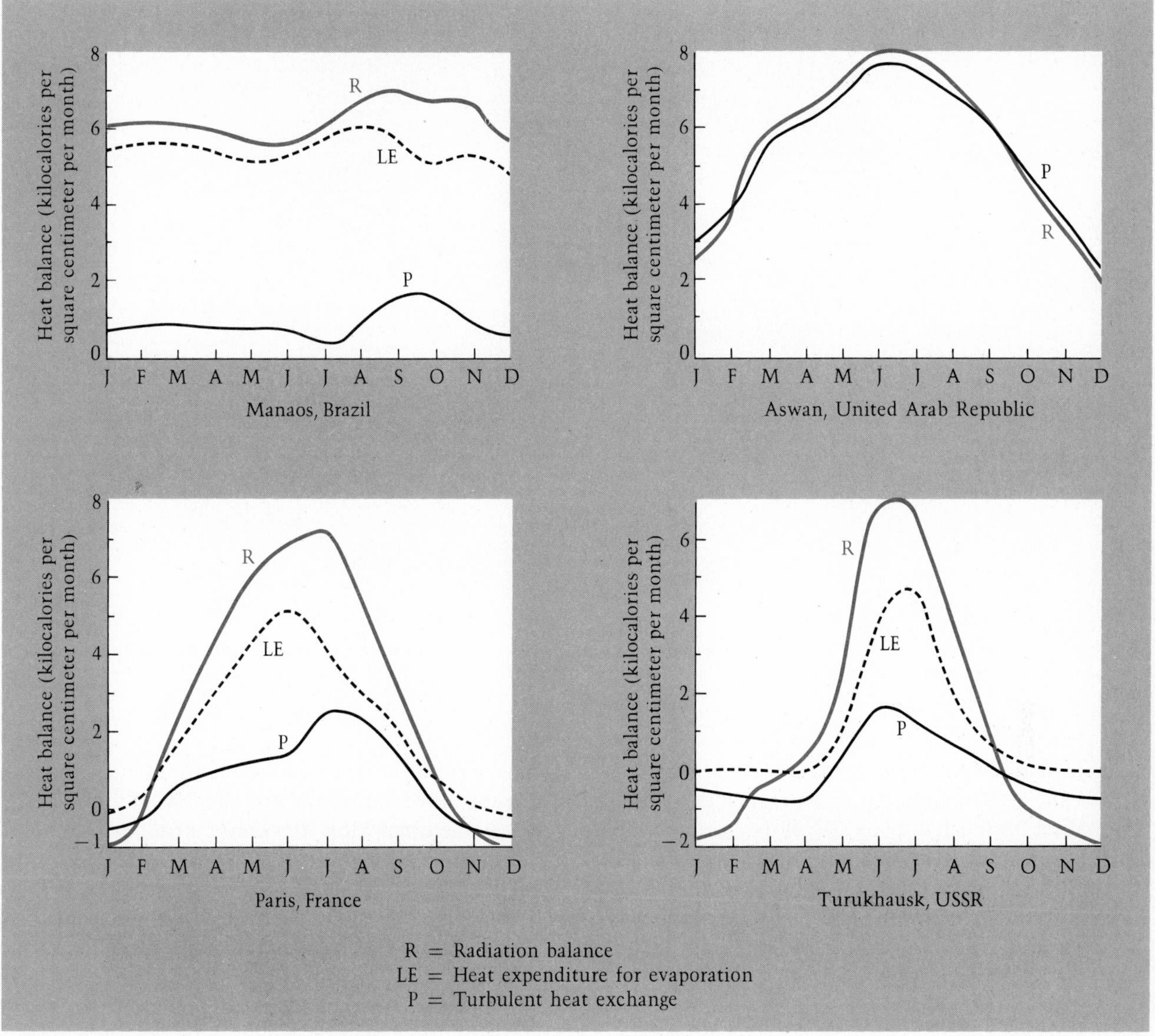

FIGURE 4-13

The heat balance. Values are in thousands of calories per square centimeter per month. a) Manáos, Brazil; b) Aswan, United Arab Republic; c) Paris, France; d) Turukhausk, USSR.

GLOBAL DISTRIBUTION OF HEAT FLOWS

We have already examined the spatial variation of net radiation (refer back to Figure 4-12), so now we will examine the disposal of net radiation through sensible and latent heat.

To find the amount of heat lost as latent heat, we multiply the amount of water evaporated by the value of the latent heat of vaporization. Consequently, the global pattern of latent heat loss, which appears in Figure 4-14, is similar to the pattern of water lost in evaporation (turn back to Figure 4-8). Over land surfaces, the largest amount of latent heat loss occurs near the equator. Latent heat loss declines in subtropical latitudes, increases in midlatitudes, and then further declines in high latitudes. Over ocean surfaces, where water is always available for evaporation, latent heat loss is greatest in the subtropics. Here there are few clouds to reduce radiant heat input. Because of the effect of cloud cover, latent heat loss over oceans is not as great in equatorial latitudes as in subtropical latitudes. As over the land surfaces, latent heat loss is least over oceans in high altitudes.

Sensible heat loss over the land is greatest in the subtropics and less toward the poles and the equator, as Figure 4-15 shows. Over ocean surfaces, the amount of

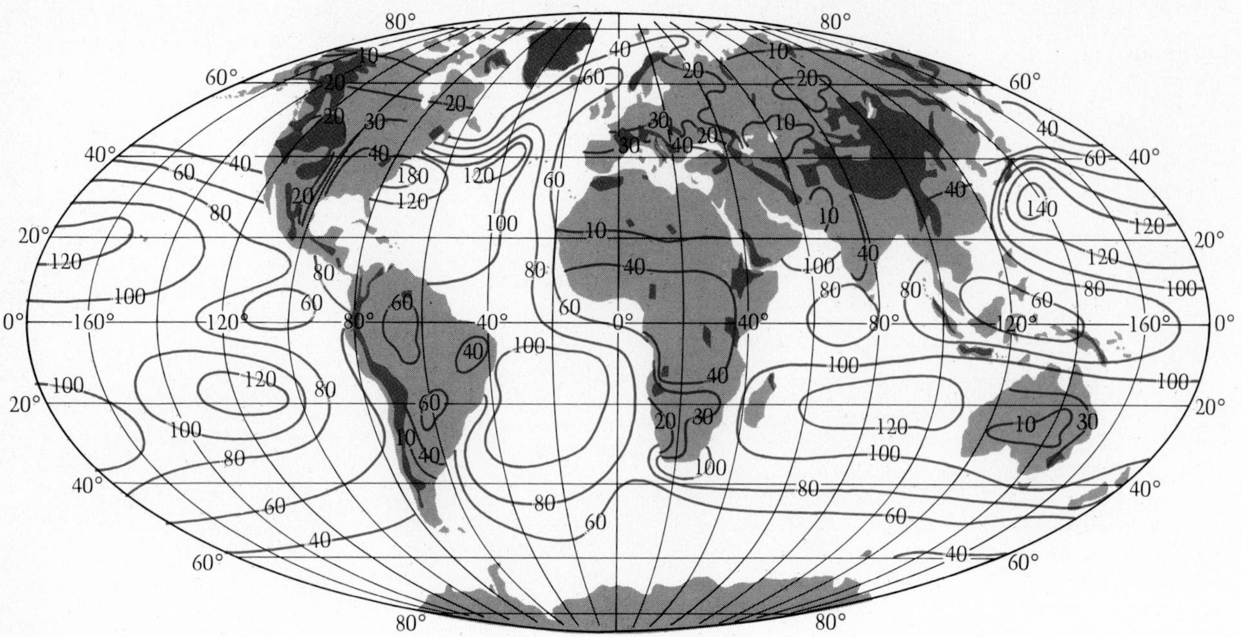

FIGURE 4-14

The global distribution of latent heat loss. The heat used in evaporation and evapotranspiration is expressed in thousands of calories per square centimeter per year.

sensible heat loss tends to increase with latitude.

Over the earth's surface as a whole, the net radiation heat gain is itself balanced by 70% of the heat being lost in the latent form and 30% being used to heat the air as sensible heat. But at any one point on the earth's surface, there is a unique interaction between the values of net radiation and evaporative and sensible heat flow. Temperature is the result of these flows.

THE DISTRIBUTION OF EARTH TEMPERATURE IN TIME AND SPACE

We should now have some clues to how the air that we feel around us actually becomes heated. The various flows of radiation resulted in a sum of net radiation that can heat the earth. At the height of our heads, the air has been principally heated by the flow of sensible heat from the earth. If there is not much water to be evaporated, most of the net radiation will be available to heat the air by sensible heat flows. Where water is evaporated, a smaller proportion of the net radiation can be used in sensible heat, and the air temperature generally is lower.

The change of temperature during a day is called the *diurnal* change. Shortly after dawn, radiation from the sun begins to exceed the radiant loss from the earth. The earth begins to heat the air, so air temperature rises. It continues to rise as the net radiation rises. But the heating of the ground and the flow of sensible heat take some time. Thus maximum air temperatures are not experienced simultaneously with the maximum net radiation peaks at solar noon, but an hour or so later. In late afternoon, net radiation and sensible heat flow decline. Therefore temperatures decline. After the sun has set, more radiation leaves the earth than arrives at the surface, which produces a negative net radiation. The surface and the air above it enter a cooling period that lasts all through the night. Temperatures are lowest near dawn, and then the whole cycle starts again.

The *annual* cycle of temperature in middle and high latitudes is rather similar to the diurnal cycle. In the spring, net radiation becomes positive, and air temperatures begin to rise. The highest temperatures do not occur at the time of the greatest net radiation, the summer solstice, but usually a month after. In autumn, decreasing net radiation leads to progressively lower temperatures. The lowest winter temperatures occur toward the end of the period of lowest, and often negative, net radiation, and when the ground has lost most of the heat it gained in summer. Then it is spring again, and net radiation once more begins to increase.

The time it takes to heat the ground in any particular place determines when the highest air temperatures will occur. The difference between land and ocean

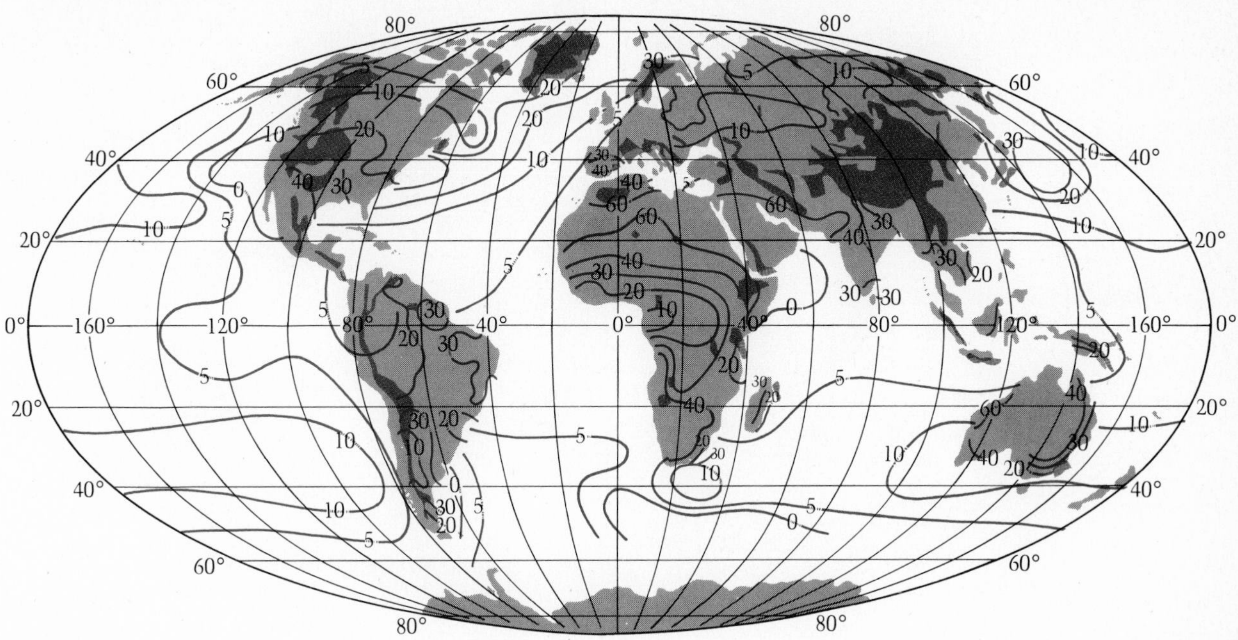

FIGURE 4-15
The global distribution of sensible heat loss. Values are in thousands of calories per square centimeter per year.

offers the most clear-cut example. Dry land heats and cools relatively rapidly, but water takes much longer to heat up and to cool down, partly because radiation can penetrate water to deeper depths. As a consequence, the air above an ocean remains cooler in summer and warmer in winter than the air over a land surface at the same latitude. In places where the ocean air is transported onto the continents, as at Paris, air temperatures do not become extremely warm or cold. This is called the *maritime effect* on climate. In the opposite case, where the ocean has a minimal moderating effect on air temperatures of the land, there is a *continental effect*.

Sometimes air from outside an area has more influence on air temperatures than do radiation and sensible and latent heat flows. For example, there might be quite a large amount of net radiation at midday during a Minnesota winter, but the air temperatures may still be very low. This is because air may come from thousands of kilometers to the north, where a completely different heat balance has been reached. Other times, the air may become heated during the condensation process or when longwave radiation heats it.

In summary, air temperature depends on the amounts of heat that make up the heat balance. Air temperatures may also be affected by the advection of air from a location with a different heat balance. The results of different heat balances and the advection of air are seen in the global distribution of air temperatures in Figure 4-16 a and b. Two things are clear in this distribution: Higher amounts of net radiation at low latitudes lead to higher temperatures, and the oceans moderate air temperatures over them and on the land surfaces near them.

RELATIONS AMONG FLOWS

The flows of energy and water that lead to the water, radiation, and heat balances, and eventually to the temperatures, of the earth's surface may be regarded as interrelated systems. Figure 4-17 places these systems side by side and shows the major links. Net radiation, the result of the radiation balance, is the kingpin of the heat balance. It causes evaporation, which plays a vital role in the water balance. There are many other links. For example, clouds, in which precipitation forms, decrease the flow of shortwave radiation to the earth and may increase the amount of longwave radiation. No latent heat can find its way into the atmosphere if the land surface is dry through lack of rain. Soil heat flow and sensible heat flow also depend in part on how much water exists at the surface. Try to think of the other linkages occurring between the three systems. The list of links among the three systems is almost endless, and Figure 4-16 would resemble the web of a spider if all the links were drawn on it.

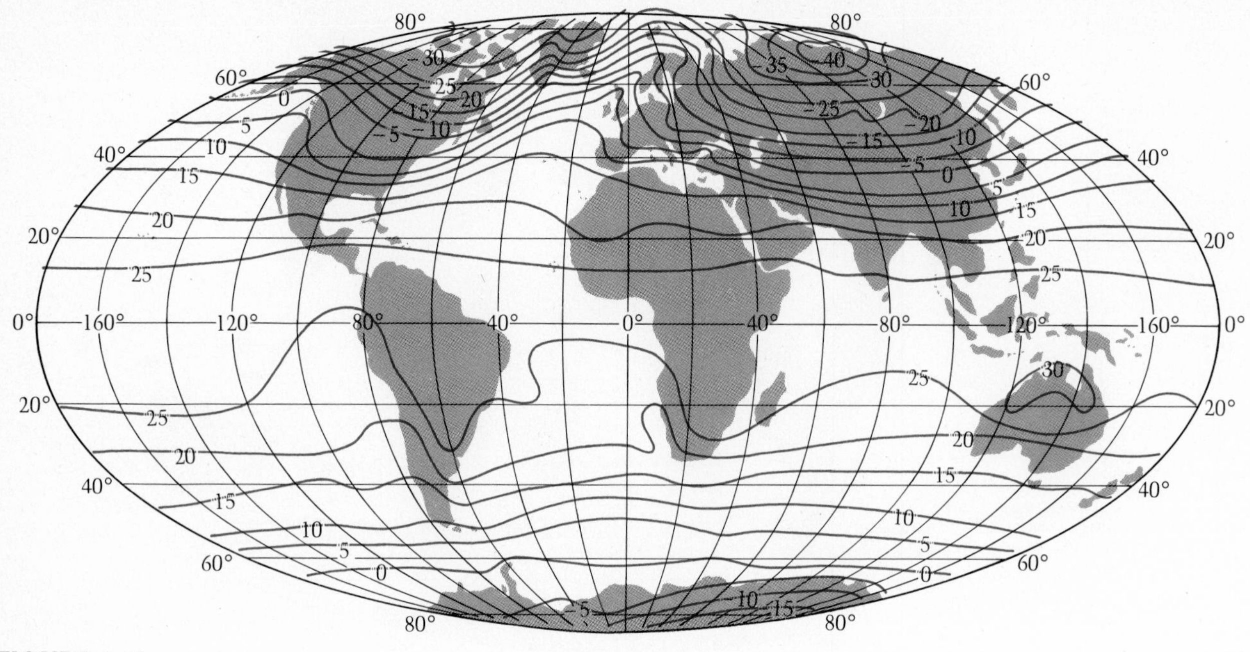

FIGURE 4-16

Mean sea level temperatures. Numbers are in degrees Celsius. a) January; b) July.

FIGURE 4-17

The major links among the water, radiation, and heat balances of the earth's surface.

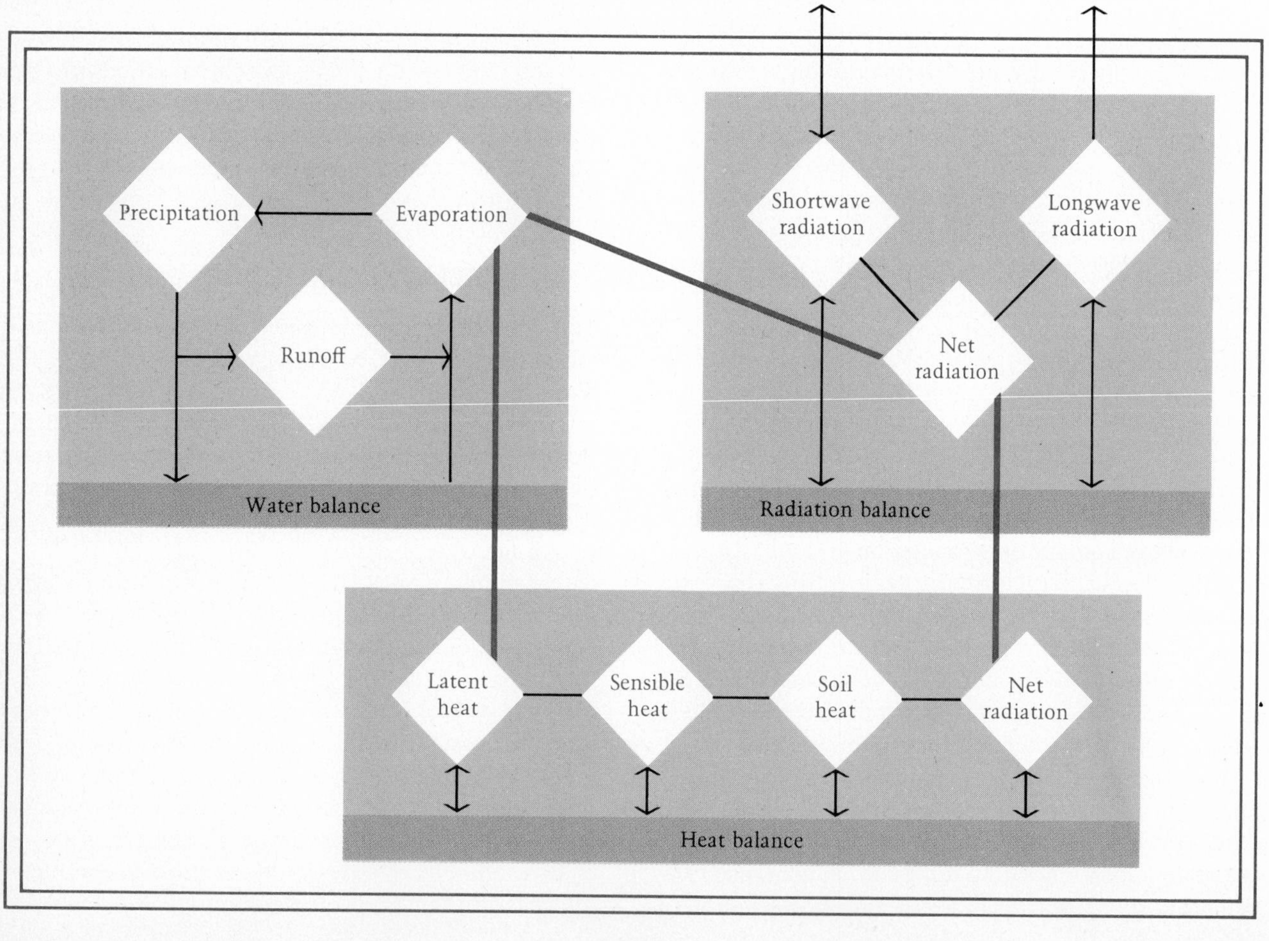

Precipitation
Evaporation
Runoff
Water balance
Shortwave radiation
Longwave radiation
Net radiation
Radiation balance
Latent heat
Sensible heat
Soil heat
Net radiation
Heat balance

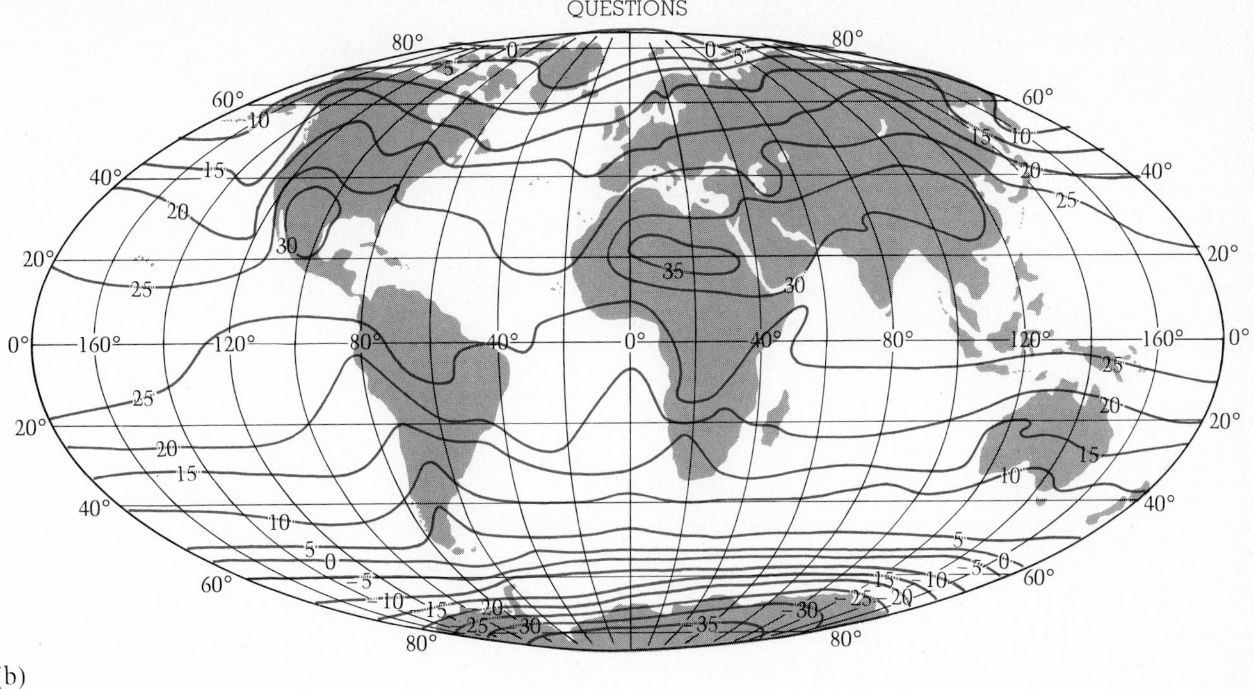

(b)

SUMMARY

The various flows of energy and water to and from the earth's surface result in three important balances at the earth's surface. The water balance is the result of flows of water in various forms, which constitute the hydrologic cycle. The hydrologic cycle depends on the ability of water to change from liquid to gas and back again and on the driving power of the sun's radiation.

The radiation balance of the earth's surface is the result of flows of shortwave and longwave radiation. Many factors affect the size of the final balance, called net radiation, including characteristics of the atmosphere, such as cloud cover, and those of the surface, such as albedo.

Net radiation is one of the terms in the heat balance of the earth's surface. The others are latent heat flow, sensible heat flow, and soil heat flow. Climates and air temperatures on the earth may be explained by the flows of the heat balance and by advection, the horizontal transport of air. The flows resulting in the water, radiation, and heat balances have many interconnecting links.

QUESTIONS

1. Fresh water is vital for human life, yet fresh water comprises a mere 3% of the earth's water, and only 11.5% of this is available for human consumption. How is this tiny amount sufficient for the huge population of the earth? What process is largely responsible for maintaining this water in a potable form?

2. Explain why the saturated adiabatic lapse rate is smaller than the dry adiabatic lapse rate. What important property of water produces this difference?

3. Cold air can hold less water in the form of vapor than warm air. Explain dew formation in terms of relative humidity and the diurnal temperature cycle.

4. Assuming that the relative humidity is not 100%, what makes a wet-bulb thermometer cooler than a dry-bulb thermometer? Why must an appropriate set of humidity tables be consulted to determine the proper relative humidity?

5. The two subsystems of the hydrologic cycle, the one concerned with precipitation and evaporation over the oceans and the other with precipitation and evapotranspiration over land areas, are connected by horizontal flows in both directions. What are some of these interconnective flows? Is the flow from ocean to land greater, or that from land to ocean?

6. The hydrologic cycle can be viewed as a closed system, with water continuously transported among various subsystems. Can the two major subsystems, the sea to land and the land to sea transport of water, be viewed as a pair of open systems in dynamic equilibrium? What major, long-term earth event might act

as a positive feedback mechanism to upset this equilibrium? Is there a negative feedback mechanism that restores the original equilibrium?

7. Evaporation can only take place when two conditions are met. Heat energy must be available at the water surface to change the liquid water to a vapor, and a vapor pressure gradient must be present. In the high latitudes, little evaporation occurs compared with the rest of the globe. Similarly, over the land surfaces in the subtropics little evaporation occurs. What conditions have not been met in each case?

8. What happens to precipitation in excess of the amount that can be handled by potential evapotranspiration? In what types of environments would you expect to find actual evapotranspiration always equal to potential evapotranspiration?

9. Occasionally a situation occurs in nature in which moisture-laden air cools and reaches a humidity greater than 100% relative humidity. At this point the air is said to be supersaturated. What factor in the process of precipitation formation allows this to happen?

10. In the ice-crystal process of precipitation, ice crystals that grow large enough to fall from the clouds are often melted on their way to earth by the warmer air that they encounter at lower altitudes. If the additional heat encountered at lower altitudes is sufficient to melt the ice crystals into water drops, why don't they entirely evaporate or sublimate into water vapor?

11. Which of the water balances described in this chapter is most like the one where you live? During which months does your area have a water surplus, and during which does it have a water deficit? Draw an approximation of the water balance in your area using the ones in Figure 4-6 as models.

12. The amount of runoff in any location is limited by the amount of precipitation received. In the Olympic rainforest of Washington, clouds obscure the sky most of the time. An immense amount of precipitation falls throughout the year, yet almost all of it contributes to the water surplus and is removed as groundwater or runoff. In the southwestern states, there is little runoff compared with the amount of precipitation received. What accounts for the differences between the two locations?

13. What factors account for the small amount of precipitation in the high latitudes? Do the same factors account for the large amounts of precipitation in the low latitudes? Why is there a large amount of precipitation over the oceans in the subtropics, but very little over the landmasses?

14. A phenomenon often noted in the country is that animals like to lie on the warm asphalt roads on cool summer nights, creating a hazard for motorists. What factors in the radiation balance system create this situation?

15. If we were to make measurements of the principal wavelength components of incoming and outgoing radiation several meters above various surfaces, what would we find above the following: A mountain snowfield on a clear, sunny day? A redwood forest near the coast on a foggy, overcast day? A wheat field on a partly cloudy day? Support your answers in terms of the radiation balance system.

16. What wavelength component of sunlight is principally responsible for sunburn, skin cancer, and eye damage? Why must mountain climbers at high altitudes and bathers at the beach be especially wary of sunburn?

17. Perihelion, the point of the earth's orbit around the sun where the two bodies are closest, occurs about January 3. However, winter temperatures are at their coldest at this time in the Northern Hemisphere. If we are closest to the sun at January 3, why aren't temperatures warmest at this time?

18. In Chapter 3 we noted that stable air conditions inhibit cloud formation and that the air over the major deserts is usually stable. The deserts receive a great deal of shortwave radiation every day, and surface soil temperatures have reached the vicinity of 96°C in Death Valley and in Algeria. Yet night temperatures in these same areas can be near the freezing point. What accounts for this immense daily variation?

19. The net radiation for the earth surface amounts to approximately one-fourth of the incoming radiation. Much of this radiation is converted into heat. How do we know that the heat balance of the earth is in a state of dynamic equilibrium when examined at a daily or yearly time scale? What components of the heat balance help to maintain this equilibrium?

20. What other factors besides net radiation over an area determine what its temperature will be? If the greatest radiation for the Northern Hemisphere is received in July, why are the warmest temperatures in August? Why are coastal areas more temperate than areas further inland, in most cases?

THE RESTLESS ATMOSPHERE AND OCEANS

CHAPTER 5

One of the largest and most beautiful of the earth's subsystems is the general circulation of the atmosphere and oceans. The most important function of these large-scale movements of air and water is to redistribute heat and moisture across the earth's surface. Were it not for the transport of heat from the equator toward the poles, vast areas of the earth would be either too cold, too hot, or too dry for human habitation. Atmospheric circulation accounts for about 80% of this heat redistribution, and that of the ocean accounts for the remainder.

On a large scale, the atmosphere moves the way it does because of the difference in the amount of radiant heat received at the equator and the poles and because the earth is spinning. On a smaller scale, several forces act on an individual particle of air or water. The combination of these forces creates distinct kinds of water and wind movement. We will examine movement in the atmosphere on both scales. We will also see how movement in the atmosphere and ocean are closely related. We must become familiar with the flow of ocean and air currents on a global basis because we depend on it for our understanding of the weather and climate of each particular location on the surface of our planet.

THE CAUSES OF MOVEMENT IN THE ATMOSPHERE

Since the days when it became common for sailing ships to make transoceanic voyages, humans have known that the winds of the world are arranged in a particular fashion. This information was vital in planning the routes of voyages that might take two or three years, but it was often of little assistance in guiding the ships through the fluctuating range of day-to-day weather. It seemed helpful to separate large-scale movement from small-scale movement, even if the two were sometimes closely related. We start by looking at the causes of the large-scale movement.

BASIC CAUSES OF ATMOSPHERIC CIRCULATION

Two simple factors explain the circulation of the atmosphere: The earth receives an unequal amount of heat at different latitudes, and it rotates.

Imbalance of heat In Chapter 4 we saw a map of the net radiation at the earth's surface (refer to Figure 4-11). If we examine the amount of incoming and outgoing radiation averaged over all longitudes, as in Figure 5-1, we find that there is a marked surplus of radiation between the equator and the thirty-fifth parallels. At higher latitudes, outgoing radiation exceeds incoming radiation. The main reason for this is that rays from the sun strike the earth's surface at higher angles, and therefore at a greater intensity, at the lower latitudes than at the higher latitudes. As a result, the equator receives about two and one-half times as much annual solar radiation as the poles do. If this latitudinal imbalance of heat were not balanced somehow, life would be possible only in a narrow zone at about 35°N and 35°S. Heat is therefore transferred toward the poles, and the amount of heat movement performed by the general circulation is also indicated in Figure 5-1. We can see from the figure that the most heat is transferred toward the poles from the midlatitude areas.

Imagine for a moment that the world is stationary and that it has no heat absorption differences between land and sea. Under these circumstances heat transfer could occur by a simple cellular movement: Warm air rises at low latitudes, travels toward the poles at high altitudes, descends, and then returns to low latitudes as a surface wind. But such a situation cannot exist because the earth spins.

The Coriolis force The earth's rotation leads to two major factors that affect how the general circulation operates. The first is through an apparent *deflective force.* Suppose you are sitting on a horse on a merry-go-round, trying to throw a ball to a friend sitting on a horse ahead of you. When you throw the ball, it travels toward the outside of the merry-go-round. Between the time when you release the ball and when it was meant to reach your friend, the merry-go-round continues spinning. Your friend is no longer in the position he or she was in when you threw the ball. To you, it appears that the ball has been deflected to the right of its path. But if an observer standing outside the merry-go-round were to mentally plot the path of the ball, he or she would see that it actually travels in a straight line.

The apparent deflecting force of a rotating body is called the *Coriolis force.* Because our earth spins, anything moving on it—from guided missiles to air particles—is subjected to the Coriolis force. In the absence of any other force, moving objects are deflected to the right in the Northern Hemisphere and to the left in the Southern Hemisphere. If we did, therefore, have a wind blowing from the North Pole it would be deflected to the right and in all probability become an easterly wind (Figure 5-2). (Winds are named according to the direction *from* which they blow). We see later how the Coriolis force affects the general circulation pattern.

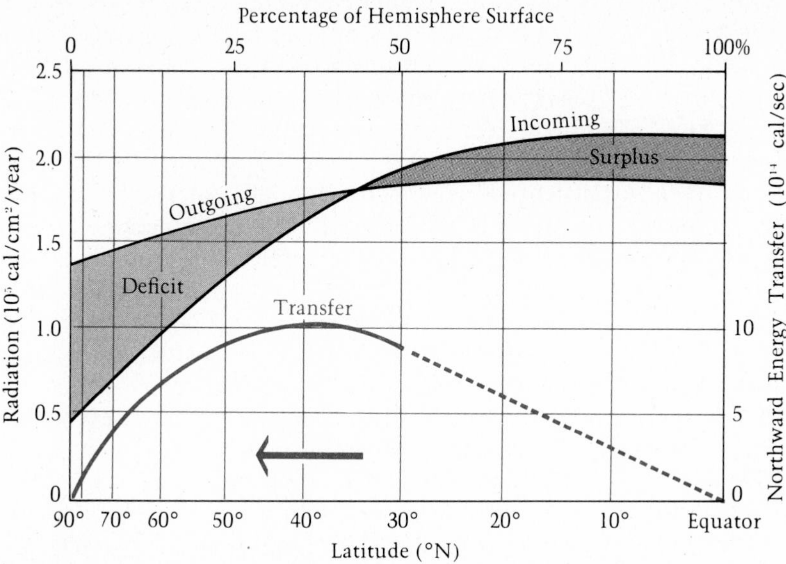

FIGURE 5-1

Latitudinal radiation balance averaged over all longitudes and the consequent poleward transfer of heat.

FIGURE 5-2

The Coriolis force in action. The North Pole of a map of the Northern Hemisphere is placed over the spindle of a record turntable, and the turntable is spun counterclockwise to simulate the earth's rotation. A line starting from the Pole and drawn along the edge of a stationary ruler describes an arc and ends up traveling toward the west.

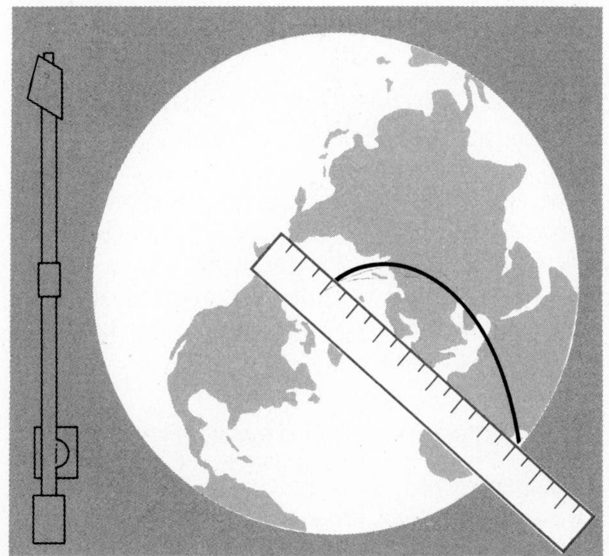

Angular momentum The fact that the earth spins may be put another way by saying that it has *angular momentum*. Angular momentum is directly proportional to the rate of spin and the distance of an object from the axis of rotation. The faster the spin and the farther the distance, the greater is the angular momentum. The angular momentum within the earth system is highest at the equator, which is the farthest point on earth from the north-south axis. Anything at the earth's equator actually moves at 1680 km per hour (1050 mi per hour). Although it is not attached to the earth's surface, the atmosphere moves with the earth around its axis of rotation.

In any rotating system, the total angular momentum must remain constant. This principle is called the *conservation of angular momentum*. Because of this principle, if any large mass of air changes its position on the earth's surface, its angular momentum also changes to keep constant the angular momentum of the total earth–air rotating system. We will see later how the need to conserve angular momentum also affects the earth's wind systems.

FORCES ON AN AIR PARTICLE

The earth's heat imbalance and its rotation may be regarded as the foremost basic causes of the general circulation of the atmosphere, but the factors that cause winds are best approached by investigating the forces that act on an individual particle of moving air.

There are four of these: the pressure gradient force, the Coriolis force, the centripetal force, and the frictional force. We start first with the relationship between wind and the pressure gradient force.

Because of gravity, air presses on the earth's surface, causing atmospheric pressure. The pressure at the bottom of two vertical tubes at two different locations on the earth's surface might not be the same, perhaps because of different temperatures, which are, in turn, due to different densities. If the two tubes were connected at the bottom, air would move from the high-pressure tube to the low-pressure tube. This happens on a larger scale in the atmosphere. The tendency for air to move from areas of relatively high pressure toward those with relatively low pressure is called the *pressure gradient force.* The pressure gradient force on an individual air particle increases as the difference in air pressure across a specified distance increases, and decreases as the air density increases.

We are already familiar with the *Coriolis force,* which acts on both large parcels of air and small particles. The magnitude of the Coriolis force depends on the speed of the earth's spin, the sine of the angle of latitude at which it is applied, and the speed of the moving air particle. The sine of the angle of latitude is explained in Figure 5-3, where for any given angle of latitude its sine is computed by dividing length *B* by length *A*. If you think about the numbers involved in the ratio *B*/*A*, you will arrive at the following fact. The sine of zero degrees of latitude (the equator) is zero because distance *B* is zero. The sine of ninety degrees of latitude (the North or South Pole) is one because distances *A* and *B* are the same. Thus, because the sine of the angle of latitude affects the strength of the Coriolis force, this force is a minimum of zero at the equator and rises to a maximum at the poles.

Another force that acts on an air particle moving in a circular fashion is the *centripetal force.* Think of a person whirling a stone on a string. The force on the whirling stone is the inward pull exerted by the string. This inward force is the centripetal force. The greater the mass and velocity of a parcel of moving air, the greater the centripetal force on an air particle. The centripetal force increases as the mass of the moving air increases. The force also increases as the square of the velocity of the air particle increases. But the centripetal force increases as the distance between the air particle and the center of the curved path decreases.

Finally, some of the motion in the atmosphere takes place near the earth's surface. Thus the individual air particle near the earth's surface is slowed by a *frictional force.* The magnitude of the frictional force depends on the velocity of the air particle, its distance from the earth's surface, and how rough the surface is at any particular location. There is less friction with movement across a smooth snow surface than across a large city.

We can now examine how the factors that we have discussed act together in the restless atmosphere. So far we have begun with the large scale and moved to the smaller. This time we will reverse the procedure.

FIGURE 5-3

The sine of the angle of latitude is the ratio of the distance *B* to the distance *A* (*B*/*A*). At the equator, *B* is zero, so the sine of zero degrees is zero, and consequently the Coriolis force is zero. At the poles, the distances *A* and *B* will be equal, so the sine of 90° is one (its highest possible), and the Coriolis force is at its maximum.

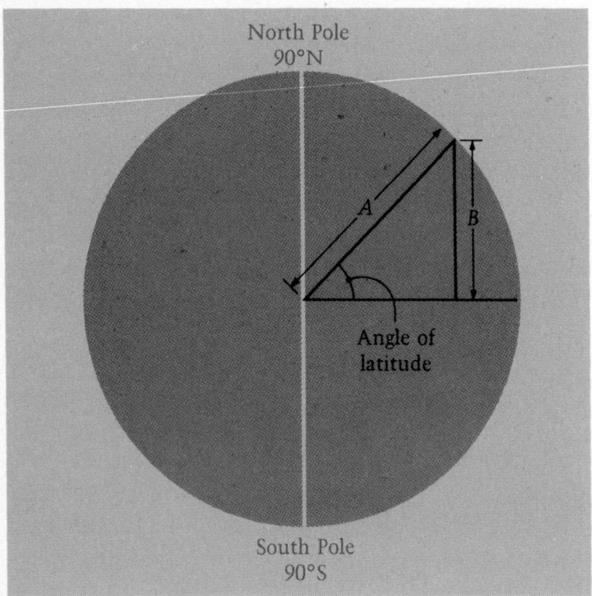

PATTERNS OF ATMOSPHERIC CIRCULATION

Except for small winds, such as a sea breeze (see Chapter 6) and winds near the equator, winds never blow directly from a high-pressure area to a low-pressure area. Once a wind starts moving, it immediately comes under the influence of the forces we have just discussed.

GEOSTROPHIC AND GRADIENT WINDS

Once a particle of air starts moving under the influence of a pressure gradient force, the Coriolis force deflects it to the right if it is in the Northern Hemisphere.

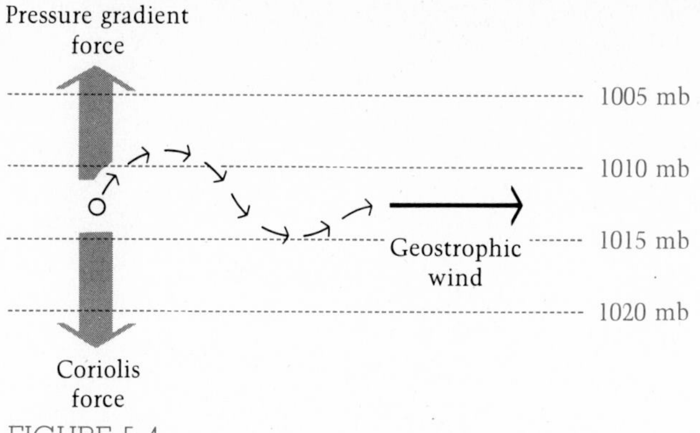

FIGURE 5-4

Formation of a geostrophic wind in the Northern Hemisphere.

Figure 5-4 diagrams the path that results. The particle may oscillate a bit, but eventually the pressure gradient force and the Coriolis force balance each other. The resultant wind, a *geostrophic wind,* is a good example of dynamic equilibrium. Geostrophic wind is common in the atmosphere some distance above the earth's surface.

Because of the balance of forces, a geostrophic wind always flows parallel to the *isobars,* imaginary lines that join areas of equal atmospheric pressure. If we have a map showing the air pressure of the upper atmosphere, we can get a rather good idea of where the winds blow. We can also predict the velocity and direction of a geostrophic wind if we know the pressure gradient force, the air density, and the latitude of the place we are interested in.

One notable by-product of the geostrophic relationship between the Coriolis force and the pressure gradient force is the idea that pressures are determined by winds just as much as winds are determined by pressures. To understand this relationship, regard the depressed surface of the water disappearing down a drain in a bath as a pressure surface. Such a surface is a response more to the whirlpoollike motion of the water than to anything else. In the atmosphere too, motion and pressure often have a two-way relationship.

In much global scale air movement, we need only concern ourselves with the geostrophic wind as a balance between the pressure gradient and Coriolis forces. But where the air movement is curved, the centripetal force comes into play. The centripetal force and the Coriolis force together balance the pressure gradient force, and the resultant wind is called the *gradient wind.*

Curved air motion is particularly common in the phenomena called cyclones and anticyclones. Recall our image of a tube of air and its relation to pressure at the earth's surface. If the tube intersected the earth's surface at an area of relatively low pressure, then that area would be called a cyclone. As Figure 5-5 shows, a *cyclone* is an area of low pressure with pressures increasing away from the center. Isobars are roughly

FIGURE 5-5

Formation of a gradient wind in the Northern Hemisphere. The pressure gradient, Coriolis, and centripetal forces balance to form a gradient wind around an anticyclone or a cyclone.

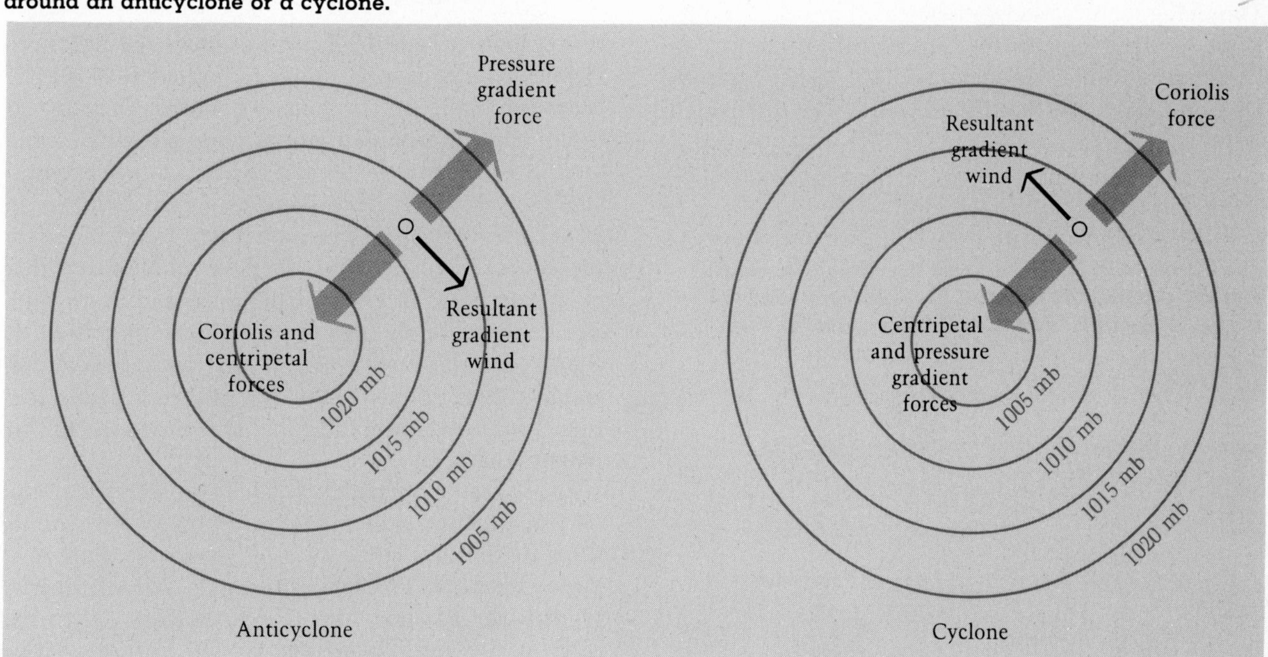

circular as is the motion of the gradient wind. The word cyclone comes from the Greek word meaning "to revolve." In the Northern Hemisphere winds move in a counterclockwise rotation around a cyclone. The opposite case, with an area of high pressure at the center and clockwise circulation of airflow in the Northern Hemisphere, is called an *anticyclone.* In the Southern Hemisphere the rotation of the airflow is reversed.

In summary, the geostrophic wind is the result of two balanced forces, whereas a gradient wind is the result of three.

Near the surface of the earth, below about 500 to 1000 m (about 1500 to 3000 ft), frictional force comes into play. It reduces the speed and alters the direction of geostrophic and gradient winds. The frictional force acts against the Coriolis force, so the winds near the surface blow across the isobars instead of parallel to them.

The action of the frictional force in causing wind to blow across isobars has two main implications. First, if an imaginary tube is set over the low-pressure area in the center of a cyclone (L in Figure 5-6), the winds at the surface converge toward it. The air has to go somewhere, so it will rise vertically through the tube. The reverse is true in an anticyclone: Air moving outward draws air downward through the tube. Thus cyclones are associated with rising air at their centers, and anticyclones are related to the downward movement of air. Chapter 6 shows how this explains the rain in cyclones and the fine weather in anticyclones.

Second, the farther a wind is from the earth's surface, the smaller the effect of friction on it. As a result, wind direction changes and velocity increases. The change of direction with height, if projected down to the surface, creates the curved pattern shown in Figure 5-7, and is called an *Ekman spiral* after V. Walfrid Ekman, the Swedish physicist who discovered it. Ekman first discovered the spiral in the ocean, where water movement becomes increasingly "geostrophic" with depth. This fact further demonstrates a link between circulation of the atmosphere and the ocean.

FIGURE 5-6

The frictional force acts against the Coriolis force and thereby disturbs the balance. The resultant winds blow across the isobars instead of being parallel to them.

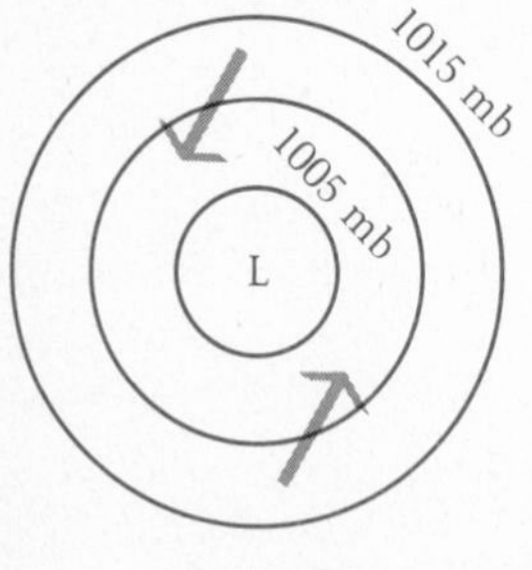

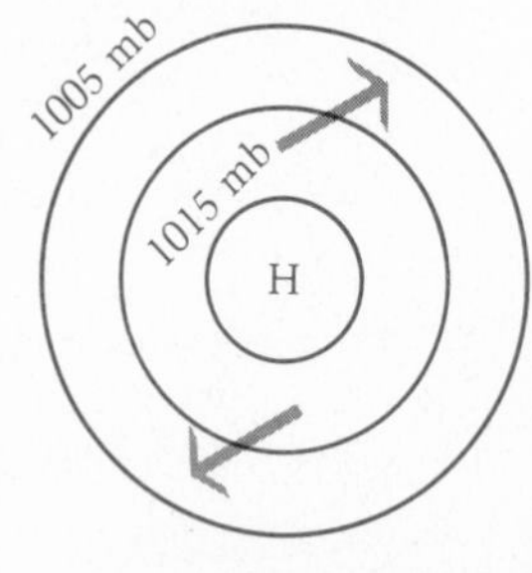

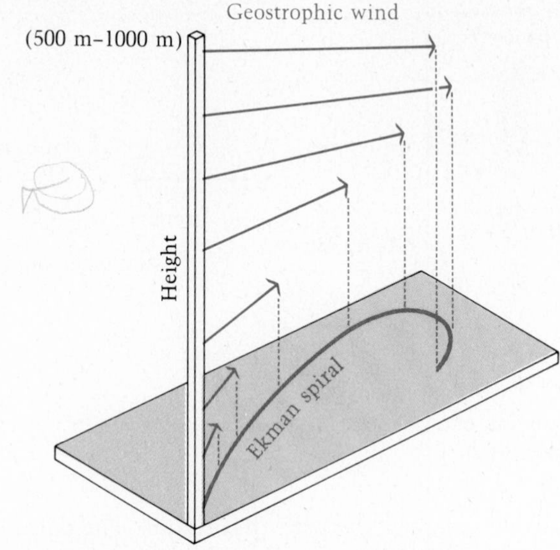

FIGURE 5-7

The Ekman spiral. In this example from the middle and higher latitudes of the Northern Hemisphere, geostrophic winds shift direction and velocity as they escape the effects of frictional force by climbing higher in the atmosphere. The longer the line representing the geostrophic wind, the greater the velocity.

MERIDIONAL CIRCULATION

If you look at Figure 5-8, which shows the Northern Hemisphere from space, you can immediately see the beautiful arrangements of cloud patterns. They appear to fall roughly into two groups. One group is a band around 10°N, and the second group is found between 40 °N and 60°N. This picture, although taken in March, is typical of the Northern Hemisphere in the summer. Recalling our tubes, we could assume that cloud bands are in low-pressure areas and that cloud-free regions are in high-pressure areas. But there is actually a vast amount of information in this photograph, and we should follow the historical path scientists took in understanding the currents of air represented here.

In 1735 George Hadley, a London philosopher and lawyer, explained the air motion observed by sailing ships in the low latitudes as follows: "The air as it moves from the Tropicks [of Cancer and Capricorn] towards the Equator, having a less Velocity than the parts of the earth it arrives at, will have a relative

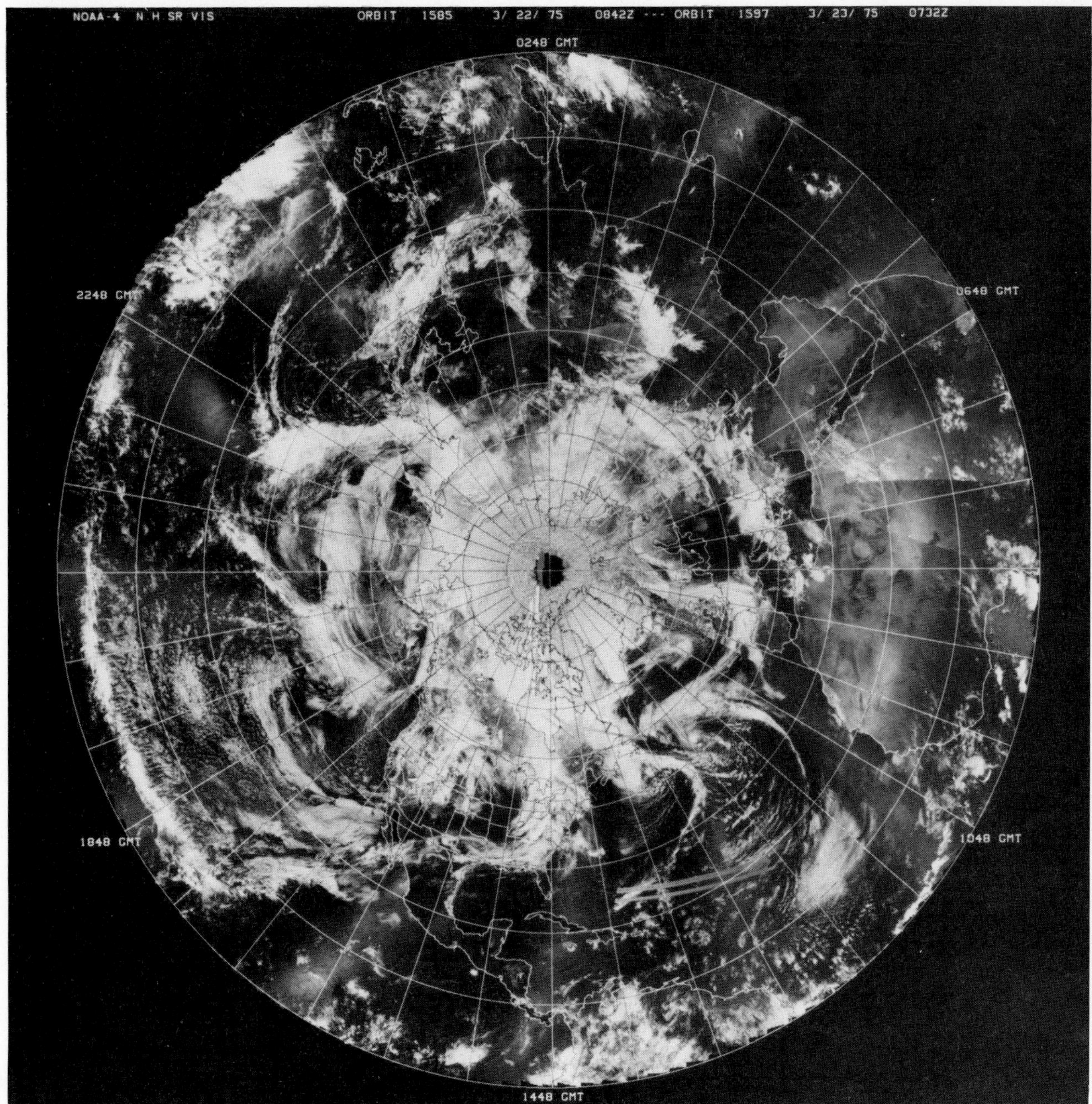

FIGURE 5-8
Mosaic of satellite pictures of the Northern Hemisphere, March 22, 1975.

Motion contrary to that of the diurnal Motion of the Earth . . . ." He was describing the large-scale action of the Coriolis force and laying the basis for beliefs in *meridional circulation,* the north–south movement of air along the meridians.

The essential idea in Hadley's time was that the heat imbalance of the earth was offset by hot air rising in the equatorial latitudes. At high altitudes, air moved toward the poles, and at the surface, air flowed toward the equator. This air movement was supposed to form a single large *cell,* a vertical current of air in which warm air rises and cool air descends. A single cell explains the northeasterly and southeasterly air flows in the tropics known as the *trade winds.* Air moving at the earth's surface toward the equator is deflected by the Coriolis force to the right in the Northern Hemi-

THE SAILOR'S LEGACY: NAMING THE WINDS

Spanish sea captains headed to the Caribbean and the Philippines in search of gold and spices. They depended on a band of steady winds to fill the sails as they traveled west. These winds were named the *trade winds*. These were the winds that blew Christopher Columbus and his flotilla to North America.

At the equator the *trade winds* converge in a zone of unpredictable breezes and calm seas. Sailors dreaded being caught in the *doldrums*. A ship stranded here might drift aimlessly for days. This was the fate of the ship in Coleridge's *The Rime of the Ancient Mariner*.

Ships also were becalmed by the light, variable winds in the subtropics at about 30°N and S. Spanish conquerors who ran afoul of the breezes in these blistering regions threw their horses overboard to save water for the crew. The trail of floating corpses caused navigators of the seventeenth century to label this area the *horse latitudes*.

In the Southern Hemisphere ships followed the strong westerly winds near 50° and 60°S. The winds were powerful but more stormy than the trades, so they became known as the *Roaring Forties* and *Screaming Fifties*. Thus many of the names given to wind belts date from the early days of sailing.

sphere, thus forming the northeasterly trade winds. In the Southern Hemisphere it is deflected to the left to form the southeasterly trade winds. However, it could not explain the westerly winds of midlatitudes. These were explained 206 years later by another model, conceived by Carl Gustav Rossby in 1941.

The Rossby model, depicted in Figure 5-9, shows a vertical cell in low latitudes. Even today, this low-latitude cell goes by the name *Hadley cell*. By using concepts of vertical tubes and the Coriolis force in Figure 5-9, we can see how the Rossby model explains high pressure in the subtropical latitudes and low pressure near the equator. We can also see why the surface winds in these latitudes blow from the northeast. In high latitudes, cold air, sinking because of its higher density, spreads out from the poles and, under the influence of the Coriolis force, forms *polar easterlies*. The high-latitude and low-latitude cells drive a midlatitude cell by friction, much as interlocking sprockets drive gears. This midlatitude cell, called the *Ferrel cell*, combines with the Coriolis force to explain the easterly winds observed at the surface in midlatitudes. Furthermore, where warm air from the south meets cold air from the north, the warm air gradually rises along an area known as the *polar front*, causing low pressure at midlatitudes. In many ways this model seems to explain how heat can be moved from the equator to the poles and how cold air can be transported from the poles. But the development of high-flying aircraft and other observational techniques during World War II revealed additional phenomena that are not explained by Rossby's model.

First, no easterlies were found at midlatitudes high above the earth's surface. A Ferrel cell would have

FIGURE 5-9

The three-cell model of the Northern Hemisphere meridional circulation published by Rossby in 1941, and its relation to surface wind and pressure zones. Black arrows indicate vertical circulation of air. Where air descends, high pressure is created at the surface. Where air ascends, low pressure develops. Latitudinal belts of high and low pressure are indicated. Color arrows are the surface winds that result from the interaction of the surface part of the cellular flow and the deflective Coriolis force.

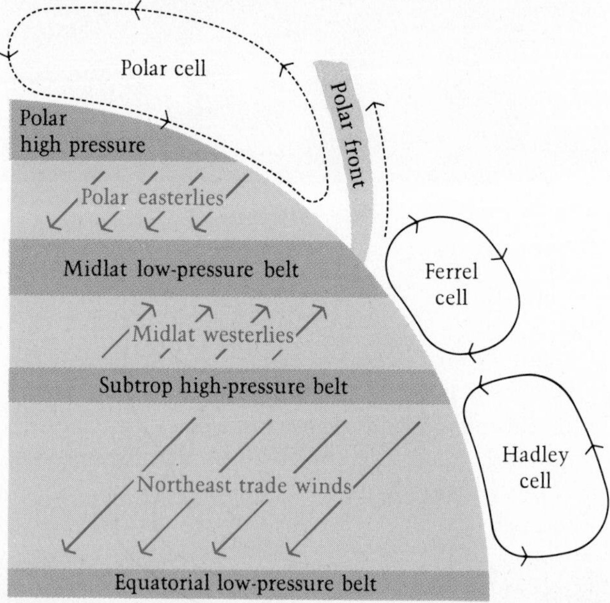

indicated easterly winds at high altitudes because of the Coriolis force. Instead, westerly winds extend to the tropopause in midlatitudes. Second, there are two fast-flowing currents of air called *jet streams,* one above the subtropical high-pressure zone and one above the polar front. Rossby's model could not explain these. Third, the rates of circulation in the Ferrel and Hadley cells were measured. Researchers found that it takes a particle of air 80 days to circulate around the Hadley cell and 120 days to move around the Ferrel cell. This is far too slow to compensate for the earth's heat imbalance.

Ten years after Rossby's model, E. Palmén of the Academy of Finland developed what may be the most useful model of meridional circulation, which is shown in Figure 5-10. He incorporated the Hadley cell and a smaller version of the Ferrel cell. The Hadley cell is in fact a useful, real feature of the atmosphere. But Palmén gave more emphasis to the role of the polar front and subtropical jet streams. This emphasis was not inappropriate, because by the mid-1950s, atmospheric scientists had discovered the importance of zonal motion in the general circulation.

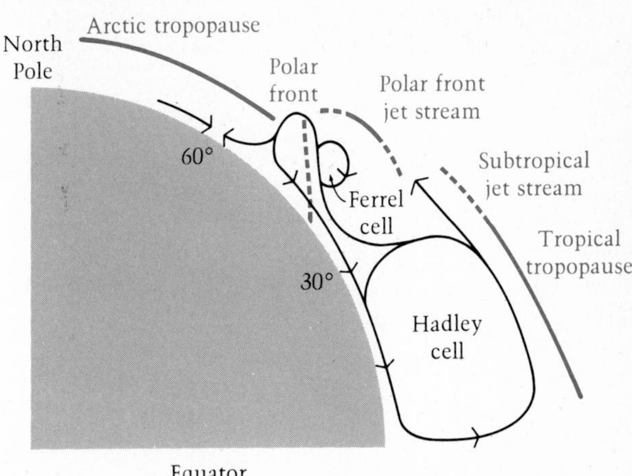

FIGURE 5-10
Palmén's meridional model, 1951. The Hadley cell is the only direct cell.

ZONAL CIRCULATION

The great increase in observations of the upper atmosphere in the past 35 years has given us a much clearer idea of how the winds actually move. It became apparent that zonal motion, that is, west-east and east-west movement along parallels, was more important to the general circulation than meridional motion. Figure 5-11 shows that, in midlatitudes and high latitudes, westerly winds predominate. In low latitudes a belt of easterly winds is most noticeable. As one hemisphere changes from summer to winter, the broad belt of westerly winds expands at the expense of the easterlies. In addition, the westerly winds are fastest between 30 ° and 60° latitude. The role of the polar easterlies is not so great as Rossby's meridional model suggested.

Another way of examining the atmospheric circulation is to look at a map of pressure in the upper atmosphere, such as that in Figure 5-12. A map showing isobars can be drawn for the surface or for any point in the upper atmosphere. As Chapter 3 explains, atmospheric pressure decreases with height. Because of this relationship between pressure and height, meteorologists often talk about a pressure instead of a height. Instead of drawing a map for pressure at the height of, say, 5500 m, they draw a map of the height of the 500 mb pressure surface. Either way, the areas of

URBAN DUST DOMES

A giant dome of dust buries Chicago! It's not the plot of a new horror movie. Many of our cities lie beneath a dust dome. The brownish haze stands out against the blue sky.

Before Chicago spawned sprawling suburbs, its dust dome was clearly defined. During World War II, the wall of dust began near Midway Airport, about 12 km (7.4 mi) from downtown Chicago. In the countryside visibility might be 25 km (15 mi), but within the dome it was only .5 to .75 km (.3 to .5 mi). Los Angeles too is beneath a thick layer of pollution. But because the city lies in a basin, its dust dome is shaped more like soup in a bowl. Why doesn't the pollution simply blow away?

The movement of dust over cities shows that heat generated in urban areas forms a small circulation cell. Air currents capture the dust and mold it into a dome. Dust and pollution particles rise in air currents at the center of the city where the temperature is warmest. As they move upward the air cools. The particles gradually drift toward the edges of the city and settle downward. Near the ground they are drawn into the center of the city to complete the circular motion. Temperature inversions over the city prevent escape upward, and the particles tend to remain trapped in this continuous cycle of air movement. We will talk about dust domes further in Chapter 8.

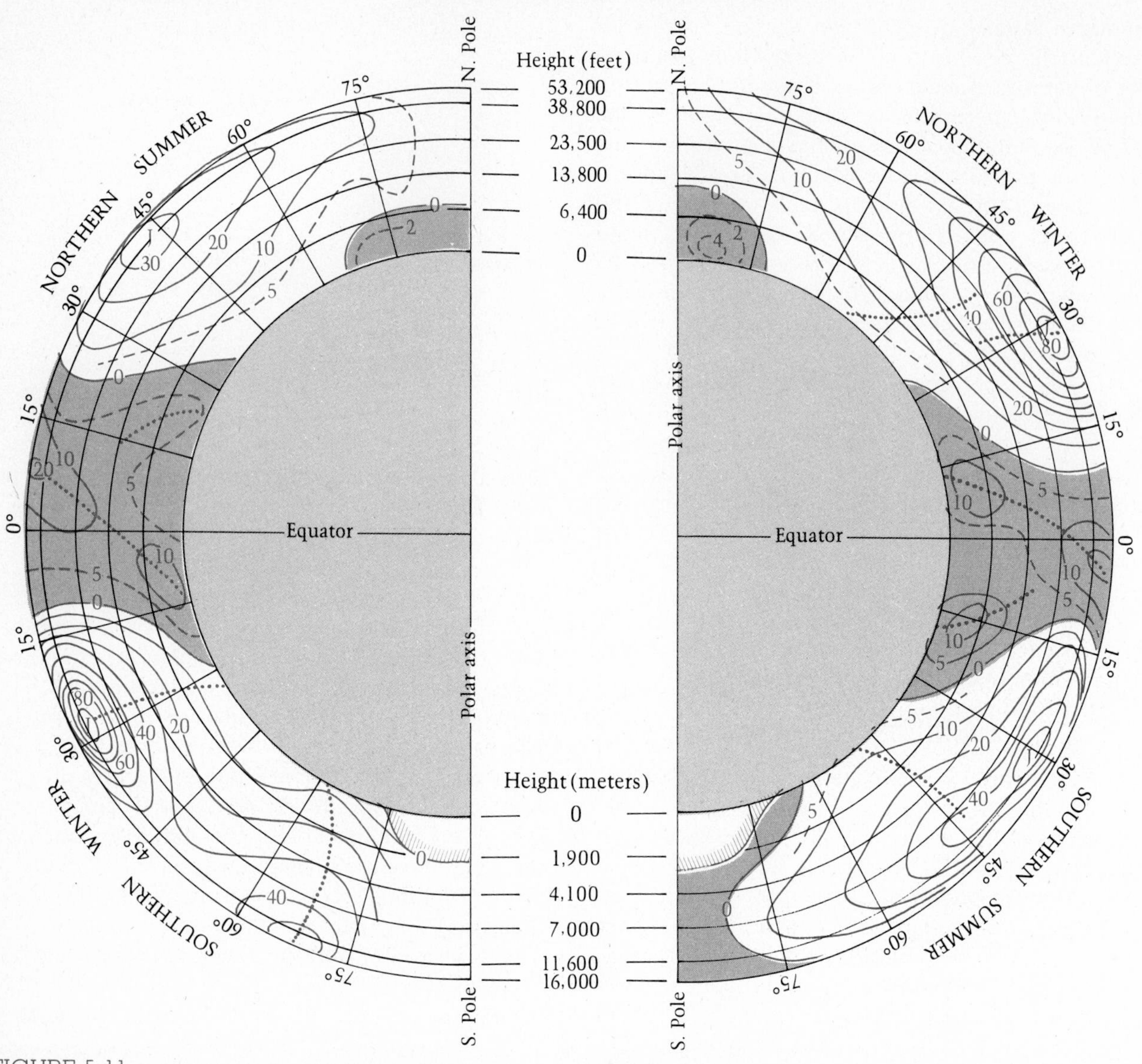

FIGURE 5-11

The average speed of zonal winds in summer and winter. Isotachs (lines of equal wind speed) are in meters per second. Regions of easterly winds are shaded.

high and low pressure—or high and low areas of one particular pressure surface—are the same. Figure 5-12 shows the height of the 500 mb pressure surface. Applying what we know about geostrophic winds, we can tell we are looking at a great vortex of air circulating counterclockwise. Where the lines are closest together, the wind is flowing fastest.

Rivers of air Figure 5-12 shows the great westerly whirl of air, which does not flow directly along lines of latitude but meanders like a large river. The meanders may be regarded as waves, called *Rossby* or *planetary long waves.* They consist of *troughs* of relatively low pressure and *ridges* of relatively high pressure. The fastest-flowing air currents, in the midlatitudes, comprise the *polar front jet stream.* The jet stream is not shown at its full speed in Figure 5-11, which shows only east-west motion, because the jet stream oscillates like a snake. Thus there is not a direct high-speed westerly wind.

The discovery of the meandering westerlies and the jet streams revolutionized scientists' thinking on the general circulation. They quickly established that heat was taken poleward and cold air moved southward not by vertical meridional currents but by horizontal movement of two kinds: the horizontal movement of

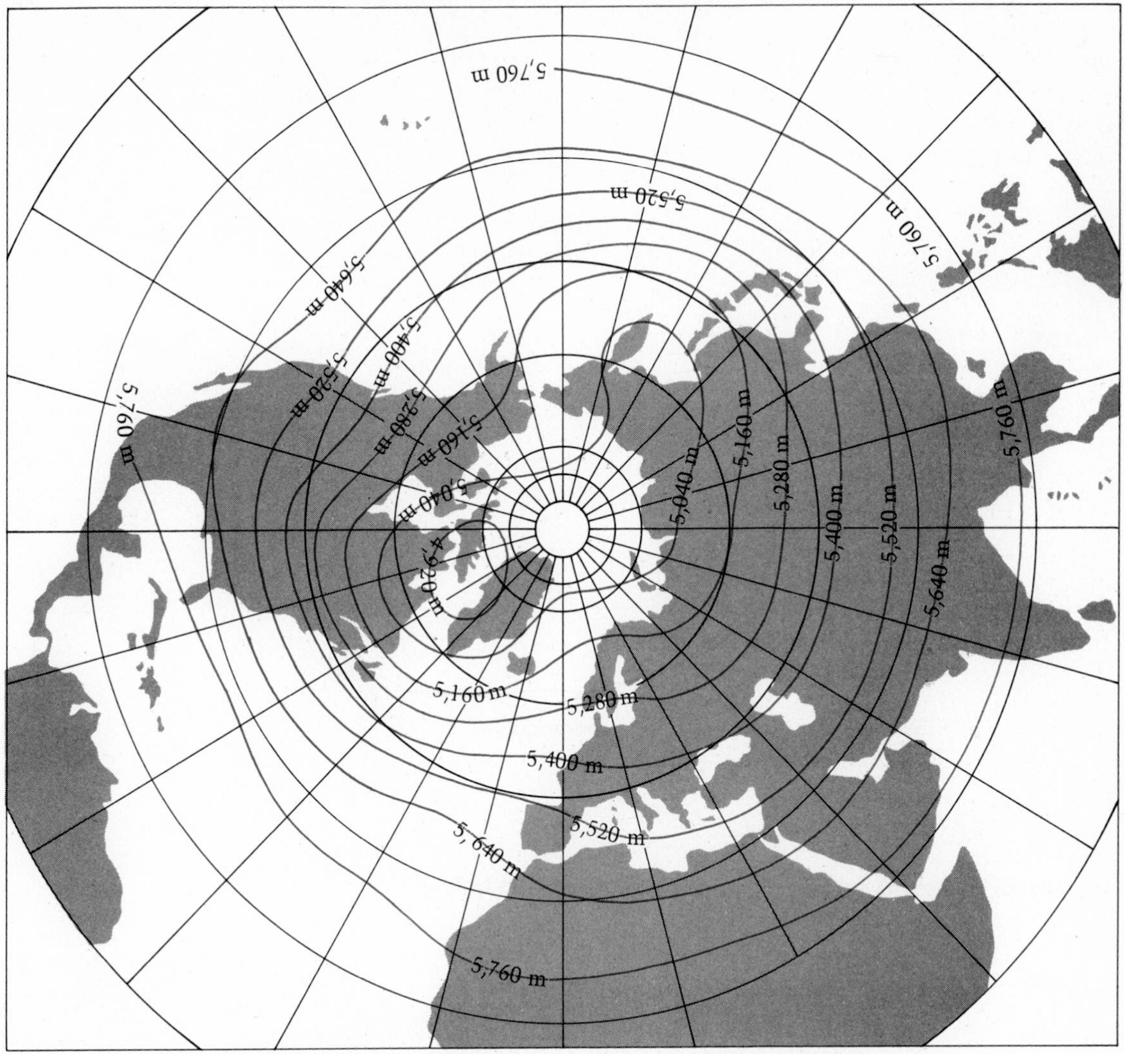

FIGURE 5-12

The Northern Hemisphere westerly winds as shown by the height of the 500-mb pressure surface in January.

the snakelike upper westerlies and the horizontal movement around the cells of high and low pressure, the anticyclones and cyclones.

Heat redistribution for the upper westerlies is best seen in the phenomenon called the *index cycle.* Over an irregular period—from three to eight weeks—the upper westerlies often change from small waves to large waves. They may even exhibit individual horizontal circulation in cells. Figure 5-13 shows how large quantities of warm and cold air can be redistributed over the globe in this fashion. The same method of redistribution can be seen at work in Figure 5-12.

Anticyclones and cyclones also redistribute heat. Imagine that the anticyclone depicted in Figure 5-6 is located off the western coast of the United States. As air circulates around the anticyclone, cold air from the north is drawn down its eastern side, and warm air from the south is pushed poleward on the western side. The circulation of air around a cyclone is similar but reversed.

The anticyclones and cyclones that commonly travel from west to east in the midlatitudes are related to the wavelike motion of the upper westerlies. If you were on a bridge overlooking a river, you might notice that the main current meandered from bank to bank. Small circular eddy currents, whirlpools, would be carried along by the main current. Traveling cyclones and anticyclones are the eddy currents in the main current of the upper westerlies, and they have their counterparts in the ocean's circulation. In modern

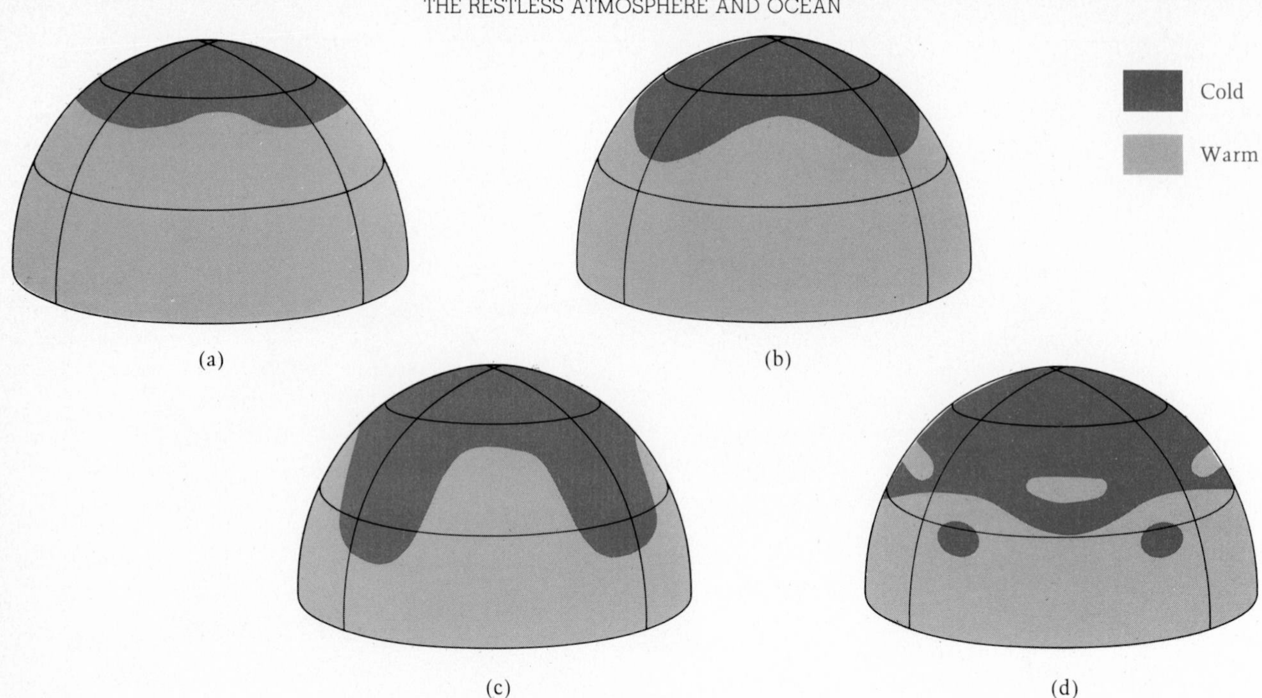

FIGURE 5-13
The zonal index cycle. Black indicates cold air; color, warm air.

FIGURE 5-14
Surface pressure and winds across the globe in a) January and b) July.

(a)

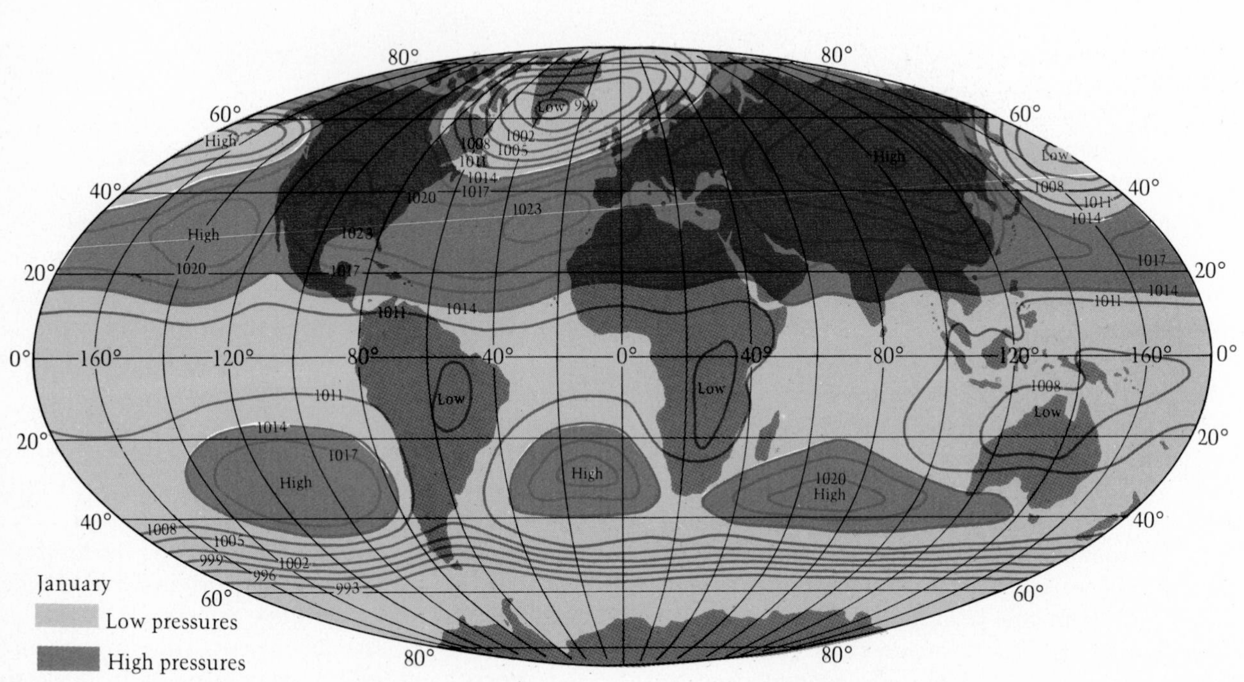

thinking, then, a large part of the heat redistribution is accomplished not by meridional circulations, but by traveling waves and cyclonic and anticyclonic eddies. This is certain for the midlatitudes but possibly insufficient by itself for the low latitudes, where the Hadley cell still maintains its 200-year-old position in atmospheric theory.

The riverlike nature of zonal flow can also account for the redistribution of angular momentum in the general circulation. This is a complicated matter, but the answer again lies with the wavelike westerlies and their traveling eddies. In certain situations, the upper air waves slope away from the ideal north–south alignment. But most of the transfer is accomplished by the eddies themselves as they move both south–north and north–south.

Detecting zones The major features in the satellite mosaic in Figure 5-8 may be explained by the concepts we have just encountered. The band of clouds around 10°N represents the upward moving limb of the Hadley cell. Surface northeasterly and southeasterly winds meet along this band to form the *intertropical convergence zone* (see Chapter 6 for further details).

The downward limb of the Hadley cell produces clear weather in *subtropical high-pressure zones.* These high-pressure zones probably build up when air moving toward the poles is deflected eastward and angular momentum is conserved. Or perhaps the subtropical jet stream throws off anticyclonic whirls that form the upper parts of the subtropical high-pressure zones, just as the main current of a river throws off smaller eddies.

Poleward of the high-pressure zones are a band of midlatitude cyclones, or depressions, that produce the spiraling cloud bands. They are carried along by the great westerly wind current, and their centers are to the west of where the polar front jet stream plunges toward the equator.

In high latitudes the snowfields of northern Canada and Russia and the Greenland icecaps are visible. The lack of clouds is a clue to the presence of a *polar high-pressure zone.* Although this zone does not exist in the upper air, it is rather well marked at the surface.

HIGHS AND LOWS

To understand the actual climates at the earth's surface, we must look more closely at the distribution of pressure patterns and their relationship to the motion of the atmosphere. The actual distribution of surface pressures and winds in January and July, shown in Figure 5-14, is much more complex than the schematic circulation system shown in Figure 5-9. The real pat-

(b)

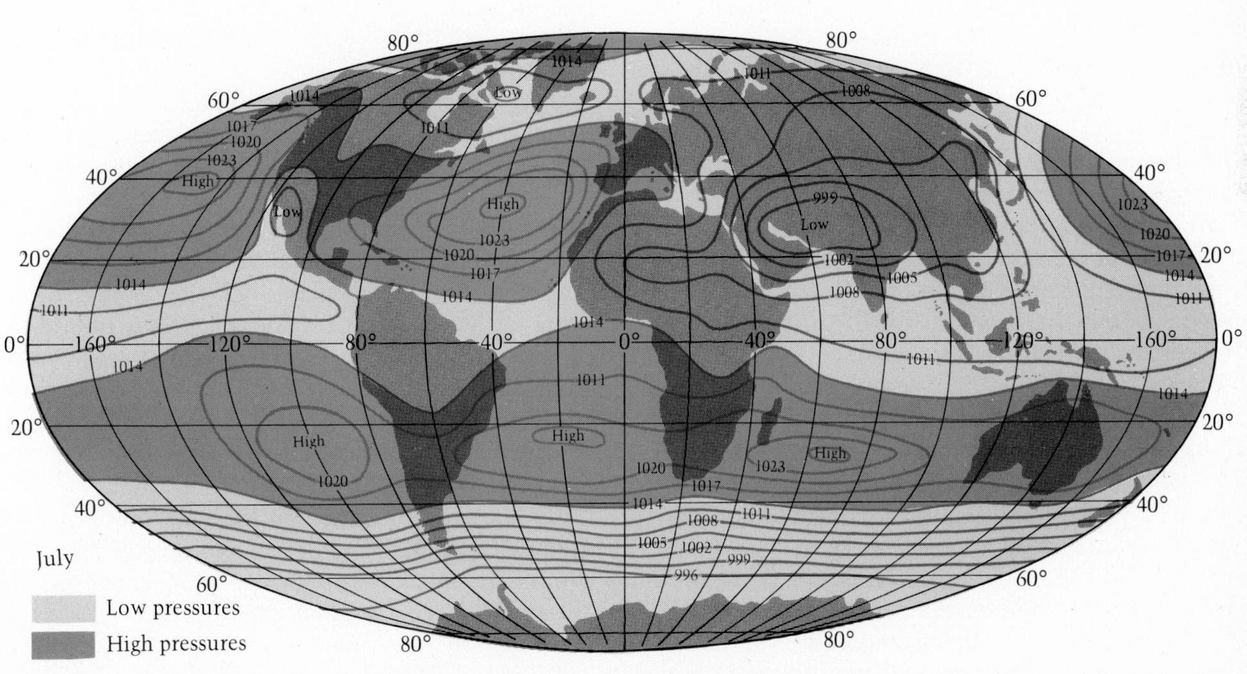

tern is made up of pressure cells rather than simple belts.

The pattern of cells at the surface is amplified by the different heating and cooling effects of land and sea. Land and sea have different effects on surface pressure because sea surfaces take longer to cool down and warm up than do land surfaces, as Chapter 4 explains. Nowhere is this more clearly demonstrated than in a comparison of the Northern and Southern hemispheres. As you can see in Figure 5-14, the pressure distribution in the Southern Hemisphere is much more closely related than that of the Northern Hemisphere to the theoretical pressure distribution of Figure 5-9. Only in winter do the relatively small land masses of South America, Africa, and Australia break up the subtropical high-pressure belt. In contrast, the large land masses in the Northern Hemisphere markedly disturb the orderly pattern of pressure zones. For example, Figure 5-14 shows that in January the *Hawaiian high* in the Pacific Ocean and the *Bermudan high* in the Atlantic Ocean tend to link up. In July continental heating, especially in the southwestern United States, creates rising air and a lower-pressure area that separates these two highs. The same phenomenon can be seen to some extent in the Southern Hemisphere and over the Asian continent.

The effect of surface heating and cooling is also apparent in the pressure cells of midlatitudes. The surface *Siberian high,* also shown in Figure 5-14, is quite pronounced in January because cold air sinks toward the earth's surface. But this high-pressure cell occurs at latitudes where we would expect a low-pressure cell. The Siberian high is actually a *cold anticyclone* that exists only to about 2500 m (8000 ft). Above this level, it is replaced by a low-pressure cell. The winter Siberian high is matched by a similar feature over North America.

In contrast, subtropical highs are *warm anticyclones* created when the air within them subsides. The subtropical high-pressure cells, one of the most important features of the cell system, exist at all times of the year, whereas the midlatitude low-pressure cells are statistical phenomena created by averaging many eastward-moving cyclones. Besides being arranged in cells in the horizontal sense, the subtropical highs are tilted so that their eastern half is closer to the earth's surface than their western half. The high-level parts of the cells are probably related to the subtropical jet stream.

The midlatitude surface pressure in July (refer again to Figure 5-14) is noteworthy for the *Icelandic lows* and *Aleutian lows.* To understand these, you should refer back to the chart of upper-air circulation in Figure 5-12. If you compare the distribution of pressure at the surface and in the upper air, you can see that the low-pressure areas of the surface lie to the east of the upper-air troughs. Chapter 6 shows that this arrangement of surface and upper-air lows is no accident. But at this point, we need only realize that the upper air does tend to form fixed troughs and ridges. This is thought to reflect the influence of such mountain ranges as the Rockies and the Himalayas and the particular distribution of land and sea in the Northern Hemisphere.

You are by now undoubtedly aware that the general circulation of the atmosphere is a complex system. Scientists still do not understand many of its interactions. But we have seen that there is a close relationship between wind movement and pressure, namely, the geostrophic wind relationship. This shows itself on a small scale by determining the speed and direction of winds around cyclones and anticyclones. On a large scale we could use the relationship to predict the location of the polar front jet stream in Figure 5-12 by finding areas where the isobars were close together. If you take one more look at Figure 5-8, you may now realize that there is more going on than at first meets the eye.

THE CIRCULATION OF THE OCEAN

The ocean, which covers 71% of the earth's surface, plays an integral part in adjusting the earth's heat imbalance. It does so principally through the circulation of water in large ocean currents. Warm currents travel toward the poles, and cold currents move toward the equator.

Ocean currents can be started in several ways. Sometimes water piles up at one coastline, such as at Brazil, yielding a higher sea level. Gravity then forces the water back, forming narrow equatorial countercurrents. Another source of ocean movement is differences in density. Density differences can come from temperature differences, as when the chilled surface water of the Arctic area sinks and spreads toward the equator, or from salinity differences. The ocean under an equatorial rain belt is less saline than that under the subtropical high-pressure zones. The saline water, being heavier than the nonsalty water, tends to sink and give way to a current of lesser salinity.

But the greatest reason for oceanic motion is the frictional drag on the water surface set up by prevailing winds. Frictional drag transfers kinetic energy, the energy of movement, from the air to the water. Once set in motion, the water immediately succumbs to the deflective Coriolis force, just as the air does. In a pattern also similar to the atmospheric circulation, friction within the oceans leads to an Ekman spiral

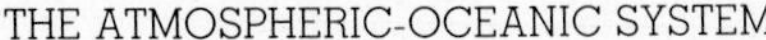

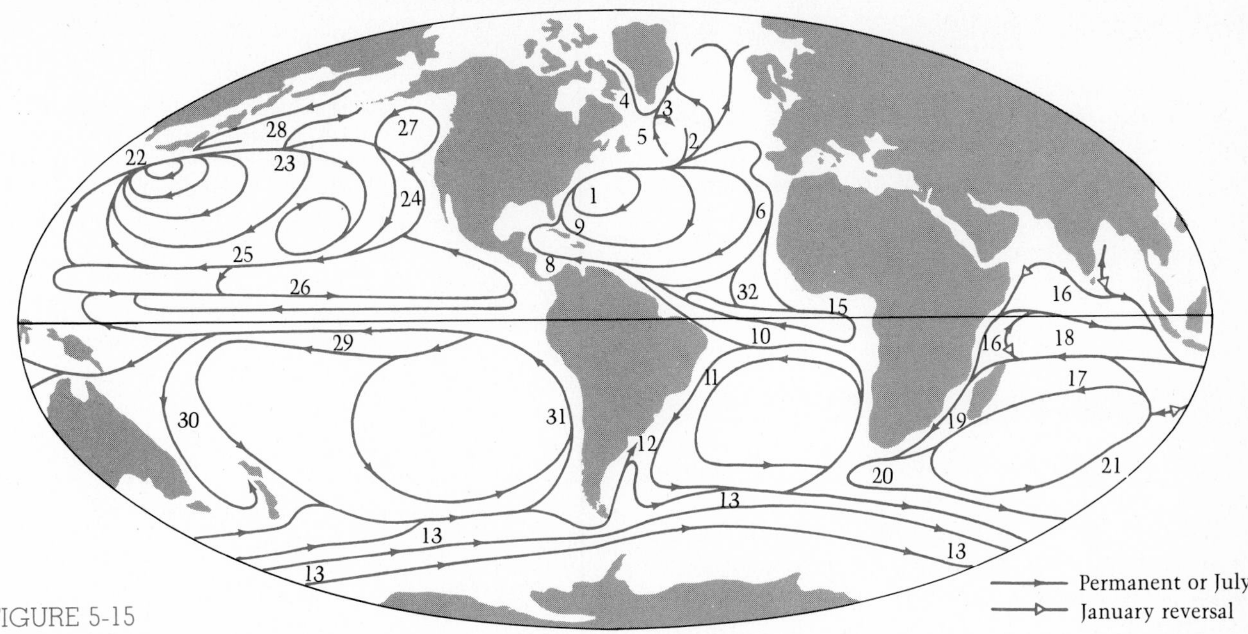

FIGURE 5-15

Major global ocean currents.

that describes the change in speed and direction of the water movement with increasing depth.

Prevailing winds, the Coriolis force, and sometimes the configuration of land masses often combine to send ocean currents into circulations that resemble large cyclones and anticyclones. In the oceans, these are called *gyres*, a name used for both cyclonic and anticyclonic circulations. Gyres are so large that they fill a whole ocean, as Figure 5-15 indicates. Scientists have recently discovered that smaller whirls, with diameters of several hundred kilometers, may also play a role in redistributing heat energy. Figure 5-15 diagrams the distribution of major ocean currents.

If you compare Figures 5-14 and 5-15, you will find many similarities, a fine demonstration of the relationship between the oceanic and atmospheric systems. But this is a two-way relationship; the atmosphere creates currents in the ocean, and ocean currents affect the atmosphere. For example, under the influence of the Hawaiian high and its related winds, part of the *North Pacific drift current* turns southward at the North American coast to form the *California current*. Coming from high latitudes, the California current plays a large role in cooling the coastline along which it flows. But by the time the current reaches California, the Ekman effect causes the surface water to move westward away from the coast. The surface water is replaced by even colder water rising from depths of about 100 to 300 m (300 to 1000 ft), and the Ekman effect spreads this cold water westward too. Air crossing this water can be chilled considerably, and its water vapor quickly condenses. Thus we have an explanation of not only why it is too cold to swim off the northern California beaches, but also why San Francisco has its famous fogs.

The need to be aware of the interaction of ocean and atmosphere has been demonstrated dramatically by a Russian proposal to remove ice from the Arctic Ocean. In essence, the Russians have proposed damming the Bering Strait to block cold arctic water from the warm Pacific. Consequently, Russia would have ice-free ports on the Pacific Ocean all year round. They also proposed to pump warm Pacific water into the Arctic Ocean, thereby warming arctic coastlands and melting the arctic ice pack. However, this maneuver would send arctic water across the northern shores of North America, further chilling the already icy *Labrador current* and seriously distressing the eastern coast of Canada. Furthermore, the melting ice might raise the sea level and submerge the very ports the Russians were attempting to open, as well as many of the earth's major coastal cities. We must find out more about the complex interaction of the atmosphere and oceans if we are to prevent such potential disasters.

THE ATMOSPHERIC-OCEANIC SYSTEM

Let us look again at the amount of heat moved from the equator to the poles by the combined forces of the atmosphere and the oceans. Calculations by the Russian scientist Mikhail Budyko, reproduced in Table 5-1, show that the atmosphere accounts for 87% of the total heat moved by these systems and that the oceans are responsible for 13%. More recent estimates indicate that the oceans may be responsible for as much as 25%

TABLE 5-1

The annual average heat flow at the earth's surface in units of 10^{14} calories per second

LATITUDE	TOTAL	OCEAN	ATMOSPHERE
60°	7.6	0.7	6.9
50°	10.5	1.3	6.9
40°	12.0	1.8	10.2
30°	11.0	2.1	8.9
20°	8.4	1.3	7.1
10°	4.6	-0.3	4.9
TOTALS	54.1	7.2	47.2

of the heat movement in low latitudes and midlatitudes.

In general, the atmosphere carries more heat because it moves faster than the oceans; the movement in the atmosphere occurs in two ways. Much of the heat is transferred as sensible heat, as when the northward flowing part of the polar front jet stream carries warm air toward the poles. In addition, a great deal of water evaporates from the low-latitude oceans because of the radiation surplus there. This water vapor will transport its latent heat until the heat is released by condensation. The latent heat of the water vapor may be carried toward the poles by the general circulation, especially by the Hadley cell. Figure 5-16 shows the relative amounts of heat carried toward the poles in the form of sensible and latent heat.

Figure 5-16 includes values of *potential energy,* the energy an object has by virtue of its position with respect to another object. When you lift a blackboard eraser, you work against the earth's gravitational field and give the eraser potential energy. If you drop the eraser and let it fall, the potential energy becomes the kinetic energy of movement. Radiant energy from the sun cannot directly power the general circulation of the atmosphere. Instead, the sun warms the earth, the earth warms the air, and then the air rises, gaining potential energy as it moves away from the earth.

In 1906 Max Margules, a scientist working in Vienna, pointed out that, whenever cool air sinks and warm air rises in an atmospheric system or subsystem, potential energy is released in the form of movement (kinetic energy). This very phenomenon occurs in the Hadley cell and in midlatitude cyclones, whence the general circulation gets its driving kinetic energy. The kinetic energy is continually dispersed by friction with land surfaces and ocean surfaces, thus forming the ocean currents. Without a conversion from potential to kinetic energy, the atmosphere would virtually come to a complete standstill within two weeks.

Now we can understand Figure 5-17, a simplified version of the atmospheric-oceanic circulation system. Here we can see the circulations acting like a giant heat engine. Solar radiation comes into the system, and although much of it leaves as outgoing radiation, some of it gives the atmosphere potential energy. This potential energy is converted into kinetic energy. The kinetic energy is used in redistributing heat across the globe through the oceans and the atmosphere. Finally, the kinetic energy is dissipated by friction and ends in the form of heat energy. A small amount of heat energy can serve as positive feedback to form more potential energy, although most of it is used to increase the entropy of the larger sun-earth system. An engineer would not consider this a very efficient machine, because it uses less than 1% of all the energy coming to the earth from the sun. But from our point of view, it performs perfectly—it allows human occupation of large parts of the globe.

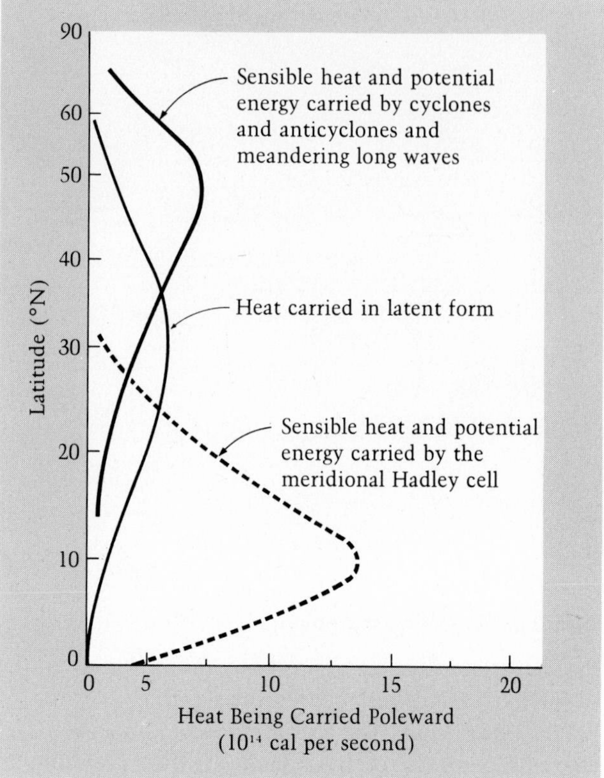

FIGURE 5-16

Relative amounts of heat carried poleward by different processes.

SUMMARY

The basic causes of the large-scale movement of air in the atmosphere, known as the general circulation, are

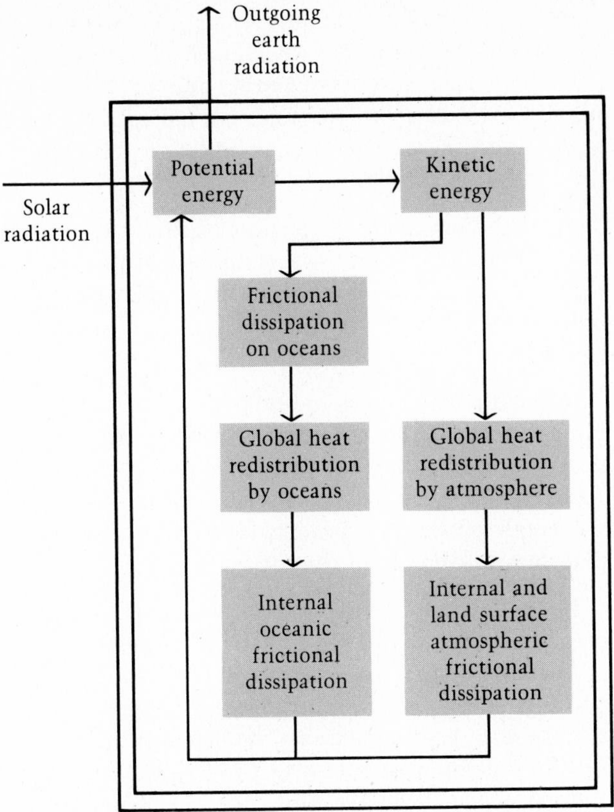

FIGURE 5-17
The atmospheric-oceanic general circulation system and its energy exchanges.

the unequal amounts of heat energy at the equator and the poles and the rotation of the earth. On a smaller scale, air movement results from the interaction of the pressure gradient, Coriolis, centripetal, and frictional forces.

The pressure gradient and Coriolis forces combine to form a wind called the geostrophic wind, and these forces and centripetal force are responsible for the gradient wind. The winds of the general atmospheric circulation can be described by examining separately meridional and zonal movement. Much of the work of redressing the heat imbalance is performed by traveling waves and cyclonic and anticyclonic eddies in midlatitudes and by the Hadley cell in low latitudes. These elements affect atmospheric pressure and engender surface winds.

Surface winds are one of the main driving forces for the currents existing in the oceans. The circulation of the oceans resembles, and is related to, atmospheric circulation.

Finally, the atmosphere and ocean may be regarded as a single system. This is most apparent when energy considerations are taken into account.

QUESTIONS

1. Using Figure 5-8, explain the patterns of clouds at the following latitudes: 10°N; 20°N; 50°N.

2. In Figure 5-5, how do we know that the gradient winds shown are forming in the Northern Hemisphere? What would be different if they were forming in the Southern Hemisphere?

3. Why is it difficult to tell a geostrophic wind from a gradient wind below about 500 m? What causes the differences in their behavior that makes it easier to distinguish them above 1000 m, and what are their distinguishing features?

4. Large-scale circulation of the earth's atmosphere is distinguished from small-scale movement. What is primarily responsible for large-scale movements of air on the earth? What are the forces controlling an individual parcel of air's movement?

5. The polar-front jet stream, a geostrophic wind, is found in the Northern and Southern hemispheres high above the surface of the earth, usually between 6 and 11 km (4 to 7 mi) in altitude. Explain why it is not closer to the earth, and what controls its speed and direction.

6. In Figure 5-7, what causes the velocity and direction changes of the wind with height?

7. Although Hadley's explanation of the trade winds was a great step forward for atmospheric science, it failed to explain the midlatitude westerly winds. While he understood the working of the Coriolis force in terms of the direction of propagation of the trade winds, Hadley underestimated the strength of the force, believing that the rising air from the equatorial cell traveled all the way to the polar regions. What part would this actual Coriolis force, stronger than Hadley anticipated, play in the global circulation system as he conceived it?

8. According to the text and Figure 5-9, what is the driving force for Rossby's midlatitude cell, the Ferrel cell? How does it explain the westerly winds at the surface in the midlatitudes?

9. Palmén's model of meridional circulation (Figure 5-10) differs from the Rossby model mainly from the midlatitudes poleward. What is the principal difference between the models, and what advantage does Palmén's model have over Rossby's?

10. How has the concept of zonal flow of heat in the earth's atmosphere revolutionized thinking about atmospheric heat transfer?

11. Describe the process of formation of Rossby waves (Figure 5-13) in terms of the wind types developed during the index cycle. Do geostrophic winds ever entirely disappear from the system?

12. Why is zonal circulation considered insufficient to explain the circulation of heat from the equator to about 30°N and S, but a sufficient explanation of the circulation in the midlatitudes and polar regions?

13. Why are the midlatitude zones of subsidence in the form of cells rather than a worldwide belt? Why do they tend to be more beltlike in the Southern Hemisphere than in the Northern?

14. What is the principal difference between the midlatitude high-pressure zones, such as the Siberian high, and the subtropical highs, such as the Azores high or the Hawaiian high? What causes the high-pressure cells to form in the midlatitudes, where we would normally expect low pressure?

15. Winds give ocean currents their motion through frictional drag on the water surface. Is this the only reason that ocean currents tend to have patterns analogous to those of the large cyclones and anticyclones in the atmosphere?

16. San Francisco is famous for its summertime fog and cold waters, while Los Angeles, only 725 km south, is known as a surfer's paradise due to the warm summertime water. What causes the radical difference in summertime climate along these two stretches of the California coast?

17. If the oceans contain most of the water in the earth's hydrosphere, and water has one of the highest heat capacities known, why do the oceans transport only about 15 to 25% of the heat from the equator to the poles?

18. Figure 5-16 shows the relative amounts of heat carried poleward in the form of latent and sensible heat. Why is the flow of latent heat shown only up to about 40°N? What accounts for the two "humps" in the graph of sensible heat transfer, and why is the bottom one larger than the top one?

19. In Chapter 4 we saw that only a small amount of the radiation from the sun actually reaches the surface of the earth, due to a variety of atmospheric phenomena. Even less of this energy is used by the atmosphere-ocean "machine." What eventually happens to the potential energy gained by the atmosphere and the oceans from the sun?

WEATHER SYSTEMS

CHAPTER 6

On Sunday June 5, 1944, 1300 aircraft, 5000 ships, and 200,000 allied troops were poised to leave the shores of Britain. Their mission: to invade France and bring the European part of World War II to a close. But they did not leave on that day. One of the thousands of weather systems in the atmosphere prevented the release of the greatest assemblage of military power the world had ever known. The passage of the same weather system made possible the historic victory of the following day, D-day. And so it is that humans must constantly yield to the changes of the restless atmosphere.

Weather systems are organized phenomena of the atmosphere—with inputs and outputs and with changes of energy and moisture. They can be as large as a hurricane or as small as a thunderstorm. Such weather systems, together with the constant flows of moisture, radiation, and heat energy, make up our daily weather. A description of our average weather, and its departures from average, over a long period of time, say 30 years, constitutes climate.

Chapter 5 describes a distinction in the general circulation between the easterly winds and Hadley cell of the low latitudes and the westerly "rivers" of air above about 30° latitude. This chapter is organized around this distinction. We begin our look at weather systems with the high-energy systems of low latitudes, typically demonstrated by violent rains. Then we will move on to the changing patterns of the higher latitudes. We will end our investigation by looking at some of the smaller weather systems, which are often caused by isolated features of the earth's surface. In this chapter, then, we are examining the major types of systems that make up our weather.

LOW-LATITUDE WEATHER SYSTEMS

In the low latitudes, heat is a constant feature of climate. The heat provides the energy for driving winds as well as the energy to evaporate water that is later carried by the winds. Convergence of the winds and cooling of the vapors causes the rainfall so common in the tropics. The largest of the tropical weather systems is found where trade winds from the two hemispheres come together.

THE INTERTROPICAL CONVERGENCE ZONE

Chapter 5 explains how the Hadley cell is manifested at the earth's surface as a flow of winds called the

FIGURE 6-1

Formation, movement, and rainfall of the intertropical convergence zone. a) The formation of the ITCZ where trade winds converge. b) The idealized seasonal track of the ITCZ. c) The seasonal rainfall distributions ideally resulting from the movement of the ITCZ.

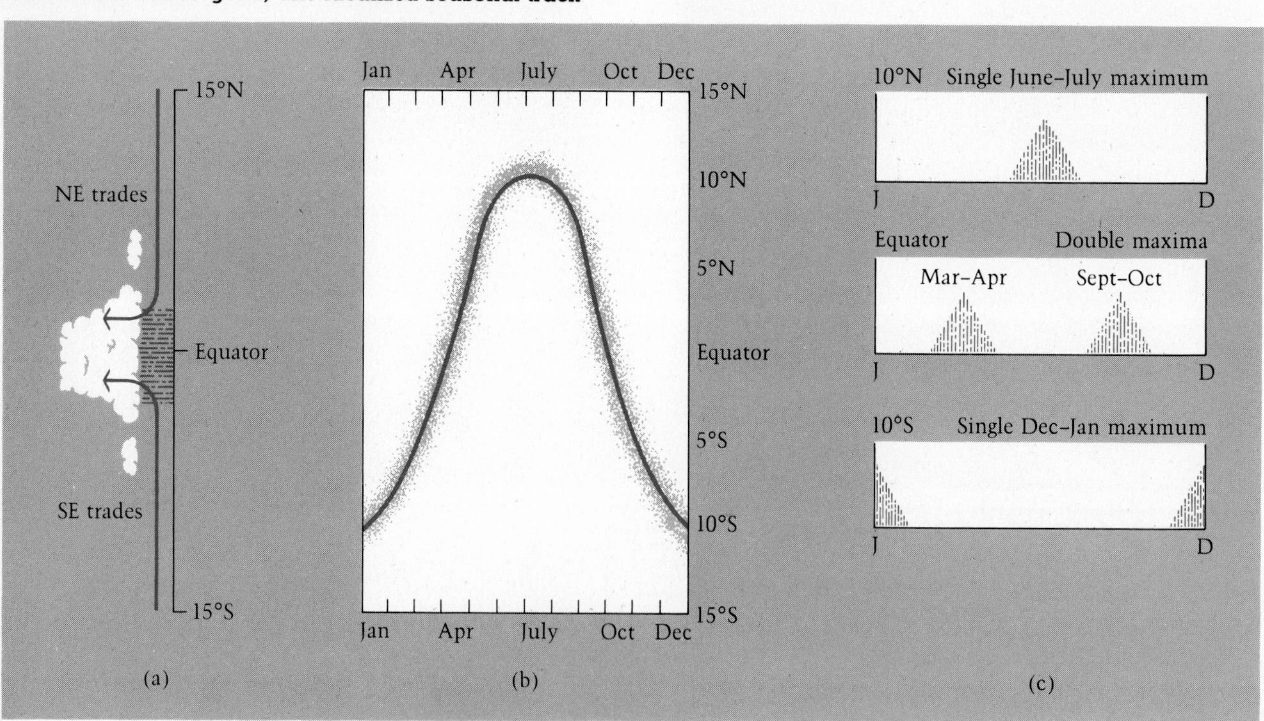

northeasterly and southeasterly trade winds. The two trade-wind systems converge in the equatorial trough of low pressure, and the rising air is responsible for clouds and rain. The clouds of the *intertropical convergence zone* (ITCZ) can be seen quite clearly in the satellite picture in Figure 5-8. The ITCZ occurs at low latitudes all around the earth, but it is most clearly marked over the oceans, particularly the Pacific Ocean. But the rainfall its clouds deliver over the land is vital to many inhabitants of tropical areas.

The ITCZ changes its location throughout the year, mainly following the areas of maximum solar heating. Thus as Figure 6-1 demonstrates, in July the ITCZ lies at about 10°N, but in December it is found around 10°S. The ITCZ passes twice over the equator, giving the greatest amount of rainfall during March and April and again in September and October.

Although the ITCZ owes its origin to the position of

A BIBLICAL PLAGUE: UPDATED

Clouds of locusts have harassed farmers and frightened city people since Biblical times. Moses called for a plague of locusts to destroy the crops of Egypt. The first crop of the Mormon settlers in Utah was nearly completely devoured by the insects. Until the 1950s, locusts were especially destructive in Africa. They descended on fields and quickly destroyed the precious food supply. Agricultural specialists were helpless, but atmospheric scientists unraveled part of the problem. They discovered links between locusts and the intertropical convergence zone.

The first breakthrough came when they found that the locust breeding grounds coincide with the rains of the ITCZ in its northernmost position. Spraying in the right area at the right time now wipes out many of the pests before they reach the sky. Next, investigators realized that the flight path of the insects is guided by winds flowing toward the ITCZ. Thus the swarms can be traced and attacked in the air. Finally, scientists recognized that locusts gather in regions of low-level air convergence, which can be located by radar. Spraying is more efficient now that it can be directed where locusts congregate.

The outcome of this story is a happy one: no serious infestations of locusts have occurred in Africa in the last twenty years.

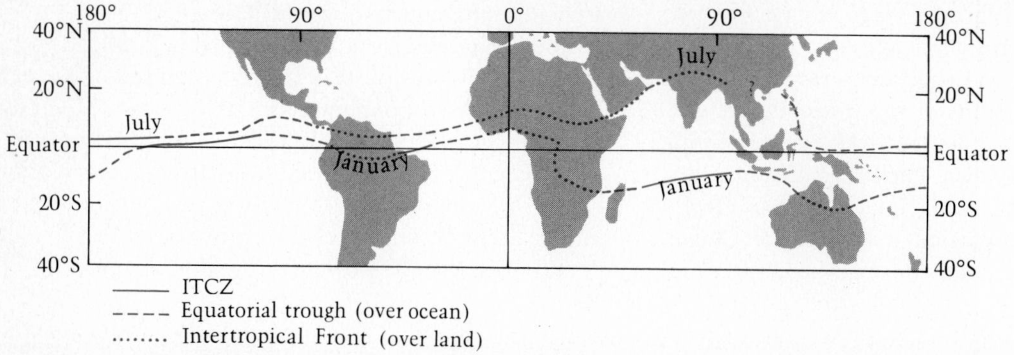

—— ITCZ
- - - - Equatorial trough (over ocean)
.... Intertropical Front (over land)

FIGURE 6-2
The actual average location of the intertropical convergence zone in January and July.

the overhead sun, this is not the only factor that controls its location. The distribution of land and sea and the flow of the tropical atmosphere are also important. For these reasons, at any one tine the ITCZ may not be where it should be, according to simple theory. Figure 6-2 shows the ITCZ at two extremes. In July, the zone does not appear in the Southern Hemisphere. Yet in January it ranges from 20°S, over Australia, to within the Northern Hemisphere in the eastern Pacific and Atlantic oceans.

Despite these fluctuations, the ITCZ may be the single most important weather system in the tropics. Its only rival is a large-scale seasonal wind reversal known as the monsoon.

MONSOONS

The Arabic word for "season" is *mausim*, and from this name is derived the word *monsoon*. Indeed, in some places certain seasons are called monsoon seasons. The climate of southern Asia is dominated by the twin influence of the northeastern monsoon of the colder months and the rain-carrying southwestern monsoon of summer. These winds override the expected pattern of the large-scale general circulation and yet are still a part of it. To explain this paradox, we must once more consider the upper air, the atmosphere at about 10,000 m (32,800 ft) above the earth's surface.

Monsoons of southern Asia primarily result from the influence of the land masses on the seasonal movement of pressure and winds. Over the sea, the seasonal shifts of the heat and pressure zones are rather small, in keeping with the similarly small annual temperature changes. But over land, where temperature variations are larger, the movement of the heat and pressure zones is exaggerated. This shows up in the summer and winter movement of the ITCZ, diagramed in Figure 6-2. Over India, in summer, the ITCZ is more than 30° away from the equator. This extreme position is associated with the large degree of heat at the land surface, which creates rising air and low pressure. In summer, therefore, the lower pressure over India draws the ITCZ away from the equator. This produces a southwesterly current of air where we

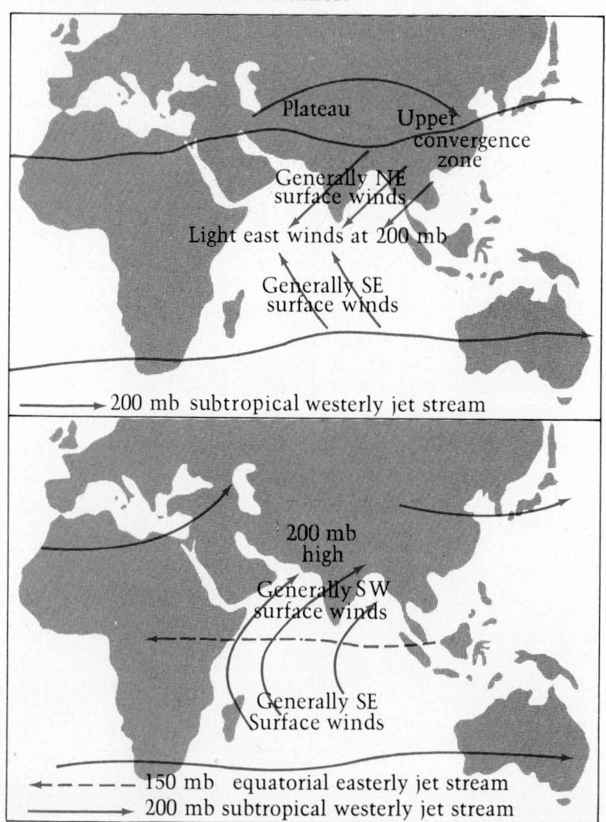

FIGURE 6-3
Major features of the Asian monsoon. Upper-northern winter. Lower-northern summer.

would expect the low-latitude easterlies, which in this case exist only in winter.

This explanation, dealing mainly with the surface winds and shifting surface pressure belts, must be supplemented with a description of other actions in the upper air. In the northern winter, the subtropical jet stream continues its westerly flow. But it is divided over India, as Figure 6-3 indicates, possibly because of the high Tibetan Plateau. One part of the jet stream flows to the north and one to the south of the plateau. Beneath the southern branch of the jet stream, the air subsides, pushing northerly and northeasterly winds across India and Pakistan. In May or June the southern branch of the subtropical jet stream weakens and disappears. At the same time, a warm anticyclone forms over the Tibetan Plateau, and air from the southern side of the anticyclone contributes to a new, but easterly, jet stream near the equator, as Figure 6-3 shows.

These events in the upper air herald the onset of the southwestern monsoons on the earth's surface. To 600 million Indians, these monsoons are a matter of life or death. The season of the northeastern monsoon runs from January to May. A fine, sunny "winter" season in January and February merges into very hot weather in April and May, when little rain falls and the ground becomes parched. In June, after the upper air flow has changed, the southwestern monsoon begins. Life-giving rains burst upon the land, and grass grows again. The situation gradually reverses in September and October, when the southwestern monsoons begin to retreat and the skies begin to clear. The temperature rises before the sun moves south and "winter" begins once again.

The southwestern monsoons are composed of two main branches, as Figure 6-4 shows. One branch crosses the Bay of Bengal to Burma and northeastern India. A second branch, with its own tendency to split into two parts, arrives from the Arabian Sea. The rains from the two branches gradually spread across the subcontinent and fill the wells. But the monsoon rains are not continuous. They depend on low-pressure areas within the southwesterly air flows. These "depressions" lift the air and enhance the formation of rain.

Monsoon rains are not limited to southern Asia. Indonesia, northern Australia, and parts of western Africa also depend on them for much of their moisture. Yet it is only over India that the subtropical westerly jet stream readily develops in the upper atmosphere.

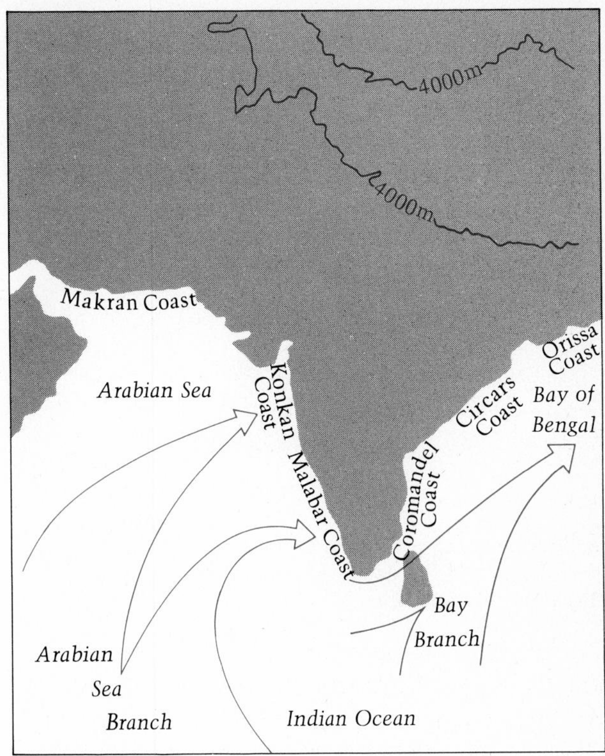

FIGURE 6-4
The main branches of the southwestern monsoon current.

EASTERLY WAVES

For centuries humans have known about the trade winds, the constant flow of tropical air. Only within the past two decades has it been discovered that wavelike phenomena carried on this flow. These waves give rise to distinctive weather systems.

Chapter 5 explains that geostrophic winds parallel isobars. If we apply this relationship to Figure 6-5, we can see that the isobars indicate the easterly trade winds. Imagine that the isobars are skipping ropes attached between western Africa and America. If some great being stood on the African continent and moved the ropes up and down, a wave would travel along them toward the west. This is directly akin to what actually happens to the isobars when a traveling easterly wave is formed.

The concept of conservation of angular momentum is also introduced in Chapter 5. A particular form of this concept is called the law of *conservation of absolute vorticity*, which again deals with conserving the total amount of spin in a system. Imagine a column of air spinning like a top and (remember our vertical tube) traveling across the spinning earth. In this case, the sum of the spin of the air column and the spin of the earth, divided by the depth of the air column, must be a constant number.

As our air column moves toward the west along an easterly limb of the wave, it moves to higher latitudes

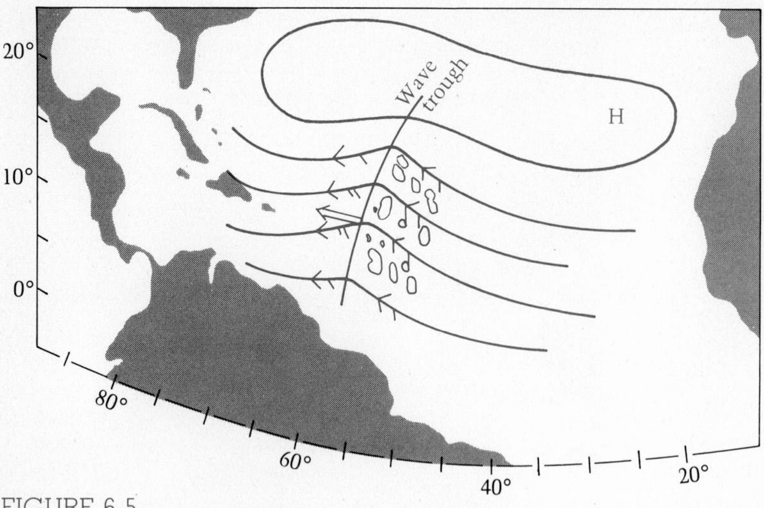

FIGURE 6-5

Isobars showing a traveling easterly wave in the Caribbean and the development of towering rain clouds in relation to it.

and therefore to an area of greater earth spin. In addition, rotation in the same sense as the earth's rotation—counterclockwise in the Northern Hemisphere—is defined as being positive. So the air column is also moving to an area of higher positive air spin, as shown by the counterclockwise curvature of the isobars. To keep the total spin of the system constant, in accordance with the conservation of absolute vorticity, the depth of the air column must also increase. (Meteorologists talk about depth when they mean height in the atmosphere.) Therefore, our column of air increases in height, and the air rises. This in turn creates clouds and often rainfall. On the eastern side of an easterly wave, we may expect towering, puffy clouds and rain. On the western side, the argument is reversed. Clear weather results because the air column decreases in depth.

Weather systems associated with easterly waves are most frequent in the Caribbean Sea, but similar phenomena are often observed in the west-central Pacific Ocean and in the northern China Sea. They travel toward the west at about 5 to 7 m per second (11 to 16

TYPHOONS: A CONTRAST OF SOLUTIONS

Despite the enormous destruction caused by cyclones, the Japanese people refuse to use new techniques that might tame the storms. About four typhoons batter Japan every year, and the property damage amounts to over $100 million. But a quarter of Japan's rainfall arrives with the typhoons, so Japanese farmers know this ill wind blows them a good harvest.

The United States has experimented with cloud seeding to subdue cyclones. In the 1960s the meteorologists of Project Stormfury dropped silver iodide crystals into hurricanes to soften their punch. But the Japanese insist that American seeding operations in the Pacific only tamper with storms at sea. Cyclones headed for the islands of Japan must be allowed to crash into the coast with their full force and their full load of rain.

The Japanese would rather invest their money and expertise in warning systems and rescue efforts. A radar station atop Mount Fuji sights typhoons 50 miles offshore. A network of 16 other radar installations plots the movement of approaching storms. Residents along the coast are alerted and helped with preparations for the coming heavy weather. The war on cyclones illustrates two conflicting attitudes toward the environment. Besides developing sophisticated warning systems, Americans look for technological ways to control nature, but the Japanese look for ways to adapt to natural conditions.

FIGURE 6-6

An Apollo 7 view of Hurricane Gladys, a tropical cyclone in the Caribbean. The picture, looking toward the southwest, shows Cuba in the background. Winds of 40 m per second (90 mi per hour) were reported during this storm.

mi per hour) and are relatively predictable in their motion. This cannot be said for the next tropical system to be described.

TROPICAL CYCLONES

The tropical cyclone is one of the most fascinating and terrifying features of the atmosphere. It is the *hurricane* of the southern North Atlantic and Indian oceans, the *typhoon* of the China Sea, and the *willy-willy* of western Australia.

The tropical cyclone is a moving low-pressure area, normally originating in the warm, moist air of the low-latitude atmosphere. It has distinctly circular wind and pressure fields. In Figure 6-6, a portrait of Hurricane Gladys, the circular wind pattern extends vertically, and the clouds associated with the center of the system take on a shape much like a doughnut.

The World Meteorological Organization describes the tropical cyclone as having wind speeds greater than 32 m per second (71½ mi per hour) and a central surface pressure less than 900 mb. Wind speeds can, however, exceed 90 m per second (200 m per hour) and lead to great destruction. The strongest winds are found near the center of the storm but not actually at the center.

The middle of the system contains an "eye," the hole in the doughnut, from which spiral bands of cloud spread out. In the eye, winds are light, and there is little or no rain. But the surrounding clouds are responsible for torrential rain. The rain, like the wind, is heaviest near the center and sometimes exceeds 500 mm (20 in) per day.

Temperatures remain the same throughout the area of a tropical cyclone, which has an average diameter of about 500 km (310 mi). But the diameter can vary from 80 km (50 mi) to about 2500 km (1550 mi). In height, cyclones range from 8 km (5 mi) to 17.5 km (11 mi). The greater the amounts of energy and moisture involved in a system, the larger the cyclone will be.

Tropical cyclones form most often in late summer and autumn over warm sea surfaces with temperatures greater than 27°C (80°F). Their origin appears to be related to the equatorial trough of low pressure. They form when latent heat warms the center of a preexisting storm and helps to intensify an anticyclone present in the upper troposphere. Once the storm has developed, it becomes self-sustaining. Vast quantities of heat energy are siphoned from the warm ocean and transported into the atmosphere as latent heat. This energy is released as sensible heat when clouds form. The sensible heat provides the system with potential energy, which is partially converted into kinetic energy, causing the violent winds of the cyclone.

Because of the conditions necessary for their formation, tropical cyclones originate between 5° and 20° latitude in all tropical oceans except the South Atlantic Ocean and the southeastern Pacific Ocean, where the ITCZ seldom occurs. Once formed, their movement is erratic. In general, they first travel westward and then to the northwest in the Northern Hemisphere before curving around to the east, where they come under the influence of the westerly winds of the middle latitudes. The cyclones usually move with speeds of 16 to 24 km per hour (10 to 15 mi per hour). If they pass over a large body of land, their energy source—the warm ocean—is cut off, and they gradually die away.

Before these storms die, however, they often cause widespread damage on land. Most of the damage is done by high winds and heavy rains, but near coastlines, waves and tides rise to destructive levels. Such havoc can be understood when you consider that the amount of energy unleashed by a tropical storm in one hour equals all the electric power generated in the United States in one year. The tropical cyclone is the most dangerous of all atmospheric weather systems. In the United States, these storms have cost more than 1500 lives in this century. In other countries it is not unusual to count in the thousands the death toll of one storm. Figure 6-7 contains satellite photographs of one tropical cyclone that smashed into Bangladesh.

Violence and drama mark the energy-laden air of low latitudes. We have not considered all the weather

November 11, 1970

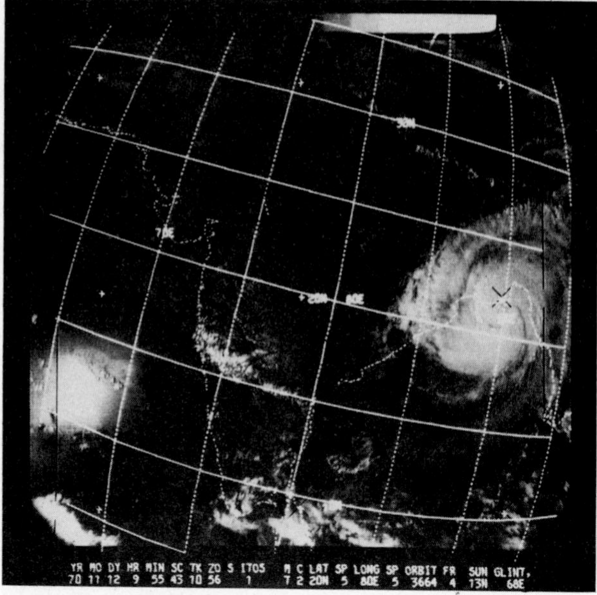

November 12, 1970

FIGURE 6-7

Satellite pictures taken on 11-12 November, 1970, showing the passage of a tropical cyclone across the Bay of Bengal and into Bangladesh. This cyclone resulted in a storm surge with a height between 5 and 10 m (16 and 33 ft). The wave of water swept over the land, killing hundreds of people.

systems that develop in the tropics. Individual thunderstorms, for example, are influential and frequent, as we will learn later. Weather systems in the higher latitudes are often less dramatic but seldom less important.

WEATHER SYSTEMS IN HIGHER LATITUDES

The general circulation of middle and higher latitudes has some basic differences from that of tropical latitudes. The principal difference is that the Coriolis force is more influential above about 30° latitude. Thus the Hadley cell dominates tropical regions; but in the higher latitudes, high- and low-pressure areas moving eastward are carried along on the majestic westerly winds of the upper air. Within midlatitudes, air of different origins—cold, warm, moist, dry—mixes constantly, and fast-flowing jet streams are associated with sharp differences in temperature.

THE POLAR FRONT JET STREAM

You have probably seen your local television weather forecaster show the jet stream on a chart while discussing your regional weather. The increasing frequency of this practice demonstrates our growing awareness that the location of the jet stream is one of the most important influences on midlatitude weather.

Chapter 5 explains how the fast-moving, upper-air current of the jet stream is related to the process of moving cold air from the high latitudes and warm air from the low. The jet stream might also be thought of as the boundary between the large areas of cold and warm air. The polar front jet stream, although not in itself a weather system, controls the most important midlatitude weather systems.

Figure 6-8 shows the lines of constant temperature sloping from the equator to the pole. In the middle latitudes, this temperature gradient drops sharply, and warm and cold air face each other horizontally. Whenever warm and cold air are opposed like this, the area or zone of contact is called a *front.* The name was inspired by the fronts formed by opposing armies in World War I. The polar front thus separates relatively cold polar air from relatively warm tropical air, and the jet stream associated with this sharp temperature contrast is therefore called the polar front jet stream.

At the core of the wavelike westerly winds, the polar front jet stream snakes its way around the globe in large meanders. These meanders may be explained through the law of conservation of absolute vorticity. Because we are concerned here with global distances, we may regard the depth of a column of spinning air as unchanging and concentrate on the spin of the air column and of the earth.

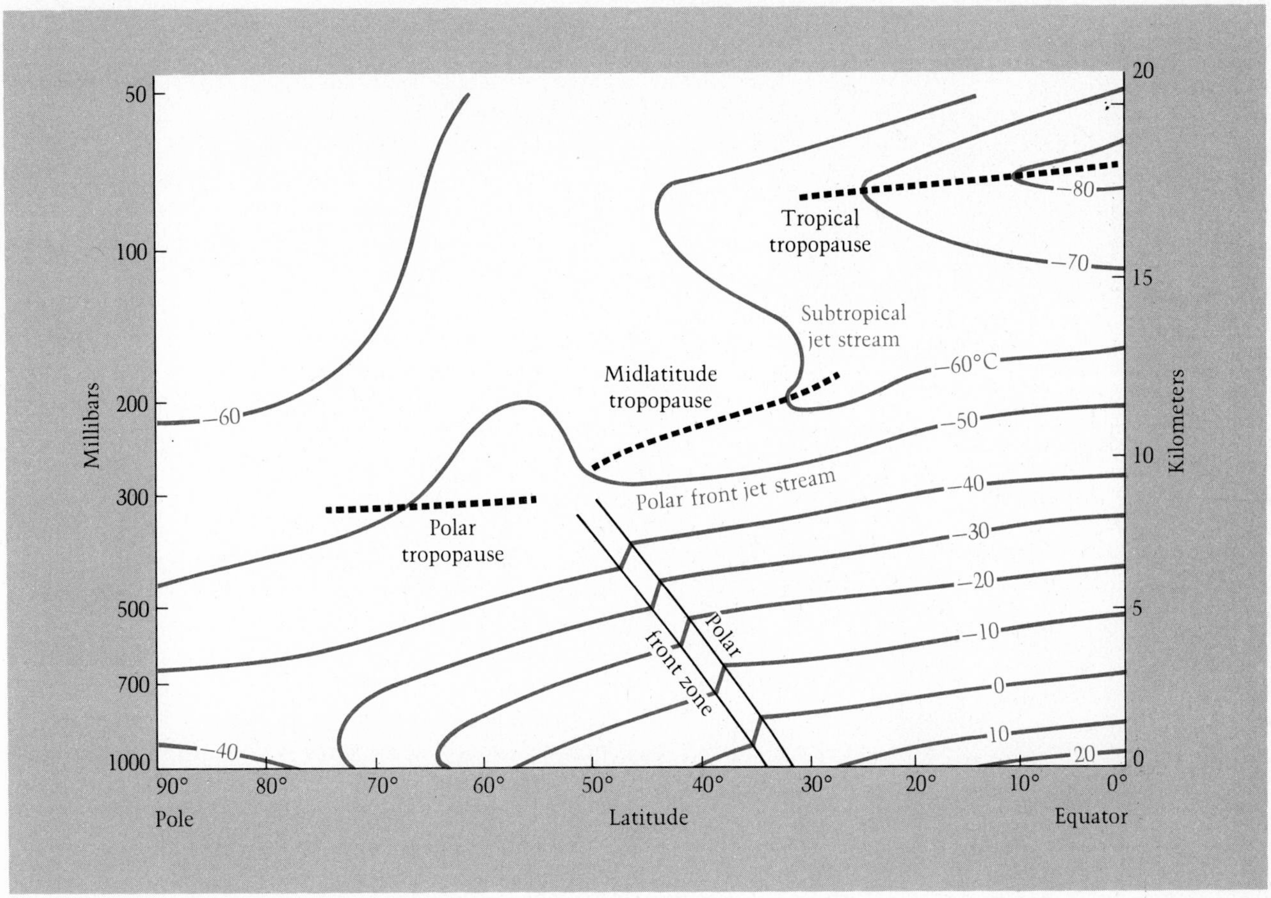

FIGURE 6-8
The temperature structure of the atmosphere, showing location of the polar front and jet stream.

The spin of a column of air is known as *relative vorticity*. In the Northern Hemisphere, a counterclockwise spin is considered positive. The spin of the earth is its *planetary vorticity*, which is always zero at the equator and a maximum at the poles. *Absolute vorticity*, the sum of relative and planetary vorticity, must be conserved and must remain constant. Therefore, if either relative or planetary vorticity increases, the other must decrease proportionally.

Let us assume that a column of air is traveling on the earth's surface from *A* to *B*, as indicated in Figure 6-9. As the column moves toward the pole, planetary vorticity, the spin of the earth, increases in effect. Consequently, the relative vorticity of the column must decrease, and it does so by spinning in a clockwise or more negative direction. By the time the air column reaches *B*, this new direction of spin affects the column's overall direction of movement. As a result, the column again travels toward the equator. The air traveling from *B* to *C*, toward the equator, is now moving to an area of decreasing planetary vorticity and compensates by gaining relative vorticity. It does so by again spinning counterclockwise. At *C*, the newly positive relative vorticity of the air column changes its overall path once more, and the wave motion continues. Thus the polar front jet stream and the plane-

FIGURE 6-9
Global wave motion in the polar front jet stream and planetary waves of the atmosphere.

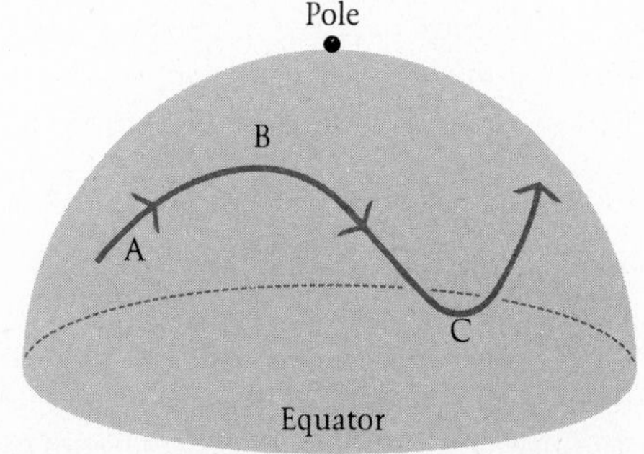

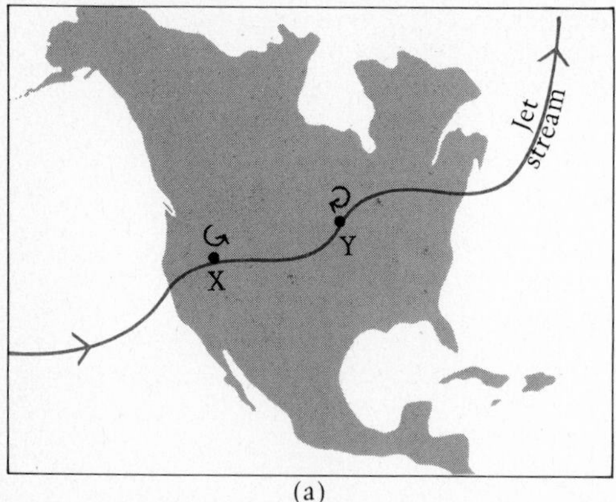

(a)

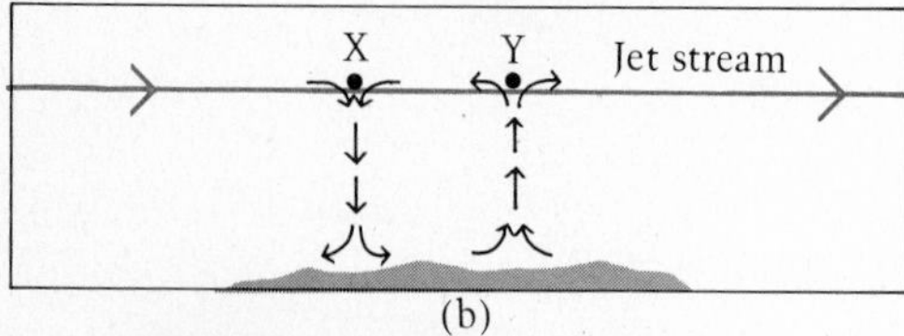

(b)

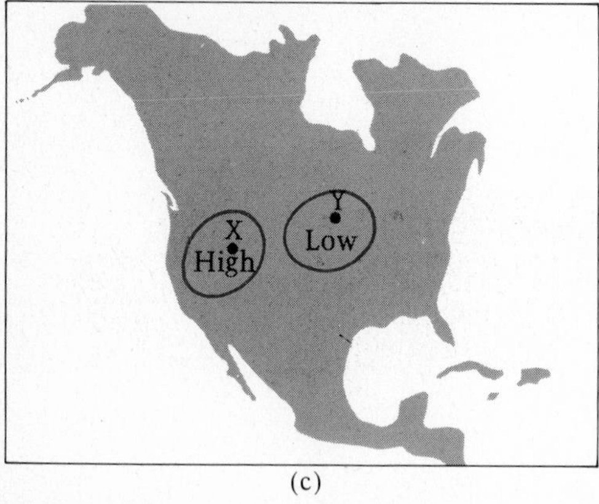

(c)

FIGURE 6-10
The relationship between the polar front jet stream and surface pressure. a) Location of the jet stream 9000 meters above North America; b) cross-sectional view of the associated air movement; c) the resulting surface pressure.

tary waves of midlatitude air continue on their large meandering path around the planet.

Spin or vorticity in the atmosphere is important because the air acts just like a flowing liquid in many ways. The riverlike nature of the streams in the atmosphere explains not only the planetary waves in the jet stream on a worldwide scale, but also where we might expect storms.

MIDLATITUDE CYCLONES

The weather at the earth's surface in midlatitudes is largely determined by the position of the polar front jet stream. The reason for this is yet another expression of the law of conservation of absolute vorticity but on a smaller scale. Smaller waves are superimposed on the larger planetary waves somewhat like small ripples on an ocean swell.

In Figure 6-10a, the jet stream, with its smaller waves, passes across North America 9000 m (29,500 ft) above the ground. At point *X* the relative vorticity becomes more positive, and the air at the top of the air column rotates counterclockwise. In such a cyclonic circulation, air converges toward the center. Figure 6-10b, a cross-sectional view, shows this through air-movement arrows at the level of the jet stream at point *X*. The air converging at this point cannot rise because the tropopause forms a barrier, so it descends. You may imagine the total air column becoming shorter, or decreasing in depth. The descending air causes high pressure at the surface, as shown in Figure 6-10c, and spreads out, or diverges, from the surface anticyclone. At point *X*, therefore, we could expect good weather because of the descending air.

Meanwhile, farther "downriver" in the jet stream, at point Y, the relative vorticity becomes more negative. High above the earth, a clockwise anticyclonic circulation develops, and air spreads outward. The diverging air aloft must be replaced, and because the tropopause is still a barrier, new air is drawn up from below. Once more you may imagine the rising air increasing the depth of the total column. As the air is drawn up the

FIGURE 6-11
The spiral cloud bands of midlatitudes in the Northern Hemisphere indicating midlatitude cyclones in different stages of their life cycle.

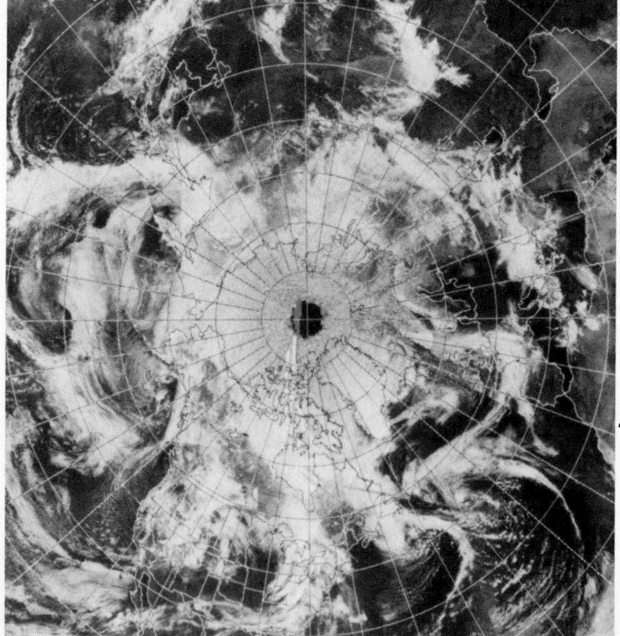

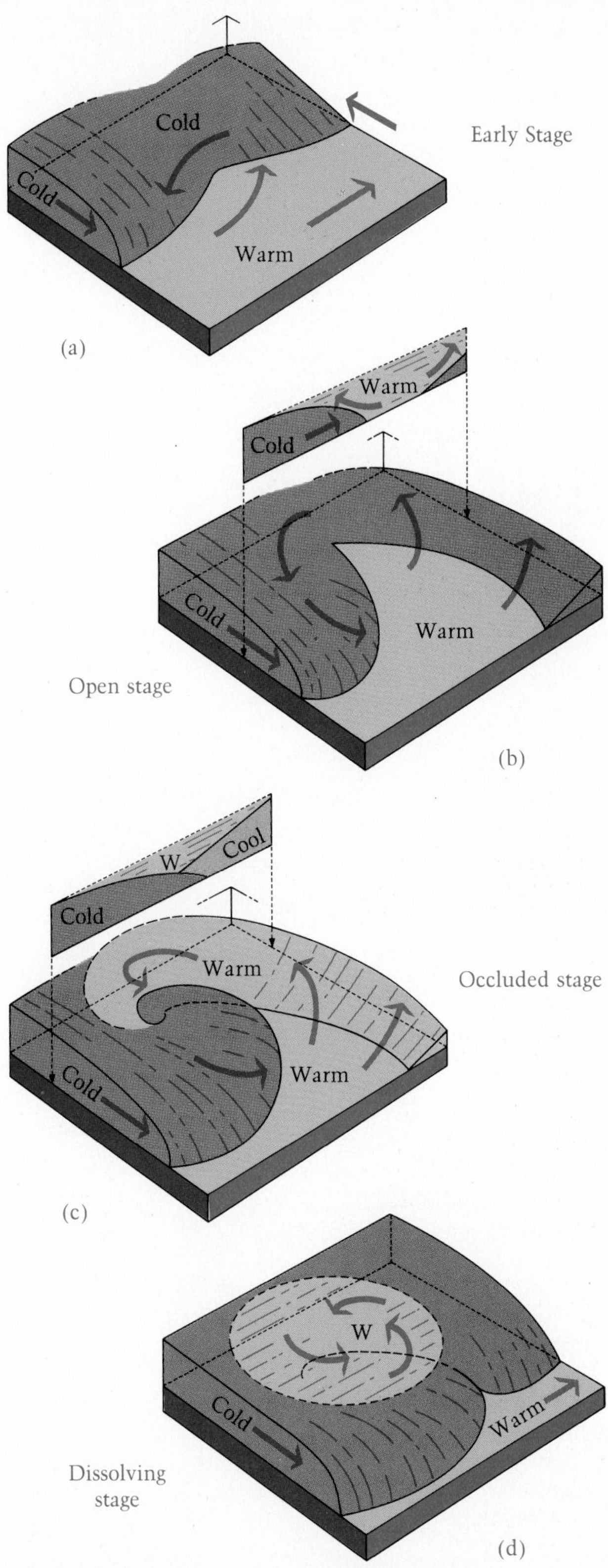

FIGURE 6-12

The four stages in the life cycle of a midlatitude cyclone. W indicates warm air; C indicates cold. a) Early stage; b) open stage; c) occluded stage; d) dissolving stage.

column, a converging cyclonic circulation develops at the surface, with its characteristic low pressure. The rising air cools and forms clouds and rainfall. In this manner the jet stream becomes the primary cause of good weather when it flows equatorward and bad when it flows poleward.

Life cycle of a midlatitude cyclone Every day, spiral cloud bands, like so many galaxies, sprawl across the middle latitudes of the globe. The satellite photograph in Figure 6-11 indicates the presence of 20 or 30 low-pressure areas accompanying midlatitude cyclones. These are the most common large-scale weather systems found outside the tropics. The midlatitude cyclone, like its low-latitude counterpart, has several names. It is often called a depression or an extratropical (meaning "outside of") cyclone. A midlatitude cyclone is characterized by its circular wind and low pressure field as well as the interaction of air of different properties. As with many atmospheric weather systems, it has a complete life cycle.

Figure 6-12a shows the earliest stage in the development of a midlatitude cyclone. A mass of cold air and a mass of warm air lie side by side at what is called a *stationary front.* Anticyclonic circulation above the surface causes a slight cyclonic motion at the surface. A small kink appears in the stationary front.

As cyclonic motion develops, the kink grows larger and becomes an *open wave,* as Figure 6-12b shows. In this open-wave stage, the warm and cold air interact in distinct ways. On the eastern side of the cyclone, warm air glides up over the cold air mass along a surface called a *warm front.* To the west, the cold air tunnels under the warm air, in an action resembling that of a snowplow. The surface where the cold air forces the warm air to rise is known as a *cold front.* The open-wave stage represents maturity in the life cycle of a midlatitude cyclone.

With further growth, old age sets in. Figure 6-12c shows how the cold front travels faster than the warm, overtaking it first at the center of the cyclone. When this happens, the snowplowlike cold front lifts the warm air entirely off the ground, causing an *occluded front* to form.

In time, the entire section of warm air is lifted to the colder altitudes above the earth, and the cyclone dies away, as Figure 6-12d indicates. But it does not die where it was born. The polar front jet stream continually steers midlatitude cyclones eastward.

This wave model of the cyclone was first put forward by Norwegian weather forecasters at the time of World War I. It is a useful forecasting tool, and we can see why if we look more closely at the open-wave stage.

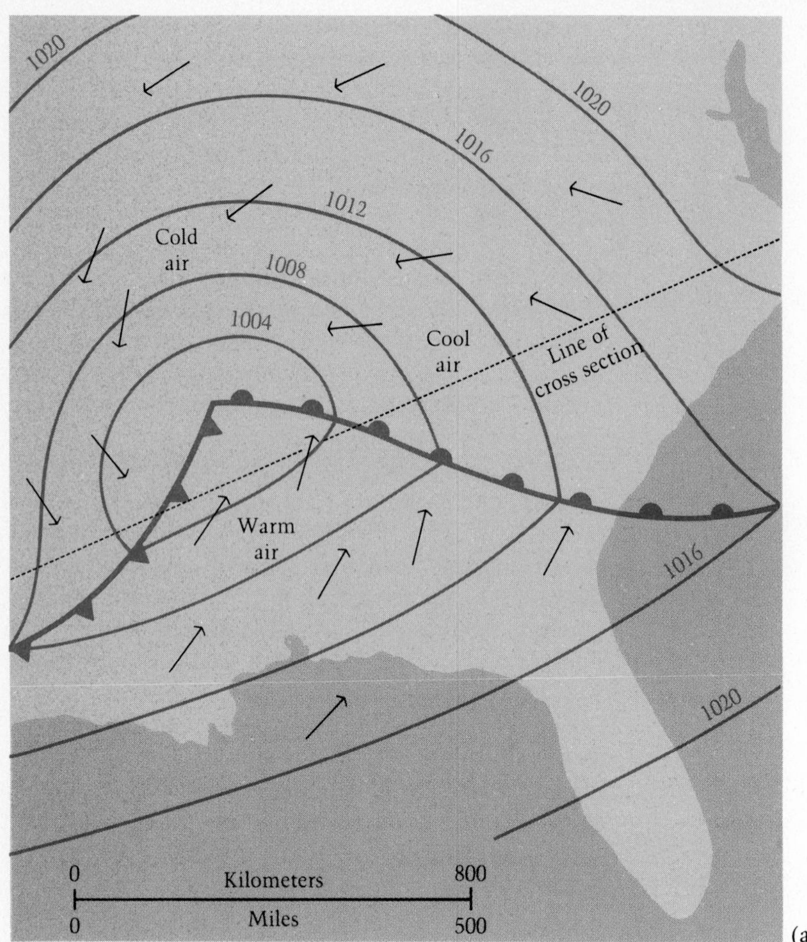

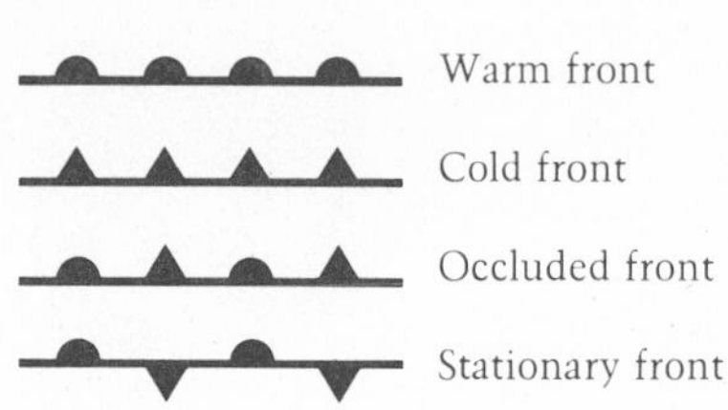

(a)

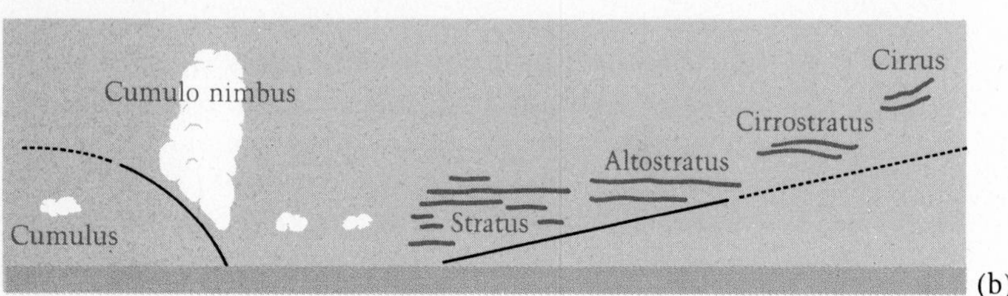

(b)

Post cold front	Warm sector	Pre warm front	
Heavy rain	Showers	Rain and fog	Precipitation
↘ ↘	↗ ↗	↖ ↖	Wind direction
Rising	Low steady	Falling	Pressure
Low	High	Low	Temperature

(c)

◀FIGURE 6-13

The open-wave stage of the midlatitude cyclone. a) Positions of pressure fields, wind fields, and fronts over the southeastern United States; b) cross-sectional view; c) weather conditions at the surface along the line of the cross section.

Weather forecasting and the open-wave stage The most obvious feature of Figure 6-13a, which shows the southeastern United States, is the roughly circular nature of the isobars and the position of the fronts. Although the isobars are circular for the most part, they tend to be straight within the area enclosed by the fronts. This region of warm air is called the *warm sector.* Also noteworthy is the distinct angle the isobars form at the fronts themselves.

The next thing to notice is the wind arrows. They tend to cross the isobars at a slight angle, just as they should when friction at the earth's surface partially upsets the geostrophic balance. The wind arrows indicate a large whirl of air gradually accumulating at the center of the low-pressure area. This converging air rises and cools, especially when it lifts at the fronts, and its water vapor condenses to form precipitation.

In the cross-sectional view in Figure 6-13b, the warm front has a gentle slope of about ½° to 1° (although it is exaggerated in the figure). As the warm air moves up this slope, it cools and condenses to form clouds. The highest clouds to form are thin, wispy ones known as *cirrus.* These are followed by high layered clouds called *cirrostratus.* Lower, below 6 km (20,000 ft), similar clouds are named *altostratus,* and even lower, below 3 km (10,000 ft), similar layered cloud sheets are simply named *stratus.* The rain of the warm front comes from the stratus and the altostratus.

The cold front is much steeper than the warm front, having a slope of about 2°. This forces the warm air to rise much more quickly, creating towering, puffy clouds called *cumulus.* If rain comes from them, they are called *cumulonimbus.* The passage of a cold front such as this caused the postponement of the World War II invasion of France. In the warm sector, and behind the cold front, clear skies prevail, except for a few small cumulus clouds.

Although no two depressions are ever the same, there are enough similarities for an observer on the surface to predict the pattern of atmospheric events outlined above. These are summarized in Figure 6-13c. As the midlatitude cyclone moves along the line of the cross section (as shown in Figure 6-13a), clouds thicken, steady rain falls, and pressure drops. As it passes the warm front, temperatures rise, wind direction changes, pressure remains steady, and the rain clears or turns to occasional showers. Then, at the violent cold front, there is heavy rain, another wind change, and an abrupt drop in temperatures. The whole system usually moves eastward, in the direction of the isobars in the warm sector, usually roughly parallel to the upper jet stream. Thus a weather forecaster must predict how fast the system will move and when it will occlude. But these two predictions are difficult and account for most of the forecasts that go wrong. Because it is so influential and prevalent, the midlatitude cyclone is a dominant climatic control outside the tropics.

These systems follow paths that move toward the poles in the summer and toward the equator in the winter, under the influence of the polar front and its jet stream. Their general eastward movement varies in speed from 0 to 60 km per hour (35 mi per hour) in winter. In summer, the velocities range from 0 to 40 km per hour (25 mi per hour). The depressions may be anywhere from 300 to 3000 km (200 to 2000 mi) in diameter and range from 8 to 11 km (5 to 7 mi) in height. They provide the greatest source of rain in the midlatitudes, and they generate kinetic energy that helps power the general circulation of the atmosphere.

Much of this kinetic energy of the midlatitude cyclone comes from the mixing of cold and warm air masses existing on either side of the polar front. It is to these that we now turn.

AIR MASSES IN THE ATMOSPHERE

In large areas of the world, such as the snow-covered arctic wastes and the warm tropical oceans, masses of air remain stationary for several days. Here they obtain the properties of the underlying surface. Extensive geographic areas with relatively uniform characteristics of temperature and moisture form the *source regions* where air masses can be produced. The air over a source region reaches an equilibrium with the temperature and moisture condition of the source region. An *air mass* is therefore an extensive portion of air in the lower troposphere with relatively uniform qualities of temperature and moisture in the horizontal dimension. Air masses are given code letters according to whether their source region is maritime (m), continental (c), tropical (T), or polar (P).

In a warm source region, such as the Caribbean Sea, the equilibrium of temperature and moisture is established in two or three days. Unstable air in either a maritime tropical air mass (mT) or a continental tropical air mass (cT) is warmed to about 3000 m (10,000 ft). In contrast, cold air masses can take a week or longer to achieve equilibrium. Only a relatively shallow layer, up to about 900 m (3000 ft), cools in continental or maritime polar air masses (cP or mP). Of course, the air above is cold but for reasons other than contact with the surface. It takes a relatively long time for a cold air mass to become established because cooling at its base stabilizes lapse rates, thus preventing vertical mixing and prohibiting efficient heat exchange.

After the first four types of air mass (mT, cT, mP, and cP) were recognized, two more were detected in the highest latitudes. These are called continental arctic (cA) and maritime arctic (mA), and paradoxically, they form at higher latitudes than the polar air

FIGURE 6-14
The most likely paths of the air masses affecting the United States.

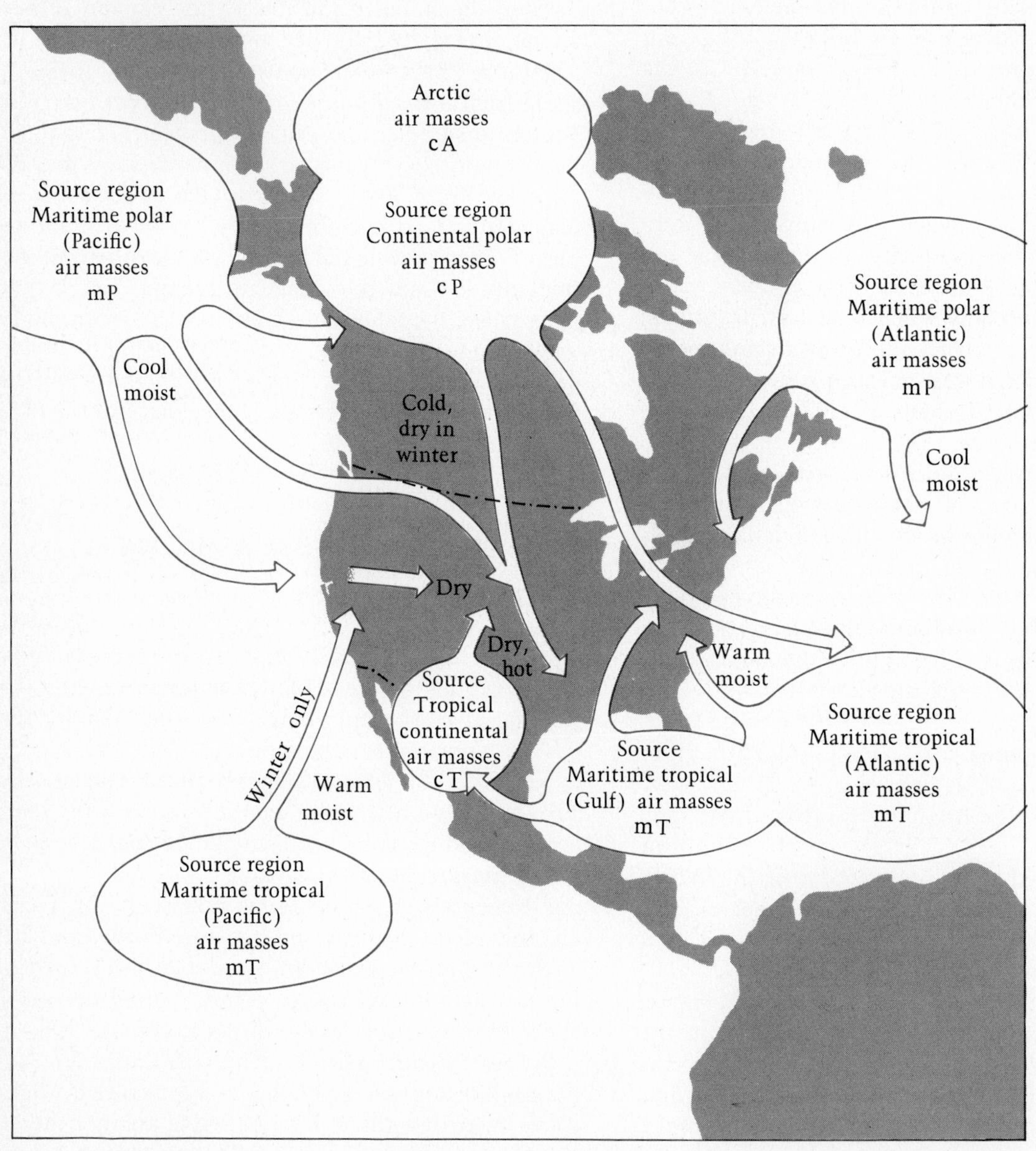

m – Maritime
c – Continental
T – Tropical
P – Polar

masses do. The letters used to label air masses are sometimes supplemented by letters indicating whether the mass is colder (k) or warmer (w) than the underlying surface and whether the air is stable (s) or unstable (u).

If you live in the Midwest of the United States, you can testify to the bitterness of the polar air that streams from Canada and the oppressiveness of the warm, humid air from the Gulf of Mexico. You can see that the air masses affect not only their source areas, but also the places they flow over. As an air mass moves, it may be modified by contact with the ground or by changes within the air, but many of its original characteristics remain identifiable far from its source region. This is a boon to weather forecasters, because they can predict weather conditions if they know the characteristics of the air mass and its rate and direction of movement.

The principal air masses affecting North America are shown in Figure 6-14. The arrows represent the most common paths of movement. If this diagram looks like the plan for a battle, that may be because opposing air masses battle for supremacy throughout the year. Some arrows representing the different types of air that interact in the midlatitude cyclone might well have come from Figure 6-13a. Maritime tropical air (mTw), warmer than the underlying surface, would lie in the warm sector surrounded to the northwest by continental polar air (cPk), colder than the underlying surface. Losses and gains of territory in this battle are controlled by one important factor, the polar front and its associated jet stream. The main current of a river does not let eddies pass from one bank to another, and so it is with the jet stream and polar front providing an unseen barrier to air masses. This fact is a key to understanding weather outside the tropics.

SMALLER-SCALE WEATHER SYSTEMS

The large-scale weather systems of the tropics and higher latitudes are fundamental to an understanding of the climates of these areas. Yet full appreciation depends on a grasp of detail as well. Medium-scale and small-scale systems make up the texture of the atmosphere. The first is common to both low and midlatitudes, and has long commanded the respect and fear of human beings and other animals.

THUNDERSTORMS

Like the midlatitude cyclone, low-latitude and midlatitude thunderstorms have a distinct life cycle, but it lasts for only a few hours. Figure 6-15 is a portrait of one of these thunderstorms. The cycle begins with a warm parcel of air in an unstable atmosphere. The parcel rises, and as condensation occurs, large quantities of latent heat are released. This heat energy makes the air much warmer than its surroundings, and the resulting updrafts of warm air attain speeds of 10 m per second (22 mi per hour). They may sometimes move as fast as 30 m per second (67 mi per hour) in this early stage of development, as Figure 6-16 indicates. Raindrops and ice crystals may form at this stage, but they do not reach the ground because of the updrafts.

FIGURE 6-15
An Apollo 9 view of the top of a thundercloud over South America.

When the middle, or mature, stage is reached, enough raindrops fall to cause downdrafts of cold air. Evaporation from the falling drops accentuates the cooling. Thus heavy rain begins to fall from the bottom of the cloud, and cold air spreads out in a wedge formation, sometimes developing into a small-scale cold front.

In time, the moisture in the storm is used up. The dissipation (dispersal) stage is reached when the latent heat source starts to fail. Downdrafts gradually predominate over updrafts, and this situation continues until the storm dies away.

Thunderstorms produce towering cumulonimbus clouds. These are sometimes drawn out by upper-air winds to form an *anvil top*, a name derived from the shape of a blacksmith's anvil. Another feature that often accompanies a thunderstorm is the formation of

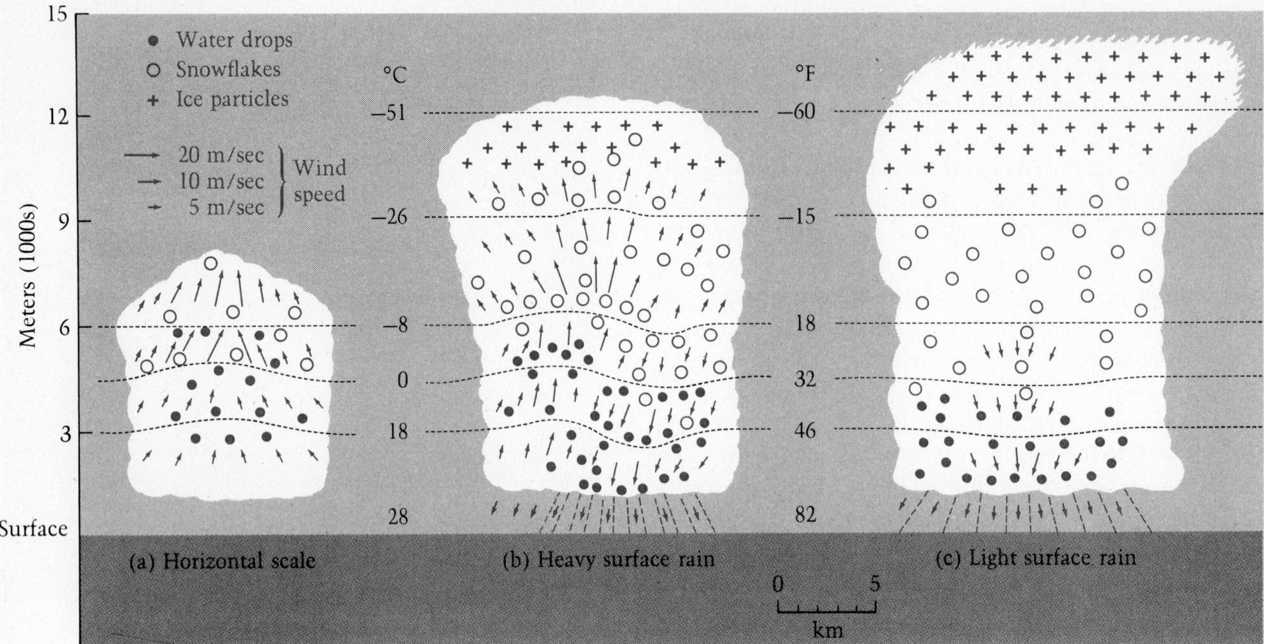

FIGURE 6-16

The cycle of thunderstorm development. The arrows indicate the direction and speed of the vertical air currents. a) Developing stage of the initial updraft; b) mature stage, with updrafts and downdrafts; c) dissipating stage, dominated by cool downdrafts.

hail. Figure 6-16 shows that, in the mature stage of the storm, an ice crystal might be caught in a circulation that continually moves it above and below the freezing level. This circulation can create concentric rings of ice in the hailstone. In some storms, such as in a *squall-line storm,* these circulatory systems can be exaggerated. Figure 6-17 shows how falling hailstones can be scooped back into the main cloud by a small low-level jet stream moving into the system. Thus hailstones may have several journeys through the freezing level before falling to earth. Many a farmer's crop has been destroyed because of conditions such as these. In extreme cases, hailstones the size of baseballs pound the earth.

The thunderstorm is a good example of a convectional cloud. *Convection* is the name given to vertical movement in the atmosphere. Because upward movement of air usually leads to condensation, subsequent rainfall is often termed *convectional* rainfall.

Thunderstorms do not usually exist in single cells like the simplified thunderstorm in Figure 6-16. Radar studies have shown that thunderstorms actually consist of several cells arranged in clusters, 2 to 8 km (1¼ to 5 mi) in diameter. As the storms move, new cells develop on the right flank to replace old ones dissipating on the left. If the downdrafts of two cells meet at the surface, a new updraft and a new cell may be initiated. This may lead to the development of another notable phenomenon.

TORNADOES

The interiors of large continents are subject, especially in spring and early summer, to a small but vicious storm known as the tornado. The tornado is a small vortex of air about 100 to 200 m (330 to 660 ft) in diameter. The pressure in the center of the funnel might be 100 to 150 mb below the pressure in the surrounding air. The difference between the interior pressure of a sealed house and the low pressure of the funnel sometimes causes the house to explode. Major damage is also done by the high winds of tornadoes, which can reach speeds between 50 and 200 m per second (110 and 450 mi per hour). Tornadoes follow rather straight paths at a speed and direction determined by a low-level jet stream. Their awesome dark color is caused by the vegetation, loose soil, and other movable objects sucked into the tube of this giant vacuum cleaner (Figure 6-18).

In the United States, tornadoes usually develop from squall-line thunderstorms over the Great Plains. Dry air from the high western plateaus moves eastward over maritime tropical air. Sometimes the moist air from the south is forced northward by a low-level jet

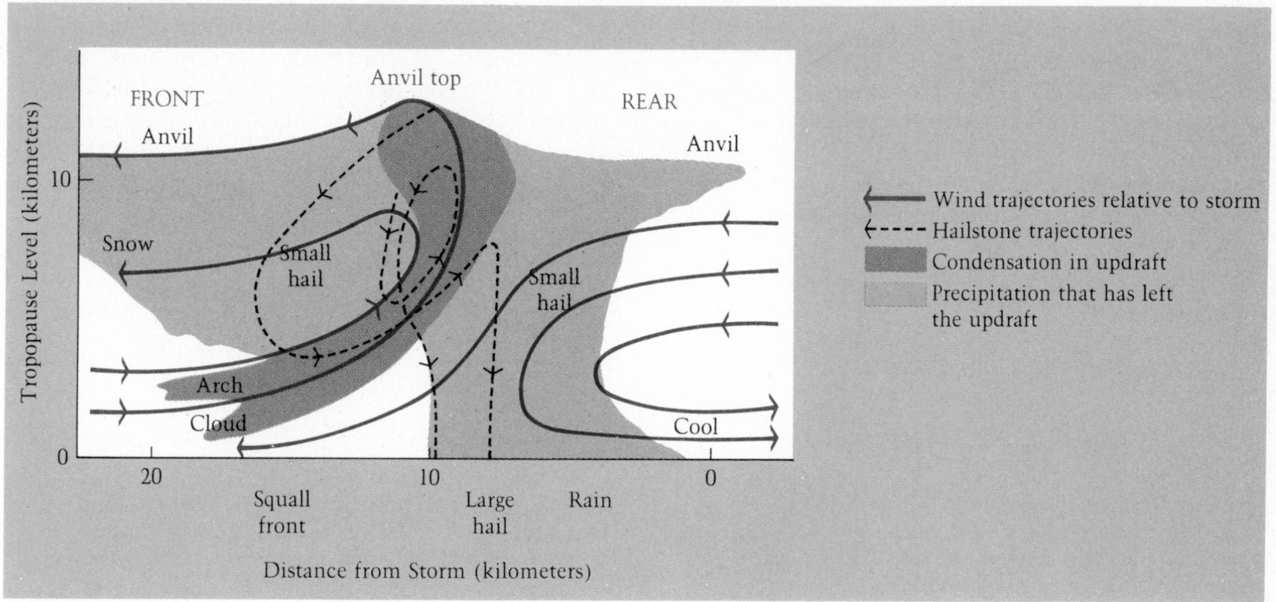

FIGURE 6-17

Model of a thunderstorm producing large hailstones. This squall-line storm advances from right to left.

stream. If the tongue of moist air intrudes beneath warm, dry air in the presence of a midlatitude cyclone, convergence and ascending air can trigger extreme instability. Tornadoes can develop in the large cumulus clouds that result. But a midlatitude cyclone is not always necessary; it appears that almost any vertical movement of air can trigger tornadoes.

Many Midwesterners have tragic tales of tornadoes. Among them are the people of Xenia, Ohio. On the afternoon of April 3, 1974, a deadly double-spiraled tornado swept through the western part of Xenia. It killed 34 persons and injured 1150 others. By chance the tornado struck Xenia after school had let out, otherwise hundreds of children would have been killed. Fortunately, not every atmospheric weather system is so severe.

FIGURE 6-18

A tornado near Manhattan, Kansas, May 31, 1949.

LAND AND SEA BREEZES

Brighton is a seaside resort on the southern coast of Britain. At 3 PM on a warm summer day, a cool breeze usually blows from the sea, often discouraging vacationers from removing their shirts. By 9 at night, the wind brings warmth from the land, and once again the stoic Briton curses the climate for its perversity. These land and sea breezes are common to most coastlines all year round, although they are most dramatic in summer.

During the day, the land heats up, and the warm air

gradually ascends. It is replaced by cool air from the sea, which takes longer to heat. At night, the land cools faster than the sea. Air rising over the ocean is replaced by air from the land, and the land breeze forms. Sea breezes have the greater velocity, which commonly ranges from 4 to 7 m per second (9 to 15½ mi per hour). Land breezes seldom exceed 2 m per second (4½ mi per hour).

Differential heating of land and sea is not always enough to create land breezes. Usually unstable air must develop over the land. Land breezes are often accompanied by a small cold front, where the cool sea air encroaches into the warm air of the land. Seagulls and glider pilots, who both look for ascending air currents, commonly share the uplift of air at this front. During the day, the front can move inland about 50 km (30 mi), although incursions of up to 100 km (60 mi) have been recorded. The sea breeze can travel far enough to come under the influence of the Coriolis force and eventually blow parallel to the coast. The sea-breeze front may be accompanied by some cumulus clouds, but usually no rain develops. After it passes,

FIGURE 6-19

The formation of mountain and valley winds. a) Sunrise: valley cold, plains warm; b) forenoon: valley temperature same as plains; c) noon and early afternoon: valley warmer than plains. d) late afternoon: valley warmer than plains; e) evening: valley only slightly warmer than plains; f) early night: valley temperature same as plains; g) middle of night: valley colder than plains; h) late night: valley much colder than plains.

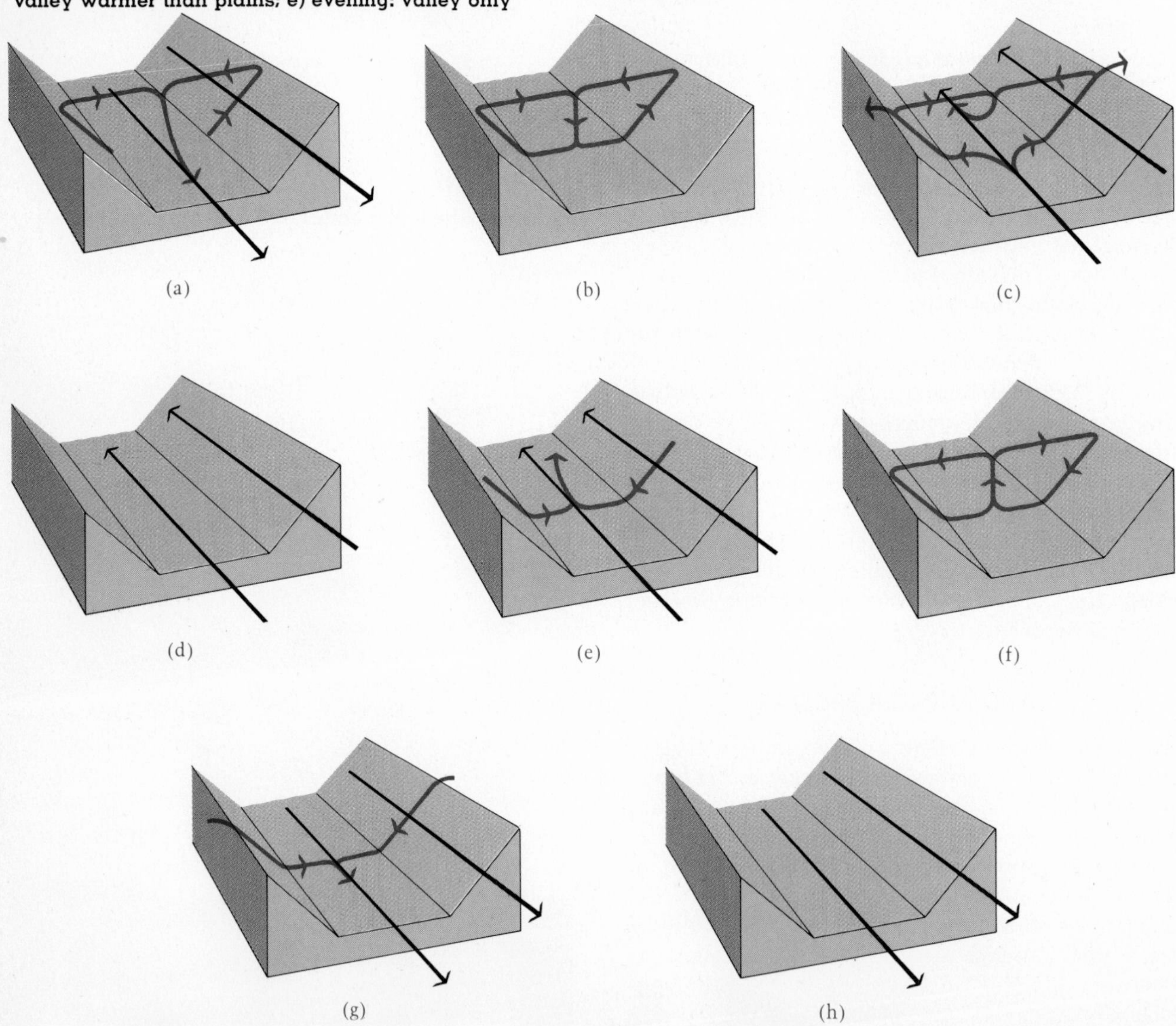

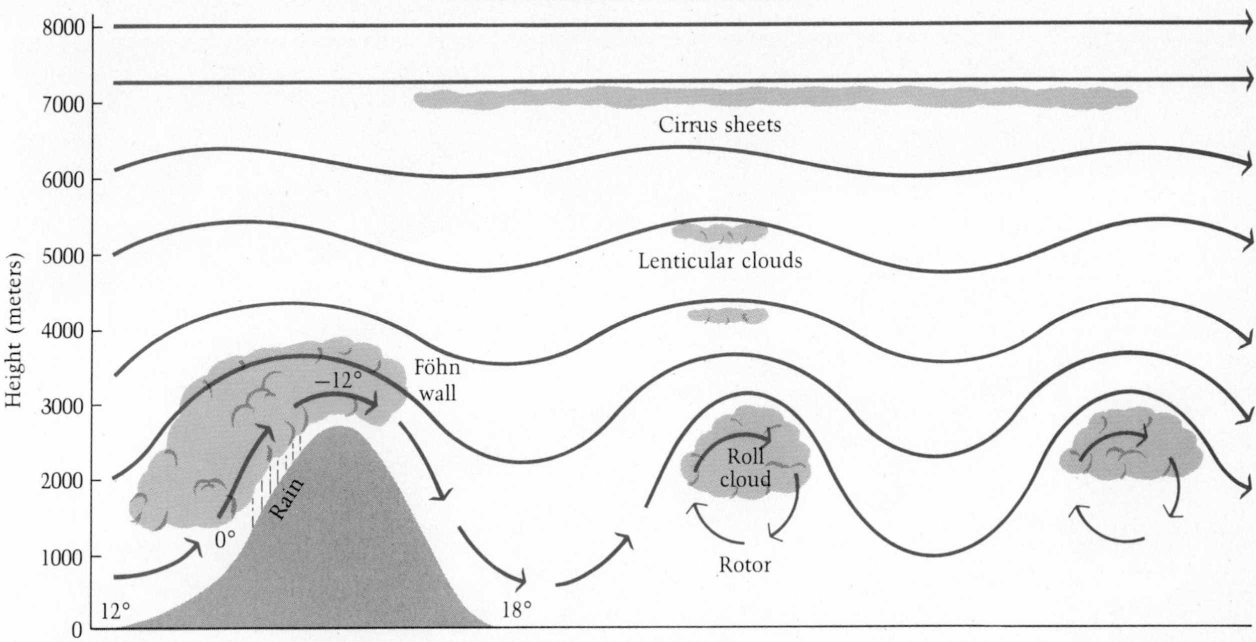

FIGURE 6-20

Airflow over a mountain range. The mountain affects temperature, rainfall, and wave and cloud development.

temperatures drop, and relative humidity generally increases.

MOUNTAIN AND VALLEY WINDS

Chilton Valley is a beautiful alpine valley deep in the mountains of New Zealand's South Island. On a summer afternoon when no strong large-scale winds are blowing, hikers can eat their lunch in a breeze that comes up the valley. The breeze that bends the tall grass toward the peak of the mountain is a valley wind.

As the day matures, the valley sides become warm, and the rising hot air is replaced by cooler air from below. At night the valley sides become cool, and they cool the air, which then forms a mountain wind that sinks like heavy syrup to the bottom of the hill.

The mountain and valley wind system, which develops on most calm days in mountainous terrain, is actually two wind systems, as Figure 6-19 indicates. The mountain and valley winds blow along the length of the valley, and the slope winds act on the valley sides.

Air blowing up a slope as a result of warming is called an *anabatic* wind. Air moving down a slope because it has cooled is called a *katabatic* wind. Katabatic winds are sometimes called *cold-air drainage.* When katabatic winds create pools of cold air at the bottom of a slope, orchard farmers in lowlands fear damage to their fruit. The collection of cold air can often produce condensation, so people sitting on the side of Chilton Valley at night can sometimes overlook a sea of clouds.

AIR FLOW OVER MOUNTAINS

Mountains, and highlands in general, have many effects on the atmosphere. On the global scale, the Rocky Mountains and the Himalayas help to anchor planetary waves in the westerly winds. Air flowing over mountains is affected on both its forced ascent and its descent.

Imagine a parcel of air with a temperature of 12°C (53.6°F) arriving at the sea-level base of a mountain 3000 m (9800 ft) high. Figure 6-20 diagrams this situation. The parcel of air is forced to rise up the mountain, and it starts to cool at the DALR (1°C/100 m). (Lapse rates are discussed in Chapter 3.) At this rate, the air parcel cools to 0°C (32°F) by the time it reaches 1200 m (3900 ft). Assume that the moisture conditions of the air parcel are such that condensation occurs at this freezing point. Therefore clouds form, usually leading to heavy rainfall on the side of the mountain the wind hits, the windward side. Rainfall caused by mountains in this manner is called *orographic rainfall* (*oros* is the Greek word for "mountain"). The air continues to rise, but because condensation has begun, it cools at the SALR. (Let us assume

FIGURE 6-21

A symphony of clouds in the lee of the Sierra Nevada. Both rotor clouds and cirrus sheets are visible, as is dust from the valley floor picked up by the warm, dry descending wind.

that the SALR is 0.65°C/100 m.) By the time it reaches the mountaintop, in this case 1800 m (5900 ft) later, its temperature is some 12°C (21.6°F) lower, or -12°C (10.4°F).

Now the air parcel begins to descend to the warmer temperatures of the lower atmosphere. There is now no cause for condensation, so the air parcel warms at the DALR. When it reaches the bottom of the 3000 m (9800 ft) mountain, its temperature will have risen 30°C (54°F), to 18°C (64.4°F). Thus the air parcel is much warmer on the lee side of the mountain, the side away from the wind, than on the windward side, where it started. Because the air has lost much of its moisture on the windward side of the mountain and the air is warmer on the lee side, the leeward side often has very little rain. The lack of rain caused this way is called a *rain shadow* effect. The mountain has caused both rainfall, on the windward side, and air heating, on the lee side.

The lee side of mountains is often characterized by warm, dry air. The warm wind is most commonly called a *föhn* wind. As Chapter 1 explains, the Santa Ana wind of California and chinook wind of the western plains of Canada and the United States are good examples. The warm chinook wind is sometimes known as the "snow eater." Just before the cloud in the descending air disappears, it looks like a wall, and this is called a *föhn wall cloud.*

As Figure 6-20 shows, the mountain also imposes a vertical wave motion on the air's general flow. The wave motion is most dramatic in the lower atmosphere and becomes less noticeable at higher altitudes. Sometimes the air ascending on the other side of the mountain gives rise to a *roll* or *rotor cloud* at about the height of the mountaintop. Wave motion at higher altitudes may cause clouds often perceived as flying saucers, called *lenticular clouds.* At even higher levels, a continuous sheet of cirrus cloud may form. When seen together, as in Figure 6-21, all these clouds present an atmospheric symphony of great splendor.

ENERGY AND MOISTURE WITHIN WEATHER SYSTEMS

Satellite photographs have shown, and your own experience probably confirms, that every day a multitude of weather systems sweep across the earth's surface. Each one, although distinctive in detail, has a characteristic internal organization. In each, warm air is transformed into faster-moving air, and moisture becomes precipitation. The processes of such changes form a unique system, diagramed in Figure 6-22.

There are two inputs into this system: warm air carrying the energy of heat and moisture in the form of water vapor. The heat energy of the rising warm air adds potential energy to the system, and the system's center of gravity lifts higher off the ground. At the same time, the moisture condenses, which has two effects. First, the latent heat that is released helps to swell the store of potential energy. Second, the condensed moisture appears as clouds. The growing store of potential energy changes to kinetic energy, and air movement increases. The kinetic energy leaves the system in the form of winds. Meanwhile, the droplets in the clouds grow to precipitable size. They too are exported from the system in the various forms of precipitation that bathe the earth.

FIGURE 6-22

Moisture and energy transformations within a weather system.

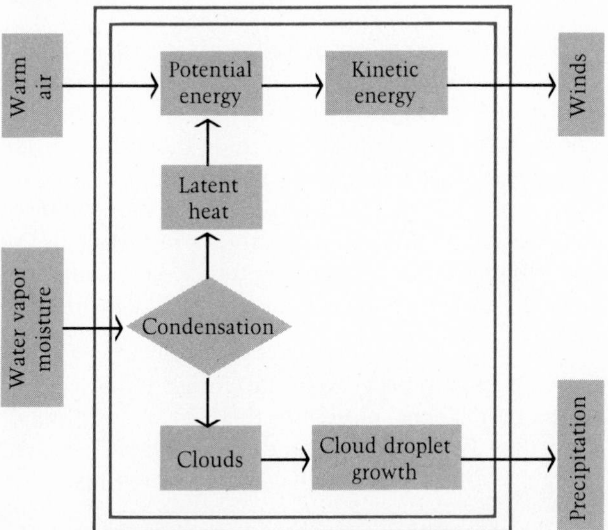

These events occur in the clouds of the ITCZ, in the tropical easterly waves, and in the wet monsoons. You may be able to envision them best in the tropical cyclones and thunderstorms. Yet whether in a tornado or a midlatitude cyclone, these events take place in a similar way. Because of the inputs and outputs of energy and moisture to many of these phenomena, we have considerable justification for calling them *weather systems.* In the next chapter we see how these systems combine together with the general circulation to form the climates of the earth.

SUMMARY

Weather systems, organized phenomena of the atmosphere, are characterized by inputs and outputs and by changes in energy and moisture. On a large scale, different systems exist in low latitudes and higher latitudes. The ITCZ, monsoons, easterly waves, and tropical cyclones are most typical of low latitudes, where the effect of the Coriolis force is minimal.

An understanding of midlatitude and high-latitude weather depends on an understanding of the polar front jet stream and of how midlatitude cyclones form on different parts of it. In contrast to tropical cyclones, midlatitude cyclones involve the interaction of air masses with different properties.

There are also many small-scale weather systems throughout the atmosphere. The ones discussed in this chapter are thunderstorms, tornadoes, land and sea breezes, mountain and valley winds, and air flow over mountains. All these systems, both large and small, make up the fabric of the weather that affects our daily lives.

QUESTIONS

1. The location of the intertropical convergence zone is dependent on the operation of the Coriolis force, which forces air to pile up and subside at 30° N and S, completing the Hadley cell. If the earth rotated in the opposite direction but with the same speed, how would this affect the location of the ITCZ?

2. Figure 6-1 shows that the equator has two periods of rainfall maxima each year, one in March–April and one in September–October. Why does the equator have two maxima per year, while latitudes 10° N and S have only one each? Why don't the maxima occur in the months February–March, August–September,

May–June, and November–December instead of when they do?

3. Figure 6-2 shows the ITCZ moving northward over India and Asia during the summer and southward during the winter. Keeping in mind the configuration of landmasses and oceans, the seasonal heating differences north and south of the equator, and the directions of the subtropical jet streams, explain why the ITCZ doesn't follow the same pattern over North and South America.

4. How is the wet summer monsoon in India related to the changes in the circulation of the upper atmosphere? Why do the winds of the summer monsoon not continue as southeasterly winds and bring the monsoon rains to the east coast of Africa and the Saudi Arabian peninsula?

5. The law of conservation of absolute vorticity states that the sum of the spin of an air column and the spin of the earth, divided by the depth of the air column, is a constant. Using this concept, explain the easterly wave phenomena found in the Caribbean, including the production of clouds and rainfall on the eastern side of the easterly wave, and the clear weather on the western side, in terms of a system attempting to maintain dynamic equilibrium through a negative feedback mechanism.

6. Why do tropical cyclones form in the latitudes between 5° and 20° N and S? How is the path that tropical cyclones usually follow in the Caribbean/Gulf Coast area related to the method of their formation?

7. Why are tropical cyclones potentially the most destructive of all weather phenomena?

8. In the formation of midlatitude cyclonic storms according to the concept of the conservation of absolute vorticity and the action of the meandering polar front jet stream, what is the principal cause of the development of surface air movement patterns? In Figure 6-10 can we assume that the swirling anticyclonic pattern which would develop at point *Y* in the jet stream accurately reflects the currents at the earth's surface?

9. Why do we only see the midlatitude cyclones and not the anticyclones in the mosaic in Figure 6-11? What accounts for their regular spacing around the earth's parallels?

10. How do the differing shapes of the warm and cold fronts in a midlatitude cyclonic storm affect cloud shape and formation? In Figure 6-13a, where would we find cirrostratus clouds? Cumulus clouds? Cirrus clouds?

11. Assume that the midlatitude cyclone in Figure 6-13 is forming over your town in Tennessee, and you have to deliver the evening weather forecast. Your prediction yesterday of a drop in barometric pressure, rain, and fog today was correct. What factors must you take into consideration for today's weather report? What will your prediction be?

12. Mount Rainier (4800 m) sometimes gets 30 m of snow in a year, even though temperatures in nearby Seattle rarely get below -10°C during the winter. In Duluth, Minnesota at an elevation of 376 m, -15°C winter weather is the norm, but snowfall is only about 3.3 m per year. Looking at the map of the air masses that affect North America (Figure 6-14), what effect do you think they have on this situation?

13. If the path of polar front jet stream moves north in the summer and south in the winter, what effect does this have on the location of the air masses affecting the weather in the continental United States?

14. It is not uncommon for mountain climbers to be caught in brief hailstorms at high altitudes. What conditions would produce such a phenomenon on an otherwise sunny day?

15. Tornadoes exhibit more kinetic energy in concentrated form than any other weather phenomenon. Where does this energy come from, and why are tornadoes such dangerous events?

16. Why do sea breezes generally have a much greater velocity than land breezes in coastal areas?

17. Why do sea breezes rarely go more than 50 km inland?

18. The Central Valley of California is flanked by the Coast Mountains on the west and the Sierra Nevada on the east. A dense nighttime radiation fog often develops in the valley, known as thule fog. It has caused huge freeway accidents as drivers slam into one another, unaware of the limited visibility and danger ahead of them. What causes this fog to collect in the low-lying portions of the Central Valley, since we normally expect condensation to occur at higher altitudes?

19. The northwest coast of the United States receives the highest average rainfall in the country, yet inland about 350 km in Oregon and Washington is desert. What accounts for this radical change in climate?

CLIMATE: VARIATION OVER SPACE

CHAPTER 7

Climate is the day-to-day weather each of us expects in a particular place. If we live in Hawaii, we expect to wear light clothes for most of the year. If we live in Alaska, we expect to wear heavy jackets. Climate is a synthesis of the succession of weather events we have learned to expect in any particular location.

Because there is a degree of organization in the heat and water exchanges at the earth's surface and in the general circulation of the atmosphere, there is a broadly simple pattern of climates across the globe. But this pattern is altered in detail by such factors as the location of land and sea, ocean currents, and highland areas. More specifically, the climate of a place may be defined as the average values of weather elements, such as temperature and precipitation, over a 30-year period and the important departures from the average values. In this chapter we will examine the way broad pattern and fine detail create distinctly different climates across the earth.

Climatology is the study of climates, whereas meteorology is the study of the physics, and specifically the weather, of the atmosphere. To understand the distinction, we may use the rule that weather happens now, but climate goes on all the time.

CLASSIFYING CLIMATE

The ideal climate classification system would achieve five objectives. First, it would clearly differentiate among all the major types of climates found on the globe. Second, it would show the relationships among these types. Third, it would apply to the whole world, and fourth, it would provide a framework for further subdivision to cover specific locales. Finally, the perfect climate classification system could demonstrate the factors that cause any particular climate.

In the same way that no map projection can satisfy all our requirements at the same time, no climate classification system can achieve all the above objectives at once. There are two reasons. First, there are so many factors that contribute to climate that we must compromise between simplicity and complexity. We have values for radiation, temperature, precipitation, evaporation, wind speed, and so on. We could use one variable, as the Greeks used latitude, and have a classification system that is too simple to be very useful, or we could use all the values and gain infinite detail but overwhelming complexity. Second, earth climates form a continuum. There are seldom sharp boundaries between the major types of climate.

When confronted with these perplexities, climatologists center their compromises around a single rule. The development, or choice, of a climatic classification must be determined by the particular use for which the scheme is intended. With so many potential uses, there are many different classification systems. This chapter concentrates on two of them. The Köppen system, which we will use most, provides us with a description of world climates that brilliantly treads the tightrope between complexity and simplicity. The second system, developed by Werner Terjung, helps us determine what a particular climate will feel like. To Terjung, the key to the classification of climate was the human body. To Wladimir Köppen, the key was plant life.

THE KÖPPEN CLASSIFICATION SCHEME

In 1874 a Swiss botanist, Alphonse de Candolle, produced a classification of world vegetation based on the internal functions of plant organs. It is to Wladimir Köppen's everlasting credit that he recognized that the plant, or assemblages of plants, at a particular place represent a synthesis of the many variations of the climate parameters experienced there. He therefore looked to de Candolle's classification of vegetation to solve the puzzle of the organization of climates.

Köppen compared the distribution of vegetation with the distribution of temperature and precipitation and used the correlations between atmosphere and biosphere to distinguish one climate from another. For example, he noted that, in high latitudes, the boundary marking the presence or absence of trees closely coincides with the presence or absence of at least one month in the year with an average temperature of 10°C (50°F). He noted many other correlations and put forward the first classification scheme at the turn of the century. Since that time, the scheme has been modified by Köppen and other workers to become the most widely used climatic classification system.

Figure 7-1 presents a simplified version of the Köppen system. Just as code letters are used to describe air masses, the Köppen classification uses a shorthand notation of letters to distinguish different characteristics of the major climates. The six major divisions are labeled *A* through *E* and *H*. The major tropical (*A*), mesothermal (*C*), microthermal (*D*), and polar (*E*) climates are differentiated by temperature. "Mesothermal" implies a moderate amount of heat, and "microthermal" implies a small amount. The major dry (*B*) climates are distinguished by potential evapotranspiration in excess of precipitation. Finally, the variable highland climates (*H*) are grouped into another major category.

The major groups are further subdivided in terms of heat or moisture by the use of a second and, in some cases, a third letter. The letters *f*, *m*, *w*, and *s* tell us

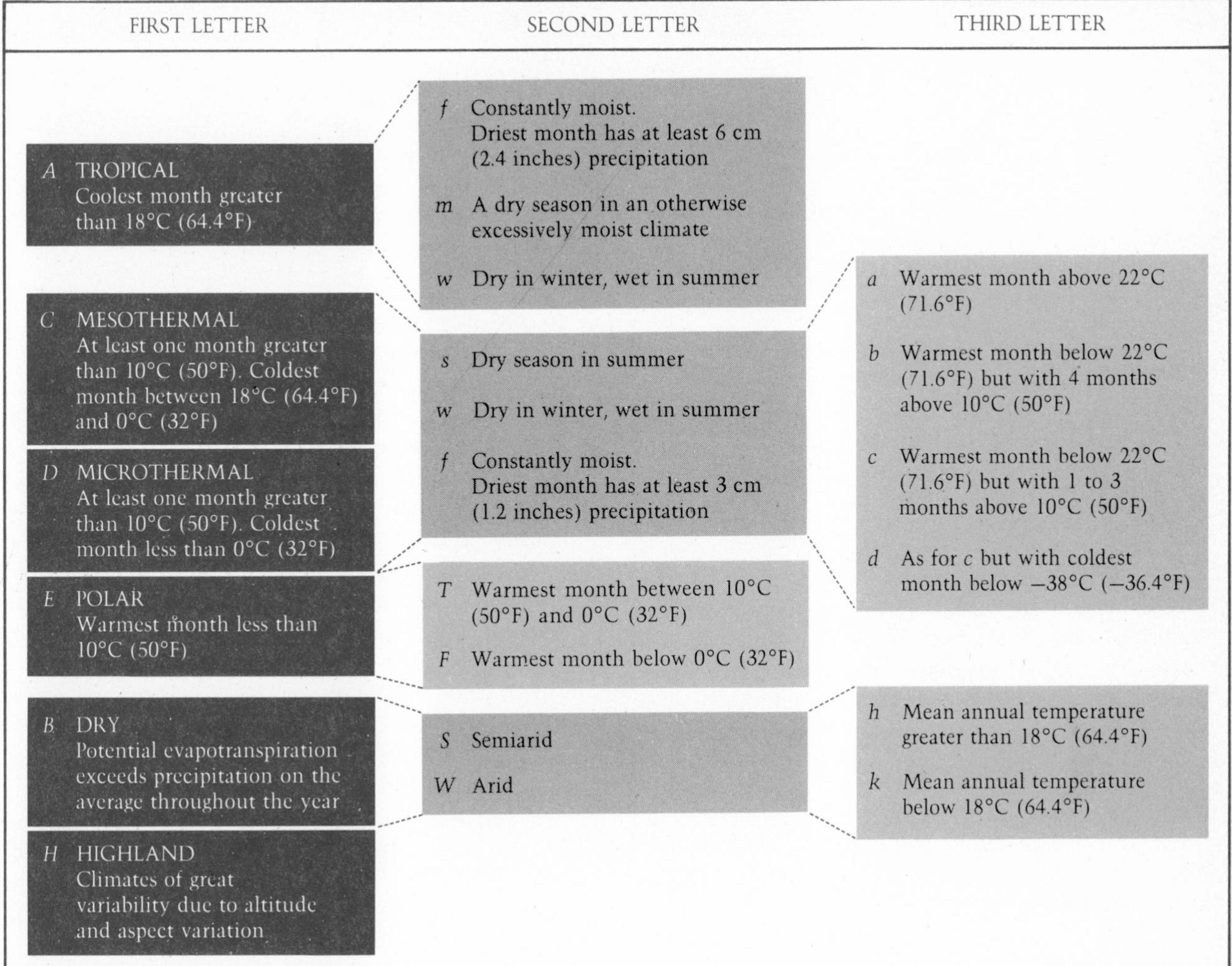

FIGURE 7-1

A simplified version of the modern Köppen climatic classification system.

when precipitation occurs during the year and are applicable to *A*, *C*, and *D* climates. *S* and *W* indicate the degree of aridity in dry (*B*) climates. Letters referring to moisture conditions are shown in the gray boxes in Figure 7-1. The heat modifiers *a*, *b*, *c*, and *d* provide details on the temperatures of *C* and *D* climates. The letters *h* and *k* do the same for *B* climates, and *T* and *F* subdivide the temperatures in polar (*E*) climates. Letters referring to heat are shown in the red boxes in Figure 7-1.

We can define many different climates with this shorthand code. Later in this chapter we will find examples of combinations of the code letters, and you will be able to determine their meaning by referring to Figure 7-1. For example, an *Af* climate is a tropical climate where every month the average temperature exceeds 18°C (64.4°F) and total precipitation exceeds 6 cm (2.4 in).

The classification system developed by Köppen has been criticized because it does not consider the causes of climate and because some of its vegetation-climate links are not very strong. Yet it remains the most often-used classification. Obviously it has proven itself.

THE COMFORT-INDEX CLASSIFICATION

In the twentieth century, scientists have worked hard to relate human comfort to factors in the climatic environment. Werner Terjung, a professor of geography at UCLA, synthesized many findings by the use of a psychrometric chart. This chart, in Figure 7-2, simply plots wet bulb temperature against dry bulb temperature and values of relative humidity. (These parameters are discussed in Chapter 4.) The combination of

TALES OF DISTANT PLACES

1. "The butter when stabbed with a knife flew like very brittle toffee. The lower skirts of the inner tent are solid with ice. All our [sleeping] bags were so saturated with water that they froze too stiff to bend with safety, so we packed them one on the other full length, like coffins, on the sledge."

2. "The rain poured steadily down, turning the little patch of reclaimed ground on which his house stood back into swamp again. The window of his room blew to and fro: at some time during the night, the catch had been broken by a squall of wind. Now the rain had blown in, his dressing table was soaking wet, and there was a pool of water on the floor."

3. "It was a breathless wind, with the furnace taste sometimes known in Egypt when a khamsin came, and, as the day went on and the sun rose in the sky it grew stronger, more filled with the dust of Nefudh, the great sand desert of Northern Arabia, close by us over there, but invisible through the haze."

4. "The spring came richly, and the hills lay asleep in grass—emerald green, the rank thick grass; the slopes were sleek and fat with it. The stock, sensing a great quantity of food shooting up on the sidehills, increased the bearing of the young. When April came, and warm grass-scented days, the flower burdened the hills with color, the poppies gold and the lupines blue, in spreads and comforters."

What do these rather poetic excerpts have in common? They attempt to convey a sense of climate, the impact a particular climate—especially a demanding and harsh climate—has on people who attempt to explore and live in it. Description of climate is one of the basic elements of the literature of travel and exploration. Perhaps nothing conveys a sense of an environment as efficiently and dramatically as notation of its winds, rains, and temperatures. And as these skillful writers show, climate tells you a great deal about the "feel" of a place.

To relieve your curiosity: Excerpt 1 is from Robert Falcon Scott's memoirs of his explorations in the Arctic, published in 1914; excerpt 2 is by Graham Greene, who after being stationed in West Africa during World War II, wrote *The Heart of the Matter;* excerpt 3 is by T.E. Lawrence (of Arabia), the British soldier who organized the Arab nations against the Turks during World War I; and excerpt 4 is by John Steinbeck writing in *To a God Unknown,* about the Mediterranean climate of northern California.

humidity and temperature is a more realistic clue to human comfort than temperature alone. You may be able to recall feeling uncomfortable on a hot day with high humidity and feeling much more at ease on an equally hot day when the air was dry. Terjung used the results of many other studies to divide the psychrometric chart into eleven regions of human comfort.

Seven of the regions are indicated in Figure 7-2, and the limits of the other four are indicated at the top left of the diagram. The comfort zones are divided by lines of effective temperature and wet bulb temperature. *Effective temperature* is the temperature of still air saturated with water vapor at which a certain percentage of humans experience a subjectively equivalent sensation of comfort. For example, tests on thousands of people in the United States showed 98% of them to be comfortable at an effective temperature of 21.7°C (71°F). On the psychrometric chart, lines of effective temperature correspond to saturation lines drawn between points of equal dry bulb, wet bulb, and dew-point temperatures.

Each of the comfort zones is assigned a number ranging from +4 through 0, the most comfortable, to −6. These numbers are the comfort indexes. The indexes for day and night conditions are combined into the final symbols, each with subscript numbers, of comfort for a place. In Table 7-1 the larger the subscript number, the larger the change of conditions between night and day.

Although climate is essentially an abstract idea, the Köppen and comfort-index systems give us complementary views of the climates existing on the earth. Köppen deals with temperatures and rainfall. These are related to vegetation and hence the food we eat. The comfort index is concerned with temperatures and humidities. It tells us how we will feel. Both systems work for different purposes, and both are necessary to

TABLE 7-1

Day and night combination of comfort index. Capital letters indicates day conditions; subscript indicates diurnal variation. First numeral refers to day, second numeral to night conditions.

EXTREMELY HOT	SULTRY	HOT	WARM
$+4/+3EH_1$	$+3/+3S_1$	$+2/+2H_1$	$+1/+1W_1$
$+4/+2EH_2$	$+3/+2S_2$	$+2/+1H_2$	$+1/\ \ 0W_2$
$+4/+1EH_3$	$+3/+1S_3$	$+2/\ \ 0H_3$	$+1/-1W_3$
etc.	etc.	etc.	etc.
MILD	**COOL**	**KEEN**	**COLD**
$0/0\ M_1$	$-1/-1C_1$	$-2/-2K_1$	$-3/-3CD_1$
$0/-1M_2$	$-1/-2C_2$	$-2/-3K_2$	$-3/-4CD_2$
$0/-2M_3$	$-1/-3C_3$	$-2/-4K_3$	$-3/-5CD_3$
etc.	etc.	etc.	etc.
VERY COLD	**EXTREMELY COLD**	**ULTRA COLD**	
$-4/-4VC_1$	$-5/-5EC_1$	$-6/-6UC_1$	
$-4/-5VC_2$	$-5/-6EC_2$		

FIGURE 7-2

Terjung's comfort index as derived from the psychrometric chart. The symbols of the index: -6, ultra cold; -5, extremely cold; -4, very cold; -3, cold; -2, keen; -1, cool; 0, mild; 1, warm; 2, hot; 3, sultry; 4, extremely hot.

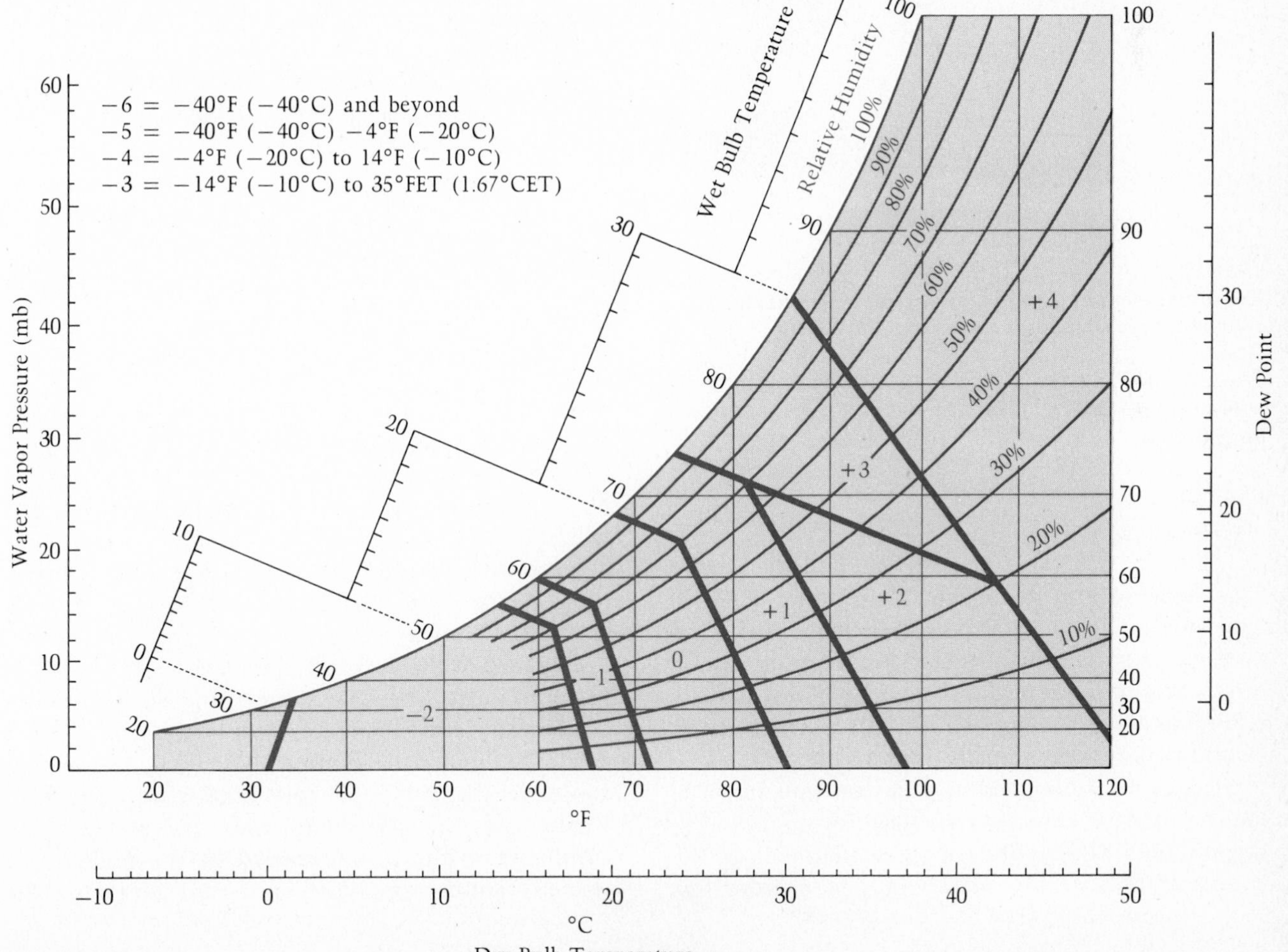

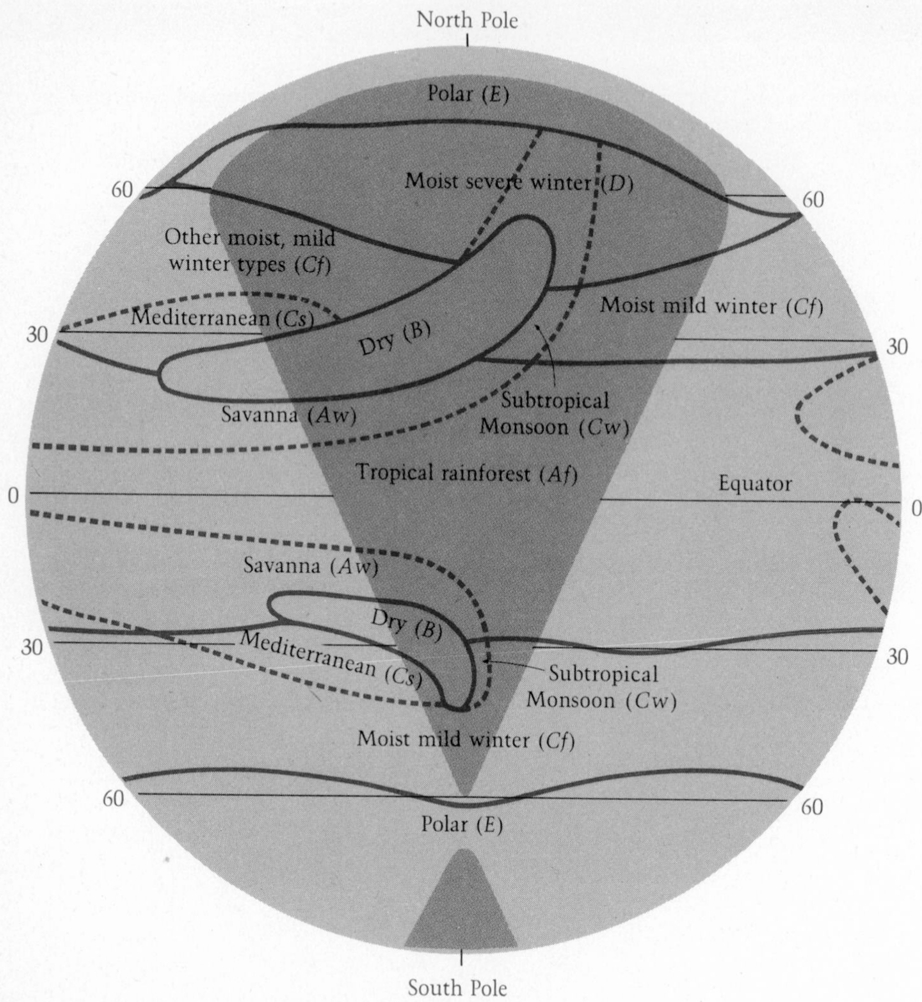

FIGURE 7-3
The distribution of climatic types over a hypothetical continent of low uniform elevation.

give a clear picture of climate. We look now at climate distribution.

THE WORLD DISTRIBUTION OF CLIMATE

The easiest way to picture the distribution of world climates is to first imagine how they would be laid out on a hypothetical continent of low and uniform elevation. The continent outlined in Figure 7-3 is vaguely reminiscent of the Euro-Afro-Asiatic and American "world islands" that actually exist on the globe.

An area of *A* climates straddles the equator. The moister of these climates (*Af*) nurtures tropical rainforest. The area is widest on the eastern side of the continent because the trade winds blow from this direction. The subtropical high-pressure zones, whose influence is greatest on the eastern side of oceans, lead to the dry *B* climates intruding into the continent in both the Northern and Southern hemispheres. Between the *A* and *B* climates lies a belt of transitional climate with a winter dry season, *Aw*. This is sometimes called a savanna climate because of the tall grasses that grow there.

Poleward of the dry *B* climatic areas, small zones of Mediterranean climate (*Cs*) are found on the western coast, where the subtropical high-pressure zone has a drying influence in summer. Poleward from these zones, the climates with moist mild winters (*Cf*) owe their existence, especially on the western side of the continent, to the storms carried by the midlatitude westerly winds. *Cf* climates are also found on the eastern side of the continent, where they obtain moisture from humid air on the western side of the ocean's subtropical high-pressure zone.

In the Northern Hemisphere, *D* climates, with a moist severe winter, exist poleward of the *C* climates. There are no large land masses in the Southern Hemisphere, so here latitude and the effect of land do not give rise to the cold winters of the *D* climates. In both hemispheres, however, land areas above the Arctic and Antarctic circles, with deficits of net radiation, have polar (*E*) climates.

Figure 7-4 shows the actual distribution of climates. It rather closely resembles the overall pattern outlined in Figure 7-3, although there are differences. First, the actual distribution of land and sea varies markedly from that of the hypothetical continent. This is important, because oceans moderate the climates of land masses, especially where winds blow from the ocean onto the land. A second difference lies in the large areas of highland actually found on the globe. Mountains produce lower temperatures and affect the pattern of rainfall.

The point made by examining the distribution of climates in Köppen's terms, as we have just done, is even better substantiated by examining the different climates in terms of the comfort index. Figure 7-5 shows that many climates are transitional. In any one place, several comfort categories may exist throughout the year. Where one category predominates, only one capital letter is used, and a small letter indicates the number of months in the year characterized by that category. What we feel, therefore, usually changes throughout the year.

Very few parts of the world fall into the ideal, mild category. The Northern Hemisphere has extensive areas of extremely hot (*EH*) and keen (*K*) and colder regions. Hot (*H*) and extremely hot climates appear in the low-latitude deserts, and the cyclonic storm belt is most often represented by keen climates. "Keen" is the comfort index class between "cold" and "cool." During keen conditions, people sometimes talk of there being a "nip" in the air. Mild (*M*) climates are found only near coasts with offshore cold currents and in low-latitude highlands. This broad climatic distribution testifies to the magnificent adaptability of the widespread human species.

TROPICAL CLIMATES (*A*)

In a continuous belt astride the equator, 20° to 40° wide, is the warmth and moisture of the tropical climates. Warmth is derived from their closeness to the equator, and moisture comes from the rains of the ITCZ, easterly waves, cyclones, and monsoons. In these climates Alfred Russel Wallace became entranced by the great variety of plant and animal species and developed his contribution to the theory of evolution. Tropical climates have three major subdivisions: the savanna, the monsoon rainforest, and the tropical rainforest.

The *savanna* (*Aw*) areas are found in a transitional zone between the subtropical high-pressure and equatorial low-pressure regions. Although these areas often receive between 750 and 1500 mm (30 and 60in) of rainfall per year, there is a distinct dry season in the part of the year when the sun is lowest. The dry season can cause moisture deficits in the soil. The season's length and severity are proportional to the area's distance from the equator. In eastern Africa, the Masai tribe (see Figure 7-6) traditionally move their herds of cattle north and south in search of new grass sprouted by the rains. This area grows seasonal coarse grass, clumps of trees or individual trees, and thorn bushes.There are extensive areas of savanna climate in the world, including large parts of South America, Africa, and India. Rainfall is so unreliable on the poleward sides of these climates that parts of the savanna lands are marginal to economic development. These areas are also highly susceptible to minor changes of long-term climate, as Chapter 9 explains.

The *monsoon rainforest* (*Am*) is restricted to tropical coasts that are often backed by highlands. It too has a dry season, but a short one. A compensating season of heavy rains generally prevents any soil moisture deficits. In some places, such as northeastern India and Burma, the highlands add an orographic effect to the monsoonal rains. Thus at Baguio, on the slopes of the central ranges on Luzon, in the Philippines, 1168 mm (46 in) of rain fell on one July day in 1911. The vegetation in this climate consists mainly of evergreen trees with occasional grasslands. The trees, however, are not so dense as those found in the areas of tropical rainforest.

The *tropical rainforest* (*Af*) shows the greatest effects of heat and moisture. Tall trees, green all year round, completely cover the land surface, as Figure 7-7 shows. Virgin forest has little undergrowth, because the trees let through such a small amount of light. But where the trees have been cleared, a dense jungle develops.

São Gabriel is a small Brazilian town high upstream on the Río Negro, a tributary of the Amazon. It is

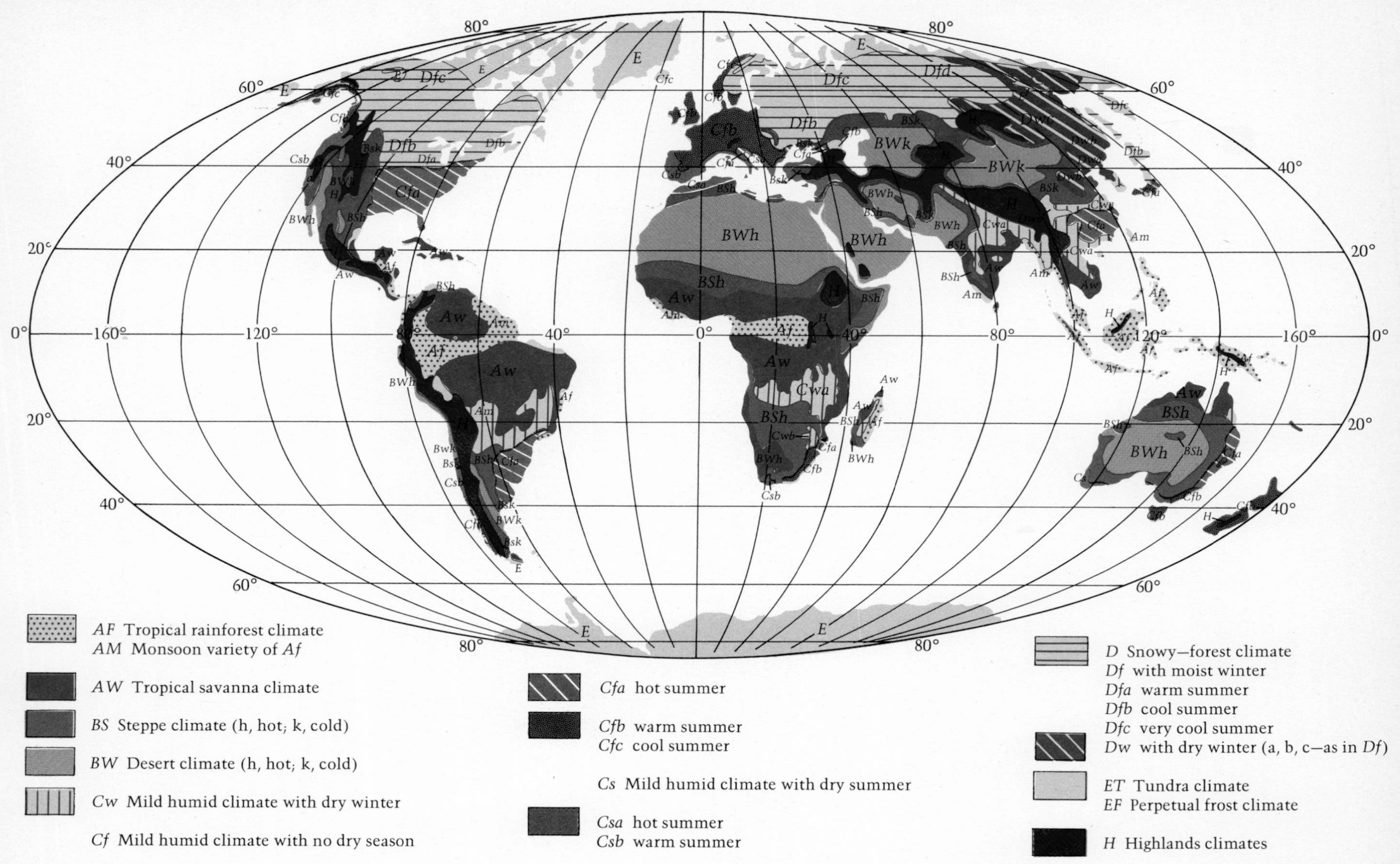

FIGURE 7-4

The global distribution of climates according to the modern Köppen classification system.

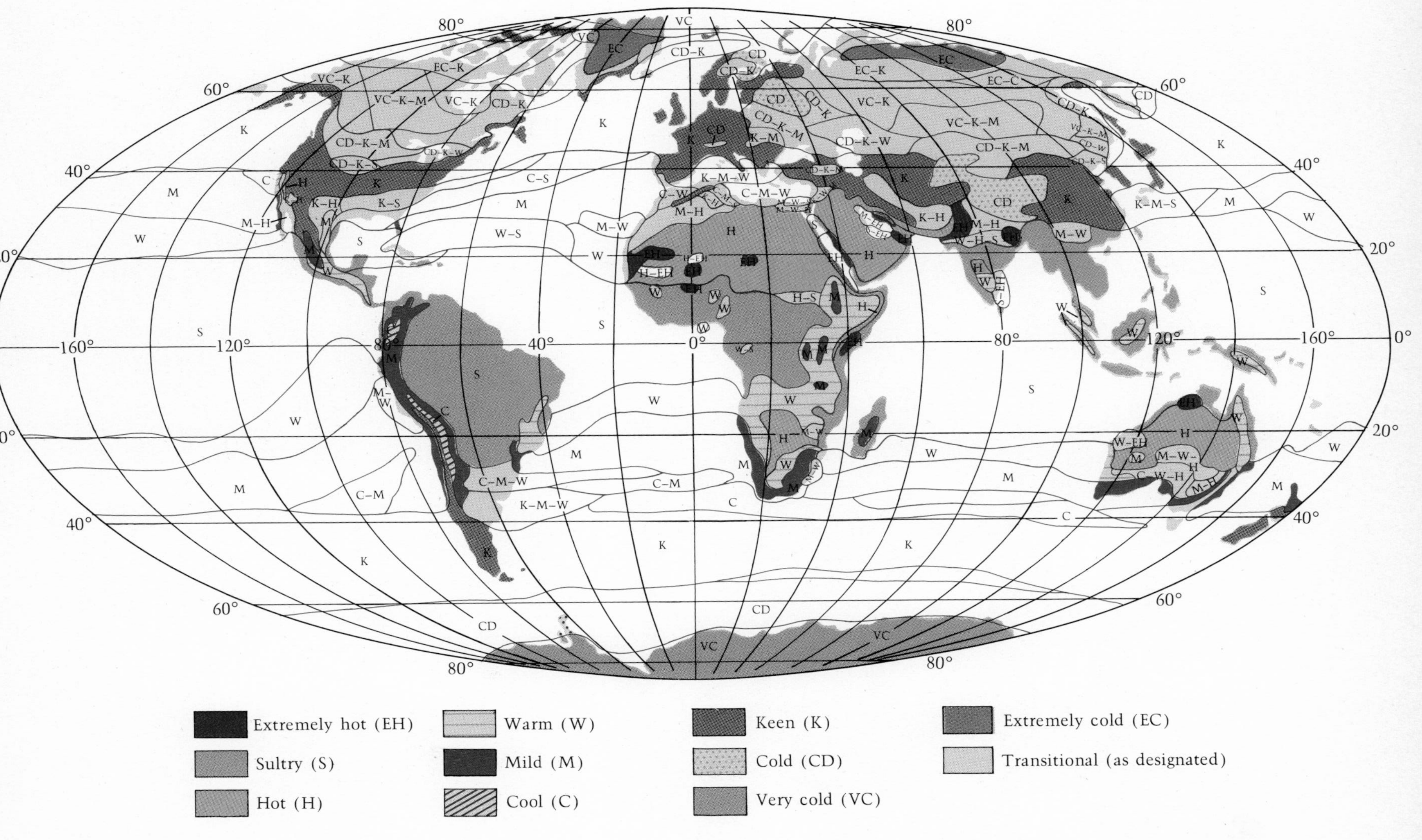

FIGURE 7-5

The global distribution of climates according to the Terjung comfort index.

FIGURE 7-6
The cattle of the Masai in the savanna (*Aw*) climate of Kenya.

located almost exactly on the equator. Among other things, Figure 7-8 shows that temperatures are consistently around 26°C (79°F) throughout the year. In fact, the daily temperature may change as much as 6°C (11°F), an amount greater than the annual range of 1.6°C (3°F). Precipitation also tends to fall in a diurnal rather than seasonal rhythm. A normal daily pattern consists of relatively clear skies in the morning and a gradual build-up of convectional clouds, from the vertical movement of air, with the increasing heat of the day. In early afternoon, the storms burst forth with torrential rain. Such a pattern is repeated day after day.

In São Gabriel, no month receives less than 130 mm (5.1 in) of rainfall. In May, about 10 mm (0.39 in) of rain falls per day. The total rainfall for the year is 2800 mm (109 in). Such high amounts of rainfall keep the water balance in perpetual surplus, as Figure 7-8 indicates. Evapotranspiration in this area constantly occurs at its potential rate, and the surplus water runs off into the Río Negro and from there into the Amazon. As a result, most of the net radiation is used in evaporative heat (also shown in Figure 7-8), and so the moisture in

FIGURE 7-7
The dense foliage of the tropical rainforest completely covers the surface of the earth.

Af – Tropical Rainforest Climate

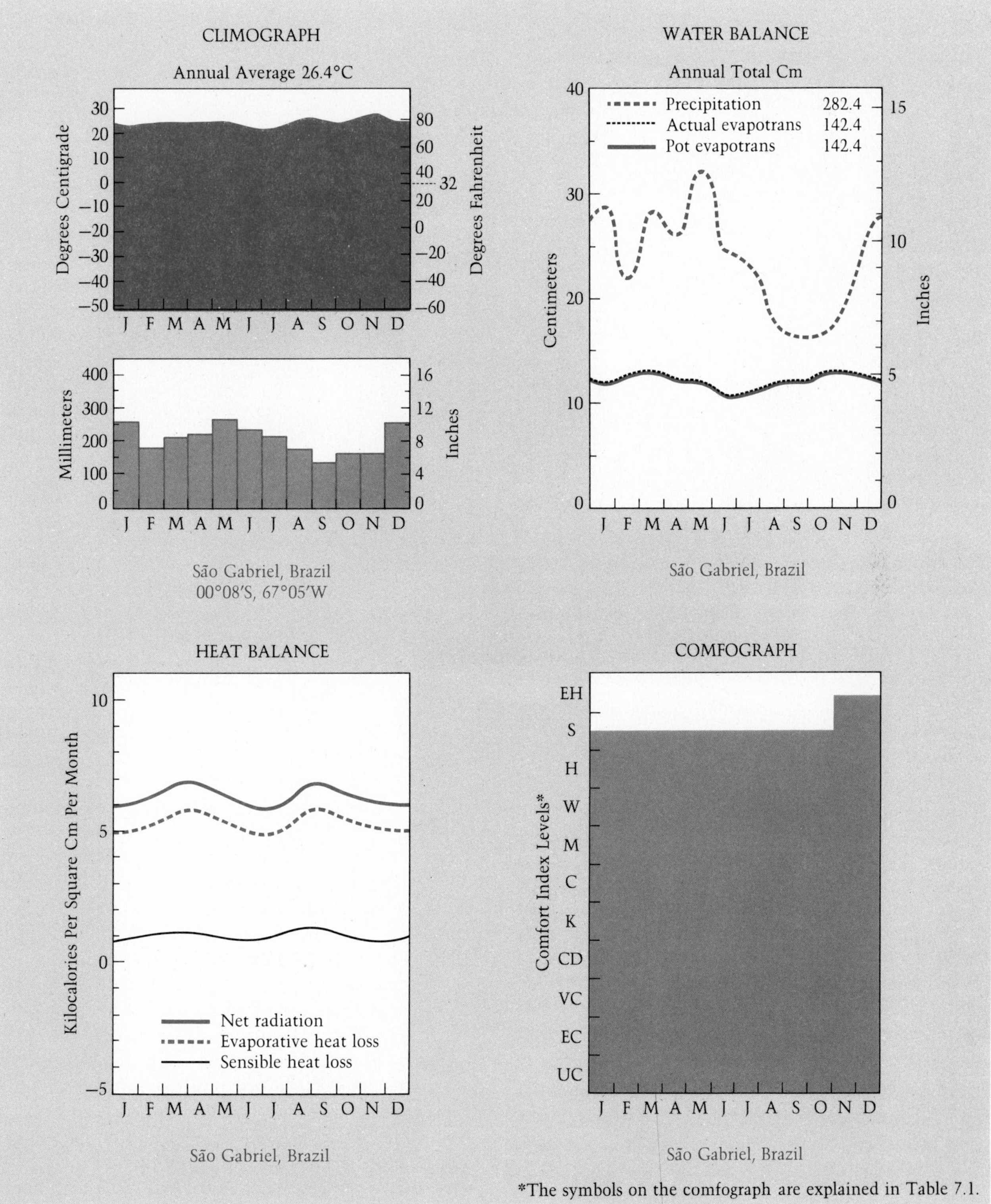

FIGURE 7-8
Details of a tropical rainforest (*Af*) climate.

FIGURE 7-9
The Sahara, a dry (*B*) climate.

the air is continuously replaced. When the moisture condenses, latent heat adds to the monotonous warmth of the atmosphere. When the sun passes directly overhead at the equinox, the rainfall values are not affected, because they are always high. But the heat balance is dramatically affected, as demonstrated by the high values of net radiation and evaporative heat loss in Figure 7-8. The continually hot and highly humid air makes such a climate unpleasant for people unaccustomed to it. This is shown in the comfort index— the comfograph in Figure 7-8. All months of the year, except for November and December, are sultry. November and December are worse.

DRY CLIMATES (*B*)

Dry *B* climates (see Figure 7-9) provide a striking contrast to the *A* climates. In these areas, where no permanent streams originate, potential evapotranspiration always exceeds the moisture from precipitation.

Dry climates are found in two general locations. The first group consists of two interrupted zonal belts near 30°N and 30°S, where the subtropical high-pressure areas are dominant. The largest areas of this type are the Sahara and deserts of the Middle East and Australia. The driest of these areas are often located near the western coasts of continents. For example, Iquique, a seaport in Chile, once had no rain for 14 consecutive years. The second group of dry climates can be attributed to two different factors. First, a continental interior, remote from any moisture source, with cold high-pressure zones in winter, can have an arid climate. The southern USSR is a case in point. Second, rain shadow areas in the lee of mountain ranges often give rise to cold deserts, as in Argentina and North America.

Average annual temperatures in dry climates are usually typical of those that might be expected at any given latitude—high in low latitudes and low in higher latitudes. However, the annual range of temperature is often higher than might be expected. In the Sahara, for example, a range of 17°C (30°F) at 25°N is partly the result of a continental effect. The daily range of temperature can also be large, averaging between 14°C and 25°C (25°F and 45°F). On one day in Tripoli, Libya, the temperature went from below freezing, −0.5°C (31°F), at dawn to 37°C (99°F) in the afternoon. Rocks expanding through such changes of heat sometimes break with a sharp crack. North African soldiers in World War II occasionally mistook the breaking rocks for rifle shots.

Dry climates of deserts are not necessarily dry all the time. Their precipitation can range from near 0 to about 750 mm (30 in) per year. The temperature and precipitation boundaries for the *desert* (*BW*) and the more moist *steppe* (*BS*) climates are shown in Figure 7-10. In summer, the margins of subtropical *B* climates nearest the equator are sometimes affected by the rains of the ITCZ as it moves toward the poles. In winter, the margins of the dry areas nearest the poles may be subject to midlatitude cyclonic rains. There are also rare thunderstorms, which sometimes lead to flash floods. At Helwan in the United Arab Republic, seven such storms in 20 years gave a total of 780 mm (31 in) of rain. Furthermore, along coasts with cold ocean currents, rather frequent fogs provide a little moisture. Swakopmund in South-West Africa experiences 150 days of fog per year. But for the most part, extreme dryness is most common in *B* areas.

The dryness is exemplified by the American state where people take their troubled sinuses—Arizona. At Yuma, Arizona the annual average temperature is 23.5°C (74.3°F), with an annual range of some 23°C (41°F). The climate here is *BWh*, a hot, dry desert. The small amount of rainfall shown in Figure 7-11, 89 mm (3.4 in), comes from winter storms or from convectional clouds of the summer.

The water balance in the figure shows a marked deficit throughout the year; potential evapotranspiration is far in excess of precipitation. Most water that falls quickly evaporates back into the dry atmosphere. But there is little water for evaporation, so the evaporative heat loss is negligible. Consequently, most of the heat input from net radiation, which is markedly seasonal, is expended as sensible heat. The sensible heat adds warmth to the already hot air.

You may get an impression of what it would feel like to live in Yuma from the comfograph in Figure 7-11. Surprisingly, January is actually cool, and four months are mild. No months are sultry, but July, August, and September rate extremely hot.

The world's dry climates are usually inhospitable to human life. Settlement is restricted to the rare water

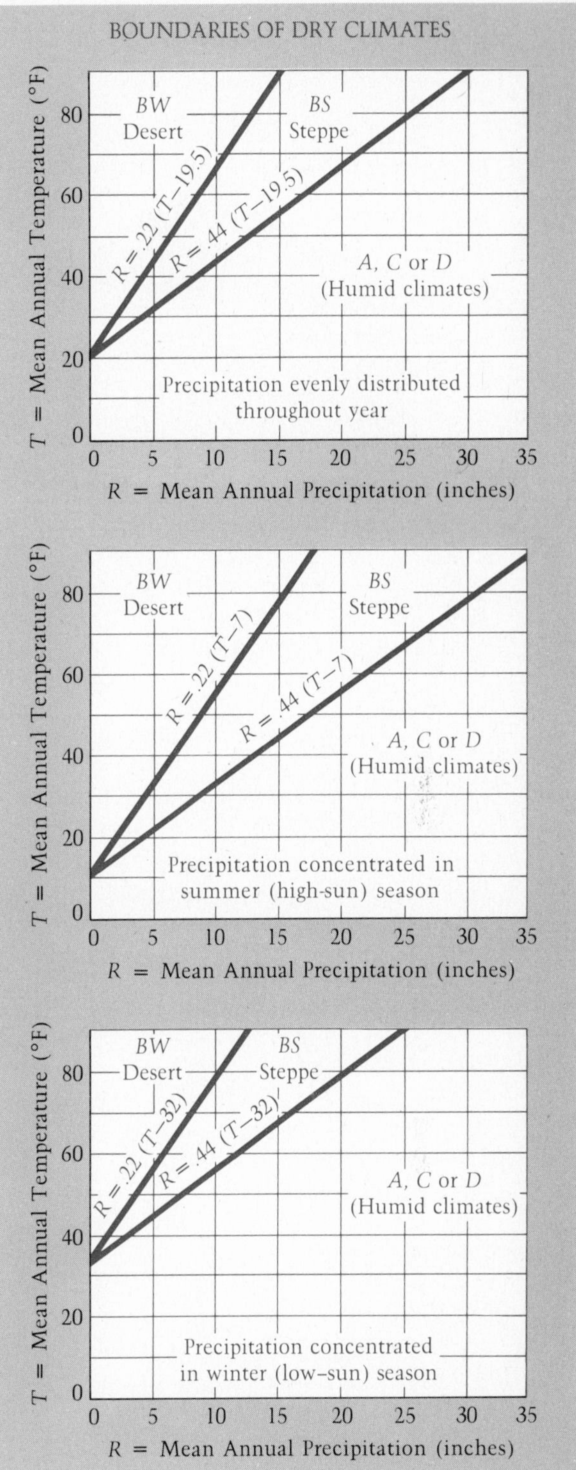

FIGURE 7-10

Temperature and precipitation boundaries of dry (*B*) climates.

BWh— Hot Dry Desert

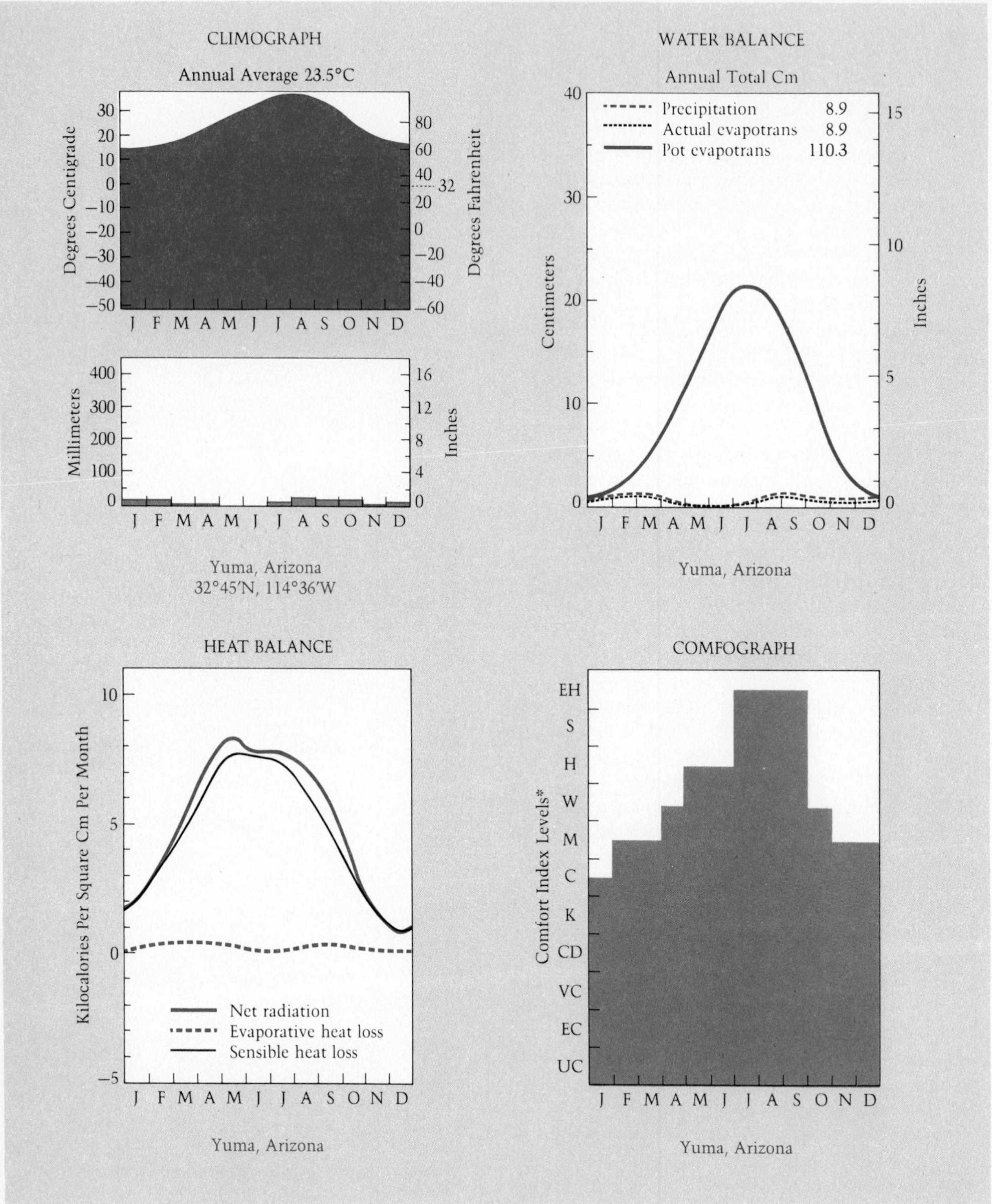

FIGURE 7-11
Details of a hot dry desert (*BWh*) climate.

sources of oases or to rivers that originate outside the area. In some cases, irrigation can bring life. But it is all too easy for irrigation water to wash important minerals from the soil. Scientists are constantly made aware that the balance between the earth's systems in dry areas is extremely fragile.

HUMID MESOTHERMAL CLIMATES (C)

If the importance of climates is to be assessed by the numbers of people who live in them, the warm, temperate, rainy climates of the midlatitudes reign supreme. Three factors characterize these climates: the westerly winds and their jet streams, the continual battle of conflicting air masses, and the steady rhythm of seasonal change. The weather changes by the day, necessitating frequent changes of clothing style and offering a perpetual topic of conversation.

The humid mesothermal *C* climates are found in two midlatitude belts, one in the Northern Hemisphere and one in the Southern Hemisphere. In the Northern Hemisphere, the belt tends to be wider on the western sides of continents, because of the moderating influence that westerly winds bring from the oceans. At the same time, the belt of westerlies shifts from north to south and back again as the seasons change.

We may distinquish three important types of humid mesothermal climate according to their pattern of precipitation occurrence. The *Cf* climates are perpetually moist. The *Cs* climate has a dry season in the summer, and the *Cw* climate is dry in the winter.

PERPETUALLY MOIST AREAS (*Cf*)

The perpetually moist mesothermal climates (*Cf*) are found in two major groups of areas, both near a source of water. The first group, the coasts caressed by the westerlies, is occasionally called a *marine west coast* type of climate. Figure 7-12 shows one such area; others are found in the northwestern United States and western Canada and in southern Chile, southeastern Australia, and New Zealand. In all these locations, storms from midlatitude cyclones bring moist temperate air to the land. There are seldom any extremes of temperature. Average monthly temperatures never exceed 22°C (71.6°F), and there are always four or more months with mean temperatures exceeding 10°C (50°F) in the *Cfb* areas and somewhat less in the *Cfc* regions.

Cfb climates are located exactly in the middle of the midlatitude westerlies. The cooler *Cfc* climates are found only in the Northern Hemisphere in high latitudes, where strong, warm ocean currents sweep poleward along western shores. Western civilization moved to the *Cfb* climates of northwestern Europe after leaving its Mediterranean cradle. The adequate year-round precipitation naturally produces a forest of evergreen conifers and broad-leaved trees that shed their leaves in winter. Although droughts are not unknown in these areas, the soil rarely has a moisture deficit for long periods.

A warmer, moist mesothermal climate, sometimes called a *humid subtropical climate* (*Cfa*), is found in the land areas to the southeast of the five major continents. It owes its existence to the effects of the warm moist air traveling northwestward (in the North-

FORECASTS FROM THE WEATHER BUREAU

Each morning before she starts her laundry a Connecticut woman calls the local Weather Bureau office to check the forecasts on cloud ceilings. For months the weather forecasters were baffled. A request for rain predictions would have made sense—but the height of the cloud cover? Their curiosity finally overwhelmed them.

The answer? "When the clouds are high I can hang out my wash, but when the ceiling is low the sea gulls fly over my backyard. That's no time to have clean clothes hanging on the line."

Millions of other people also rely on the forecasts of the United States Weather Bureau. Each year the Bureau makes about 1.3 million general weather forecasts and 1.7 million aviation forecasts. Thousands more special warnings alert Americans to tornadoes, hurricanes, floods, hail, lightning, high winds, blizzards, fog, and frost. No other government agency except the Postal service reaches more people every day.

In addition to fresh laundry, the services of the Weather Bureau save Americans over a billion dollars each year. Storm warnings are especially valuable. The Bureau estimates that these special forecasts prevent about $220 million in property damage and save about 2440 lives per year.

FIGURE 7-12
The verdant landscape of England, in a marine west coast (*Cf*) climate.

ern Hemisphere) around the western portions of the oceanic subtropical high-pressure zones. Additional moisture comes from the movement of midlatitude cyclones toward the equator in winter and from tropical cyclones in summer and autumn.

Miyazaki, on the southern island of Japan, has a humid subtropical climate similar to that of the southeastern United States. The graphs in Figure 7-13 show a strong seasonal effect. Temperatures range from 6.8°C (44°F) in January to 26.7°C (80°F) in a hot humid August, and the average is 16.7°C (67°F). Rainfall for the year almost reaches that of the tropical rainforest. The total is 2560 mm (100 ins), falling mainly in summer. The water balance at Miyazaki is never at a deficit, not even in the relatively dry winter. Throughout the year, actual evapotranspiration always equals the potential evapotranspiration.

The heat balance too demonstrates the humid nature of the climate. Most of the net radiation is used to evaporate water, and a relatively small proportion passes into the air as sensible heat. While the people of Miyazaki go about their daily business, they are subject to atmospheric comfort conditions that vary from three keen months in the winter to three sultry months in the summer, as Figure 7-13 shows. The seasonal character of midlatitudinal climates is shown not only by curves in the climograph and the heat and water balances, but also by the transition through no less than six comfort categories as the year progresses.

AREAS WITH DRY WINTERS (*Cw*)

The *Cw* climate, another type of *C* climate, is little different from the savanna climate (*Aw*). The only departures are the amounts of rainfall and a distinct cool season with the average temperature of one month falling below 15°C (64.4°F). The *Cw* climate is the only division of the *C* climates found extensively

Cfa – Humid Subtropical Climate

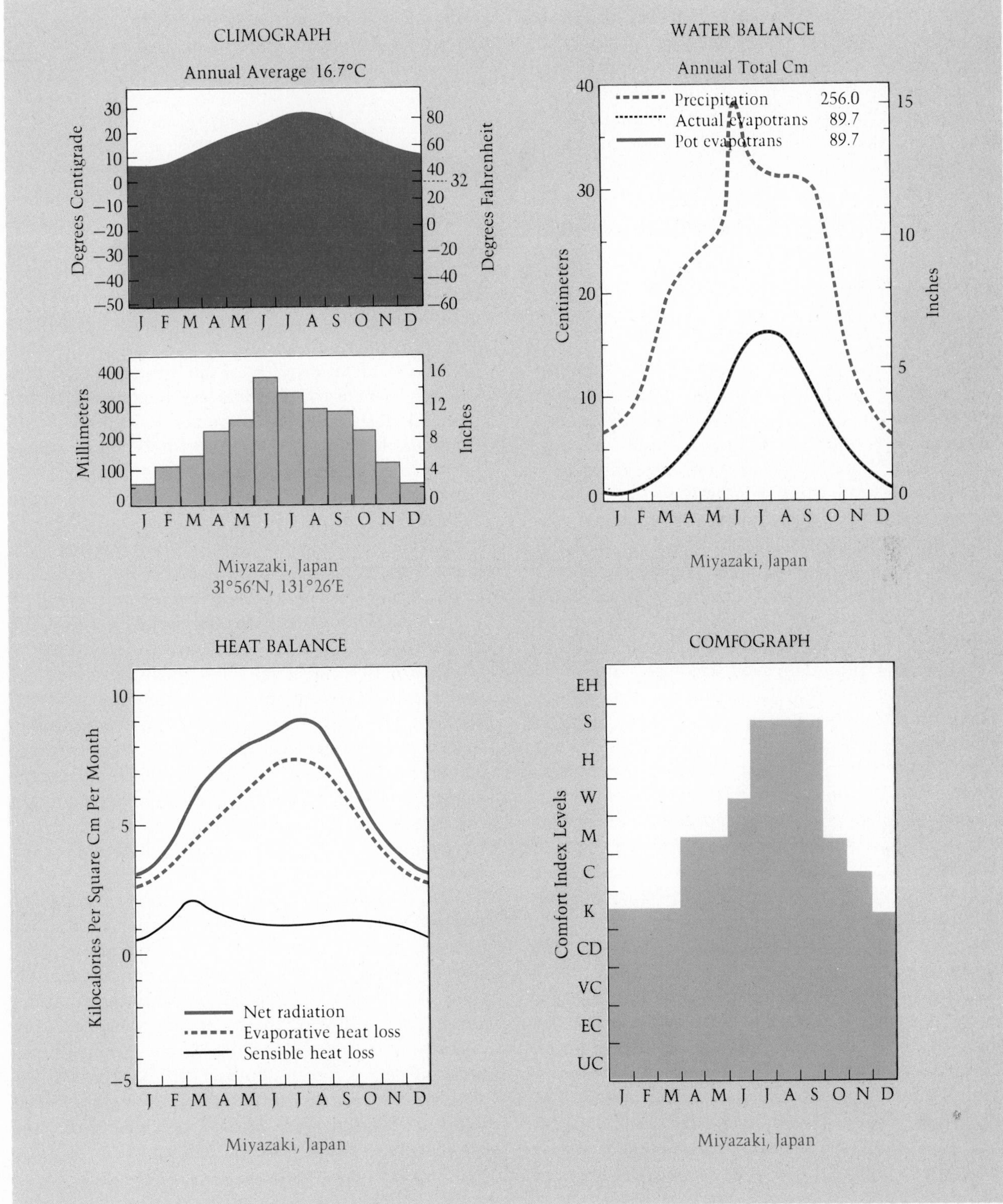

FIGURE 7-13
Details of a humid subtropical (*Cfa*) climate.

in the tropics. For this reason it often shares with the *Aw* climates the designation *tropical wet-dry climate.*

In many cases the winter dry season of *Cw* climates comes from the offshore winds accompanying a winter monsoon. In Southeast Asia and northeastern Australia, rainfall is provided by the wet summer monsoon. In other cases, such as in south-central Africa, the winter dry season occurs because the rains of the ITCZ travel into the Northern Hemisphere with the sun.

AREAS WITH DRY SUMMERS (*Cs*)

The other *C* climate is the *Mediterranean climate* (*Cs*). It is frequently described, by people of European descent, as the most desirable climate on earth. The climate that attracted film makers to Hollywood is famous for its clear light and dependable sunshine. The warmer type of Mediterranean climate (*Csa*) is found in the Mediterranean itself and in other interior locations. The cooler type (*Csb*) is found on coasts near cool offshore ocean currents. Coastal areas of California, central Chile, southern Australia, South Africa, and the Iberian Peninsula also enjoy this vacation climate.

In the *Cs* climate, rainfall arrives in the cool season. Annual totals are moderate, ranging between about 400 mm (16 in) and 650 mm (25 in). The rains are the result of the winter storms from midlatitude cyclones. The long dry summers, such as those enjoyed by Californians, result from the dominance of the subsiding air on the eastern part of the oceanic subtropical high-pressure areas.

Some aspects of the "perfect" climate can be seen in Figure 7-14. The cool offshore current, with its frequent fogs, keeps the average monthly temperatures almost constant throughout the year. At San Francisco, they range only 6.6°C (12°F) around an annual average of 13.6°C (56.5°F). The water balance values for San Francisco indicate annual rainfall of 551 mm (21 in). The rains usually arrive with winter storms steered southward by the jet stream.

Although in winter actual evapotranspiration reaches the potential amount, in summer it falls far short. The effect of the dry summer is reflected in the heat balance estimates for Sacramento, California. In winter much of the net radiation is expended as evaporative heat loss. But in the summer, soil moisture deficits require that a large proportion of the net radiation be used in the form of sensible heat to warm the air.

Perhaps the most important proof of the pleasure experienced in *Cs* regions is shown by the comfort index. In Santa Monica, southern California, which is not cooled quite so much as San Francisco is by the effect of offshore currents, seven months of the year fall into the ideal mild comfort category; the others are either warm or cool. Millions of Americans have moved to California, which may well exemplify the best of all possible worlds—at least as far as climate goes.

HUMID MICROTHERMAL CLIMATES (*D*)

In the Northern Hemisphere, continental land masses extend east and west for thousands of kilometers. For example, the great-circle distance between Bergen, Norway and Okhotsk, on the eastern coast of the USSR is about 4800 km (3000 mi). Consequently, vast areas of land in the middle and high latitudes are far away from the moderating effect of the oceans. The result is a climate where the seasonal rhythms of the higher latitudes are carried to extremes. Distinctly warm summers are balanced by frigid winters. These are the humid microthermal climates (*D*), distinguished by a warm month, or months, when temperatures are above 10°C (50°F), and a period, averaging longer than a month, when temperatures are below freezing. The people who live in *D* climates need life styles as varied as their wardrobes to cope with such conditions.

The cold-winter *D* climates are mainly found in the midlatitude and high-latitude continental expanses of North America and Asia. These areas are sometimes punctuated by regions of the completely summerless *E* climates, especially in northeastern Asia and Alaska. The *D* climates are also found in certain Northern Hemisphere midlatitude highlands and on the eastern sides of continental masses. These climates represent the epitome of the continental effect. Whereas they cover only 7% of the surface of the whole globe, they cover 21% of its land area.

The *D* climates are often divided into two groups. At the lower latitudes, where more heat is available, we find the *humid continental climates* (*Da* and *Db*). Farther toward the poles, where the summer net radiation cannot raise the average monthly temperatures above 22°C (71.6°F), there are the *subarctic taiga climates* (*Dc* and *Dd*). *Taiga* is the Russian word meaning "snow forest." Both groups of climates usually receive enough precipitation to be classified as moist all year round (*f*). But in eastern Asia, the cold air of winter cannot hold enough moisture, so a dry winter season (*w*) results.

All *D* climates show singular extremes of temperature throughout the year. Chapter 4 shows the heat balance at the Siberian town of Turukhausk. In July the temperature here is a pleasant 14.7°C (58.5°F), but

Csb – Mediterranean Climate

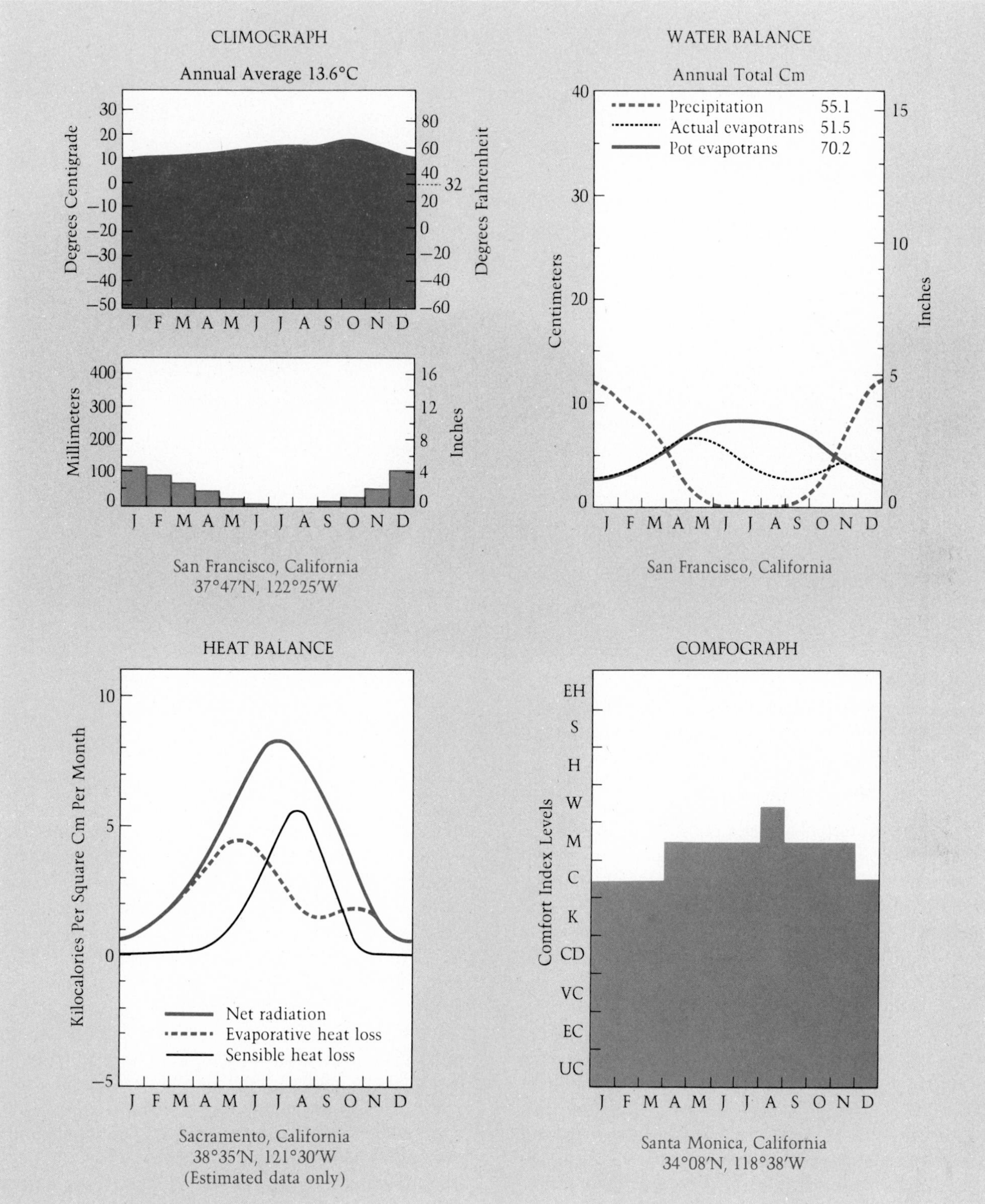

FIGURE 7-14
Details of a Mediterranean (*Cs*) climate.

FIGURE 7-15
An area where railroad fill caused the permafrost to thaw. This in turn led to subsidence and a rather peculiar railroad.

the January mean temperature is −31°C (−24.5° F). In Verkhoyansk, also in Siberia, the January temperature plunges to −50°C (−58.5°F). The ranges of temperature—the difference between the highest and lowest monthly averages—are 45.7°C (83°F) in Turukhausk and 64°C (115°F) in Verkhoyansk. These are some of the largest temperature ranges found anywhere on the earth's surface.

Such temperatures have a far-reaching effect. There is a short, intense growing season, because the long summer days permit large quantities of heat and light to fall on the earth. Special quick-growing strains of vegetables and wheat have been developed to make full use of this short period of warmth and moisture. On the Chukotski Peninsula, one of the easternmost points of Asia, one variety of cucumbers usually grows to full size within 40 days.

But agriculture is severly hampered by the thin soil (most was removed by the passing of extensive ice sheets thousands of years ago) and permafrost. *Permafrost* is a permanently frozen layer of the subsoil that sometimes exceeds 300 m (1000 ft) in depth. Although the top layer of soil thaws in summer, the lower ice presents a barrier that water cannot permeate. Thus the surface is often poorly drained. Furthermore, the annual freeze–thaw cycle expands and contracts the soil, making construction of any kind difficult, as Figure 7-15 demonstrates. Because of the soil's instability, the Alaskan oil pipeline must be placed above ground, on pedestals, for 25% of its length. This prevents the heat of oil in the pipe from melting the permafrost and causing land instability that might damage the pipe.

Long-lasting snow cover has other effects. It reflects most of the small amount of radiation that reaches it, so that little is absorbed. It cools the air and contributes to the production of areas of high atmospheric pressure. It presents difficulties for human transport and

Dfb — Humid Continental Climate

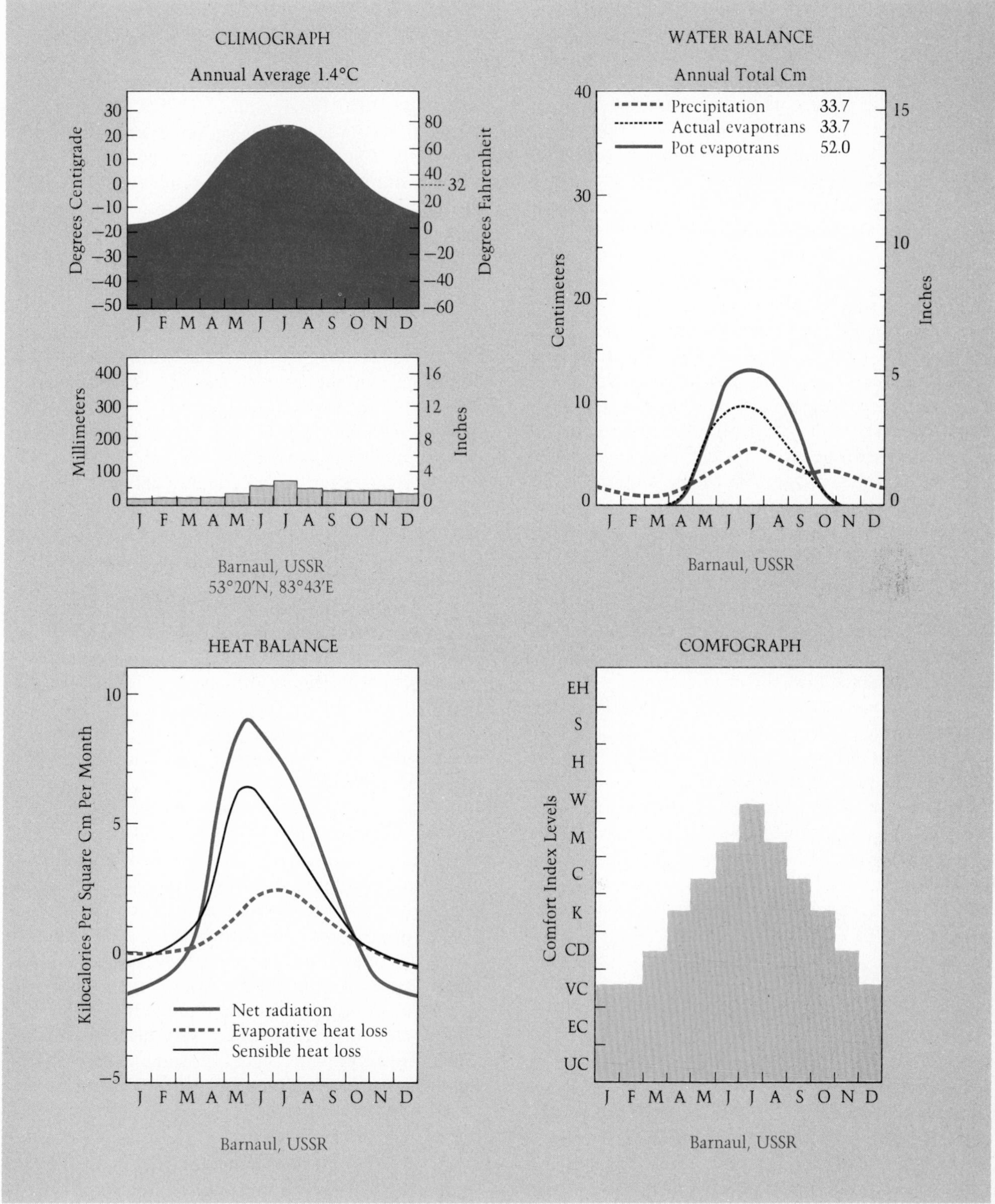

FIGURE 7-16
Details of a humid continental (*Dfb*) climate.

other activities, but paradoxically, it does keep the soil and dormant plants warm. Snow is a poor conductor of heat. Some measurements in Leningrad have shown that temperatures below a snow cover can be as high as −2.8° C (27°F)when the air on top of the cover is a bitter −41°C (−39°F). Humans sometimes use snow caves to survive these cold air temperatures, and the plants too make use of the principle every winter.

In the humid microthermal climates, most of the precipitation comes in the warmer months. This is not just because the warmer air can hold more moisture. In winter, large anticyclones develop in the lower layers of the atmosphere. Within the anticyclones, the air is stable, and the anticyclones tend to block midlatitude cyclones. In summer, convection in the unstable warmer air creates storms, so midlatitude cyclones can pass through *D* climates more frequently then. In some places there is also a summer monsoon season.

The features of this severe type of climate appear in the climatic values for Barnaul. Barnaul, in central Russia, is located where the western Siberian lowlands meet the foothills of the Mongolian Mountains. In every part of Figure 7-16, the extreme seasonal change is apparent. Average monthly temperatures range from 20°C (68°F) in July to −17.7°C (0°F) in January. The yearly average is 1.4°C (34.5°F). Most of the precipitation falls in summer, but it is not enough to satisfy the potential evapotranspiration rate.

The heat balance diagram shows the large amount of net radiation that allows the intense summer growing season. Lack of naturally available surface water means that only a little of this energy is used in evaporation in summer. Most of it warms the air as sensible heat. As in Turukhausk, the low winter temperatures are the result of net radiation deficits that are not balanced by a significant flow of sensible heat to the surface. The comfograph supports the remark about large wardrobes: Warm, mild, cool, keen, cold, and very cold conditions are experienced in this highly seasonal climate.

POLAR CLIMATES (*E*)

Beyond the Arctic and Antarctic circles, summer and winter become synonymous with night and day. Near the poles, there are six months of daylight in summer, when the monthly average temperature might "leap" to −22°C (−9°F). In winter, six months' darkness and continual outgoing radiation lead to the lowest temperatures and most extensive ice fields of the planet (see Figure 7-17). The lowest temperature ever published— −88° C (−127°F)—is for a chilly day at one of the highest stations in central Antarctica, the Russian station Vostok.

Polar climates (*E*) are defined as those where the average temperature of the warmest month is less than 10°C (50°F). There are two major kinds of polar climate. If the warmest average monthly temperature is between 10°C (50°F) and 0° (32°F), the climate is called *tundra* (*ET*), after its vegetation of mosses, lichens, and stunted trees. Practically all of the *ET* climates are found in the Northern Hemisphere; here the continental effect gives long bitter winters. Many of the *ET* areas border the Arctic Ocean, which provides a moisture source for frequent fogs when it is not frozen. In summer, poor drainage leads to stagnant water, the breeding grounds for flies and mosquitoes. Some arctic researchers fear mosquitoes more than cold. Precipitation, mainly from frontal midlatitude depressions in the warmer months, seldom exceeds 300mm (12 in) for the year.

In *icecap climates* (*EF*), we find the lowest temperatures throughout the year. In a year, about 90 mm (3.5 in) of precipitation, usually snow, falls onto the icy wastes. The inhospitable climate has made it difficult to collect data from these areas, but in recent years, international cooperation has made some data available.

The Russian station Mirnyi is in Antarctica just inside the Antarctic Circle. Its relatively low latitude hosts temperatures that are rather warm for icecap climates, but the average annual temperature of −11°C (12°F) is not high, and monthly averages rarely rise above freezing, as Figure 7-18 suggests. Year-round snow makes it almost impossible to obtain accurate water balance data.

But the Russians have measured the heat balance, and this vividly explains the frigid climate. During only about four months of the year is there significant positive net radiation at Mirnyi. More radiation leaves the earth than enters during the other months. One might imagine that the situation is even worse at higher latitudes. Sensible heat flow throughout the year is directed from the air toward the ground, the final result of the general circulation of the atmosphere moving heat toward the poles. Sensible heat and net radiation can provide energy for some evaporation in the warmer months, but in the winter, condensation of moisture onto the surface provides only a minor source of heat.

The comfort values in Figure 7-18 come from a station in the middle of the Greenland icecap that was called Eismitte by the hardy Scandinavians responsible for its establishment. The graph would be better designated an uncomfograph, because as you can see, the categories never rise above the cold level. During most of the year, Eismitte is extremely cold. The ultracold category of February requires, by definition, more than seven layers of clothing for protection. Scientists are now living in these climates, searching for oil and other secrets of the earth. But every one of them needs artificial heating, food, and water; without them human life could not exist in the coldest climate of the planet.

HIGHLAND CLIMATES (*H*)

This description of the mosaic of climates existing on the earth's surface would not be complete without some mention of the climates of highland areas. Highland areas reach into the lower temperatures and pressures of the troposphere. So as one moves up into a highland area, changes in climate mimic those of a latitudinal movement from equator to pole. Highland areas also possess other distinct characteristics.

One of the outstanding features of highland climates (*H*) is their distinct zonation by altitude. Nowhere has this been better shown than in the tropical mountain ranges of South America. Figure 7-19 combines the characteristics of many of these mountains. The foothills of the Andes lie in tropical rainforest climates in the east and dry climates tempered by cool ocean currents in the west. Above 1200 m (4000 ft), tropical climates give way to the subtropical zone. At 2400 m (8000 ft), the mesothermal climates appear, with vegetation reminiscent of that found in Mediterranean climatic regions. These in turn give way to microthermal climates at 3600 m (12,000 ft), and above 4800 m (16,000 ft), permanent ice and snow creates a climate like that of polar areas. It is sometimes said that, if one misses a bend at the top of an Andean highway, the car and driver will pass through four different climates before hitting the bottom.

The altitudinal zones mainly reflect the decrease of temperatures with height. But wind speeds tend to increase with height, as do precipitation, fog, and

FIGURE 7-17
A view of polar (*E*) wastes and the Admiralty Mountain Range in Antarctica.

EF — Ice Cap Climate

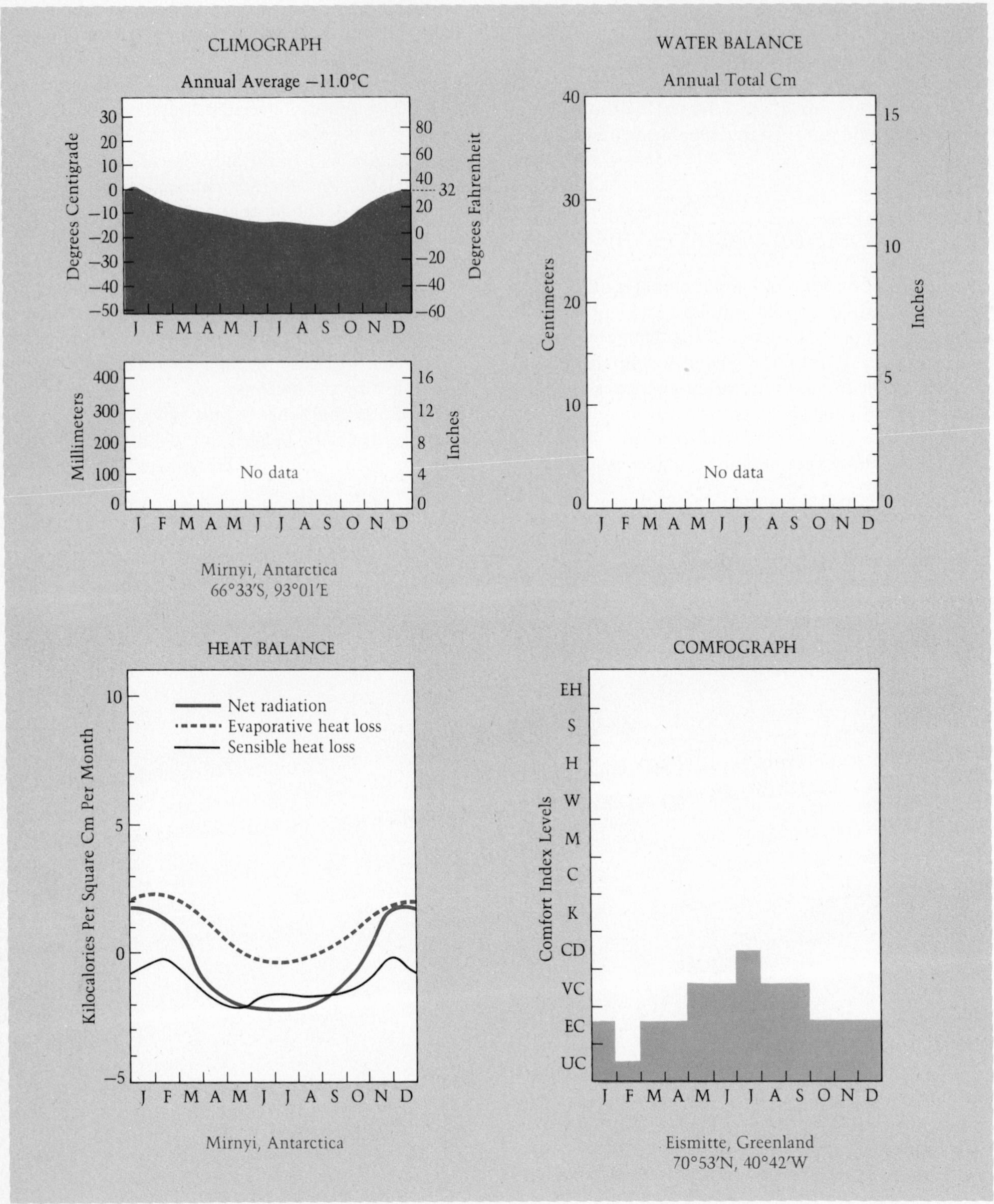

FIGURE 7-18
Details of an icecap (*EF*) climate.

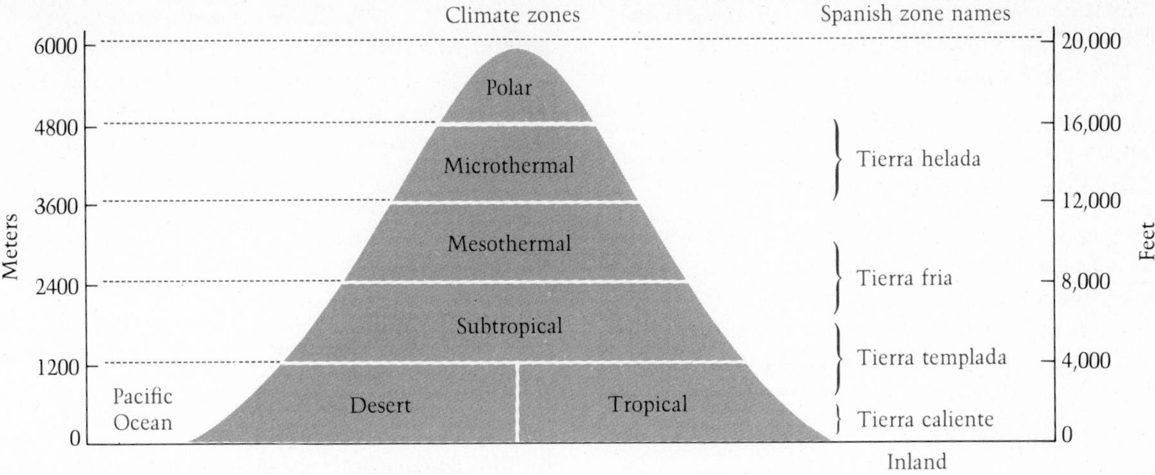

FIGURE 7-19

A hypothetical South American mountain, showing the changes of climatic zones with height and their corresponding Spanish names.

cloud cover in many cases. The radiation balance is markedly altered by altitude. Because less shortwave radiation is absorbed by the atmosphere in higher altitudes, greater values are recorded. This is particularly true of the ultraviolet radiation responsible for snow blindness and suntans of mountaineers and skiers. Where there are snow surfaces, much of the incoming radiation is reflected and not absorbed, also keeping the temperatures low.

The directions in which slopes face are called *aspects.* In highland areas the many different aspects receive variable amounts of radiation. Some slopes receive high values, whereas others obtain no direct shortwave radiation at all.

Other features contribute to the unique climate of every part of each highland or mountain area, including the general orientation of the mountains and the detailed topography of the hills. The vegetation covering the surface is also important. Finally, the climate of a particular highland area depends on its location with respect to the large-scale, global factors of climate, such as the general circulation of the atmosphere.

SUMMARY

A broadly simple pattern of climates occurs because of the heat and water exchanges at the earth's surface and because of the organization of the general circulation of the atmosphere. This pattern is altered in detail by the specific location of land and sea areas, ocean currents, and highland regions.

The resulting mosaic of climates may be classified in different ways. We have mainly followed the classification of Köppen, who divided the climates of the earth into six major types and a number of subtypes. Additional information has come from a human comfort classification.

The major climates are classified as tropical, dry, mesothermal, microthermal, polar, and highland types. In each, both vegetation and human activity adapt to the climate.

QUESTIONS

1. Climate is the overall pattern of weather expected in a particular place. Why does the formal definition of climate require the averaging of values of weather elements over a 30-year period? In this respect, how is climatology different from meteorology?

2. The problems in developing an adequate worldwide climatic classification scheme are very similar to those encountered in producing adequate map projections: Some compromise, based on the ultimate use of the classification, is called for. In map projections, the factors to be balanced are scale, area, and shape. What trade-offs must be made for a climate classification scheme?

3. Köppen's climatic classification system is based on the relationship between the types of plants at a particular place and the climatic characteristics of the place. What are the strengths and weaknesses of such an approach? In what kinds of studies would you expect the Köppen system to be most useful?

4. Using Köppen's climatic classification (Figure 7-1) how would you classify the following areas of the

United States: The Arizona desert? The northern Great Plains states? Southern Florida? The Olympic Peninsula of Washington? Massachusetts?

5. Do the same criticisms that were leveled against the Köppen classification system apply to the comfort index classification of climates? Why or why not?

6. How might you use the two classification systems presented in this chapter to predict the areas of greatest concentration of human beings in the world? What factors in the Köppen system would we expect to correlate with large populations? What factors on the comfort index scale? Are the two sets of factors related?

7. In the hypothetical continent of low and uniform elevation (Figure 7-3), what accounts for the *D* climates with moist, severe winters which occur in the Northern Hemisphere only?

8. What are the primary causes of the differences between the generalized picture of the world-wide climates in the uniform continent example and the actual distribution of climates on the globe?

9. In tropical rainforest climates, such as the one found at São Gabriel, Brazil, what factors influence the almost continually sultry climate? How would the heat balance change there if the water balance showed a continual deficit instead of a continual surplus?

10. How can we account for the extremes of temperature in the desert belts at 30° N and S on a daily and a yearly scale? Why are these extremes greater than those of the tropical rainforest areas?

11. We saw that in São Gabriel, Brazil, the comfograph was largely a product of the interaction of a continual water surplus and a huge amount of insolation. In the heat balance and water balance diagrams for Yuma, Arizona, how can you explain the high sensible heat loss curve? What is the low evaporative heat loss curve related to? What primarily controls the shape of the comfograph for Yuma?

12. What is the significance of humid mesothermal climates? Why are these climates especially suited for human habitation?

13. Describe the principal factors controlling the humid mesothermal climates, using the concepts of the variations in the jet stream, the influence of maritime winds, and seasonal variations in precipitation and insolation.

14. In the perpetually moist subtropical climate around Miyazaki, Japan, precipitation exceeds actual and potential evapotranspiration throughout the year. There is also a marked seasonal variation in insolation. Why is the sudden "spike" of precipitation in the summer months not compensated for by a decreased sensible heat loss curve and an increased evaporative heat loss curve? What is the controlling factor in the comfograph here?

15. What factors make the Mediterranean climate the "ideal" climate for people of European descent?

16. Although they receive large amounts of insolation during the long summer days, the high-latitude humid continental climates (*Dc* and *Dd*) are not particularly well suited to human beings. Why?

17. How are the winter low-lying cold air masses, with their associated high pressures, related to the microthermal climates found in Siberia and northern Canada? What effect would these winter anticyclones have on the precipitation at Barnaul (Figure 7-16)?

18. Explain the heat balance curves for Mirnyi, Antarctica based on the general circulation of the earth and what you know about the radiation balance in this area (Figure 7-18).

19. The north and south rims of the Grand Canyon in Arizona exhibit markedly different climates. The South Rim, most visited by tourists, is a full 350 m lower in elevation than the North Rim. The South Rim could be described as a desertlike chaparral landscape, with little precipitation, and moderate to hot temperatures. The North Rim, between eight and fifteen miles away across the Canyon, has an appearance more like alpine topography. What accounts for the radically different climates at these two locations that are actually right next to each other?

20. Imagine that a mountain with a shape like a four-sided pyramid sits in the Cascade Mountains in Oregon. One side of the mountain faces directly south, the other three face north, east, and west, completing the rectangular shape of the base. On which face of the mountain would you expect the tree line to be highest? Which side of the mountain would receive the most precipitation? If the mountain is snow-covered, which faces will have the greatest amount of snow left in July? Will these factors affect the climatic zones on each slope?

CLIMATE, PEOPLE, AND CITIES

CHAPTER 8

Over 75% of the inhabitants of the United States live in cities. In this chapter we must be concerned with the effect of climate on people and with the effect that people and cities have on climate.

This chapter begins by briefly specifying the relationship of the human body to its immediate surroundings. We will examine the concept of comfort and how it relates to elements of climate. To a large degree, our comfort depends on having adequate shelter. Humans have obtained shelter in small controlled-climate spaces ranging from caves to the modern house.

Modern houses tend to be grouped together in cities, and the cities have developed climates of their own. One of the most detrimental aspects of city climate is the concentration of the wrong things in the wrong place at the wrong time, a phenomenon called pollution. We will consider those aspects of pollution that occur in the atmosphere. In many cases, the air of urban areas can affect the surrounding countryside as well. When many cities are close together, their climates and their pollution can have a global effect.

We will move, therefore, from the individual person to the whole globe and inquire at the end what has been done about some of our self-made problems. A good way to start on this journey is to examine the human body as a heating and cooling system.

THE HEAT BALANCE OF THE HUMAN BODY

Chapter 4 introduces the concept of the heat balance with respect to the surface of the earth. The idea of examining the flows of heat energy to and from an object can usefully be applied to the human body. It is essential that the internal temperature of the body remain at about 37°C (98.6°F). Depending on the person, temperature fluctuations exceeding 3°C to 6°C (5.4°F to 10.8°F) result in death. The body therefore has to be kept within a limited range of temperature by regulating the flows of heat to and from it. Four types of heat flows can be altered: radiant, metabolic, evaporative, and convective. Figure 8-1 diagrams these flows.

First, we all live in a radiation environment. We receive shortwave radiation from the sun and longwave radiation from our surroundings— clothes, walls, the earth's surface, and so on. We also emit longwave radiation. Our radiation balance, the sum of ingoing and outgoing radiation, can be positive or negative according to the environment we are in. More often than not it is positive. So net radiation is usually a heat gain for us, especially during the day.

Another heat gain is the heat our bodies produce, called *metabolic heat.* Our bodies produce metabolic heat by changing the chemical energy in the food we

FIGURE 8-1
Energy flows to and from the human body.

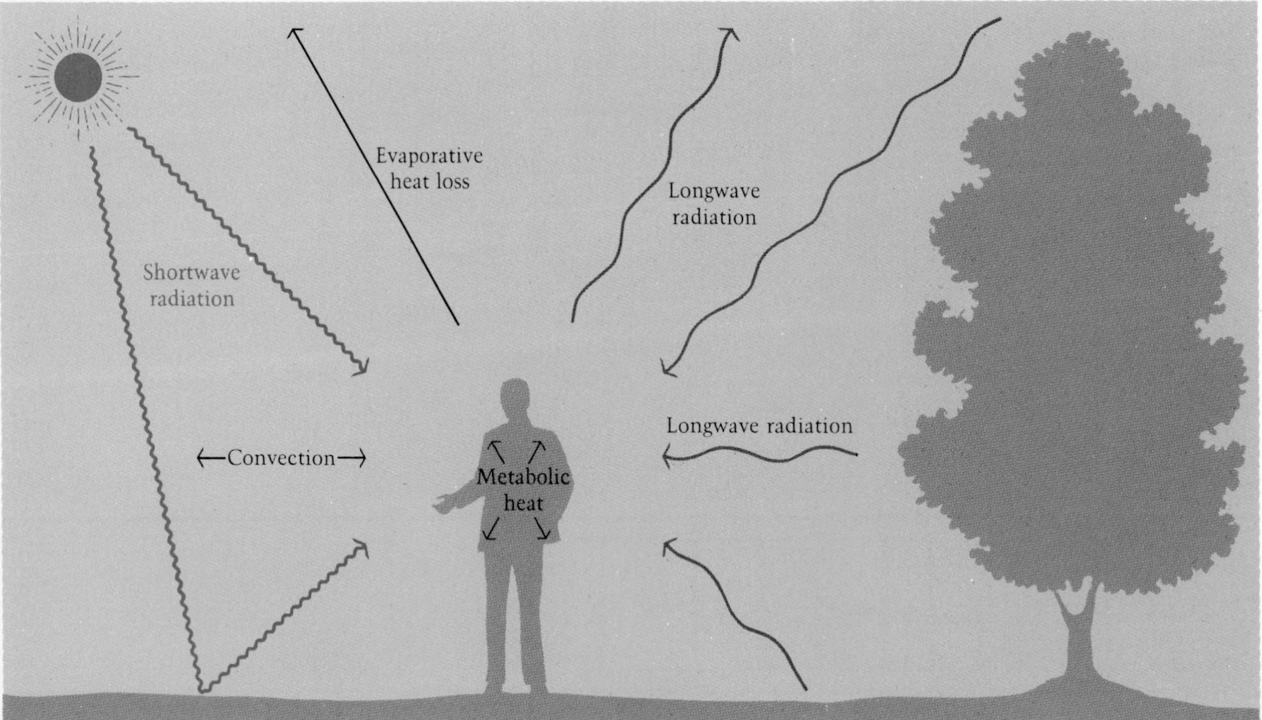

COLDS AND THE CALENDAR

Despite much evidence to the contrary, we still cling to the idea that temperature somehow relates to catching cold.

At the height of the recent energy crisis, some doctors predicted better health for Americans. They explained that lower temperatures in homes and stores would mean fewer colds for people moving from heated buildings to cooler outdoor settings.

A company marketing cold tablets pointed out our belief that temperature affects colds. They said, "a summer cold is a different animal." They defended the claim to the Federal Trade Commission by reporting that two-thirds of the people they surveyed believed summer colds are less serious but more persistent than winter colds. Therefore, they claimed, their advertisement was true.

In the Northern Hemisphere, colds increase from September to March. Doesn't this prove that colds correspond to cool temperatures? Perhaps we have tried to oversimplify the situation. Cold temperatures do stress the human body and may lower resistance to infections, but this does not automatically lead to the sniffles. Temperature changes and dry air in warm buildings also stress membranes in the body. In cold weather people spend more time indoors in groups. So the viruses have more opportunity to spread and thrive. Doctors still do not understand the common cold, but we may hypothesize that a combination of conditions including cool weather may lead to a ripe situation for an epidemic of colds.

assimilate into heat energy. The amount of metabolic heat we produce depends on, among other things, our age, activity, and environmental temperature. An older person at rest produces metabolic heat equal to that used to run a 75-watt light bulb. A five-year-old child produces the equivalent of 120 watts, and an active adult, about 260 watts. This rate could double if the adult were playing a game of tennis. We produce even more metabolic heat in colder temperatures. At 33°C (92° F) an adult creates 3100 cal per day, whereas at 0°C (32°F) an adult's metabolism creates 3930 cal per day.

Metabolic heat is always a heat source to the human body. The evaporation of water from the skin through perspiration is always a heat loss. Perspiration is a vital function, because some of the heat used in evaporating the perspired water is taken from the body and thus cools it down. One of the factors determining how much evaporation can take place is the amount of water vapor already in the surrounding air, as measured by the relative humidity. As we all know, we can feel cool on a hot day if the relative humidity is low, but we can feel distinctly uncomfortable, even at a lower temperature, if the relative humidity is high. You can see in Figure 8-2 that, if other heat losses and gains are kept constant, relative humidity acts as an index of how efficient our evaporative cooling system is. It could become a deciding factor between life and death.

Finally, a flow of sensible heat can act either as a cooling or a heating mechanism, depending on the relative temperatures of the body and air, by means of a *convective* flow. Hot air blowing onto our bodies makes us gain heat, whereas a cold wind leads to a rapid heat loss. When we breathe, we pass air to and from our lungs. This air may be warmer or colder than our lungs. For convenience you may regard the loss or gain of heat by breathing as a convective heat flow.

Our bodies consciously and unconsciously regulate these four types of heat flow so that body temperature stays within the narrow vital range. For example, if our environmental temperature changes over a short period of time from 40°C (104°F) to 10°C (50°F), the metabolic heat production might stay the same as

FIGURE 8-2

The limitation imposed on our evaporative cooling system by relative humidity.

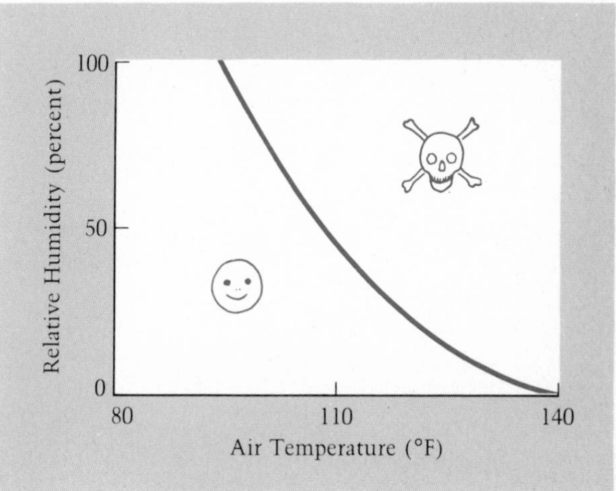

Figure 8-3 shows. But radiant and convectional heat flows change from being a slight heat gain to being a marked heat loss. Evaporative cooling of the body moves from a high level to a relatively low level in the cooler environment. Many of our feelings of comfort or discomfort and almost all of our artificial adaptions to the environment, such as clothing and shelter, are related to the balance of these energy flows that maintain our constant internal temperature.

COMFORT AND CLIMATE

The measurement of heat flows to and from the body is a difficult procedure. Consequently, scientists have attempted to develop simpler indexes relating our sensations of comfort to the more usually measured weather elements, such as temperature and humidity. Chapter 7 shows how these are used in a comfort index.

Another measure, called the *wind-chill index*, gives us an idea of how cold we would feel in given conditions of wind speed and air temperature. Although this index does not take into account evaporative heat loss and the amount of protective clothing we wear, the wind-chill index is closely related to the occurrence of frostbite. The index thus applies mainly to convectional heat loss. You can see from Table 8-1 that an equal amount of cold can be experienced at different combinations of wind speeds and air temperatures. You would begin to feel very cold if the temperature were −15°F (−26°C) in calm conditions or if the temperature were 35°F (2°C) with a wind of 15 mi per hour (6.7 m/sec).

CLIMATE, BEHAVIOR, AND HEALTH

Climate affects our behavior and health, as well as our comfort. It is not difficult to find cases where short-term atmospheric events definitely affected important decisions and subsequent behavior. During the critical stages of the American Revolution, in early January 1777, George Washington was frustrated because an unseasonable thaw had turned the roads to mud, preventing him from moving his army against the British at Princeton. On January 2 the wind shifted to the northwest, and within a few hours the roads were frozen. Washington immediately decided to take advantage of this, moved the troops, and won a vital victory by catching the British off guard.

Nor is it hard to find statistical correlations between the way people act and the changing weather. In one experiment the mistakes of wireless telegraph operators were recorded. At 28°C (85°F) the average number of mistakes per operator per hour was 12.0. At 30°C (87°F) the number was 15.3, at 38°C (100°F) it was 17.3, and at 41°C (105°F) the number soared to 94.7.

FIGURE 8-3

The adjustment of heat flows under changing environmental temperatures to maintain a constant internal body temperature.

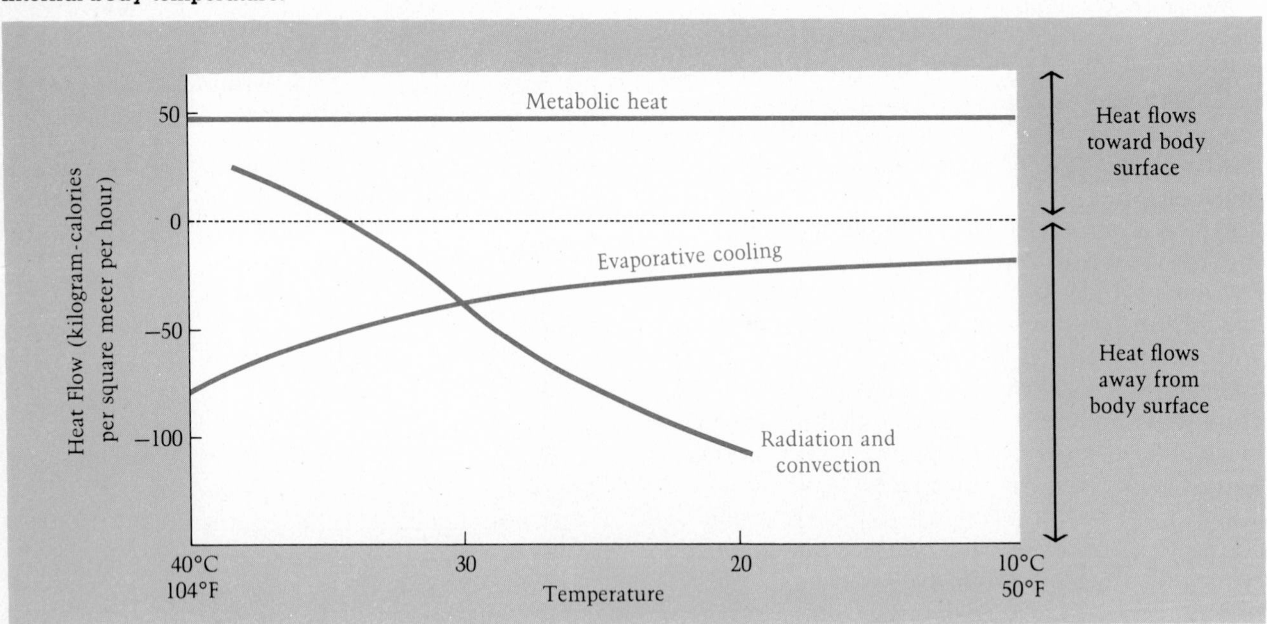

MPH	Dry-Bulb Temperature (°F) 35	30	25	20	15	10	5	0	-5	-10	-15	-20	-25	-30	-35	-40	-45
	WIND-CHILL INDEX (°F)																
Calm	35	30	25	20	15	10	5	0	-5	-10	-15	-20	-25	-30	-35	-40	-45
5	33	27	21	16	12	7	1	-6	-11	-15	-20	-26	-31	-35	-41	-47	-54
10	21	16	9	2	-2	-9	-15	-22	-27	-31	-38	-45	-52	-58	-64	-70	-77
15	16	11	1	-6	-11	-18	-25	-33	-40	-45	-51	-60	-65	-70	-78	-85	-90
20	12	3	-4	-9	-17	-24	-32	-40	-46	-52	-60	-68	-76	-81	-88	-96	-103
25	7	0	-7	-15	-22	-29	-37	-45	-52	-58	-67	-75	-83	-89	-96	-104	-112
30	5	-2	-11	-18	-26	-33	-41	-49	-56	-63	-70	-78	-87	-94	-101	-109	-117
35	3	-4	-13	-20	-27	-35	-43	-52	-60	-67	-72	-83	-90	-98	-105	-113	-123
40	1	-4	-15	-22	-29	-36	-45	-54	-62	-69	-76	-87	-94	-101	-108	-118	-128
45	1	-6	-17	-24	-31	-38	-46	-54	-63	-70	-78	-87	-94	-101	-108	-118	-128
50	0	-7	-17	-24	-31	-38	-47	-56	-63	-70	-79	-88	-96	-103	-110	-120	-128

Very cold · Bitterly cold · Extremely cold

TABLE 8-1
The wind-chill index, a measure of cooling power on exposed flesh. Wind speeds over 40 miles per hour have little additional chilling effect.

Each operator's efficiency was drastically reduced by rising temperatures.

In another experiment, the records of British Navy sick lists showed that, on days when the average temperature was 16°C (60°F), some 3% of the British sailors were reported sick. At 27°C (80°F) 5% reported sick, and at 32°C (90°F), 9%. After air conditioning was widely installed, the sick lists shrank dramatically. However, in these experiments, especially the latter, social influences may also be at work.

Behavior is also affected by the passage of a midlatitude cyclone. Studies have shown that, as pressure falls, a whole range of abnormal behavior occurs. General irritability increases. Violent acts and accidents of all kinds are more frequent. Reaction times become slower, and industrial production falls. Forgetfulness and irresponsibility rise. The rate of attempted suicide increases. There are more headaches, fainting spells, nosebleeds, and cases of appendicitis, depression, and nervous disorder.

One of the more spectacular new discoveries along these lines is the effect of ions on humans. The formation of ions begins in the air when enough energy acts on a gas molecule to eject the part of an atom called an electron. The displaced electron attaches itself to an adjacent molecule, which then carries an electrical charge and becomes a negative ion. The original molecule remains a positive ion. People living in cities inevitably breathe ion-depleted air for large portions of their lives. There is increasing evidence to show that this leads to discomfort, enervation, and the loss of mental and physical efficiency. An overabundance of positive ions can also have negative physiological and psychological effects, and this is what happens in the case of the föhn-type winds such as the Santa Ana.

Today scientists are attempting to replace mere speculation and statistical correlation with a complete knowledge of the actual links between weather and the human body. When astronaut Richard Gordon made his space walk from Gemini 11, the temperature inside his suit rose to 43° C (109°F), and humidity reached

100%. His pulse rate went to 180, compared to 120 for a healthy adult. Gordon had to be pulled back because he became unable to perform even the simplest tasks. Problems like this are forcing the National Aeronautics and Space Administration to inquire still further into the effect of climatic environment on people.

SHELTER, HOUSES, AND BUILDINGS

Ever since humans first felt the effects of sensible heat loss by winds or an increased evaporative cooling when they were rained upon, they have sought some form of shelter. The form of shelter depended to some degree on the most prevalent features of the climate and on the available building materials (see Figure 8-4). In low-latitude climates, prehistoric people used the bones of animals or branches thatched with leaves to form huts. In midlatitudes, caves often provided the handiest shelters, and at high latitudes the Eskimos, even to this day, have developed their homes from the building material most readily available. The first settlers on the Great Plains protected themselves against the severe plains winters in dwellings dug out of the ground and roofed with grass sods. These houses were not too different from those built by Eskimos in northern Alaska. Climate still plays a large role in some of the design features of our dwellings. In hot climates, where there is a need to promote convectional cooling, Arabian nomads roll up the side walls of their tents. In India, screens are aligned to catch the effect of any cooling breeze.

Now let us consider the houses with which you are probably more familiar. In dealing with the human body, we used the concept of heat balance. The same concept may be extended to the average modern house. As in the case of the body, there are four flows of heat to and from the roof and walls of a house. Instead of metabolism, the house contains an *artificial heating system*. This often takes the form of open fires, but in more-developed countries, it is usually a central heating system. An artificial heating system usually channels heat from the central interior toward the walls and roof of the house. The *evaporative cooling* part of the standard heat balance is of least significance, because a house is designed with a sloping, roof that removes surface water as quickly as possible. Most houses have the important function of reducing the amount of *convective cooling* by wind. They provide shelter from the wind, and facilitate control of the inside climate by the heating system.

Radiant heating and cooling applies to the outside of a house much as it applies to the body, and architects take it into consideration in several ways. If they want to use radiation for heating the house, by means of the greenhouse effect (see Chapter 4), they can ensure that the windows of the buildings are open to the direct rays of the sun. Sometimes they use a partial shade to cut out direct rays coming from high sun angles in the summer, when heating is not required. Rays from low sun angles bypass the shade and provide winter heating. In addition, commercially available solar heating systems can provide hot water and hot-water radiant heat for the entire house.

By adjusting their designs to the operation of the four factors of the heat balance system, architects can provide our houses with a small environment perfectly adjusted to the needs of our bodies. When portability

FIGURE 8-4

Some human shelters. a) African natives in leaf-thatched wooden huts.

b) An Eskimo igloo.

and social convention require, our clothes meet similar needs. In many ways housing can be thought of as an extension of our clothing.

CITY CLIMATES

The marvels of modern architectural engineering and design work well, for the most part, as small, individual open systems. But in today's world, where at least 25% of the population lives in cities, individual buildings do not function as single entities but as a group.

MASS, ENERGY, AND HEAT IN THE CITY

The city is an extraordinary processor of mass and energy. It has its own metabolism. A daily input of water, food, fuels, and energy of various kinds is matched by an output of sewage, solid refuse, air pollutants, and energy and materials that have been changed in some way. The quantities involved are enormous. Every day, directly or indirectly, the average city dweller uses about 600 ℓ (156 gal) of water, 2 kg (4.4 lb) of food, and 8 kg (17.6 lb) of fossil fuel. This is converted into roughly 500 ℓ (130 gal) of sewage, 2 kg (4.4 lb) of refuse, and almost a kilogram of air pollutants. Multiply these figures by the population of your town, and you will get some idea of what a large processor it is, even without taking into account the consumption and output of its industry. Many aspects of this energy use affect the atmosphere of the city, particularly in the production of heat.

In winter the values of heat produced by a town can approach, or in some cases surpass, the amount of heat available from the sun. All the heat that warms a house or other building eventually diffuses to the surrounding air. The process is quickest where houses are poorly insulated. But an automobile produces enough heat to warm an average house in winter. And if a house were perfectly insulated, a human would also produce more than enough heat to warm it. Therefore, apart from any industrial production of heat, an urban area tends to be warmer than the surrounding countryside.

The burning of fuel such as in an automobile is called *combustion*. The increase in heat is not the result of combustion alone. Two other factors are at play. The first is the heat capacity of the materials used in city buildings. During the day, heat from the sun can be conducted into this material and stored—to be released at night. This contributes to cities' higher overall temperatures. In the surrounding countryside, materials have a relatively low heat capacity, and a grass or vegetative blanket keeps heat from easily flowing into and out of the soil.

A second factor is that radiant heat coming into the city from the sun is trapped in two ways. It is trapped by a continuing series of reflections among and within the large number of vertical surfaces that the buildings present. It is also trapped by the pollution dome that most cities acquire. Just as in the greenhouse effect, shortwave radiation from the sun passes through the dome more easily than outgoing longwave radiation does. Longwave radiation is absorbed by the constituents of the dome and reradiated back to the city.

There are therefore numerous reasons why the city will be warmer than its surrounding area. These reasons all together lead to a phenomenon known as the *urban heat island*.

c) Eskimo sod house, northern Alaska.

d) The wind scoops of Hyderabad, India.

FIGURE 8-5
The Queen Elizabeth Arboretum in the foreground, with the Vancouver heat island in the background.

URBAN HEAT ISLANDS

If we regard isotherms (lines of equal temperature) as analogous to contour lines on a map, then in certain conditions the distribution of temperatures within a city gives the impression of an area of high land or an island of higher temperatures set on a more homogeneous plain. The *urban heat island* is a concept whose existence has been established for most large cities and many smaller cities of the world.

Heat islands develop best under the light wind conditions associated with anticyclones. But in large cities, they can form at almost any time. Temperatures are usually highest in the downtown area of a city and where there is heavy industry. Temperatures decrease toward the suburbs and often decline sharply where built-up land changes to rural land. The precise form of the islands depends on several factors. The island can be elongated away from the prevailing wind. Pools of cold air can be found over any extensive parkland within the city (Figure 8-5). Sometimes tongues of cold air follow the course of major rivers or river valleys. The heat island extends vertically as well. In San Francisco increased temperatures were found to extend about three times the height of the buildings.

When the island is well-developed, local variations can be extreme. In winter, busy streets in most cities can be 1.7°C (3°F) warmer than the side streets. The areas near stoplights can also be 1.7°C (3°F) warmer than the areas between them, because of the effect of idling cars. Spot temperature ranges of up to 26°C (47°F) have been measured within the winter environment of Madison, Wisconsin. In small towns the existence of something like a large dairy factory can also be detected in the form of higher temperatures.

The maximum difference in temperature between a matched urban and rural environment is called the *heat-island intensity* for that region. In general, the larger the city, the greater its heat-island intensity. The actual size of the intensity depends on such factors as the population density, structure, and function of the cities. For example, statistics averaged over a long period show that the average annual temperature in London is 1.4°C (2.5°F) greater than that of the surrounding countryside.

The consequences of higher temperatures in urban areas should not be neglected. During July 1966, there was a heat wave in parts of the United States. This was felt most in the cities. In St. Louis, deaths attributed to heat rose dramatically in temperatures above 32.2°C (90°F). When the temperature reached 32.8°C (91°F), 11 people died, but the figure rose to 73 when the temperature reached 35°C (95°F). The total death toll from heat in this heat wave reached 500 in St. Louis and over 1100 each in New York City and the state of Illinois. City parks could have lowered the temperatures by an apparently critical three degrees. Urban planners are beginning to realize the importance of parks for reasons that are not purely social.

CITY SURFACE

The presence or absence of moisture is affected by the unique features of the city surface. Furthermore, in most cities about 50% of the surface is impenetrable by water. In downtown Detroit, 47% of the land surface is devoted to roads alone. Roofs, streets, and parking areas all cause even gentle rain to run off almost immediately. Rapid runoff leaves many city surfaces dry for most of the time between rainfalls, and little water is available for the cooling process of evaporation. The city thus presents somewhat of a desert in an area where soil moisture may be plentiful. Consequently, city air is drier, and relative humidities are usually lower.

Wind is another item that is affected by the city surface. Buildings increase the amount of friction on air flowing over a town. Wind speeds are 5% lower in central London than in the rural surroundings. Furthermore, 30 m (100 ft) above the ground, the wind in a city may be as much as 80% slower. Winds are therefore less efficient in dispersing pollutants when they travel over cities. Friction also alters the wind direction so that geostrophic or gradient winds (see Chapter 5) occur almost twice as high over a city as over the surrounding area. Although wind speeds in the city are generally reduced, in some cases local turbulence increases. High-velocity wind eddies on street corners and increased wind speeds from the channeling effects of streets and tall buildings are common experiences.

DUST DOMES

Other peculiarities of city climates originate in the modified air over the urban landscape. There are large quantities of dust and gaseous pollutants in the city atmosphere, so much so that the city is often covered by a *dust dome*. Within the dome, dust is sometimes caught in the wind circulation shown in Figure 8-6. Dust domes have notable effects on both the radiation and the atmospheric moisture of cities.

FIGURE 8-6
Wind circulation in an urban dust dome.

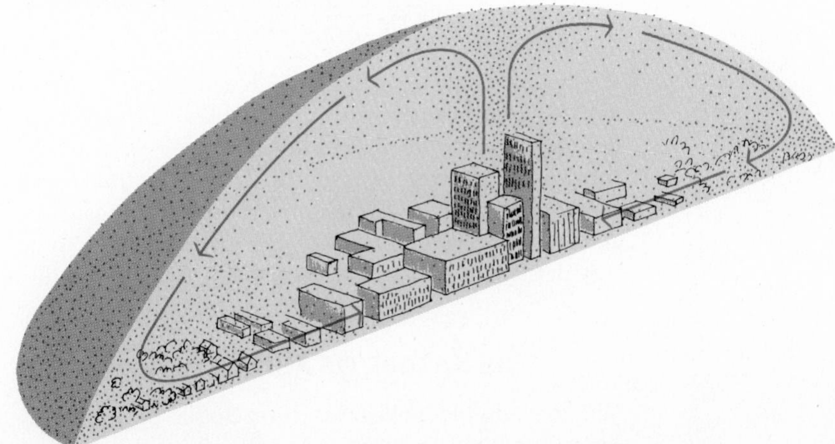

All parts of the radiation balance discussed in Chapter 4 are altered in the urban environment. The interception of incoming shortwave radiation is most apparent during the winter when the pollution is worse and the sun's rays are striking at a lower angle. In London, 8.5% of the direct solar radiation is lost when the sun's elevation is 30°, but 12.8% is lost with an elevation of 14°. On very cloudy days, 90% of this radiation can be lost. Consequently, the number of hours with bright sunshine is reduced. Outside London, the sun shines an average of 4.33 hours per day, and it shines 3.6 hours per day in the center of the city.

In addition, the loss of radiation is not evenly distributed through the solar spectrum. More ultraviolet light is lost than light of other wavelengths. It is the ultraviolet light in sunshine that tans us, so the Caucasian urbanite is always going to appear pale in comparison with a person from the countryside.

How much of the incoming shortwave radiation is absorbed by the city surface depends on the albedo of the surface. This in turn depends on the actual materials used in construction and varies from city to city. English cities are usually made of dark materials or ones that have been blackened by smoke. They have a lower albedo (17%) than agricultural land does (22% to 24%) and therefore absorb more radiation. In contrast, downtown areas in Los Angeles have lighter colors and therefore show a higher albedo than the more vegetated residential zones.

Scientists know little about the longwave radiation in cities, but preliminary studies show that both upward and downward longwave radiation increase. The downward flow increases more than the upward flow; increased longwave radiation can offset the decreased shortwave radiation. The result is a net radiation that is not too different from surrounding country areas. Of the net radiation arriving at the city surface, some studies suggest that about 80% of it is lost as sensible heat warming the city air and that the rest mainly acts as "soil heat flow" to warm the city materials. Very little heat is used in evaporative cooling.

We must also investigate the modified moisture content of the urban atmosphere. In the urban dust dome, there are many more particles that can act as condensation nuclei than there are in rural areas. Although little water vapor rises from the city surface, horizontal flow in the atmosphere brings just as much moisture to the city as to the country. Because of the greater number of condensation particles over a city, there is a greater propensity for condensation and the formation of small particles of liquid mist, fog, and cloud. As a result, fog is more frequent in cities than in surrounding areas. In Manchester, England, winter fogs are twice as frequent as in the surrounding country. Visibility in towns thus decreases. These effects increase when there are local moisture sources, such as rivers or lakes, within the city. Rainfall can be increased by the presence of a city. Most towns seem to have about 10% more rainfall than the surrounding area, but some, such as the urban area of Champaign-Urbana, Illinois, show larger increases, as Figure 8-7 shows. In St. Louis, where we find the same effect, the cause appears to be partly the greater turbulence in the urban atmosphere caused by hot air rising from the urban surface.

The most important climate changes in urban areas are summarized in Table 8-2. Many of these changes are directly attributable to the pollutants that cities discharge into the air.

FIGURE 8-7

Average yearly precipitation (in inches) in Champaign-Urbana, Illinois, 1949–1967.

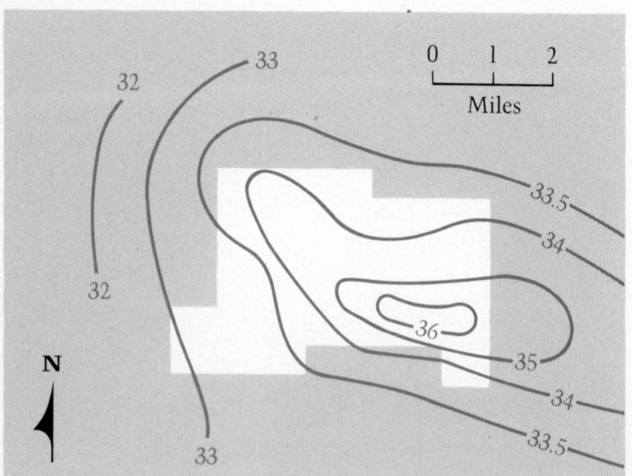

AIR POLLUTION

We may say that air is polluted when its composition departs significantly from its natural composition, defined in Chapter 3, of such gases as oxygen and nitrogen. However, we might call cigarette smoke pollution but not the smell of a charbroiled steak. We are therefore concerned with factors that are in some way detrimental to, or uncomfortable for, human life.

THE NATURE OF AIR POLLUTION

We may divide pollutants into two types. *Primary pollutants* may be gaseous or solid. They come from

TABLE 8-2
The effect of cities on climatic elements.

ELEMENT	COMPARISON WITH RURAL ENVIRONMENT
Radiation	
Global	15% to 20% less
Ultraviolet, winter	30% less
Ultraviolet, summer	5% less
Sunshine duration	5% to 15% less
Temperature	
Annual average	0.5°C to 1.0°C more
Winter low (average)	1°C to 2°C more
Heating degree days	10% less
Contaminants	
Condensation nuclei and particles	10 times more
Gaseous mixtures	5 to 25 times more
Wind speed	
Annual average	20% to 30% less
Extreme gusts	10% to 20% less
Calms	5% to 20% more
Precipitation	
Totals	5% to 10% more
Days with less than 5 mm	10% more
Snowfall	5% less
Cloudiness	
Cover	5% to 10% more
Fog, winter	100% more
Fog, summer	30% more
Relative humidity	
Winter	2% less
Summer	8% less

industrial and domestic sources and the internal-combustion engine. The main gaseous primary pollutants are carbon dioxide, water vapor, hydrocarbons, carbon monoxide, and oxides of sulfur and nitrogen, especially sulfur dioxide and nitrogen dioxide. The pollutants toward the end of this list are sometimes called "status-symbol pollutants," because they are particularly associated with the industrially developed countries. The principal solid primary pollutants are iron, manganese, titanium, lead, benzene, nickel, copper, and suspended coal or smoke particles. Except where coal or wood is burned in homes, these pollutants come mainly from industrial sources.

Secondary pollutants are produced in the air by interaction of two or more primary pollutants or by reaction with normal atmospheric constituents. There are two kinds of secondary pollution. The first is the *reducing type*. An example of this is when sulfur dioxide changes to sulfur trioxide in combustion or in the atmosphere. The sulfur trioxide then joins with atmospheric water to form drops of sulfuric acid. Sulfuric acid is corrosive, irritating, and can attract water, thus enhancing the development of rain droplets. Sulfuric acid droplets are also noted for their tendency to attract an oily film.

A more recently recognized type of secondary pollution is called the *oxidation type*. In this type, photochemical effects, the effects of sunlight, play a role. Nitrogen dioxide, which gives the brown color to urban dust domes, reacts with sunlight to form nitrogen monoxide and one odd oxygen atom. The oxygen atom can then join normal oxygen (O_2) to form ozone (O_3) that acts as an irritant.

The initial vertical and horizontal distribution of pollutants depends on the location of their source. Any further spread of air pollution is associated with two main factors. The first is the stability of the air and its

RESCUING THE AIR OF LOS ANGELES

One gasping resident of Los Angeles has suggested blowing away the smog with giant fans on the surrounding hills. Southern Californians have concocted grandiose schemes to free the city from its smoggy veil. Desperate situations require desperate solutions, but would any of the proposals work?

People who hope to blow away the smog apparently do not realize that operating the fans would require one-sixth of the United States' electrical power for one year. What about destroying the temperature inversion so the pollution could escape? Some inventive Californians would do this by using heaters to create hot air currents. The catch is that the heaters would consume energy equivalent to all the crude oil processed by the city's refineries in 12 days. Other people believe that the inversion could be dissolved with air turbulence generated by helicopters. How many helicopters would be needed to hover over the city? 40,000. The only low energy proposal calls for a reduction of incoming sunlight. Cutting down the incoming rays would decrease the photochemical changes that produce more pollutants. But how can the sunlight be screened? One citizen suggested covering the city with a layer of white oil smoke. For Los Angeles, the cure may be worse than the disease.

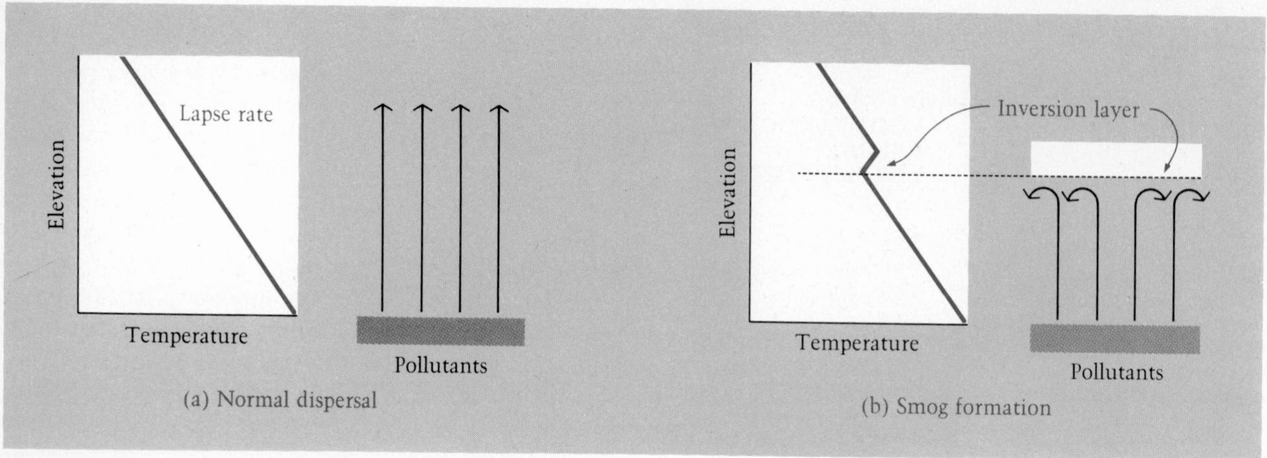

FIGURE 8-8
The effect of a temperature inversion on the vertical dispersal of atmospheric pollutants. a) A normal dispersal; b) the formation of smog.

power to allow vertical mixing. The second combines air stability and the flushing effect of the horizontal wind.

Chapter 3 introduces the concept of atmospheric stability. An individual parcel of hot air rises as long as its temperature is higher than that of the surrounding air. When the parcel reaches a place where its temperature is less than that of the surrounding air, it stops and often returns to the surface. Figure 8-8 shows how the temperature structure affects the vertical mixing of pollutants.

Because of the nightly cooling of the earth's surface and the atmosphere near the ground, it is common for temperature inversions—warm air over cold air—to occur in early morning over both city and countryside. These inversions, which form an atmospheric "lid," can be broken down by rapid heating of the earth's surface or by windy conditions. Without these conditions, air pollutants are trapped in the urban environment. Cold-air drainage at night and large-scale subsidence of air can also contribute to pollution trapping, as residents of Los Angeles should know. Cold air drains toward Los Angeles from the surrounding hills, and high-level inversions at 1000 m (3300 ft) are typical of the eastern end of the Hawaiian anticyclone, which affects Los Angeles weather.

Horizontal flushing by winds can help relieve air pollution. *Air-pollution potential* can be estimated by calculating the vertical range of vigorously mixed pollutants and the average wind speed through the mixing layer. Figure 8-9 shows the average number of days per year in the United States when inversions and light wind conditions create a high air-pollution potential. On the average, it appears that ventilation conditions are best in the Northeast, Florida, and the Midwest and poorest in the eastern and western states. Air-pollution potential alone might be regarded as a matter of climatological luck. But the frequency of pollution sources also determines the level of air pollution.

Air pollution is an old, prevalent, and costly problem. An early treatise on London's air pollution was written in the seventeenth century. An oft-quoted death toll of 4000 in London in December 1952 and similar tragedies in Belgium and Donora, Pennsylvania are evidence of air pollution's dangerous impact. New figures published almost daily indicate the large increases in respiratory disease that accompany air pollution. The costs of air pollution are difficult to estimate accurately but are assuredly astronomical. Some years ago it cost $2 million to reface the New York City Hall, and the cost of deterioration of the city's public and private property due to air pollution has been estimated at $520 million per year.

GLOBAL AIR POLLUTION

The problem of air pollution is even more complex. Just as individual houses tend to interact in a city, so cities have effects beyond their immediate vicinity, contributing to pollution on a world scale. We have already become familiar with urban dust domes (see Figure 8-10). When the wind is greater than about 13 km per hour (8 mi per hour), the dome streams out from the city across the surrounding countryside and becomes a *plume*. The plume beyond Chicago has been observed to stretch to Madison, Wisconsin under a southerly wind, a distance of 240 km (150 mi). Cities

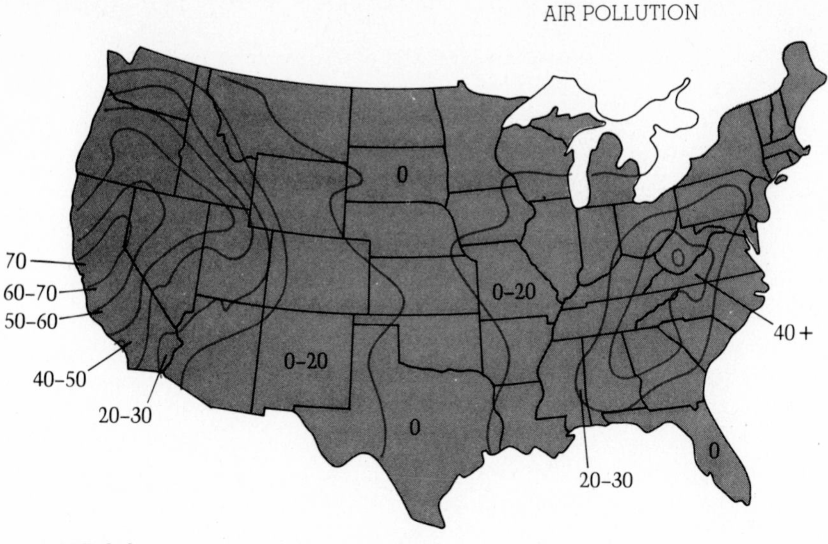

FIGURE 8-9

Air-pollution potential in the contiguous United States, in terms of the number of days per year with light winds and inversion conditions.

FIGURE 8-10

The polluted urban dust dome of Denver, Colorado.

around the world—from Fargo, North Dakota to Hong Kong—are affected by the urban plumes from other cities.

The plume brings with it many of the climatic effects associated with cities. The amount of short-wave solar radiation received on a clear day in the environs of Toronto depends markedly on which way the wind is blowing. Plumes affect more than the pollution and radiation levels of the immediate surroundings. La Porte, Indiana is 48 km (30 mi) downwind from the Chicago-Gary industrial area. The year-to-year variation of precipitation at La Porte appears to be related to the production of steel and the number of smoke-haze days in Chicago-Gary. The amount of precipitation and the frequency of thunderstorms and hail at La Porte has risen dramatically since 1925.

Another factor complicating the problem of global pollution is that many of our cities are concentrated into certain geographical areas. The cities of the eastern seaboard of North America and some of those in western Europe are so close that their urban plumes act together rather than individually. These areas are so large that they act as pollution sources on a world scale. Thus the release of sulfur dioxide over Britain leads to an increase in the acidity of rainfall in Sweden.

Unfortunately, cities are not the only sources of pollution. Many industrial activities result in both isolated and widespread pollution sources away from urban areas. Quarrying, mining, and pulp milling pollute from the location of their raw materials. Because of the danger of fallout, a particularly hazardous form of air pollution, the testing of nuclear weapons must be performed in remote areas. Pesticides, widely distributed by the farming industry, often pollute the air. Pesticides appearing in the West Indies seem to have traveled from Europe or North America. The vapor trails from jet aircraft have been accused of causing widespread cirrus clouds in some areas. Our communities have accentuated natural pollutants by concentrating fires, pollen, bacteria, and blowing dust. An increase in pollution over northwestern India and Pakistan during a time of local warfare has been attributed to tank traffic over the dry, dusty landscape. Air pollutants are everywhere.

Nobody is sure that the atmosphere can cleanse itself of these new sources of pollution in addition to the natural sources, such as volcanoes, that it has always handled. So we have become more concerned with helping the atmosphere. We have given the greatest attention by far to lowering urban air pollution. One common measure is to filter pollutants at their source. Traps are used in industrial smokestacks, and various methods of recycling or filtering exhaust from automobile engines have been developed. Industrial activities are carefully scheduled to avoid the emission of smoke and other pollutants when the atmosphere is stable and has a low dispersal potential. Legislatures are trying to ensure, among other things, a greater use of "clean" fuels. As a result of the clean-air act in London, the frequency of medium thickness fogs [visibilities in the range of 200 to 1000 m (650 to 3300 ft)] has become much less. We are also learning more about the natural ways the atmosphere cleans itself. These include the normal fallout process under gravity, where the size of the pollutant particles is critical, and the process called *scavenging*, the washout effect achieved by precipitation.

The effects associated with the urban plume have attracted a large amount of interest. Teams of US scientists are intensively concerned with identifying the effects and physical processes of the urban atmosphere and its plume. But on the global scale, little has been done so far. There have been suggestions for global monitoring of pollutants, and there is a certain degree of international cooperation in regard to atmospheric nuclear testing. In this chapter we have not dealt with all types of pollution in the atmosphere, and we have not touched on pollution problems in the other major spheres of our planet. In view of this, global pollution becomes an even greater problem.

SUMMARY

This chapter started with a discussion of the four flows of heat to and from the human body. These are the radiant, metabolic, evaporative, and convective flows.

The human body attempts to adjust these flows for optimum comfort. But optimum comfort is elusive, and people's behavior and health is often affected by changing weather conditions.

Shelters, houses, and buildings act in the same ways as clothing to adjust heat flows. They provide a small space that is comfortable to the human body.

When many houses and buildings are clustered together, as in a city, a special climate may develop. The city climate varies in many ways from that of the country. Urban heat islands and air pollution are cases in point. Air pollution from cities is one of the major causes of pollution of the whole atmosphere.

QUESTIONS

1. In what ways can the human body's heat balance be compared to that of the planet earth as a whole? Are there aspects of one system that are not found in the other?

2. The human body maintains a normal body temperature of 98.2°F. How does it compensate for low environmental temperatures? For high environmental temperatures?

3. How do nomadic tribes survive in the scorching desert environments of northern Africa? Could inhabitants of the tropical rainforest in Brazil survive if temperatures reached the extreme of 136°F found in the Libyan desert? Why or why not?

4. January and February mean temperatures for St. Paul, Minnesota are 12°F and 15°F respectively. People living in this area, however, often wear winter clothing as if temperatures were 10 or 15 degrees below 0°F. How would you explain this degree of overcompensation?

5. In many Latin American and Mediterranean cultures, the midafternoon siesta, during which commercial and social life halts while people rest, is an important feature of daily life. What physical factors explain the popularity of the siesta in many different cultures? Aside from historical and political reasons, why hasn't the traditional siesta become a feature of northern European and American life?

6. Shelters are needed in almost all of the climates found on the earth at some time during the year. The form of the shelter usually depends on two factors—the availability of building materials and the dominant climatic elements found in a particular location. From this basis, explain the significance of the Eskimo's igloo and the sod hut of the settler on the Great Plains. What is the major difference between these two dwellings? How do their different shapes and different materials help in dealing with the climates found in each location?

7. As an architect, you have been asked to design a house appropriate for a semiarid climate in which there are fairly large diurnal changes in temperature, from a low of 40°F at night, to a high of 105°F during the daytime in the summer. Using the four flows of heat corresponding to those of the human body, explain how your concern for the climatic factors will influence your choice of roofing material, window area, artificial heating system, roof shape, and house color.

8. On hot summer nights it used to be common in the 1920s for city dwellers to drive into the countryside to seek relief from the heat. What factors made the countryside, which then was much closer to urban centers, so much cooler than the cities?

9. Chicago and the southern shore of Lake Michigan are one of the major American "strip cities." The area from Gary, Indiana to South Chicago is almost entirely filled with industrial activity, including steel mills, dock facilities, light and heavy manufacturing, rail transportation, and freeways carrying Chicago-bound commuters and heavy truck traffic. The area north of Chicago includes some industry but is mainly suburban residential. Describe the effect these factors would probably have on the shape of the heat island of this region. Would the shape change at different times of the day? If so, how?

10. There are many practical reasons to place conventional coal and oil-fired power-producing generators near the major cities they serve. Access to fuel is often more economical, since major transportation networks already exist. The transport of electricity over short distances is efficient. With all these good reasons to place power plants near cities, what characteristics of the urban heat island make such placement really impractical?

11. In spite of the fact that cities have an average of 10% more rainfall than the surrounding rural areas, relative humidities in the cities tend to be lower. How would you explain this contradiction?

12. How does the shape of the dust dome found above many cities correspond to the shape of the heat island of the city?

13. The mean duration of sunshine is 4.33 hours per day outside the city of London, compared with 3.6 hours per day in the center of the city in the winter months. Given your background from this chapter, why is the surrounding countryside not warmer than the city?

14. Why is the amount of ultraviolet radiation received in the center of Chicago radically different in summer and winter? Would we expect to find this much of a seasonal difference in Mexico City? Why or why not?

15. Large cities in areas of the world that are not heavily industrialized can still have severe pollution problems. What differences would you expect to find in the type of pollution in Cleveland, Ohio and that in Nairobi, Kenya?

16. Pollution is not only an urban problem. The Central Valley of California, flanked by the Sierra Nevada on the east and the Coast Mountains on the west, develops dense radiation fogs at night, which settle in the low-lying parts of the valley due to cold-air drainage. The valley also has severe pollution problems, and agricultural burning is only allowed on

certain days during the year. What causes the pollution in the valley? What climatic conditions are required on days when burning is allowed?

17. According to Figure 8-9, San Francisco has a higher air pollution potential than Los Angeles, yet Los Angeles is probably the smoggiest city in the country. What accounts for this paradox?

18. What health and economic effects does air pollution have? Would we expect to find different effects of air pollution in industrial and nonindustrial areas?

19. What effects does the interaction of prevailing winds and urban dust domes have on local weather patterns? On global weather patterns?

20. Is pollution less of a long-term problem for those countries that are not highly industrialized? Why or why not?

CLIMATE:
VARIATION
OVER
TIME

CHAPTER 9

A courageous Viking known as Erik the Red discovered the coast of Greenland in AD 982. Three years later, he founded the Norse colony of Osterbygd there. This colony and others flourished at first, yet a little over 450 years later, the last colonist died. The disappearance of Osterbygd was the only recorded instance of a well-developed European outpost being completely extinguished. Was it because of inbreeding? Was it because of plague transferred to Greenland by pirates? Were the Norsemen murdered by pirates? Or was it because the climate became so cold that the brave Norsemen were pushed beyond the margin of survival? We may never know for certain. Yet we do know that the demise of the Norse colony of Greenland coincided with the coldest temperatures in the area within the past 1400 years.

The disappearance of this Viking community is not an isolated incident stranded in the past. An average cooling of only about 0.4°C (0.7°F) in the Northern Hemisphere since 1940 has already reduced the growing season in England by a week or two. We do not know if this trend will continue, but we can be sure that people—lay-persons and scientists alike—will be asking such questions for several decades to come.

In this chapter we will try to discover just what is, and what is not, known about the changing climate. Climatic change is an urgent issue, with implications for our future, yet it is normal for the climate of the earth to change. The importance of today's changes can only be fully assessed when viewed against a background of past changes.

CHANGING CLIMATE IN HUMAN TERMS

On May 3, 1976 the Central Intelligence Agency of the United States issued a report warning of disastrous world-wide food shortages resulting from recent climatic changes. The report argued that weather patterns indicate a return to a climate much more like that of the last century. More specifically, the CIA report predicted a cooling in some significant agricultural areas and a drought in others.

A 1°C (1.8°F) drop in the average annual temperature of the Northern Hemisphere would eventually lead to a major drought in India every four years. The first drought might be responsible for the death of 150 million people. China would have a major famine every five years and would require a supplement of 50 million metric tons (55 million tons) of grain. The Soviet Union would lose grain produced in the Kazakhstan region and thus would require 48 million metric tons (53 million tons) of grain. Canada would lose 50% of its grain-production capability and 75% of its exporting capabilities.

Such food shortages, famine, and death would not pass without global political and economic upheaval. An earlier CIA report noted that grave food shortages would prompt powerful but increasingly desperate nations to get grain any way they could. The CIA also predicted massive migrations, sometimes backed by force.

Political and economic speculations are always dangerous, because human affairs are much harder to predict than the events of the physical world are. But we cannot lightly dismiss the possibility of such shortages and subsequent events. Instead, we must look at the topic of climatic change as a whole so we may be able to make a balanced judgment. Newspapers usually select only the most newsworthy facts from scientific reports.

DOES THE CLIMATE CHANGE?

Climate is sometimes defined as the average values of the various weather elements taken over a 30-year period. The most widely accepted 30-year period is that between 1931 and 1960. Average values of many weather parameters during this period, called *climatic normals,* have been published by the World Meteorological Organization for hundreds of places on the earth.

An even more specific definition is used for studies of possible changes of climate. A *climatic state* is defined as the average (together with the variability and other statistics) of the complete set of atmospheric, hydrospheric, and cryospheric (ice) variables over a specified period of time in a specified domain of the earth-atmosphere system. It is therefore possible to have monthly, seasonal, annual, or decadal climatic states. From now on, we will use the word "climate" as an abbreviation for climatic state.

There are three kinds of evidence that might show a change in climate.

1. There is *direct evidence,* as in the daily weather records gathered all over the earth. These are extremely valuable but have been available for only about the past two centuries or less.

2. There is *historical climatic data,* consisting of such historical sources as written records and such qualitative observations as figures on crop yields and drought. These data have been kept, in some cases, for thousands of years. Ancient Egyptians kept a record of the water level of the Nile for hundreds of years.

3. There is evidence to be found all over the natural

THE EARLY SIGNS OF CLIMATE CHANGE

"FAST MOVING GLACIER PROBLEM
What is perhaps the largest glacier in mainland Canada has begun advancing rapidly . . . serious threat of floods downstream . . ."

Albuquerque Tribune
December 6, 1973
(Ottawa Enterprise Science News)

"GROWING ICE CAP MAY BE TREND
Weather Satellites sweeping across the Northern Hemisphere have come up with a surprise, the permanent snow and ice cap has increased sharply.

. . . increased by 12 per cent in the Northern Hemisphere in 1971 and has remained at the new level."

Albuquerque Journal
May 15, 1974 (AP)

During the past million years or so, the earth has experienced several phases of extreme coldness and, as a result, expansion of the polar ice sheets. This period of earth history is called the Pleistocene, and it was after the most recent withdrawal of the glaciers—just 10,000 years ago—that the earth's human population multiplied and grew to its present size.

Now a truly chilling reality faces us: the most recent withdrawal of the ice may not be the end of the ice age, and it appears that the ice sheets are getting ready to come back. The signs include icebergs in polar waters, larger and longer-lasting than they have been in some periods of the past. Arctic lands once free of snow in the summer lie under permanent snow today. Temperature averages in the Northern Hemisphere are declining. Wind belts are shifting, and disastrous droughts are occurring in many places in the first half of the 1970s.

The last time the glaciers advanced, they covered the United States as far south as the Ohio River. Think of the huge population which now lives *north* of that border. If we are indeed only between ice ages, how would we be able to provide for the displaced population forced south by advancing ice sheets? The scattered news headlines might be the first faint clues of a major disaster.

Source for headlines: Nels Winkless III and Iben Browning, *Climate and the Affairs of Men* (New York: Harper's Magazine Press, 1975).

world, from the sediments of the sea to the rings in trees. This kind of evidence has been called *proxy climatic data* because it is indirect. Table 9-1 summarizes some proxy data.

The methods of gathering proxy climatic data are much more complicated than those used in gathering direct data. Let us look at some examples.

SEDIMENTARY EVIDENCE

Sediments accumulated at the bottom of the oceans offer data on sea-surface temperature, the amount of ice on the globe, and other factors. Samples of these sediments are tested for accumulations of ash and sand, the fossils of minute sea creatures, and mineral composition.

Nuclear particles within the atoms of these substances present especially valuable information. The nucleus of an atom is made up of positively charged protons and uncharged neutrons. Two atoms of one element, such as oxygen, may have different numbers of neutrons, although they will always have the same number of protons. The related forms of the element, differentiated by the number of neutrons each has, are called *isotopes.* The concentration and ratio of different isotopes in samples, or *cores,* of ocean sediment (and other related factors) give information on past temperatures and the volume of polar ice (see Figure 9-1).

Sediments at the bottom of lakes are often useful in researching past temperature and precipitation. This is particularly true of lakes that at one time existed at the edge of ice sheets or glaciers. Each year, spring and summer runoff carries both fine and coarse sediment to the lake. The coarse sediment, because of its weight, is deposited on the lake bed first. Then, in the still water conditions under the ice of the winter, the fine sediment is gradually deposited. Thus, the passing of each year is accompanied by a layer composed of coarse sediment at the bottom and fine sediment on the top.

These alternating layers of sediments are called *varves.* Their testimony to melting and runoff conditions, indicative of past temperatures and precipitation, may sometimes be traced back for 5000 years.

Longer but discontinuous records of hydrologic and temperature conditions may also be derived from lakes in closed drainage basins. Old shorelines in such lakes show past levels of their long-time hydrologic equi-

TABLE 9-1
Characteristics of Proxy Data Sources

PROXY DATA SOURCE	CONTINUITY OF EVIDENCE	MAXIMUM PERIOD OPEN TO STUDY (Years before Present)	CLIMATIC INFERENCE
Ocean sediments	Continuous	1,000,000	Sea surface temperature, salinity, sea ice extent, global ice volume, bottom water flow, water chemistry.
Lake sediments	Continuous	5,000	Temperature, precipitation
Closed basin lakes	Episodic	50,000	Evaporation, runoff, precipitation, temperature
Marine shorelines	Episodic	400,000	Sea level, ice volume
Mountain glaciers	Episodic	40,000	Glacial extent
Ice sheets	Episodic	1,000,000	Ice sheet area
Layered ice cores	Continuous	100,000	Temperature
Fossil pollen	Continuous	200,000	Temperature, precipitation, soil moisture
Tree rings	Continuous	8,000	Temperature, runoff, precipitation, soil moisture
Ancient soils	Episodic	100,000	Temperature, precipitation, drainage
Cave deposits	Continuous		Temperature
Sedimentary rocks	Episodic	600,000,000	Temperature, precipitation, wind direction

FIGURE 9-1
Scientists can measure the isotopic content of small marine fossils to ascertain the amount of ice on the globe in past periods.

librium. Thus data on evaporation, temperature, runoff, and precipitation are integrated. The same is true, but on a much larger scale, of ocean shorelines of past times. Ancient marine shorelines, which may be either above or below the present sea level, indicate past sea levels and volumes of ice on the globe.

EVIDENCE AT THE EARTH'S SURFACE

Ice on the earth was the first phenomenon to set scientists' minds thinking about climatic change. Between 1800 and 1830, investigators in Switzerland and Norway put forward the idea that the landscapes in their mountain areas could only be explained by a great extension of glaciers in the past. Since then, hundreds of people have studied landscapes carved out of moving ice to determine the previous extent of glaciers and ice sheets. Figure 9-2 shows one example. Recently the ice has yielded some amazing secrets. Cores of ice taken from the Greenland ice sheet, when examined isotopically, have provided a detailed temperature record for the past 100,000 years.

Biological features on the land surface can also tell investigators details of past climates. The types of fossil pollen in bogs give clues to past climates and their changes. Trees are sometimes good indicators too. The

(a)

(b)

(c)

(d)

FIGURE 9-2

Rhone Glacier, Switzerland. Changes in the glacier's extent tell us of changes of climate in the present century. a) About 1856. b) 1870. c) 1931. d) 1970.

width of their annual growth rings, as shown in Figure 9-3, may be used to determine details of the hydrologic and thermal conditions prevailing prior to and during the growth of the rings. Studies of bristlecone pines in the White Mountains of eastern California have yielded a climatic record of the past 8200 years. Most tree-ring records are much shorter than this but are still valuable.

LITHOSPHERIC EVIDENCE

Scientists also look for clues within the lithosphere, the earth's crust. Chapter 10 shows that climate helps determine the type of soil to develop in an area. Ancient soils that have been buried and later exposed can therefore indicate the climates that prevailed when they were formed.

A different source of evidence comes from limestone caves. Slow-forming deposits of calcite made the iciclelike stalactites and stalagmites that grace Figure 9-4. Through isotopic analysis, these deposits can indicate past temperatures.

The rocks of the earth also contain the story of past climates. Sometimes the rock type evinces the climatic conditions under which it was formed. Other times certain features of the rock tell a story. For example, fossilized sand dunes demonstrate not only desert conditions but also the prevailing wind direction during formation.

The most detailed rock record of past climates comes from the fossil plants and animals found in rocks. These fossils, when interpreted, can give much information on past life forms and therefore past climatic environment.

These proxy data and other evidence are the pieces of a giant jigsaw puzzle. When put together, they give us information on past climates. But each form of evidence tells about a limited time span. Furthermore,

FIGURE 9-3
Tree rings. Their width indicates the climatic conditions prevailing when they were formed.

the geographical extent of any particular piece of evidence is often small. Therefore, not all of the pieces of the jigsaw are available. Even so, there is no doubt that the climate does change. The next thing we must find out is how it has changed.

THE CLIMATIC HISTORY OF THE EARTH

Most history books start with the oldest times and work their way forward. In contrast, most descriptions of past climates start with the most recent times and work backward. There is much more information available on the more recent times, which allows statements to be made with greater certainty. Another advantage of such an approach is that it allows investigators to use certain climatic events as signposts as they go further into the more obscure past. The following account will also begin with the more recent periods, and we will use certain events for guides as we change time scales within the past.

THE PAST HUNDRED YEARS

Anyone with any doubt about the changeability of climate need only look at the temperature changes of the past hundred years, in Figure 9-5 to have his or her doubts removed. The most marked feature of the average annual temperature of the Northern Hemisphere in the past hundred years is a warming trend beginning in the 1880s and a cooling trend since about 1940. This pattern is plotted in Figure 9-6a.

Different parts of the world have experienced this change to a greater or lesser degree. The Atlantic sector of the Arctic Circle, for example, has cooled the most since 1940, with average winter temperatures in some locations almost 3°C (5.6°F) cooler. On the other hand, no cooling at all has been detected in parts of the Southern Hemisphere, such as New Zealand, where the warming trend since the late nineteenth century is still apparent. This warming trend is in fact a pronounced feature of the climate of the past thousand years.

THE PAST THOUSAND YEARS

The index of winter severity in eastern Europe, shown in Figure 9-6b, may be taken as representative of the thermal conditions of the past thousand years. Such evidence as tree rings in California and cave deposits in New Zealand confirm the general pattern shown for Europe.

The first part of the last millennium—between AD 1100 and 1400—is sometimes called the Middle Ages warm epoch. This was the time of Norse settlement in Greenland, yet was not as warm as the first part of the present century. Between AD 1430 and 1850, cooler temperatures prevailed, giving rise to what is known as the *Little Ice Age.* During this time, glaciers in most parts of the world expanded greatly. Grain could no longer be grown in Iceland, and English literature abounded with references to the cold—from the snow

FIGURE 9-4
Stalactites and stalagmites in a limestone cave in South Africa. Isotopic analysis of the slowly forming deposits of calcite gives a continuous record of past temperature.

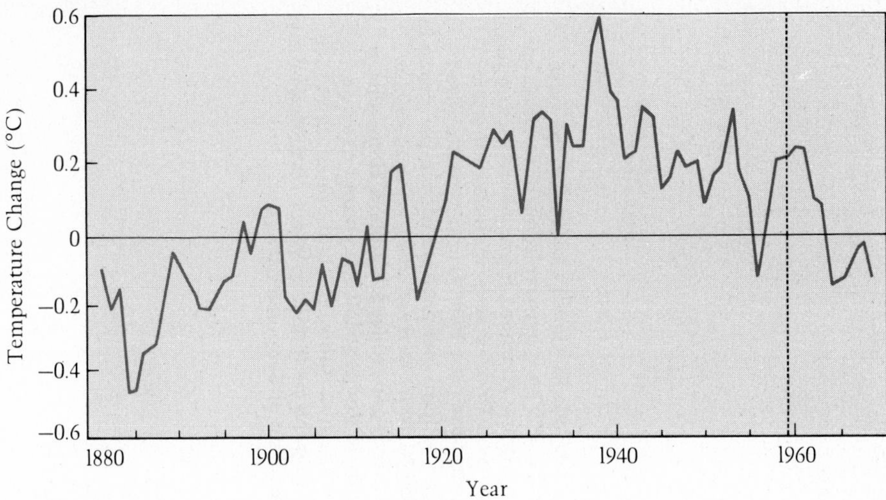

FIGURE 9-5
Recorded changes in average annual temperatures of the Northern Hemisphere in the past 90 years.

scene of a Charles Dicken's Christmas to Shakespeare's milk coming "frozen home in pail." The Little Ice Age, when average temperatures were about 1.5°C (3.7°F) lower than in the 1940s, also appears in an extended time scale.

THE PAST TWENTY-FIVE THOUSAND YEARS

Within this period, the Little Ice Age is a minor yet still significant feature of the temperature record of the midlatitudes of the Northern Hemisphere, as Figure 9-6c indicates. It is one of three periods of depressed temperatures occurring at about 2500-year intervals in the past 6000 years. There are, however, other more marked features in the thermal record of the past 25,000 years. This period is most noteworthy for a long warm interval and for the last stages of the most recent major full glacial period.

Between 22,000 and 18,000 years ago, the ice sheets of northeastern America and Scandinavia extended to 90% of their greatest extent of the last million years, diagramed in Figure 9-7. Around 14,000 years ago, the ice sheet surmounting the mountain ranges of western North America reached its maximum limit. Then rather quickly, all the ice sheets began to retreat. By 10,000 years ago, the mountain ice sheet had melted, and by 7000 years ago, ice sheets in all parts of the world had shrunk to the approximate areas they now occupy. The retreat of the ice sheets was followed by the *Holocene* interglacial period that has essentially lasted to the present time. The maximum warmth within the interglacial was experienced between 7000 and 5000 years ago, a period sometimes called the *thermal maximum* or the *climatic optimum*.

Even during the time when the northern European ice sheet was retreating and the climate was gradually becoming warmer, there were still some sharp returns to glacial conditions. One of these, known as the *Younger Dryas* event, occurred between 10,800 and 10,100 years ago. Virtually full glacial conditions were reestablished within 100 years, and many northern forest areas were destroyed. Such events as this in the climatic record lend fuel to those who prophesy "instant" glaciation in the near future. Yet coming where it does on the overall temperature curve (refer to Figure 9-6c), the Younger Dryas event is more remarkable for its quick onset than for its severity. Speedy changes appear to mark the end of cold as well as warm periods, as we can see on the next time scale.

THE PAST HUNDRED-AND-FIFTY THOUSAND YEARS

The past 150,000 years is a period of alternating cold and warmth, with temperatures mostly below those of the present day. The most noteworthy factors of the temperature record for this time period, which appears in Figure 9-6d, are the two interglacial intervals. These are the Holocene period, which we now live in, and an equally warm interglacial interval called the *Eemian* that occurred about 125,000 years ago and lasted for about 10,000 years. Both the Eemian and the Holocene interglacials started with extremely fast disappearances of glaciers and rapidly warming temperatures.

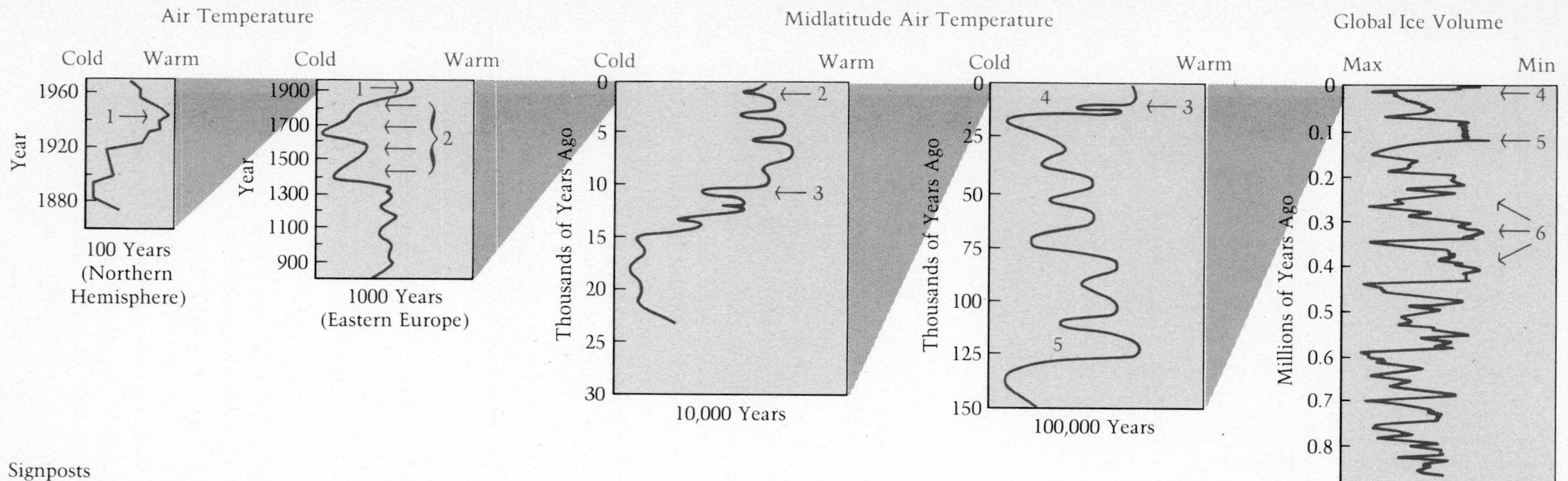

Signposts

1. Thermal maximum of 1940s
2. Little ice age
3. Younger dryas cold interval
4. Present interglacial (Holocene)
5. Last previous interglacial (Eemian)
6. Earlier pleistocene interglacials

FIGURE 9-6

Generalized trends in global climates in the past million years. a) Five-year averages of average air temperature in the Northern Hemisphere during the past 100 years; b) winter severity index for eastern Europe during the past 1000 years; c) generalized midlatitude Northern Hemisphere air temperatures in the past 25,000 years derived from various proxy climatic data; d) generalized Northern Hemisphere temperatures in the past 150,000 years based on various proxy data; and e) fluctuations of global ice volume during the past million years as indicated by isotopic analysis of deep-sea cores.

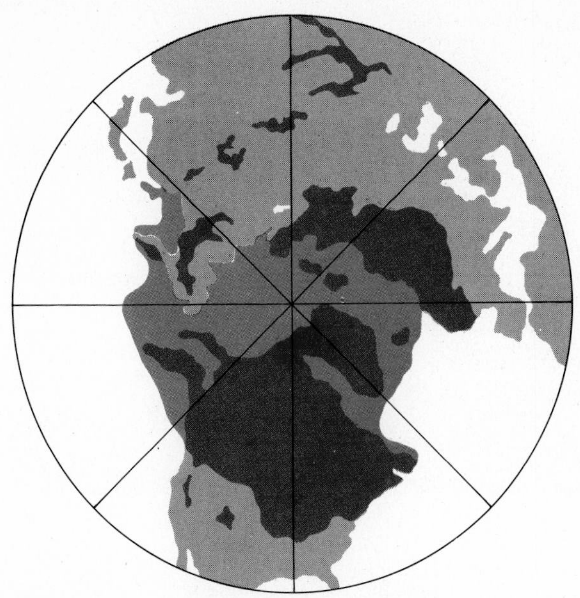

FIGURE 9-7
Maximum extent of the Northern Hemisphere ice sheets during the Pleistocene glaciation.

THE PAST MILLION YEARS

The present interglacial and the Eemian warm periods are well marked on the climatic record of the past million years, as indicated by the volume of ice present on the globe (see Figure 9-6e). Three other prominent interglacials are apparent in the last half of the period, each one coming at intervals of about 100,000 years. Other interglacials occurred in the first half of the period, but they were not so regularly spaced, and proxy records do not so well demonstrate their presence. Whether or not glacial and interglacial periods occur regularly, the pattern of the climatic fluctuations of the past million years is quite clear. Cold and warm conditions seem to alternate constantly. The complete million-year suite of changeovers may together be called the *Pleistocene Ice Age.* The Pleistocene epoch is the geological epoch that contained most of the past ice age.

The Pleistocene geological epoch lasted much longer than a million years, especially if it is defined in climatic terms. Most often it is thought to have lasted 2½ million years, although there is some evidence of glacial events somewhat older than this. It is sobering to think that just 15 years ago the Pleistocene was considered to have lasted only 1 million years. The "recent" ice age was a larger-scale event than scientists previously believed. Indeed, for some general descriptive purposes, the Pleistocene Ice Age may be said to have begun 10 million years ago.

THE PAST THOUSAND MILLION YEARS

From a broad overview of all geologic time, the great surges of ice and frigid temperatures of the Pleistocene Ice Age appear unusual in the climatic history of the earth. Most evidence points to a warm earth climate, with temperatures about 5°C (9°F) higher than in the present. It has been suggested that such balmy conditions have existed for more than 90% of the past 500 million years, at least at latitudes north of 40°N. It is known for certain that the "normal" warmth was punctuated twice before the Pleistocene Ice Age. One world-wide ice age occurred about 300 million years ago. Another one took place some 600 million years ago. Evidence for these glaciations comes from geological deposits left by ice moving over the earth's surface. Some researchers have suggested as many as five major glaciations previous to 600 million years ago, but they lack definite proof.

The picture of past climates and their variations is even today only a hazy one. It becomes clearer as more recent times are reached, there is still a need for more facts. One fact will be apparent to you by now: Climates do change. Perhaps you have already anticipated the next question: How?

THE MECHANISMS OF CLIMATIC CHANGE

Earlier we considered the evidence for climatic change as pieces of a gigantic jigsaw puzzle, many of which are missing. It is not therefore possible to be sure that the picture of climatic change that has to be explained is the real picture. This is the first difficulty in searching for the causes of climatic change. Another difficulty is that there are many probable causes, all acting on different time and space scales. A further complicating factor is that many phenomena affecting climate are linked and interact together, so that changes in one cause changes in others. As if this were not enough, there still are gaps in our knowledge of the exact way the atmosphere and oceans operate, both individually and together. There may even be variables that have not yet been considered.

Given these difficulties, we will nevertheless build a framework for thoughts and facts about the causes of climatic change. First we must specify the system within which climatic changes take place. This system cannot deal with the atmosphere alone. It must also account for the ice of the world, known as the *cryosphere,* and the oceanic part of the hydrosphere.

Such an interlinked atmosphere-ocean-ice-earth system is shown in Figure 9-8. The atmospheric processes should be familiar to you (see Chapter 4). Within the

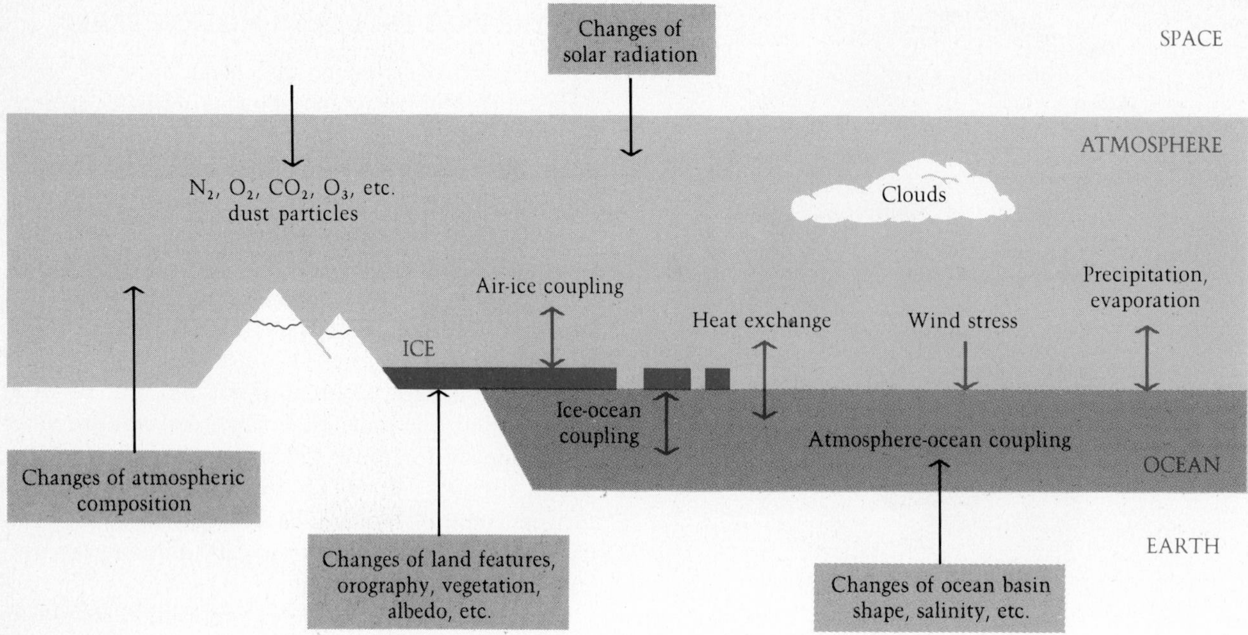

FIGURE 9-8

The atmosphere-ocean-ice-earth climatic system. Dark arrows denote external processes; light arrows indicate internal processes.

system, both the heat energy system and the hydrologic cycle are at work. In addition, the figure diagrams the interactions of wind, ocean currents, ice, and sea. Outside the climatic system, a number of boxes list items, such as a change in radiation coming from the sun, that have the power to alter the system externally. A change in any one or more of these processes creates a climatic change.

EXTERNAL PROCESSES

At night, any particular earth location receives markedly less radiation from the sun than it receives during the day. In a similar way, solar radiation changing over a longer period of time affects the earth's climate. There are both short-term and long-term changes in the behavior of the sun. Short-term changes take place when relatively cool storm areas occur on the surface of the sun. These are called sunspots. A sunspot cycle occurs so that about every 11 years there is a high frequency of spots on the sun. Figure 9-9 shows the spots at a maximum and a minimum. There is some evidence that sunspots have some effect on the earth's weather, but measurements are not yet available to indicate whether sunspots affect the amount of solar radiation received at the earth's surface.

Longer-term changes also occur, because of three specific peculiarities in the earth's orbit and axis. The first is a distortion, or stretch, of the earth's orbit around the sun, as shown in Figure 9-10a. A change of this kind occurs about every 90,000 to 100,000 years.

Every 41,000 years, there is a variation in the obliqueness of the earth's axis. In other words, the earth rolls like a ship, as in Figure 9-10b, so that the angle between the axis and the plane of the ecliptic changes from 21.8° to 24.4°. Its present angle of 23.5° is therefore not constant.

A third motion might best be described as a wobble (see Figure 9-10c). Like a spinning top, the earth's axis swivels once every 21,000 years. When these cycles are put together, the amount of solar radiation estimated to reach the earth is strikingly parallel to the waxing and waning of the global ice cover in the past million years. Yet the orbital and axial changes presumably also took place in the 90% of earth history when ice ages did not occur.

Many of the other external processes that might lead to climatic change are considered elsewhere. Chapters 3 and 4 mention how changes in atmospheric composition, especially of such gases as carbon dioxide, can alter the intensity of the greenhouse effect and thus the temperature of the atmosphere. Volcanic dust may also enlarge the greenhouse effect. It is also possible for the land surface to change. Most of Part Four explains how the land surface is lifted up and worn away. In addition, Chapter 15, on wandering continents, shows how the geographical arrangement of oceans and continents may alter through time. Before about 30 million

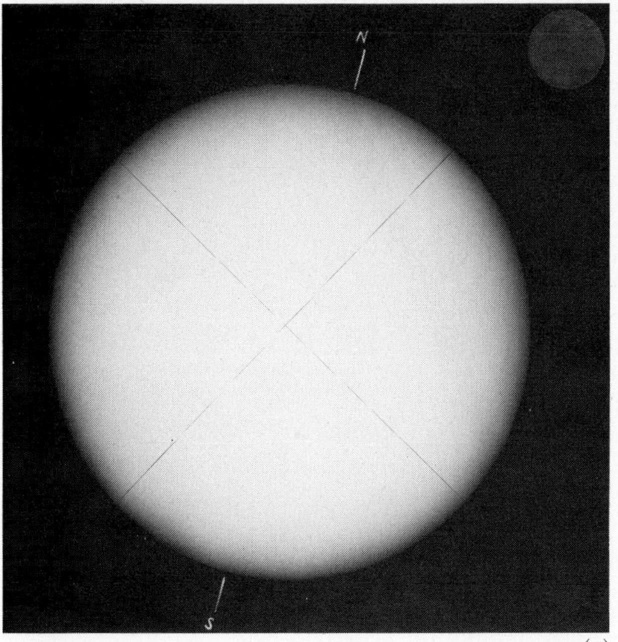

(a)

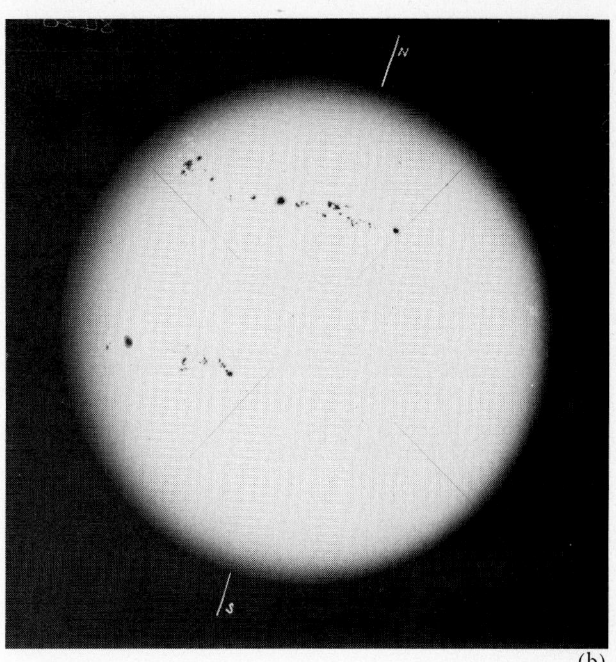

(b)

FIGURE 9-9

Minimum and maximum numbers of sunspots on the surface of the sun.

WEATHER CHANGES IN AN INDUSTRIAL TOWN

In April 1968, *Saturday Review* ran an article entitled "Home-brewed thunderstorms of La Porte, Indiana." Two weeks later, *Newsweek* picked up the same story. What catapulted the weather of this midwestern town into the national news media?

Climatologist Stanley Changnon, Jr., writing in the *Bulletin of the American Meteorological Society* in 1968, presented some of the startling facts behind the headlines. Beginning about 1925, La Porte experienced a dramatic increase in rain, thunderstorms, and hail. Nearby weather stations in other towns did not show evidence of the same changes. Some people charged that pollution from steel plants 30 miles away in Chicago and Gary caused the increase in storms. Others blamed errors by the observers recording the data.

But the changes were real, and meteorologists believe they are related to the heat, pollutants, and vapor from the industrial complex to the west of La Porte. During the past 50 years, La Porte has been drenched by a 30–40% increase in precipitation. Especially heavy rainfall occurred in years of increased steel production in Chicago. Since 1949, La Porte has experienced 38% more thunderstorms than the surrounding towns. Since 1920, La Porte has averaged 59% more hail than surrounding communities; from 1951 to 1965, the figure was 246% more hail.

This strikingly altered weather pattern is probably due to, according to Changnon, "an interaction between two major problem areas of meteorology: weather modification and atmospheric pollution . . . La Porte is located downwind of a major industrial complex capable of sizeable increase in heating, moisture content, and condensation nuclei and freezing nuclei, and the temporal fluctuations in smoke days and steel production compare favorably with those of the precipitation conditions."

La Porte stands out as the prime example of weather modification by human activities, a phenomenon associated to some degree with most heavily industrialized areas.

Source: *Bulletin of the American Meteorological Society*, 49(1): 4-11, 1968.

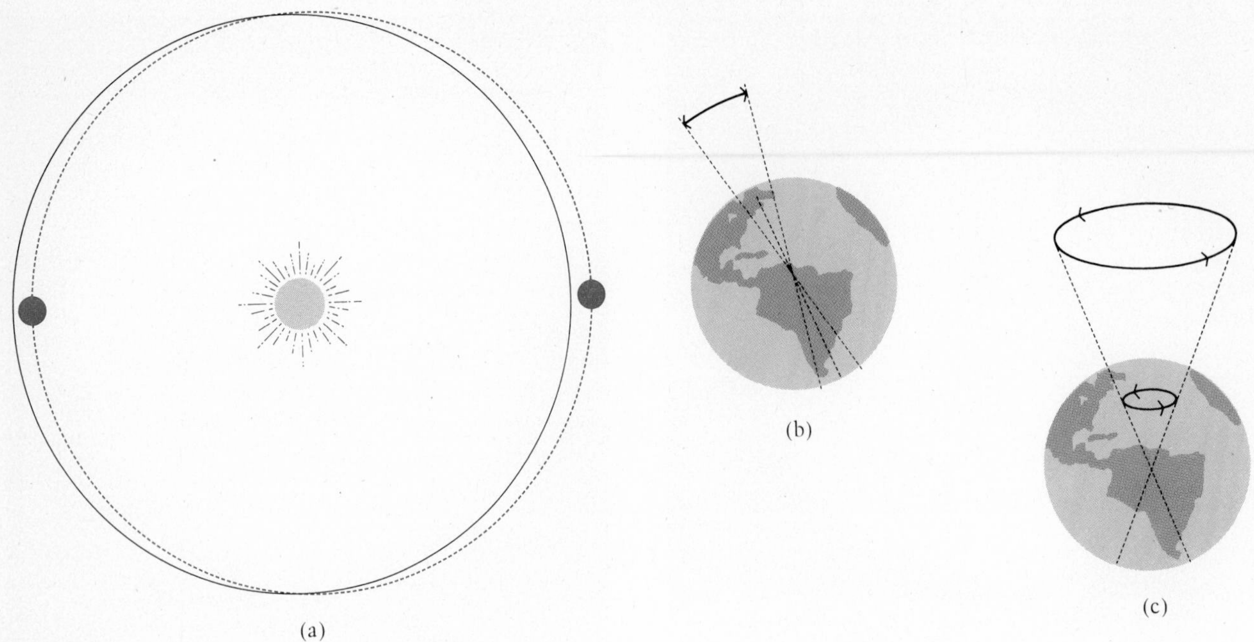

FIGURE 9-10
Changes in the earth's orbit and axis: a) "stretch," b) "roll," and c) "wobble."

years ago, both South America and Australia were connected to Antarctica. Their separation allowed a circumpolar ocean current to form in the Southern Hemisphere, an event of crucial importance to world climate.

INTERNAL PROCESSES

Chapters 4, 5, and 6 describe the water and heat exchanges within the atmosphere, the way that the general circulation distributes heat and moisture across wide areas of the earth, and the weather systems forming the fine grain of atmospheric movement and operation. All these are internal processes, and many of them work as systems by themselves. Changes in any one of them could lead to climatic variation. But they are linked by feedback mechanisms.

An increase in unusual variations in ocean temperatures provides an example of positive feedback. A change in the temperature of the ocean surface may modify the amount of sensible heat transferred to the atmosphere. This alters the atmospheric circulation and cloudiness. Variations in radiation, wind-driven mixing of ocean water, and other factors may in turn affect the temperature of the original ocean surface. In the Northern Pacific Ocean, the ocean-surface temperature has increased for several years at a time because of positive feedback mechanisms like those just described.

An example of negative feedback is when a snow-covered land or ice surface reduces atmospheric temperatures. The cooler atmosphere can hold less water vapor, and thus less snow falls. The low precipitation levels in Antarctica are partly a result of such negative feedback.

Many other examples of both positive and negative feedback exist. Thus it is highly unlikely that climatic changes could be explained by any one isolated factor. It is more likely that different features operate on different time scales.

THE PULSES OF THE EARTH

As far as climatic change is concerned, the earth has many pulses, and they are not all regular. In some ways ice ages could be described as times when the climatic system becomes very excited. To make sense of all the possible causes of climatic change, we must specify the lengths of time over which they act, as done in Figure 9-11. The diagram shows that changes in the amount of volcanic dust and the composition of the atmosphere have acted on climate over time periods ranging from less than a year up to a thousand million years. In contrast, changes in climate resulting from the alteration of the earth's surface by humans has been significant only in the past thousand years. Thus humans could not have been responsible for the last ice age, because their large-scale alterations of the earth system are recent. But changes in the atmospheric composition from erupting volcanoes might have been a cause.

Although humans were not responsible for the last ice age, our ancestors were certainly affected by it. Many migrations became necessary because of the advancing ice. Changes in climate still affect us. We will now examine one of the more recent changes in the climatic system, one that has caused great suffering for large numbers of our species.

RECENT CLIMATIC CHANGES

Erik the Red gave the name "Greenland" to the land where he established the Norse colony because the land was green when he arrived. He arrived there at the end of a warm period that lasted longer than any that have occurred since. Yet Greenland's climate was marginal for survival, and a deterioration of climate, most noticeable in marginal areas, was probably responsible for the failure of the colony. On the map of world climates in Figure 7-4, every boundary is a climatic margin. The world is full of them. Some are not only climatically marginal but also climatically marginal for human survival. The distinction is vital.

Sahel is the Arabic word meaning "shore." The southern shore of the Sahara, the Sahel has been the scene of immense human suffering in recent years. The suffering is directly attributed to a change of climate. Figure 9-12 shows the average annual rainfall

FIGURE 9-11

Time scales of characteristic events that affect the climatic system and that may cause global climatic change.

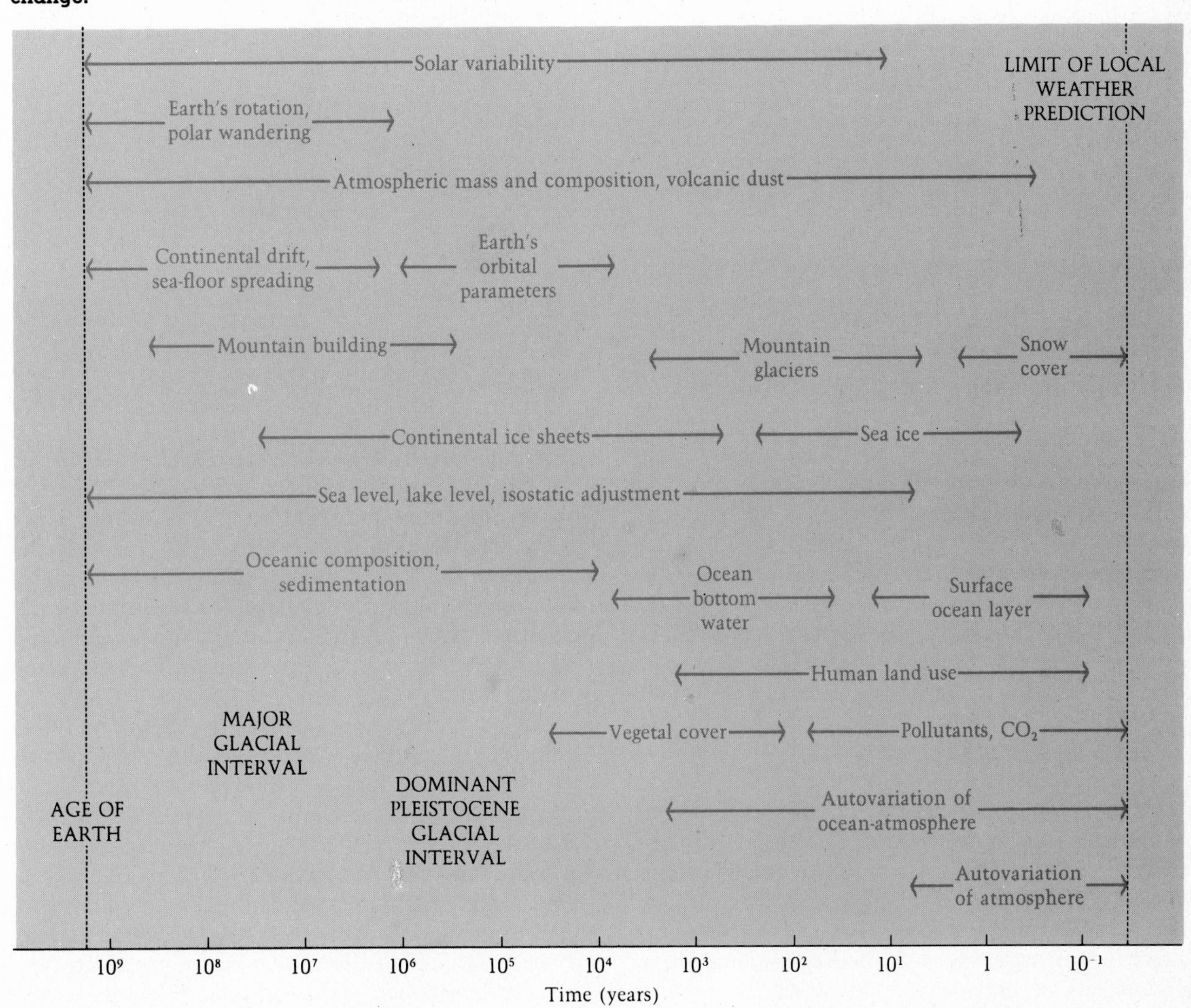

FIGURE 9-12

Average annual rainfall in the Sahel. a) Locations of rainfall stations in the Sahel area of West Africa. b) Average annual rainfall for five stations in the Sahel expressed as a percentage of the 1931-1960 average.

(a)

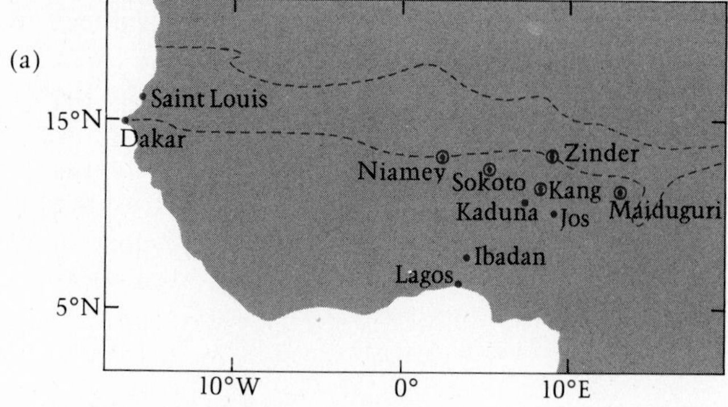

(b)

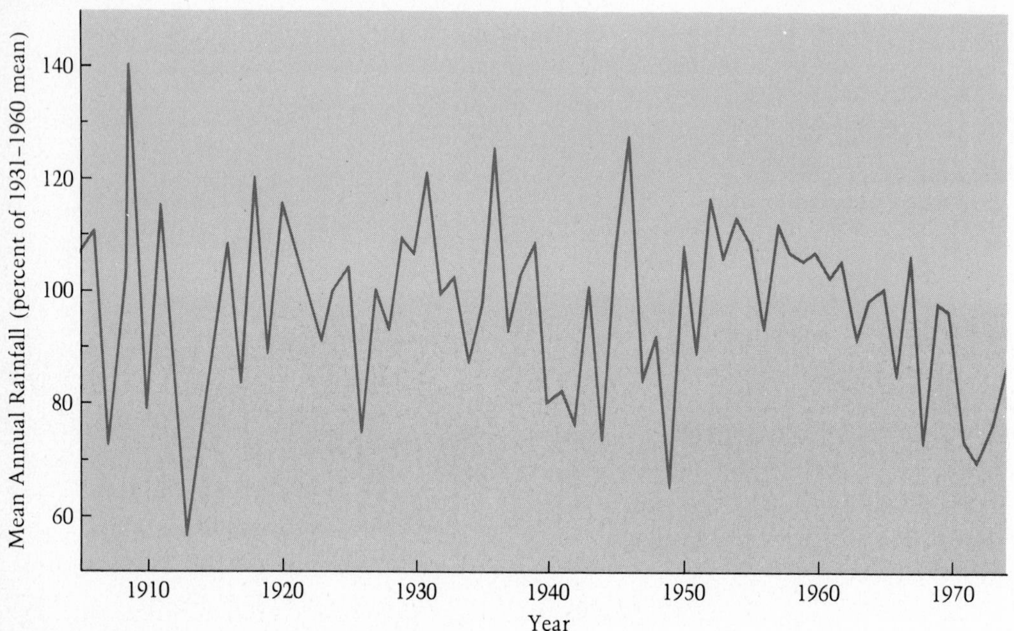

for five rainfall stations in the Nigerian part of the Sahel. The rainfall values are expressed as a percentage of the 1931–1960 average yearly rainfall. Between 1905 and the early 1950s, annual rainfall values varied considerably: In some years rainfall was well above the average, but in others it was markedly below the average. Since the early 1950s, there has been an obvious downward trend in rainfall values. In fact, since the mid-1960s, only one year has had rainfall exceeding the 1931–1960 average. Beyond any doubt, in the mid-1970s a drought gripped the southern shore of the Sahara.

Different climatologists have different explanations of how the drought came about. One has pointed out that decreasing rainfall *south* of the Sahara has accompanied increased rainfall to the *north.* The drought might be the product of an expansion of the great westerly swirl of winds of the Northern Hemisphere. The expansion brings rain-bearing storm belts to the northern desert and pushes the subsiding air of the subtropical high-pressure area southward. Therefore, the summer monsoon rains of the Sahel cannot penetrate as far northward as in 1931–1960. Another climatologist has argued that the Sahel's drought results from a general warming of the earth's surface with increased carbon dioxide in the atmosphere and from an associated increase of temperature differences from north to south.

Nor is there any agreement as to whether the drought will continue. One investigator predicts that the trend will continue until about AD 2030. Other scientists have pointed out that when the known rainfall values of the Sahel are subjected to every possible statistical test, no established trends can be detected. They demonstrate that the recent succession of drought years falls within statistical expectation. Their message is that the available past records provide no basis for predicting the future in this case.

While climatologists argue, the vultures on the southern shore of the Sahara become fat. Between 30% and 70% of the cattle have died in recent years (see Figure 9-13), and crop failures are widespread. At one point, 3000 to 4000 Ethiopians were dying each week, basically from hunger. Relief through international agencies, although vital, is only an interim measure. But long-term planning depends on guesses about further changes of climate. This question is of consequence to the Sahel and to places as widely scattered as India and the high plains of the United States. On a larger scale, the whole world, and every person on it, is affected. So what of the future?

THE FUTURE CLIMATE

Climatologists are not arguing about rainfall in the Sahel because they enjoy argument. They are almost desperate to find the secrets of the climatic system so that the Sahel and future humankind may be saved. In addition, the climatic system is complex, and a change of climate might result from one or a combination of a wide variety of internal and external causes. The overall product is a spectrum of climatic predictions for the world—from no immediate change to an imminent "instant" ice age. A select committee of the National Academy of Sciences of the United States has

FIGURE 9-13

The face of death in the Senegal region of the Sahel.

recently put forward a balanced opinion on future climate. Much of what you have just read is based on their work.

The committee goes to great lengths to point up the uncertainties in making climatic predictions. In addition to those we have already discovered, there are three more. The first is the possibility that a significant part of the climatic system acts in a completely random way. If this is true, climate could never be predicted absolutely, and future predictions could only be given in probabilities. Second, it may be natural for the atmosphere to have not one "normal" state but two or more. In such a case the alternate states might change rapidly. A system operating in such a manner is described as being *almost transitive*. The third uncertainty we will save for a moment.

Given the past climatic record of the earth and the uncertainty about how the climatic system works, the American select committee was able to say the following about the future climate of the earth:

> There seems little doubt that the present period of unusual warmth will eventually give way to a time of colder climate, but there is no consensus with regard to either the magnitude or rapidity of the transition. The onset of this climatic decline could be several thousand years in the future, although there is a finite probability that a serious worldwide cooling could befall the earth within the next hundred years.

This is as much as can be said with the facts scientists possess at present. Yet whatever certainty there is in this statement, and it is a remarkably strong statement for a highly respected body to make, may be destroyed by the third area of uncertainty—human interference. Increased atmospheric carbon dioxide, atmospheric contamination by areosols, and waste heat from human activity, as well as factors that may not even have been recognized yet, all could alter the unknown natural pattern of climatic change. Only one thing is certain. As stated by the National Academy, "the clear need is for greatly increased research on both the nature and the causes of climatic variation." The answers must be found. No doubt the spirit of Erik the Red would echo the statement were he ever to discover the fate of the Greenland colony.

SUMMARY

Climatic changes on the earth have marked effects on human activity. Their impact is more widely felt in an increasingly populated world. The full importance of recent changes can be fully assessed only when viewed against a background of changes in the past.

Evidence of climatic change is of three kinds. Direct evidence comes from past weather records. Historical climatic data can be used. Indirect proxy evidence is available in sediments, surface features, and the lithosphere.

All this evidence shows that the climate of the earth has been warmer than it is at the present time. However, oscillations between cold and warm periods of various lengths are normal. Three major ice ages have occurred in the past 600 million year.

The causes of climatic change cannot be definitely distinguished, but they are probably related to processes both internal and external to the earth's climatic system. Whatever the causes, recent climatic changes have engendered severe human suffering.

QUESTIONS

1. Why are climatic changes of very small magnitude more noticeable now than they perhaps were at the time of the decline of Østerbygd? What factors make a decrease in mean annual temperature of 1°C in the Northern Hemisphere more of a threat today than might have been the case as recently as 200 years ago?

2. How is a climatic state different from a climatic normal? Why do we have both terms?

3. How would you compare the three kinds of evidence of climatic change in terms of their potential accuracy? What would be the primary factor determining which kind of evidence was examined for a particular study of climatic change?

4. In the late nineteenth and early twentieth centuries it was common for scientists of the US Geologic Survey to request core soil samples from farmers digging wells in the Midwest. What type of analysis would probably be performed on these core samples, and what climatic information would the scientists expect to find?

5. How would old ocean shorelines indicate the extent of past glaciation? How would this evidence tell us the amount of ice on the globe? How would the calculations be made, and what factors would have to be taken into consideration?

6. How might we compute the changing amount of energy necessary to create an ice age if we know the volume of water that was lost to the ice sheets from old ocean shoreline evidence? Would sublimation affect this calculation?

7. What are the disadvantages of using tree ring data for studying past climatic regimes? How does the

reliability of these data vary around the world?

8. Considering the large number of types of proxy climatic data, should we be able to write detailed climatic histories for all areas of the earth for the last 100,000 years? Why or why not?

9. When we speak of the temperature trend of the last 100 years, what is the shortest time span we are interested in, as depicted in Figure 9-5? If, instead of looking at the mean temperature for each year, we look at the mean temperature for each five-year period, starting in 1880, would the graph have the same general appearance? How about ten-year means?

10. Much of the evidence for the climate during the last 1000 years comes from historic climatic data, such as literary descriptions and records of the type and amount of grain grown in certain areas. When interpreting such data, what factors must be taken into consideration? Is is valid to rely on this data alone for an accurate characterization of the climatic history of the last 1000 years? Why or why not?

11. In Figure 9-6d, "Midlatitude air temperature over the last 100,000 years," where would you expect to find the maximum extent of glaciation in the last glacial period? In Figure 9-6c, where is the thermal maximum of the Holocene interglacial found?

12. From your examination of the graphs in Figure 9-6, what generalizations can we make about glacial activity on the earth?

13. Why is the existence of the Pleistocene Ice Age so well documented, while the previous ice ages, approximately 300 and 600 million years ago, are known only in the most approximate terms? How is the evidence for these older ice ages different from that for the Pleistocene Ice Age?

14. Climate change is a complex topic, usually studied by teams of scientists from different disciplines. Why would a phenomenon such as an ice age require an interdisciplinary approach?

15. What effects do the various earth-sun positions have on the amount of radiation received from the sun? Why are these factors thought to have contributed to the formation of ice ages on the earth? What arguments can be used to counter the position that earth-sun relationships are largely responsible for ice age formation?

16. Briefly describe the major external causes scientists have considered as possible events leading to ice age formation. Are they independent, or might they work together to trigger formation of an ice age?

17. What are some examples of positive and negative feedback mechanisms in the earth's climatic system which might contribute to the formation and ending of ice ages?

18. How can the process of long-term climate change be viewed as a gigantic system, with external and internal processes affecting it? Why is this approach preferable to one that considers changes in the atmosphere, the hydrosphere, and the cryosphere alone?

19. Why do boundaries between climatic regions exhibit the effects of climatic change much more rapidly and drastically than do homogeneous areas? Is it unusual for these boundaries to change their positions over time? What implications does this have for the distribution of human habitation over the globe?

20. Why is it difficult to predict future climatic change with any degree of certainty? Why is so much importance attached to finding out the implications and meanings of patterns of change exhibited in climate today, such as in the Sahel? If humans were able to survive the much greater changes of the early Pleistocene ice ages, why do we need to concern ourselves with relatively minor fluctuations such as those in the Sahel?

SUPPLEMENTARY READING FOR PART TWO

Many sources on the atmosphere cover much of the material considered in this part. Two good books are *Atmosphere, Weather, and Climate* by R.G. Barry and R.J. Chorley (the third edition published by Methuen in 1976 is recommended), and *World Climatology: An Environmental Approach* by J.G. Lockwood, published by Edward Arnold in 1974. The earth's heat balance is eloquently described by M.I. Budyko in an article in *Soviet Geography,* 1962, vol 3, number 5, pages 3–15. Details on the spatial variation of climate can be obtained as volumes become available from the series *World Survey of Climatology,* edited by H.E. Landsberg and published by Elsevier. If you would like more details on human comfort across the world, see the article "World Patterns of the Distribution of the Monthly Comfort Index" by W.H. Terjung, published in the *International Journal of Biometeorology,* 1968, vol 12, number 2, pages 119–151. Topics relating climate to people and other applications are treated quite fully by K. Smith in *Applied Climatology,* published by McGraw-Hill in 1975. A balanced view of the topic of climatic change is given in *Understanding Climatic Change: A Program for Action* published by the National Academy of Sciences in 1975. Finally, for light reading, we recommend the *Time-Life* book called *Weather* by P.D. Thompson and R. O'Brien, published in 1967. This goes a long way toward capturing the excitement of the atmosphere.

FROM THE GEOGRAPHER'S NOTEBOOK

THE CITY AS LABORATORY

Three out of every four Americans live in cities. So, the odds are that you are a city dweller. If so, then for 24 hours a day and often for 12 months a year, the air you breathe is city air, the temperatures you feel are city temperatures, and the rainfalls and humidities you experience are those of the urban atmosphere. In 1853 it was written of London: "the idea of the immense area which is covered by this gigantic town may be approximately realised from the fact that many learned physicians discuss the climate differences of many parts of the town exactly as if they were comparing the climates of Italy and Germany." The city climate has its own characteristics and its own variations. What are the special features of the climate of the city you live in, or the one nearest you? What are some simple ways to observe the urban atmosphere?

We can design a field trip that can be divided into several parts. We will start inside the building where you live, then move outside, and finally have a brief look at the city as a whole. Only a few aspects of the city atmosphere will be inspected, but this inspection will probably stimulate you to carry out further investigations.

THE INDOOR ENVIRONMENT

Much of your time is spent indoors. Some parts of this living environment are well-designed; others are poorly designed. One aspect of a well-designed living space is an even distribution of air with similar temperatures throughout it. Pick the room in which you spend most of your time and see whether it has hot spots and cold spots or whether the air is fairly evenly heated.

Do this by taking ten or twenty ice cubes from an ice tray that produces uniform-size cubes. Place the cubes on saucers that will catch the water when it melts. Place the saucers around all parts of your room. Don't forget to test the vertical variation of temperature by putting some ice cubes in high places and others on the floor.

After setting the cubes out, look at them again every 30 minutes. If the room is evenly heated, the cubes will all be melting at the same rate. If not, then some cubes will be smaller than others and you will be able to identify the warm and cold spots of the room from the sizes of the ice cubes. See if you can draw a map of the temperatures in your room showing their horizontal and vertical distribution. You can repeat the ice cube experiment outside the building. See how much faster the cubes melt on the sunny side than on the shady side of any structure.

TRACKING THE SUN

Another way the effects of the sun can be quantified with respect to buildings is by the use of a solar path diagram such as shown in Figure 1. Get four copies of the diagram made. Here is how to use them.

The base of the diagram is a series of concentric circles. These are marked from 0° to 80°, and 90° is the point in the middle. Imagine you are standing in the middle of the diagram—then these circles represent the angles above your head. The 0° circle represents the horizon, and the center point (90°) is the position directly above your head. Around the edge of the diagram are the compass bearings showing the direction in the horizontal out from the center of the diagram (or where you are standing). Superimposed on the concentric circles and radial direction lines are the paths that the sun takes across the sky at different times of the year. These are marked by the angular distance of the sun north or south of the equator. This varies from 23°27′on June 21 to −23°27′on December 22. Note that the summer paths are much longer than the winter paths. Also marked on the diagram are the times when the sun is in a particular position along the path. The hours shown on the diagram are in "sun" time where noon occurs when the sun is highest in the sky. "Sun" or solar time will differ from your local time unless you live on the meridian on which your time zone is based. See Chapter 2 for a further explanation of this.

We want to add to the diagram the amount of sky that is shaded by buildings. Select a place on the north side of your building to take your measurements. You also need a compass or a protractor unless you wish to estimate the angles by eye. Mount the protractor on a stick fitted with a plumb bob as shown in Figure 2. Mark on your diagram the angle of the local horizon to the north of where you stand. The horizon might be a tree or a building or a fence. Whatever it is, mark its angular height on the solar path diagram at the north position. Now continue this procedure marking the

FROM THE GEOGRAPHER'S NOTEBOOK

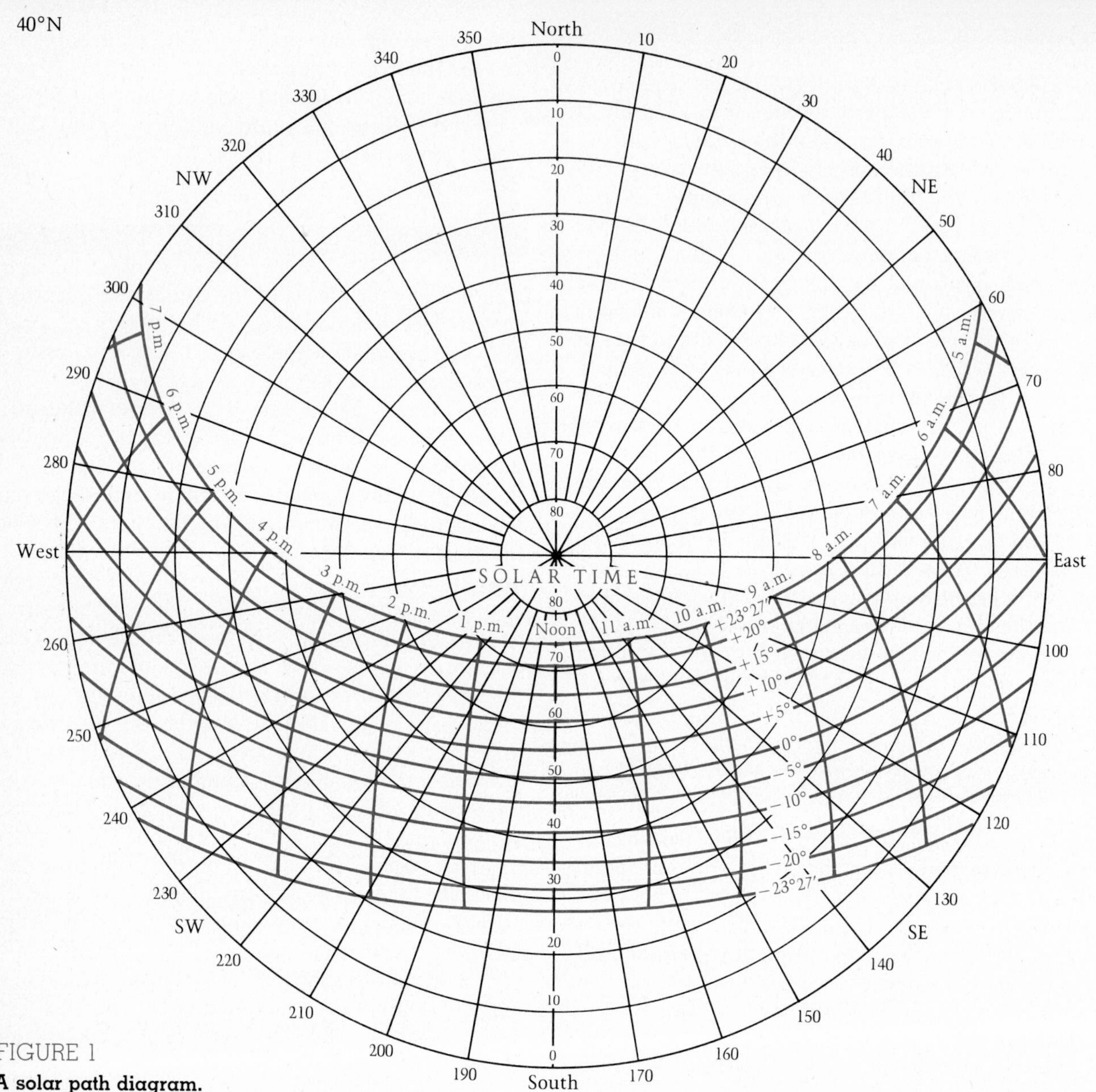

FIGURE 1
A solar path diagram.

horizon elevation at 10° compass intervals all the way around the point where you are standing. You can then fill in the horizon on the diagram. In the example shown in Figure 3 there is a large building shading the sky to the south.

Now, just by looking at your diagram you can make predictions of when the sun is going to appear and set at the point where the measurements were made. You only have to select the sun path appropriate to the time of year and assume that the sky will be cloudless. Repeat the procedure and construct a solar path diagram for a point on the south side of your building. You will notice a great difference. By adding the hours along the solar paths, you can make predictions of the potential amounts of sunshine (in hours) that will be received near the two sides of the building. The difference easily explains the temperature differences on both sides that were previously indicated by the melting ice cubes.

Before finishing with the solar path diagrams, it is

FROM THE GEOGRAPHER'S NOTEBOOK

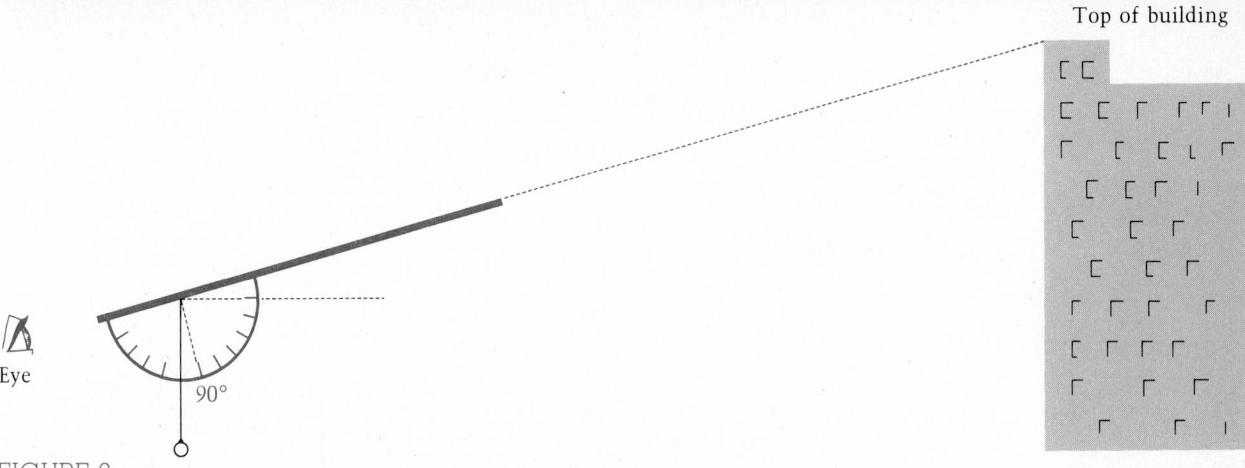

FIGURE 2

A protractor with a plumb bob mounted on a stick measures the angular height of a building. Note that the required angle is 90° minus the angle read by the string of the plumb bob.

instructive to draw two more horizons—one for a typical suburban area and one for the central business district of your town. The one for the central business district will probably dramatically illustrate how little sunshine is received at street level in the central parts of some towns. The sidewalks of Wall Street in New York City receive hardly any sunlight at all and are a good example of a dark "urban canyon."

THE URBAN HEAT PROFILE

Now let us examine the horizontal temperature differences across town. There are at least two ways you can do this—one is by using records that others have collected. The other is by collecting records yourself if you can obtain a thermometer.

First, try to obtain temperature records collected by other people. Look up the National Weather Service or the National Oceanic and Atmospheric Administration in your telephone book and contact them. Ideally, you require two sets of information. For both sets you require air temperatures for every recording station in and around your city. The first set would be minimum temperatures for any day you choose that has clear skies and low wind speeds. The second set would be average minimum temperatures for a winter month—say January.

Having obtained these temperatures, find or construct a map of your city and plot them in the appropriate places on the map. It is probably best to use two maps—one for each set of temperatures. Try to construct isotherms (lines of equal temperatures) on the maps. These will show areas of high and low temperatures like those shown in Figure 4. Try to explain these areas of high and low temperatures using the principles outlined in Chapter 8.

Rather than asking the weather service for records, it is far more fun to obtain your own. Cheap thermometers can be obtained from mail order stores—sometimes for the price of a long playing record. Such thermometers often come in the form of psychrometers or hygrometers that give humidity readings as well. It is also sometimes possible to borrow a thermometer from your school equipment.

THE PORTABLE GEOGRAPHER

If you can get a thermometer, then you can make measurements of the heat island of your city. Using bicycle, car, or bus, travel from the built-up area to the rural area surrounding the city. Take temperature measurements along your path at points that can be noted on a map. If your city is small, start in the downtown area and work outward. If possible, include an industrial area in your route. The greatest dif-

FROM THE GEOGRAPHER'S NOTEBOOK

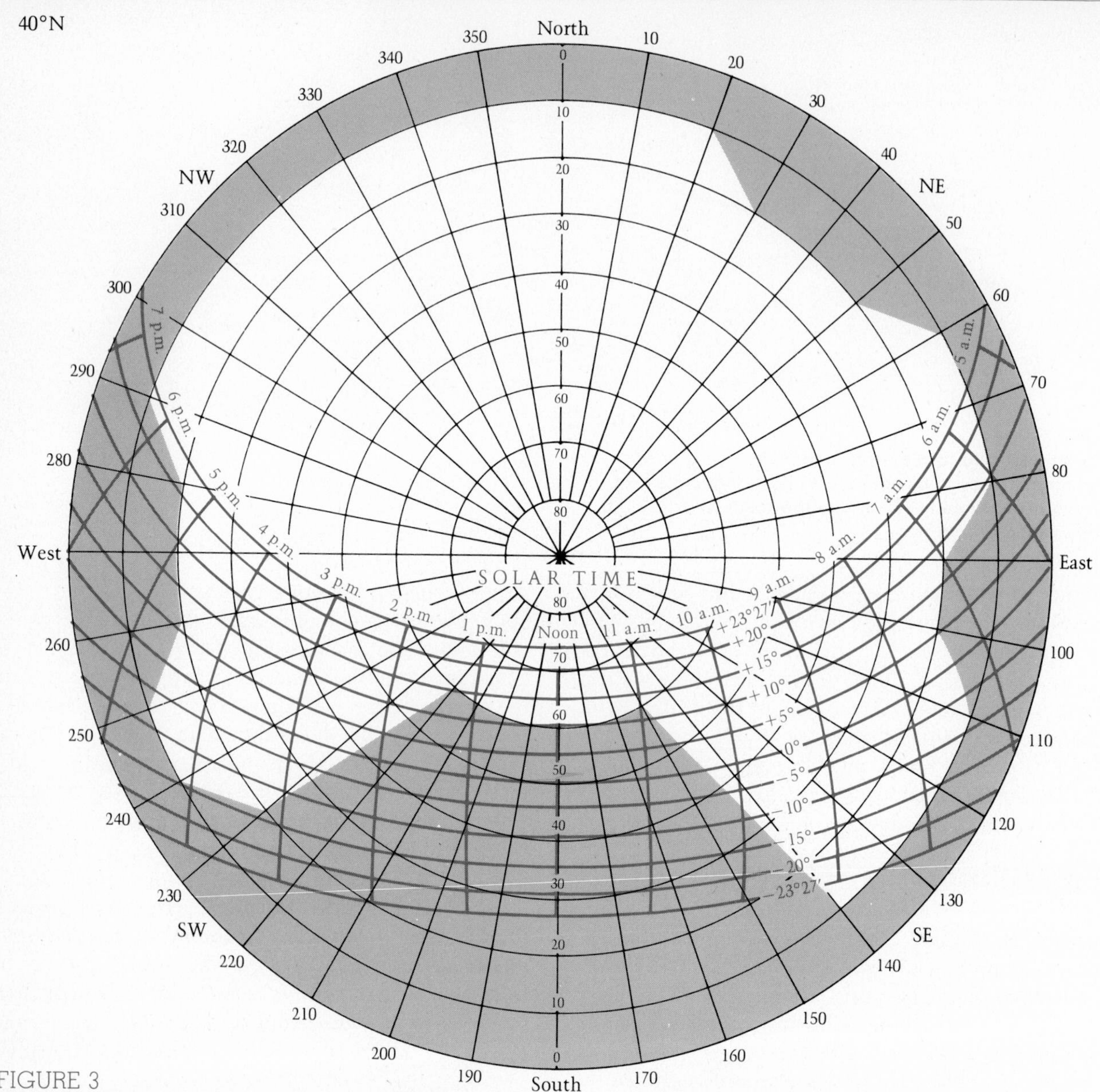

FIGURE 3

A solar path diagram showing the local horizon. A large building lies to the south of the observation point.

ferences in temperature should be found on a clear night when there are no winds.

If your town is small you may have time to make two journeys at right angles across the town. Take measurements about every 400 m (about ¼ mi). Plot the recorded temperatures on a map and draw isotherms (lines of equal temperature) of the city heat island for the time of your transects (see Figure 4). Try to make your measurements as quickly as possible because the normal daily change of temperature with time will probably be taking place simultaneously, and you really require spatial and not temporal temperature differences. One way to try to counter the effect of the time change of temperatures is to return over the same route repeating the measurements and for every observation point average the two temperatures that you recorded. These observations should again enable you to identify warm and cool areas of your town.

FROM THE GEOGRAPHER'S NOTEBOOK

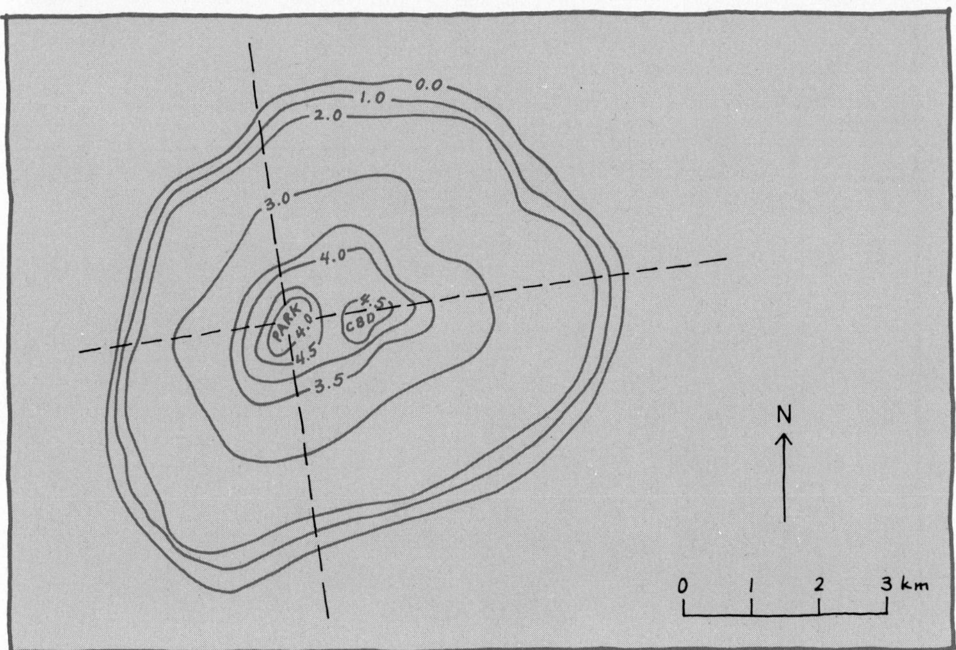

FIGURE 4

A sample sketch map of a city heat island. Dashed lines show the route of transect. Temperatures are in °C. Note the cooler area of city park and the warmer area of central business district (CBD).

In cases where you have time for one transect only, the temperature recordings can be made into a graph by plotting temperature on the vertical axis and distance along the transect on the horizontal axis. Also mark on the horizontal axis important features such as factories, parks, dense housing, and open country. This will help you to interpret your results.

There are many other observations you can make as part of your field work in urban climatology. Your local weather service may be able to provide you with rainfall information. If there are enough stations, you can construct a rainfall map of the urban area. If you manage to get a psychrometer, you can make measurements of humidity as well as temperature and map the results in the same way.

Many observations can be made quite casually as you go about your daily business. If your town is large, chances are it will have a pollution dome on certain days. Note its shape and size. Sketch it. See how it is affected by wind coming from different directions. Look around you. See if you can see any evidence of the way your city is adapted to its climatic environment. Are there salt boxes and snow-clearing equipment? Or is each building studded with air conditioners? Or both? By observing such factors and by doing all or some of the suggestions made here, you will gain a much better appreciation of the urban atmosphere.

Solar path diagrams for latitudes other than 40°N are given in *Smithsonian MeteorologicalTables*, edited by R.J. List (Washington, DC: Smithsonian Institution, 1958). However, the one provided here may be used as a first approximation for the purposes of the field trip in most parts of the USA and Canada.

PART THREE

THE BIOSPHERE

Several billion years ago life appeared on our planet. Through the various processes of evolution, a wide variety of animals and plants gradually colonized the earth. Humans developed in the last twentieth of one percent of all earth history. Life forms, both human and nonhuman, have spread to all the other spheres of the earth. Wherever there is life on earth, that place is part of the biosphere. Thus the biosphere includes all parts of the earth containing life—plants and animals and all other organisms in water and in the air.

We start our discussion of the biosphere with an examination of soils. Soil provides the home for many plants and animals, although as a subject in itself, it belongs more technically in the lithosphere. Soil is actually an intimate cohesion of all the earth spheres. Both its history of formation and its behavior as a system are fascinating subjects.

We look also at plants and animals—how they developed, the importance of photosynthesis, and how the energy captured by it flows through ecological systems. The energetics of ecology and the influences of the evolutionary process help us to understand the present-day distributions in the layer of life.

In the last chapter of this part we describe the distribution of soils and the major vegetation types on the planet. Finally, we consider some of the many spatial relationships among vegetation, soils, and climate. The close connections between all parts of the physical world are demonstrated in the biosphere more clearly than in any of the other spheres.

THE SOIL LAYER

CHAPTER 10

Chapters 3 through 9 are almost totally concerned with the atmospheric layer. It is now time to look at some of the others. Between the vegetation layer and the rock layer lies the soil layer, an intimate cohesion of the lithosphere, biosphere, hydrosphere, and atmosphere. The soil layer is usually considered part of the lithosphere, but it is closely linked with the other three spheres of our physical world. Like so many parts of the physical world, the soil is vital to our existence.

The Greek word for "ground" is *pedon.* Hence the science that deals with soils is called *pedology.* In this chapter we will start to become soil scientists, or *pedologists,* and examine the nature, place, and value of the soil layer. Next we will investigate the origin and formation of soil. We can consider this detective work as unique as anything Sherlock Holmes ever did. Where did the soil come from, and why does it have the characteristics we observe today? Physical properties are often an important clue to the operation of a system. This principle holds true for the soil layer. Finally, the soil in people's gardens is just one of many types of soil. Each soil has a unique story of development, the details of which can be detected by close examination of the soil system.

THE NATURE, PLACE, AND VALUE OF SOIL

Soil can effectively be compared to a living system. It breathes, uses water, has inputs and outputs, and has a life cycle just like a living creature. Using more formal terms, the United States Soil Conservation Service defines *soil* as "a mixture of fragmented and weathered grains of minerals and rocks with variable proportions of air and water; the mixture has a fairly distinct layering; and its development is influenced by climate and living organisms." Weathering, the chemical alteration and physical disintegration of earth materials by the action of air, water, and organisms, is more closely examined in Chapter 18. Here we will examine the rest.

The term "soil" actually means different things to different people. There are some people in the world who actually eat soil for medicinal purposes. More conventionally, the agriculturalist regards soil as the few top layers of weathered material that plants root and grow in. Geologists use the term to refer to all materials that are produced by weathering at a particular site. Thus soils that were produced thousands of years ago, and are now covered by layers of other material, are still considered soil even though it is impossible to grow plants in them. Civil engineers look upon soil as something to build on, something that clogs up their dams, and anything that doesn't have to be blasted away. Physical geographers prefer the viewpoint of the geologists.

Soil is obviously located on the earth's surface. It is not, except in small quantities, rightfully found in the air, although when dry or mismanaged, soil is subject to the action of wind. Nor should it be found, except in small amounts, in rivers or behind engineers' dams, although the action of water may put it there. The rightful location is at the interface of the atmosphere, hydrosphere, biosphere, and lithosphere. The soil layer in fact presents one of the most active interfaces among all four of the earth's main spheres.

The value of soil is incalculable. The soil provides a medium for the plant growth that supplies our food (see Figure 10-1). It is little wonder that one of the first tasks of the Mars probe was to scratch into the soil of the moon in search of any possible forms of life. Benjamin Franklin once said that, if you want to know the value of money, try to borrow some. The same holds true for soil. Oklahoma was settled by whites in

FIGURE 10-1

A rich, fertile soil excellent for growing crops.

FIGURE 10-2
Fine-grained soil particles blown away in Colorado in a scene similar to that of the Dust Bowl of the 1930s.

the last years of the nineteenth century. About 35 to 40 years later, a dark cloud hung over the city of New York and the surrounding coast. The cloud, on its way to the ocean, was the dust from the topsoil of Oklahoma and other nearby areas. Figure 10-2 shows a similar event in Colorado. The so-called Dust Bowl of the 1930s had a devastating influence on American agriculture. The mismanagement of the precious soil was a mistake that must not be repeated.

THE FORMATION OF SOIL

At one time in earth history there was no soil. Bare rock and material freshly thrown from volcanic processes dominated the earth's surface. Gradually, with the formation of the atmosphere, the soil began to form, a process that is still going on.

SOIL COMPONENTS

If you pick up a handful of soil, you can see that it contains four items. Figure 10-3 depicts the rock particles made of minerals, the organic matter of various types, the water that usually clings to the surfaces of the rock particles, and the air in the intervening gaps. Let us examine these items in more detail.

A *mineral* is a naturally occurring chemical element or compound with a crystalline structure formed of matter that is neither animal nor vegetable. Almost 100% of the earth's crust is composed of oxygen, silicon, aluminum, iron, calcium, sodium, potassium,

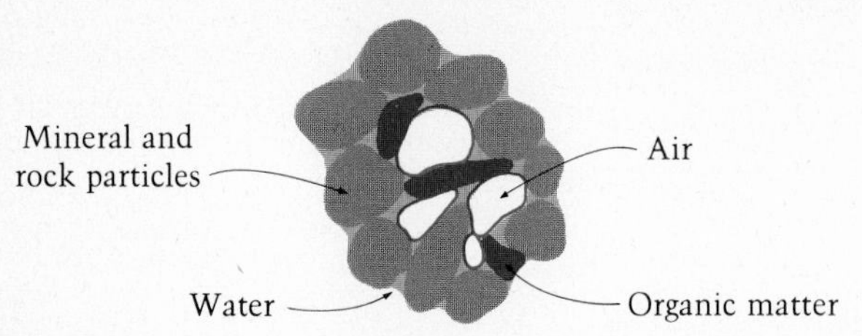

FIGURE 10-3

The four major components of soil—minerals, organic matter, water, and air.

and magnesium. These elements combine in various ways to form minerals. There are thousands of combinations and therefore thousands of minerals. However, minerals tend to fall into the six groups shown in Figure 10-4. Some of the mineral groups have familiar uses. Feldspars, for example, are widely used in the ceramic and pottery industry. Quartz is used in making glass. Olivines include the garnets used in semiprecious jewelry. Pyroxenes, amphiboles, and a group that includes the micas, chlorites, and clays are all associated with particular types of rock (see Chapter 14).

The next soil component—*organic matter*—is material that forms from living matter. In the upper layers of the soil there is an accumulation of decaying and decayed remains of the leaves, stems, and roots of plants. There is also waste matter from worms, insects, and other animals. The decay processes are carried out by an astronomical number of microorganisms, such as bacteria and fungi. All of these contribute to the organic matter of the soil.

Soil also contains life-sustaining *water*. There is an electrical attraction between the mineral particles and the water molecules surrounding them. Normally water fills much of the space between the minerals. But even in very dry soils, the attraction is so persistent that there may be a thin film of water, possibly one or two molecules thick, around the mineral particles. The water is not pure but exists as a weak solution of the many chemicals found in the soil. Without water, the many chemical changes occurring in the soil could not take place.

Air fills the spaces among the mineral particles, organic matter, and water. It is not exactly the kind of air we know in the atmosphere. Soil air contains more carbon dioxide and less oxygen and nitrogen than atmospheric air does.

It is simple enough to deduce that these components came from the other earth spheres. The more interesting point is to discover how. Five important factors will direct us to the solution.

FACTORS IN THE FORMATION OF SOIL

Some of the factors in soil formation are obvious, whereas others are not so apparent. The rocks of the earth are the parents of soil, and so we give them, and the deposits formed from them, the name *parent material.* The atmosphere provides the water and air for the soil layer, but the processes of the atmosphere vary across the earth's surface. We must therefore look at *climate* as a distinct soil-forming factor. Because of the presence of organic matter, we might also expect *vegetation* and other biological agents to be a factor. Only close observation tells us that soils seem to vary

FIGURE 10-4

The way the eight elements of the crust combine into six major mineral groups.

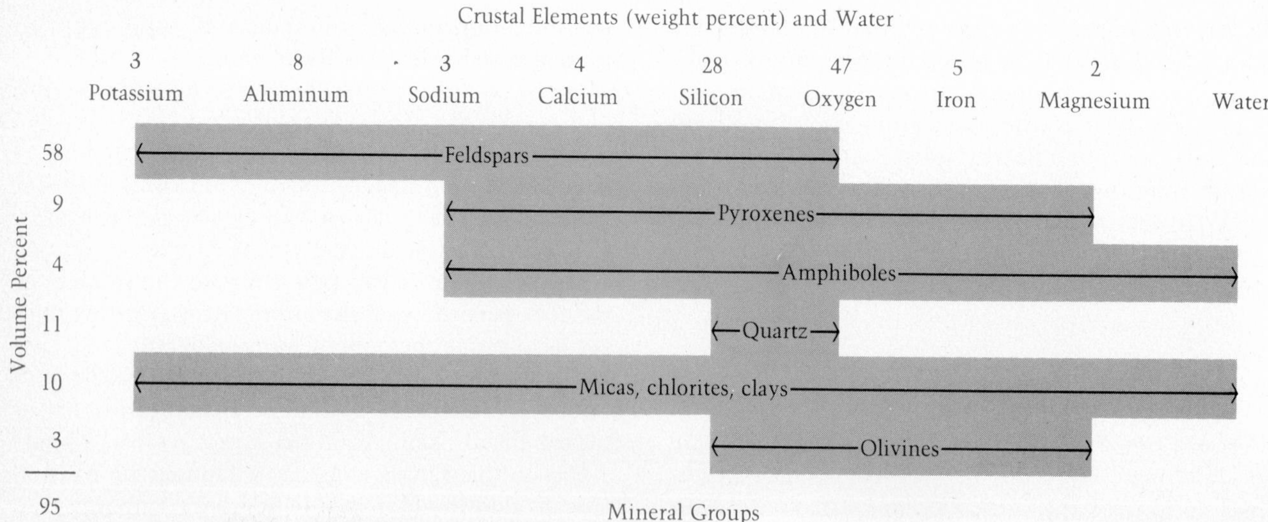

with the ruggedness of the landscape. Thus the *topography* is not so obvious a factor in soil formation. Neither is the last factor. Only by stopping to think could we conclude that the formation of soils must depend on *time*. Let us see how all these factors come into play.

Parent material When a soil forms directly from underlying rock, the soil minerals and type may bear a direct relationship to the original rock. This, the simplest kind of soil formation, gives rise to what is known as a *residual soil*. Thus on the Atlantic coast of the United States, thick red soils contain the insoluble residues of the iron oxides and aluminum silicates from the original rock. Such cases are quite common throughout the world. Yet even on the East Coast of the United States there may also be soil differences resulting from the variation of climate along the coast.

In a second category of soil, known as *transported soil*, the soil may be totally independent of the underlying solid rock because the parent material has been brought from somewhere else. During the most recent ice age, large quantities of soils and other materials were transported thousands of kilometers and deposited in new areas. These materials then formed the parent material for new soil formation. Many of the eastern and midwestern states have soils formed of such parent material.

The sediment laid down in streams and river valleys is another source of parent material. This type of parent material often creates fertile soils, as in the case of the lands bordering the Mississippi River.

In still other instances, the wind of the atmosphere carries and eventually deposits thick blankets of fine matter that form the parent materials of new soil. Such wind-blown material is called *loess*. Much of the Great Plains of the United States are underlain by loess deposits that are sometimes 30 m (100 ft) thick.

Climate In the Piedmont areas of Georgia and Maryland, the soil has developed from a rock called granite (refer to Chapter 14). Yet the resulting soils are not the same, because the two states have different climates. The warmer, moister climate of the Deep South has engendered a much more complete pattern of chemical change in the Georgian soils.

As far as soils are concerned, the important elements of climate are moisture, temperature, and wind. The amount of soil moisture is determined by the amount of precipitation and evaporation at a particular location. Both moisture and higher temperatures accelerate the chemical reactions of the soil. Thus we could expect to find thick, well-developed soils in the lower, warmer latitudes. Wind is another factor in the formation of some soil parent materials, such as loess. Elsewhere, as in the Dust Bowl of the Great Plains, wind is responsible for the removal of soil.

Vegetation Climate determines the types and amounts of vegetation that grow in an area. Outside the tropics usually the greater the amount of vegetation on the soil, the greater the amount of organic matter in the soil. Partially decomposed organic matter, called *humus*, forms a dark layer at the top of the soil. The most fertile soils are rich in humus. In many respects humus and vegetation form a closed system.

The substances that circulate through this system are the foods of plants—such as nitrogen compounds, phosphates, and potassium—collectively known as *plant nutrients*. We would need a magnifying glass to see the small bacteria and fungi that are the key to this circulation. "Decomposing" microorganisms continually change the nutrients into simpler compounds that can enter plants through their roots. Other bacteria "fix" atmospheric nitrogen so that it too can be absorbed. The plants use these nutrients in growth. When the plants die the decomposing bacteria return the nutrients to the soil to continue the cycle. This system turns into an open system when humans or animals remove the vegetation. Then the soil must often be balanced with artificial nutrients or fertilizers.

Several macroorganisms living in the soil also act as biological agents of formation. While Sherlock Holmes was performing his detective work in London, Charles Darwin was doing some detective work of his own. He was watching earthworms in a pot. His observations led him to conclude that worms can turn up to 6.5 metric tons per hectare (18.1 tons per acre) of soil every year. This mixing activity greatly helps soil formation and fertility. You may have seen advertisements for earthworms. Worms are a great aid to a healthy garden. Ants, termites, gophers, and scores of other burrowing animals perform the same function but usually not with such efficiency.

Topography Another factor that affects the formation of a soil is its location with respect to some of the earth's features. In the case of mountain climates (see Chapter 7), the aspect of a slope partially determines its receipt of radiation and therefore the amount of moisture evaporating from it. Slopes facing in different directions also receive varying amounts of precipitation. Their steepness affects runoff and thus the amount of moisture that penetrates to the lower layers of the soil. These phenomena to a large degree control the amounts of moisture and heat in a soil-forming area. A hillside might have a relatively thin layer of soil because of its efficient drainage. If the hillside faces away from the sun, this is even more likely, because both heat and moisture are minimized. In contrast, a less well-drained valley bottom, receiving a large amount of heat and moisture, is an optimal location for the chemical processes of soil formation. We could thus expect a deep soil layer in a valley (Figure 10-5).

Time Whether a soil is deep or shallow, it still needs a long time to form. The soil does not mix with the speed of the atmosphere. Thus time becomes an important factor in soil formation.

An example of relatively rapid soil formation comes from the volcanic island of Krakatoa. In its tropical climate, 35 cm (13.5 in) of soil developed on newly deposited materials within 45 years. The same process often takes much longer in colder climates. Some of the organic matter in arctic soils at Point Barrow, Alaska is still not thoroughly decomposed, even though it is around 2900 years old. The soils on the Jutland Peninsula, the Danish mainland, probably started forming 16,000 years ago.

Now we know what factors affect the formation of soil. Yet we still have not discovered exactly how soil is formed. Let us dig a hole in the soil in one location and see what we find. Pick up your shovel, because it is impossible to study soils without getting your hands dirty.

THE SOIL PROFILE

After about an hour of vigorous digging, we might create a hole, or soil pit, about 2 m (6.5 ft) deep. We can now perceive a major clue: The soil consists of a series of layers. Each of these layers is a *soil horizon* with one or more distinctive characteristics. All the soil hori-

FIGURE 10-5

Both slope and vegetation affect the soil on this hill in Greece.

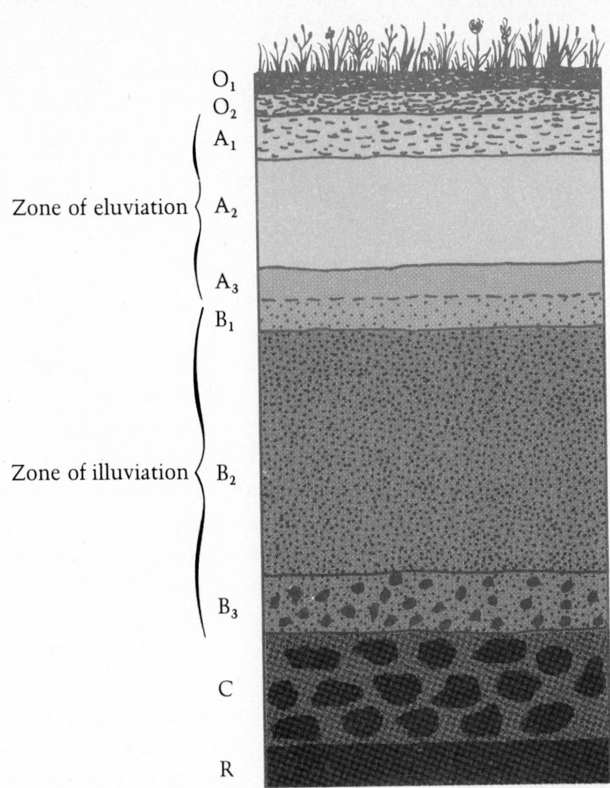

FIGURE 10-6
A soil profile showing the soil horizons.

zons together are known as a *soil profile.* A soil profile is to the soil scientist what a fingerprint is to a detective.

A typical soil profile in humid midlatitudes might look like that in Figure 10-6. The uppermost layer, rich in fresh organic material and humus from plant and animal remains, is called the 0_1 horizon. This lies above a slightly less dark layer, the 0_2 horizon, where the humus is partly decomposed. The next horizon, even less dark but still obviously containing organic material, is the A_1 horizon. The main part of the A horizon, A_2, is much lighter. Water continuously percolates through this light zone, dissolving and washing away many of the soil's mineral substances in a process called *leaching.*

Between the A_2 horizon and the B_2 layer is a transitional zone, with a gradation from lighter to darker material. This is designated A_3 at the top and B_1 at the bottom. The main part of the B horizon, B_2, is darker because many of the materials leached from the A layer are deposited in the B layer. The layers the materials are washed from are called the zone of *eluviation.* The letter *e* before a word often means "out of" or "from." The layers where minerals accumulate are called the zone of *illuviation.* Another transitional zone, B_3, lies between the B_2 horizon and the C layer. In the C layer, rocks and other substances of the parent material do not completely break down through weathering processes. The residual rocks make digging difficult. Digging often comes to a stop altogether at the R horizon of the parent material, especially if it happens to be solid rock.

This one diagram shows two of the most important soil-forming processes. The first, the weathering of the parent material, is quite clear at the bottom of the pit. The second is the downward percolation of water and the chemical effects associated with the percolation. The water transports oxygen and carbon dioxide, together with acids and nutrients, from higher to lower parts of the soil profile, and this accounts for the layered nature of the soil.

THE EXCHANGE OF IONS

Several other events take place in the soil, but we would need a highly powerful microscope to observe them. Particles within the soil are sometimes less than 0.0001 mm (0.000004 in) in diameter. Such finely subdivided substances are called *colloids.* The particles within them, called colloidal particles, may be formed of minute mineral or organic materials. Colloidal particles are sometimes electrically charged and sometimes electrically neutral. When they are charged, they attract and hold ions, the stray charged particles of chemical substances.

Within the soil, ions are continually shuttled between mineral surfaces and the surrounding liquid, between adjacent mineral surfaces, or between roots and mineral surfaces. Ions of elements valuable to plants, such as calcium, magnesium, hydrogen, aluminum, and potassium, transfer from the colloidal particles to the plants. Minerals containing these chemicals are referred to in soil science as *bases,* a term that has a more specific meaning in chemistry. The release of bases to the plant is known as *base exchange.* Other ions released in the exchange later form acids, alkalis, and salts that are dissolved and leached away. This exchange continues in all soils, although the presence of water is usually a necessary condition.

SOIL-FORMING PROCESSES

The excavation of one soil pit is not enough to tell us of all the processes occurring in the soil. Other pits and other profiles might look markedly different from the hypothetical profile we have described. So we will again pick up our shovel and magnifying glass and examine three different processes found in the soil. Climate, to a large extent, determines the differences.

CALCIFICATION

In the grasslands, the steppes, and the semidry areas of the world, a soil profile is likely to show a dark brown color in the A_2 horizon and a whitish B_2 horizon, as in Figure 10-7. This profile implies a process called calcification. In areas where the average evaporation exceeds rainfall, there is little leaching. Calcium and magnesium remain in the soil, in the A horizon, to be used by the grasses growing on top of it.

Meanwhile, in the B and C horizons, the air gaps between the soil particles act like drinking straws. If the hole of a straw is very fine, water moves up it because molecules at the water surface are pulling together. The rise of water up the straw, or through the small holes in the soil, as a result of the tension between the water molecules is called *capillary action.* In soils where calcification is occurring, water containing calcium carbonate is drawn from below to the B_2 horizon. Here the water evaporates, leaving the white calcium carbonate in the form of nodules or slabs.

FIGURE 10-7

A soil developed under the influence of the calcification process.

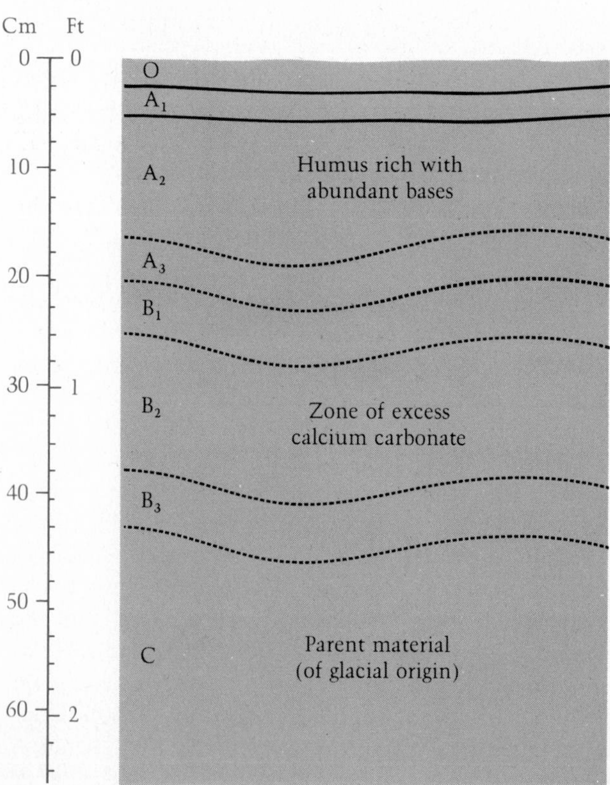

Sometimes it is left as dense stony layers called *caliche*. In Spanish, *caliche* means "a pebble in a brick."

Calcified soils are also noteworthy for the large amount of humus in both the A and B horizons. In the drier climates, microorganisms do not efficiently decompose humus. This is also true in the next process to be described, but for another reason.

PODSOLIZATION

In cooler climates, bacterial action decreases. This is typical of a soil-forming process named from the Russian word *podzol,* which means "ash." The soils resulting from podsolization look like ashes from a fire.

The most characteristic feature of podsolization is the leaching of the A horizon by water action. Humus accumulates in the O_1 and O_2 layers and produces acids. When the acids combine with rain water, they remove the bases and colloidal substances, together with iron and aluminum, from the A horizon. Most of these items end up in the B horizon, leaving the A horizon with only silica. Figure 10-8 shows the characteristic light gray color at the top of soil formed by podsolization and its dark color at the bottom.

FIGURE 10-8
A soil developed under the influence of the podsolization process.

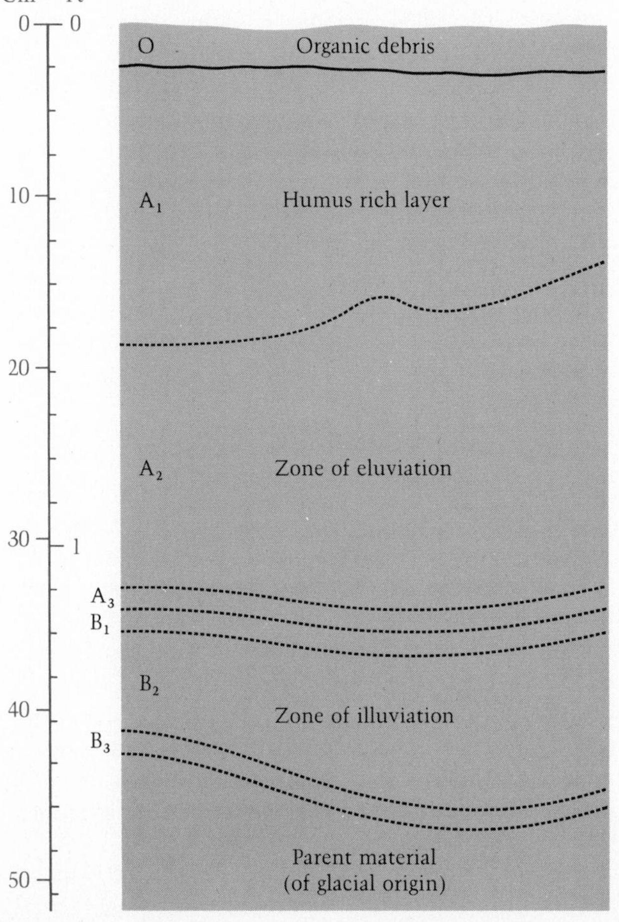

FIGURE 10-9
A localized cup podsol.

Sometimes leaching is limited to small areas, as when the rain water runs down tree trunks. Local *cup podsols,* like those of Figure 10-9, then form. It is common to find podsolization in the humid parts of midlatitudes and higher latitudes, and it is best developed in climatic areas where there are evergreen forests of fir, spruce, and pine.

LATERIZATION

In the high-rainfall areas of the tropics, large amounts of water combine with the heat to give three major results. First, the heat encourages energetic bacterial action. So organic matter is destroyed quickly, leaving little or no humus. Second, because there is little humus, there are few acids produced from it, and iron and aluminum are oxidized (combined with oxygen) and left to accumulate in the soil. The accumulations form a soil layer called an *oxic* horizon. This horizon and the process of laterization characterize the type of soils called *oxisols.* Sometimes the accumulations are so hard that they interfere with tropical agriculture. Third, silica leaches out, and residual iron and aluminum are found throughout the soil profile. Thus, with the exception of the oxic horizon, it would be difficult to pick out distinct layers in a soil pit dug in the tropics as Figure 10-10 indicates. If in fact we came across a hard oxic layer, we might not be able to dig far at all.

The three processes—calcification, podsolization, and laterization—are listed here in increasing order of leaching present. The major residual compound of the calcification process is calcium (Ca), and soils developed through this process are sometimes called

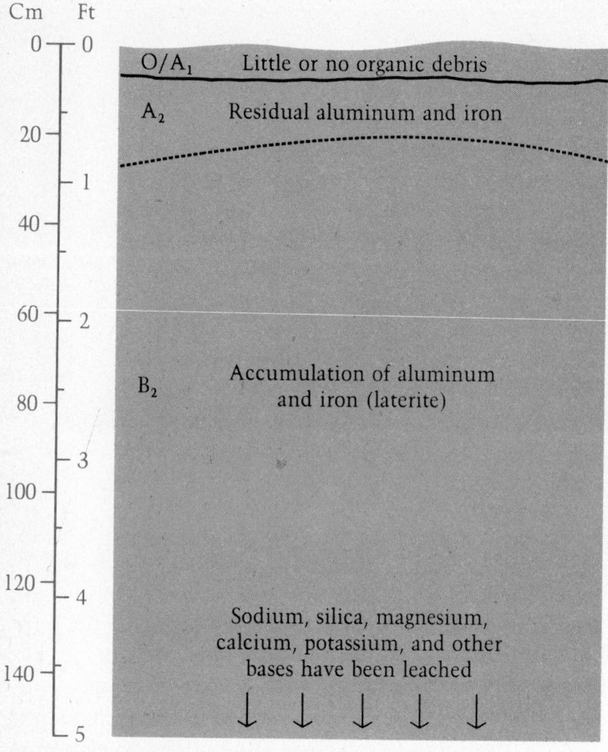

FIGURE 10-10

A soil developed under the influence of laterization. Chemical weathering, from high temperatures and rainfall, is so far advanced that parent material is not reached even at a depth of 140 cm (4.6 ft).

pedocals. Aluminum (Al) and iron (Fe) result from podsolization, and so the resulting soils are also known as *pedalfers.* The majority of soils are either pedocals, pedalfers, or oxisols (the result of laterization).

PHYSICAL PROPERTIES OF SOIL

Dr. Watson, listing Sherlock Holmes's attributes, remarked: "Knowledge of Geology—practical but limited. Tells at a glance different soils from each other. After walks has shown splashes upon his trousers, and told me by their colour and consistence in what part of London he had received them." If there are many different soils just in London, there must be an infinite number in the world. The large variety stems in part from the physical properties of the soil: texture, structure, color, and acidity-alkalinity. One of the most variable of these physical properties is the texture.

SOIL TEXTURE

Texture refers to the size of the mineral particles that compose the soil. The main categories of texture are gravel, sand, silt, and clay. According to the US Department of Agriculture, gravel has particles greater than 1 mm (0.04 in) in diameter. Particles with diameters between 0.05 and 1.0 mm (0.002 and 0.04 in) are sands. Silt particles have diameters between 0.002 and 0.05 mm (0.00008 and 0.002 in). Particles less than 0.002 mm (0.00008 in) across are called clays, the smallest of which are colloidal particles.

Soils usually contain one or more of the three smallest grain-size groups. Most soils tend to be coarser in the A horizon than in the B horizon, because the finer particles are most likely to be washed to the lower layers. Figure 10-11 shows the proportions of sand, silt, and clay in the various descriptive texture categories. A soil that contains all three categories is a *loam,* and a loam that is dominated by one of the categories bears the name of that grain size, as in "silty loam."

The textural names are important in the description of soils but are fundamental in indicating a soil's potential for holding water. Sandy soils hold very little water and thus, even in areas of high rainfall, barely support vegetation. Silts retain an intermediate amount of water, whereas clays hold so much water that they are often too waterlogged for most plants. An extreme ability to hold water is found in soils with large quantities of colloidal particles. For agricultural purposes, the ideal mixture is loam. Loam permits reasonable drainage, so air can circulate in the soil, and still retains enough water to support good plant growth.

The ability to support buildings also depends on the texture of the soil. Fine-grained soils, because of their water-holding ability, are more easily compressed and compacted than larger-grained soils. Gravel, for example, provides a strong base, because the individual particles touch and support one another. Sand mixed with gravel is even stronger, because the smaller sand particles fill the spaces between gravel particles, giving additional support.

SOIL STRUCTURE

The large-scale arrangement of soil materials commonly falls into four categories depicted in Figure 10-12. *Platy* structures have a flakelike appearance, with the soil particles arranged in overlapping horizontal planes. A *prismatic* structure has soil particles arranged in columns. This structure is often characteristic of loess soils. The prisms are usually about 0.5 to 10 cm (0.2 to 4 in) long. Another common structure, called *blocky,* is composed of irregularly shaped aggregates of earth that tend to have straight edges. Finally, many soils tend to have a *granular* structure that looks like layers of bread crumbs.

The structure of a soil depends on its mineral composition and texture. Structure is therefore related

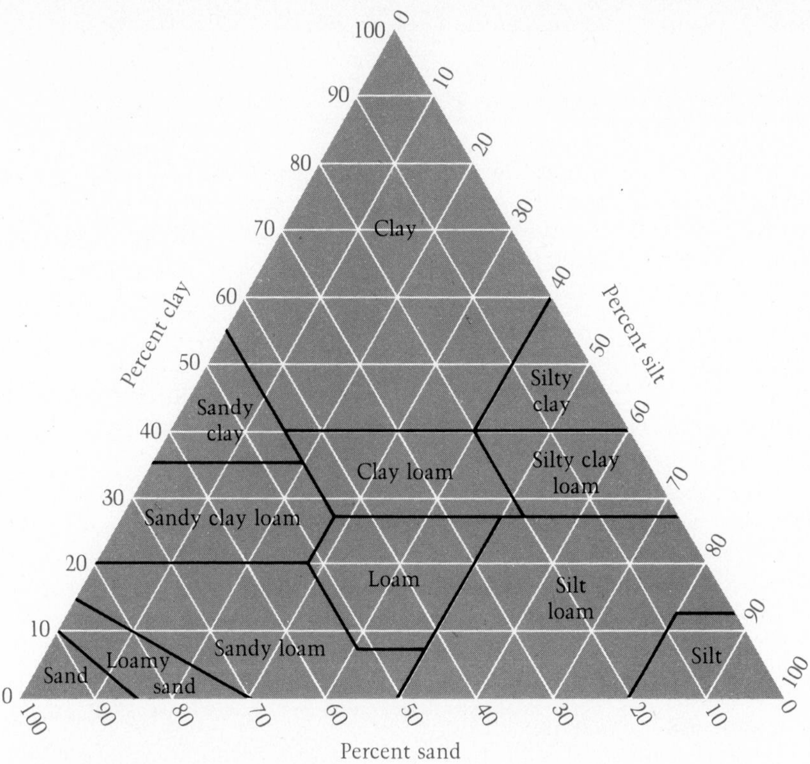

FIGURE 10-11
Texture categories of soils, defined by the percentage of sand, silt, and clay found in a soil sample.

FIGURE 10-12
Four principal kinds of soil structure.

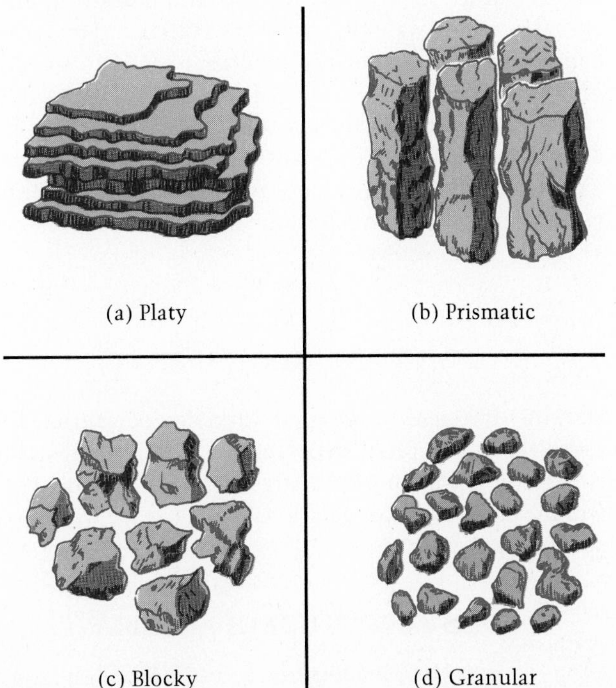

to some of the other properties of soil, such as its water-retaining ability, its ease of cultivation, and its susceptibility to erosion. In many areas, the original structure of a soil has been destroyed by mechanical and chemical agricultural practices. Sometimes when the soil structure is destroyed the soil is said to be *puddled.*

SOIL COLOR

In some traditional societies, colored soils provide artists with painting materials. The color of a soil, although not necessarily important in itself, gives the most obvious clues to the processes involved in its development. For example, the presence of humus usually accounts for the darkness of the upper layers. With decreasing humus content, soils may range in color from black through brown to white. Darker soils, with large amounts of organic material, are likely to be the most fertile, because the humus stimulates many of the chemical and biological activities occurring in the soil.

Iron compounds also give rise to a range of colors. Soil is red when the iron has been oxidized (rusted), as in the red clays of Georgia. Similar iron compounds give a yellow color to soils with plentiful moisture.

ARCHITECTURE ON THE EARTH

All man-made construction—houses, skyscrapers, factories, highways, airports, dams, and others—need a supporting foundation. Hard, consolidated rock obviously provides the strongest foundation. Yet surface distribution of hard bedrock is more common in regions of steep topography and high relief, where construction is more difficult and costly. In areas of gentle topography and low relief, ground surface is mostly loose, unconsolidated materials, either the weathered products of hard rock or transported sediments that are glacial, alluvial, or from mass wasting. Although several physical properties determine a soil's suitability for construction, one of the most important is its *bearing strength*, or ability to support heavy loads.

Soil texture largely determines its ability to support a heavy load. Fine-grained soils are more porous and therefore more easily compressed and compacted when loaded than coarse-grained soils. Gravelly textured soils have good bearing strength since the individual particles of hard rock are in contact with and support each other. Mixtures of sand and gravel are stronger still because the sand particles fill the pore space of the gravel particles and provide additional support.

Bearing Capacities of Earth Materials

MATERIAL	BEARING CAPACITY (metric ton/0.1m²)
Medium-soft clay	1.5
Loose fine sand	2
Loose gravel or compact coarse sand	4
Compact sand-gravel mixture	6
Sedimentary rocks, such as hard shales, sandstones, limestones, and siltstones	15
Massive bedrock, such as granite, gneiss, and basalt	100

Abridged and adapted from Léo Laporte, *Encounter with the Earth: Resources* (San Francisco: Canfield Press, 1975), pp. 133–136.

When the iron is reduced, the opposite of oxidized, grays or blues may occur. This often happens in poorly drained soils.

You may have noticed that soils also change their color according to their degree of wetness. With so many factors, the range of hues is almost infinite. Soil scientists use the standard Munsell Soil Color Chart to describe soil color objectively. The chart contains hundreds of different possible shades, each with a number and letter code. Thus soil detectives have to be observant to determine the source location of a soil by its color alone.

SOIL ACIDITY AND ALKALINITY

As we have discovered, soil colloids are associated with the presence of ions in the soil. Hydrogen ions are very common, and their presence in the soil solution defines an acid condition. Conversely, a relative absence of hydrogen ions makes the soil alkaline (or "basic") in nature. The acidity of soil is measured by its pH value. A pH value of 7.0 is considered neutral. Lower values (normally between 4.0 and 7.0) indicate acid soils, whereas higher values (7.0 to 11.0) indicate basic soils.

The acidity of a soil is closely related to its fertility, because acids are necessary to make nutrients available to plants. However, extreme acidity or alkalinity is detrimental to a soil. The best way to clear a soil of excess alkalinity is to flush it with large quantities of water. Unfortunately, alkaline soils often occur in dry climates, where water is not always available. The most common treatment for too much acidity is to spread lime on the soil or to drop it from the air. Different plants and microorganisms are adapted to varying degrees of acidity. The variation of acidity often bears a relationship to both climate and parent material and is associated with the different soil-forming processes, as Figure 10-13 shows.

SOILS AND GEOGRAPHY

Despite the large number of variables pertaining to soils, there are often geographical relationships between soil types, on both a large and a small scale. We will examine two such relationships. First we will look at the large scale.

SOILS ACROSS NORTH AMERICA

Suppose that we take a journey across North America, from the southwestern United States to Newfound-

WINE COUNTRY

European wine growers pass up the most fertile soil. Ironically, the best wine comes from difficult soil. Fertile land encourages each vine to support too many bunches of grapes rather than nourishing a few more carefully. Soil in central California can produce 1000 cases of ordinary wine per acre, yet some of the most distinguished vineyards in France yield only 150 cases per acre. In the Bordeaux region of France the richest soil lies along the riverbanks, but the wine laws prohibit planting on this land. These productive strips cannot grow the high-quality grapes associated with the region.

Certain varieties of grapes grow best in certain types of soil. In Europe the matching process took hundreds of years of experimentation. Today regions are famous for particular types of grapes. Cabernet Sauvignon grapes flourish in the Bordeaux region. Pinot Noir and Chardonnay thrive in the Burgundy region of France. Riesling grapes are cultivated along the Rhine River.

Each plot of soil creates slightly different wine. Traditionally, the small vineyards in France are called *climats*, which suggests the English word, microclimates. The owners realize that the soil, subsoil, drainage, and exposure to the sun make each slope unique. One chateau may bottle common wine while on the adjoining land the same techniques result in a spectacular vintage. Experts can identify the source region of the grapes by the taste of the wine.

pH	Acidity	Lime Requirements	Frequency of Occurrence	Soil Forming Process
4.0	Very highly acid	Required except for crops requiring acid soil	Seldom	
4.5			Often	Podsolization
5.0	Highly acid	Required except for crops tolerating acid soil	Very often found in cultivated soils in humid climates	
5.5	Moderately acid			
6.5	Slightly acid	Not usually required	Very often found in arid and subhumid climates	Laterization
	Neutral	Not required		
7.0	Slightly alkaline			Calcification
8.0	Moderately alkaline			
9.0	Highly alkaline		Sometimes found in desert areas	
10.0				
11.0	Very highly alkaline			

FIGURE 10-13
The pH scale of soil acidity and its relationship to the major soil-forming conditions.

land (see Figure 10-14) and carry with us our spades, magnifying glasses, and acidity-testing equipment. In Arizona and New Mexico, we find thin desert soils with little humus and some accumulations of salts and calcium carbonate.

Farther to the northeast, the soils become thicker and progressively brown and *chernozem* (the Russian word for "black") in color. There are still accumulations of calcium carbonate in the B horizon, but organic material becomes much more marked in the higher layers. The calcium carbonate becomes much less apparent in the prairie soils of northeastern Oklahoma, eastern Kansas, and Iowa.

Farther along, in the Ohio Valley, the carbonates are replaced by accumulations of clay and iron in the B horizon. Leaching of the A horizon is apparent from the pale colors found below the noticeably dark humus layers near the surface. Podsolization has therefore replaced calcification as the dominant soil-forming process, and it becomes more marked the farther we proceed to the northeast.

By the time we reach Newfoundland, podsolization is clearly dominant, but because of the cold conditions, chemical alteration of minerals and organic material is slow. Thus the depth of the soils decreases once more. In some cases, these so-called tundra soils do not have distinctive profiles but consist of thin layers of sandy clay and raw, highly acid humus.

A few minutes' study of Figure 10-14 indicates the close geographic relationships among the soil depth,

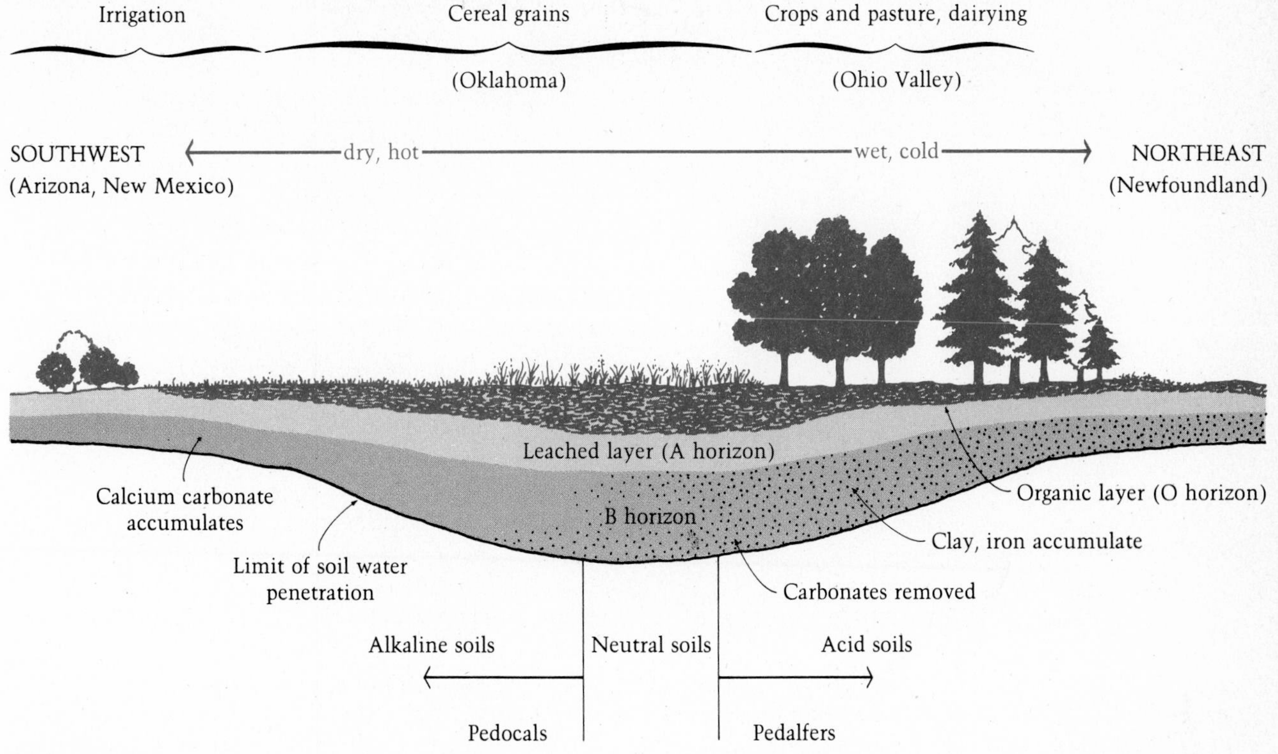

FIGURE 10-14

The soil across North America, from Arizona to Newfoundland.

acidity, natural vegetation, possible agricultural use, and of course, the climate that has played such a large part in its formation. Analogous relationships can also be found on a smaller scale.

SOILS OF HILLS AND VALLEYS

Soil formation depends on topography, and this is often demonstrated in the spatial arrangement of soil on single slopes or in groups of landforms. An example of the different soils found on varying landforms is diagramed in Figure 10-15. The soil profile of this undulating landscape looks much like the first profile we examined in Figure 10-6. In the flat upland, the soil is noteworthy for the thick B horizon. This horizon collects most of the materials that had previously been leached to form a dense clay layer. Hilly topography is usually associated with thin soil, a result of the rapid drainage of water on the steeper slopes. In poorly drained meadows and bogs, it is often difficult to distinguish A from B horizons. Therefore, if you knew the prevalent type of landforms, you might be able to make a reasonable guess of at least the depth of soil and to some extent the type of soil.

On hillsides in the southern Sudan, and in many other parts of the world, soils have a characteristic arrangement from the top to the bottom of the slopes. As Figure 10-16 shows, lateritic soils and a capping of hard laterite (the name sometimes given to a very hard oxic horizon) are found on the top of the hillside. Some of the soil particles washed downhill by water come to rest farther down the slope. They form a material known as *colluvium,* and the soils that develop from this are called colluvial soils. Most of the material brought down slopes by water ends up as *alluvium* on the valley floor, and alluvial soils may develop from it. These are often poorly drained. Thus different soil types are arranged along the hillside.

Such a series of soils is called a soil *catena,* from the Latin word meaning "chain" or "series." A catena is usually defined as a sequence of soil profiles appearing in regular succession on landform features of uniform rock type. It is important to make sure that the rock type, or parent material, is homogeneous before deciding that a catena exists. The Thames River basin near London is distinguished by a number of different parent materials, and so Mr. S. Holmes would not have been able to use the catena concept in remembering his soil types.

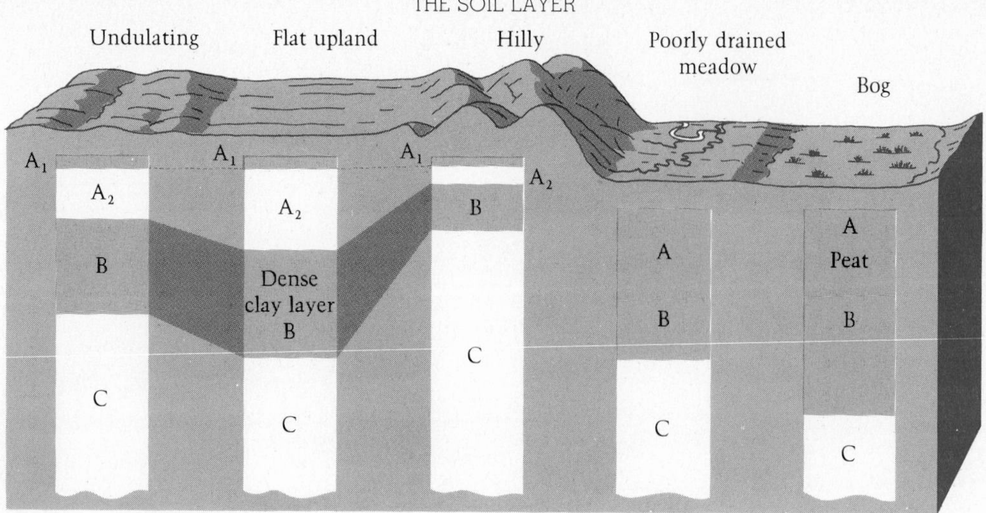

FIGURE 10-15
The soils of different landforms.

THE SYSTEM OF SOIL DEVELOPMENT

We are now close to putting together all the facts in the mysterious case of the development of the soil. We can conveniently place them into a coherent picture if we regard the total process and the soil profile in terms of a system, as Figure 10-17 does. An overall view of this system reinforces the statement about the role of the soil as an interface among the atmosphere, biosphere, lithosphere, and hydrosphere.

The four main spheres are the starting point of the soil development system. The atmosphere and hydrosphere provide heat and moisture, while the lithosphere and biosphere furnish the materials. Then a variety of chemical and physical weathering processes break down and transform the parent material and organic matter. The biosphere provides not only organic material, but when this material is broken down, it also gives new chemicals, especially acids.

FIGURE 10-16
A common type of soil *catena* in the southern Sudan.

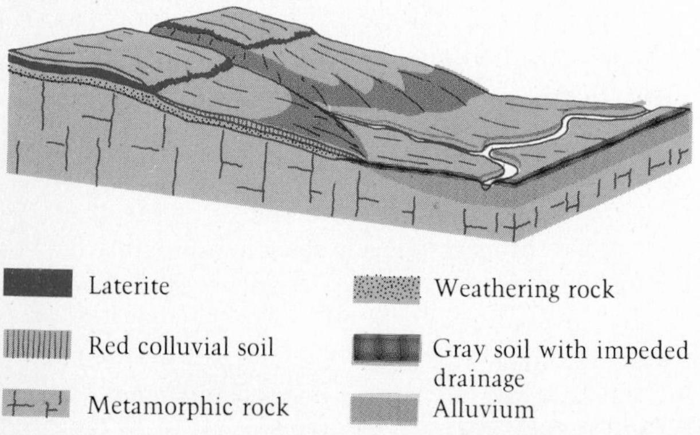

These can contribute to further weathering of the inorganic matter of the lithosphere. At the same time, the chemical and physical weathering processes can release nutrients to the plants of the biosphere. The interaction between the biosphere and the soil of the lithosphere is therefore both vital and reciprocal.

As a result of the weathering process and the decay of the plants of the biosphere, the soil system ends up with four items. First, there is organic matter. Second, there is the resistant residue that cannot be altered in any way by the weathering processes. This often takes the form of silica, such as the quartz particles we find in most soils, particularly sandy soils. Third, there is a whole host of newly altered chemical compounds, such as oxides and carbonates. These include the various clay minerals, which by ion exchange react with the fourth item, the soil solution. The soil solution contains many of the minerals extracted from the original parent material.

These four items are then subjected to various processes of dispersion, migration, and aggregation. Some of the more important of these processes are the flow of water through the soil under the action of gravity, the upward movement of water by capillary action, and the evaporation of water from the soil. Water is clearly important. Although the diagram does not include such events as the carbon dioxide and nitrogen cycles, the groundwater part of the hydrologic cycle is illustrated.

The final result of these, and many more, processes is usually the soil profile, with its distinctive horizons. We need all our powers of observation and deduction to gain some understanding of soils, because all the parts of the system often operate simultaneously. Soils are a dynamic component of the physical world.

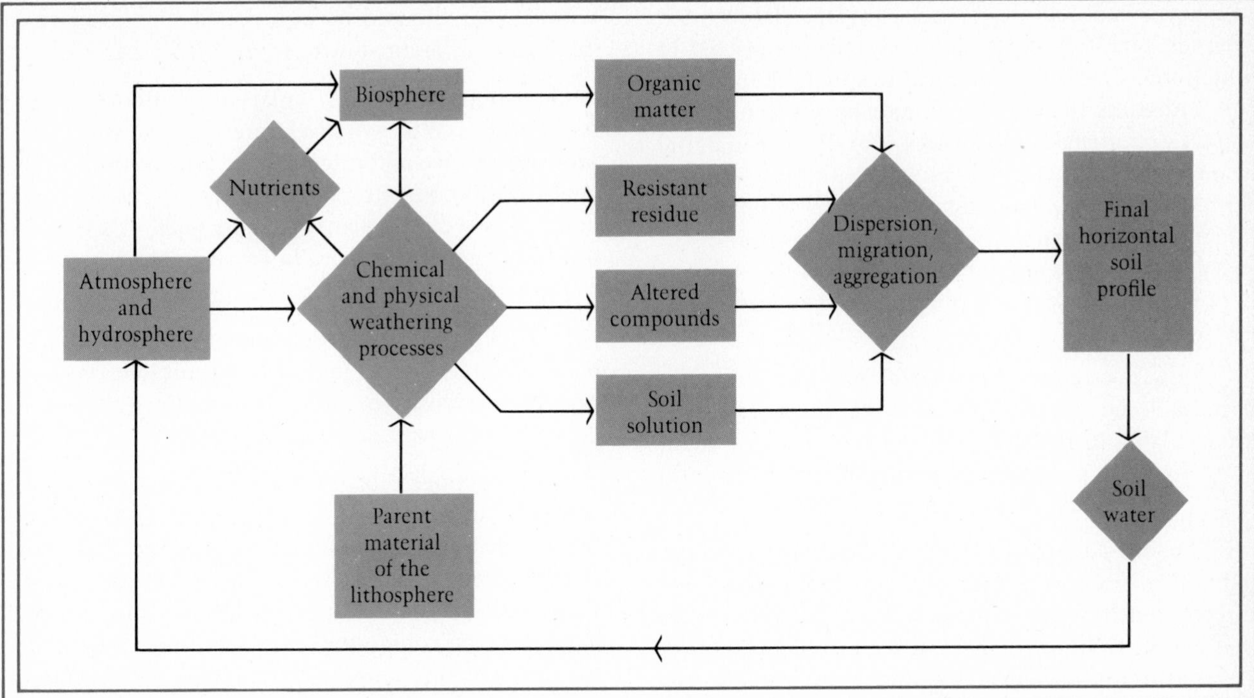

FIGURE 10-17
A simplified system of soil development.

SUMMARY

Soils are most important in providing the medium for plant growth and, subsequently, our food. Soils are composed of minerals, organic matter, water, and air.

The major factors in the formation of soils are parent material, climate, vegetation, topography, and time. Weathering of parent material at the bottom of a soil and the transportation of materials down through the soil by water often lead to a layering of the soil—called the soil profile. Different soil profiles develop in different climates. Common soil-formation processes include calcification, podsolization, and laterization. The most important physical properties of soils are texture, structure, color, and acidity.

Many factors and processes are at work in forming soils. To a certain extent, these features show a geographic organization on both a large and small scale. The factors and processes may also be regarded as forming one soil system.

QUESTIONS

1. Although soil is intimately linked with the biosphere, the atmosphere, and the hydrosphere, it is usually considered part of the lithosphere. How would you explain this classification?

2. Why is soil considered important enough to be studied as a scientific subject in its own right? What types of historical events have pointed up the importance of soil for human survival?

3. According to the chart in Figure 10-4, which group of minerals is the most complex? Which group is the most important in terms of the percentage of the earth's crust it composes? Which two elements together compose more than 75% of the earth's crust? How is this reflected in the mineral content of the earth's crust?

4. In what way does the soil layer appear to be a synthesis of the four major spheres of the earth? In what ways are the components of the soil similar to their counterparts in each of the four spheres, and in what ways are they different?

5. What type of soil, residual or transported, would you expect to find in upper Michigan? Why? What factors would have to be taken into account before we could actually determine that the soil is residual or transported? What changes might become apparent with increasing depth?

6. Three climatic factors are important for soil formation: moisture, temperature, and wind. Large,

well-developed deposits of soil correlate closely with high temperatures and high moisture content, both of which help speed the chemical reactions involved in soil formation. In addition, climate largely determines the vegetation present in an area, and thus the material available for humus development. With these ideas in mind, what are the major climatic factors which would affect soil development in (a) a tropical rainforest? (b) in a desert, such as might be found at latitude 30° N or S? (c) in a tundra region in the very high latitudes of North America?

7. How does commercial farming interrupt an important closed system in the soil? What remedial steps are available to make up for this loss in the system? How does the entire cycle of the system occur, and what part does soil play in the process?

8. If a hill having the shape of a regular four-sided pyramid is oriented so that each of the four sides faces directly north, south, east, or west, which side would you expect to have the best developed soil? Can you answer this question from the information given, or is soil formation dependent on other, additional topographical information? If so, what else?

9. Time is an intangible but crucial variable in soil formation. In the examples given in this chapter, the tropical climate of Krakatoa and the arctic climates of Point Barrow and Jutland, what factors influence the amount of time needed for soil formation? Upon what factors in general might the time of soil formation depend?

10. What categorizing principle is used to designate the horizons of the soil profile?

11. In the calcification process found in the world's grassland, steppe, and semidry region soils, what processes keep calcium and magnesium in the top layers of the soil? Which of these phenomena and processes contribute directly to this effect and which indirectly? How does climate affect the calcification process?

12. How is the characteristic color of podsols related to their formation? In what ways is podsolization related to climatic factors? Why are podsols common in areas of evergreen forests of fir, spruce, and pine trees?

13. Generally high levels of moisture are found associated with both podsolization and laterization. What are the climatic, biological, and chemical factors that cause different soils to form in each case?

14. Soil texture is an important attribute for the study of soils. Why would a farmer be interested in this property of the soil on his or her farm? Why would a civil engineer be interested in the soil texture near a dam? Why would a plant biologist need to know the soil texture characteristics in an area before he or she could explain the plant regimes found there?

15. How does soil color relate to other soil properties? If we know the parent material of a soil in terms of its mineral composition, do we automatically know its color? Why or why not?

16. It was once thought that merely by introducing sufficient quantities of water to arid and semiarid regions that agriculture could proceed much as it did in the more temperate climates, with their rich, fertile soils. This was later found to be overly optimistic. What factors would still limit agriculture after water was brought into an arid region? Would the same be true of soils beneath northern pine forests?

17. The Great Dust Bowl of the Great Plains states was a disaster for more reasons than most people realize. When that region was opened up for settlement in the mid-nineteenth century, farmers reported fertile, humus-laden soil to depths up to 30 feet. What accounted for the great depth of the topsoil in this area before it was lost by poor agricultural practices?

18. Why is long-term, successful agriculture usually found in relatively flat areas? What factors have been changed to permit intensive agriculture on the mountainsides in southern China? Where is most American agriculture located, and why?

19. What are the "ultimate" products of the process of soil formation, and how are they eventually distributed? What keeps the process of soil formation going?

20. How does the interrelationship of soil, climate, and plant life alter your thinking on the broad validity of the Köppen system of climate classification? Remember that the Köppen system classifies climatic areas largely on the basis of the plants found in those particular areas, correlating plant regimes with temperature and moisture regimes. What drawbacks can you see to such an approach?

THE LAYER OF LIFE

CHAPTER 11

Life on the earth's surface exists almost exclusively in a shallow layer, the biosphere, which can be regarded as a transition zone between the top of the lithosphere and the bottom of the atmosphere. The most visually dramatic evidence of the biosphere is the many plants on the planet. Only slightly less obvious are the multitude of animals. Biologists give the name *species* to a group of organisms that has the potential to interbreed and produce fertile offspring. The current estimate for our planet is 550,000 species of green plants and fungi and about 1,300,000 species of animals. Animal life is closely tied to the vegetation layer, and so in this chapter we will deal with both plants (*flora*) and animals (*fauna*) whenever it is appropriate.

Physical geographers are interested in the biosphere for three reasons. They want to know how it interacts with other parts of the physical world. They want to know the role of the species *Homo sapiens* (us) within the biosphere. Finally, they want to know the spatial distributions of the components of the biosphere, and how these distributions relate to those in the other spheres. We will emphasize the first two aspects in this chapter. The question of distribution is the subject of Chapter 12.

In this chapter, we will first examine how the vegetation layer formed, a series of events intimately related to the development of a process called photosynthesis. This vital process is the initial input of energy into the biosphere. We are recipients of the energy first captured by the plants. Beyond the question of energy, we will relate the development of plant and animal life to certain evolutionary processes. These processes help to explain both the spread of species and many aspects of their structure and life style. Finally, we will discuss the controls acting on the spread of vegetation species. But first we turn to how it all began.

THE BIRTH OF THE BIOSPHERE

The story of the development of life on earth parallels that of the formation of the atmosphere. The primitive atmosphere was rich in methane, ammonia, carbon dioxide, and water vapor. Around 1950, an American scientist named Stanley Miller put these components into a flask. He then subjected them to electrical discharges like lightning and to continual boiling. The result was amino acids, the building blocks of proteins, which are, in turn, the constituents of all living things. Another investigator froze a water solution of some of the constituents of the primitive atmosphere. Again, organic material formed, including one of the four components of DNA (deoxyribonucleic acid), the information carrier of life. Scientists suggest that these experiments repeat what actually took place in the physical world about 3 billion years ago to form complex organic molecules in the primitive ocean. The first simple forms of life—single-celled algae and bacteria—developed from these molecules.

Photosynthesis, as we will soon discover, releases oxygen to the atmosphere. This oxygen is necessary for breathing and for forming ozone (0_3). Ozone blocks the sun's lethal ultraviolet rays, and its formation allowed life forms to forsake the protection of the sea and emerge onto the land. As Figure 11-1 shows, this invasion of the land took place about 400 million years ago. Plants and insects led the invasion, to be followed later by amphibians and reptiles. Plant life, in many forms, spread throughout all the land that was neither too dry nor too cold. About 200 million years ago, the first mammals appeared, and 2 million years ago, the human species entered the scene. Humans arrived in the biosphere in the last twentieth of 1% of all earth history.

In this manner, a variety of life forms colonized the earth. There were many prerequisites, but one in particular continues to be crucial to life. This is photosynthesis.

PHOTOSYNTHESIS

The process of photosynthesis seems disarmingly simple. In the leaves of plants, solar energy combines with carbon dioxide (CO_2) and water (H_2O) to form carbohydrates and oxygen (O_2). Photosynthesis can be described in the following terms:

$$CO_2 + H_2O \xrightarrow{\text{solar energy}} \text{carbohydrate} + O_2$$

A fundamental requirement for photosynthesis is a green pigment, *chlorophyll*, at the surface of the plant. Figure 11-2 offers a close-up view of chlorophyll. The color of this pigment in part ensures that the correct wavelength of light from the sun may be absorbed for photosynthesis. It is no accident, therefore, that plants are green. If photosynthesis required another wavelength of light, our landscape might appear blue or orange.

The apparent simplicity of photosynthesis belies its many implications for the living and physical world. For one thing, it removes carbon dioxide and substitutes oxygen. This was critical in the atmosphere's formative years, and we are still thankful for the removal of carbon dioxide. If it were not removed, the temperature of the atmosphere would rise. Photosynthesis also produces carbohydrates, the food substances of plants. The natural production of organic food substances would cease entirely without photosynthesis. All our fossil fuels—coal, oil, and gas—have also been produced from carbohydrates originally formed in photosynthesis.

Water has a vital role in the production of food, even though photosynthesis itself uses little water. However, carbon dioxide from the atmosphere is not available for photosynthesis until it dissolves in water at the plant surface. Thus continued photosynthesis requires moisture. Small holes in the leaf surface, called *stomata*, are portholes for the water that arrives from the roots and stem of the plant. The more water there is at the stomata, the more carbon dioxide that is dissolved and the greater the production of plant food. Evaporation may also increase. Thus there is an intimate relationship between the amount of water lost in evapotranspiration and the production of organic plant matter, sometimes called *biomass* or *phytomass* (plant mass). Some of the results of this relationship appear in Figure 11-3 which shows potential food production on a global scale.

Of course, photosynthesis also depends on radiant energy from the sun. The production of one gram molecule of glucose takes 686,000 cal of radiant energy.

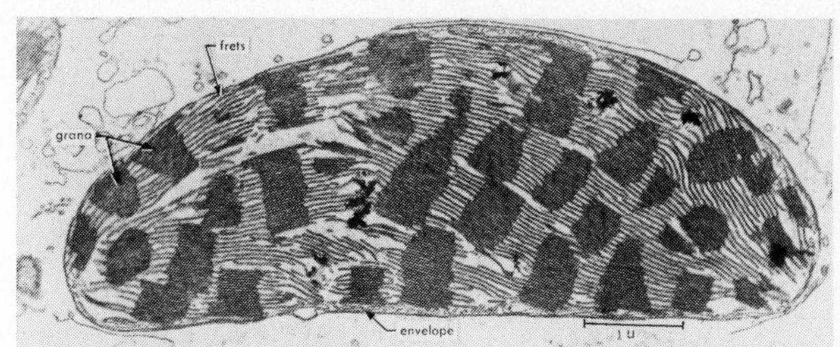

FIGURE 11-2
An electron micrograph of a corn chloroplast, the location of the chlorophyll necessary for photosynthesis. The scale at the bottom right measures a millionth of a meter.

FIGURE 11-1
The development of the biosphere on earth, in millions of years.

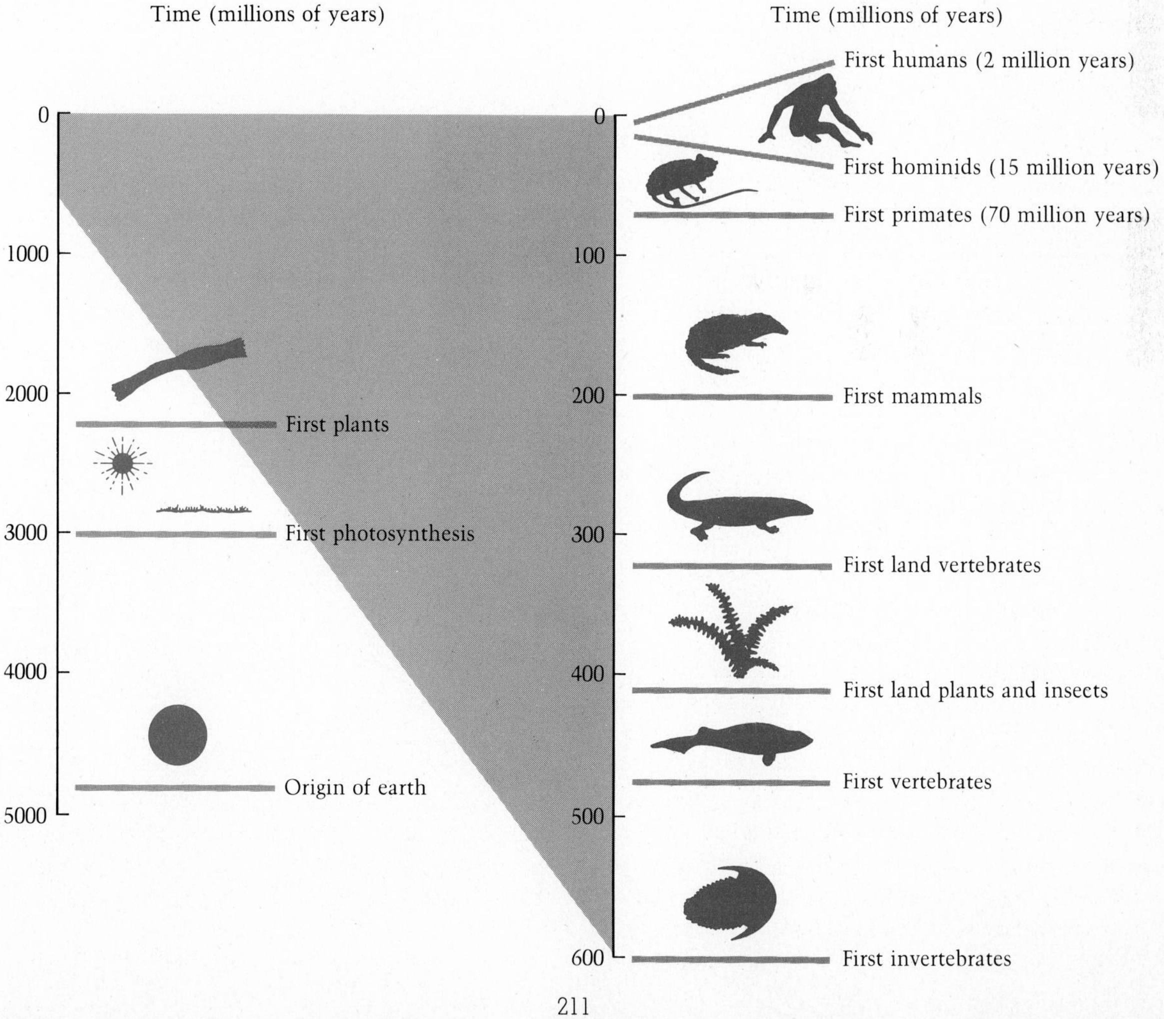

Time (millions of years)

0
1000
2000
First plants
3000
First photosynthesis
4000
Origin of earth
5000

Time (millions of years)

First humans (2 million years)
0
First hominids (15 million years)
First primates (70 million years)
100
200
First mammals
300
First land vertebrates
400
First land plants and insects
First vertebrates
500
600
First invertebrates

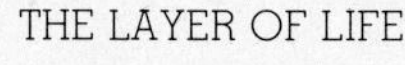

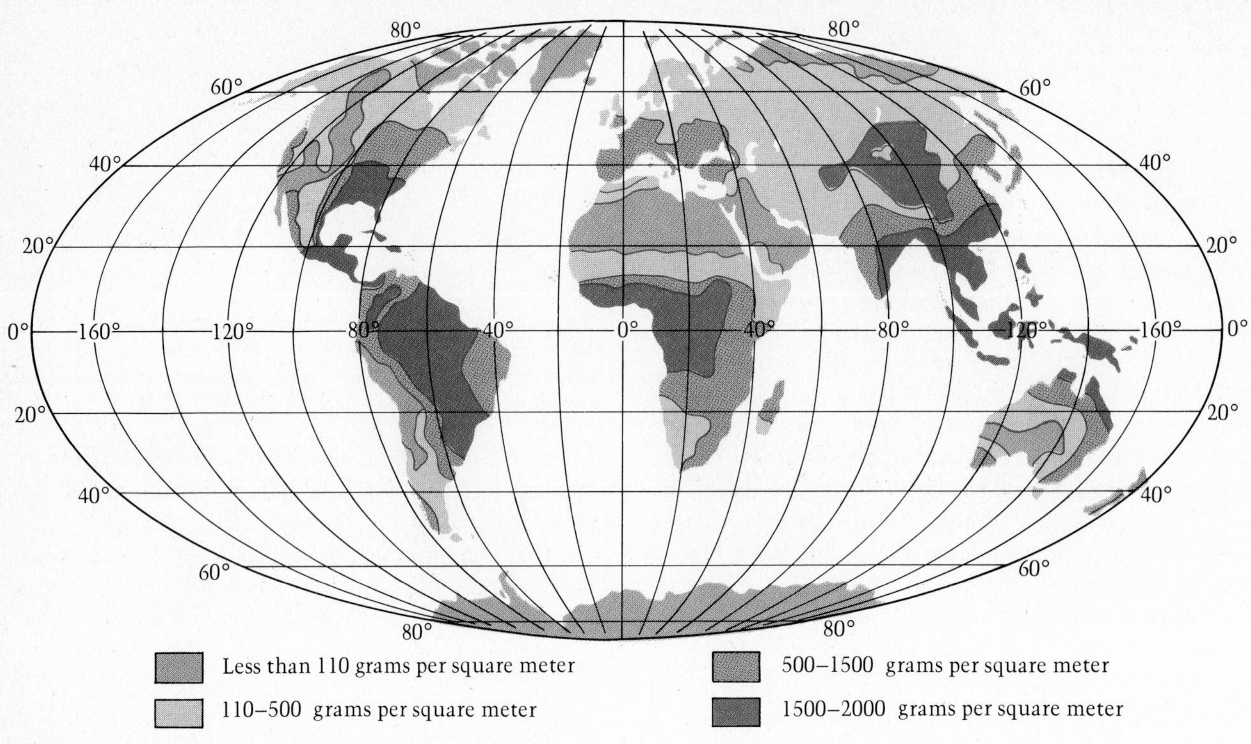

FIGURE 11-3

The global distribution of annual net productivity of organic matter in grams per square meter.

The Russian scientist A. Drozdov has correlated values of net radiation and the net production of organic matter on a global scale. Figure 11-4 plots his data. On this large scale, production increases equatorward at the rate of approximately 1 metric ton per hectare (1.1 tons per 2.47 acres) for every 2000 cal of additional net radiation. [1000 cal is called a kilocalorie (kcal).]

Photosynthesis is the key that lets solar energy into the life systems of the biosphere; it is the key that turns sunlight into our breakfast cornflakes. This is, as you might imagine, somewhat of a simplification, and we would do better to follow the paths of the energy through the life systems.

ECOLOGICAL ENERGETICS

In an *ecosystem*, plants and animals are linked to their environment through a series of feedback loops. It is an open system as far as energy is concerned: Solar energy is taken in, and chemical and heat energy are lost in several ways.

One such ecosystem is to be found in Silver Springs, a pond in Florida. In many ways this is a typical small pond. Energy from sunlight is taken up in photosynthesis by microscopic green plants, called *phytoplankton*, that in turn produce food. Organisms and all green plants that manufacture their own organic materials from inorganic chemicals are called *autotrophs.* The phytoplankton provide food for small larvae and other microlife forms collectively called *zooplankton.* These are eaten by small fish that are subsequently devoured by larger fish. Meanwhile, plants and animals die and decay, and this process releases chemicals back into the lake water to be used

FIGURE 11-4

Relationship between the production of organic matter by photosynthesis and the net radiation received at a location.

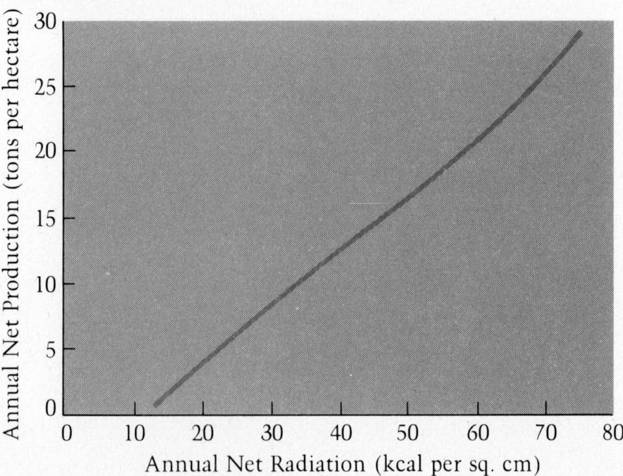

once more by autotrophs in the production of food. Thus the process by which food energy passes through an ecosystem is a continuous chain with numerous links—a *food chain*. Such a food chain exists in the ecosystem of Silver Springs.

THE FLOW OF ENERGY

The only unusual fact about Silver Springs is that the energy in its food chain has actually been measured. Such measurements are difficult to make, and it is worth our while to examine the results. The receipt and outflow of energy at the first stage, where autotrophs produce their own food, is shown in Figure 11-5. You can see that little of the incoming solar radiation, about 5%, is actually incorporated into the ecosystem. The phytoplankton, the autotrophs in this case, produce 20,810 kcal of energy per square meter per year. Only 3368 kcal is passed on to the zooplankton.

What happens to the remaining 17,442 kcal of energy produced by the phytoplankton of Silver Springs? Some 11,977 kcal are used in a process known as *respiration*, the oxidation of organic substances within cells that is accompanied by the release of energy. In other words, the autotroph, or any other living thing, burns its food to release chemical energy. Respiration is also the process by which our lungs take in oxygen, but in ecology we use the word in a more general sense.

An undetermined amount of energy is not used at all. This energy remains in the chemical bonds of the autotroph's molecules. When the autotroph dies, it falls to the bottom of the lake to add to the accumulating sediments. This kind of energy can be released hundreds of millions of years later when the sediments, in the form of coal, are burned in a fire.

Another unknown quantity of energy is used in the decomposition of the autotroph when it dies. The release of heat energy associated with the action of decomposing bacteria sometimes causes spontaneous combustion in compost heaps and other accumulations of organic litter.

ECOLOGICAL EFFICIENCY

Although autotrophic producers can easily fix energy from the sun in the form of food, there are many calls on this energy. Only part of it is passed on to the *herbivores*, the zooplankton and other animals that live on plants and other autotrophs. This is true all the way through the food chain of the ecosystem.

In a food chain, herbivores feed on the autotrophic producers, carnivores feed on the herbivores, and sometimes the carnivores are themselves eaten by so-called top carnivores. Thus in Silver Springs the phytoplankton are the producers, the zooplankton are the herbivores, the small fish are the carnivores, and the larger fish are the top carnivores. Each of these stages along the food chain is called a *trophic level*. *Trophe* means "food" in Greek. The producers are the first trophic level, the plant eaters are the second, and so on.

Central to the way the biosphere operates is the fact that, at each trophic level, chemical food energy is lost in respiration, decomposition, and random dispersal, so that only a small part of it is passed on to the next trophic level. In the Silver Springs food chain, diagramed in Figure 11-6, only 16% of the food energy

FIGURE 11-5

Energy flow in the Silver Springs ecosystem at the autotrophic production stage, in kilocalories per square meter per year.

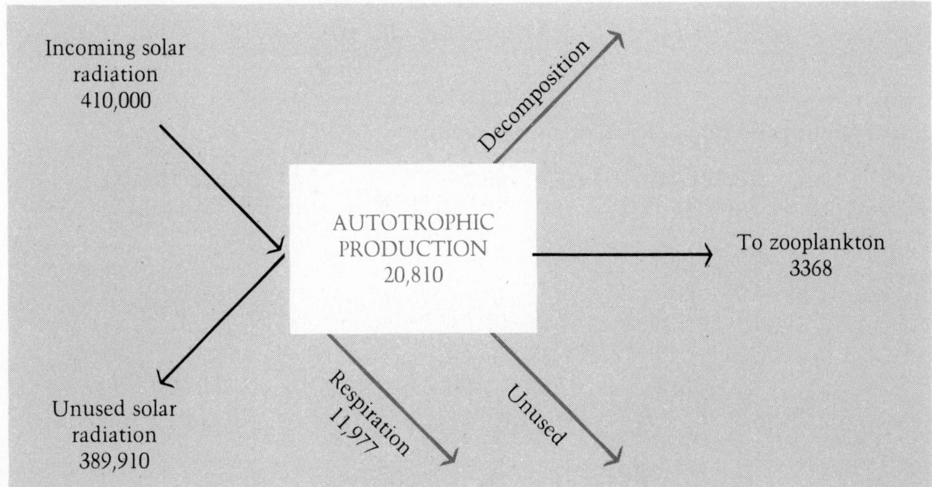

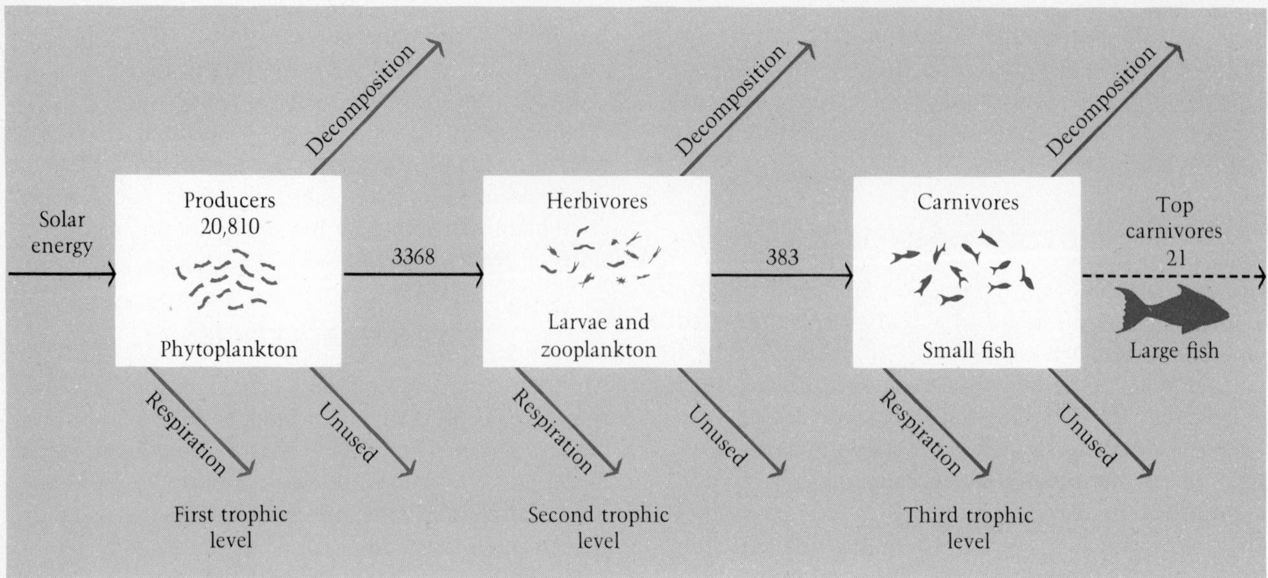

FIGURE 11-6

Energy flow through the trophic levels of the food chain at Silver Springs.

captured by the producers is passed on to the herbivores. Only 11% of the energy of the herbivores reaches the carnivores, and only 5% of this goes to the larger fish (top carnivores) when they eat the smaller ones. These percentages, or measures of the ability of organisms in a food chain to convert chemical food energy into living matter, indicate the *ecological efficiency* of the various trophic levels.

Different ecosystems have different degrees of ecological efficiency, as Table 11-1 shows. In this table, *net productivity* indicates the increase of living tissue in a given time period. The ecological efficiency of the producer stage, expressed as a percentage, is the amount of net productivity divided by the net radiation. The efficiency at the producer stage is usually less than 4%, but it varies with climate. Desert ecosystems have the lowest efficiency with values of less than 0.1%. Reedswamps in tropical climates are most efficient, with values ranging up to 4%. In land ecosystems, ecological efficiency at the producer stage seems to increase with rising values of net radiation and available moisture.

Efficiency also varies within the food chain. It goes from 16% at the second trophic level to 11% at the third level in Silver Springs. A pasture-steer-human ecosystem in North America has an efficiency of only

TABLE 11-1

Ecological efficiency at the producer (autotrophic) stage of various ecosystems

ECOSYSTEM	CLIMATE	NET PRODUCTIVITY (kcal/cm^2/yr)	NET RADIATION (kcal/cm^2/yr)	ECOLOGICAL EFFICIENCY (%)
Desert	Arid	0.04 ± 0.02	50–70	<0.1
Ocean		0.08 ± 0.04	35–125	0.1–0.2
Deciduous forest	Temperate	0.48 ± 0.12	40–70	0.7–1.2
Coniferous forest	Temperate	1.12 ± 0.28	30–60	1.8–3.7
Rainforest	Tropical	2.00 ± 0.40	70–80	2.5–3.0
Reedswamp	Tropical	3.00 ± 0.45	75–85	3.5–4.0

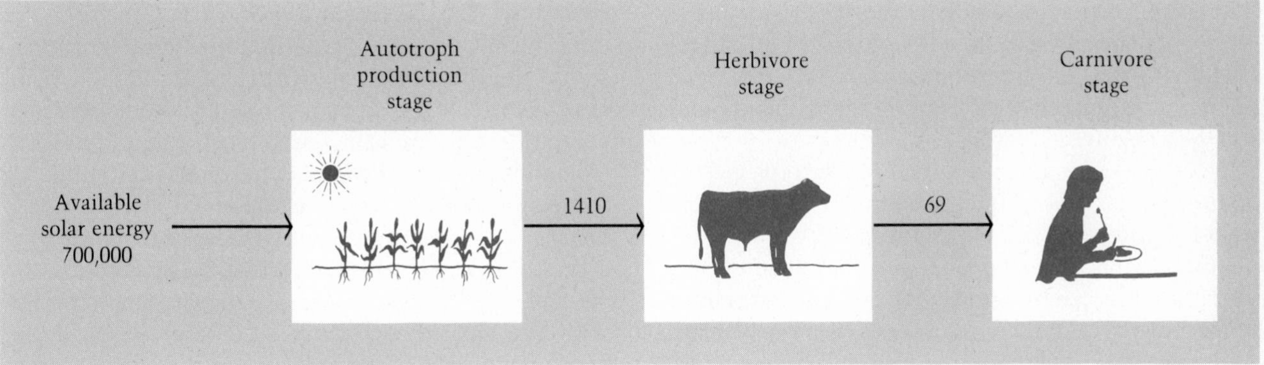

FIGURE 11-7

The transfer of energy through the food chain of a pasture-steer-human ecosystem. Units are in kilocalories per square meter per year.

4.8% between the producer and the herbivore stages, as Figure 11-7 illustrates. The ecological efficiency of energy transfer between trophic levels within ecosystems usually averages about 10%.

Several important consequences arise from the various efficiencies in ecosystems. First, there must always be a large number of producers to support smaller quantities of herbivores and even fewer carnivores. The masses of living material at each trophic level therefore stack up like the pyramid in Figure 11-8. Because only about 10% of the energy produced in the form of food is passed from one stage to another, to obtain enough food the animals at the higher trophic levels must have large territorial areas that provide enough of the species at the lower trophic level. This explains why large carnivores, such as lions, require a large territorial range.

Another consequence is that, because food and energy move along a chain in only one direction, the whole system collapses if earlier links of the chain are broken. Thus the removal of the autotrophs (grass) by too many rabbits in Australia caused the breakdown of a whole ecosystem.

A related consequence of the chain structure is the fact that undesirable materials can be passed along and concentrated by the ecosystem. The insecticide DDD was applied in Clear Lake, California to kill gnats. It was sprayed onto the water at a density of 0.02 parts per million. The DDD density was 5 ppm in the plankton, 15 ppm in the herbivores feeding on the plankton, 100 ppm in the fish, and 1600 ppm in the grebes (birds) that ate the fish. The grebes died.

It is therefore important to understand the nature of food energy flow through ecological systems. We turn now to look at a vegetation system that changes, and find once again that energy is crucial.

PLANT SUCCESSIONS

The flow of energy through a food chain illustrates the dynamic nature of the biosphere. Indeed, the vegetation layer is as dynamic as the atmosphere, soil, or crust of the earth. Some of the changes occur within a stable ecosystem, but sometimes one type of vegetation is replaced by another. This is called a *plant succession.*

There are three kinds of plant succession. A *linear autogenic* succession is when the plants themselves initiate changes in the land surface that cause consequent vegetational changes. "Linear" indicates that the order of succession in any one place is not normally repeated. Figure 11-9 shows how the growth of vegetation on an area that has been a lake is part of a linear autogenic succession. As the lake gradually fills with sediments, the water becomes chemically enriched.

FIGURE 11-8

Mass of living materials per unit of area in different trophic levels of an ecosystem.

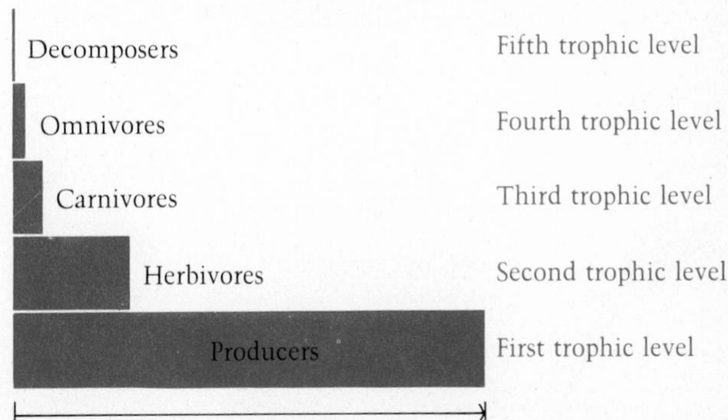

Plant productivity increases in the lake, and other plants encroach around its edges. Mosses and sedges and floating rafts of vegetation build up. After the lake has completely filled with organic debris, plants and trees may finally take over.

Sometimes one kind of vegetation is replaced by another, which is in turn replaced by the first. Possibly the original vegetation follows a series of two or three others. This is a *cyclic autogenic* succession. An example of this can be found at the northernmost limit of tree growth in Alaska, where permafrost lies beneath tundra vegetation of grasses, sedges, and bare ground. The permafrost melts to a sufficient depth in summer to allow colonization by willow scrub and later by spruce trees. Gradually, the forest becomes denser and forms a layer of litter. The permafrost, thus insulated, gradually rebuilds. The forest degenerates and eventually gives way to the original tundra vegetation. The cycle is then complete and ready for another turn.

A third type of succession occurs where vegetation changes because of some outside environmental force. Devastation through nuclear radiation might be one such force, but disease is more common. An epidemic of chestnut blight in the eastern United States created oak and oak-hickory forests where oak-chestnut forests had once existed. This kind of succession is termed an *allogenic* succession, because the agent of change comes from outside the plant's immediate environment.

Linear autogenic successions may start on completely bare ground, such as might result from the passage of ice, the formation of a river delta, or the eruption of a volcano. These are called *primary* successions. The successions starting from abandoned farmland, burned-over forest, or some other previous vegetation type are called *secondary* successions.

In all types of successions, the vegetation builds up through a series of stages, as indicated in Figure 11-9. Except in cyclic successions, these stages are called *seral stages* or *seres*. When the final sere is reached, the vegetation and its ecosystem are in complete harmony with the soil, the climate, and other parts of the environment. This balance is called a *climax community* of vegetation. The major vegetation types, described in Chapter 12, all represent climax communities.

The climax community is characterized by an ecosystem with a stable amount of accumulated energy. Solar energy is taken in and energy is lost through respiration and other processes, but the stored energy of the biomass is relatively constant. But in a plant succession, the amount of energy stored in the biomass increases as Figure 11-10 shows. Input of energy must exceed losses for the plant succession to develop.

Energy considerations are pertinent to all parts of the biosphere and have far-reaching implications. The evolutionary discoveries of Charles Darwin and Alfred Wallace, made earlier than the findings on energy flow, have had an equally profound effect on our understanding of the biosphere. Both discoveries show clearly that the living world is a world of movement and change.

FIGURE 11-9

An idealized sequence of a linear autogenic plant succession, by which a lake is eventually colonized by shrubs and trees.

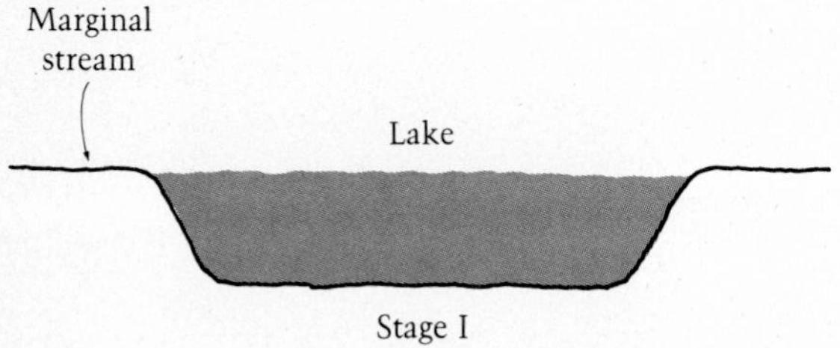

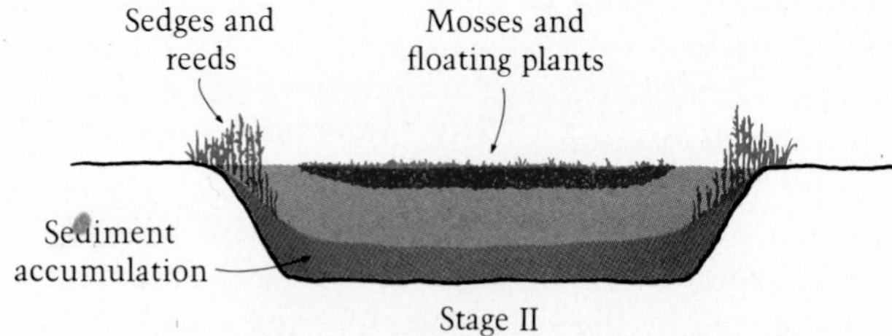

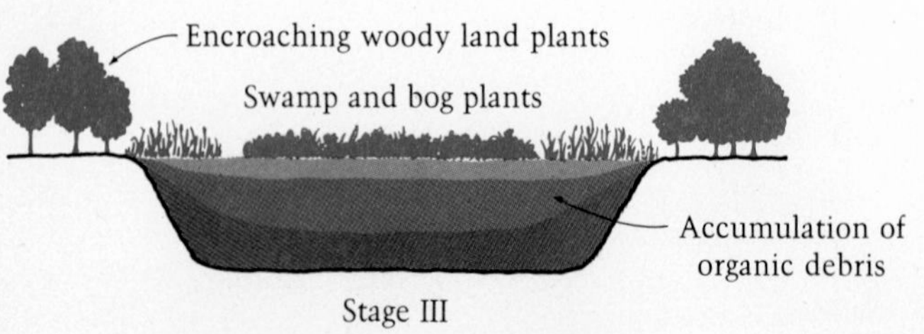

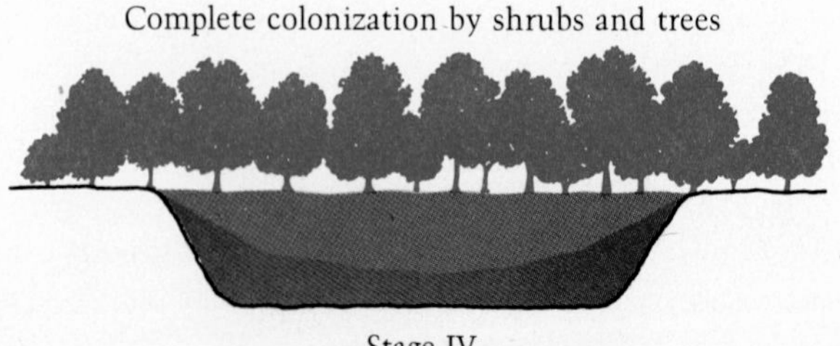

THE SCHOOLYARD AS AN ECOSYSTEM

One weekend a bulldozer ripped up the asphalt at a school playground in Berkeley, California. The school board exploded. But it was too late. The asphalt was gone, and plants had begun to grow. The children, with some help from a few landscape architects, created a new play space for learning about the environment.

In the past few years the children and some volunteers have established plant communities in the yard. In one corner grows a cluster of pine trees, and beside it, a struggling meadow with wild flowers. On a small ridge a stand of tiny redwoods has taken root. Each tree is claimed by one class. The class names their tree, measures its growth, and tends to its needs. Nearby, the chaparral thrives. The sturdy manzanita and wild mountain lilac bushes withstand the Mediterranean climate, as well as the jostling of young children. A small stream fills a pond that is a favorite spot for many of the children. They spend recess tossing in bits of paper and fishing them out, and they also get acquainted with the miniature marsh. One small plot is set aside for planting vegetables and flowers.

In just a few seasons the children have been able to watch the succession process. New grasses and wild flowers have crept in to fill bare spots. Paths have compacted from heavy use and lost their green cover. The Washington Schoolyard is now a special ecosystem where children can study and participate in its growth and change.

THE PROCESSES OF EVOLUTION

Up to the middle of the nineteenth century, students of the biosphere were confronted by a great problem: There appeared to be an infinite variety of plants and animals on earth, but the porcesses of nature appeared to act in some kind of orderly fashion. The formulation of the theory of evolution resolved this paradox by showing how both variety and order are essential components of life on the planet.

EVOLUTIONARY THEORY

The idea that there is natural variation within a species is basic to evolutionary theory. This natural variation stems mainly from the sexual reproductive process. Information carriers, called *genes*, from both parents join in such a way as to combine a degree of randomness, or luck, with a high amount of specification and continuity. Thus a child may have blue or green eyes but will most likely have only one head. The mechan-

FIGURE 11-10

The increase in stored energy of the biomass in a typical secondary autogenic succession. In this case, a deciduous hardwood forest takes over from an abandoned field.

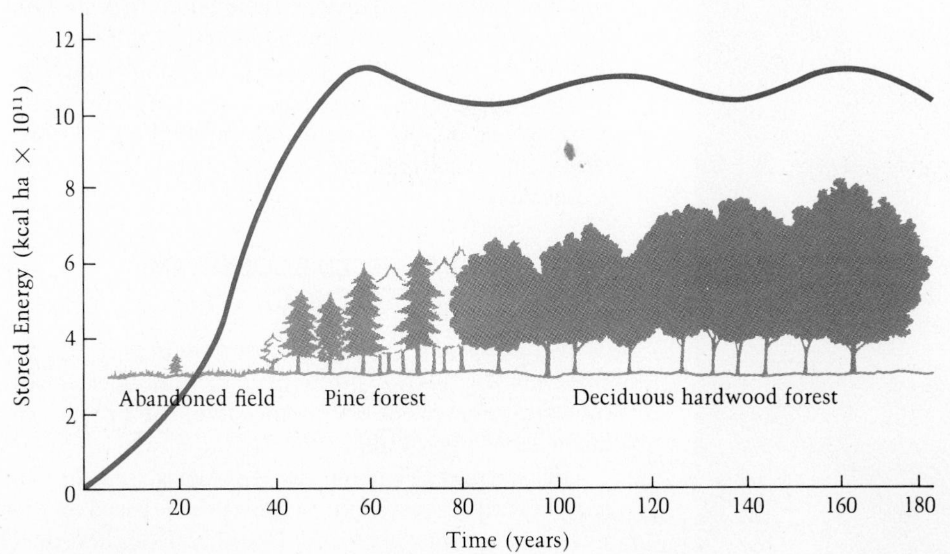

ics of continuity occasionally break down, and so the exact message of heredity is not always passed on. New species might originate from these *mutations*.

Another idea in evolutionary theory is that a species produces more offspring than will survive to reproduce. In all species, many immature young die by accident, predation, and disease. The pyramid shape of food chains is compatible with the idea that premature death is the norm. For example, many autotrophs must be eaten by the herbivores, and many herbivores are consumed by the carnivores. But the individual is not important to evolution; it is the population that evolves. An individual carries only a small proportion of the information from the gene pool of the entire population.

Each species also carves for itself an *ecological niche*, a way to make its living in nature. The environmental space in which an organism operates most efficiently in the niche can be very large, such as the large portion of earth occupied by *Homo sapiens*, or very small. The niche of the species *Myxotricha parodoxa* is in the digestive tract of Australian termites. A specialized niche reduces competition from other species but increases the risk of total annihilation. A wide niche overlaps those of other species but is more likely to survive a major environmental change. If humans depleted the resources of one area, they were, in the past, able to move to another.

FIGURE 11-11

The dark and light forms of the peppered moth on a soot-covered oak trunk in an industrial area. The light form is much more susceptible to being eaten in this particular environment.

The ecological niche is essentially an abstract concept, but a somewhat related idea is often used in practical cases of biogeographical study. The *habitat* of a species is the environment it normally occupies within its geographic range. Often the distributions of organisms can conveniently be considered in terms of vegetational type. Thus we may talk of a woodland, grassland, or seashore habitat. Each habitat may contain hundreds of ecological niches.

Another idea in evolutionary theory is that some offspring are better able to adapt to their environment, habitat, or niche than others. One example of this idea is the evolution of the giraffe's long neck. We may assume that neck length varied in early giraffe populations. The animals with the longest necks could reach highest into the trees and obtain more food than those with shorter necks could. Shorter-necked animals had to compete for food near the ground with several other species and were therefore less successful. Today giraffes have long necks. Another example is the peppered moths in Great Britain. Two varieties have developed. The dark form has evolved in areas of air pollution. You can see in Figure 11-11 that it is better hidden against a soot-covered oak tree than its light-colored companion is. The dark form's newly developed camouflage is a fine example of adaption to the environment.

Finally, the core of evolutionary theory is the idea that better-adapted organisms are more likely to survive and reproduce. Through their genes, they pass on the favorable aspects of their adaptation to their successors. As Figure 11-12 suggests, the long-necked giraffes were more successful in surviving and reproducing, so the genetic information that a long neck helps in getting food was passed on to subsequent generations.

A BALANCED ECOSYSTEM: THE SERENGETI PLAIN

Individual species develop and find their own niches, although the niches often overlap. But because of ecological energetics, the species live and interact in one balanced ecosystem.

The Serengeti Plain of eastern Africa, one of the most beautiful ecosystems in the world, is based on the grazing of savanna grassland. Figure 11-13 shows some

FIGURE 11-12

Three stages in the evolution of the giraffe's long neck. The earliest giraffes had a variety of neck lengths, but the ones with longer necks could reach more food in trees. These prosperous giraffes produced more offspring, who passed the genetic information concerning the long neck to later generations.

FIGURE 11-13

The balanced ecosystem of the Serengeti Plains, eastern Africa.

of the animals that live in the Serengeti. The herbivores are so finely adapted that they use different portions of the vegetation at different times. During part of the year, mixed herds graze on the short grass, satisfying their protein requirements without requiring them to use too much energy in respiration while obtaining the food. Eventually the short grass becomes overgrazed. Then the largest animals, the zebras and buffaloes, move into areas of mixed vegetation—tall grasses, short grasses, and herbs. Zebras and buffaloes eat the stems and tops of the taller grasses and trample and soften the lower vegetation. The wildebeests, a little like the American buffalo in appearance, can then graze the middle level of vegetation, trampling the level below. The Thompson's gazelles then move into the softened area to eat the low leaves, herbs, and fallen fruit at ground level. The pastoral indigenous people of the Serengeti Plains have long been a part of this well-balanced ecosystem. Some scientists believe that their regular use of fire has helped to maintain the short grass best adapted to the native animals.

The Serengeti ecosystem perfectly illustrates the arrangement of plants and animals that has developed under the rules governing both energy flow and evolution. An understanding of these rules can explain both the nature of species and their geographic spread throughout the biosphere.

GEOGRAPHIC DISPERSAL

Geographers are interested in the spatial distribution of the species of the biosphere, and that is the topic of Chapter 12. But here we will investigate the factors that determine the spread and geographical limits of any particular group of organisms. The limiting factors may be either physical or biotic.

Each species has a *range of optimum* where it can survive and maintain a large, healthy population, as Figure 11-14 shows. Beyond this range, the species suffers *zones of increasing physiological stress.* Although it can still survive in these zones, population is small. When conditions become even more extreme, in the *zones of intolerance,* the species is absent altogether, except possibly for short intermittent periods.

PHYSICAL FACTORS

Temperature is a common limiting factor for both plants and animals. Chapter 7 mentions the correspondence, suggested by Köppen, of the northern tree line and certain temperature conditions. Another example of temperature's effect on distribution was discovered by the English ecologist Sir Edward Salisbury in the 1920s. One species he studied was a creeping woody plant in scrubby country, known as the wild madder, *Rubia peregrina.* He found that the northern boundary of the wild madder in Europe coincides closely with the January 4.5°C (40°F) isotherm. This temperature is critical, because in January the plant forms new shoots, and lower temperatures would inhibit their development and subsequent growth.

Temperatures play such a large role in determining plant distributions that plants are sometimes classified according to their propensity for heat. Thus plants adapted to heat are called *megatherms,* those that can withstand low temperatures are designated *micro-*

FIGURE 11-14

A model of population abundance in relation to the physical factors in a species' environment.

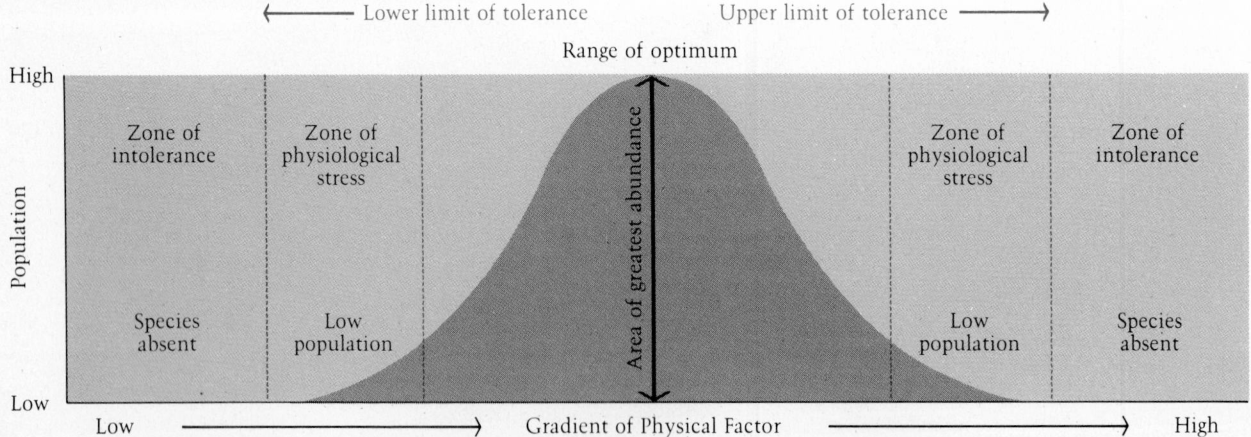

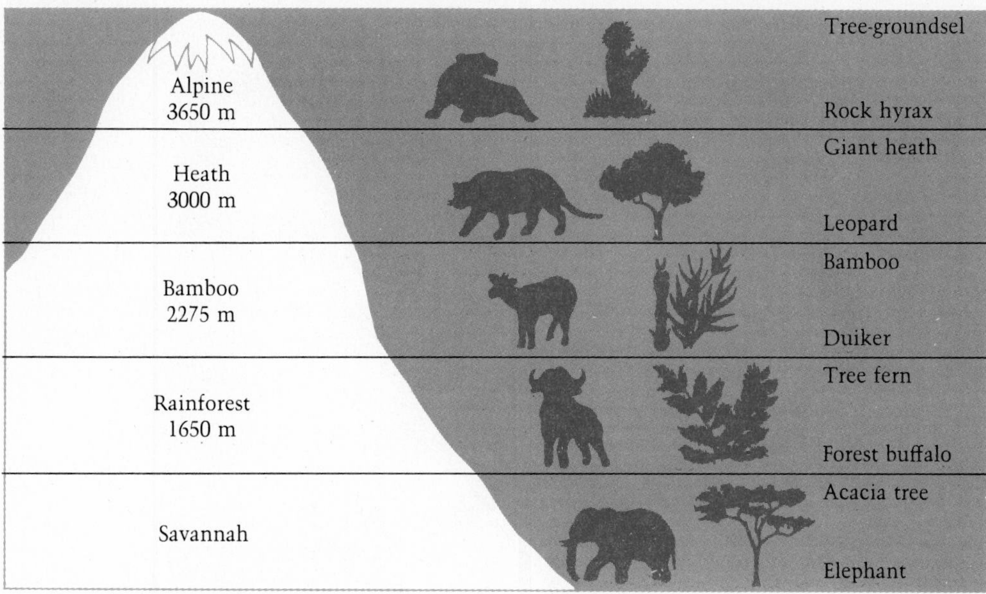

FIGURE 11-15

Zones of vegetation and animal life on the flanks of Mt. Kenya in eastern Africa.

therms, and those with a preference for intermediate temperatures are called *mesotherms*.

Availability of water is another vital factor limiting the spread of plants and animals throughout the physical world. Water is essential in photosynthesis and in other functions of plants and animals. Several plant classifications take water availability into account.

Plants that are adapted to dry areas are called *xerophytes*. Desert plants have evolved many fine adjustments. Stomata are deeply sunken into the leaf surface to reduce water loss by evapotranspiration. Roots often reach 5 m (16 ft) or more into the ground in search of water or, more commonly, spread horizontally for great distances.

The plants that live in wet environments are classified as *hygrophytes*. Swamps, marshes, lakes, and bogs are the habitats of this vegetation. The aquatic buttercup, a curious example of a hygrophyte, produces two kinds of leaves. It develops finely dissected leaves when in water and simple, entire leaves when exposed to the air.

Plants that develop in areas of neither extreme humidity nor extreme aridity are called *mesophytes*. Most plants growing in regions of plentiful rainfall and well-drained topography fall into this category.

In tropical climates with a dry season, flowering trees and plants have developed the habit of dropping their leaves to reduce water loss during the dry season. This habit spread to plants of higher latitudes, where the formation of ice in winter sometimes causes a water shortage. Trees and plants that seasonally drop their leaves are called *deciduous*, and those that keep their leaves all year round are called *evergreen*.

Other climatic factors play a role in the dispersal of plants. These include the availability of light, the winds, and the length of snow cover. The position of a species within a habitat shared by other species determines the amount of light available to it. In deciduous forests of temperate latitudes, many low shrubs grow intensely in spring before the leaves of taller trees cut out light. The amount of available light is further determined by latitude and associated length of daylight. Growth in the short warm season of humid microthermal (*D*) climates is enhanced by the long daylight hours of summer. Wind can effect the spread of plants in several ways. It can limit growth or even destroy plants and trees in extreme situations. It also spreads pollen and the seeds of some species.

The *distribution of soils* is another factor affecting plant distribution. Factors concerned with the soil are known as *edaphic* factors. The most important edaphic factors are soil structure and texture, the presence of nutrients, and the quantities of air and water. Soil structure and texture affect a plant's ability to root. Nutrients determine, to some extent, the type of vegetation. Grasses, for example, need large quantities of calcium, so they are more likely to be found on pedocals than pedalfers. The presence of soil water depends not only on precipitation but also on the porosity of the soil. Thus given the same rainfall, a grassland might exist over porous soils and a forest over nonporous soils.

A final physical factor is *landforms.* Landforms control vegetation distribution in many ways. On a large scale, vegetation changes with altitude, as Chapter 7 explains. On the flanks of Mount Kenya in eastern Africa (Figure 11-15 on page 221), there is a transition from savanna grassland below 1650 m (5400 ft) to alpine vegetation above 3650 m (12,000 ft). Landforms have a small-scale effect as well. Steep slopes foster rapid drainage and a possible lack of soil water. Aspect of the slopes (also discussed in Chapter 7) controls the amount of incoming radiation and determines shelter from the wind. Thus it too helps determine the distribution of plant and animal species.

BIOTIC FACTORS

As might be expected from the interactions in any one ecosystem, many biotic, as well as physical, factors affect the distribution of plants and animals. The biosphere is seldom static. Individual species may compete with one another, be suppressed by other species, be predators or prey, or live in intimate cooperation with other species. *Competition* for food and space plays a strong part in plant and animal distribution. Sometimes new species compete for resources so well that they eliminate old species. In the intermountain region of the United States such as Utah, native bluebunch wheatgrass (*Agrophyron spicatum*) has been almost totally replaced by European cheatgrass (*Bromus tectorum*). One reason for the success of the cheatgrass is that it produces 65 to 200 times as many seeds as the native species does.

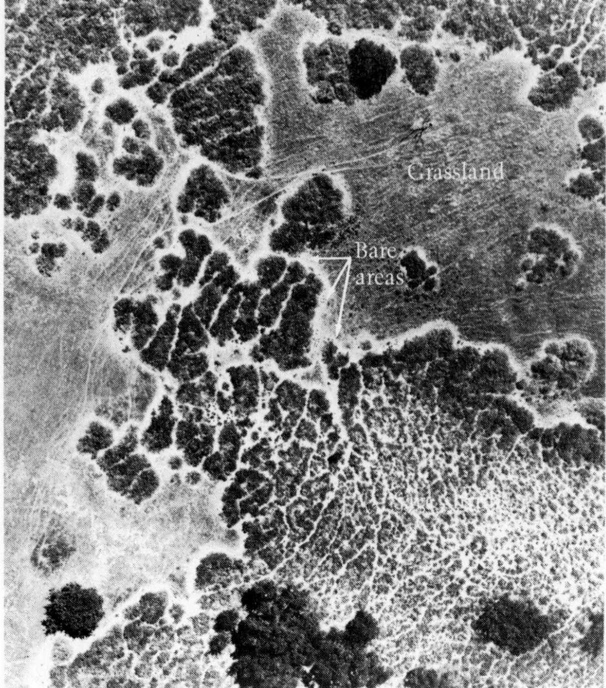

FIGURE 11-17
Aerial photograph of sage shrubs suppressing the growth of grass in grasslands. The light areas around the sage are bare of grass.

FIGURE 11-16
The spread of the starling across North America between 1905 and 1955.

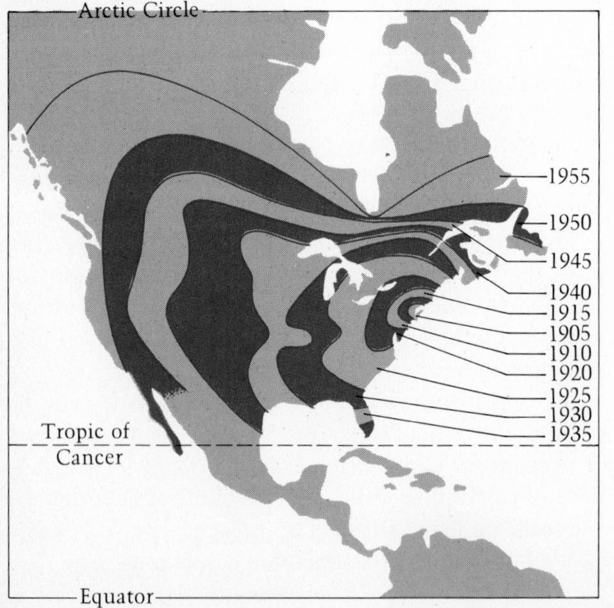

Another example of the displacement of a native species by an invader is the fast spread of the European starling (*Sturnus vulgaris*) to the detriment of the bluebird (*Sialia sialis*) and the yellow-shafted flicker (*Colaptes auratus*). The starlings are more efficient at occupying and holding nesting holes. Their rapid spread across North America since their introduction into Central Park, New York in 1891, mapped in Figure 11-16, is quite astounding.

Another form of biological interaction, the inhibition of one species by another, is called *amensalism.* In the coastal hills of southern California, sage shrubs grow on slopes, and grasses inhabit the valleys. Sometimes, however, the sage shrubs occur in the grassland. As Figure 11-17 shows, the shrubs are usually surrounded by grass-free bare ground. Foraging by birds and small animals clears some of the ground, but a major cause is the cineole, a liquid with camphorlike odor, and camphor oil emitted by the sage. Both are toxic to grass seedlings. The sage therefore suppresses the growth of the grass.

THE RUBBER SEED CAPER

During the mid-nineteenth century, rubber trees thrived on plantations in Brazil. Profits flowed into the country, and the Brazilian government actively maintained a monopoly on this valuable crop. But the British government wanted a share of the treasure, and began a covert plot to end Brazil's corner on the market.

They suspected that rubber trees would grow in their colony of Ceylon (now Sri Lanka), but they needed some seeds. The government commissioned H. A. Wickham to smuggle the goods out of Brazil. His pay was about $50.00 for each thousand seeds, packed in baskets and labeled "Botanical Specimens for Her Majesty's Garden at Kew." Thanks to casual security, the treasure slipped through. The seeds sprouted in London and were then shipped to Ceylon. Today 95% of the world's natural rubber comes from the Far East, and Brazil was left far behind in the international competition.

Today, governments have elaborate means of protecting their cash crops. Hawaii, for example, rigorously controls shipment of any sugar cane sprouts beyond its borders. It would be more difficult to duplicate Wickham's feat today. In addition to his $3500.00 in payment, he was named knight of the British Empire.

Anyone who has seen the ravaged landscapes of Australia before rabbit control began cannot doubt the efficiency of *predation* as a factor in the distribution of vegetation. But examples of one species eating all members of another species tend to be rare and artificial. It is in the best interest of predators in balanced ecosystems to rely on a number of prey species so that their food will never be exhausted. In this more natural situation, predation affects plant distribution mainly by reducing the pressure of competition among prey species. In general, the presence of predators tends to increase the number of species in a given ecosystem. Darwin suggested that ungrazed pasture in southern England was dominated by fast-growing, tall grasses that kept out light. Consequently, the ungrazed areas contained only about 11 species, whereas the grazed lands possessed as many as 20.

Yet another biological interaction is termed *mutual-*

THE AFRICAN STOWAWAYS

About 1929 a few African mosquitoes arrived in Brazil. They had probably stowed away aboard a high-speed French destroyer in the African city of Dakar. Once in Brazil, the immigrants established a colony in a marsh along the coast. Although the residents of a nearby town suffered from an outbreak of malaria, no one seemed to notice the foreign mosquitoes for a long time.

The insects settled comfortably in their new environment, and during the next few years they spread about 320 kilometers (200 mi) along the coast. Then in 1938 a malaria epidemic swept northeast Brazil. In 1939 the disease continued to rage through the region. Hundreds of thousands of people fell sick, and nearly 20,000 died.

Brazil always had malaria-carrying mosquitoes, but none quite like the new African variety. The native mosquitoes tended to stay in the forest. But the foreign pests could breed in sunny ponds outside the forest, and they made a habit of flying into houses to find people to bite. The Rockefeller Foundation and the Brazilian government hired over 3000 people and spent over $2 million to attack the invaders. After studying the ecological characteristics of the enemy, they sprayed houses and ponds. Within three years the battle wiped out the African mosquitoes in South America. Brazil also began a quarantine and inspection program for incoming airplanes to keep out unwanted stowaways.

High-speed transportation developed by humans has inadvertently carried many plants and animals to other continents. In the new environment the immigrants may react in three ways. They may languish and die; they may fit into the existing ecosystem; or they may tick away like a time bomb and eventually explode like the African mosquitoes.

Source: Charles S. Elton, "The Invaders," in *Man's Impact on Environment*, Thomas R. Detwyler, ed. (New York: McGraw-Hill, 1971), pp. 447–458.

ism, or sometimes *symbiosis,* the coexistence of two or more species because one (or both) is absolutely necessary to the other. A most obvious example are the bees, moths, and beetles that cross-pollinate flowers and thereby ensure more flowers. Another example is the hermit crab that inhabits an empty mollusk shell. Clearly, the species involved in symbiosis must have similar spatial distributions.

No species has been more responsible for affecting the distribution of plants and animals than *Homo sapiens.* It is fitting that we should end by illuminating some of the ways humans have altered the biosphere's spatial relationships.

VEGETATION AND HUMANS

If we define natural vegetation as vegetation that has never experienced any effects of the activities of humans, then there is probably no natural vegetation now existing on the planet. Air pollution reaches monitoring stations as far afield as Alaska and the South Pole, and "acid" rains from distant sources fall more and more frequently in Scandinavia. Despite this widespread nature of "accidental" tampering with the biosphere, it is a drop in the bucket compared to the deliberate changes made by agricultural technology.

The United Nations Food and Agriculture Organization estimates that, during the 1970s, 10% of the planet's land area is used as cropland, and 19% is used as pasture and meadow. This 29% of the land's surface has mainly been taken from natural woodlands in eastern North America, Europe, and eastern Asia and from the monsoon woodlands of southern Asia. Large areas of natural grassland have been replaced by new cultivated grassland areas.

A striking example of the removal of woodlands is provided in Figure 11-18. In Cadiz township in Wisconsin, in a little over 100 years, 96% of the wooded land was cleared for cultivation. The change had widespread effects on plants and animals alike.

The effect of humans has been not only to remove original species but to replace them with new ones. Sometimes the new species are brought from different parts of the world, as when the American potato was introduced into Europe. In other cases, humans develop entirely new strains. New varieties of rice called miracle rice have been developed in the Philippines. These produce up to seven times more rice, are disease resistant, and are less sensitive to the length of the day, so that two or three crops may be grown each year.

Many species are entirely eliminated through the action of humans. Since 1600, approximately 250

FIGURE 11-18
The removal of woodland in Cadiz township, Wisconsin since the beginning of European settlement.

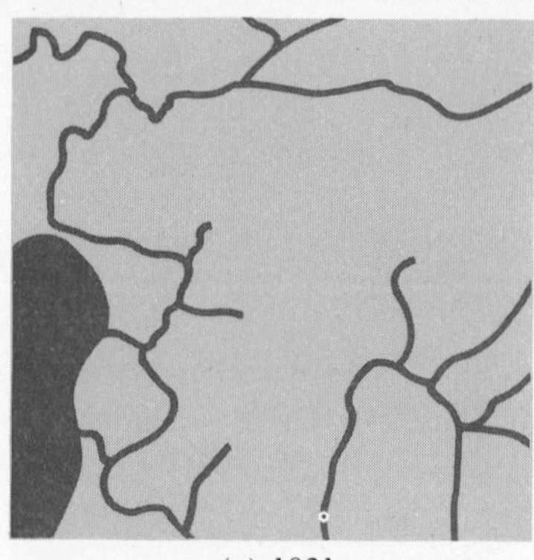
(a) 1831

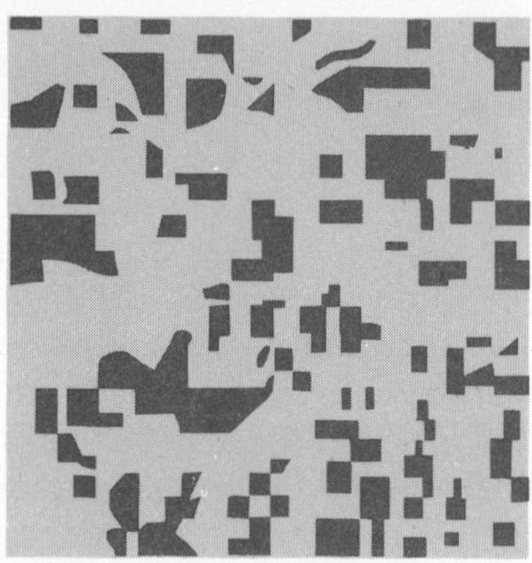
(b) 1882

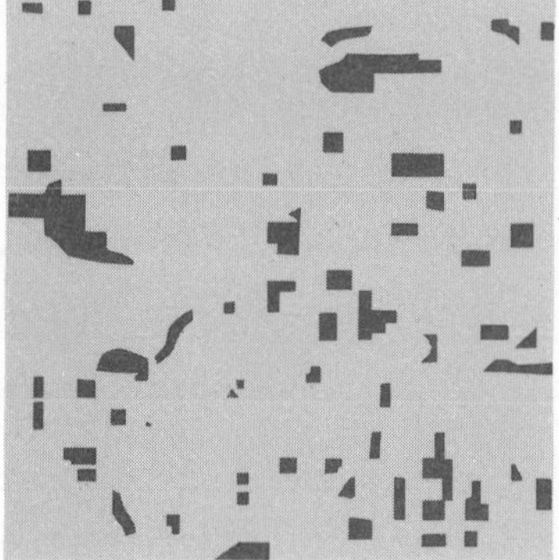
(c) 1902

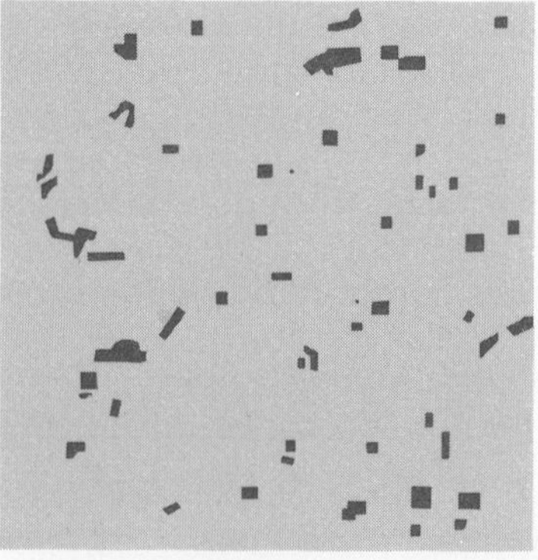
(d) 1950

species of animals have become extinct. In addition, 817 varieties of birds and animals are severely threatened. When accompanied by similar losses of flora, these are sad statistics. Evolution suffers, because the information in the gene pool of these species is lost to the biosphere forever. Ecological energetics suffers too. *Homo sapiens* is just another species, subject to the laws of ecological energy flow and food chain links just as other species are. Yet modern agriculture opts for the development of just a few species without determining whether solar energy can be best harvested in a vegetation layer of a single species or in one of greater diversity. The exploding human population requires sound and rapid decisions on most questions of the biosphere.

SUMMARY

The vegetation layer began to form when gases of the atmosphere developed into amino acids and when photosynthesis evolved. Photosynthesis is the key by which solar energy is taken into the life systems of the biosphere.

Autotrophs, the basic food producers, convert solar energy into chemical energy. The chemical energy is lost at various stages of the food chain or passed on to higher trophic levels. An understanding of the way energy passes through the various stages of ecological systems is important because humans belong to some of these systems and interact with others.

The biosphere is never static. One example of its dynamic behavior is in plant successions. Another is the processes of evolution. Principles of energy flow through systems and of evolution are helpful in understanding the ways that plants and animals have dispersed across the earth's surface.

QUESTIONS

1. In what ways was the primitive atmosphere of the earth different from that which exists today? Could we reproduce the experiments in which the elements of the primitive atmosphere were boiled and subjected to electrical discharges, or frozen, to produce primitive organic compounds using the components of the present atmosphere? Why or why not?

2. If life began about 3 billion years ago on the earth, why was it confined to the sea for such a large proportion of the time, until about 400 million years ago? How is the formation of the atmosphere related to this invasion of the land?

3. What major effects does the process of photosynthesis have on the physical composition of the atmosphere? Apart from the obvious importance of the production of oxygen, which allows animal life to exist on the earth, what physical changes in the earth's surface can be traced to the continuing changes in the atmosphere wrought by photosynthesis over immense periods of time?

4. How do the factors that control photosynthesis affect the actual amount of production of new biomass? How are these factors related to aspects of climate that we have previously examined?

5. Figure 11-4 shows a simple relationship between values of net radiation and the net production of organic matter on a global scale. What other factors must be taken into account to find the actual global distribution of plant matter production?

6. How are animal and plant life interdependent? If plant life had existed alone on the earth, what would have been the limiting factor in the duration of life?

7. How does the food chain found in the pond ecosystem in Silver Springs, Florida form a complete cycle? Is this system a closed or an open system? Why?

8. If all of the incoming radiation from the sun were used to produce biomass, we would have many times more plant life on the planet than presently exists. What happens to the remaining 95% of the incoming solar radiation? If only about 16% of the 5% of the incoming radiation incorporated into the earth ecosystem is used for carbohydrate production, what happens to the rest of it?

9. How are human beings different from most of the other consumers in the food chain? Are we efficient users of the energy of the global ecosystem? How is the efficiency of the human level increased in some societies, and decreased in others? Are the eating habits of most Americans an efficient or inefficient use of energy, and why?

10. Why did the DDD insecticide applied in Clear Lake, California become more concentrated as it traveled up the food chain? Assuming that almost all of the DDD that entered the system of any organism in this food chain remained in that organism until it was incorporated into the system of some higher level organism, what information can be gained about the relative efficiency of each of the organisms from the concentration figures of DDD?

11. What has been the most widespread agent of change in allogenic successions in the United States? In Europe? What type of plant successions would you

expect in areas reclaimed from stripmining activity, as required by recent environmental laws? Can the land really be returned to its original appearance, as the laws require, after disruptive events of this magnitude?

12. In systems terms what factors characterize a climax community?

13. Combining the idea of the pyramid-shaped food chain with the concept of species producing more offspring than survive to reproduce, which animals in the food chain would you expect to have the highest number of offspring, based solely on feeding habits? What natural controls inherent in the food chain keep any species from excessively dominating all others in the chain?

14. Using the related concepts of the ecological niche and adaptability to environmental change, Darwin came up with the beginnings of evolutionary theory. The success of any species is a measure of its ability to exploit and to extend its ecological niche to the fullest extent possible. For the following animals, which is the most important factor in the species' success or failure—the adaptation to a diverse habitat or the exploitation of a single, specialized niche: (a) the dark peppered moth? (b) the brontosaurus? (c) human beings?

15. Temperature was found by Köppen and Salisbury to be a limiting factor in the northward spread of certain trees and plants. How could the habitat of the wild madder, which Salisbury found to be limited by the January 4.5°C isotherm, possibly affect that of other species for which this isotherm is well within the zone of increasing physiological stress or even within the range of optimum in temperature terms? What implications does this have for plant communities and their boundaries?

16. We now have a mechanism to explain the variation in plant types which occurs in different climates. Explain briefly how plants have evolved to be sensitive to both temperature and water availability, as well as seasonal change.

17. If occupation of a wide and diverse ecological niche implies competition among species, does this negate the idea of a stable ecosystem? Does the introduction of a new species better adapted to an environment mean the permanent end of a stable ecosystem in an area? Why or why not?

18. Human beings have on several occasions introduced predatory species to control other species considered pests, usually for economic reasons. What are the dangers inherent in this type of amensalistic control of species? What other amensalistic role have we had?

19. In his famous essay, "Homestead and Community on the Middle Border," Carl Sauer writes of the differences in farming and eating habits between the time of his childhood, when farm families were largely self-sufficient for their food, and the changes that came with agricultural specialization in the early twentieth century:

> The farm orchards now are largely gone, and the gardens are going. Many varieties of fruits that were familiar and appreciated have been lost. A family orchard was stocked with diverse sorts of apple trees for early and midsummer sauce, for making apple butter and cider in the fall, for laying down in cool bins in the cellar to be used, one kind after another, until the russet closed out the season late in winter. The agricultural bulletins and yearbooks of the past century invited attention to new kinds of fruits and vegetables that might be added to the home orchard and garden, with diversification, not standardization, in view.*

What advantages and disadvantages are associated with such agricultural specialization? What have the genetic effects of this agricultural specialization been?

20. New advances in agriculture, especially the artificial crossbreeding of new species and varieties such as the miracle rices IR8 and C4-63, have allowed greater crop yields per acre, more resistance to disease, and less sensitivity to factors such as moisture availability and day length. In addition, modern agriculture can now control soil conditions through fertilization and the addition of missing textures and materials, as well as through irrigation on a massive scale. Can today's huge crop yields properly be considered the result of having achieved climax community vegetation? Why or why not?

Landscape, vol. 12, no. 1, 1962, pp. 3–7.

SOILS AND VEGETATION: VARIATION OVER SPACE

CHAPTER 12

The processes that determine the development of soils and vegetation depend, to a great extent, on climatic factors. Thus soils and vegetation differ from place to place just as climate does. If we are to discover exactly how soils and vegetation are distributed, we must take a two-part approach. First, we must classify the soils and vegetation in a manner suitable for our purposes. Second, we must describe the mosaic of soils and vegetation that actually exists. Once we determine the connection between the distributions of soils and vegetation and the climate, we can turn our attention to the grand design at the interface of the earth's spheres.

Classifications often reflect the state of a science, and the history of soil classifications speaks well of the advances made by twentieth-century soil scientists. Following a description of the soils of our planet, we will turn our attention to vegetation. We will find that, in any spot on the globe, vegetation, soils, and climate are related. An example is at work in northern Canada, where the boreal forests meet the arctic wastes. We will then move to larger-scale global interactions, where relationships at the interface are even more apparent.

CLASSIFYING SOILS

Two countries that span many climatic boundaries are primarily responsible for the soil classification systems in use today. These are Russia and the United States. The story of soil classification begins in Russia toward the end of the last century.

A HISTORY OF SOIL CLASSIFICATION

Up to about 1850, soils were thought of as weathered versions of the original underlying hard rock. Around 1870, the Russian scientist V. V. Dokuchaiev realized that soil was independent, with distinctive characteristics and unique forms of origin. Dokuchaiev and his colleagues developed several soil classification schemes based on the origin of soils and their relationship to

FIGURE 12-1

Rice terraces in the Philippines. The soil has been severely altered by this agricultural practice.

TABLE 12-1
Organization of the US Comprehensive Soil Classification System

LEVEL	DESCRIPTION
Order	The most general class. Soils of a given order have a similar degree of horizon development, degree of weathering or leaching, gross composition, and presence or absence of specific diagnostic horizons, such as an oxic horizon in an oxisol
Suborder	Distinguished by the chemical and physical properties of the soil. At this level formative and environmental factors are taken into account
Great group	Distinguished by the kind, array, or absence of diagnostic horizons
Subgroup	Determined by the extent of development or deviation from the major characteristics of the great groups
Family	Soil texture, mineral composition, temperature, and chemistry are distinguishing features
Series	A collection of individual soils that might vary only in such items as slope, stoniness, and depth to bedrock.

climate and vegetation. Once translated into English, the Russian work had a great effect on soil scientists in the United States.

C. F. Marbutt, who was chief of the US Soil Survey, was particularly influential. He incorporated Dokuchaiev's ideas into his own and produced a soil classification in 1927 that was used, with some refinements, for several decades. New classifications put out by the US Soil Conservation Service in 1935 and 1949 contained major revisions of Marbutt's ideas.

Soil scientists were still troubled over a number of points. Many soils have been altered from their natural state either by agricultural practices (see Figure 12-1) or by other events of the physical world. Knowing this, soil scientists felt that too much emphasis was being given in the classifications to the original forming factors and processes. In the last analysis, the soils, and not their formation factors, were to be classified. Thus definitions had to be stated in terms of the soils and not related factors, such as climate. They also argued that the origin of a soil is sometimes unknown, and that the soil is therefore impossible to classify by formation factors. To solve problems of this kind, a completely new classification system was developed and published in 1960. Although it takes formation factors into account, the new system focuses on the properties of the soils as they exist now and not on the properties they would have had under their formative conditions.

THE COMPREHENSIVE SOIL CLASSIFICATION SYSTEM

The US Comprehensive Soil Classification System has six levels of organization, all described in Table 12-1. From the highest level to the lowest level, soils become more difficult to subdivide. For simplicity, we will only deal with the ten major orders and some of their suborders. See Table 12-2 for a summary of their characteristics.

The first three orders—entisols, histosols, and vertisols—are soils that have not, for some reason, developed the kind of distinct horizons we expect most soils to possess. They may also contain a peculiar property that makes them quite distinct.

Entisols Soils that have no horizons, except for possibly a plowed layer, are entisols. Horizons may be absent for any reason, including a recent deposit of river sediments to form alluvium or a large amount of quartz sands that will not weather. Entisols can be found in a number of different locations. They include the alluvial soils of river valleys, the thin, stony soils of mountain regions, and dune sands like those in Figure 12-2. Entisols also include the poorly developed soils found in microthermal (*D*) climates and the clay soils with low humus content in areas of poor drainage. Because there are many reasons for the absence of well-developed horizons, entisols are found in many different earth environments. The suborders of entisols, listed in Table 12-2, classify their environmental conditions.

Histosols Another order of soils having no horizons are those associated with bogs, called histosols. These soils are typically accompanied by clay and peat deposits. They are unique among the major orders because they can be completely destroyed by time or altered by drainage. It is hard to generalize about the geographic

TABLE 12-2
Suborders of soil classifications

ORDER	SUBORDER	CHARACTERISTICS
Entisol	Aquent	Show evidence of saturation at some season
	Arent	Lack horizons because of plowing or other human activity
	Fluvent	Formed in recent water-deposited sediments, as in flood plains
	Orthent	Found on recent erosional surfaces, such as high mountains
	Psamment	Found in sandy areas, such as sand dunes
Histosol		Found in bogs and poorly drained areas
Vertisol	Torrert	Found in arid climates
	Udert	Found in humid climates
	Ustert	Found in monsoon climates
	Xerert	Found in Mediterranean climates
Inceptisol	Andept	Contain a high percentage of volcanic ash or a clay mineral, developed from ash, called allophane
	Aquept	Wet with poor drainage
	Ochrept	Freely drained, light in color
	Plaggept	With a surface layer more than 50 cm thick resulting from human activity, such as manuring
	Tropept	Freely drained, brownish to reddish, found in the tropics
	Umbrept	Dark reddish or brownish, acid, freely drained, organically rich
Aridisol	Argid	With an illuvial horizon where clays have accumulated to a significant extent
	Orthid	With an altered horizon, a hard layer (called a hardpan or duripan), or an illuvial horizon of water soluble material
Mollisol	Alboll	With a surface layer that covers a white horizon from which clay and iron oxides have been removed and with a layer of clay accumulation below
	Aquoll	Saturated with water at some time during the year
	Boroll	Cool or cold, relatively freely drained
	Rendoll	Found in humid climates, developed from parent materials rich in calcium carbonate
	Udoll	Not dry for as much as 60 consecutive or 90 cumulative days per year
	Xeroll	Found in Mediterranean climates
Alfisol	Aqualf	Periodically saturated with water
	Boralf	Freely drained, found in cool places
	Udalf	Brownish to reddish, freely drained
	Ustalf	Partly or completely dry for periods longer than 3 months
	Xeralf	Found in dry climates
Spodosol	Aquod	Associated with wetness
	Humod	With a humus-enriched sesquioxide horizon
	Orthod	With significant amounts of humus and iron in the sesquioxide horizon
	Ferrod	With an iron-enriched sesquioxide horizon
Ultisol	Aquult	Found in wet places
	Humult	Freely drained, rich in humus
	Udult	Freely drained but poor in humus
	Ustult	Found in warm regions with high rainfall but a marked dry season
	Xerult	Freely drained, found in Mediterranean climates
Oxisol	Aquox	Formed under the influence of water
	Humox	Always moist, with a high humus content
	Orthox	Found in places with a short or no dry season, other than aquoxes
	Torrox	Found in arid climates, may have formed under a different climate from that now existing in their location
	Ustox	Found in humid climates with at least 60 consecutive dry days per year

FIGURE 12-2

An entisol; dune sand in White Sands National Monument.

distribution of histosols because bogs are so widespread.

Sometimes histosols are not used because of their poor drainage. But where drainage is practical, intensive cultivation of such crops as cabbage, carrots, potatoes, and other root crops is possible.

Vertisols You may have seen areas where clay soils develop large cracks in the dry season and swell with moisture when rain returns. Under the comprehensive classification system, these soils, once called grumusols, are now called vertisols. More than 35% of their content is clay particles. The clay causes the swelling and cracking. The clay particles are normally derived from the parent materials, so vertisols are found where the clay-producing materials are available—in mesothermal or tropical climates with periodic dry and moist seasons. Their suborders are closely related to climatic divisions, as Table 12-2 indicates. They are most extensive in Australia, India, and the Sudan.

Vertisols are hard to use for most human purposes, particularly construction. When they shrink and crack, fences and telephone poles may be thrown out of line (see Figure 12-3). Pavements, building foundations, and pipelines may all be damaged by the movement of these turning soils.

Inceptisols These soils take their name from the Latin word for "beginning." They are believed to have formed rather quickly, so any horizons that may be apparent have not developed by the common processes of illuviation and eluviation. But inceptisols do contain a significant amount of organic matter and/or evidence that the parent material has been weathered to a certain extent. Inceptisols are generally found in humid climates, but they may be found from the arctic to the tropics and are often found in alpine areas. If there is natural irrigation, some inceptisols may be found in arid areas. These soils most frequently develop under a forest cover but can be found under tundra or grass. Once more, the suborders in Table 12-2 indicate their environmental conditions.

None of these four major soil orders can be specified geographically with precision. They are principally related to earth spheres other than the atmosphere, as Figure 12-4 explains. But the remaining orders are definitely associated with particular climates. This is clear in the soils of dry areas of the world.

FIGURE 12-3

Fenceposts thrown out of line by swelling and "walking" of vertisols.

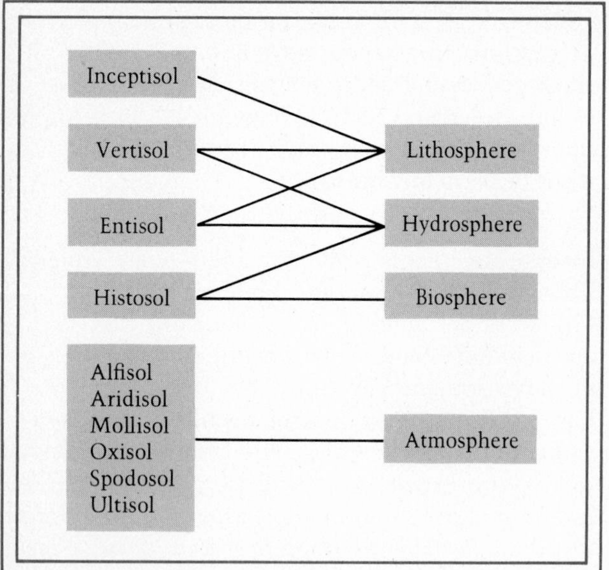

FIGURE 12-4

Primary relationships among the four earth spheres and the soil orders of the Comprehensive Classification System.

Aridisols These soils cover a larger area of the world's land surface (19.2%) than any other soil. These soils are usually dry unless they are irrigated—naturally or artificially. They usually have a thin light-colored horizon at the surface that is low in organic carbon. They often contain horizons rich in calcium, clay, gypsum, or salt minerals as shown in Figure 12-5. Table 12-2 describes the two major suborders of aridisols. Large expanses of aridisols are found in deserts such as the Sahara, the Sonoran, and the Gobi deserts.

Areas with these soils can be used for grazing or intensive production of crops with the aid of irrigation. Desert shrubs and grasses are the main vegetation, and overgrazing of the land is often a serious problem. The soil-forming process of calcification (see Chapter 10) is common in aridisols but even more pronounced in the next order.

Mollisols You can find mollisols in climates that normally have dry seasons, but temperatures can range from microthermal to tropical. Rainfall may be sufficient to leach these soils, but calcification is far more common. Mollisols are the soils of the steppes, the grass-covered plains of central Russia that lend their name to the *BS* climatic zone. The central attribute of a mollisol is a thick, dark surface layer, high in bases (alkalis) and saturated by positive ions. The layer has a ratio of carbon to nitrogen of less than 17% (13% if cultivated) and a moderate to strong stable structure. Occasionally mollisols are found under water-loving plants or deciduous hardwood forests, but the vast majority are found under tall or short grasslands.

Mollisols are often associated with large-scale commercial grain production and livestock grazing. Corn is the predominant grain when precipitation is sufficient, but sorghum is grown more and more on these soils in the southwestern United States. Drought is the most common problem facing the agriculturalist located on mollisols.

Alfisols These soils are found in moister, less continental climatic areas than the mollisols are. They are high in mineral content and usually moist. As Figure 12-6 indicates, alfisols lack the dark surface horizon of the mollisols, and show no marked signs of horizons where oxidation has occurred. But they do have a noteworthy lower horizon of clay accumulation. Alfisols are usually found under high-latitude forests or deciduous forests but occasionally can be found in

FIGURE 12-5

An aridisol (argid) where a light-colored calcium-rich horizon has developed. The scale is in units of 15 cm (6 in).

FIGURE 12-6

An alfisol (ustalf) with a massive hard horizon forming overhanging shelves in road cuts.

areas of vegetation adapted to dryness. Climatic factors are apparent in the suborders of alfisols in Table 12-2.

Areas of alfisols are noteworthy for some of the most intensive forms of agriculture. Part of the Corn Belt of the United States is found on these soils. Oats, soybeans, and alfalfa are also widely grown. In some areas of the United States, a technique called *minimum tillage*, involving chemical weed control and minimum plowing, is being used on alfisols to maintain their organic content and fertility.

Spodosols Soils, formed primarily under the podsolization process, like those in Figure 12-7 are spodosols. Consequently, they are characterized by a horizon with an illuvial accumulation of sesquioxides. *Sesquioxides* are oxides with 1½ oxygen atoms to every metallic atom. Usually this horizon shows rounded or subangular black or very dark brown pellets the size of silt. The characteristic ash-gray horizon of podsolized soil is often found in spodosols but is not a defining feature. Spodosols are found only in humid regions but from high to low latitudes. Most have a coniferous forest cover. There are four suborders listed in Table 12-2.

Because of the association of forest cover with spodosols, the timber industry is one of the most important human activities on these soils. But limestone can offset the acidity of the soil. Thus corn, oats, wheat, and hay are now being grown in the state of New York and other places where these soils occur. When artificial lime is not applied, root crops, especially potatoes and sugar beets, may be cultivated.

Ultisols Warmer, wetter climatic areas host a soil distinguished by a horizon of strong clay accumulation. The native vegetation of these ultisols may have been forest, savanna grassland, or even swampland. Ultisols have at least a few minerals that may be subject to weathering. They may well be alfisols that have been subjected to greater weathering. In Northern Hemisphere midlatitudes, ultisols lie to the south of alfisols, in a humid subtropical climate.

There are five suborders of ultisols, as Table 12-2 indicates, and they are highly favorable to cultivation.

FERTILIZER

The population explosion raises a painful question—can we feed all of these people? Our limited amount of earth must produce more and more food. One way to meet the challenge is to use fertilizer. Fertilizer enriches the existing soil and allows much larger harvests.

The American Indians pampered the earth by planting a fish with each seed. In other cultures farmers have applied organic fertilizers—primarily manure. But organic fertilizer cannot meet our needs. Today 30 million tons of nutrients are added to the soil each year. Chemical fertilizers supply the primary nutrients: nitrogen, phosphorus, and potassium.

Although pesticides, new hybrid seeds, and mechanization contribute to higher crop yields, fertilizers pay the most substantial rewards. For each kilogram of fertilizer a farmer spreads on the soil, he may harvest 10 extra kilograms of grain. By using the proper combination of nutrients to complement the land, a farmer may grow two or three times as many crops.

By the year 2000, we may need 120 million tons of nutrients to support 6 billion people. This means one 100-pound bag of chemical fertilizer will be used for each person. Understanding the composition of the soil will allow us to fertilize more efficiently and with greater safety.

FIGURE 12-7
A spodosol (aquod) where a thin duripan has developed at a depth of 30 to 38 cm (12 to 15 in).

Ultisols are often associated with the growth of cotton and peanuts.

Oxisols The final soil order is restricted to tropical areas, with high rainfall, where the laterization process is common. It is hard to distinguish the horizons in laterized soil. Thus oxisols are characterized only by a horizon with a large part of the silica, previously combined with iron and aluminum, removed or altered by weathering. This is called the *oxic* horizon. Oxisols also have a concentration of clay-sized minerals, mainly sesquioxides. The five suborders of oxisols are listed in Table 12-2. The natural vegetation on most oxisols is tropical rainforest, and the soils tend to be of low fertility except where they come from rivers or volcanoes.

The low fertility of oxisols has given rise to a pattern of *shifting agriculture:* The land is cultivated for a while and then left for many years to naturally renew its nutrients. Population pressures in some parts of the world, such as Nigeria, have altered this agricultural system, and the rapid deterioration of soil fertility is often the result.

THE SPATIAL DISTRIBUTION OF SOILS

The spatial distribution of climates across the earth's surface was, in Chapter 7, best understood through the concept of a hypothetical continent. We may begin to understand the spatial distribution of soils in the same way. Six of the orders of soil in the comprehensive classification are arranged geographically according to climate. Figure 12-8 shows how these relate to conditions of moisture and heat, as well as to one another. In addition, the major soil-forming factors—podsolization, calcification, and laterization—are related to specific groups of soil orders. The spatial arrangement in relation to climate is quite clear for most orders. Spodosols are found mostly in cold wet climates;

FIGURE 12-8
Distribution of soil orders in relation to climatic factors and major soil-forming processes.

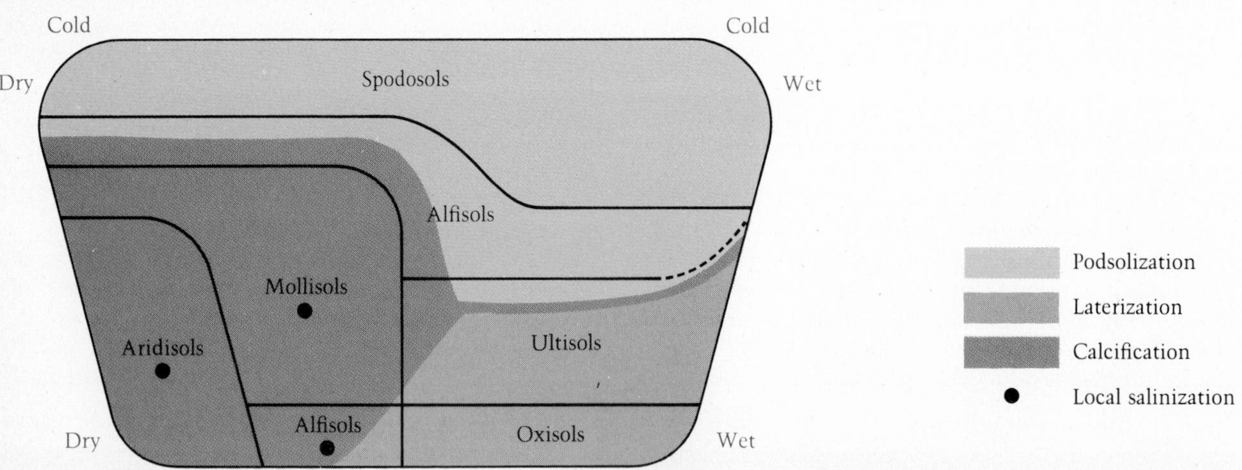

WHAT'S IN (THIS PARTICULAR) NAME?

Science is based, in part, on classification schemes. Biologists organize the living world into kingdoms. Chemists classify types of changes in matter and energy. Anthropologists identify types of kinship systems and family structures. New systems are constantly being put forward, yet only some survive. What defines the difference between a useful system and one that does not stand the test of time?

A useful system is flexible enough to accommodate new data and phenomena, yet rigid enough so that its underlying principles can be applied again and again. Because of the increasingly international aspect of science, its names should have some significance in a variety of languages. And finally, it should organize its information in ways that show useful relationships.

How does the comprehensive soil classification system, which we are surveying in this chapter, measure up according to these standards? When the U.S. Department of Agriculture first proposed the system in 1960, its major goals were (1) to replace entirely the terms used in older soil name systems; (2) to produce words that could be remembered easily and that would suggest some of the properties of each kind of soil; and (3) to provide names that would have some meaning in all languages derived from either Latin or Greek, the majority of modern European languages. In fact, the Department called upon classic scholars from the University of Ghent, Belgium and the University of Illinois to provide the Latin and Greek roots to match the soil characteristics.

ORDER NAME	DERIVATION	ROOT/KEY
Entisol		rec**ent**
Vertisol	verto, Latin for "turn"	in**vert**
Inceptisol	inceptum, Latin for "beginning"	**incept**ion
Aridisol	aridus, Latin for "dry"	**arid**
Mollisol	mollis, Latin for "soft"	**molli**fy
Spodosol	spodos, Greek for "wood ash"	**Pod**zol
Alfisol	al from aluminum, ferrum, Latin for "iron"	ped**alf**er
Ultisol	ultimus, Latin for "last"	**ulti**mate
Oxisol	oxide, French for "containing oxygen"	**oxi**de
Histosol	histos, Greek for "tissue"	**histo**logy

oxisols, in warm wet regions; and aridisols, in hot, dry areas. Certain orders, in particular alfisols, may be found in a variety of climatic conditions.

The distribution of soils is in reality more complicated. Figure 12-9 shows the pattern for the United States. The climatic relationships are quite clear with such soils as spodosols and aridosols. But complications arise because some soils, such as the inceptisols, are caused by different factors. Thus Appalachian soils, because of their mountain environment, are poorly developed, whereas the incipient stage of soils in the Mississippi Valley is a result of their youth. Terrain can complicate the actual pattern even more. The Rocky Mountain area, for example, is a complex mosaic of aridisols, entisols, mollisols, and alfisols. It is difficult to place all the soil orders that are not directly related to climate on a map of this scale. Yet all soil orders exist in the United States.

On a global scale, there are many distinct relationships between soils and climate, but there are also some differences in detail, as Figure 12-10 demonstrates. The relationship between soils and climates is clear in the western and central USSR. Here soil types vary from south to north, changing from aridisols to mollisols, then to alfisols, and finally to spodosols. These variations parallel changes in climate from desert through steppe and microthermal to polar climates. A similar progression may be seen in parts of North America. Perhaps the most obvious relationship on the global scale is that between the desert climates and the aridisols and entisols. Notice that, in classifying soils on a global scale, mountain areas and ice fields have been grouped into separate categories.

Despite the clear relationships between climate and soil types on this scale, similar climates do not always correspond with similar soil types. The humid subtropical climate (*Cfa*) is a case in point. This climate may be found in the southeastern parts of the five

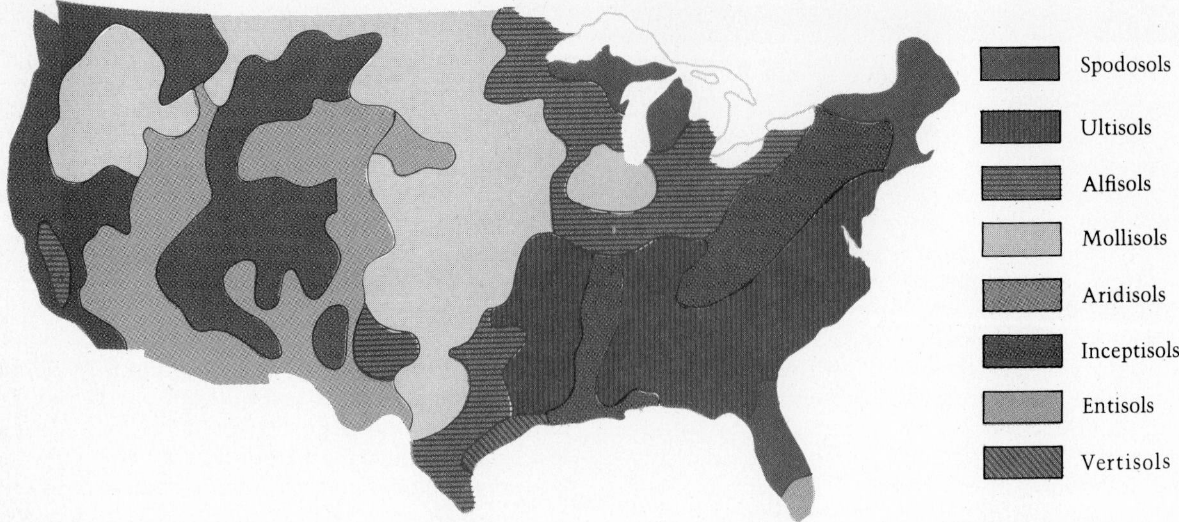

FIGURE 12-9
Simplified spatial distribution of soils across the United States.

major continents (refer to Figure 7-4), but the soil types in these areas vary. Only in the southeastern United States and southeastern China do we find ultisols. Southeastern parts of South America, around the Rio de la Plata, exhibit mollisols, whereas southeastern Africa presents alfisols. Australia, an area of humid subtropical climate, has alfisols, mountain soils, and vertisols. If you compare Figures 7-4 and 12-10, you will be able to pick out other discrepancies between climate and soils. Other factors in soil formation, such as parent material and the way soil is used by humans, help determine the prevailing type of soil.

CLASSIFYING VEGETATION

It is not only between climate and soils that we find spatial coincidence. The vegetation layer intermediates between climate and soil, and it would be surprising if similar spatial correspondences were not apparent in the vegetation. But before we set out to look for these, we must learn to distinguish one kind of vegetation from another.

SOME APPROACHES TO PLANT CLASSIFICATION

The Royal Botanic Gardens in Kew, England contain the largest collection of plant specimens in the world, more than 6½ million of them. Try to imagine the task of classifying these plants. Where would you start?

The Greek naturalist Theophrastus started (around 300 BC) to classify plants according to their *structural* characteristics, such as life form (tree, herb, shrub), length of life, and the positions of petals and reproductive organs. Up to about AD 1750, plants, except for common food-producing varieties, were mainly used as medicines. Many medicinal books, called herbals, identified them by structure and gave directions for their use.

In 1753 Swedish botanist Carolus Linnaeus developed a classification system based on the structure of plants and the number of stamens and pistils, reproductive organs, in their flowers. His Latin nomenclature describing groups of species and one other characteristic of the plant is still in use and is the basis of the *floristic* approach to plant classification.

After Darwin and Wallace put forward the theory of evolution, botanists concerned with classifying plants emphasized *evolutionary* aspects in their classifications. They tried to make the classification reflect the heredity of the plants in question. At the present time, the botanical classifier takes many different approaches to classifying plants. One approach uses the structural (morphological) characteristics of the plant. Others use the anatomy, embryology, biochemistry, and biological barriers of plants. Still others put all this evidence together to make numerical classifications.

The geographer is not as concerned with tracing evolutionary development of species as with tracing

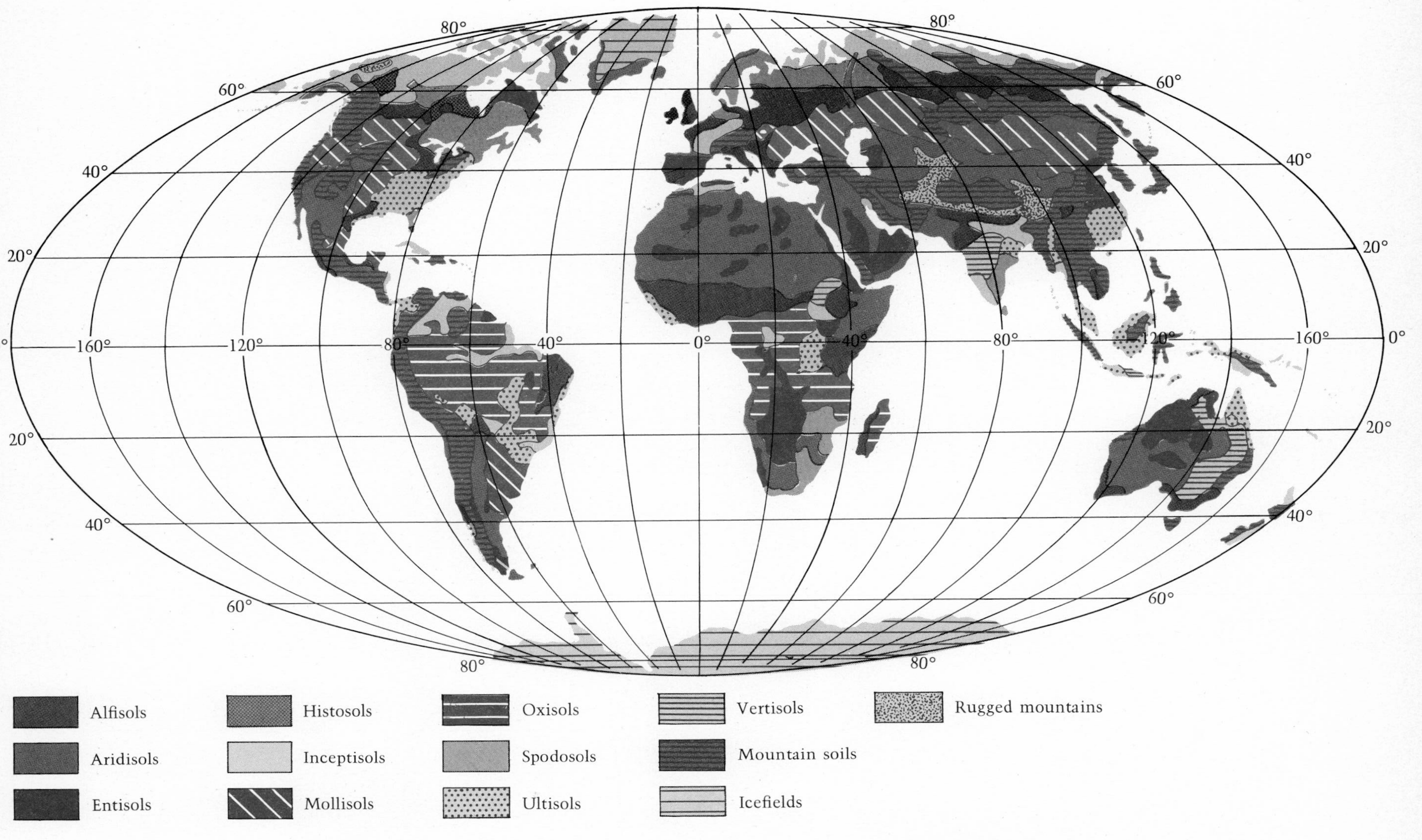

FIGURE 12-10

The global distribution of soils.

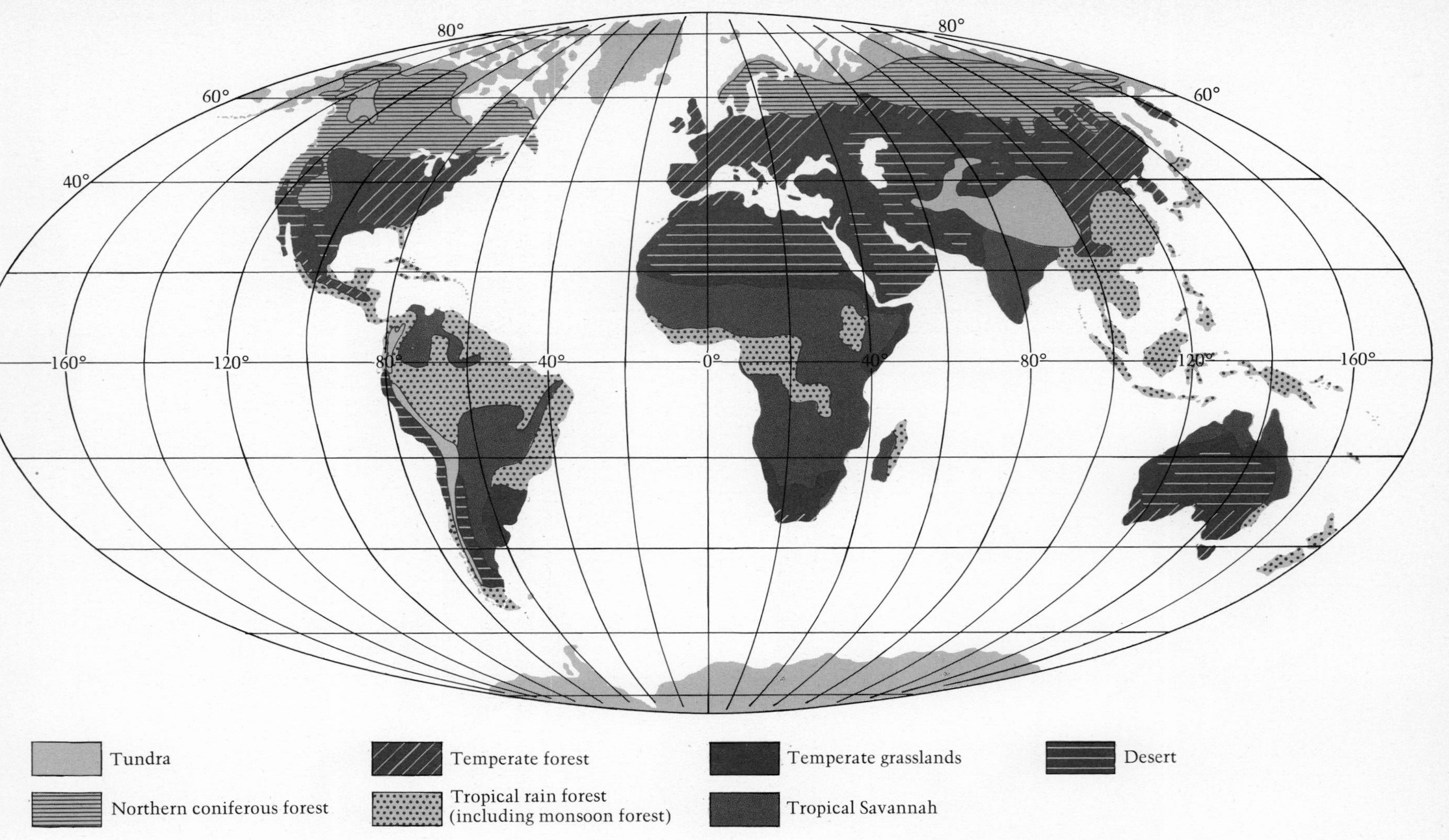

FIGURE 12-11
Global distribution of the principal terrestrial biomes.

their distribution and interaction with the physical environment. The geographer therefore takes an *ecological* approach and, as a first order of classification, examines major groupings of ecosystems. These groupings, defined on the basis of their overall appearance and climatic characteristics, are called *biomes.* A biome is thus a climatically controlled group of plants and animals, with a unique appearance, that occurs over a wide geographical area. Although the biome is one of the largest geographical and ecological units, often occupying sizable portions of continents, it is a useful unit of classification.

TERRESTRIAL BIOMES

All the earth's vegetation can be classified as one of four major types: forest, grassland, desert, or tundra. Subdivisions of these types form the terrestrial biomes. Apart from the tundra and the northern coniferous forest, which are rather continuous, most of the terrestrial biomes are found in a patchwork across the globe, as Figure 12-11 indicates. The distribution of climatic types is obviously related to that of the major biomes.

Biologists distinguish nine biomes, but we will not discuss the marine and freshwater biomes here. We will instead concentrate on the seven land biomes. The description of the plant life within them is structural rather than floristic, because the approach of Theophrastus is more appropriate than that of Linnaeus for beginning students of the earth's surface.

Tropical rainforest Wallace wrote of the tropical rainforest in 1853: "What we may fairly allow of tropical vegetation is that there is a much greater number of species, and a greater variety of forms, than in temperate zones." This was an understatement. We now know that more species of plants and animals live in tropical rainforests than in all the other world biomes combined. Parts of the Brazilian rainforest contain 300 species of trees in 2 km^2 (0.76 mi^2).

True climax tropical rainforest lets in little light. The crowns of the trees are so close together that sometimes only 1% of the light above the forest reaches the ground (see Figure 12-12). As a result, only a few shade-tolerant plants can live on the forest floor. The trees are large, often reaching 40 to 60 m (130 to 200 feet). Their roots are usually shallow, and so the bases of the trees are supported by buttresses. Another feature is the frequent presence of epiphytes and lianas. *Epiphytes* are plants that use the trees for support, but they are not parasites. Lianas are vines rooted in the ground with leaves and flowers in the canopy, the top parts of the trees. What organic matter there is decomposes rapidly, so there is little accumulation of litter on the rainforest floor.

Although it is easy to walk through the true climax tropical rainforest, many areas contain a thick, impenetrable undergrowth. This growth springs up where humans have destroyed the original forest. The areas bordering oceans, where shifting agricultural practices are common, are especially likely to have second growth. It has been estimated that most of the true tropical rainforest may disappear by the end of this century. In addition, where natural vegetation is cleared from the oxisols of this biome, hardpans frequently develop through the laterization process. Agriculture is difficult in such untillable soil.

Monsoon rainforests are included in this biome, even though they differ slightly from tropical rainforests. Monsoon rainforests are established in areas with a dry season. These areas have less variety in species. The vegetation is lower and less dense. Furthermore, the vegetation grows in layers or tiers composed of species adjusted to various light intensities.

Tropical rainforests presently cover about half the forested area of the earth. The largest expanses are in the Amazon River basin in South America and the basin of the Zäire River of Central Africa. A third area includes regions of southeastern Asia, northeastern Australia, and the island territories between them.

FIGURE 12-12
The Caribbean National Forest in Puerto Rico.

FIGURE 12-13
The savanna in Tanzania.

Savanna The savanna biome (see Figure 12-13) is a transition stage between the tropical rainforest and the desert. Savanna is tropical grassland with widely spaced trees. Bulbous plants are abundant, but thorn forests, characterized by dense, spiny, low trees, are more obvious. The most common trees of the savanna are deciduous, losing their leaves in the dry season. These include the acacia and the curious water storing, fat-trunked baobab tree. The grasses in the savanna are usually tall, sometimes growing to 5 m (16 ft), and have stiff, coarse blades.

The savanna vegetation has primarily arisen because of climate (*Aw*) in large areas of Africa, South America, northern Australia, and India and in parts of southeastern Asia. But periodic burning plays a significant role in limiting tree growth. In some places the grasses form a highly flammable straw mat in the dry season. This may be ignited through natural causes, but as Chapter 11 explains in relation to the Serengeti ecosystem, humans may play an integral role. Typical of the food chain of this biome are the large herds of grazing animals.

Desert The desert biome is characterized by sparse vegetation or even its complete absence (see Figure 12-14). Whereas the grasses of the savanna are *perennials*, persisting from year to year, many of the desert plants are *annuals*, completing their entire life cycle in a single growing season. These annuals, such as the heliotrope (*Heliotropium convolvulaceum*) of western Texas, often grow quickly after the seasonal rains, covering open, sandy areas in a spectacular display. The seeds of these annuals often lie in the soils for many years and then germinate rapidly after a rainstorm. The perennial plants in the desert biome, such as cactuses and euphorbias (spurges), are dormant much of the year. Some, with fleshy, water-storing leaves or stems, are known as *succulents*. Others have small leathery leaves or a deciduous habit. Woody plants have very long roots or are restricted to localized areas of water.

The sparse vegetation of desert ecosystems can support only small creatures of the higher trophic levels. These creatures are well adapted to the arid conditions. Rodents live in cool burrows. Insects and reptiles have waterproof skins that help them retain water. They excrete almost dry, crystalline urine.

Of course, the desert biome coincides with areas of arid (*B*) climates. Aridisols and entisols are the most common soil orders in the desert biome.

Temperate grassland This biome generally occurs over large areas of continental interiors. Perennial and sod-forming grasses like those in Figure 12-15 are dominant. Perennial herbs leave little space for annual herbs, which are seldom seen in this biome. The temperate grassland biome, like the savanna, is inhabited by herds of grazing animals and their predators.

In North America the short-grass prairie of the Great Plains gradually gives way to the moister, richer, tall-

FIGURE 12-14
Desert vegetation in Tonto National Forest, Arizona.

grass prairie of what is now the Corn Belt. The transition in vegetation is accompanied by one in the soil layer from mollisols to alfisols.

This biome has been highly susceptible to human influence. Large areas have been turned over to agriculture or domestic grazing. This is true of the interior areas of North America, the Pampas grasslands of Argentina, and the steppes of Russia, where the biome is most widespread. Temperate grasslands maintain a delicate ecological balance, and mismanagement or climatic change quickly turns then into temperate forests or deserts.

Temperate forest There are several varieties of temperate forest. In the eastern United States, Europe, and eastern China, *temperate deciduous* forests occur (see Figure 12-16). The forests are shared by herbaceous plants, most profuse in spring before the growth of new leaves on the trees. An outstanding characteristic of temperate deciduous forests is the similarity of plants found in their three locations in the Northern Hemisphere. Oak, beech, birch, hickory, walnut, maple, elm, ash, and chestnut are all common. As with the grasslands, large areas of this forest type have been

FIGURE 12-16

Temperate deciduous forest in the mountains of North Carolina.

FIGURE 12-15

Bison grazing on grass prairie (temperate grassland) in a wildlife refuge in eastern Oklahoma.

FIGURE 12-17
A redwood stand in Del Norte County, California.

turned over to agricultural purposes.

On western coasts in temperate latitudes where plentiful precipitation is the norm, *temperate evergreen* forests are found. In the Northern Hemisphere, they take the form of needle-leaf forests. The coastal redwoods (*Sequoia sempervirens*) (see Figure 12-17) and Douglas firs (*Pseudotsuga menziesii*) of the northwestern coast of North America are representative. The podcarps of the temperate evergreen rainforest of the western coast of New Zealand exemplify the broadleaf and small-leaf evergreen forests of the Southern Hemisphere.

A different sort of vegetation has developed in areas of Mediterranean climate (*Csb, Csa*). We will use the North American name *chaparral* for it. The justification for including chaparral vegetation in the temperate forest biome is the former existence of such trees as oaks and pines on the now-barren hills of Greece, the almost legendary cedars of Lebanon, the California live oaks, and the cork oaks still found in both the Eastern and Western hemispheres. But hard-leaf *scrub*, low-growing woody plants, are more typical of chaparral vegetation. They are usually evergreen with thick, waxy leaves well adapted to the long dry summers. Thus such shrubs as wild lilac, manzanita, poison oak, and juniper are now found where there may once have been forest. Chaparral vegetation is also found in the Mediterranean region, where it is called *maquis*, and in Chile, where it is known as *mattoral*. Examples also exist in southern Africa and Australia. Most of these areas now grow vines and have a flourishing wine industry. The chaparral once had a varied fauna—with such herbivores as ground squirrels, deer, and elk and with such predators as mountain lions and wolves—but the fauna is now much reduced.

Northern coniferous forest This biome goes by many different names. In North America it takes a Latin name to become the *boreal forest*. In the USSR it is called the snow forest, or in Russian, *taiga*.

The most common *coniferous* (cone-bearing) trees in this biome are spruce, hemlock, fir, and pine. These needle-leaf trees can withstand the periodic drought resulting from long periods of freezing conditions. The trees are slender and short, between 12 and 18 m (40 and 60 ft) tall. They generally live fewer than 300 years but grow quite densely.

Depressions, bogs, and lakes hide among the trees. In such areas, low-growing bushes with leathery leaves, mosses, and grasses rise out of the waterlogged soil. These forms, combined with stunted and peculiarly shaped trees, are known as *muskeg*.

All these biomes could be differentiated even more precisely. For example, the boreal forest of Canada could be divided into three subzones. The main boreal forest is characterized by the meeting of the crowns of the trees. Second, patches where the trees are broken up by open spaces of grass or muskeg are called the open boreal woodland. Finally, at the northern limits, a mixture of woodland in the valleys and tundra vegetation on the ridges is called the forest tundra zone (see Figure 12-18).

Tundra This is the most continuous of biomes; it is found almost unbroken around the northern continents. It is also found in subantarctic islands and in alpine environments above the tree line in mountains.

Only cold-tolerant plants can survive in tundra. The most common are mosses, lichens, sedges, and sometimes dwarf trees. Annual plants are rare. The perennial shrubs are pruned back by the ice-carrying winter winds and seldom reach their maximum height. Nor can plant roots be extensive in this biome, because permafrost (refer to Chapter 7) is seldom less than a meter (3.28 ft) from the surface. The permafrost also prevents good surface drainage.

In the short summer, shallow pools of water at the surface become the home of a large insect population.

In the Northern Hemisphere, birds migrate from the south to feed on the insects. The fauna is surprisingly varied considering the small biomass available. It consists of such large animals as reindeer, caribou, and musk ox and such small herbivores as hares, lemmings, and voles. Carnivores include foxes, wolves, hawks, falcons, owls, and humans.

CLIMATE, SOIL, AND VEGETATION

Let us return to the northern limit of the boreal forest to begin our discussion of the relationship among the atmosphere, lithosphere, and biosphere. This area is mentioned in Chapter 11 as a location of cyclic autogenic succession, in which soil, climate, and vegetation are intimately related.

Climate has a great impact on vegetation. The northern limit of the boreal forest is the site of an isotherm and a tree-line boundary used by Köppen. However, recent studies have shown this Köppen relationship to be rather imprecise. Rather than following the isotherm, the arctic tree line appears to follow a line of equal net radiation. More precisely, the arctic tree line coincides with lines of equal net radiation in the growing season and lines of length of season with average daily temperatures above 0°C (32°F). Snow cover also plays a key role at the northern edge of the boreal forest, especially with respect to the flora. On the tundra, in winter, the snow is thin, heavily drifted, and dense, permitting the atmosphere to freeze the dormant vegetation. But in the forest tundra and boreal woodland, winter snow cover is deep, less dense, and more uniform, thus providing better insulation for underlying flora.

Vegetation also affects the climate. The distribution and physical properties (such as density) of the snow are governed largely by the structure of the vegetation. The forest tundra, for example, helps to prevent snow from drifting and thus keeps a fairly even insulating blanket of snow on the ground. Furthermore, net radiation values are influenced by the albedo of the surface, which in turn depends on the composition and structure of the vegetation. There is therefore a

FIGURE 12-18

Boreal forest (northern coniferous forest) and tundra near the northern limit of tree growth in the Northwest Territories of Canada.

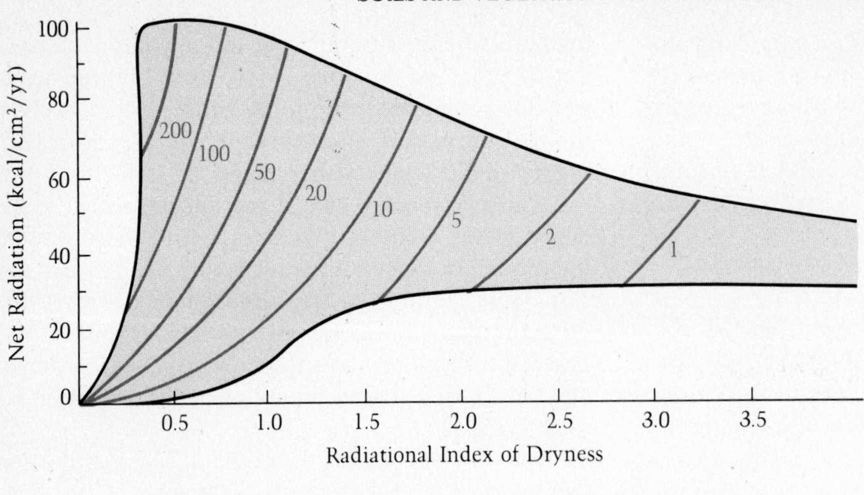

(a)

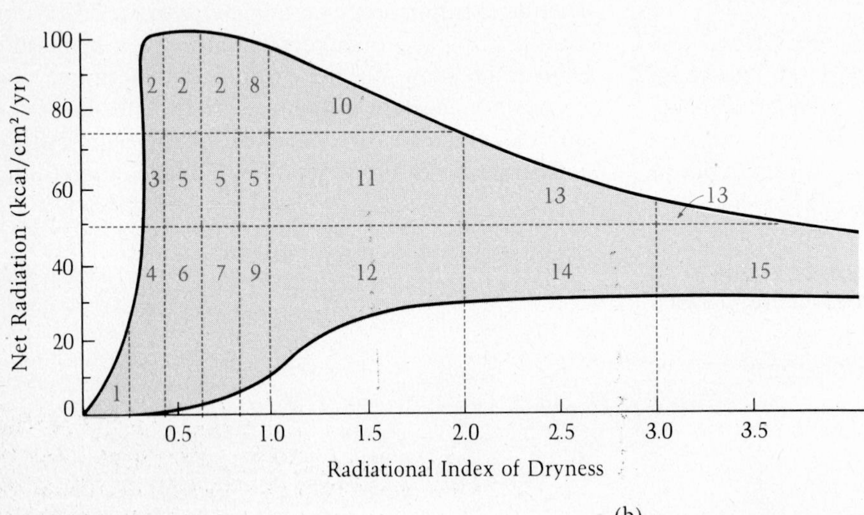

(b)

Key

1. Polar desert
2. Tropical rainforest
3. Subtropical swampland
4. Tundra and forest tundra
5. Subtropical rainforest
6. Northern coniferous forest
7. Temperate coniferous and deciduous forest
8. Wooded savanna
9. Deciduous forest and forest steppe
10. Dry savanna
11. Subtropical forest and scrub
12. Steppe
13. Subtropical semi desert
14. Temperate semi desert
15. Desert

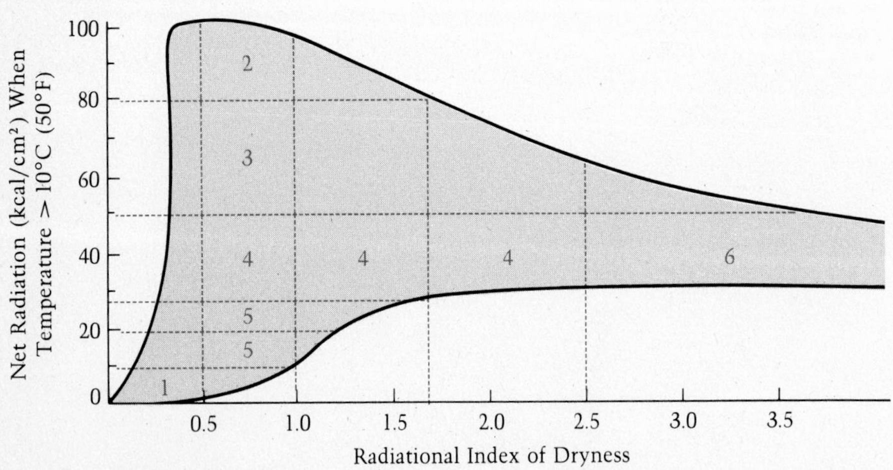

(c)

Key

1. Inceptisols
2. Oxisols
3. Ultisols
4. Mollisols
5. Spodosols
6. Aridisols

FIGURE 12-19

Uses of the radiational index of dryness for determining geographic zones. a) Runoff values in cm per year; b) vegetation types; c) soil types.

two-way link between climate and vegetation in this location. When you consider that soil development is affected by both climate and vegetation, one example of the relationships in the physical world becomes apparent.

The Russian geographers A. Z. Grigoriyev, I. P. Gerasimov, and Mikhail I. Budyko have spent decades attempting to discern order in the relationships of the spheres of the physical world. Budyko has pointed out that

> the heat and water balance of the earth's surface is . . . as a rule, the main mechanism that determines the intensity and character of all the other forms of exchange of energy and matter between . . . the climatic, hydrologic, soil forming, biologic and other phenomena occurring on the earth's surface. . . .

Budyko has suggested the use of a parameter called the *radiational index of dryness* to understand the relationship of climate, soil, and vegetation. This is the net radiation (Q^*) received at a location during a year divided by the total annual precipitation (P), which is multiplied by the latent heat of vaporization (L):

$$\text{Radiational index of dryness} = \frac{Q^*}{PL}$$

The radiational index of dryness is therefore a ratio between net radiation and the heat required to evaporate the precipitation at a place. The radiational index of dryness can be used with net radiation values to distinguish physical geographical zones on a global basis.

A simple example of the index's use is in determining zones of runoff. In Figure 12-19a the hydrologic cycle is represented by net radiation and the radiational index of dryness. Grigoriyev pointed up the relationship between these two measures of energy and the distribution of vegetation types, as indicated in Figure 12-19b. Gerasimov, using net radiation in the warmer season only, instead of for the entire year, pointed up the energetic and hydrologic relationships among soil types, as shown in Figure 12-19c.

To make these relationships more clear, let us take the example of deserts that appear on the right-hand side of these diagrams. In Figure 12-19b you can see that deserts occur in areas where the net radiation is mainly between 25 and 50 kcal per square cm per year, and where the radiational index of dryness is greater than 3.0. If Q^*/PL is greater than 3.0, the value of net radiation (Q^*) is at least three times larger than the amount of heat needed to evaporate precipitation (PL). Obviously, for deserts, the reason for the large value of Q^*/PL is the small precipitation value (P). This fact also explains the low runoff value. As you can see in Figure 12-19a, runoff in the part of the diagram coinciding with deserts is less than 1 cm per year. In addition, Figure 12-19c shows that similar values of net radiation and the radiational index of dryness (both with respect to the growing season only) give rise to the aridisols of some desert areas.

These syntheses do not describe scientific laws, but they clearly show that relationships exist among the hydrosphere, atmosphere, biosphere, and lithosphere. These relationships among energy, water, and other matter translate into spatial relationships vividly apparent when we compare the world maps of climate, soils, and vegetation (Figures 7-4, 12-10, and 12-11). It is the task of the physical geographer to bring these relationships more clearly into focus.

SUMMARY

Russian and American scientists have played important roles in the development of soil classifications. The most recent of their efforts is the Comprehensive Soil Classification System of the US Department of Agriculture. This scheme has ten major soil orders, each related to a certain spatial distribution.

There are also many ways to classify vegetation. An ecological approach relies on the concept of biomes, and seven of the principal terrestrial biomes are described in this chapter.

There are many relationships among climate, soils, and vegetation, and these relationships exist on different scales. Of course, the physical geographer is most interested in the spatial relationships.

QUESTIONS

1. Why was it particularly appropriate that the first soil classification schemes came from Russia and the United States? What major advances in geographers' thinking about soils did these early classification schemes represent?

2. What is the principal difference between the soil classification systems of Dokuchaiev and Marbutt, and those that came before them? What further problems with these classification systems led to the revisions incorporated in the 1960 Comprehensive Soil Classification System?

3. What basic principle is reflected in the levels of organization of the U.S. Comprehensive Soil Classification System? Where would the classification systems of Dokuchaiev and Marbutt have fit in this scheme?

4. Entisols are soils that for a variety of reasons have no horizons. Where in the United States would you be likely to find the following suborders of entisols: fluents? psamments? orthents? What formative factors largely control the location of each?

5. Histosols have no worked horizons and are associated with bogs. What features of their formation and location explain their lack of horizons?

6. Vertisols are soils with a high clay content, and they tend to swell and crack with changing moisture and heat conditions. Would you expect vertisols to create more problems for human activities in the Brazilian rainforest or on the Australian plateau, an area characterized by alternating dry and moist periods? Why?

7. What characteristics connect the major soil orders of histosols, entisols, vertisols, and inceptisols? How are they alike, and how are they different?

8. Figure 12-4 shows how the ten major soil orders interact with the four major spheres of the earth. In what ways does each sphere influence the formation and composition of all soils?

9. Give an example of the effect of the biosphere on latisols.

10. Why do aridisols cover a larger area of the world's land surface than any other soil? Are their features connected with factors dependent on their location, and if so, how?

11. Both mollisols and alfisols are important in commercial grain production, yet different farming techniques are used on both. What factors explain the difference? Are the techniques dependent on climate?

12. For many years, the timber industry was the only human activity to make widespread use of spodosols, since they were closely associated with coniferous forest cover. Through what process are these forests and spodosols related, and what climatic factors seem to connect the two?

13. What is the principal characteristic of oxisols? How are these soils dependent on climate, and where are they usually found?

14. How is the distribution of soils related to the distribution of climates on the hypothetical continent in Figure 7-3? What might complicate the actual distribution of soils?

15. Figure 12-9 presents a simplified spatial distribution of soils across the United States. To what landscape features do the inceptisols seem to be related? Which soil order is not present on this map, and why?

16. The relationship between soil types and climates on a world-wide scale is best seen in the USSR and in the United States, where the soil types vary from south to north, paralleling changes in climate. Soils change from aridisols successively to mollisols, while climates progress from desert through steppe. Why do these two countries exhibit fairly regular progressions, closely related to those of our hypothetical continent?

17. How have the various plant classification schemes been related to the uses to which plants have been put or the ways people have thought about them throughout history? Which classification systems are most important today, and why? Which do we most often use in geography?

18. The tropical rainforest biome, though not the largest in area, contains more species of plants and animals than in all the other biomes combined. In spite of the incredibly rich gene pool this represents, most of the true tropical rainforest may disappear by the end of this century. Why?

19. How have plants evolved differently to survive in the savannah and desert biomes? What different climatic factors do they have to compensate for in each case?

20. The marginal areas of the northern limit of the boreal forest biome and the southern limit of the tundra biome are excellent examples of the interrelationships among climate, soil, and vegetation. What are the relationships among the four spheres in this area? What are the negative and positive feedback mechanisms that maintain each biome's geographical distribution?

21. What are the advantages to using Budyko's radiational index of dryness as the measuring parameter for "the intensity and character of all the other forms of exchange of energy and matter between . . . the climatic, hydrologic, soil forming, biologic and other phenomena occurring on the earth's surface"? What are the differences in using this index with net radiation values to distinguish physical geographical zones rather than the Köppen classification factors of moisture and heat?

SUPPLEMENTARY READING FOR PART THREE

One of the best books concerning soil-forming processes and other related matters is S.R. Eyre's *Vegetation and Soils*, published by Aldine in 1963. A more up-to-date publication is *The Geography of Soils* by D. Steila published by Prentice-Hall in 1976. This is one of the first books to use the United States comprehensive soil classification system in physical geography, and it is recommended for its clarity. All biology textbooks naturally deal with the biosphere. Among the most useful are *Ecology: The Experimental Analysis of Distribution and Abundance* by C.J. Krebs, published by Harper & Row in 1972; *Biogeography* by C.B. Cox and others, published by Blackwells in 1973; and *Biology of Plants* by P.H. Raven, R.F. Evert, and H. Curtis, second edition, a Worth publication of 1976. Also of interest are *Principles of Environmental Science* by K.F. Watt, published by McGraw-Hill in 1972, and Theodore C. Foin's book *Ecological Systems and the Environment*, published in 1976 by Houghton Mifflin. For light but informative reading we recommend the *Time-Life* book *Ecology* by P. Farb, revised in 1969.

FROM THE GEOGRAPHER'S NOTEBOOK

PLANT SUCCESSION: THE CONSTANCY OF CHANGE

Much of the world's vegetation changes with time. The changes may be very slow such as those in response to a major change of climate, or they may be relatively fast such as might accompany a volcanic eruption or other catastrophic event. We are not concerned here with the slowest of vegetational changes because rather specialized methods are employed to detect them. But the changes that take place over a couple of centuries or even faster can be easily observed on the landscape, and this is the subject matter of this field trip.

Chapter 11 explains how changes in vegetation can occur in a plant succession and how there are several types of plant succession. It is just possible that there is not evidence for a plant succession near where you live. But they are more common than you might expect. Let us first see the types of places where we might find a plant succession in progress.

FINDING A PLANT SUCCESSION

The best place to look is any newly exposed land surface. River beds or deltas are likely places. When the water level is low, plants immediately start to colonize the exposed surface. The same happens on the side of a dammed lake when the water level is allowed to drop and on new road cuts. More advanced successions can be seen on burned-over forest land or abandoned farm land, both of which have been influenced by previous vegetation. Look also on the edges of forests. Can you see any evidence of the forest edge advancing or retreating? For example, new seedlings out beyond the forest edge might show that the forest is attempting to advance.

Sand dunes, estuaries, and lakes provide likely environments for a succession. So also do areas where landslides or slumps have exposed new ground. More exotic sources are regions where volcanic debris has been deposited or glacial moraines (see Chapter 21). If possible, seek the help of a biologist to confirm that you have found a succession because although these are the best places to look, it cannot be guaranteed that there is a true succession present.

Assuming that a plant succession has been identified, it is often possible to find a time sequence of colonizing plants from the very youngest to the oldest ones, and sometimes even to those plants representing a climax vegetation. The sequence of plants may occur in a line—such as from a water's edge to an area of completely dry land or perpendicular to a forest edge. In other places it may not be possible to find a linear sequence in space, but surfaces of successively greater age may be found in scattered points around a locality. In this field trip the first thing you have to do is to find an example of a plant succession. The next thing to be done is to describe the succession. Let us suppose that you have found a linear succession such as that which develops around a pond. Here are some methods that can be used to describe, and therefore observe and study the succession. Although we are talking about a pond plant succession here, the same observation techniques can be applied to any type of succession.

WHAT TO LOOK FOR

First you must organize an objective way of observing the succession. Either mark out plots of land, say about 1 m (1 yd) square, each representing different parts of the succession, or define a line, a transect, along which you can make observations at 5 m (5 yd) intervals. The actual distances and areas used will depend on the total area covered by the succession, and you will have to use your judgment in each case. Arrange your sampling interval so that you include at least one sample of each different vegetation type. Alternatively, make sure you study at least five sample sites in detail. You may find it helpful to take photographs of the sites you study so that you can refer back to the sites and include the photos in your report.

Every plant community has six major characteristics that can be observed, measured, and studied. We will assume that your sample sites are representative plant communities. So for each sample plot or point on the transect, you should make an attempt to say something about each characteristic.

CHARACTERISTICS OF A COMMUNITY

The first characteristic is the physical environment. Make estimates of the amount of light reaching the ground. You could use a camera light meter to make relative measurements of this. Also estimate the

FROM THE GEOGRAPHER'S NOTEBOOK

amount of moisture in the soil. The more you can find out about the soil the better. The best way to do this is to dig a pit and describe the soil profile. Alternatively, obtain an instrument called a soil auger. Your school or local hardware store may have one. An auger is like a large corkscrew that can be screwed into the ground and then drawn out with a small sample of soil. These instruments are invaluable for taking soil samples quickly. In addition, samples at different depths can be easily obtained. If you do not have time to dig pits and cannot obtain an auger, you will have to do your best to estimate the soil depth and determine its type from the upper horizons.

Another physical factor that you should note is the microclimate of the location. Is the site shaded? Is it sheltered from wind and rain? Is it likely to be warmer, colder, wetter, or drier than the other sites you observe?

The second characteristic of the community is the species diversity. Examine your sample plot or point on the transect, and count the number of different plants you can see. You do not need to identify them—just make sure they are different from one another and not just different sizes of the same species. In a linear succession the number of plant species present at the various seral stages tends to increase as the succession develops, with a maximum occurring in the stage immediately before the climax stage.

The third characteristic to be recorded is the growth form and structure of the plants. The major categories of growth form are trees, shrubs, herbs, and mosses. Further differentiation can be made into such divisions as broad-leaved trees, needle-leaved trees, and so on. Some of the terminology found in Chapters 11 and 12 should help you here. The different growth forms affect the amount of light the lower plants receive. See if you can detect any layering in the community. For example, the beech and maple forest that forms the climax of the pond succession in North America is often layered. At the top are the crowns of the beeches and maples. An intermediate layer would be formed by shrubs and their foliage, while on the forest floor might be a bottom layer of small plants.

The next characteristic to look for at the point you are studying is the dominance factor. Many species may be present, but a few (one, two, or more) might exert a major controlling influence because of size or number. Thus, referring again to the climax stage of the pond succession, the beech and maple are the dominant species. They determine to a considerable extent the conditions under which the other species must grow. Try to assess which characteristics of the dominant species are most influential in affecting other plants.

The fifth characteristic is a concept closely related to the idea of dominance—the idea of relative abundance. For the major species in the sample plot or along the transect, try to estimate the relative number present. Although beech and maple might dominate the pond succession climax, these trees might make up, say only 20% of the species present; small shrubs might account for about 40%; and grasses, mosses, and lichen might contribute another 40%. It would take a long time to obtain really accurate information on this, but even a rough estimate will help differentiate the different stages of the succession.

The final characteristic of a plant community is its trophic structure. Try to identify who eats whom. In some cases this will be easy. You might be able to see fish that eat small insects that fly over the pond or plankton within the pond. But in most cases the identification of the trophic structure (particularly in its entirety) will be difficult. However, you should be able to say something about the most obvious features of the structure, and give examples of producers, herbivores, and possibly carnivores.

PUTTING YOUR DATA TOGETHER

Now return to your home or school and write up the observations you have made. One way of organizing this material is shown in Figure 1. We have included samples of the kind of notes you should take for each stage. Obviously, the more details you can fit in the better. When you have filled in data on all of the stages of the succession and added photographs and any sketches you may have made, you will have a much better idea of the processes involved in the succession and of the dynamic nature of the succession itself.

It might just be that no matter how hard you try, you cannot find an example of a plant succession suitable for study nearby. In this case an alternative would be to create your own. Try to find an area of wasteland that will not be disturbed for a number of years. Mark a number of equal-sized plots, say 1 m (1 yd) square, with pegs. Use one plot as a control plot and leave this one in its natural condition. Dig the other plots so that all of the vegetation is removed and then leave all these

TROPHIC STRUCTURE		Producers = Phytoplankton Herbivores = Zooplankton Carnivores = Fish			
RELATIVE ABUNDANCE					Beech and maple 20% Small shrubs 40% Grasses, mosses, and lichens 40%
DOMINANT SPECIES				Cattails and bullrushes	
GROWTH FORM AND STRUCTURE			Hydrophytes attached to pond bottom also forming surface vegetation layer of leaves and flowers. Bottom dwellers being overshadowed.		
SPECIES DIVERSITY Number of different species counted	0	8	12	25	20
PHYSICAL FACTORS	Area completely submerged. Light penetrates to pond bottom. No soil development.	Ooze developing on pond bottom.			
RELEVANT PHOTOS					
STAGE OF SUCCESSION	 Bare bottom	 Submerged vegetation	 Emerging vegetation	 Temporary pond and meadow	 Beech and maple forest

FIGURE 1
Putting your data together.

plots for a year.

The open ground will gradually be recolonized by plants. The next year redig and clear all of the plots except one. Repeat this procedure year after year. Within two or three years a complete sequence of regrowth stages will be present. If growth is rapid in your locality, it may be possible to dig the plots at more frequent intervals. It is a good idea to try to perform this experiment at your school so that many people will benefit by observing the results of it. See if you can enlist the help of your professors.

Whichever method you choose to study the dynamics of plant successions, you will gain a greatly increased understanding of the way the living world operates. Occasionally, successions can take place quickly. One scientist studied cow droppings as ecological units and found that within the 30-day period it took for a fresh cow pat to be broken down the pat is sequentially occupied by 40 to 60 different species of insects. More often when we deal with plant successions we may expect the time periods to be spread over tens, hundreds, or thousands of years. Despite the long times involved, you will now realize that the biosphere is a dynamic sphere, and like the rest of the physical world, is subject to continual change.

PART FOUR

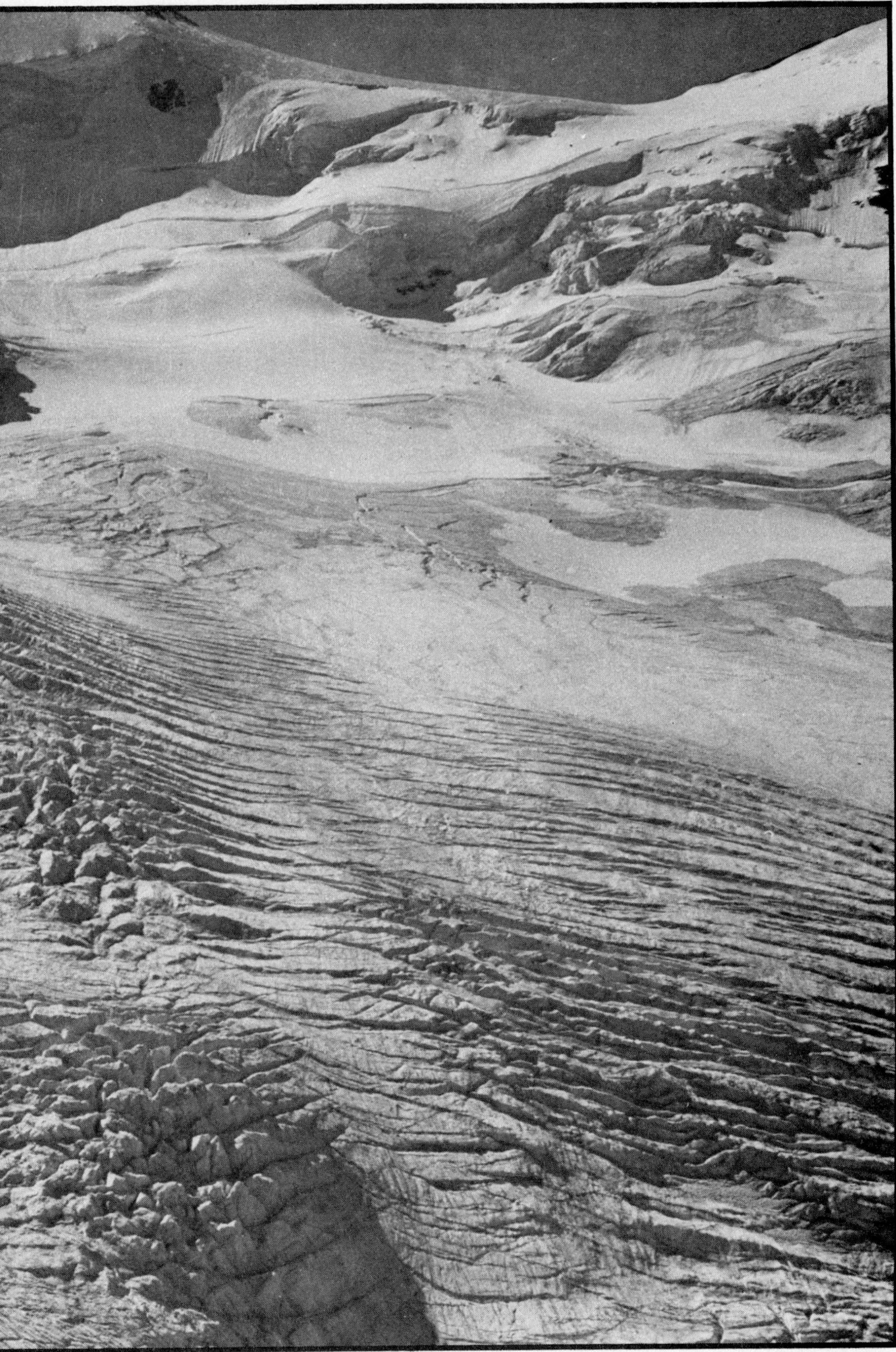

THE LITHOSPHERE

Processes that create rocks and wear them away make the lithosphere as dynamic as any of the other spheres that we have explored so far. But the processes of the lithosphere often act on a far longer time scale than others of the physical world. A story of continual stresses and strains, formation and decay, uplift and erosion is etched in the rocks of the crust of the earth. Slow but inexorable change is so much the keynote of the lithosphere that we repeat geologist James Hutton's comment that we can see "no vestige of a beginning—no prospect of an end."

In this part we examine the crust of the earth, its larger forms, and earthquakes. We discuss the rocks of the crust together with the stories and evidences they contain concerning past environments. Next we describe how large sections of the earth's crust, called plates, actually change their positions slowly through geologic time. Continents wander across the earth. Plate movement and other geologic forces give rise to the uplift of land surfaces. Then the other part of the story begins. The minute a land surface rises above the sea, it is attacked by the forces of weathering and erosion. Water plays a large part in most of these processes, helping to carve away the land surface, yet providing humans with one of their most important resources.

The shaping of the earth's surface presents us with many questions, not the least of which is how slopes are formed. We can often find answers by observing present-day events. We can see ice molding landscapes, we can watch waves attack cliffs, and we can feel the wind as it carries along small particles of sand. Landforms are being created and changed before our eyes. A battle is in progress—a battle between uplift and erosion. The results of the battle are seen in the shape—the morphology—of the landforms. We discuss this morphology both in profile and in its distribution across the face of the earth.

THE CRUST OF THE EARTH

CHAPTER 13

Other chapters in this book describe the outermost "shells" of our layered earth: the atmosphere, the biosphere, and the soils of the lithosphere. But they barely scratch the surface, the crust that sustains all this. Soils, as Chapter 10 explains, develop from the rocks that make up the earth's crust. Soils sustain vegetation cover that could not exist without moisture and warmth—the same factors that help transform rocks into soils. Soils, then, are at the *interface* between crust and atmosphere. It is appropriate now to turn from one side of this plane of interaction to the other and to learn what makes the continents and the ocean basins, the plains and the mountains.

About 70% of the earth's landscape lies *not* on the continents but under the waters of the oceans and seas. Even on the continents themselves, we need to know what lies below the surface in order to understand what is on the surface. Deep mines and even deeper boreholes help, but they afford only glimpses of a world that remains largely hidden from us. So we are in for constant surprises, and we are, in many ways, still exploring much as the old explorers did in unknown continents. It was only natural, for example, to assume that the terrain of the ocean floors constitutes an extension of the landscape we see around us. And then, just a few years ago, oceanographic research began to prove that the ocean basins are radically different from the continents. Ridges on the ocean floors, it turned out, are not simply submerged mountains like those we see on land. Undoubtedly we will be surprised by other discoveries in the future. Dead rocks provide us with some very live scientific issues!

This chapter provides an introduction to the crust of the earth. We consider its landmasses and ocean basins together with the flat areas and rugged regions of both. We briefly survey the interior structure of the earth and consider the age of our planet. Much can be learned about the earth's interior by studying earthquakes, so at the end of this chapter we explore the trembling crust.

LANDMASSES AND OCEAN BASINS

Geographers, we have become aware, are especially interested in the distribution of phenomena. As physical geographers, we would look at a globe and immediately observe some spatial qualities of its general layout that seem to demand an explanation. A prominent example: An examination of the globe shows that the earth has a land hemisphere centered on Africa and a sea hemisphere, the bulk of which is the Pacific Ocean. Why are the continents clustered mainly on one half of our planet?

Our inspection of the globe produces some additional questions. Is it an accident that three of the five major landmasses of the earth are triangular in shape, the long axes of the triangles pointing in essentially the same direction? Is it an accident that the bulges and indentations on opposite sides of the Atlantic Ocean seem to fit one another? Are all these questions somehow related, and would the answers help us to understand the complexities of continental terrains, the formation of mountains and escarpments, and the occurrence of earthquakes and volcanic eruptions? Surprisingly, the answer, as we see in the coming chapters, is—yes.

RELIEF OF THE LAND

Let us, for the moment, proceed systematically and consider more closely the *relief* of the continents, the landmasses that protrude above sea level. The term *relief* refers to the vertical difference between the highest and lowest elevations in a given area. Thus the Rocky Mountains of North America constitute an area of high relief, whereas the Great Plains form an area of low relief. Relief, then, differs from *elevation* or altitude. A nearly flat plateau with an elevation of 3600 m (12,000 ft) may have a lower relief than a mountain range whose summits reach only 2750 m (9000 ft) but whose valleys go down to 900 m (about 3000 ft). All the continents have areas of low relief as well as areas of comparatively high relief. Only Australia, smallest of the continents, lacks the high-relief zones of the others.

Perhaps it is best to start our discussion of relief with the simplest distinction. The difference between a plain and a mountain is obvious enough, but exact definitions of terms in common usage are sometimes quite difficult to establish. A *plain* is an area of low relief, and plains often lie at low elevations, as the coastal plain of the southeastern United States does. But the Great Plains of the North American interior lie as high as 1400 to 1500 m (4590 to 4920 ft)—higher than the tops of some mountains.

What is a *mountain*, as compared to a hill? Will F. Thompson suggested in 1964 that a real mountain is high enough to display climatic (and therefore vegetative) variation at different levels. But where such vegetative change cannot be observed, "it can . . . be argued that 2000 feet of local relief is a good rule-of-thumb separation value distinguishing hills from mountains." In other words, if the top of a hill is more than 600 m (2000 ft) from the bottom, the hill can be considered a mountain.

The earth's landmasses—the continents—consist of two basic geological components: shields and orogenic belts. The *shields,* often expressed as plains, are the oldest parts of the continents. They may represent the earliest examples where the molten surface began to cool enough to solidify into hard rock. Thus the shields are perhaps the nuclei that the landmasses eventually built around. So old and stable are these shield areas (they produce the oldest rocks found at the surface of the crust) that they have been worn down to a very low relief. Much of Canada east of the Rocky Mountains is such a shield zone, and as Figure 13-1 suggests, shield areas also exist in South America, Africa, Eurasia, and Australia.

Not far from the margins of the shield areas lie the *orogenic belts,* sites of the present and former mountain ranges of the earth. The term derives from the ancient Greek word *oros,* meaning "mountain." (Chapter 6 describes *oro*graphic rainfall.) Chapter 16 explains in detail, the earth over its 5-billion-year life span has experienced several periods of active mountain building, or *orogeny.* "Older" mountains, such as the Appalachians, have been worn down by 200 million years of erosion or more but have not been rebuilt by orogenic forces. Other "young" mountain ranges remain more active and, as in the case of the Andes and the Himalayas, still experience uplift.

During an orogenic period, mountain ranges undergo stages of building and enlargement separated by stages of quiescence and inactivity. During the quiet phases, erosion lowers the relief, reducing the mountains to rolling countryside and producing an *erosion surface.* Then the mountain building resumes, and the erosion surface is broken up and elevated by the new orogeny. But pieces of the erosion surface survive, and in the Rocky Mountains, the Alps, and other mountainous areas we can use these remnants to interpret orogenic history.

At present, "young" mountain ranges such as the Andes, Rocky Mountains, Alps, and Himalayas stand high above their surroundings, but they too will be reduced by erosion until new orogenies revive them. Other mountain ranges are in the making: Some geologists believe that a huge new mountain range will develop along the eastern coast of Asia, for example. We may see why they think so in Chapter 15.

The earth's plains Many of the earth's people make their homes on the plains. Plains often have fertile soils and easy communications among people. Exposure to

FIGURE 13-1

Shield areas of the earth, representing materials cooled from the earliest molten surface of the earth.

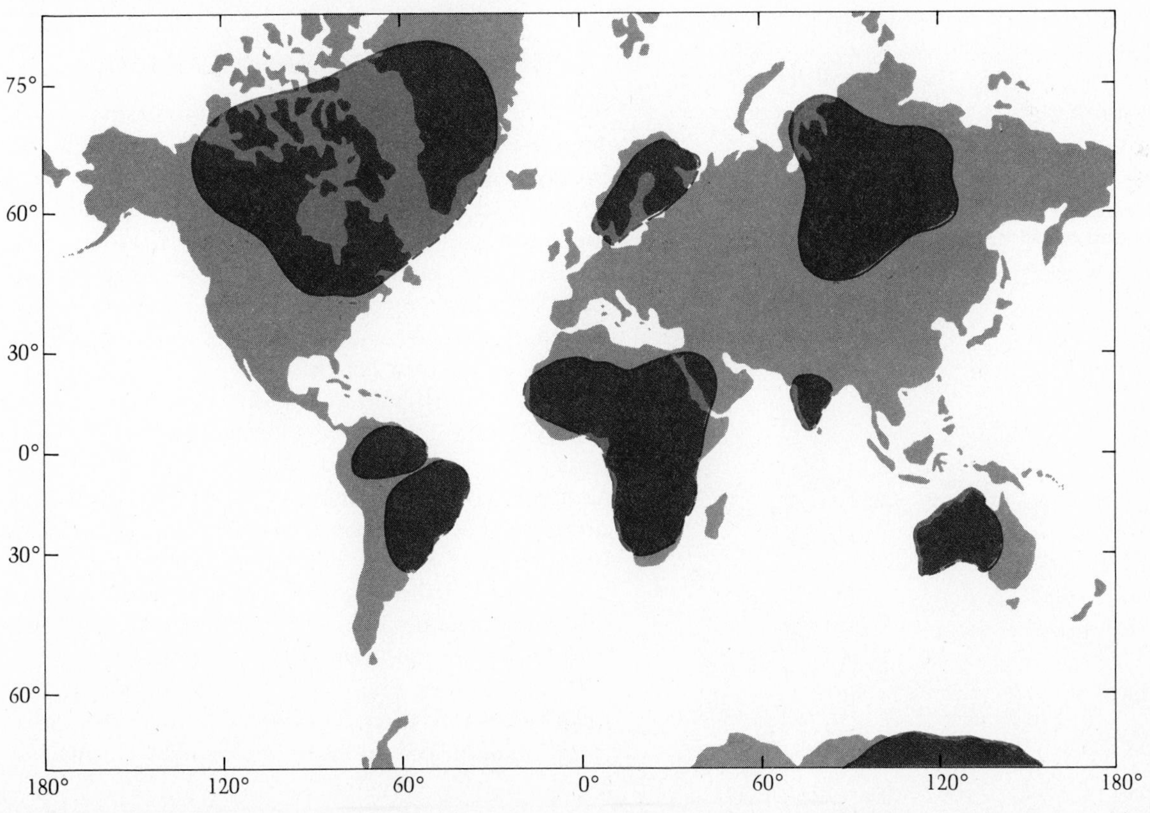

and rapid spread of innovations are among factors contributing to the clustering of peoples. Most of the world's great cities lie in plainlands.

Plains can result from long periods of erosion, when higher areas are worn down and depressions are filled in. Sometimes plains result from deposits of sedimentary layers. Often the formation of a plain is promoted by the horizontal layering of rocks. Softer overlying material is soon eroded, but a particularly hard, resistant rock layer slows erosion and allows a structural plain to develop.

An important aspect of plains on the margins of the continental landmasses (coastal plains) is that they continue beneath the ocean, sometimes for dozens of kilometers. The same rocks and structures that exist on the mainland occur for some distance from shore below the water, where the plain is called a *continental shelf.* A continental shelf is depicted in Figure 13-2. Minerals that occur on the continents also occur in their continental shelves. In the comparatively shallow waters above the continental shelves lie the largest commercial fishing grounds.

Another kind of plain, at higher elevations, is referred to as a *plateau.* A plateau, like a plain, is an area of low relief; the difference is in the elevation of the surface. A familiar example is the Colorado Plateau of the United States. Rivers are now cutting deeply into the Colorado Plateau, exposing a whole sequence of rocks in numerous localities—but most strikingly in the Grand Canyon. In East Africa, an extensive plateau is sustained by the same kinds of rocks that underlie the Canadian shield—rocks that crystallized from a molten state quite early in the history of our planet.

FIGURE 13-2

The continental shelf off the coast of the northeastern United States and the maritime provinces of Canada.

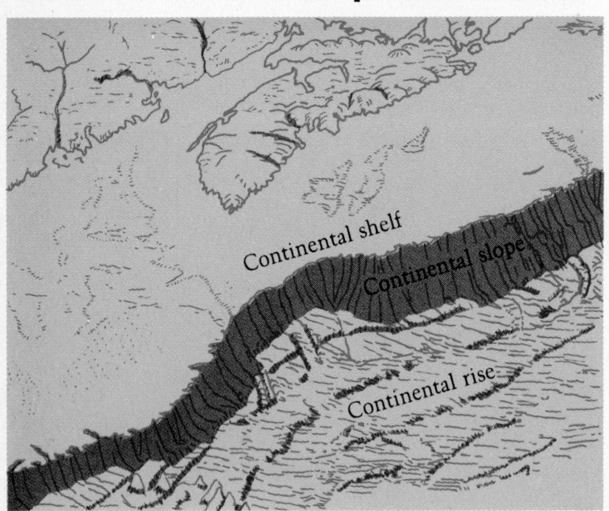

The great mountain ranges In contrast to the plains, mountain ranges are noteworthy for their sparse populations. With the notable exception of Africa, all the continents are spanned by lengthy, comparatively narrow mountainous belts. South America's Andes mountains extend from its southern tip to the Caribbean Sea; in North America the Rocky Mountains extend all the way into Alaska. A huge mountain chain crosses Eurasia, from the Alps in Europe through the Caucasus range to the Himalayas in Asia and beyond. Africa does have mountain ranges, but not of dimensions comparable to those of the Americas or Eurasia.

Topographically, the world's mountain belts are characterized by great vertical relief and by their enormous elongation. The term *chain* is appropriate, because the length and comparative narrowness of these mountains are distinguishing features. Frequently they consist of a series of parallel ridges separated by long, deep valleys. From afar—from a spacecraft, for example—these ranges look like crumpled sections of the earth's crust, as though the rock layers were a sheet or a towel that had been squeezed into giant folds. In Chapter 16, where mountain ranges are examined in more detail, we may find that impression to be quite accurate. The question that arises: Why does this happen, and what will happen next to these spectacular areas?

TOPOGRAPHY OF THE OCEAN FLOOR

The floors of the oceans—the ocean basins—cover about 70% of the solid surface of the earth. Some submarine areas are quite shallow, and fishermen drag their nets across the sea floor on the continental shelf. Midocean islands prove that the floor rises and falls and has relief. But until quite recently, maps of the ocean bottom were generalized and therefore rather inaccurate, and no geological samples were available to interpret submerged structures.

We now know that submarine topography is extremely varied, even more so than that of the continents. There are extensive plains, sheer cliffs, huge mountains, and deep trenches. Volcanic eruptions and earthquakes occur on the ocean floor, just as on land. Although there are no rivers and no wind, the floors of the oceans nevertheless undergo topographic change as drifts and currents combine with internal forces of the earth to modify the underwater scenery.

In the most general terms, we can subdivide the ocean basins into three topographical units: the continental margins, the midocean ridges, and the marine

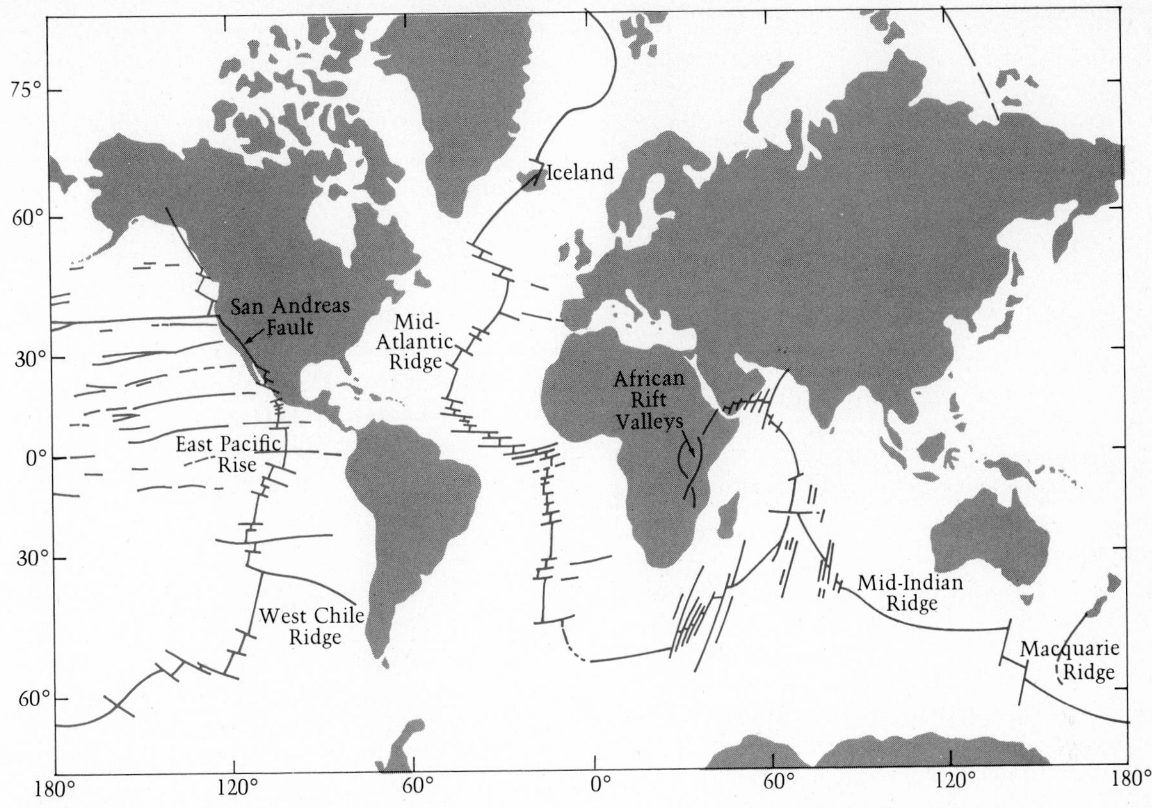

FIGURE 13-3
Midocean ridges in the major oceans of the world.

plains. As with the landscapes of the continents, numerous varieties mark each of these units.

Continental margins The rim of the ocean basins is also the margin of the continents. We have already noted that the continental shelf extends from the landmasses beneath the ocean water, sloping gently outward several dozen—or in places, as much as several hundred—kilometers. Some areas, such as eastern North America, have continental shelves of substantial width, but other places have little or practically no shelf (off the coast of Peru, for example).

The continental shelf continues to a depth of between about 120 and 180 m (400 and 600 ft) below sea level. (The 100-fathom or 600-ft contour line is usually charted as the outer edge of the shelf.) Then the surface drops quite markedly in a submarine slope called the *continental slope*, which marks the "real" margin of the continental landmasses. This slope carries the ocean floor down to much greater depths, about 1800 m (6000 ft or 1000 fathoms) on the average. There its slant becomes less, and the *continental rise*, at the foot of the continental slope, leads gently to the ocean floor itself. Remember that we are referring to the ocean floor here, so the "rise" actually involves a drop in elevation.

Midocean ridges Oceanographers for a long time have known about the submarine mountain range that traverses the Atlantic Ocean from north to south, approximately in the middle. The Mid-Atlantic Ridge is mapped in Figure 13-3. In several places this ridge emerges above the water to form islands (the Azores in the North Atlantic and Ascension Island in the South Atlantic, for example).

What was not realized until quite recently is that the Mid-Atlantic Ridge is only part of a whole system of such ridges that extends through all the oceans. Nor is this system of ridges simply a submarine equivalent of continental mountain ranges, as once believed. Rather, it marks a global fracture zone, and there is evidence that the crust here is being pulled apart by earth forces powerful enough to move whole continents. (Further explained in Chapter 15.) These midocean ridges are the key to our interpretation of the processes whereby the crust is formed.

A fault scarp 4000 meters below sea level, in the northeast Atlantic.

Marine plains Between the continental margins and the midocean ridges lie the marine plains—some nearly flat and featureless, others dotted by hills and rises, still others raised to levels we would call plateaus if they were on the continents. Where the floor is quite flat, it is probable that the accumulation of sand, silt, and mud for millions of years has buried the hard-rock topography below. These flat areas are called the *abyssal plains* of the ocean floor, and they lie at depths of as much as 4570 to 6100 m (15,000 to 20,000 ft) below sea level. We still have but limited knowledge of these deep plains; at those depths the pressure of the overlying water is enormous, making exploration most difficult. It is dark and quiet. There may be whole groups of fauna that we know as little about as we do life on other planets.

The abyssal plain 5000 meters below sea level, in the north Atlantic.

The abyssal plains are broken by *seamounts,* isolated or clustered submarine mountains that may be of volcanic origin, and by deep *trenches,* mapped in Figure 13-4. These trenches are of special interest, because their distribution suggests an association with the continental landmasses. Such trenches, or *foredeeps,* reach maximum depths of more than 9150 m (30,000 ft) below sea level. They are especially prevalent in the western Pacific Ocean, off the island arcs of Asia. Island arcs such as Japan and the Philippines are exposed sections of submarine extensions of continental mountain ranges: They are the crumpled, folded-up edges of the landmasses.

EARTH STRUCTURE

Any understanding of what we see on the earth's surface must begin with what lies below. The continental landmasses and ocean floors constitute the crust of the earth. This crust is quite thin and, in places, unstable. Below the crust, our planet takes on a different character. Rocks are solid and extremely dense near the crust, but deeper down they become viscous (sticky), so great is the heat and so high the pressure. All this subcrustal material moves in slow motion, agitated by heat generated in processes of chemical change. In the context of the total mass of the earth and the forces active within it, the crust that we depend on so greatly seems precariously vulnerable—as people living in earthquake belts or near volcanic mountains will readily attest.

The unseen, inner earth is known to be made of a series of layers, so that the whole planet resembles an orange with several skins, as in Figure 13-5. At the center of the earth is a ball of extremely heavy, dense material identified as the earth's *inner core.* Its radius is less than 1250 km (800 mi), or about one-fifth the earth's radius, and it is surrounded by a layer of similarly heavy but liquid material forming the *outer core.* Almost 2240 km (1400 mi) thick, this outer core appears to be made of metallic matter, perhaps nickel and iron, under unimaginably high pressure and at enormously high temperatures, many times greater than anything yet created artificially by scientists.

A characteristic of the earth's core is its great weight and high specific gravity. *Specific gravity* is the ratio of the weight of a given volume of a substance to the weight of the same volume of water. The earth's crust is comparatively light; the specific gravity of rocks in the crust averages about 3.0. But calculations of the total mass of the earth, based on its role in the balance of the solar system, produce an overall specific gravity of approximately 5.5. If the crust averages but 3.0 and the whole planet 5.5, then the earth's deep interior must be much heavier than the crustal rocks. To compensate for the crust's lightness, the inner earth must have a specific gravity of as much as 10.0 or even more. Only a few substances in nature, including nickel and iron, are so heavy. Hence scientists conclude that these are major elements in the inner and outer core.

Between the core and the crust is the earth's *mantle,'* about 2900 km (1800 mi) thick. By volume, the mantle constitutes the bulk of the earth, and the crust rests upon it. New interpretations of the behavior of earthquake waves in the mantle have begun to unlock its secrets. For example, we now know that earthquakes originate deep within the mantle, some more than 650 km (400 mi) below the surface, as well as relatively near the surface. This means, among other things, that the mantle is not composed of liquid or viscous material alone. There may be such pockets in it, but much of the mantle appears to be quite rigid.

Another discovery is that the crust does not gradually merge into the mantle. On the contrary, there is a

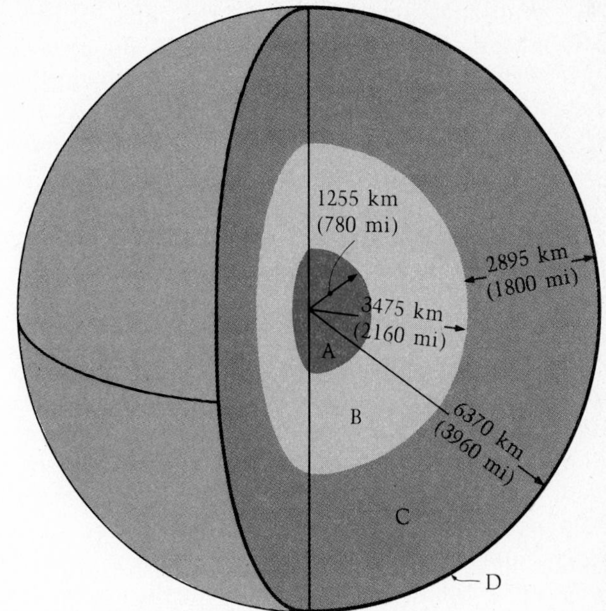

FIGURE 13-5
The principal layers of the inner earth.

FIGURE 13-4
The earth's major ocean trenches.

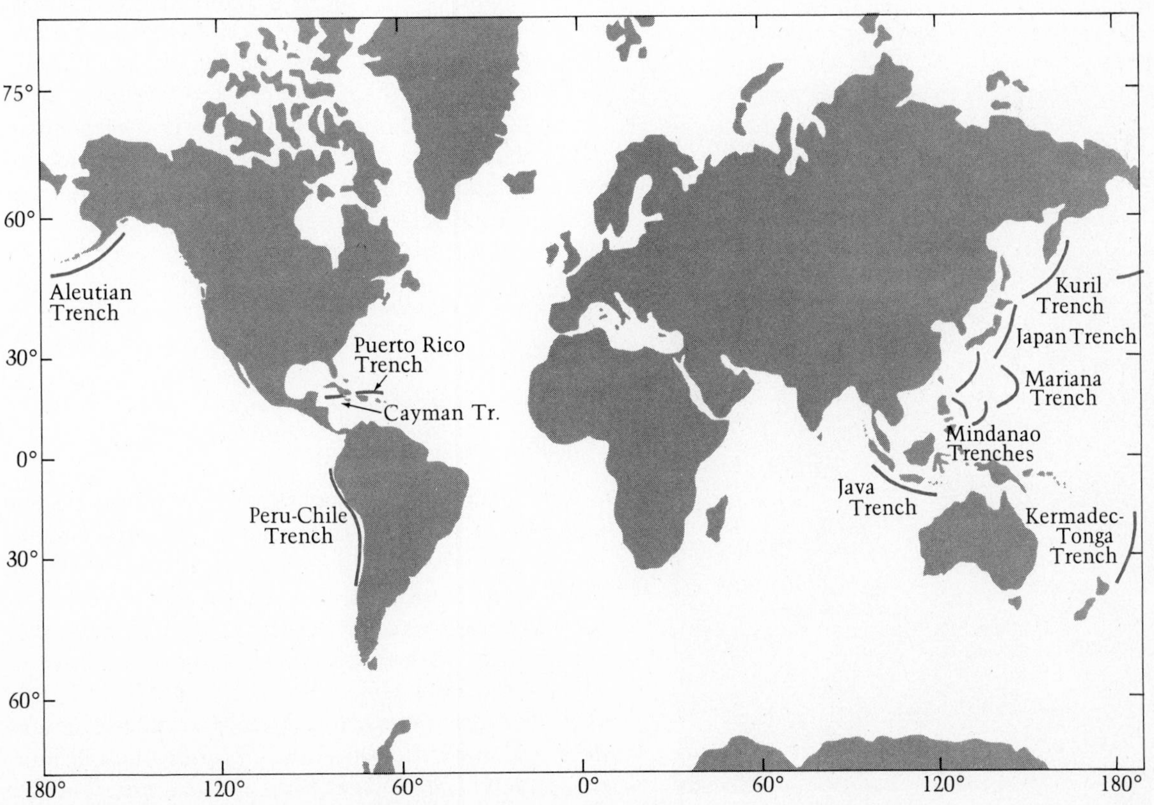

sharp break between the two shells. The Yugoslav seismologist Andrija Mohorovičić was the first to recognize the existence of this boundary, which then became known as the Mohorovičić discontinuity. The discontinuity acts as a reflector of some earthquake shock waves. Not unreasonably, it has come to be called the Moho discontinuity.

The outermost solid shell of the earth, the *crust,* is by far the thinnest, a mere 16 to 40 km (10 to 25 mi) thick. Near the upper surface of the crust are the lightest of earth rocks, the granitic rocks. Down below, the crustal rocks become heavier. There is an easy way to remember their main characteristics: In the upper layer, the granitic rocks contain mainly *al*uminum-rich *si*licates, the first two pairs of letters forming the word *sial.* Below, *si*lica is still prevalent, though less than in the granitic rocks, and a heavy substance called *ma*gnesium replaces aluminum (the mantle below also seems to be rich in magnesium). Again, the first letters form a term: *sima.* So the crust has an upper zone of sial and a lower layer of sima, except for the ocean floors. Beyond the margins of the continental shelves, the sial is absent, and only sima is present. At the base of the sima layer is the Moho discontinuity.

It would be fascinating to pursue our study of the earth's structure below the crust in greater detail, but this is really more the province of geology. Our interest is in the upper surface of the crust, in the processes that carve the landscapes we see around us. Here at the interface, the forces of weathering and erosion operating from above interact with the forces of crustal deformation from the earth below. Moisture in the soil helps rock layers below decay. A period of rainfall and consequent runoff moves loose particles from one place to another. And although we cannot be aware of it, entire regions move—perhaps only centimeters per year, but moving nevertheless. One year from now the building you are sitting in could be up to 15 cm (6 in) from where it is today (with respect to latitude and longitude). You will not feel it move, but if you multiply a few centimeters by thousands and then millions of years, you can see how significant this process is.

FIGURE 13-6

Fossil trilobites. Fossils helped establish the geologic time scale—they can be dated and they can be matched to other fossils.

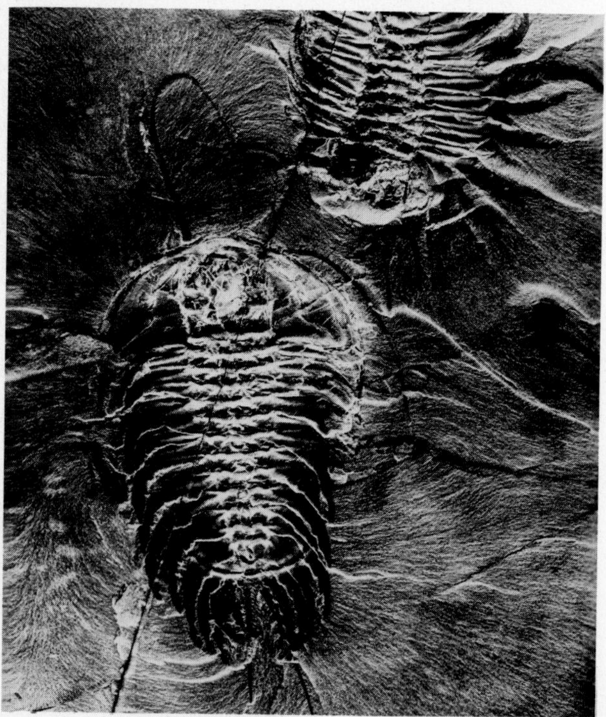

THE LIFE SPAN OF THE EARTH

When geologists began to realize that there is a degree of order in the massive accumulations of rocks that make up the earth's crust, it became necessary to establish a timetable for permanently labeling newly identified rocks. The *geologic time scale,* a chronicle of earth history, was developed more than a century and a half ago by British geologists. They could see how a geological cross-section of England involved several separate phases of accumulation and erosion of rocks. They called the oldest sequence of rocks the *Primary* and the next two the *Secondary* and *Tertiary.* Later, the youngest rocks, such as sediments now accumulating in river valleys and deltas, were separately identified as *Quaternary.* If we used these original terms, our geologic time scale would look like this (with the oldest rocks at the bottom, of course):

4. Quaternary (recent)
3. Tertiary (young)
2. Secondary (intermediate)
1. Primary (oldest)

It soon became necessary to subdivide these original major units. Bit by bit the geologic time scale became more exact—and more complicated. Scientists learned more about fossils, traces of plants and animals preserved in the rocks (see Figure 13-6), and they made correlations with discoveries in distant countries. Their discoveries about the evolution of life largely determined the names in the geologic time scale now in use, shown in Figure 13-7. Primary has become *Paleozoic* (era of ancient life), and Secondary is now

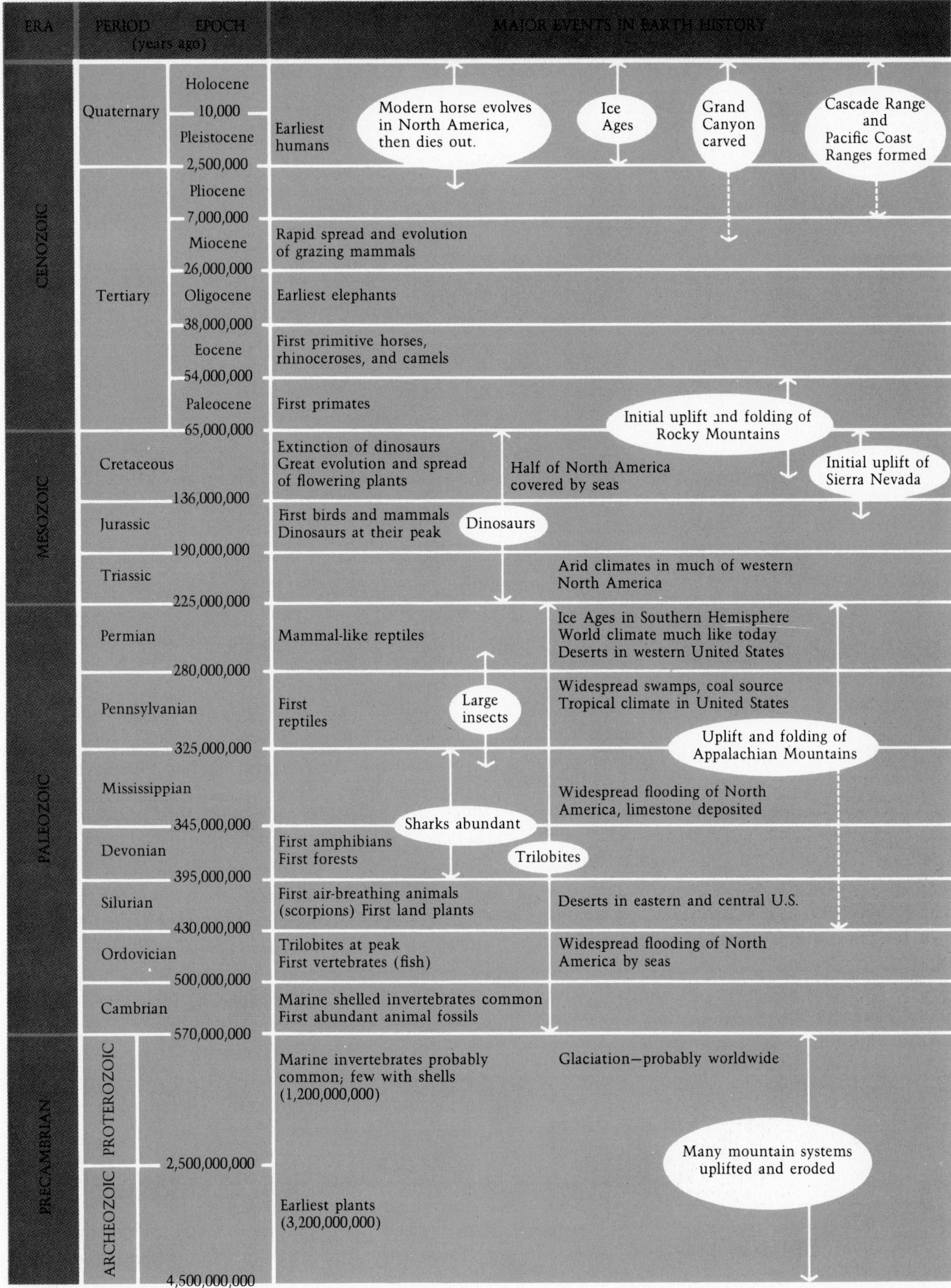

FIGURE 13-7

The geologic time scale, showing important events in the history of the earth.

EARTH'S LIFE SPAN AND YOUR AGE

If you are 20 years old or near 20, and we take as the age of the earth the time of formation of the oldest rocks we can find (about 4 billion years), consider the following:

One year of your life equals 200 million years of earth's. That puts you in the middle of the Mesozoic just one year ago.

One month of your life equals just under 17 million years of earth's. The Rocky Mountains formed just 4 months ago.

One week of your life equals about 4 million years of earth's. The Pleistocene Ice Age began yesterday.

One day of your life equals about 550,000 years of earth's. Human evolution was still in early stages just yesterday at this time of day.

One hour of your life equals some 23,000 years of earth's. In that one hour, human population grew from a few tens of thousands to four billion, and the major civilizations developed.

Where will we be one hour from now?

Mesozoic (era of medieval life). We can still find Tertiary and Quaternary in the modern classification, but they are now subdivisions of a third great era, the *Cenozoic* (era of recent life). The time scale in Figure 13-7 shows the three major *eras* divided into *periods*, which, in the case of the Tertiary and Quaternary, are further subdivided into *epochs*. These names refer to the evolution of all life on earth, not human life alone. Human communities did not appear until the latest phase of the Cenozoic era—the Quaternary. There are also rocks predating the Paleozoic era (older than about 600 million years). This era is called the *Precambrian*, because it is older than the oldest period of the Paleozoic era. The shield zones of the earth are in this oldest of age categories.

The simplified geologic time scale in Figure 13-7 shows that more is known about relatively recent geological times than about older periods. Subdivisions at the top of the chart are more numerous, and they cover shorter time intervals. This is not surprising, because the youngest rocks tend to be closest to the surface. Layers left in the northern United States and in Canada during a recent glaciation are fresh and have hardly changed, geologically speaking. Other rocks are being deposited at this very moment. We can extrapolate from the processes we see to the rocks buried deep down, which must have been laid down under similar circumstances.

We will have occasion to refer to the time scale much as we refer to the years and months of our own lifetimes. Although our interest is in the landscapes of the present-day world, it is as important to understand the sequence of events that contributed to their development as it is to comprehend the nature of rock structures and forces below the surface of the crust. Periods of mountain building, glaciation, and other events in earth history must be placed in a temporal context, or we will lose track of the whole sequence.

THE TREMBLING CRUST

In many parts of the world, including Central America, earthquakes are almost a daily part of life. In those areas tremors occur frequently, punctuated by devastating earthquakes that can destroy whole cities and claim tens of thousands of lives in a few moments of calamitous upheaval. In just the last few years severe earthquakes have struck in China, Alaska, southern California, Mexico, Nicaragua, Peru, Guatemala, Italy, and Iran.

Such earthquakes can topple substantial buildings and dislocate roads, railroads, water mains, and other facilities. Wide fissures open in the ground, and people, cars, even houses disappear into them. Earthquakes often cause destructive landslides. Finally, an earthquake at sea may cause a huge wave to roll coastward, swamping towns and villages under a wall of water. Though rare, such *tsunamis* have done severe damage along Pacific coastlines. The Philippines suffered large tsunami damage in 1976.

DISTRIBUTION OF EARTHQUAKES

From the short list of countries, just mentioned, that are frequently afflicted by severe earthquakes, it appears that the earth's younger mountain ranges are especially associated with such shocks. As Figure 13-8 shows, that impression is correct. The earth's most earthquake-prone belts lie around the margins of the Pacific and along the Alpine mountains across Eurasia.

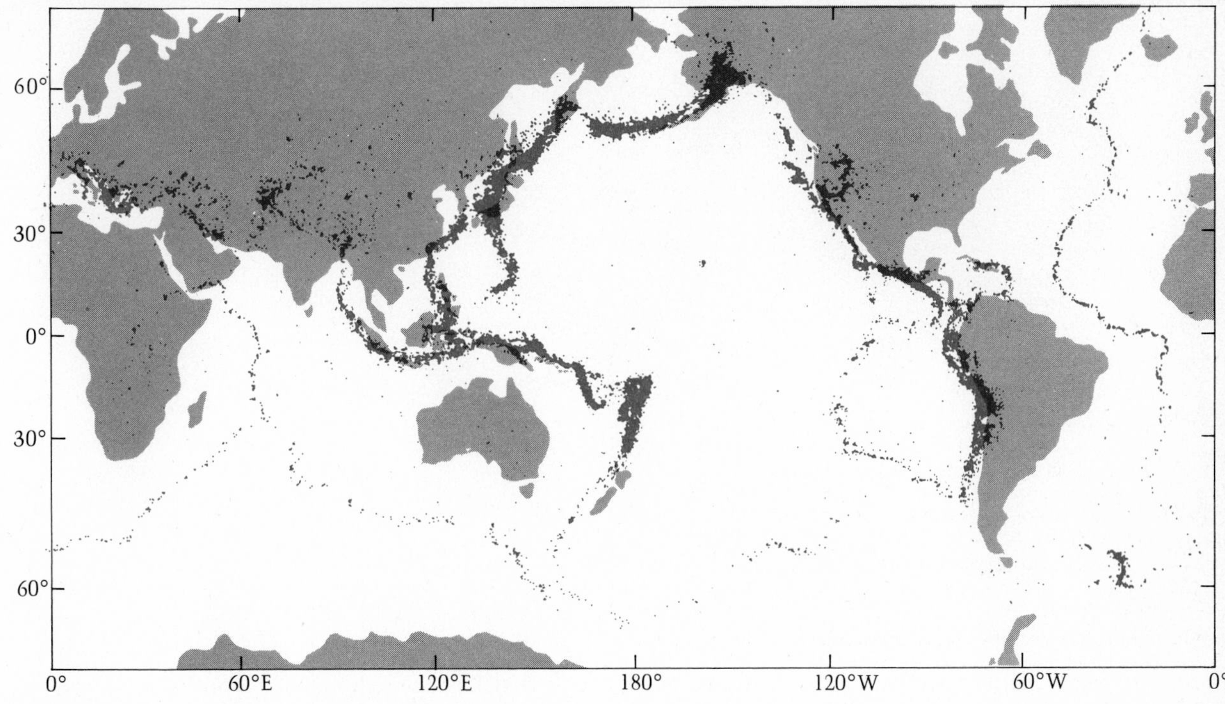

FIGURE 13-8

World-wide distribution of earthquake epicenters from 1961 to 1967.

Note, also, that the midocean ridges are marked by a high frequency of earthquakes.

Areas that experience relatively few earthquakes include the shield regions discussed earlier in this chapter. Earthquakes are rare in northern Canada, eastern South America, northern Eurasia, Africa, and Australia. Thus we believe that the shield areas are the earth's stable zones and that the younger, high mountain ranges are the unstable belts.

Figure 13-8 also reveals why earthquakes are so often destructive of human life. Especially in eastern Asia, but also in North and South America and in parts of Eurasia (Turkey and Iran, for example), the earthquake belt passes through zones of dense population. Figure 13-9 shows earthquake (or *seismic*) risk in the United States. If you reside in any of the zones of major

MANAGUA, 1972

If you look on the map of Central America, you will find the city of Managua, capital of the Republic of Nicaragua, positioned at the southern end of Lake Managua. Nicaragua has two large interior lakes, of which Lake Managua is the smaller. Should you fly over this area you will see volcanoes and lava flows, rugged mountains and large escarpments. People here are accustomed to rumbling and shaking in the earth, for this is a zone of crustal instability and high earthquake incidence.

Indeed, Managua had been hit by disastrous earthquakes twice—in 1885 and in 1931—and as Christmas 1972 approached there were people who could recall what happened in 1931, when the capital was turned into a hell of devastation and fire. But even such memories fade, and on December 23, 1972 the festive holiday season was in full swing.

On that day a series of earthquakes struck, the strongest measuring 6.25 on the Richter scale, and in minutes Managua was in ruins. Thousands of people lay buried beneath the rubble, fires swept the area unchecked. There was no water, no electricity. Aftershocks continued. Disease threatened, and on orders of the government, what remained of Managua was evacuated, the wreckage leveled by bulldozers and covered with lime. An entire city had been swept off the map.

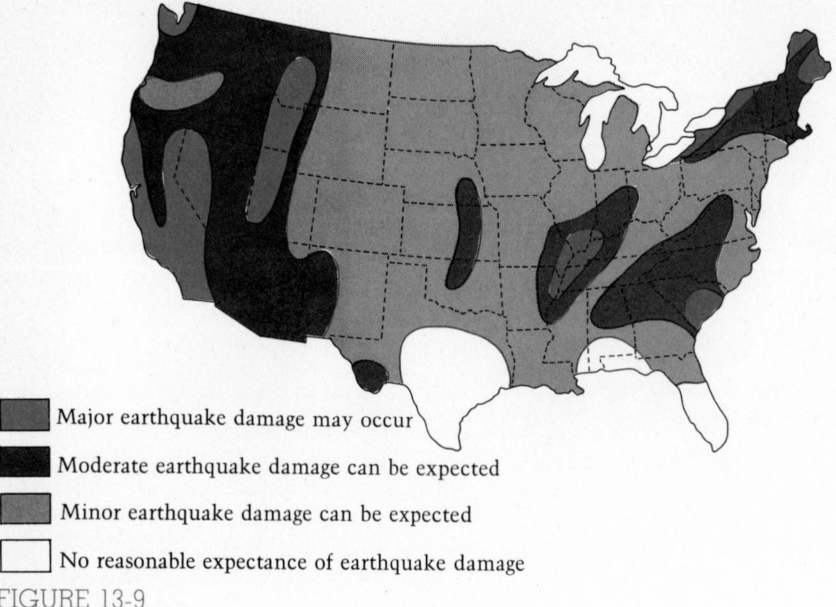

FIGURE 13-9
Seismic risk in the continental United States.

expected damage, chances are that your area will be struck by a severe earthquake. But this does not deter people from living in high-risk areas. Some people moved away from Los Angeles after the February 1971 earthquake, but most people stayed.

Earthquakes are distributed in time as well as space. If there were just one station monitoring the whole Pacific coastline from southern Chile to New Zealand, that station would record a constant, continuing succession of earthquakes, large and small. Many thousands of earthquakes, ranging in strength from mere tremors to severe shocks, occur on the earth every year, proving the instability and changeability of the crust. It has been calculated that each century between 200 and 300 of the severest earthquakes occur, between 1000 and 2000 very strong ones and hundreds of thousands of minor shocks and tremors.

ORIGIN OF EARTHQUAKES

Although we experience the manifestations of an earthquake at the surface—a rumbling roar, a shaking of the ground—earthquakes originate within the crust and sometimes deeper down. The point of origin of an earthquake is its *focus*, which may range from near the surface to far into the mantle. But most earthquakes originate within only a few kilometers of the surface. The place directly above the focus at the surface of the crust is the *epicenter*. Figure 13-8 shows earthquake epicenters for some recent years.

Earthquakes result from the sudden movement of rock that has been subjected to prolonged stress. The same forces that cause the crumpling of sections of the earth's crust into major mountain chains cause rocks that cannot yield by bending or folding to fracture. Such fractures in the crust are called *faults*, and some of these, such as the San Andreas Fault in California, have become quite famous as generators of recurrent earthquakes.

The San Andreas Fault, shown in Figure 13-10, extends for more than 960 km (600 mi) from near Eureka, along the northern California coast, through San Francisco, and inland from Santa Barbara and Los Angeles to a spot near the Mexican border. A movement along this fault produced the 1906 earthquake that destroyed much of San Francisco. The damage in and near Los Angeles in 1971 was the result of an earthquake generated by a fault near the San Andreas Fault, certainly part of the zone dominated by the San Andreas. The stresses are building up again along this fault, and another shock may not be far in the future.

The contribution of earthquakes in the generation of landscapes is often rather minor. Tremors and minor earthquakes have little or no effect on the landscape. But stronger earthquakes can and do have an impact. The displacement along faults can create cliffs called *scarps*. Although scarps can be formed in other ways as well, repeated movement along faults may produce major *escarpments*, long cliffs that withstand the forces of erosion for extended periods. Additionally, earthquakes can cause landslides that block and dam streams, and tsunamis resulting from earthquakes can modify coastlines.

MEASURING EARTHQUAKES

So far we have spoken of earthquakes as "severe," "strong," or "weak"—without attaching any precision to those descriptive terms. Still it is necessary to measure the force of individual earthquakes, if only to permit comparisons of the damage they do.

In 1935 Charles F. Richter, a famous seismologist at the California Institute of Technology, devised a scale of earthquake magnitudes that is still in use. Numbers from 0 (the smallest recordable tremor) to 9 indicate the calculated energy released at the earthquake focus. Earthquakes in the 0 to 4 range are minor; from 4 to 6 they can cause local to regional damage. From 7 upward, the earthquake is a major one whose shock waves are recorded over the entire globe. The San Francisco earthquake of 1906 is estimated to have had a magnitude of 7.8. Although the Richter scale goes up to 9, the greatest magnitudes ever recorded are in the vicinity of 8.6. The 1964 Alaska earthquake had a magnitude ranging from 8.4 to 8.6 and was one of the strongest earthquakes in recorded history. Although the epicenter of this earthquake was 120 km (75 mi) away, Anchorage suffered severe damage. In a wide surrounding region, landslides, avalanches, and changes in the level of the land occurred. The sea floor also underwent change, and tsunamis struck the coast. In 1976 a major earthquake was recorded near Peking, China. The major shock was 8.4 on the Richter scale, and the first aftershock was 7.9. The earthquake hit an industrial area, and it was reported that the nation's steel production was considerably set back as a result.

FIGURE 13-10

The San Andreas Fault, California.

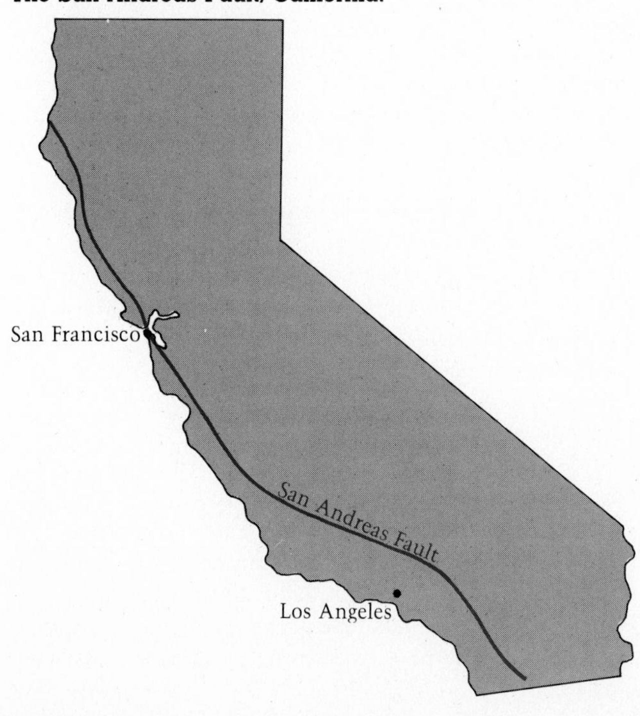

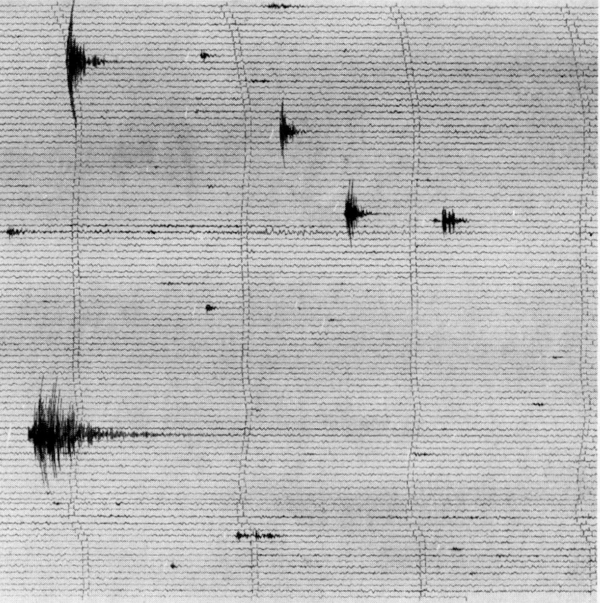

FIGURE 13-11

A seismogram, the record of a seismograph. A major earthquake is indicated by the wide arcs traced by the pen—in the summit region of Kilauea volcano in Hawaii, February 15, 1966.

MAKING USE OF EARTHQUAKES

Destructive as earthquakes are, they do have one positive side. Their records, taken at seismographs around the world, have revealed a great deal about the interior of the earth.

An earthquake sends out shock waves in all directions from its focus, as a pebble thrown in the water of a quiet pond sends out rings. In the case of the earthquake, however, the waves do not travel only along the surface, or the crust. They also penetrate downward into the mantle and through the earth's core. A distant seismograph, then, receives vibrations resulting from waves that have traveled different courses, and these different paths cause the waves to arrive at slightly different times. The *seismogram* (See Figure 13-11)—the record of an earthquake at a seismograph—shows which waves have traveled through the crust only, which have traversed the mantle, and which have penetrated the core.

Another important characteristic of earthquakes is that they generate different types of waves. The jarring motion of the rocks can push nearby rocks away to rebound again in a kind of push-pull action, or the surrounding rocks can shake in a sideways action. The seismograph shows these different types of waves, and the evidence is used to interpret the structures and materials that the earthquake's waves traveled through.

Earthquake waves, just as sound waves, travel faster through solids than through liquids (and some types of earthquake waves are not transmitted by liquids at all). The push-pull or *primary* waves generated by an earthquake, called *P* waves, can be heard as a rumble or a deep roar as the earthquake vibrations pass. If the earthquake is strong enough, these waves are recorded by all seismographs everywhere. Figure 13-12 shows how the *P* waves penetrate through the mantle and even the earth's core and travel along the crust as well. *P* waves traveling through the mantle reach seismographs first and give warning that stronger shocks are in the offing.

The waves that cause a sideways action are the *secondary* or *S* waves. You can simulate their effect by tying a rope to a fence or a pole and holding the other end of the rope in your hand. If you move your hand vigorously up and down, the resulting motions run through the rope to the point of its attachment. These *transverse* waves move at right angles to the direction of wave propagation. At the earth's surface, secondary waves can do enormous damage to structures in towns and cities.

FIGURE 13-12

Cross-section of the earth showing the paths of *P* waves, *S* waves, and surface waves.

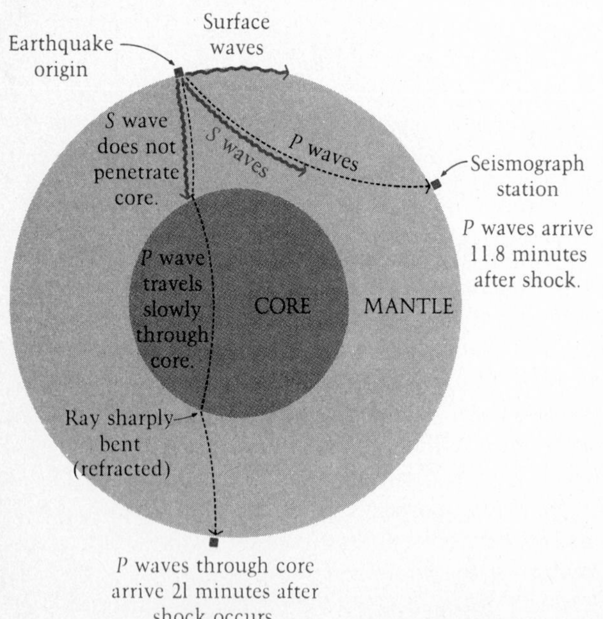

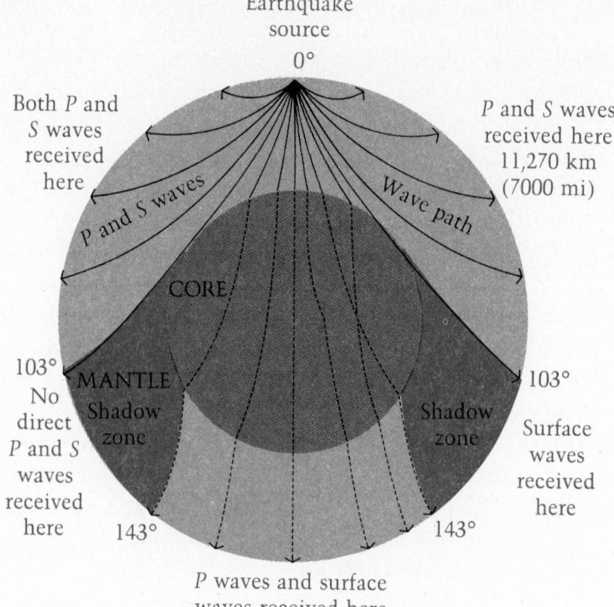

FIGURE 13-13

The possible paths of earthquake waves through the earth's interior.

S waves cannot travel through liquids. This is of vital importance in our use of seismographic records to interpret the earth's internal structure. If seismographs anywhere do not record the arrival of certain *S* waves, then there must be a liquid obstruction to their travel through the interior of the planet.

When an earthquake occurs, seismographs for about 11,000 km (7000 mi) in all directions record all *P* and *S* waves, whether they travel through the interior or along the crust, as Figure 13-13 indicates. Then for some 4300 km (2700 mi), no direct *P* and *S* waves are received at all, only surface waves. This is the so-called *shadow zone*, apparently caused by the refraction of the waves somewhere in the interior of the earth. Beyond the shadow zone, in the lower part of the diagram, the *P* waves can be recorded again, but *S* waves cannot. The *P* waves are refracted by the same material that fails to transmit the *S* waves. From global earthquake records it has been possible to deduce the existence of the earth's liquid outer core and even its fairly exact dimensions.

Many other deductions regarding the earth's interior structure have been made possible by the ever-growing seismic record and the greater precision of the recording instruments. Seismic prospecting, involving the creation of tiny earthquakes by the detonation of buried explosives, has revealed the existence of many geological structures containing resources, especially petroleum.

PREDICTION AND PLANNING

Every year earthquakes take a toll in human lives, especially in urban areas where buildings prove inadequate and collapse around their inhabitants. We can respond to this danger in two principal areas: in regulations for the construction of stronger buildings in cities and in the search for a way to predict severe earthquakes.

The 1971 earthquake in southern California came at a time when environmental concerns were predominant. Although not an especially strong earthquake (6.6 on the Richter scale) and with an epicenter a considerable distance from central Los Angeles, there was much structural damage, especially to older buildings. The Veterans Administration Hospital at Sylmar suffered severe damage: A wing collapsed, and 47 patients died in the rubble. Investigative panels later concluded that many of the buildings in the Los Angeles area, especially the older ones, were unsafe in the event of an earthquake, and that existing building codes were not sufficiently strong to safeguard occupants. Public buildings, it was urged, should be built as "earthquake-proof" as possible. New high-rise buildings towering over Los Angeles and San Francisco are constructed to bend rather than break with earthquake shocks. Of course, no matter how well designed, these buildings would be severely damaged and even toppled if an earthquake over 8.0 on the Richter scale were to strike near the central city. There is probably nothing that can be done to protect against earthquakes of that strength.

The alternative, of course, is to know that a severe earthquake is due and to evacuate the threatened area. Research into the prediction of earthquakes is in its infancy, but it is attracting the attention of an increasing number of scientists. They follow two directions. By one method, seismic records of all earthquakes, even the most minute tremors, are used in attempts to determine patterns that may normally precede major earthquakes. If there is evidence that stresses build up, then a network of recording stations might be capable of giving warning. The other method concentrates on the strain on rocks in earthquake-prone areas. If this strain can be measured and breaking points predicted, then major earthquakes might be predicted. If we had a record of the level of strain on rocks near the epicenter of the 1971 southern California earthquake and noted similar strain levels being approached elsewhere, we might expect shocks in those other areas.

A remarkable event occurred near the Hawaiian Islands in 1973, when recording devices in aircraft above the Pacific Ocean noted a change in normal atmospheric conditions approximately an hour before a major submarine earthquake. Whether this was merely a coincidence or a glimpse of a predictable pattern is not yet certain, but the matter is under investigation. An hour's warning in a major city could save thousands of lives, even it were not enough to permit wholesale evacuation. Thus it is well worth our while to learn about earthquakes.

SUMMARY

The surface of the earth may be divided into landmasses and ocean basins. Both have various kinds of relief. The land area may be divided into areas of plains and mountains and into shields and orogenic belts. The ocean floor provides as varied a topography as the land surface does.

The interior of the earth is composed of several layers of different materials. These layers include the inner and outer cores, the mantle, and the crust, which is constituted of materials sometimes called sial and sima. Geologic forces continually uplift and erode the crust.

Earthquakes experienced at the surface give important clues to the structure of the earth's interior. Earthquakes concentrate near areas of younger mountains and originate within the crust or mantle. Because they often occur in populous regions of the earth, great efforts are being made to predict them and plan for them.

QUESTIONS

1. We distinguish relief, the difference between the highest and lowest elevations in a given area, from elevation, the height above or below sea level of a given feature or area. Different combinations of relief and elevation can be found in various landforms and landscapes. Where would you expect to find both low relief and low elevation? High relief and low elevation? High relief and high elevation?

2. Why is the statement that plains are land areas of low relief and low elevation an inadequate definition? In Connecticut the problem with defining these terms can be seen in the names of several landscape features. Meshomasic Mountain, in the Connecticut River Valley, has an elevation of 260 meters, while Cream Hill, in the Taconic Range, has an elevation of 450 meters. What might account for this seeming paradox?

3. How can we use the concept of relief as a rough measure of the relative relief of the Canadian Shield,

the Appalachian Mountains, and the Rocky Mountains?

4. Plains can result from several different processes, and they may exist in areas of radically different elevation, from more than a kilometer high to several hundred meters below sea level. How are plains formed? What is their economic significance, depending on where they are formed?

5. What are the general distinguishing features of the earth's mountain ranges, in terms of appearance and location?

6. Does the point at which ocean and land meet represent the true edge of a continent? What different physical shapes can this interface take? Why?

7. Can the submarine features of the ocean floor be directly compared to features of the continents? Which features have terrestrial counterparts that look the same but were formed differently? Which have no terrestrial counterparts?

8. How has the concept of specific gravity, the ratio of the weight of a given volume of a material to the weight of the same volume of water, been used to determine the probable substances of which the earth's core is composed?

9. What differences have been inferred between the inner and outer core of the earth and the mantle? What method has primarily been used to make these approximations of the interior structure of the earth?

10. How do the specific gravity and the mineral composition of the various layers of the earth change with depth? Does this variation appear to follow a regular pattern, or does it seem random?

11. The first geologic time scale, which had four ordered divisions, was gradually replaced by a much more detailed classification system. What new factors were considered in making up the new classification system, and how does it differ from the earliest four-part scale?

12. There are several interesting aspects to the simplified geologic time scale in Figure 13-7. One of the most noticeable is the frequency of the divisions in the most recent era, which is much shorter than earlier eras. Does the fact that we have more subdivisions in the Cenozoic era mean that the earth is changing more rapidly in this period than in earlier ones? Why or why not?

13. How do the zones of greatest and least earthquake activity correlate with the oldest and youngest geologic formations of the earth? How can you explain such a relationship?

14. What features directly associated with earthquakes make them such violent physical phenomena? What factor associated with their distribution makes them exceptionally destructive of human life?

15. Earthquakes occur almost continuously, and even strong and severe quakes occur several thousand times in a century. Looking at Figure 13-8, can you explain why major earthquakes are considered rare by most people?

16. The Richter scale expresses the calculated energy released at the earthquake focus in a standard, numerical form. According to the scale, earthquakes in the 4 to 6 range are capable of causing local to regional damage, while from 7 upward, the shock waves are recorded over the entire globe. The San Francisco earthquake of 1906 had a magnitude of 7.8, while the Anchorage earthquake of 1964 had a magnitude between 8.4 and 8.6. Yet the damage to San Francisco was much more severe than that to Anchorage. What factors explain this?

17. How are *S* waves used to help determine the properties and dimensions of the earth's core? What properties of *S* waves make them useful for this specific problem?

18. How do *P* waves differ from *S* waves? Why are both types used in interpreting the interior structure of the earth?

19. What two major directions do plans to minimize earthquake damage in the United States follow? What are the limits to each type of protection?

20. Project Moho was an ambitious attempt to drill through the earth's crust to the mantle. After a great expenditure of time and money, the project was given up as technologically impossible at the present time, although much had been learned about the ocean floor where drilling was carried out. Why did the Moho researchers choose ocean drilling, far more difficult than drilling on land?

ROCKS
OF THE
EARTH'S
CRUST

CHAPTER 14

The surface of the earth's crust displays the results of the interaction of many agents and processes. The varied scenery of the continents is produced by internal geological forces as well as by water, air, heat from the sun, gravity, and other agents. Individual products of this assemblage of causes—an individual hill, a single ridge, or any other discrete shape on the surface—are referred to as *landforms*. The total scenery, made up of numerous landforms, is called the *landscape*.

Landforms and landscapes are sustained by the rocks of the crust. Under the cover of vegetation and beneath the thickest mantle of soil are rocks whose character has much to do with the landscape we observe. Hence it is essential to study the nature of rocks if the landscape is to be interpreted. Such a study is the aim of this chapter.

We begin by examining the three major kinds of rocks. Igneous rocks result directly from the solidification of molten material. Sedimentary rocks form when organic and inorganic sediments are deposited and accumulate. Metamorphic rocks occur when preexisting rocks are changed because of great pressures, high temperatures, or both. We will see that different rock types are responsible for different types of landforms, and that we can learn much about the history of the earth by studying its rocks.

THE NATURE OF ROCKS

A rock is a natural mass that forms part of the earth's solid crust. A more restrictive definition would leave out some materials properly considered rocks. When we think of rocks we tend to think of consolidated masses of "hard" rocks, such as granite and limestone, but even loose materials such as sand and gravel are often technically rocks. The variety of rocks is enormous. In everyday language we refer to such common rocks as shale, sandstone, and marble, but there are hundreds of other rocks whose names are not so familiar. Yet some of these less familiar rocks are important landscape builders.

Rocks are made of minerals. Chapter 10 defines these crystalline particles. The groups of minerals most commonly found on the earth are feldspars, quartz, olivines, pyroxenes, amphiboles, and the micas, chlorites, and clays. We are familiar with some minerals, such as copper and sulfur (*elements*, the simplest of chemical substances) or quartz and bauxite (*compounds*, combinations of two or more elements). Rocks are combinations of minerals such as these.

Mineral particles differ in their strength or weakness and in the tightness with which they are "packed" together. The quartz crystals in granite are welded so tightly to the other mineral crystals that it is difficult to pry one loose, as Figure 14-1 shows. But the grains of quartz in sandstone may come loose quite easily. In an area of moist climate, granite resists weathering better than limestone does. Granite, which is often used for tombstones, has a great deal of quartz, a very hard mineral. But the calcite, or calcium carbonate, in limestone can be dissolved in water if the water contains carbon dioxide.

Rocks develop in the crust in a variety of ways. We can observe one of the processes when the lava from a volcano cools and solidifies. Many rocks were created when a liquid mass of molten earth material cooled and hardened. Other rocks form when oceans and rivers deposit layers of sediment (such as sand and silt) that later become hard and dry. Still other kinds of rocks are produced when existing rocks are subjected to high temperatures, often accompanied by high pressures. Thus a sandstone covered by intensely hot lava from a nearby volcano is baked and cemented into something much harder—quartzite.

Rocks, then, because they are formed by different kinds of processes and of different combinations of minerals, have widely varying qualities of hardness. Sandstone, for example, may consist of quartz grains cemented together by a substance much softer than the quartz grains themselves. Weathering and erosion

FIGURE 14-1

The structure of granite, as viewed through a microscope. It is composed of the minerals quartz (Q), feldspar (F), biotite (B), and hornblende (H).

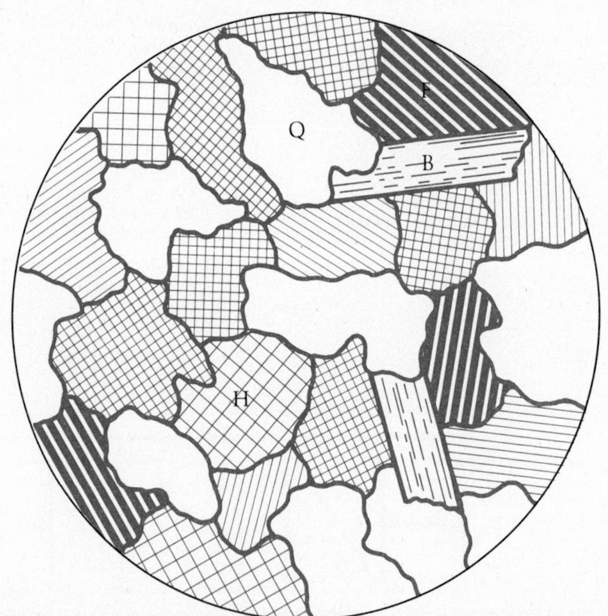

THE MOHS SCALE

One of the most useful things a "rockhound" can determine about a specimen is its hardness, for its hardness is a first clue to many other facts about that particular rock. The hardness of rocks, especially igneous and sedimentary rocks, is related to the hardness of the minerals of which they are made and the strength of the bond between the mineral particles. This latter factor relates directly to the rock's method of formation.

How do we measure the hardness of a rock? In 1822, a German mineralogist named Friedrich Mohs noticed that common minerals can be distinguished by their ability (or lack of it) to scratch other rocks—a harder rock will scratch a softer one, but not vice versa. Diamond, the hardest mineral of all, will scratch all other natural mineral surfaces, but cannot be scratched by any of the others. So Mohs classified diamond as the hardest—10 on his scale. Talc, the base material in talcum powder, was the softest mineral Mohs identified, and he numbered it 1 on the scale. His complete array: (1) talc; (2) gypsum; (3) calcite; (4) fluorite; (5) apatite; (6) orthoclase feldspar; (7) quartz; (8) topaz; (9) corundum; and (10) diamond.

We can apply Mohs numbers to other substances. For example, ordinary glass has a hardness between 5 and 6, so if you find a shiny mineral sample and you cannot decide whether it is calcite or quartz, try it on a window. Calcite will not scratch glass, but the harder quartz will.

In recent years the Mohs scale has been updated to provide greater accuracy in the harder range of the scale. The revised version is as follows: (1) talc; (2) gypsum; (3) calcite; (4) fluorite; (5) apatite; (6) orthoclase feldspar; (7) vitreous pure silica; (8) quartz; (9) topaz; (10) garnet; (11) fused zirconia; (12) fused alumina; (13) silicon carbide; (14) boron carbide; and (15) diamond.

attack this weak component. Quartzite, however, is a much harder rock. It, too, consists mostly of quartz, but the quartz is baked into a rather homogeneous, tightly packed mass. Long after the sandstone nearby has worn down, perhaps even been made into a valley, the quartzite is likely to remain standing high above the countryside. Engineers have to be aware of these properties of underlying rocks to plan the foundations of buildings and dams. Geographers are interested in rocks because they underlie the landscapes that they study.

ROCK TYPES

In the broadest possible sense, rocks form through three processes: solidification from a molten state, accumulation by deposition, and alteration by temperature and pressure. *Igneous* rocks are those that result directly from the hardening of molten material. Actually, it may be more appropriate to speak of crystallization, because various sets of minerals crystallize out of the molten mass at different stages in the cooling process. In granite, the feldspar crystals form first, and then the mica crystals, and finally the quartz crystals are forced into the niches still remaining. When the cooling rate is slow, the crystals have a lot of time to form, and they grow large and often quite well shaped, as in Figure 14-2. But sometimes the molten material cools rapidly. Obsidian, one type of igneous rock, cools so fast that no crystals form at all, and a kind of natural glass results.

Cooling rates can be affected by numerous circumstances. Rocks that solidify at some depth in the crust

FIGURE 14-2

Quartz crystals that have developed as a result of slow cooling of the original material.

cool more slowly, for example, than does the lava that pours out of a volcano. Below the surface, heat is conducted away more slowly. Liquid material that squeezes between existing rock layers cools more rapidly than a large, bulky mass does. Therefore, we can conclude that an igneous rock with large, well-formed crystals cooled slowly and at a considerable depth below the surface of the crust.

The *sedimentary* rocks result from deposition and accumulation. *Sedimentum* is a Latin word meaning "settling." Of course, sedimentary rocks go through another stage, the stage of cementation or compaction. River alluvium and desert sand are technically sedimentary rocks, but we usually think of sandstone and shale as examples of this type. Sandstone was loose sand at one time, but some agent caused the loose grains to stick together. This agent may have been water filtering between the grains, depositing a cement such as calcium carbonate in the process. Or perhaps the gathering weight of rocks accumulating above (not accompanied by any significant rise in temperature) may have aided in the compaction process.

The materials of sedimentary rocks are derived from other rocks and sometimes from biological materials, as in the formation of coral, limestone, and coal. Weathering processes pry the crystals from a granite surface. Rivers carry the grains to lakes and seas, grinding them down as they go. In river valleys, on shores of lake and sea, on the ocean bottom the sediments accumulate. Then the compacting and cementing processes take over, and sediments change to sedimentary rock layers, sometimes called *strata,* or beds.

Metamorphic rocks often have the most complex origins of the three types. The word *metamorphic* comes from a Greek word meaning "change." These rocks are made from other, preexisting rocks, by modification through increases in temperature and pressure and by the invasion of permeating fluids and gases, which sometimes penetrate upward through parts of the crust from deep below. The metamorphic process can repeat itself, so that a rock, once altered, may be changed again. It is sometimes difficult to unravel the history of such a rock, because its original igneous or sedimentary (or earlier metamorphic) characteristics are practically wiped out. By comparison, the shale-to-slate and sandstone-to-quartzite metamorphoses are relatively simple and do not transform the original beyond recognition.

All rocks fall into these three classes. Each group has particular qualities that are expressed in the landscape. As geographers, we should look at the three classes in more detail.

IGNEOUS ROCKS

The name of this group of rocks is a misnomer in a way, because it means "origin by fire." The actual process of formation is one of cooling and crystallization. It is true, however, that igneous rocks come from material that once was very hot—a liquid, molten mass called *magma.* Magma, made predominantly of silicates, also contains many gases, especially water vapor. It exists beneath the crust and pushes upward, sometimes forcing itself into and through some of the layers of the upper crust. When its upward advance eventually ceases, it cools and forms *intrusive* rock. At other times the magma escapes through fissures (cracks in the crust) or volcanic vents to flow onto the surface as lava. This eventually cools to form *extrusive* rock. The processes that form igneous rocks can also create concentrations of such valuable minerals as gold and silver.

When we study the behavior of igneous rocks under the attack of weathering and erosion, we note that igneous rocks are quite resistant but not invariably so. In a coarse-grained granite, the feldspar minerals are much softer than the quartz is, and the dark minerals, such as the mica specks, are removed early. When we use a microscope to view a section of weathered granite, we note that the planes of contact between the various crystals are especially susceptible to decay. The biotite is soon removed, leaving a pitted surface. Along the contacts between the quartz and feldspar grains,

FIGURE 14-3

Hexagonal jointing of basalt forming the Giant's Causeway, Northern Ireland.

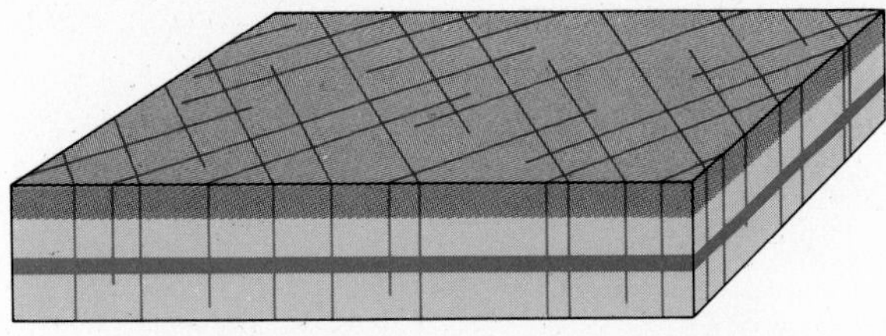

(a) Two sets of joints cutting horizontal beds

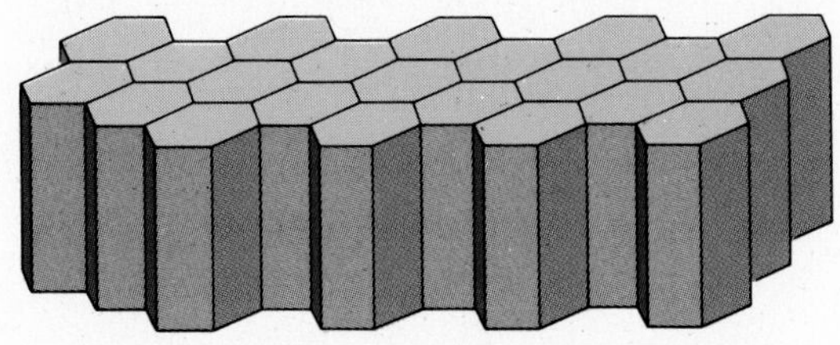

(b) Columnar joints in basalt flow

FIGURE 14-4
Jointing in horizontal sedimentary strata and in basaltic igneous rocks.

the loosening process advances.

Of course, the intrinsic hardness and resistance of a rock mean less when its large-scale structure makes it susceptible to erosion. An igneous rock that formed as a bulky mass may resist weathering and erosion for a long time. Nearby, a comparatively thin layer of igneous rock, underlain by softer sediments, may be worn down quite quickly. As the supporting sediments are exposed and removed, the igneous layer breaks up and collapses. No matter how strong the rock itself, it cannot withstand undercutting. This is why wave action is so potent in forming coastal cliffs.

Jointing and exfoliation Igneous rocks such as granite and basalt possess a property that is probably related to the cooling of magma. *Jointing,* a great aid to weathering and erosion, is the tendency for rocks to develop surfaces of fracture (breaks) without any obvious movement. The next time you pass a quarry or road cut, note that the freshly broken rock displays lines that seem to fragment the rock face into squares or rectangles. Those breaks were not made by machines—they were made by nature. Granite often breaks up into great cubes, and basalt sometimes displays a remarkable joint system that produces six-sided columns like those in Figure 14-3.

Jointing in igneous rocks appears to be related to the cooling process (the contraction of the material produces planes of weakness and separation, the *joint planes*), but jointing is not confined to igneous rocks. Sedimentary rocks, too, display joint patterns, as in Figure 14-4. These seem to result from the drying and compaction of wet sedimentary material. Joints also form when layered sedimentary rocks are bent, even slightly.

A special kind of jointing, in certain kinds of granite, produces a joint pattern resembling a series of concentric shells cut by rays—not unlike a wheel and spokes. The outer layers, or shells, peel away progressively, leaving lower layers exposed. In this way granite sometimes appears in the landscape in the form of domes, such as those in the Sierra Nevada in California. Half Dome in Yosemite National Park, shown in Figure 14-5, is just such a feature. Exactly what causes this phenomenon, called *exfoliation,* is not certain. It may be related to the intrusive character of the granite, which at one time was covered and weighted down by perhaps thousands of meters of overlying crust. As erosion removed these overlying sediments and other rocks, the pressure became less, and dilation (swelling) occurred. The outer shells were unable to contain the expanding mass and cracked and peeled along hidden joint planes. But shell-like weathering is also seen in small boulders, so the process could be external, induced by chemical decay and disintegration by weathering.

Whatever the causes of jointing and exfoliation, the joint planes are natural areas of weakness in rocks, where weathering penetrates early and begins to pry the rocks apart. Moisture trapped in the joint plane promotes the decay of individual grains in the rock, and it helps loosen grains from one another. The roots of trees and other plants grow into joint planes and, as they expand, wedge the rock apart. The joint system in rocks may affect and even control the river courses that develop, because the joint lines may be the easiest places for streams to create valleys in otherwise hard rock. From an airplane, you may be able to see rivers that follow straight lines and then make abrupt, 90° turns.

Natural forms The way igneous rock forms has a great bearing on landforms and landscapes. Magma tends to take a number of shapes that occur time and again, shapes that may eventually be exposed as erosion removes the overlying rocks. These shapes are diagramed in Figure 14-6.

Magma may insert itself in a thin layer between beds of sedimentary rock without disturbing preexisting layers to any great extent. Such an intrusion is called a *sill.* Much later, erosion may expose the sill, as in Figure 14-7, because it is harder and more resistant to further erosion than the rock around it. A sill, like any

FIGURE 14-5

The northeast side of Half Dome in Yosemite National Park, California. The outer layers of the granite mass are progressively peeling away—exfoliation on a gigantic scale. Can you find the climbers resting on a ledge halfway up?

FIGURE 14-6

Forms of igneous rocks.

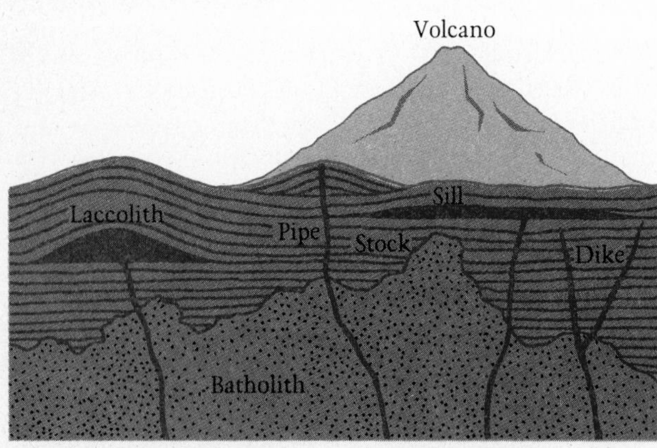

igneous intrusions that do not significantly disturb the rocks around them, is an *accordant* (or *concordant*) intrusion.

An intrusion that does disturb the older rocks, perhaps by cutting through and across them, is a *discordant* intrusion. For example, magma may penetrate rock layers vertically by squeezing along joint planes and by melting its way through those strata with no ready-made channels. When it finally hardens, it is a *dike*, like that in Figure 14-8. A dike, like a sill, may be just a few meters thick or even less, but its influence on the landscape can also be substantial. A sill protects softer rocks underneath, often showing up as the cap rock of a mesa or a butte, but a dike forms a prominent ridge, like a wall, when softer surrounding materials have been removed.

FIGURE 14-7

A basalt sill with columnar jointing intruded into sedimentary rocks, Yellowstone National Park.

Although sills and dikes are normally rather thin, they sometimes attain considerable thickness. The Palisades, overlooking the Hudson River between New York and New Jersey, are made of a sill whose thickness approaches 300 m (1000 ft) (Figure 14-9). Perhaps the most famous of all dikes is Rhodesia's Great Dike, whose length exceeds 480 km (300 mi) and whose width in places is 10 km (6 mi). The Great Dike stands above the Rhodesian plateau as a nearly continuous ridge, hundreds of meters above the surrounding countryside.

Sills and dikes are formed of magma fed to them from the main chamber, identified as the *batholith* in Figure 14-6. Batholiths cover enormous areas. From the coarse-grained quality of their granite, we may assume that they formed at great depth and were subsequently exposed by erosion. A small batholith, say approximately 100 km^2 (40 mi^2) or less, is a *stock*, an offshoot of a batholith. Although no one has ever seen all of a batholith or a stock, it seems clear that they broaden downward and that their relationships with preexisting rocks are discordant. (After broadening downward, they thin out again, according to data derived from earthquake waves.)

Figure 14-6 displays one other major underground form of igneous intrusion, the *laccolith*. This is an important feature, with origins in magma's loss of fluidity. The magma of sills and dikes is fluid and thin. But other magma, richer in silica, may lose its capacity to flow. Then on its journey upward it begins to coagulate and collect between sedimentary layers. Not being able to flow outward and continuing to be fed from below, the magma begins to push the overlying sediments upward until a lenticular (lens-shaped) structure is created. The strata below the laccolith remain more or less horizontal. If you flew over the area of a laccolith, you would see a large mound, circular or somewhat elliptical, perhaps as much as 30 km (20 mi) in diameter. It can be thousands of meters thick in the center but only a few at its edges. In the landscape, an exposed laccolith has a characteristic topography. The softer overlying sediments are soon eroded away but may remain as slight ridges encircling the exposed granite dome.

Volcanism Magma also erupts at the surface of the crust, creating volcanic mountains, lava flows, and other landforms associated with volcanism. Vulcan was the Roman god of fire. We are unable to observe the intrusive forms of igneous rock being created, but we can watch volcanic extrusions taking place. The lava pours out and hardens into igneous rock.

Volcanic eruptions rank with earthquakes in the devastation they can cause. Although volcanic activity may be more localized, it is also more spectacular. Great volcanic mountains stand silently for decades, and human settlements and farms nestle on the fertile

FIGURE 14-8

A dike of granite cutting through metamorphic rock in Colorado.

FIGURE 14-9
The Palisades on the Hudson River, seen from Yonkers, New York.

slopes derived from the newly developed nutritious soils. Then the mountain comes to life, sometimes without warning, and death-dealing gases rush down from the crater in advance of gushes of red or white-hot molten lava. Vesuvius, perhaps the most famous of all active volcanoes, lies just 11 km (7 mi) from the Italian city of Naples and has erupted disastrously 18 times since the first century AD (see Figure 14-10). In the twentieth century, Vesuvius became active in 1906, 1929, and 1944, in each instance causing destruction and death. In most cases, when the eruption is over, people return to farm the fertile soil. Volcanic soils in Java, Indonesia, sustain one of the world's densest populations.

The world distribution of volcanoes resembles the earthquake distribution shown in Figure 13-7. The "Ring of Fire" around the edges of the Pacific Ocean is both an earthquake amd a volcanic zone, although the volcanic chain is quite discontinuous in the Americas. Europe's active volcanoes are clustered in the Italian peninsula; those in Africa lie principally in the east. Western Pacific areas, from Kamchatka in the north to New Zealand in the south, form the world's strongest volcano regions.

Figure 14-11 shows the distribution of the world's *active* volcanoes, but not all volcanoes are active. A volcano is identified as active if it can be observed in action. A *dormant* volcano is one that has erupted in recorded times. A volcano is termed *extinct* if there are no recorded instances of any activity and when its slopes carry evidence of uninterrupted, long-term erosion, such as deep gulleys.

The mountains made by volcanic activity are also classified according to their shape and appearance, which in turn are related to the materials from which they are built. This material comes to the surface through a narrow conduit called a *pipe* (see Figure 14-6), at whose top is a vent. The magma, called *lava* when it reaches the surface, collects around this vent. If the lava is fluid and flows easily, the volcanic cone is lower and broader than a mountain made of pastier lava that piles up close to the vent. *Shield* volcanoes, made of fluid lava,are named for their gently curved slopes, which resemble the curve of a shield. The Hawaiian Islands are such shield volcanoes (Figure 14-12).

MONT PELÉE

On the island of Martinique, in the Caribbean Sea, lies the town of St. Pierre. Overlooking the town is Mont Pelée, a volcano whose name has been given to a type of eruption that is especially violent, with clouds of burning gas, violent explosions, and very viscous lava flows. Peléan eruptions are especially destructive, for they often come suddenly and with little warning.

In the case of Mt. Pelée, there had been rumblings and smoke before May 8, 1902, but they were discounted. A professor at the local college said on May 7 that the mountain was moody, but not dangerous. But then, shortly after 8 AM the following day, Mt. Pelée erupted with incalculable fury. Lava poured from its crater and the side of the mountain literally blew off, liberating a huge cloud of burning gas that sped downslope at about 100 km (60 mi) per hour. A wall of flame engulfed St. Pierre. People ran to the harbor and dove into the water, but it soon boiled; ships in port were pulverized in seconds. In a few minutes some 30,000 people died; there were just two survivors. All that remained of St. Pierre were the foundations of buildings and the ashes left by the inferno.

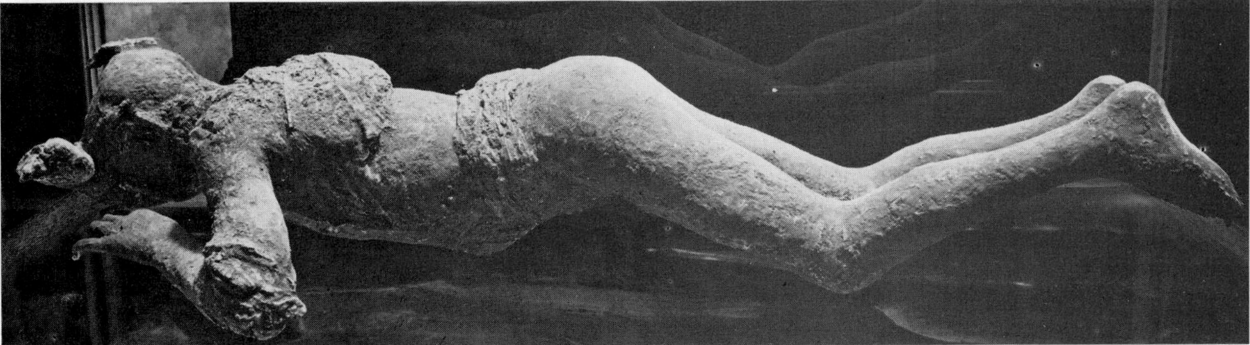

FIGURE 14-10
Plaster cast of a victim of the eruption of Pompei.

Mauna Loa rises from 4500 m (15,000 ft) below sea level to 4200 m (14,000 ft) above. Despite its great height, its slopes are very gentle and its area enormous.

Shield volcanoes emit their highly fluid lava passively, but other volcanoes intersperse such emissions with huge explosions. The more viscous, pasty magma tends to contain many pent-up gases, and these sometimes explode violently upon reaching the vent. Great quantities of debris are hurled into the air and then collect on the adjacent slopes. Lava subsequently covers these *pyroclastics.* (The term comes from the ancient Greek *pyr,* "fire," and *klastos,* "broken.") The debris cannot be hurled as far as fluid lava could flow, and the pasty lava cannot flow far either. Thus the volcanic cone tends to be steeper than that of the shield volcano. Mountains of this type, made of alternating layers of lava and pyroclastics rather than of lava alone, are identified as *composite* volcanic cones. Some of the great volcanoes of the world are in this class, exemplified in Figure 14-13: Vesuvius in Europe, Kilimanjaro in Africa, Fujiyama in Japan, Mount Hood in North America.

In a few cases the lava is so stiff when it reaches the surface that its progress cannot really be described as a flow. This lava pushes up overlying, preexisting rock (almost in the manner of a small laccolith) or penetrates and piles up in an irregular mass. The resulting mountain is a *lava dome,* a third type of volcanic landform. A similar form develops when material around the central solidified core of a volcano erodes, leaving a plug. In France, such plugs, called *puys,* often have churches built in spectacular positions on the top.

Volcanism is an awe-inspiring phenomenon, but volcanic landforms do not occupy a large part of the total area of the earth's landmasses. The process that creates volcanic landforms is a special case, a deviation from the landscape-sculpturing processes we are trying to unravel.

PREDICTING ERUPTIONS

Although the toll in human lives of volcanic eruptions is not as high as that of earthquakes, it is nevertheless estimated that perhaps a quarter of a million people have died over the past five centuries in volcanic eruptions. And in many parts of the world, people live on the slopes of volcanoes known to be dangerous. Even volcanoes classified as "dormant" may suddenly come to life. Pressures quietly build up, until the mountain breaks open and deadly gases and red-hot ash escape. Burning gas and killer ashfalls have killed more people than lava flows. Indeed, you need not be on the volcano's slope or near the lava flow itself to risk your life.

Predicting volcanic eruptions, then, is as important as predicting earthquakes—and at least as difficult. Local earth tremors, changes in the surface and shape of the mountain, temperature measurements, and gas detectors are among methods used to gauge the activity of a volcano. But the approximate moment of eruption must be predicted, especially where extremely serious eruptions occur. False warnings and apparently unnecessary evacuations will lead people to ignore future emergencies. The predictions are becoming more reliable, but they are not nearly as precise as scientists would like them to be.

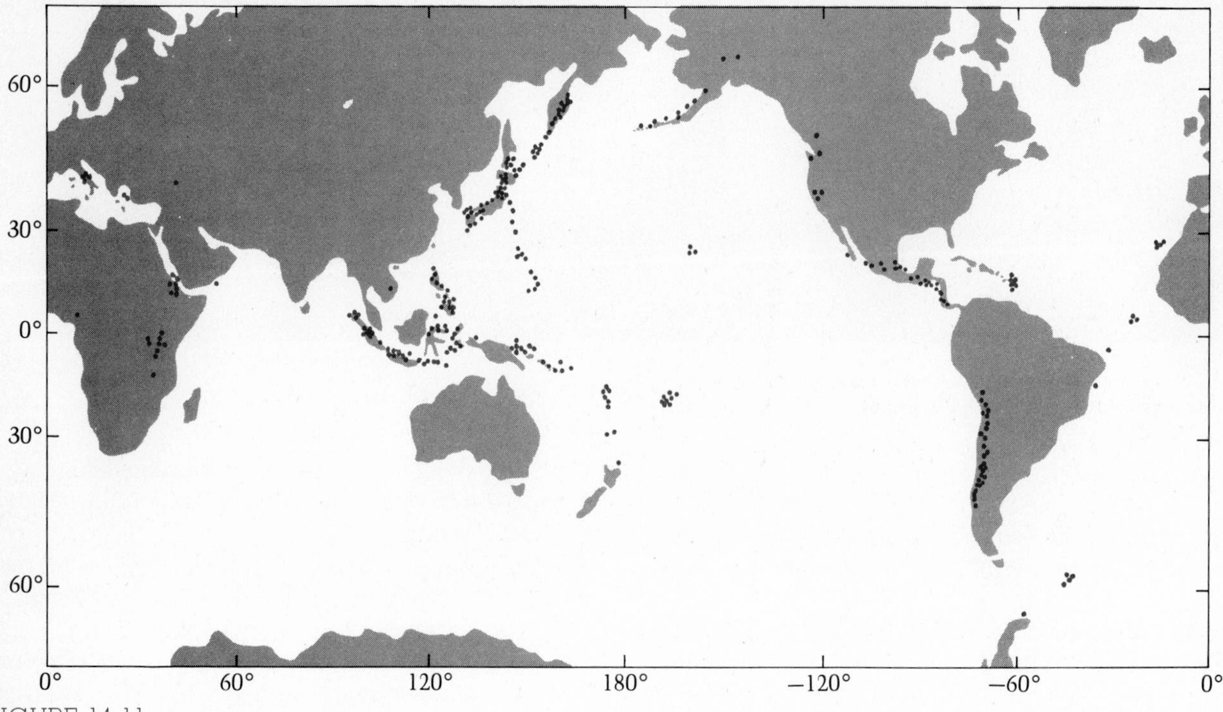

FIGURE 14-11
Global distribution of active volcanoes.

SEDIMENTARY ROCKS

There is quite a difference between lava and sedimentary rocks. Lava appears and consolidates with all the drama of fire and explosion, whereas sedimentary rocks accumulate quietly, slowly, and without fanfare. Still, sedimentary rocks cover far more of the earth's surface than lava does, because the process of sedimentation, or deposition, goes on all the time over vast areas. The sediments accumulate on ocean floors, in valleys between mountains, in river deltas, and in desert areas and eventually compact into rock layers.

FIGURE 14-12
Fluid lava erupting in the shield volcano Kilauea in Hawaii.

Clastic and nonclastic sediments So wide is the range of agents and materials that produce sedimentary rocks that the texture and structure of these rocks also vary greatly. The finest dust is blown by the wind to settle elsewhere and perhaps become part of a rock. In lakes and seas, fine clay particles accumulate and consolidate. At the other end of the size scale, glaciers and rivers move whole boulders and cobbles, which also may be cemented into permanent rock layers. These sedimentary rocks, made from particles of other rocks (whatever their size) are called *clastic* sediments. The vast majority of sedimentary rocks are clastic (again, the word derives from the Greek *klastos,* or "broken"), but other kinds of sediments do exist. Nonclastic deposits are mainly chemical, such as salt beds that form when a lake dries up. The most important nonclastic sedimentary rock is limestone, which forms when the minute shells of tiny marine organisms settle on the ocean floor.

Among the sedimentary rocks, the clastic strata are overwhelmingly the more important landscape builders, so we will concentrate on this group. All the familiar sedimentary types—sandstone, conglomerate,

FIGURE 14-13

Mount Fujiyama, a composite volcano cone.

shale—are clastic rocks. They all form when some agent cements the fragments together, in the process of *lithification* (literally "rock forming"). The strength and durability of the sedimentary rock depends greatly on the quality of the cement that binds the grains together. A silica solution seeping between the grains can deposit a strong cement, resistant to weathering. But calcium carbonate in solution produces a less effective, less lasting cement.

Examples of sedimentary rocks Clastic sedimentary rocks are most conveniently classified according to the size of their grains which range from boulders to fine sand particles and even silt. The coarsest-grained of all sedimentary rocks is *conglomerate*, shown in Figure 14-14, a rock made of pebbles and sometimes of fragments even larger than pebbles—cobbles and boulders. An important characteristic of conglomerates is that the fragments—the pebbles or boulders—tend to be quite well rounded, giving evidence that they were transported some distance, perhaps rolled down a river valley or washed back and forth across a beach. Because the pebbles are large, their origin and source area may be determined, perhaps telling us something about drainage patterns in the distant past. Sometimes the pebbles are elliptical in shape, with a significant number lying cemented with their long axes in the

FIGURE 14-14
A conglomerate of well-rounded pebbles and boulders surrounded by sand and silt particles.

same direction. This information helps reveal the orientation of a coastline that once existed where the conglomerate accumulated.

Sometimes the pebbles in a conglomerate are not rounded but are angular and jagged. Such a rock is called a *breccia*, and the sharp character of the pebbles indicates that little transport has taken place. For example, the crushing effect of a landslide and subsequent cementation of the broken fragments produce a breccia. If we find a breccia, we need only study the rock fragments it contains to conclude what rocks existed in the vicinity and if erosion has removed those rocks. Although not sedimentary, a *volcanic breccia* can also occur when pyroclastics are consolidated within volcanic ash.

Another common sedimentary rock is *sandstone*. In sandstone the grains, by definition, are sand-sized, and usually they are silica (quartz). When you look closely at a sandstone sample, you can learn something of its history. If the grains are well rounded, they have traveled farther than angular ones have. If the grains are all about the same size, the material was well sorted and deposition was relatively slow. A large range in the size of a sandstone's grains means that sorting was poor and deposition rapid. Sandstones are sometimes important for their ability to hold oil or water within the earth.

A sedimentary rock even softer than most sandstone

FIGURE 14-15
Limestone and shale alternating in a formation in Pueblo County, Colorado.

is *shale,* the finest-grained clastic sedimentary rock. Shale is compacted mud. Sandstone contains quartz particles, but shale contains clay forming minerals, such as feldspar. Shale has a tendency to split into thin layers, making this already soft rock even more susceptible to erosion. In many places, such as the Appalachian Mountains in the eastern United States, the low valleys are underlain by shale and the higher ridges by other rocks.

The last sedimentary rock to be considered here is the special case of *limestone.* Numerous kinds and varieties of limestone exist, but calcite is the chief mineral in all of them. Most limestones result from the respiration and photosynthesis of marine organisms, in which calcium carbonate is distilled from sea water. The calcium carbonate then settles on the ocean floor, where it accumulates in sometimes thick layers. Limestone may also form from the accumulation of shell fragments on the ocean floor, and in these circumstances it can be considered a clastic sedimentary rock (Figure 14-15).

Because limestone is a calcium carbonate ($CaCO_3$), it is susceptible to a solution of water and a small amount of carbon dioxide (CO_2) from the atmosphere (a mild carbonic acid): $H_2O + CO_2 = H_2CO_3$. In this form, water is an important agent in the breakdown of limestone and other rocks. The reaction of calcium carbonate with water and carbon dioxide can be represented as follows:

$$CaCO_3 + H_2O + CO_2 = Ca\ (HCO_3)_2$$

But limestone dissolves slightly in water even when no carbon dioxide is present. Where the climate is comparatively dry, then, we see limestone standing in strong, resistant ridges, but under moist conditions it yields much more rapidly. Because of its solubility, limestone forms many peculiar landscape features.

Stories from sedimentary rocks Sedimentary rocks are enormously useful in our effort to interpret the evolution of landscapes. Sedimentary rocks contain fossils, and much of what we know of earth history is based on this fossil record. The color, texture, and structure of sedimentary rocks suggest the kind of environment that existed when they were deposited, and so we can reconstruct the climatic conditions that prevailed in certain parts of the world millions of years ago. From the processes by which sediments accumulate today, we can determine whether sedimentary rocks were laid on land, in shallow water, or in deep water. In addition, we know that the sedimentary rocks now being deposited in river basins, lakes, and oceans accumulate in a horizontal position. The alignment of these sediments resembles a stack of magazines—flat. Thus we can deduce that, when we encounter sedimentary rocks in a position other than horizontal, something has disturbed the beds. Sedimentary layers not horizontal have been tilted, warped, folded, or in some other way repositioned after their deposition. There are some minor but interesting exceptions to this.

FIGURE 14-16
Ripple marks preserved in a marine sandstone that was deposited 125 million years ago in Colorado.

First, on wave-swept beaches and in wind-ridden deserts, the sands are often laid down at angles on slopes. Later, when the material is compacted, the rocks appear *cross-stratified,* or cross-bedded. The angle of repose of those strata permits us to determine the direction of currents or winds existing when the deposits were laid down. If we can find cross-stratified layers of the same age in different places, we can perhaps reconstruct the prevailing wind or current directions of millions of years past.

Second, if you have been on a beach or in an area of desert sand, you may have seen *ripple marks*—long, narrow, low ridges lying closely parallel to one another. The back-and-forth wash of the waves across a beach or the persistent blowing of wind across a sandy surface can produce ripple marks like those in Figure 14-16. Most of the time the ripples are washed away again, but sometimes they are cemented and preserved. Later, when the rock is exposed by erosion, the ripple marks show the direction of wave motion or the prevailing wind direction as it was long ago. Ripple marks thus provide another bit of the evidence that makes sedimentary rocks so valuable for reconstructing the past.

METAMORPHIC ROCKS

These are the rocks of metamorphosis, of change, of modification. Metamorphic rocks result from the alteration of other rocks by means of pressure and heat. There are probably some aspects of the process that we do not yet understand completely. Metamorphic processes can create accumulations of valuable minerals. Many of the gold rushes in history have been toward metamorphic rocks. All rocks are subject to this change. Igneous rocks, although solidified from the molten state, may be fashioned into something new. Sedimentary rocks are compressed, heated up, even liquified; when they cool down they are no longer sedimentary rocks. Even metamorphic rocks may be remade into different metamorphic rocks. Given enough time, no rocks in the crust are beyond the reach of metamorphic processes.

No one has ever seen metamorphic processes occur in the same way we can observe lava as it solidifies or sediments as they accumulate. Still we conclude that metamorphism occurs, because we find rocks of appar-

FIGURE 14-17

Contorted veined gneiss in British Columbia. The high temperatures and pressures of metamorphism allowed the rock to flow into this form.

ent igneous or sedimentary ancestry that are definitely not igneous or sedimentary rocks. One of the best examples is marble, a hard, crystallized rock made of the same substance as limestone. Only great heat and pressure could recrystallize the calcium grains in this rock, so often used by sculptors.

There are places where we can see a sedimentary layer's change into metamorphic rock. Where sedimentary strata have been subjected to folding and compression (see Chapter 16), a shale bed may change into slate or sandstone into quartzite. Elsewhere the heat from a lava flow may have begun to alter the rocks beneath, proving that heat-induced metamorphism does occur. And scientists have simulated metamorphic processes in the laboratory by subjecting pieces of igneous and sedimentary rock to high temperatures and great pressure.

The processes of metamorphism In general, sedimentary rocks subjected to metamorphism become stronger and more resistant to weathering and erosion; igneous rocks stay about the same. Granite, for example, is altered into *gneiss*, a rock of essentially the same composition as granite but newly recrystallized and looking a bit different.

In the field, gneiss may be recognized by its linear character; that is, the crystals of different minerals tend to line up in bands, as in Figure 14-17. This appears to be the result of the crystals' adjustment to severe pressure. The gneiss (or *schist*, another metamorphic rock that possesses this quality) may be most susceptible to weathering and breakdown along these bands. The property of banding in metamorphic rocks is referred to as *foliation* (not to be confused with exfoliation), and the bands of minerals are surface expressions of foliation planes that extend deep down into the rock masses. The foliation planes in gneiss are sometimes so extensive that vegetation concentrates itself along these weaker, softer lines.

This brings us to the different kinds of metamorphism that can occur and the process most important to us in our study of landscape. Sometimes lava or magma flow alters nearby rocks. This *contact* metamorphism is a localized phenomenon. Of much more importance to us is *regional* metamorphism, which affects huge rock masses. For example, when mountains are formed, enormous zones of rocks push against each other and press deeply into the crust. Metamorphic changes transform all rock types in these zones. Eventually the forces of uplift and the process of erosion expose the altered rock masses at the surface. Thus whole regions of metamorphic rocks exist, obviously created far below the surface.

Metamorphic rocks in the landscape We have already noted that granite is altered into gneiss and limestone into marble. Slate comes from the sedimentary rock shale, and in slate foliation is so complete that the rock splits along flat, smooth planes. Slate rock is quite hard, as we know—walkways and patio surfaces are often made of it. Yet the slate, when it forms the side of a hill or mountain, may erode away rather quickly, because the various processes of weathering can take hold in the cracks between the slate layers and pry them apart. In the Appalachian Mountains, for example, slate breaks down because the foliation planes become moist. Then winter freeze and thaw causes movement and loosening of the rocks. And the roots of trees find their way into the cracks and wedge the layers apart. Nevertheless, slate resists erosion better than shale, which has all the weaknesses of slate and none of its hardness and density.

A widespread metamorphic rock apparently derived from sedimentary rocks is *schist*. The original sedimentary rocks are so severely altered that it is no longer possible to determine just what they were. Schist is characterized by its foliation, but it does not have the regularity of slate. Schist breaks unevenly along the parallel foliation planes, sometimes in thick fragments, often in thin sheets. Schist is not especially resistant to weathering and erosion, but it is usually formed in huge masses and is capable of sustaining large landforms.

Quartzite, a metamorphosed sandstone, is much harder than its predecessor. It sustains prominent landforms in different parts of the world. In Wisconsin, the high hills overlooking Baraboo Lake in the Wisconsin Dells area are made of hard quartzite. The famous gold-bearing ridge in South Africa, the Witwatersrand, is underlain principally by quartzite.

THE ROCK CYCLE

The earth's crust and its rocks are in a constant state of change, a planetary example of dynamic equilibrium. Above the crust, rocks are worn away principally by the running water of the hydrologic cycle. Below the crust, such forces as the slow-moving currents of the mantle help move the crust upward. Between the upper and the lower forces, the three main types of rock are continually formed and transformed in one great cycle. The rock cycle is diagramed in Figure 14-18.

Weathering and erosion of all kinds of rock lead to the formation of sedimentary rocks. Some of these are uplifted by large-scale movements. Others are changed

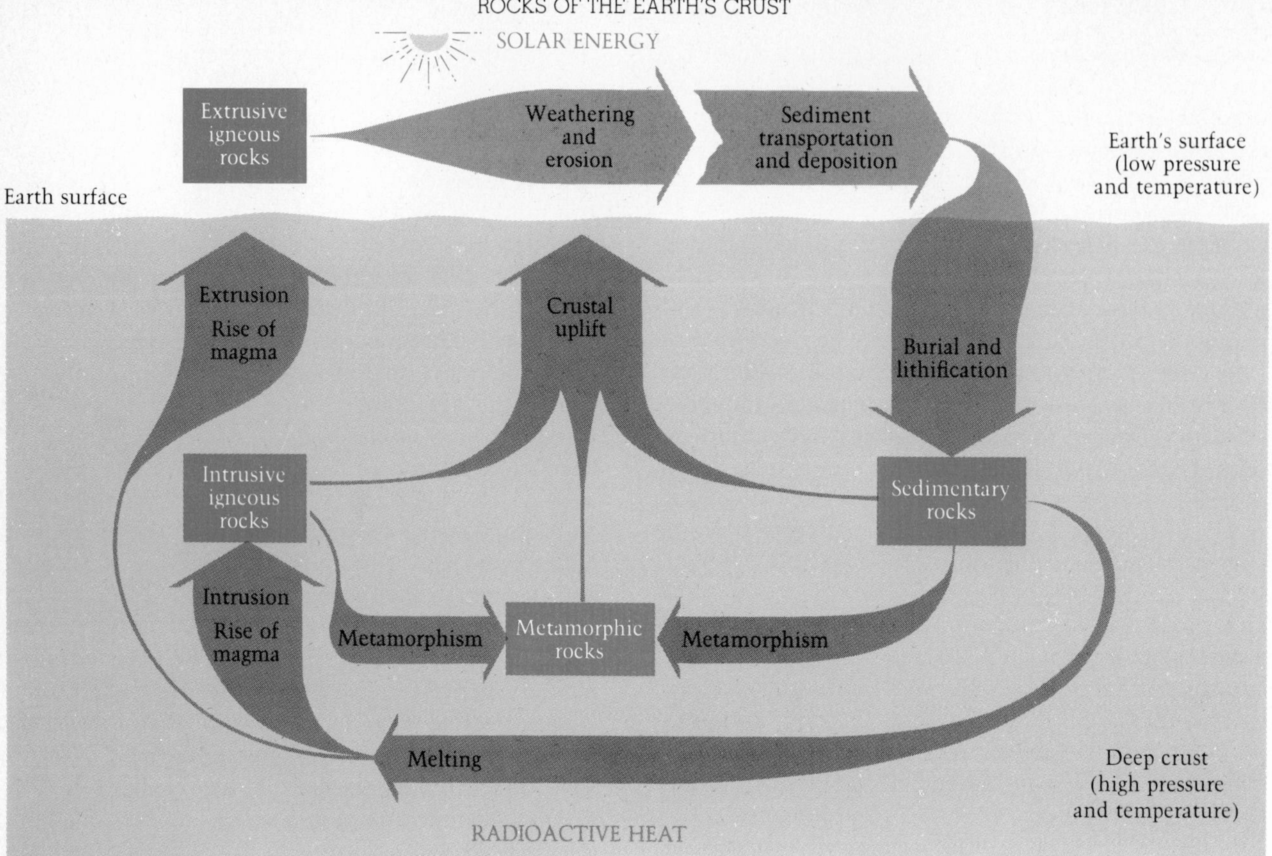

FIGURE 14-18

The rock cycle. The flow of materials within, above, and below the earth's crust continually forms and destroys igneous, sedimentary, and metamorphic rocks.

into metamorphic rocks, and still others are melted, deep in the crust, to go into the pool of igneous rocks. The igneous material may be intruded into the crust or extruded onto the surface. The resulting rocks are subject both to erosion and to metamorphism. Metamorphic rock, as well as igneous and sedimentary rock, can be uplifted with the crust.

Thus the rock cycle continues on a spatial scale as large as the earth itself and on a time scale as long as the history of our planet. It is difficult for us to imagine such distances and lengths of time. The facts of Chapter 15 are an even greater challenge to our imagination.

SUMMARY

The rocks of the earth's crust must be understood if we are to comprehend the various landforms existing at the surface. There are three main kinds of rock. Igneous rocks result directly from the solidification of molten material. Sedimentary rocks result from the deposition and accumulation of both organic and inorganic sediments. Metamorphic rocks are made from preexisting rocks of all types by means of great pressure and high temperatures.

Different rock types can be seen in the different landforms they produce. For example, igneous rocks produce a variety of intrusions and extrusions. Volcanoes are the most spectacular of the igneous extrusions. Sedimentary rocks are especially useful in giving information about the environments in which they were formed. Metamorphic rocks are often stronger than the original rocks. Often rock types express themselves in subtle ways, such as creating particular patterns of rivers. All the rock types are linked in the continual flow of the rock cycle.

QUESTIONS

1. How do we distinguish between landforms and landscapes? Yosemite Valley is a glacial feature of the Sierra Nevada. Within Yosemite is a granite dome

called Half Dome, also produced by glacial action. Can either of these features be definitely identified as a landform? Can either be exclusively identified as a landscape? Why or why not?

2. How is the scientific definition of a rock different from that in everyday use? What things not normally considered rocks can we now consider under this label?

3. Many of the tombstones in King's Chapel graveyard in Boston have lost their legibility in less than 150 years, while Cleopatra's Needle, a pre-Christian era obelisk brought from Egypt to New York, was still legible when it arrived in the United States. What factors of material and climate might cause this difference?

4. Is it strictly proper to speak of the process by which igneous rocks form as "hardening"? How, more accurately, should we refer to the process, and what factors determine the form the original magma, or molten material, will take when it finally cools?

5. What are the stages of formation of a sedimentary rock? Is the presence of sediment sufficient to produce a sedimentary rock? Why or why not?

6. Why is it more difficult to determine the origins of metamorphic rocks than those of any other rock type? Are metamorphic rocks the only type formed from other rock types? Why or why not?

7. Igneous rocks are among the most resistant found in nature, yet there is great variability in how long individual rocks or rock masses can withstand the forces of weathering and erosion. What must we consider before we can estimate the resistance of igneous rocks?

8. In areas where wet mud dries out completely, we can often see a pattern of hexagonal shapes in the shrinking mud. Another type of fracturing in nature occurs when large salt crystals are lightly tapped with a hammer—they separate into miniature grains, each a replica of the original crystal. How do these two processes—the fracturing of shrinking masses as they pass from the liquid to the solid state and the cleavage of crystals along planes in a regular fashion—relate to joint structures in rocks? What are the different processes involved in each type of fracturing, and what rock masses show jointing characterisitc of each type of fracturing?

9. Why is jointing an important factor in the study of landforms and landscapes? When does jointing become apparent, and with what other processes does it work?

10. Sills and dikes are normally very thin features, formed when magma is squeezed between layers of rock or into joints, and then cooled, deep within the earth. At what point do they become important surface features, and how may they influence the formation of other landforms? Would you expect the Palisades and Rhodesia's Great Dike to be accordant or discordant intrusion features, and why?

11. Volcanoes, along with earthquakes, are among the world's most destructive natural phenomena. Despite this fact, people continue to live near volcanoes and earthquake-prone areas. What are the two major reasons for these risky patterns of human settlement?

12. How are the different shapes of volcanoes, such as shield volcanoes, composite volcanic cones, and lava domes, related to the composition and state of the lava types from which they are formed?

13. Volcanic activity tends to be episodic, and the production of igneous rocks at the earth's surface is limited both in area and in time. The process of formation of sedimentary rocks, on the other hand, has no such limitations. Why are sedimentary rocks so much more widespread than igneous rocks?

14. Sedimentary rocks are among the most useful to scientists as sources of indirect information about processes related to lithification in the past, including weathering, erosion, plant and animal life, and the relative ages of various phenomena. What information, for example, can be gained from the following sedimentary rocks: conglomerate? sandstone?

15. Unlike most sedimentary rocks, limestone can be classified as either clastic or nonclastic. In addition, it is one of the rocks most easily dissolved. How does the composition of limestone explain each of these properties?

16. Why were sedimentary rocks of major importance in the development of the geologic time scale? How have they provided clues about major movements of the earth's crust?

17. What are the main forces involved in metamorphism? How do scientists know what these forces are, since they have never seen metamorphism in action?

18. How do sedimentary rocks change when subjected to metamorphism? Is the same true of igneous rocks? Why or why not?

19. Why do metamorphic rocks develop bands, and how are these bands important in determining landscape features? What are some important foliated

metamorphic rocks, and what are their geologic predecessors?

20. The rock cycle, shown in Figure 14-18, probably did have a definite beginning—when the first rocks were formed as the molten material of the earth cooled and crystallized. Is it accurate to view the process of formation of various types of rocks in a linear manner, since there was a definite beginning to the process? Why or why not?

OUR WANDERING CONTINENTS

CHAPTER 15

On May 6, 1976 residents of towns and villages in northeastern Italy found themselves in one of the most frightening circumstances imaginable—a powerful earthquake. The initial shock lasted 14 seconds and was accompanied by a deafening roar. When it was over, countless houses, churches, and public buildings lay in ruins, more than 1000 persons were dead (over 400 in Gemona alone), and thousands more were injured. The earthquake was felt throughout much of Europe. As far away as Brussels, Belgium, people rushed from apartment buildings and homes. In the days that followed, about 40 aftershocks made damaged buildings collapse, hampering rescue efforts. The peaceful countryside of the southern fringe of the Italian Alps was transformed into a place of terror and sudden death.

People who live in and near the earth's most active earthquake zones are familiar with the instability of parts of the crust, even though intervals of quiescence tend to dull memories and allay fears. This chapter begins by describing an earthquake because it is one of the most obvious signs of the instability of the earth's crust. Moreover, earthquakes are symptoms of the large-scale movement of the earth's crust that is the subject of this chapter. We will encounter a theory of movement called the theory of plate tectonics. It is without doubt the most important theory to develop in the earth sciences during this century. Like all good theories, it is important because it can explain many different facts in one sweep. But also like many other important theories, its development has been turbulent.

We begin this chapter with some of the recent evidence that led to the general acceptance of the notion of horizontal movement of parts of the earth's crust. Then we will examine the movement itself. Before the recent discoveries, evidence already existed to suggest that the earth's crust could travel horizontally. We will look at some of this evidence, much of it in the landscape itself. Finally, we will assess the significance of the theory of plate tectonics, both to scientists and, as a practical help, to humans. We enter this story of discovery well after the play has begun.

NEW EVIDENCE FOR WANDERING CONTINENTS

In the past two decades, scientists have made discoveries that confirm once and for all the fact that continents, and other parts of the earth, do actually move across the face of our planet. This amazing idea was finally accepted following studies of past magnetism (*paleomagnetism*) and investigations of the ocean floor. Let us look at these discoveries.

PALEOMAGNETISM

The movement of material in the earth's outer (liquid) core produces a magnetic field that encompasses the earth. This magnetic field has probably existed as long

FIGURE 15-1

Possible previous location of southern continents. Lines of paleomagnetism (*A*) point to a south pole between Africa and Antarctica.

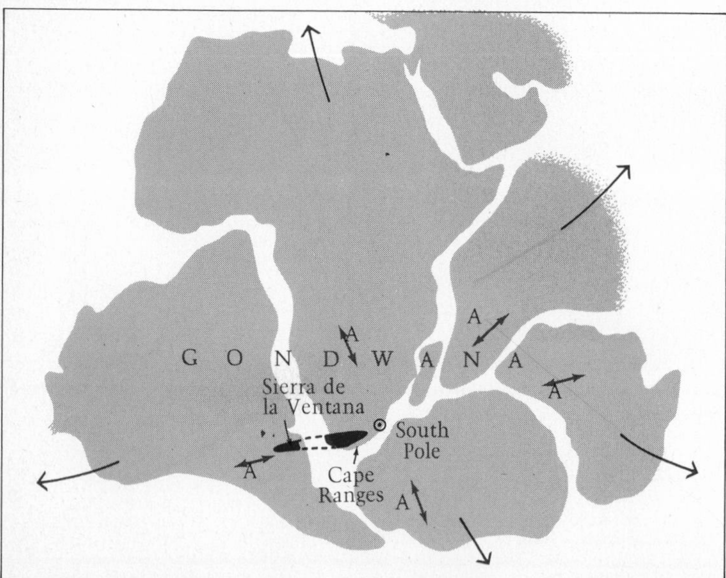

as the earth has possessed internal layers, as it does today. Because the earth's magnetic and rotational poles are always close together, we can deduce the approximate location of the rotational poles if we can find out where the magnetic poles were during past geologic periods.

Molten lava, liquid igneous rocks, and water-accumulated sedimentary rocks all contain small elongated grains of iron (and other substances that respond to magnetism). In the process of their solidification or lithification, a significant number of these grains orient themselves parallel to the direction of the magnetic field, directly toward the magnetic pole.

We will immediately confront an astounding idea. Let us suppose that the land areas of South America, Africa, India, Australia, and Antarctica were once all close together in a large continent, as in Figure 15-1, that we will call Gondwana. If this supposition is correct, rocks formed at this time would show iron grains aligned toward the south magnetic pole. The alignment of iron grains in rocks of the same age is shown in the figure by the arrows marked *A*. As you can see, they do indeed point toward a south pole—but a south pole between Africa and Antarctica, which were, at the time, lying side by side. The alignment of these iron grains in the present-day continents leads scientists to believe that not only have the continents moved outward in the directions shown in Figure 15-1 but that the south magnetic pole has also moved. Thus the frozen magnetism in the rocks suggests that the southern landmasses have relocated.

Paleomagnetism tells us something else as well. Scientists have discovered another amazing fact. The earth's magnetic field sometimes reverses itself, so the north and south magnetic poles exchange places. Geologically speaking, the reversal takes place quickly, and then the magnetic field remains stable for about half a million years. The direction of the stable field is imprinted into new rocks that are formed.

In the early 1960s, geologists mapped the magnetic directions imprinted in rocks on the ocean floor southwest of Iceland. The result was a zebra pattern of rocks, some magnetized in one direction and some in the other, as in Figure 15-2. A surprising explanation was put forward for this: Material rising from the interior at the midocean ridges slowly spread across the sea floor. It was a slow process, but inexorably new rocks formed and the ocean floor spread.

SEA-FLOOR SPREADING

The first scientist to postulate that the earth's crust is "born" of upwelling magma along the midocean ridges, and then moves away, was Harry H. Hess. He suggested in 1962 that discrete segments of the earth's crust are formed at the midocean ridges as the upwelling magma solidifies. Then, as the segments pull apart, space is made for new liquid magma to flow through the fissures and vents, replacing that which has moved away. The rate of movement is slow, perhaps 2.5 cm (1 in) per year, but it is measurable. The whole system resembles two giant conveyor belts positioned next to each other, slowly carrying their loads away from the middle, the midocean ridge, as in Figure 15-3.

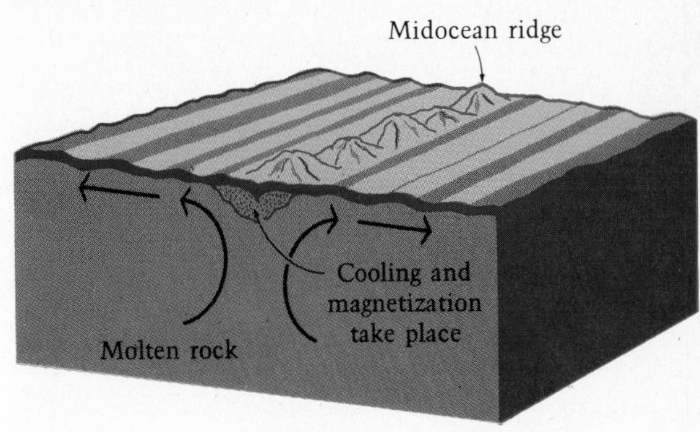

FIGURE 15-2

A map of magnetism on the ocean floor. White areas show alignment to the earth's present magnetic field. Black areas indicate a field reversed from that of the present.

One of the discoveries that led to this interpretation, besides paleomagnetic data, relates to the age of rocks in the ocean floors. When it became possible to drill holes and recover rock samples from the ocean floor, no rocks were found whose age was as great as that of the older rocks on the continents. The ocean basins are covered by comparatively young rocks, no older than, say, 125 million years. The scientists also found that, as they studied samples along cross sections of the ocean, that the rocks became younger as they approached the midocean ridges. This fits the theory that the ocean basins were created along magma-producing rifts that became progressively wider.

When geologists examined the midocean ridges, they found them to possess some unexpected qualities.

FIGURE 15-3

Hess's 1965 sketch illustrating the idea that the earth's crust is firmly attached to the upper mantle and that magmatic material reaches the crust through the midocean ridges.

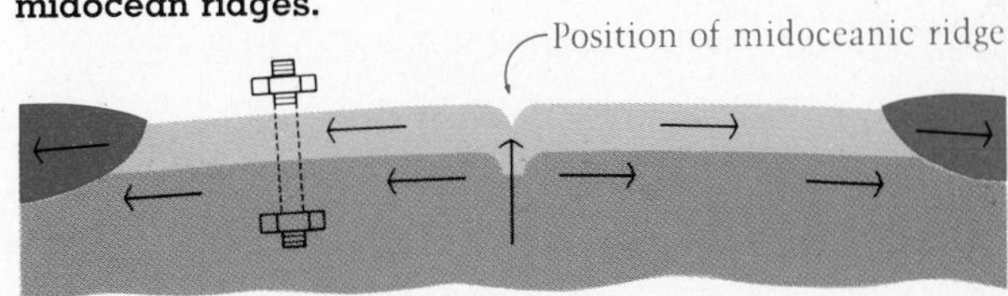

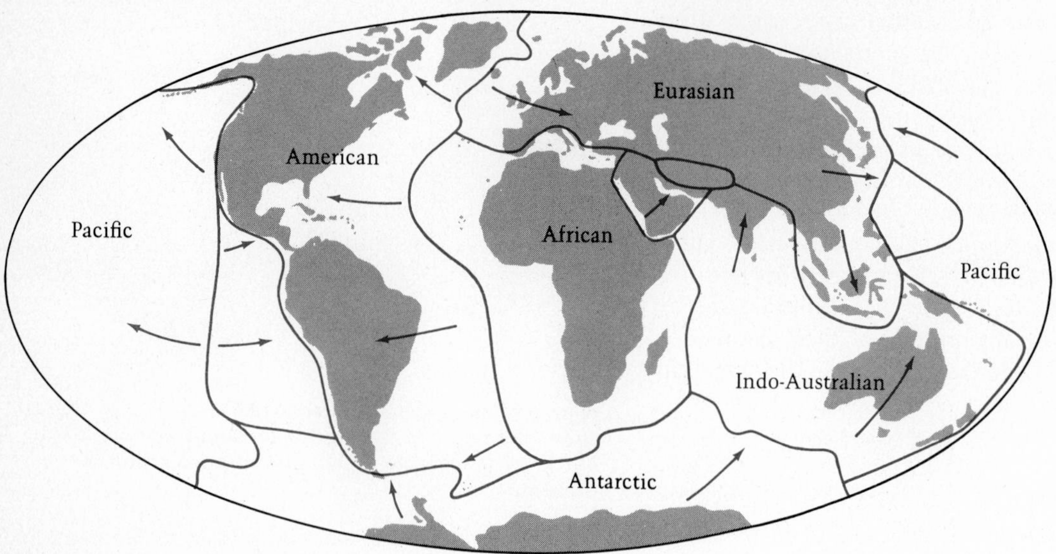

FIGURE 15-4
The six major plates of the earth and some of the minor ones. Arrows indicate the general direction of plate movement.

From deep mines and boreholes, they have learned that the earth's crust has a temperature gradient. The deeper the hole, the higher the temperature. The average rate of increase is about 9.4°C per 300 meters (17°F per 1000 feet). In the vicinity of the midocean ridges, this gradient is much higher than it is elsewhere on the ocean floor. From this and other information, geologists have concluded that the midocean ridges represent places where upwelling of the mantle's material takes place.

THE MOVING SURFACE

The great heat flow in the area of the midocean ridges is only one piece of evidence indicating the existence of subcrustal movement of magma. Studies of the distribution of the force of gravity in the vicinity of island arcs also point to its presence. Near Indonesia, for example, F. A. Vening Meinesz found gravity's force to be far below the recorded average, and he concluded that, sinking movements pull the crust down there. Where the crust was pulled down toward the center of the earth, there would be less mass and thus lower gravity.

CONTINENTAL AND OCEANIC PLATES

Studies of the earth's crust under the oceans, and especially research on the global system of midocean ridges, have generated a new view of the earth's outermost shells. It is now believed that the earth's crust and the upper portion of the underlying mantle (together called the *lithosphere*) consist of a set of rigid *plates,* whose exact number and location are not yet certain. These rigid, hard plates rest on the weaker *asthenosphere,* a plastic layer below the lithosphere. The plates appear to average about 100 km (60 mi) in thickness, and the larger plates have diameters of thousands of kilometers. For some time it was thought that the globe had but six of these enormous plates, but there are indications that the number is greater. Figure 15-4 diagrams them. *Plate tectonics* is the name given to the study of the way these plates move and interact.

Chapter 13 notes that there is a difference between continental (sialic) and oceanic (simatic) crust. Similarly, because the global plates in some cases include continental crust and in other cases are solely oceanic, a distinction is made between continental plates and oceanic plates. When a plate includes all or part of a continental landmass it is termed a *continental plate;* when it consists only of a segment of crust beneath the ocean, it is an *oceanic plate.*

As a result of the movement of magma in the asthenosphere and below, the plates of the earth's lithosphere move. They are borne along that worldwide system of ocean ridges amid a giant upwelling of basaltic magma, then spread apart and inevitably collide.

Three kinds of boundaries between plates have been recognized. The first kind is the *divergent plate boundary,* which occurs at the midocean ridges. As Figure 15-5a shows, this is where plates move away, or diverge,

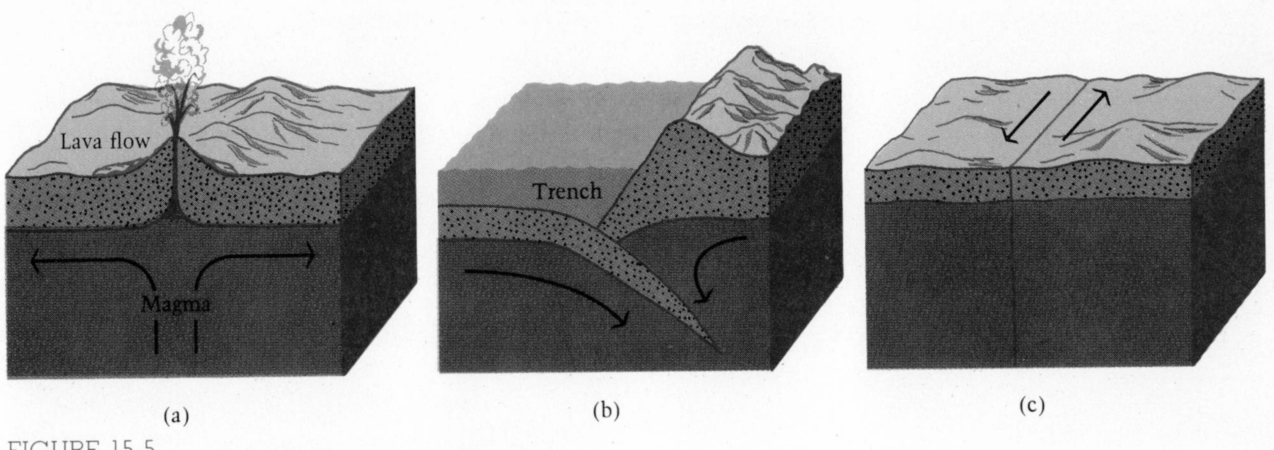

FIGURE 15-5

Three types of plate boundaries. a) Divergent plate boundary, often found at the midocean ridges; b) convergent plate boundary, where two plates collide; c) transcurrent plate boundary, where two plates in contact along a fault move sideways.

from each other. When this happens, basaltic magmas emerge to fill the gap in the ocean floor, and the ocean floor keeps spreading. Divergent boundaries are often accompanied by volcanoes that form islands in the midocean ridges. Surtsey, the island that appeared off southern Iceland on November 14, 1963 is a case in point. Figure 15-6 documents its birth.

A second kind of boundary is the *convergent plate boundary* that occurs when two plates move toward each other and collide, as in Figure 15-5b. At the collision point, one plate usually plunges underneath the other. The material of the submerging plate is gradually reincorporated into the upper mantle and crust. An attractive feature of the theory of plate tectonics is the dynamic equilibrium existing between a gain of new surface material at divergent plate boundaries and a loss of material at convergent boundaries. The area where the material is lost at the convergent boundary is called a *subduction zone.*

Two important landforms accompany a convergent boundary. Where one plate dives beneath another, a deep-sea trench is formed, such as the Peru Trench. These subduction zones may be efficient places to dispose of dangerous radioactive wastes. The other plate involved in the collision is crumpled up and folded at its edge to form a major mountain range, such as the Andes. Convergence of plates is always attended by earthquakes and sometimes by the eruption of lava from vents and fissures. The people in northern Italy live on the contact zone between the African and the Eurasian continental plates, and the 1976 earthquake resulted from the collision of those plates. That collision has already generated the giant Alpine-Himalayan mountain system, shown in Figure 15-7, that traverses the Eurasian landmass.

A third boundary is called a *transcurrent boundary.* This exists where two plates in contact along a fault in the crust tend to move sideways, as in Figure 15-5c. The San Andreas Fault, marking the contact plane between the Pacific and American plates, is an example of a transcurrent boundary (Figure 15-8). Much of the rest of the Pacific "Ring of Fire" consists of convergent boundaries, with a high incidence of earthquakes and frequent volcanism attesting to the crust's instability.

As physical geographers we are interested in the effects of plate tectonics and continental movement on the landscapes of the landmasses. But before we consider this something should be said about the mechanism of plate tectonics: What drives the plates, and what moves the sea floor?

THE GIANT CONVEYOR BELT

It is believed that magma within the mantle moves in *convection currents,* where a heated fluid rises from a heat source. As it cools, the current sinks to be reheated and then rises again, completing a cellular circulation. The Hadley cell of the atmosphere (see Chapter 5) is just such a current, and convection currents form in a pot of water when it is heated.

The idea that subcrustal convection currents are responsible for continental motion is not new. Arthur Holmes, a British geologist, mentioned the possibility as long ago as 1928. With the discovery of sea-floor spreading and the apparent confirmation that internal convection currents do exist, we are one step closer to

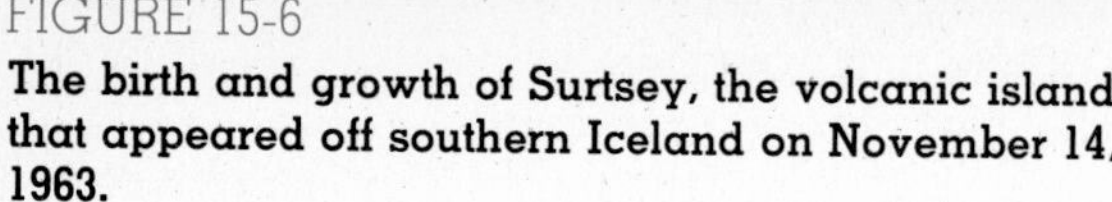

FIGURE 15-6
The birth and growth of Surtsey, the volcanic island that appeared off southern Iceland on November 14, 1963.

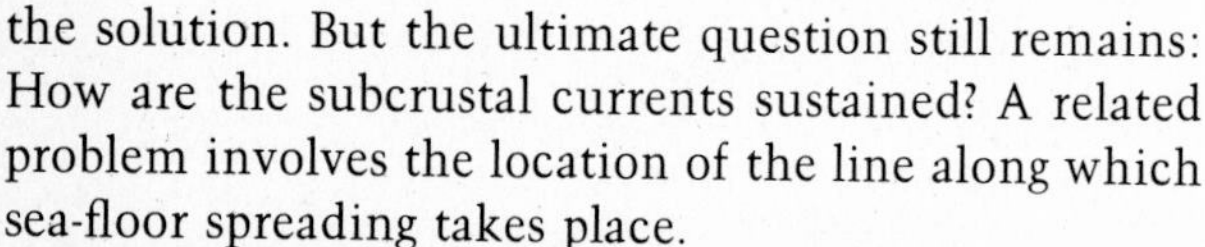

the solution. But the ultimate question still remains: How are the subcrustal currents sustained? A related problem involves the location of the line along which sea-floor spreading takes place.

These are questions to be solved by geologists, not physical geographers, but the answers are of great interest to physical geographers. Indeed, some of the world's continental landscapes may contain important evidence relating to various stages and locations of sea-floor spreading. Today, Africa and South America lie about the same distance from the Mid-Atlantic Ridge, where their separation began. Today, what keeps them moving farther apart is the convection system and its magmatic upwelling along that great midocean ridge. But there was a time when Africa and South America were united and when the spreading occurred midcontinent, not midocean. As we will note when we study continental drift, this kind of continental breakup may not be over yet. There is evidence that eastern Africa, from Ethiopia to Mozambique, is separating from the rest of the continent.

CONTINENTAL SHAPE AND DISTRIBUTION

Let us turn our attention to a question raised in Chapter 13—the form and distribution of the continental landmasses. Our globe shows the continents in locations familiar to us: The Americas lie across the wide Atlantic Ocean from Europe and Africa, Australia hangs off the southeastern end of Asia, and Antarctica straddles the South Pole. When we turn the land hemisphere to face us, Africa lies at its heart, surrounded by the other continents. On the opposite side from Africa lies the Pacific Ocean, constituting nearly all of the sea hemisphere. This distribution of the landmasses, concentrated on one side of the globe,

FIGURE 15-7

The European Alps, formed at the convergent boundary between the African and Eurasian plates.

FIGURE 15-8

Radar image of the San Francisco peninsula, with the San Andreas Fault system clearly delineated.

has been a subject for speculation as long as it has been recognized to exist. Similarly, the puzzle-like match between coastlines on opposite sides of the Atlantic Ocean attracted scientific attention very early. Is the continental concentration related to the jigsaw-puzzle fit? Or is it a matter of chance?

The development of the concept of plate tectonics and the identification of sea-floor spreading began with notions of *continental drift*. It is hard to believe that the continents have not always been where we see them today and that they will not stay in their places in the future. But there is no doubt: The ground on which we live our daily lives, the city we know well, the landscape that constitutes our surroundings—all lie on a mobile, moving landmass. Like huge rafts, the continents are constantly in motion, imperceptibly but inexorably. And sometimes the continents' movement is not so imperceptible. Where North and South America are pushing out into the Pacific Ocean, incredible forces are at work. Rock masses are crushed, bent, folded up. It happens slowly but not smoothly. We can feel the shocks: They throw the surface into turmoil and destroy buildings and sometimes lives. The earth shudders as the continents move.

It would be impossible to interpret landscapes without recognizing the effects of the mobility of continents. Whereas weathering and erosion constantly modify the continental surfaces, the process of drift, with its attendant stresses and strains, continuously deforms the crust—and its surface as a result. As the continents drift, mountains are pushed up, plateaus are tilted, plains are warped, river courses are altered, basins are formed where once there were uplands. In one way or another, directly or indirectly, practically every landscape and landform on the earth is related to continental drift.

In this chapter we face a doubly demanding task. As we will find, Africa's central position in the land hemisphere is more than a passing matter. Before the present phase of continental drift began, Africa formed the core of an ancient supercontinent that included all the dispersed landmasses of today. That continent is called Pangaea and is sometimes subdivided into Laurasia in the north and Gondwana in the south. Africa acquired certain surface features peculiar to its central position. Being largely a shield continent and thus made of hard, crystalline rocks, much of what characterized Africa's surface in those days was etched into the landscape and can still be discerned. So we must not only acquaint ourselves with the process of continental drift and its role as landscape builder; we must also become familiar with the surface properties of Africa, to us a rather unfamiliar continent. As we will discover, it is quite an exciting exploration.

OLD EVIDENCE FOR WANDERING CONTINENTS

Many ideas and concepts now accepted and used in physical geography failed to enjoy immediate approval, but we have not traced their history. But continental drift and its successor, plate tectonics, are so central to our study of landscapes, and the development of the concept is so interesting and instructive, that we will make an exception in this case. As we trace the evolution of the drift concept, we can observe the appearance of major related problems and issues and

THE RIDDLE OF MADAGASCAR

Look at a good atlas and consider the island of Madagascar, off the southeast coast of Africa. Madagascar is no ordinary island. Not only is it one of the world's largest, but its configuration and composition make it one of the most interesting.

Even the most casual look at the map of Madagascar reveals the remarkable straightness of the island's east coast. This is created by a persistent, high escarpment whose appearance and dimensions strongly resemble the Great Escarpment of Africa. Madagascar's highest elevations lie in the east, along the top of that escarpment. They decline to the west, toward Africa.

Madagascar's rock strata also resemble those of nearby southern Africa, and so does the fossil sequence. But all this similarity ends abruptly, and if you went to Madagascar today you would find none of Africa's lions or leopards, antelopes or reptiles. What you would find would be lemurs, marsupials distantly related not to African, but to southeast Asian and Australian relatives.

And by one of those accidents that make history, even the population of Madagascar is strongly influenced by Southeast Asian strains. Language and culture have Indonesian foundations. Even the place names reflect the island's unique qualities.

their impact on the search for solutions. Some of these major problems—such as why there should be convection currents in the mantle—are still not solved.

THE HISTORY OF AN IDEA

As long ago as 1619, Francis Bacon, the great naturalist, remarked that the opposite coasts of the southern Atlantic Ocean were so similar and well matched that they might at one time have been joined. No one paid much attention then nor in the nineteenth century, when some scholars began to take note of the jigsawlike character of the Atlantic and the growing list of similarities on the ocean's two sides.

Then in 1915 a German scientist named Alfred Wegener published a book amazing for its time. It contained the first systematic statement of the concept that the earth's continents were once united in a single vast landmass that broke apart to form the oceans and continents that we know today. His book, *The Origin of Continents and Oceans,* for the first time brought together virtually everything that was known to support the idea of *continental drift.* Wegener's three-stage map, in Figure 15-9, depicted the drift concept cartographically.

It was a decade before Wegener's work was translated into English, but news of his thesis preceded the translation. In many quarters Wegener's views were not merely criticized—they were ridiculed. True, Wegener had assembled an enormous mass of data suggesting that the continents were at one time united. He correlated rock types on opposite sides of the southern Atlantic, traced zones of crustal deformation across the Atlantic, recorded the distribution of fossil life that had required connections across the ocean, interpreted climatic changes common to distant continents. He argued that the jigsawlike fit of Africa and South America could not be a matter of chance. But there was one problem: Wegener could not come up with a plausible explanation for the mechanism of continental drift. How could it happen?

Wegener suggested that the earth's gravitational force, slightly lower at the equator, was sufficient to pull the continents apart. But this proposal was so inadequate that it only encouraged his critics. And so the idea of drifting continents was shelved, to be periodically taken down for inspection and shelved again.

Among such inspections were those of an American, F. B. Taylor, and of two South Africans, Alexander L. du Toit and Lester C. King. Taylor, viewing the globe and its maldistributed continents, suggested that the answer might lie in the relationship between the earth and its satellite, the moon. He theorized that the moon was captured by the earth (that is, brought into the earth's orbit) quite suddenly during the Cretaceous period. A segment of crust was then dislodged in what is now the vast Pacific Basin. Since then, he thought, the continental landmasses have been moving in directions dictated by the earth's "refilling" the Pacific gap thus created.

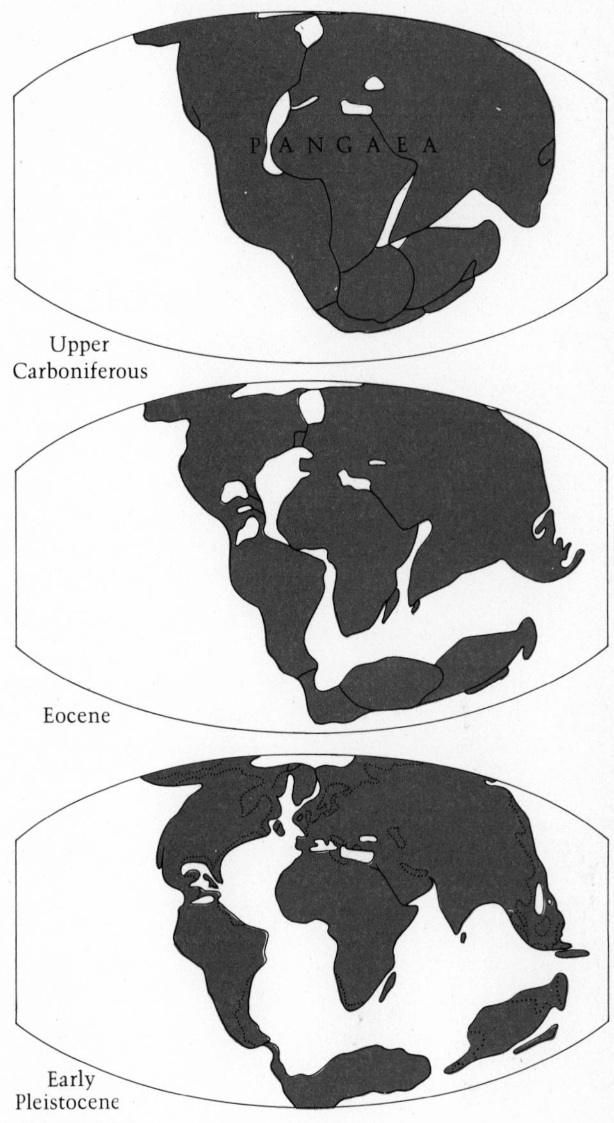

FIGURE 15-9
The original concept of the nested continents of Pangaea and their gradual separation by continental drift.

Du Toit assembled a vast amount of geologic evidence for the previous existence of a supercontinent. The geologic similarities between the southern continents, he said, could only be explained by their drifting apart. He also demonstrated that Gondwana had been

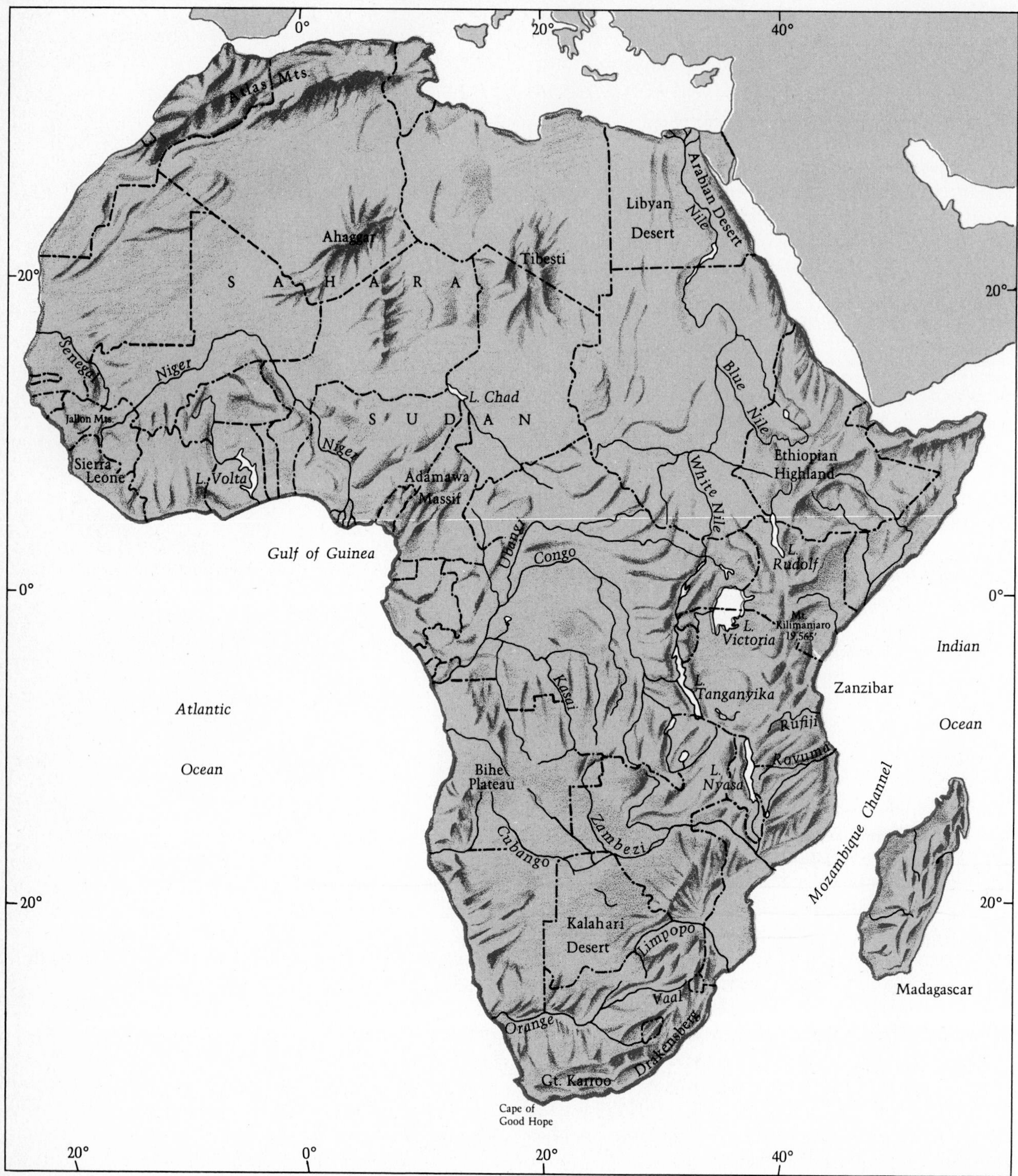

FIGURE 15-10

The physical features of Africa.

subjected to an ice age. King believed that, if the present-day southern continents had once been joined, evidence would be apparent in their landscapes.

Despite the vast amount of evidence, the concept of continental drift remained unfashionable and was often laughed at until about 1960, when studies on paleomagnetism and sea-floor spreading demanded an explanation in terms of a moving crust. The evidence on the surface of the earth was ignored by the greater part of the scientific community. As physical geographers, we are particularly interested in the earth's surface, and so it would be appropriate to examine some of this evidence in more detail.

CONTINENTAL DRIFT AND LANDSCAPE: AFRICA

When du Toit published *Our Wandering Continents* in 1937, he wrote on the title page: "Africa Forms the Key." From Africa's landscapes, du Toit said, we should be able to draw vital conclusions about continental drift. After all, Africa was the only landmass that was centrally positioned in Gondwana, surrounded on all sides by the other landmasses that have since drifted far away. If that was the case, Africa should show the evidence of its former centrality and encirclement.

African landscapes Let us examine a physical map of Africa, in Figure 15-10, to see if facts support the idea of continental drift. One of the first things we notice is that the African continent has a generally high elevation, much of it at least 600 m (2000 ft) above sea level. In addition, in many places the high elevations arise from steep coastal escarpments. Such names as *Sierra* Leone in western Africa, the old *Abyss*inia in eastern Africa, and Drakens*berg* in South Africa all imply a mountainous area of one kind or another.

Another remarkable feature of the continent is the absence of a major mountain range comparable to the Andes or the Alps. In the north, the Atlas Mountains are an extension of the Alpine system. In the far south, the Cape ranges have structures similar to those of the Andes and Alps, but in scale they are nothing like those great systems. Africa's mountainous landscapes have been created mainly by erosion and dissection of the plateau (as along the escarpments) or by volcanism (the great Kilimanjaro, for example). Africa lacks the kind of linear, folded, full-length mountain chains we find in all the other landmasses. Certainly this confirms Africa's plateaulike character.

Another aspect of Africa has the same implication. As a world map quickly confirms, continents tend to have coastal plains, extensive low-lying areas that slope gently seaward. Much of the southeastern United States is such a coastal plain, and you can identify similar, often densely populated, areas in other parts of the world. Again, Africa is markedly deficient in such low-lying surfaces. True, it has the deltas of the Nile and Niger rivers (in Egypt and Nigeria respectively),

FIGURE 15-11

The Great Escarpment in southern Africa, looking toward the Valley of Desolation.

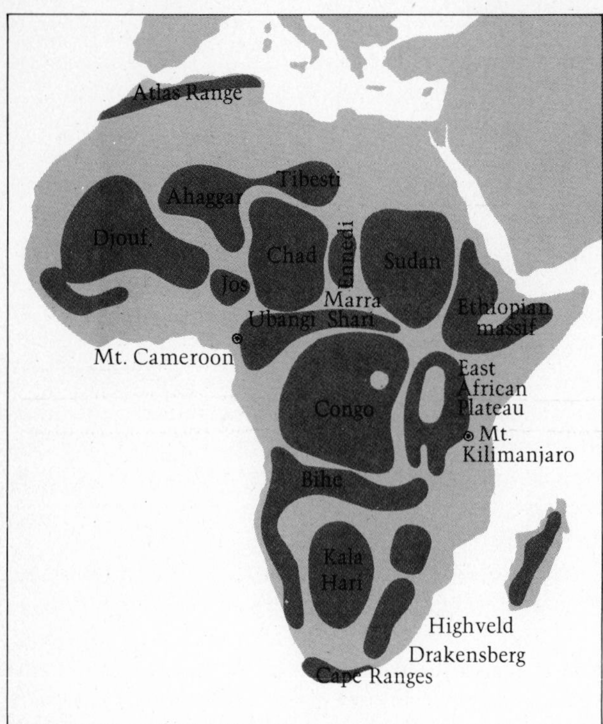

FIGURE 15-12
The major inland basins of Africa.

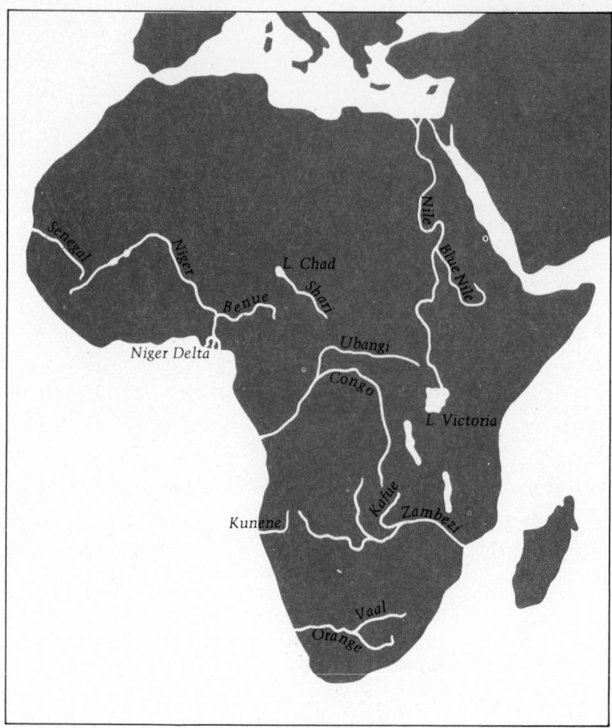

FIGURE 15-13
The major river systems of the African continent.

and it has lowlands in Somali and Mozambique. But considering Africa's bulk, coastal plains like those of the United States hardly exist. Most of Africa's coastal lowlands lie squashed between the foot of the Great Escarpment, shown in Figure 15-11, and the beaches. Nor are these lowlands flat and gently sloping. They are mostly hilly and steeply sloped.

If we look at the inner parts of the continent, two other features strike us. One is the possession of inland basins. The other concerns the rivers of the continent. The basins are shown in Figure 15-12. Some of these basins have been well known for a long time: In equatorial Africa, for instance, we know the Zäire Basin. In the northeast, the basin of the Nile River has for a century been known as the Sudan Basin. In Southern Africa, the Kalahari Desert occupies a basin of the same name. And we can identify two more such major basins: the Chad Basin, west of the Sudan Basin, and in western West Africa, the Djouf Basin. Five enormous depressions that cover well over half the African plateau surface lie separated by somewhat higher divides. Surely this is a situation we should try to explain!

Basins on the scale we now have before us are formed when something causes the crust to sag. Often the great weight of accumulated sediments has this effect—sediments brought to the basins by inflowing rivers. Perhaps Africa's river systems will reveal something about these depressions.

Africa has five major river systems—the Nile, Niger, Zäire, Zambezi, and Orange—mapped in Figure 15-13. All these river systems, with the possible exception of the Orange, have similar characteristics that seem to be related and demand an explanation. These characteristics are the location of deltas far inland, unusual course reversals, a series of waterfalls, and the fact that the waterfalls occur below the interior deltas and usually below a major change of course.

Because deltas tend to form on a coast, it is perhaps not unreasonable to think that the rivers did indeed reach coastlines long ago—not the ocean's coasts but the shores of internal seas that filled the basins. To those basins the rivers brought the sediments that now lie exposed as desert sand in the Sahara and the Kalahari Desert or buried beneath tropical vegetation in the Zäire basin. Then the interior seas were opened to the ocean and drained. The waters rushed out, carving waterfalls and widening valleys. The old rivers that once filled the basins were connected to the new ones that drained the pent-up water to the coast, establishing the present pattern. Tributaries such as the Kafue River, which had been oriented toward the interior basin, now were redirected or even reversed and abandoned their old courses. Note how the Kafue

LAKE VICTORIA

Lake Victoria, East Africa's largest lake, has none of the properties we usually find in that region's Great Lakes (shape, relative location, high-relief margins). Lake Victoria occupies an area of 62,940 sq km (24,300 sq mi) of East African plateau, about midway between the Eastern and the Western Rift Valley. Why is Lake Victoria the exception to the rules the map of East Africa so clearly reveals?

Lake Victoria has a complicated history. In further contrast to its neighbors, it is a very shallow lake—its deepest point is about 80 m (265 ft), while Lake Tanganyika reaches a depth of over 1470 m (4820 ft) and Lake Nyasa, 706 m (2316 ft). When geographers began to study the lake margins, they found that the rivers now flowing into it had been reversed—that is, they once flowed *away* from where the lake now lies! Further research suggests that Lake Victoria occupies the top of what was once a dome-shaped upland, which must have collapsed into a very shallow basin. In fact, the river patterns reveal much. Study your map carefully and note how it is possible to connect rivers flowing into Lake Victoria to rivers flowing into the Zaïre (Congo) River. So in this instance, the crust did not fracture—it dimpled and the giant hollow slowly filled with water.

elbows from a southwestward to an eastward course.

These events would all be explained if Africa had been centralized in the ancient Gondwana. Yet other features in the African landscape suggest that the breakup of the present continent is not complete.

The future of the African continent If you were to fly over the east African plateau, at least two kinds of landscapes would impress you strongly. The first of these would be the large volcanic mountains that rise above the surface. The second would undoubtedly be the lengthy, steep-walled, deep and persistent valleys that cut across the region. It would be difficult to miss these great trenches. They are hundreds of meters deep, dozens of kilometers wide, and in places filled with water—the water of eastern Africa's major lakes.

Chapter 13 refers to faults that are, in effect, fractures of the earth's outer crust (and along which earthquakes tend to occur). In eastern Africa, the land has sunk down, or has been pushed down, in strips between parallel faults. The trough or trench thus created is referred to as a *rift valley,* as shown in Figure 15-14. Africa's system of rift valleys extends for some 9700 km (6000 mi), from the northern end of the Red Sea (itself an entire rift valley), through Ethiopia and eastern Africa, to Natal in South Africa. In certain areas the rifts have filled with water, forming the great elongated lakes. Rift valleys are not static or stable. Note, on the map of earthquake incidence in Figure 13-8, how eastern Africa stands out as an active zone amid the comparative stability of the rest of the shield-dominated landmass.

The rifts are quite obviously lines of weakness, along which the African landmass is breaking up. Madagascar probably once lay along the eastern edge of a rift. Certainly the Red Sea, several million years ago, looked a great deal like the rift in which Lake Rudolf lies today. The Mozambique Channel, the Red Sea, and the Ethiopian rift valley may represent three stages in the formation of a rift valley and the fragmentation of a landmass. Africa a few million years hence may look as depicted in Figure 15-15, with a fragment larger than Madagascar lying off its new eastern coast.

Thus the evidence from our map points not only to the reality of continental drift but also to Africa's position at the heart of the ancient Gondwana. The plateau of Africa was the center of Gondwana. Its escarpments, already cut back by erosion, mark the

FIGURE 15-14

The wall of a rift valley in southeastern Africa, about 600 m (2000 ft) high. The other wall of the valley, behind the photographer, is identical.

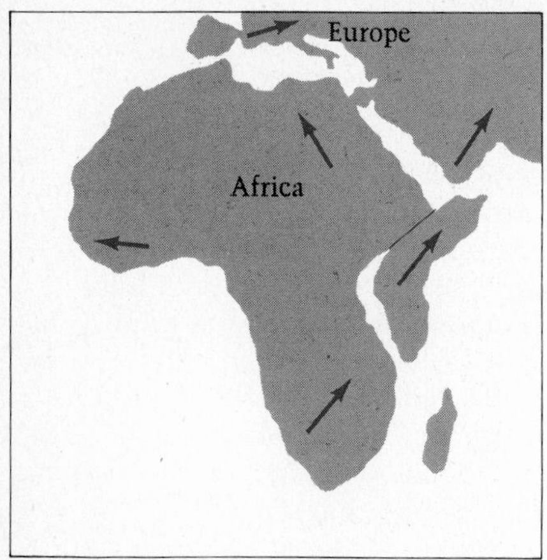

FIGURE 15-15

The African continent 50 million years from now.

fractures of the fragmenting supercontinent. While it was part of Gondwana, Africa's plateau buckled under the weight of accumulating sediments and under the waters of internal seas. Then the rupture of Gondwana produced outlets, the rift valleys became active, and the rivers that drained the internal seas formed deep valleys and enormous falls and cataracts. Basaltic lava poured out of fissures and covered much of the heart of Gondwana just as the landmass fragments began to spread apart. This was not just the beginning of the continents as we know them. Here also began the making of the Atlantic and Indian oceans, the Red Sea, the Himalayas. The earth as we know it now began to be shaped. Remember that the evidence we have just encountered came principally from a surface map, hence from the landscape.

CONTINENTAL DRIFT AND LANDSCAPE: THE WORLD

Scientists no longer doubt that Africa was once the center of a supercontinent. The parallel detail of rock successions in the Cape ranges of South Africa and the Sierra de la Ventana of Argentina; comparable fossils on both sides of the Atlantic Ocean; a similar climatic history in the southern continents—all these evidences, once neglected and scorned, are now generally regarded as indisputable evidence of our wandering continents.

The concept of plate tectonics can help explain some of the most important features of the earth's crust. Figure 15-16 shows these important features—the ocean ridges, mountain ranges, ocean trenches, volcanic zones, and earthquake zones. If you compare this figure with Figure 15-4, you will see that all these features are intimately related to the location of a plate boundary. The continents and crustal plates have drifted through different climatic zones. They have pushed and dragged, and their leading edges have crumpled up. Their surfaces, once worn flat by erosion, now lie warped and tilted. Coastal escarpments, ancient mountains, and desert dunes all have the same

Major crustal structures of the earth.

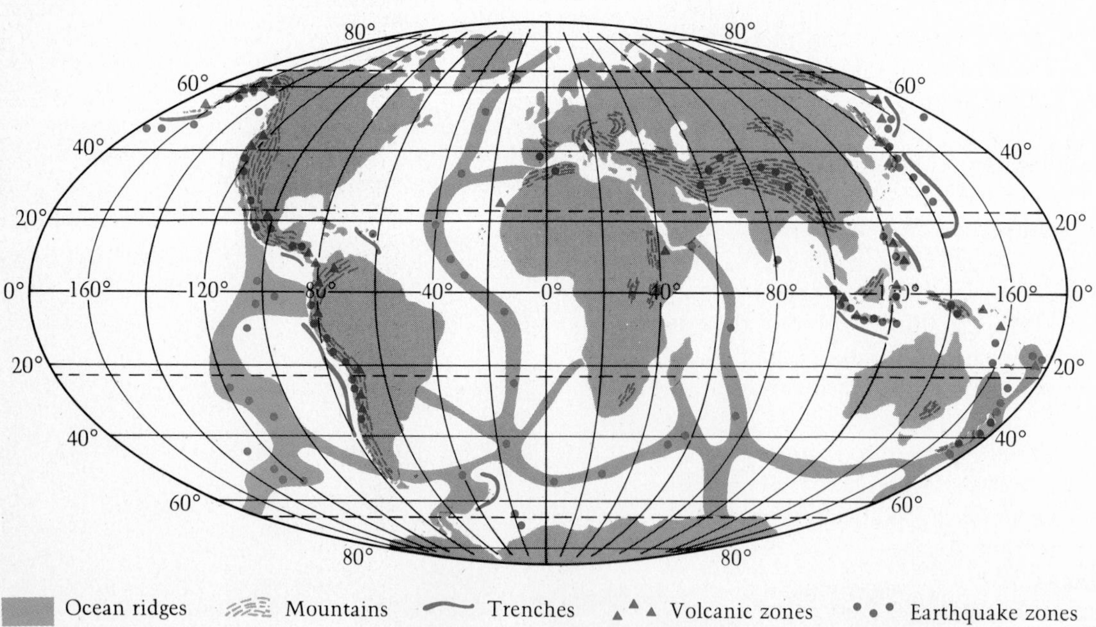

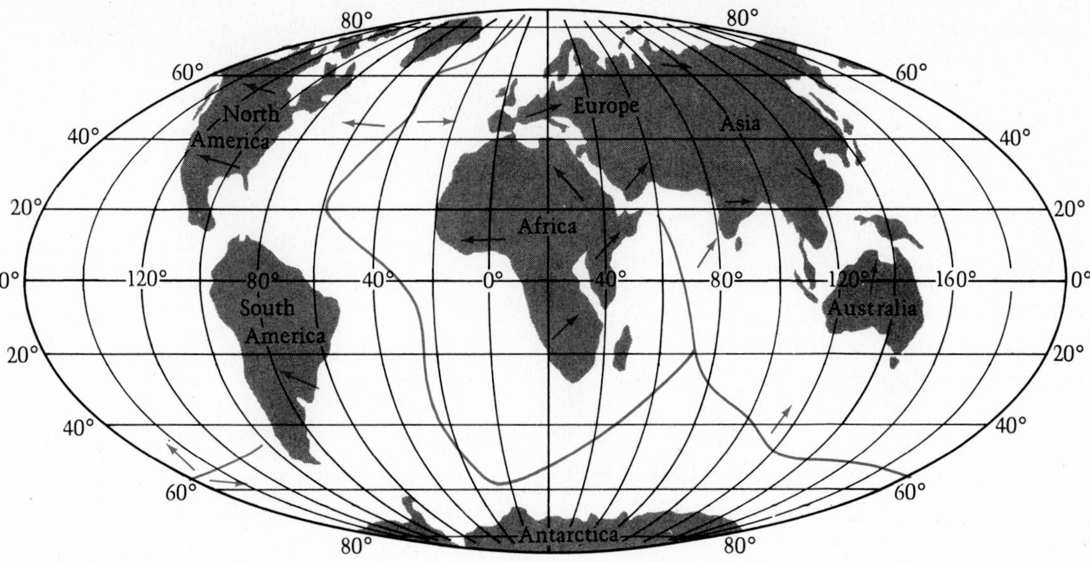

FIGURE 15-17
The world as it may look 50 million years from now, following further movement of the major plates.

relationship to the drift process. Plate tectonics is a fundamental concept for earth scientists. For physical geographers, it is an invaluable aid for interpretations of the earth's surface.

THE SIGNIFICANCE OF PLATE TECTONICS

The theory of plate tectonics has points of significance beyond the arguments that scientists may have. We can learn lessons from it, and it can directly help to improve our lives.

THE LESSONS WE LEARN

One lesson we learn is that landscapes can teach us a great deal about earth history without our delving into the rock layers below. Another is that scientists are sometimes reluctant to accept the obvious, even when confronted by powerful evidence. Still another lesson is that we sometimes forget the contributions of researchers who first formulated important ideas that finally become popularly accepted. Some of the earliest protagonists of the concept of continental drift have been all but forgotten.

But possibly the greatest lesson is that we should always try to keep an open mind. The future of physical geography may well lie substantially in the renewed interpretation of landscapes in the context of plate movement and continental drift.

THE ECONOMIC ADVANTAGE

Staggering as the idea of wandering continents may be in itself, it goes further, in its plate tectonic form to suggest considerable economic advantages for us. The greatest of these is in aiding the discovery of new mineral resources.

It has been discovered that many deposits of valuable minerals, such as copper and uranium, lie along plate boundaries. Previously the midocean ridges were neglected in mineral explorations, but now they have been found to be enriched with various metals. Much closer investigations are underway. Fitting the continents together in their predrift positions may also reveal areas of important mineral deposits. For example, diamonds now found in northern South America may have been formed from source rocks in western Africa. The gold-bearing regions in the two continents can also be matched. This jigsaw puzzle promises great rewards.

FUTURE MOVEMENT

A close study of all the evidence for plate tectonics can be used to predict the future movements of the major plates. The earth's land surfaces may be distributed as in Figure 15-17 50 million years from now. A strip of land will break off western North America. Los Angeles will move northward past San Francisco. North and South America will no longer be connected. A large part of eastern Africa will move toward Asia, and

Australia will be much closer to Asia. In 1976 a satellite named Lageos was placed into orbit and for the next 8 million years will provide a constant "spacemark" by which plate movement on earth can be measured.

The fragmentation of Pangaea raises an important question: If the earth is as old as we believe it to be, why did the continents hold together until just recently, in the Mesozoic era, when they finally drifted away? What crucial change generated continental drift so late in the earth's history?

The answer is hypothetical, but the hypothesis is a likely one. The fragmentation of Pangaea was probably only the most recent phase of continental drift. Continental landmasses may have always been mobile—to converge, fragment, unite again, and diverge once more. A record of only the latest of these cycles remains, and time has obliterated virtually all the evidence that could remain of previous ones. But it is likely that the drift process always prevailed and that it is a cyclic process.

The drift process is not slow at all, by the standards of geologic change. Calculate what 2.5 cm (1 in) per year means over the hundreds of millions of years that are etched in the geologic record. In a relatively short time, mountain ranges are thrown up and ocean trenches fold down into the crust as the earth's plates move and meet, stretch and crumble, slide and fracture. Not a centimeter of our landscape is left unaffected by this, the greatest of earth processes.

SUMMARY

Recently discovered evidence, of such phenomena as paleomagnetism and sea-floor spreading, has led to the acceptance of the idea that parts of the earth's crust move across the face of the planet. Crustal plates of the lithosphere move across the weaker and more plastic asthenosphere.

Where plates meet, there are three kinds of boundaries—divergent, convergent, and transcurrent. The plates are thought to be moved by convection currents within the magma of the mantle.

The new theory of plate tectonics has developed from the original idea of continental drift. The idea of continental drift was ridiculed for decades after it was first suggested, yet evidence was present within the rocks of the earth and visible in its landscapes. The evidence is particularly visible in the landscapes of Africa. It shows not only that Africa was once the center of a supercontinent called Pangaea but that the drift process continues.

QUESTIONS

1. How has the alignment of magnetic iron particles in ancient rocks contributed to our knowledge of the past configuration of the continents?

2. Assume that you are a geologist examining the rocks of various continents to determine the previous positions of the magnetic pole and the relationships between the continents. You have limited your examination to rocks containing aligned magnetic grains. What other information about the rocks must you have before you can directly compare the rocks and determine the positions of the poles and the continents?

3. Two factors control the zebra pattern of magnetism exhibited in Figure 15-2. What are these two factors, and how do they operate together to produce this pattern of magnetized rock? How does this phenomenon help scientists date processes of crustal movement and processes of magnetic change in the earth?

4. Evidence from bore holes drilled in the ocean floor shows that the bottoms of the ocean basins have comparatively young rocks—less than 125 million years old. Where would you expect to find the youngest rocks of the Atlantic Ocean floor, and where would you expect to find the oldest ones? Why?

5. Scientists have found that the rocks of the earth's crust are older, the further we move from the mid-oceanic ridges. These data and paleomagnetic studies are among the indications that the crust is not a single, solid mass, as was once thought. How have studies of temperature gradients and of gravity in various areas contributed to the mass of evidence that supports the theory of plate tectonics?

6. Scientists now view the earth's outermost shells as a set of rigid plates resting on a more plastic layer below, known as the asthenosphere. Some of these plates are combinations of continental and oceanic rock masses, while others are purely oceanic. We make a distinction between continental plates and oceanic plates. Is the distinction solely one of location, or are there compositional and structural differences as well?

7. By comparing the types of plate boundaries with the map of the world-wide distribution of plates, Figures 15-5 and 15-4, we can often determine the types of boundaries actually found between different plates. What type of boundary would you expect to find between the American and the African plates? Between the Pacific and the American plates? Why?

8. Each of the major plate boundary types is associated with important landform features. What features are associated with a divergent plate boundary? With a convergent plate boundary? Explain how the boundaries and their landform features are related.

9. One of the earliest problems with the theory of plate tectonics was explaining an adequate driving mechanism for the system of continually moving plates. What theory has been put forward to explain this movement? What important question remains unanswered?

10. What two factors connected with the shapes and distribution of the continents led to speculation about possible continental drift?

11. What is the importance of Africa's position and structure in the study of continental drift?

12. Alfred Wegener presented one of the first encyclopedic collections of data in support of the theory of continental drift. These data included information on correlated rock types on opposite sides of the Atlantic, distribution of fossil life that required connections across the ocean, climatic changes common to distant continents, and zones of crustal deformation that seemingly continued across the Atlantic. Why were his proposals that the continents were once a single unit ridiculed?

13. What features distinguish Africa from the other continents, and how might these be related to its former central position at the hub of Gondwanaland?

14. How are the inland basins of Africa and the present courses of the Nile, Niger, Zäire, and Zambezi rivers related to the breaking up of Gondwanaland? If it had not broken up, would these features appear as they do today? Why?

15. How does the continuing breakup of the ancient heartland of Gondwanaland affect the present-day landscape of Africa? How are the Mocambique Channel, the Red Sea, and the Ethiopian Rift Valley connected?

16. Are earthquakes and volcanoes good indicators of the presence of plate boundaries? Can either or both be associated with a specific type of plate boundary? Why or why not? (Use Figure 15-16.)

17. Two potential economic advantages, one with positive implications for humans and one with negative implications, have been derived from the theory of plate tectonics. What are these advantages, and how are they connected with our new knowledge about the crust of the earth and its movement?

18. Plate tectonics is an ongoing process, and the present distribution of the continents will change dramatically with immense amounts of time. Which continents will exhibit the greatest change, and which the least? What is the predicted impact of continued continental drift for North America?

19. We have seen two major cycles in this chapter, one connected with the movement of rock through the crust and the mantle of the earth and the other with the distribution of the earth's crustal masses over immense periods of time, including the period before the supercontinent of Pangaea existed. How do these two cycles work, and what type of evidence do we have that they actually exist?

20. The process of continental drift is a relatively rapid one in geologic terms. The rate of drift of the African and American plates is estimated to be approximately 2½ cm per year. How did scientists arrive at this figure, assuming that they did not directly measure the mid-Atlantic ridge at numerous points for one year?

DEFORMATION OF THE CRUST

CHAPTER 16

The continental landmasses of our planet are continuously moving, carried along on plates that are themselves drifting on the asthenosphere. In the process, the rocks of which the continents are made undergo dislocation and deformation. Sedimentary layers deposited horizontally are folded and crushed. The intense pressure makes metamorphic rocks out of sandstone and shale. Lava erupts from volcanic vents and fissures. Earthquakes punctuate the zones of contact between plates. Elsewhere, the effects of continental drifting are recorded more passively. In the interior of the continents, plains and plateaus change quietly, their surfaces alter imperceptibly. But everywhere there is change.

In this chapter we consider the way the landscape reflects the forces that constantly modify the continental surfaces. One important principle involves the mutual adjustment of vertical columns of the crust so each eventually has the same weight at some point within the earth itself. We also describe how mountains are built when large accumulations of sediments are uplifted, folded, and broken. In other cases land may rise or fall without noticeable distortion of its rocks.

THE FLOATING CRUST

The upper surfaces of the continents, we know, display a great degree of topographic variety. Mountain ranges rise high above surrounding plains. Plateaus and hills alternately dominate the landscape elsewhere.

Mountain ranges have mass. Because of the law of gravity, they exert a certain attraction on other objects. If we were to hang a plumb line somewhere on the flank of a mountain range, we would expect the mountains to attract the plumb line from the vertical toward the range. More than a century ago, British scientist George Everest, after whom the world's highest mountain is named, took measurements along the southern flanks of the Himalayas in India. He suspended his plumb line and did indeed find that the great Himalayas caused some attraction—but far less than his calculations, based on the mass of the mountain range, had led him to expect. Everest and his colleagues soon realized the importance of what they had discovered. If the deviation of the plumb line toward the mountains was less than calculated, there must be rocks of lesser densities extending far below the Himalayas, displacing the heavier, simatic material that would have caused greater attraction. In other words, the lighter sial appears to extend far down into the sima, and mountain ranges seem to have "roots" penetrating farthest where the elevation is greatest.

This possibility was realized as early as 1855 by Sir George Airy, whose hypothesis of mountain roots is depicted in Figure 16-1. In Figure 16-2 the sialic part of the crust is likened to blocks of copper that, because they are less dense, float in the mercury representing the sima. The higher the block stands above the dashed line representing sea level, the deeper the "root" below pushes into the simulated sima. Thus the blocks, or parts of the earth's crust, reach a kind of balance. Under the Himalayas and other major mountain ranges, the sialic part of the crust is comparatively thick. Under plateaus it is thinner, and under lowlying plains it is thinner still. Thus the relief of the continental landmasses has a mirror image below.

Since the development of plate tectonics theory, described in Chapter 15, we have come to believe that the balance is not a question of sialic "rafts" floating on a simatic "sea." Rather, the balancing movements occur at the base of the lithosphere, far below the Moho discontinuity (see Chapter 13). The vertical changes in the crust are thought to take place for two reasons. First, the lithosphere floats on the asthenosphere as the copper blocks float on mercury. Second, the lithosphere is subjected to changes of density from time to time.

This situation of sustained adjustment, as visualized by Airy (and modified by others after him), has come to be known as the principle of *isostasy*. The source of this term is not difficult to determine: *Iso* means "the same" or "equal" (isobar, isotherm), and *stasy* comes from the ancient Greek "to stand." Thus isostasy is a

FIGURE 16-1

Airy's idea of how mountain ranges have roots of sialic rock.

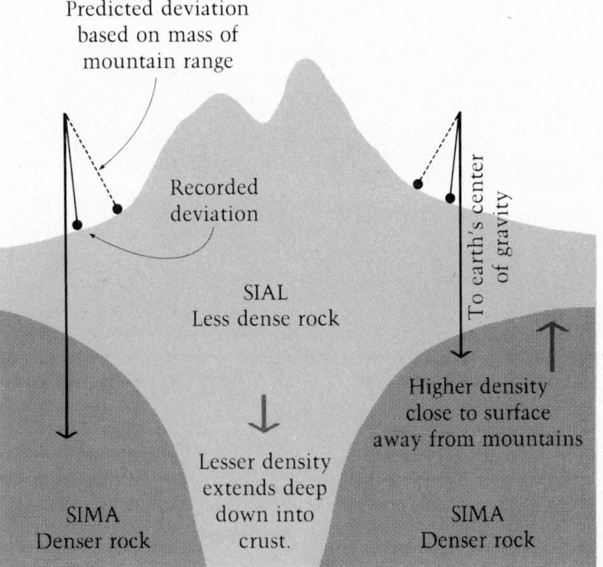

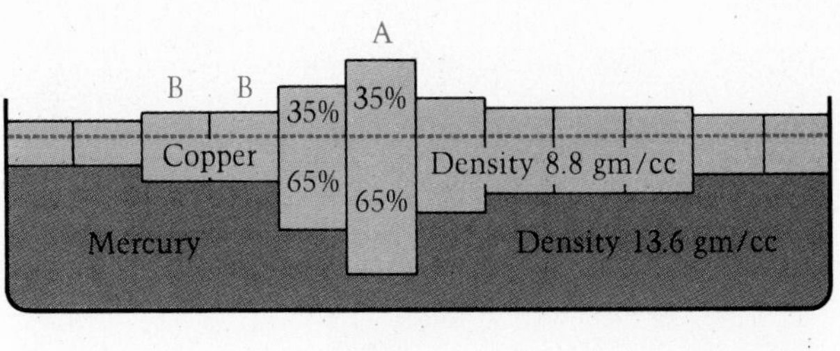

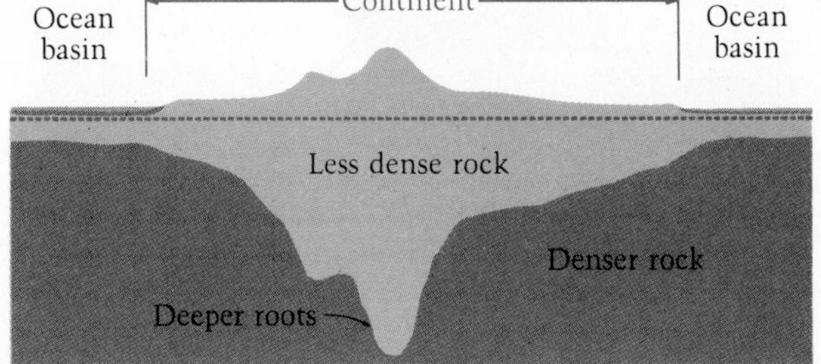

FIGURE 16-2

Isostasy. The distribution of sial and sima is like blocks of copper floating on mercury.

condition of equilibrium, maintained despite the forces that tend to change the landmasses all the time.

ISOSTASY AND EROSION

We can use the Airy model shown in Figure 16-2 to envision what would happen if a high mountain range were subjected to a lengthy period of erosion. If we were to saw off the upper 10% of the column marked *A*, we would expect that column to rise slightly—not quite to the height it was before but nearly so. If we were to place the sawed-off copper on the two columns marked *B*, they would sink slightly, and their upper surface would adjust to a slightly higher elevation than before. Thus column *A* would have a lower height and a shorter root, whereas the columns *B* would have a greater height and a deeper root.

This tendency would explain why erosional forces in the real world have not completely flattened all mountain ranges. Scientific experiments have indicated that, at present rates of erosion, the earth's mountains would be leveled in a single geological period, certainly within 50 million years. But mountains hundreds of millions of years old, such as the Appalachians, still stand above their surroundings. What seems to happen is that, as erosion removes the load from the ridges, isostatic adjustment raises the rocks to compensate. Rocks formed deep below the surface, tens of thousands of meters down, are thereby exposed to our view and to weathering and erosion.

Now we can answer some puzzling questions about the deposition of enormous thicknesses of sedimentary rocks. One sequence of sedimentary rocks found in Africa involved the accumulation of nearly 6.5 km (over 20,000 ft) of various sediments, followed by an outpouring of great quantities of lava. Other parts of the world have even thicker deposits. As we can see in Chapter 14, it is possible to deduce, from the character of the deposits, the environment under which deposition took place. In some areas such deposition took place in shallow water. Although thousands of meters of sediments collected over millions of years, the depth of the water somehow remained the same. In the accumulating sediments near the Bahamas, for example, rocks formed in shallow waters or intertidal flats are now 5500 m (18,000 ft) thick. We may conclude that some cause or combination of causes, continuously depresses the region of deposition, keeping the surface at about the same level. Such slowly accumulating sediments might be another place to

RESOURCES OF THE CONTINENTAL SHELF

If the continental landmasses extend under the ocean water to the edge of the continental shelf, then it follows that the submerged continental margins should contain various mineral deposits, just as the exposed rocks do. The total area of the continental shelves is about 5% of the whole earth's surface, or as much as one-sixth of the land area of the continents. Without the shelves, the continents constitute about 30% of the surface of the earth, but when the continental shelves are added the figure is 35%. Considering the enormous mineral yield of the exposed landmasses, the submerged areas ought to prove fertile fields for exploration.

Already, the yield is high. On the continental shelves of Europe, Asia, and the Americas stand artificial islands, oil platforms whose pipes reach through hundreds of feet of water to penetrate the oil-bearing strata below. Improving technology is making it possible to reach ever deeper oil deposits, and eventually as much as 10% of the world's oil may come from submerged fields. In Western Europe, the North Sea oil fields, including the famous pioneer Ekofisk field, produce more oil than all the mainland's oil deposits combined.

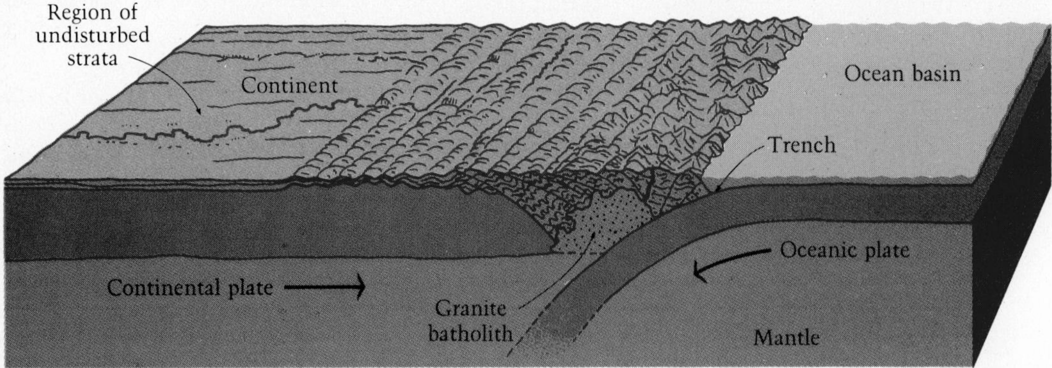

FIGURE 16-3

The formation of a mountain range and ocean trench from compression of convergent plates.

effectively dispose of waste matter produced by human activity. It might slowly sink fron sight with the sediments. Slow accumulation is now taking place in the Mississippi delta. For millions of years the great river has been pouring sediments into its delta, but these deposits have not formed a great pile nor has the river blocked itself. Isostatic adjustment lowers the material to make room for more.

ISOSTASY AND DRIFTING PLATES

If isostasy involves a condition of equilibrium, then the contact and collision of drifting plates must greatly affect that situation. When a continental plate meets an oceanic plate, the oceanic plate plunges below the continental plate, causing the enormous deformation and dislocation shown in Figure 16-3. Along the leading edge of the continental plate, rocks are crushed and folded, sediments (even those from the ocean bottom) are baked into metamorphic rocks, and magma penetrates and erupts along fissures and from volcanoes. In effect, the sialic mass increases in volume and, according to isostatic principles, rises upward. Right next to the high mountains so formed, the downward thrust of the oceanic plate often creates a deep trench, as deep as or deeper than the highest mountains are high. This is the situation along much of the Pacific's "Ring of Fire," where crustal instability and isostatic maladjustment are greatest. Earthquakes occur continuously along this zone as the plates converge and collide.

But the process does not go on forever at the same rate. We still do not know just why, but the geological record shows that the earth has gone through various rather distinct periods of mountain building and other, quieter periods. Eventually even the "Ring of Fire" will quiet down, and plate collision and mountain building may start elsewhere.

Then erosion will take over and begin to lower the mountains that have been created—but very slowly, because isostatic uplift will occur. But the first phase of erosion may in fact be rather rapid. The sialic mountains along the plate margin may have been pushed beyond the elevations justified by their roots, and isostatic readjustment does not begin until the overload has been removed. After that, however, the mountains stand no matter how severe the erosion to which they are subjected. Southern Africa's Cape ranges and the Appalachians had roots deep enough to ensure their topographic prominence for over 200 million years. Some mountain ranges have indeed been flattened by erosion, all the way down to their roots, but these are much older still.

ISOSTASY AND REGIONAL LANDSCAPES

In studying the effects of the isostatic principle, we tend to be preoccupied with mountain ranges, mountain building, plate compression, and associated phenomena. But we should not lose sight of the consequences of isostasy in areas of less prominent, less dramatic relief.

Erosion is active on the continents' plains too, and millions of tons of material are carried away by streams, rivers, and other agents. Even moving ice and wind denude or reduce land surfaces. Unlike the mountain zones, however, the plains are vast in area (consider the region drained by the Mississippi River system), and slopes are gentler. Rivers erode less spectacularly on the plains than in the mountains, there may be fewer streams per unit area than in the mountains, and there may well be less precipitation than in the mountains. All these circumstances mean that eroded material is removed from the plains at a slower rate.

The sialic crust has a certain rigidity. It does not behave, as in Airy's model, as a series of discrete columns. Therefore, isostasy affects plains and plateaus in phases. For a certain period, the amount of material removed is not enough to "trigger" isostatic readjustment, because the hardness of the crust prevents continuous uplift. But when the plain has been denuded sufficiently for the push of isostatic uplift to overcome any resistance by the crust, a change takes place. Thus, at a given moment, an area may not be in isostatic equilibrium. Instead, it awaits the time when readjustment will occur.

Scientists suggest that this periodic adjustment may also occur in mountain ranges, especially older ones. In the beginning, when the sialic root is deep, almost continuous isostatic uplift occurs. But as time goes on, the root becomes shorter, erosion continues, and comparatively more eroded material must be removed for readjustment to occur. In fact, the Appalachians were probably flattened almost completely and then rejuvenated by a recurrence of isostatic uplift. Now the old ridges are being worn down again, and the whole area may be made into a plain before another readjustment occurs.

Some other manifestations of isostatic change also are of interest. When ice sheets spread over continental areas during glacial ages, the weight of the ice (whose thickness can reach several thousand meters) causes isostatic sinking of the crust below, just as a sedimentary accumulation would. This is what happened in North America and Eurasia during the most recent glaciation, the Pleistocene, whose most recent ice sheets melted just 12,000 years ago. While the ice sheets pushed southward, across the Great Lakes area and as far as the Ohio River, the crust below was depressed by their weight. But when the ice sheets retreated, isostatic readjustment brought the crust upward.

This upward readjustment, however, could not keep pace with the relatively rapid melting of the ice. The last Pleistocene ice sheets were at their full development just 25,000 to 50,000 years ago, but they had melted by 12,000 years ago. In geologic terms, the melting removed the enormous load of the ice sheets almost instantaneously. Studies show that the ensuing isostatic readjustment is still going on; the crust has still not rebounded completely. In the heart of Scandinavia, the site of a huge Pleistocene ice sheet, the crust is rising at more than 1 m (3.28 ft) per century, a very fast rate. In some coastal parts of Norway, metal rings placed in rocks centuries ago to tie up boats are now much too high above sea level to be of use. Similarly, much of coastal California is flanked by ancient beaches tens of meters above the present-day sea level as shown in Figure 16-4.

Even human works on the surface of the earth can produce isostatic reaction. When a river dam is constructed, the weight of the trapped water may be

FIGURE 16-4

A raised beach on the California coast. The flat area, some 45 m (150 ft) above the present-day beach, is an ancient beach uplifted in the general isostatic readjustment of the land surface.

FIGURE 16-5
Aerial view of Lake Mead behind Hoover Dam. The reservoir contains 21 million acre-feet of water. Isostatic adjustment to the weight of this water has been measured.

enough to produce isostatic accommodation in the crust. Measurable readjustment of this kind has taken place in the area of Kariba Lake, the great dam on the Zambezi River in southern Africa and around Lake Mead on the Colorado River in Nevada (Figure 16-5). It will undoubtedly take place as the great Aswan Dam of Egypt, on the Nile River, fills up. We cannot see these changes with the bare eye, but scientific instruments detect them. To us, in our everyday existence, the crust below may seem permanent, unchangeable, and solid. But even our own comparatively miniscule works can disturb its equilibrium.

OROGENY: THE BIRTH OF MOUNTAINS

Chapter 13 briefly notes that the earth's landscapes are dominated by shields and orogenic belts. The shields form the heart of the continental plates. The orogenic belts are lengthy chains of intensely folded rocks, resulting from prolonged compression along the margins of colliding plates.

The earth has experienced a number of successive periods of mountain building, and each such period is known as an *orogeny.* The latest products of an orogeny are well known to us, because they stand today as the largest mountain chains of the world and include the Rocky Mountains and the Himalayas. But the earth has recorded several orogenies prior to this latest one, far back into the Precambrian era. The products of the oldest orogenies have been wiped out by erosion, but more recent mountain-building periods, such as the one that produced the Appalachians, can still be seen in the landscape.

MAJOR LINEAR MOUNTAIN CHAINS

The globe today displays three major mountain systems. In the Americas, a mountainous zone consisting of several belts extends over thousands of kilometers—from Alaska, through the Canadian and US Rocky Mountains, through Mexico and Central America, and along the western margin of South America. In Eurasia, the Alpine system has a branch, the Atlas Mountains, in North Africa and also includes the European Alps, the mountains of eastern Europe, the Caucasus range, several ranges in southwestern Asia, the great Himalayas, and even some mountains beyond. In contrast to the Rockies-Andes system, the Alpine system lies mainly east–west. The third system consists of the western half of the Pacific "Ring of Fire," and it extends along the Pacific margins of Asia and Australia. A great deal of the topography of this developing mountain chain is yet under water. The protruding peninsulas, archipelagoes, and islands of eastern Australasia are the crests of the ranges. Japan, parts of the Philippines, Indonesia, Malaysia, New Guinea, and New Zealand all are part of this third system. The mountains in these three major orogenic belts are created by compression, in contrast to the world-wide system of mid-ocean ridges, which, as we saw, results from divergence of plates.

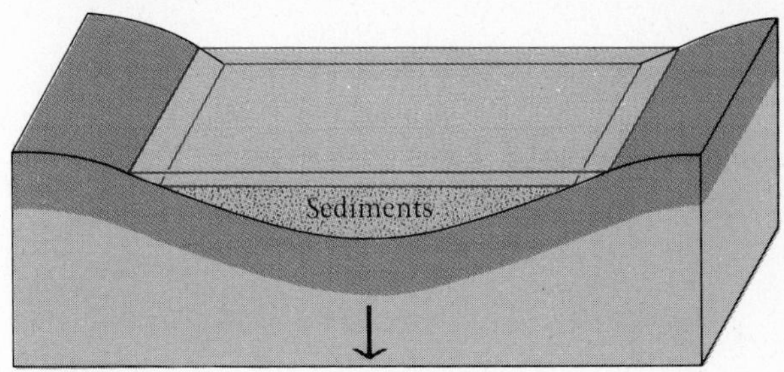

FIGURE 16-6

A geosyncline. The great depression in the earth's surface is several hundred kilometers wide. Continual deposition is accompanied by continual downwarping.

Although the world's major orogenic belts often consist of folded strata, we should not conclude that folded sedimentary rocks alone make up these ranges. The process of their formation is enormously complex. It begins quietly enough, with the deposition of sediments along the margin of a landmass or in a large basin of accumulation called a *geosyncline.* A geosyncline is a large subsiding trough of deposition, a huge but gentle depression in the crust that slowly fills with sediments from higher adjacent land areas, as illustrated in Figure 16-6. It looks like the simple depression that is sometimes called a syncline, but it is so large that the term *geosyncline* is used. Accumulation of sediments continues for a long time; thousands of meters of sediments may accumulate. Areas where geosynclines are thought to exist today include the Mediterranean Sea and the continental terraces and adjoining continental rises on the eastern and western sides of the Atlantic Ocean.

When the orogeny commences, the convergence of forces causes these neatly layered rocks to be bent and crushed. Some are pushed upward and others are thrust downward, deep into the crust. Temperatures and pressures are so high in the crust that the layered rocks melt, mix with magmatic material, and intrude upward in the form of batholiths (see Chapter 14). Thus the heart of the "folded" linear mountain chain is likely to be made of crystallized igneous and metamorphic rocks flanked by the deformed sedimentary layers. Erosion and isostatic uplift expose the deeper intrusives as time goes on.

Not all orogenic activity begins with a long period of sedimentation, however. Off eastern Asia, where a major orogenic zone is developing, sediments are scarce. The orogenic forces are expressed in the landscape mainly as volcanic mountains.

The three major orogenic belts are still active, as demonstrated by the coincidence between their distribution and that of seismic activity (refer to Figure 13-8). Older orogenies are much more stable. Compare the Appalachians, which were formed some 250 million years ago, to the young Rocky Mountains, formed 65 million years ago. There are also old orogenic zones

MOUNTAINS AND PEOPLE

Compare the location of the world's major mountain ranges (Figure 15-16) to the present world distribution of population, and you will note that, with few exceptions, mountainous areas are sparsely populated and dense, major population concentrations tend to occur in the world's plains and coastal areas.

Why should this be so? Mountains provide climatic variety, scenic beauty, and other attractions. Keep this issue in mind as you read further, and consider such items as the soils found in mountainous areas, the effect of slope on living space, the impact of high relief on communications and transportation, and the relationships between high elevation and local climate in middle- and high-latitude areas. You could make the list longer by further comparing the world distribution of major mineral deposits and the location of major mountain ranges. A few mineral concentrations do exist in the great mountainous zones, but far more lie along the foothills and in the adjacent plainlands.

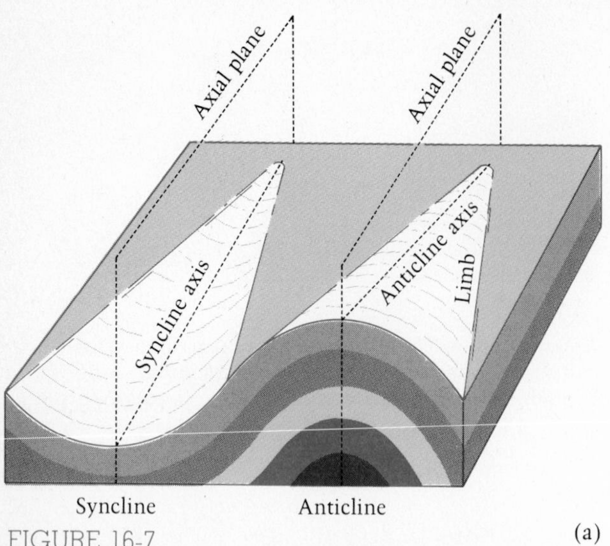

FIGURE 16-7

a) The component parts of folded structures. b) An anticline (upfold on the left) and a syncline (downfold in the middle) in deformed sedimentary rock in the French Alps.

in the Scandinavian peninsula, the British Isles, New England, eastern Australia, southern Africa, and, on a smaller scale, numerous other areas.

FOLDS IN THE EARTH: SYNCLINES AND ANTICLINES

Layered sedimentary rocks, when subjected to the stresses of severe compression associated with converging plates, change character in amazing ways. It is difficult to visualize comparatively hard, well-compacted, rocks such as sandstones and shales behaving rather like shaving cream as it comes out of a foam can and settling into the most complex forms imaginable. But the evidence that it happens is etched into the landscape where once-horizontal layers are bent into tight folds. There are also fractures large and small, and we will soon focus on the nature of these associated faults.

Folding of sedimentary strata often produces such a jumble of structures that it seems at first that nature has abandoned all order. But when we map the distribution of the rock beds and analyze the topography, we find that recognizable and recurrent structures do exist. It is useful to have some terms for identifying the various parts of folds in rocks. For example, one reason for the linear character of folded mountain ranges relates to the parallel *axes* of the folds. As Figure 16-7 shows, the axis of a fold is an imaginary line that divides the fold into two equal halves, and it lies on the *axial plane.* The two sides of a fold, the *limbs,* form the whole structure. When these limbs slope downward, so that a cross-section of the fold looks like a basin, the

FIGURE 16-8

Types of folds.

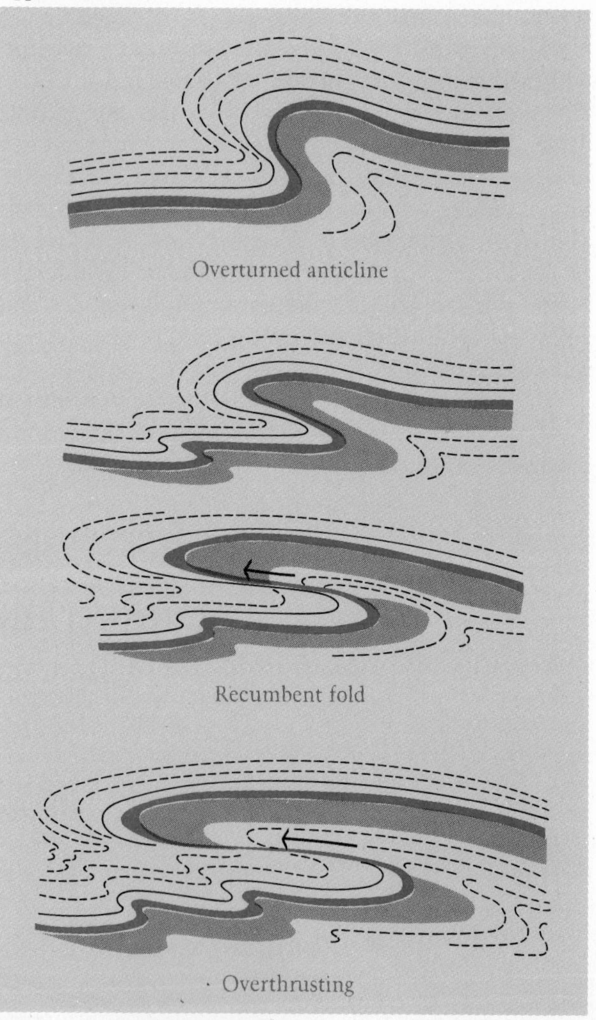

fold structure is called a *syncline.* An upward fold, whose cross-section resembles a dome, is an *anticline.*

In areas of folded mountains, such as the Andes and the Appalachians, numerous synclines and anticlines adjoin. In certain areas the folds are quite symmetrical and regular, but more often the folds have been squeezed to the point of collapse and have been overturned, as depicted in Figure 16-8. For heavily overturned folds, the term *recumbent* is used. Where faulting enters the picture, the fold is broken up by *overthrusting.*

All these structures, and many variations of them, occur in the orogenic belts of the earth. And out of the folded, faulted, intruded, and metamorphosed rocks, erosion fashions landscapes. The hard rocks in the fold structures stand up against erosion, whereas the softer ones give way faster, and the resulting landscapes allow us to interpret the structures that lie below. Correct interpretations of folded structures are important for geologists searching for precious minerals.

When the axes of adjacent anticlines and synclines lie horizontally, the anticlines may form anticlinal mountains and the synclines may create synclinal valleys. That simple situation may eventually be destroyed, however. Below the hard surface rock that sustains the anticline may be much softer layers. Once erosion has worn away the hard upper layer, the anticlinal ridge may be rapidly destroyed and converted into a valley. The syncline, initially a valley, may survive and emerge as a ridge. This is a case of topographic reversal, and examples can readily be found in the long-eroded Appalachians.

Often the axes of adjacent folds do not lie horizontally but plunge as in Figure 16-9. The resulting topography displays characteristic zigzig patterns with hardrock ridges standing out. In the case of a syncline, these ridges open up in the direction of the plunge of the axis. In an anticline, the ridges converge in the direction of plunge.

The topography of these complex structures is sculptured by nature's great excavators, the rivers and streams. They attack the softer rocks and carry them away. But the harder, more resistant strata stand up longer, revealing the structures being exhumed. River courses are to some extent controlled by structures: softer rocks make easy passages for running water, whereas harder rocks form barriers. But time is on the side of the rivers, and streams eventually cut through even the hardest of rocks. It is therefore helpful to study the drainage patterns that have developed on the fold structures. Some rivers maintain their position throughout an entire period of erosion and isostatic uplift, so we may discern where the earlier ridges and valleys were even when they can no longer be seen. This is considered in greater detail in Chapter 19.

MOVEMENT WITHOUT FOLDS: EPEIROGENY

Before we leave the topic of mountains, we should remind ourselves that there are mountainous zones

FIGURE 16-9

Sheep Mountain, Wyoming. A folded structure with a plunging axis erodes in characteristic patterns.

THE DECLINE AND FALL OF SOUTHEAST TEXAS

If you believe that such matters as crustal tilting and warping are merely textbook topics, consider what is happening in Baytown, Texas, and nearby areas. Baytown lies near Trinity Bay, an extension of Galveston Bay. With its offshore bars and lagoons (Galveston itself is located on such a bar), this would seem to be a coastline of emergence and accumulation. Instead, it is an area of subsidence, and in places very rapid subsidence. The ground is tilting downward and the water is invading the land.

The rate of decline is fast indeed—15 cm (6 inches) per year in some areas. Seawalls that once stood high above the water now barely hold back the waves. Areas that were once high and dry are now swampy and wet. Some people have already abandoned their homes and simply left, having given up the battle against the course of nature. And it is not just Baytown that is threatened. The whole Houston region is subsiding, tilting southward and downward, and what the people in Baytown are facing is only the beginning.

whose relief is not due to orogenic activity of the kind just discussed. The mountains of Ethiopia, many of them over 3000 m (10,000 ft) above sea level, are not part of one of the world's three major orogenic zones, but they are bigger than some of the folded ranges. Ethiopia was once called Abyssinia, an appropriate name indeed, because its high areas are separated by deep precipices and canyons. These mountains are the result of the erosion and fracturing of a plateau. But how did the surface of that plateau rise to an elevation over 3000 m (10,000 ft), even 4200 m (14,000 ft), above sea level?

Even before plate tectonics was beginning to be understood, scientists realized that there are at least two kinds of crustal uplift. Orogenic uplift relates to crustal compression and the thrusting up of segments of sial. But shield areas, too, are susceptible to gradual uplift, and not all of this uplift is a case of isostatic readjustment. Indeed, plains and plateaus warp downward. This kind of vertical movement in the crust is called *epeirogenic* movement, and its chief features are tilting and warping with little deformation of the rocks.

Large-scale, regional tilting and warping of the crust can have enormous topographical significance. It does not take a complicated calculation to prove that the tilting of a surface to just 1° produces huge elevations over distances of, say 2000 kilometers (1240 mi)—not an excessive distance in view of the size of shield areas. To the naked eye, the change is imperceptible. But the effects are unmistakable.

For example, the Karroo deposition in Africa occurred at relatively low elevations and in accord with isostatic principles. Today those horizontal layers sit thousands of meters above sea level. The Drakensberg range exceeds 3000 m (10,000 ft) in several places. But they did not rise as a result of orogenic deformation. Many of the strata are virtually undisturbed. Their present elevation appears to be related to a wholesale tilting of the African plate, up in the east and down toward the west. If you look again at the map of Africa in Figure 15-10, you will note an axis of high-level surfaces in the east, from Ethiopia to South Africa. That axis is ascribed to epeirogenic forces. Now the region appears to be affected by orogenic forces as crustal divergence takes place.

Erosional forces, of course, attack the highlands created by epeirogeny in the same way they assault the orogenic belts. Penetrating along fractures, seeking out the weaker rocks, erosional processes soon accentuate the vertical relief by creating contrasts and give the region a mountainous topography. But these mountains are a different sort of remnant and you might find nearly undisturbed strata near the tops of the rises.

The term *epeirogeny,* to denote regional tilting and warping of crystalline and other zones, now seems to be falling into disuse, but there is as yet no comprehensive answer to the questions posed by the results of this process. What causes areas to decline in elevation even when there is no evidence of overloading (such as deposition, ice, or water)? Why do certain areas rise or tilt without apparent isostatic incentive? Perhaps the answer lies in still-unknown properties of the continental plates or the asthenosphere below. Certainly the landscapes verify that such an answer is needed.

FAULTING: CRACKS IN THE EARTH

We have, in passing, referred to the fracturing, the breaking, of rocks along planes called *faults.* The rift valleys of eastern Africa are fault structures. Faulting is prominent in orogenic belts when the forces of compression are applied too rapidly. The rocks do not have enough time to change to a plastic state, so they break. Faulting has a considerable impact on the landscape, and some of the major escarpments (along the Sierra Nevada in California, for example) were created by faulting in the crust.

Again, we need to know some terms. The surface of the fault, along which the blocks or rocks move, is the *fault plane*. This plane, when partially exposed, creates the fault relief in the landscape. Movement along this fault plane takes place in small doses—say a few centimeters, on the averate, at a time—but it may add up to hundreds of thousands of meters, even several dozen kilometers. Movement along a fault, of course, produces the shocks we know as earthquakes and often destroys human structures.

Displacement along the fault plane relates to the nature and direction of the forces involved. A *normal fault*, diagramed in Figure 16-10a, results from tension in the crust. The fault plane is inclined in the direction of the block that has moved downward, the *downthrown block*. Tension pulls the blocks apart, and gravity lowers the downthrown block.

The opposite situation prevails when the force is compression rather than tension. In this case one block pushes over the other. The fault plane inclines toward the upthrown block, forming an overhanging escarpment. But bit by bit the overhang breaks off. Landslides are a prominent feature of the scarps of *reverse faults*, shown in Figure 16-10b.

Movement along the fault plane is vertical in both normal and reverse faults. But when the blocks move horizontally, the fault produces no major fault scarp. This occurs in the so-called *strike-slip fault*, shown in Figure 16-10c. The San Andreas Fault (see Figure 16-11) is an example. Movement along this kind of fault sometimes disconnects fences, roads, and railways.

We have already encountered overthrusting in the context of orogenic zones. The *overthrust fault*, depicted in Figure 16-10d, is a low-angle variant of the reverse fault; the fault plane lies nearly horizontal. It is quite possible that older rocks would lie on top of younger layers at an overthrust fault, as the sketch suggests.

In the field, it is sometimes possible to recognize an escarpment as a fault scarp even from far away. A fault scarp is likely to be straight and smooth, and to mark the boundary between contrasting elevations, especially if it has been formed recently. Streams may fall from the upthrown area or build an alluvial fan at the base of the escarpment, as in Figure 16-12. Sometimes, when the faulting has exposed water-bearing rock layers, springs emerge from the fault wall.

As time goes on, the escarpment first formed by faulting is driven back by erosion. The steepness of the wall is reduced, the upthrown block is *dissected*, (eroded into close valleys), and the evidence of its

FIGURE 16-10

Four basic forms of faults and how they appear in the landscape.

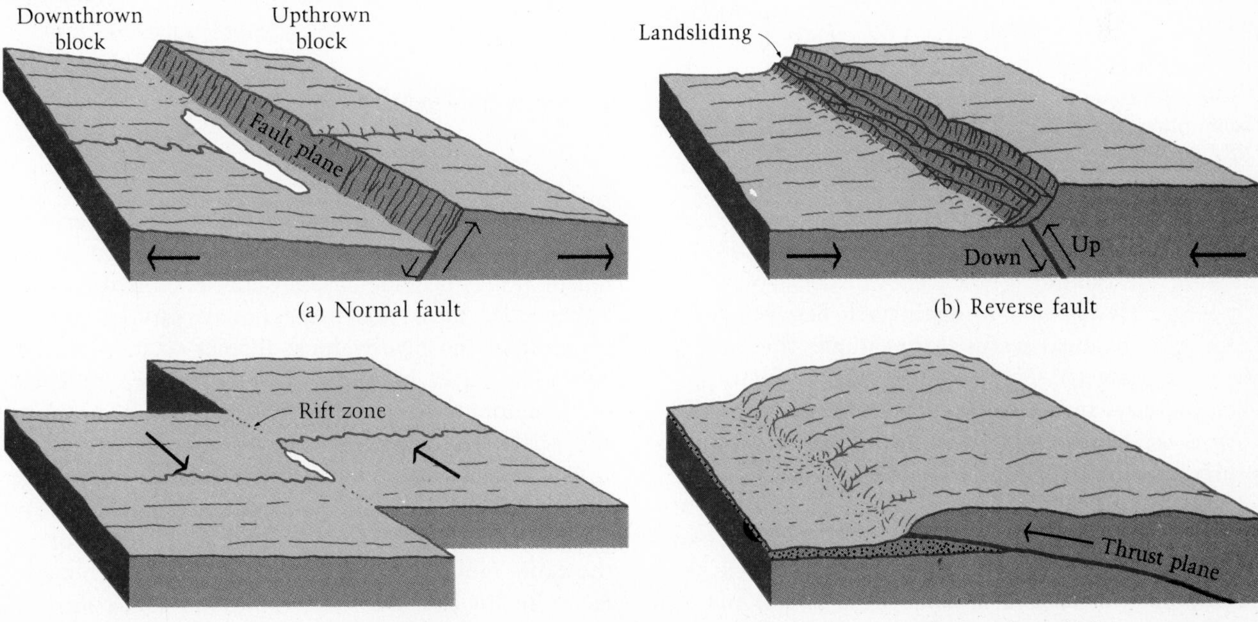

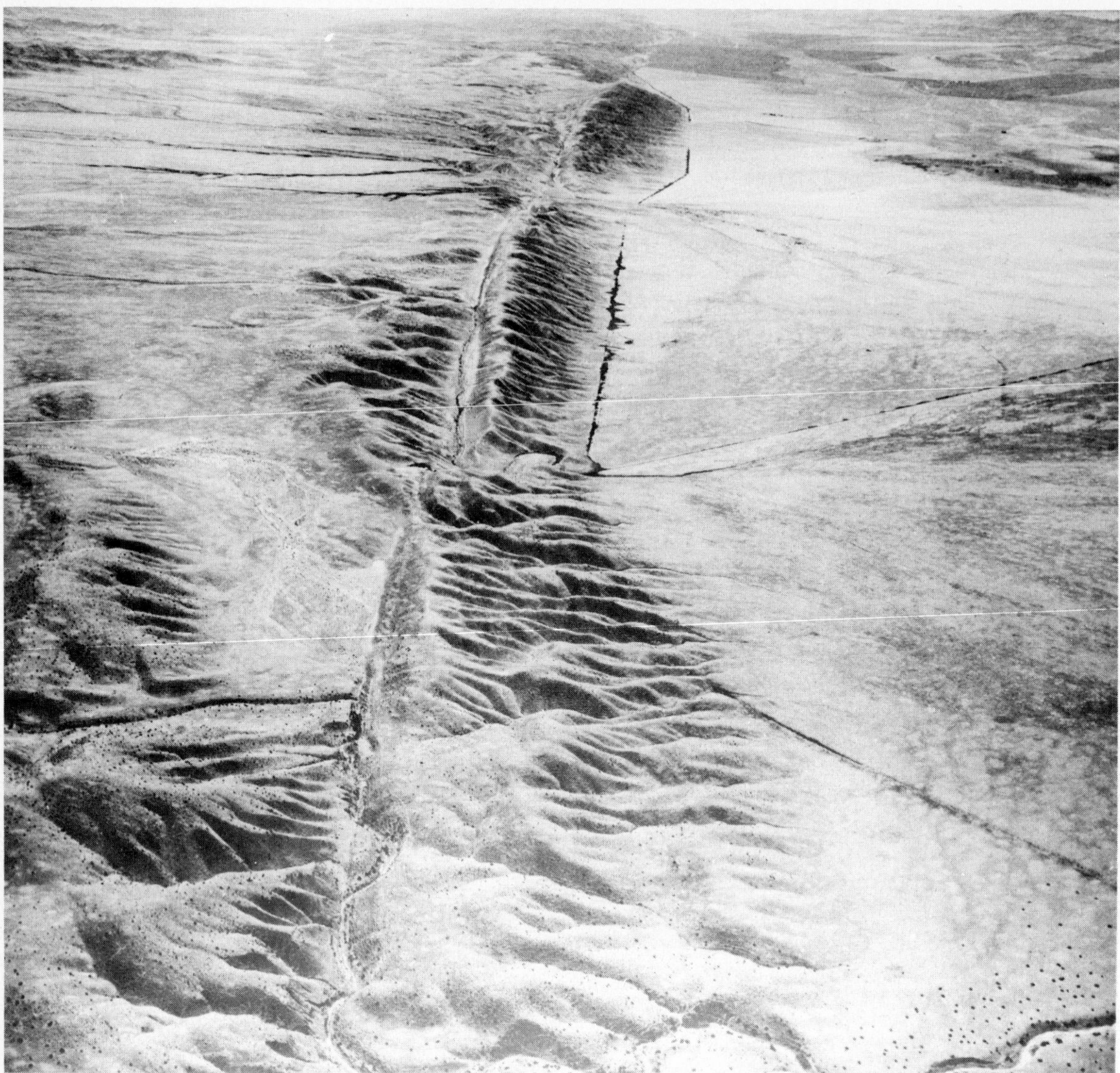

FIGURE 16-11

The San Andreas Fault, a large strike-slip fault. The Pacific plate (right) is moving northward relative to the American Plate (left). Displaced streams can be seen.

faulted origin is wiped out. Sometimes parts of the escarpment are worn away, but pieces of the upthrown block are left standing. These pieces are called *outliers*.

It is not always easy to distinguish between fault scarps and erosional scarps (for example, the eroded edge of a plateau). There is little doubt of faulting when there are springs and alluvial fans and when the scarp is as straight as a ruler. But there may be no springs, no alluvial fans, no obvious straightness.

One way to determine an escarpment's origin is to search for breccia, sure evidence of faulting. When two blocks move along a fault plane, pieces of rock are cracked off and shattered and crushed. Several meters of such broken material may lie in the fault plane after repeated movements along the fault. This material is compacted and cemented through weak metamorphism. It may become harder than the faulted rock on either side. Thus this breccia may survive erosion better than the nearby rock. It may stand in a line, rather like a dike, long after erosion has leveled the rest of the faulted area. Evidence of breccia near a scarp supports a faulted origin.

Another bit of evidence sometimes shows in the actual face of the fault scarp. Rock surfaces may become so smoothed by the sandpaperlike action of the fault movement that they resemble mirror surfaces. In this case the surfaces are known as *slickensides*.

FIGURE 16-12

Aerial view of alluvial fans along the north slope of the Avawatz Mountains in the Mojave Desert.

The discussion of eastern Africa's rift valleys briefly mentioned that sections of crust there had collapsed or were depressed between parallel sets of faults, creating a pair of fault scarps facing each other. This can result from tension (two normal faults) or compression (two overthrust or reverse faults). The German term *graben* is frequently used for such a rift valley. When a segment of crust is squeezed up between two reverse faults, a *block mountain*, or *horst*, is created. Figure 16-13 depicts both a graben and a horst. Grabens are not unique to eastern Africa; for example, the valley of the Rhine River in Germany is a graben. Block mountains are also quite common, as in the western United States.

GAPS IN TIME

So far this chapter has dealt with the things that can happen to rock masses in the crust as a result of different kinds of forces. We have looked at these phenomena as though they are always clear and obvious from the landscape. But we must be prepared to interpret landscapes of the past as well, landscapes buried deeply below layers of rock. Eventually prominent fault scarps are leveled by erosion, rift valleys are filled in and buried, active faults become dormant. If we can interpret past landscapes—"exhume" and reconstruct them—we can decipher the history of the landmasses.

FIGURE 16-13

Relative movements of crustal blocks in grabens and horsts.

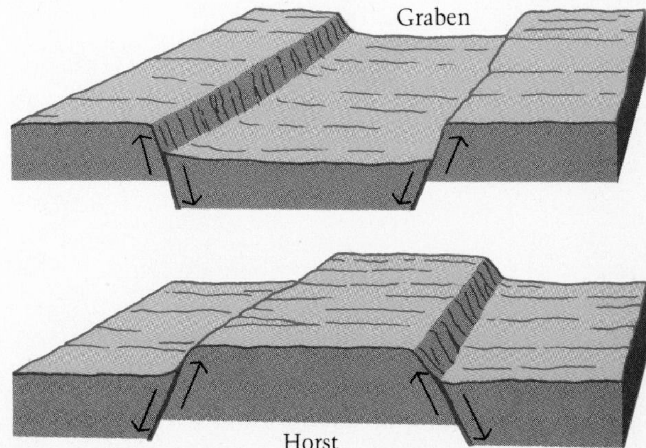

UNCONFORMITIES

When tilting, warping, or folding affects the rocks of a region, erosion soon begins to lower the ridges and other topographic prominences. If the region is unaffected by major deformation of the crust for a time, erosion may succeed in completely wiping out the relief induced by geological forces. The result may be a practically featureless plain that hides all the complex structures of geology below.

In the next phase of the region's history, several layers of sediment may accumulate on top of that erosion-created plain. Obviously these new layers will not lie at the same angle as those rocks below the erosional plain. They do not conform to the structures below, and therefore there is an *unconformity* between the new layers and what lies underneath. Figure 16-14 provides an example. An unconformity in a landscape reveals a lengthy interruption in the deformation of the crust and the accumulation of rocks, a time for erosion and denudation and a leveling off of the surface. This leveling off must be accompanied by deposition somewhere else. Material eroded from one surface is deposited on another. An unconformity, therefore, is a chapter in the history of a region. It is not really a gap in time but certainly represents a missing part of the story of the rocks.

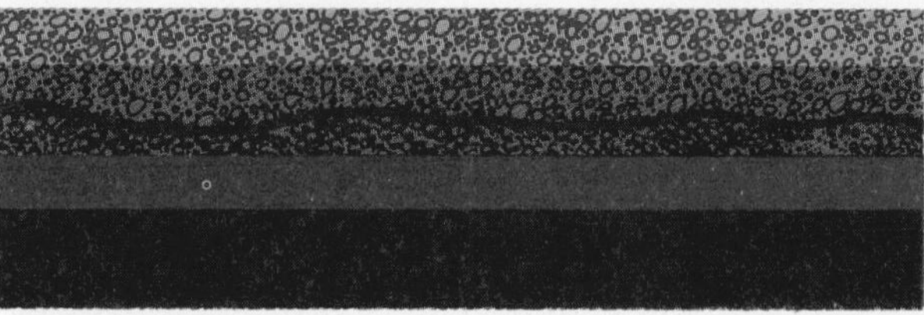

FIGURE 16-15
A disconformity.

DISCONFORMITIES

An interruption in the sequence of accumulation and other geological changes need not involve deformation. A sedimentary sequence may simply be interrupted, perhaps by sinking or rising of the land. Erosion can then attack the most recent layer. The

FIGURE 16-14
Yakataga District, Alaska. The canyon wall displays an unconformity.

erosion produces a certain relief (if the area was uplifted, the relief may be considerable), and then deposition resumes. After the hiatus, new layers are deposited at the same level as the rocks that accumulated before it. Thus there is no angular difference between them as Figure 16-15 demonstrates. Such a hiatus without any deformation is a *disconformity*. Examples of both uncomformities and disconformities exist in the exposed walls of the Grand Canyon.

SUMMARY

The earth's lithosphere acts as if it were floating on the plastic asthenosphere. Because of isostatic adjustment, parts of the earth's crust sink when additional weight is imposed and rise when the weight is removed. And mountains, made of sialic material maintain a balance with their sialic roots.

The process of building mountains is called orogeny. It often occurs at convergent plate boundaries and in areas of geosynclines. It also occurs at divergent plate boundaries, such as the midocean ridges. During orogeny, rock strata are subject to many kinds of folding and faulting. When only minimal deformation occurs, epeirogeny is said to take place.

Two important features of erosion are unconformities and disconformities. They represent a part of the earth's history that is lost to us.

QUESTIONS

1. The basic relationship between the mass of any two objects, their gravitational attraction, has been understood since the time of Isaac Newton. When George Everest and a party of surveyors in northern India found that a plumb line they were using to sight stars with was somewhat inaccurate in northern India, but less inaccurate in southern India, they hypothesized that the gravitational attraction of the Himalayas was deflecting the plumb line. On the basis of calculations, they expected greater deflection than was observed. What characteristic of the mountain range, coupled with gravitational attraction, caused the deflection of the plumb line to be less than expected?

2. What part does the concept of specific gravity play in the development of Airy's hypothesis of mountain roots, depicted in Figure 16-1?

3. An important change in the conceptualization of the form of the crust and its properties is implied in Airy's hypothesis about the floating mountains with their deep roots extending into the denser material beneath them. What was this implied assumption underlying Airy's theory, and how has it been changed with the development of plate tectonic theory?

4. How is the principle of isostasy an example of a mechanism that maintains various lithospheric systems in dynamic equilibrium?

5. An anomaly that puzzled scientists for many years was the evidence indicating that most of the igneous rocks exposed at the surface of the earth were actually formed far beneath the earth's surface. Large crystals, in granite, for instance, indicated that it cooled extremely slowly, which could only occur far below the earth surface, insulated from the colder surface temperatures. How is this anomaly now explained?

6. If the effects of isostatic adjustment were not operative, sedimentary layers could not be deeper than the oceans in which they formed. However, in the Bahamas, sedimentary layers up to 5500 m thick have been deposited, even though the animals that deposited them could survive only in very shallow waters. What can you infer from this information about the rate at which isostatic changes occur?

7. How do the theory of plate tectonics and the principle of isostasy combine to help explain landforms at the margins and boundaries of the earth's plates? Will the landforms produced by plate collisions automatically be subject to isostatic equilibrium? Why or why not?

8. How does isostatic adjustment maintain mountain ranges as prominent surface features long past the time in which erosional forces would otherwise have flattened them completely, approximately within 50 million years? Can isostatic adjustment maintain mountains as surface features indefinitely? What are the limiting factors, if any?

9. Why does isostatic adjustment function somewhat differently in areas of low relief, such as plains, than it does in areas of great relief, such as mountain ranges? What does this imply about the formation and maintenance of regional landscapes of plain areas?

10. In coastal parts of Norway, metal rings placed in rocks 500 years ago to tie up boats are now much too high above sea level to be of use. Before concluding that sea level has dropped by some four or five meters, what factors connected with the geological history of this area and the concept of isostasy should we consider?

11. Major periods of mountain building are known as orogenies. What mountain chains in North America

are products of the latest orogeny? Which chain is the product of an earlier orogeny? What indicates the relative age of the orogeny to which a mountain chain belongs?

12. Periods of mountain building tend to begin with the deformation of layers of sedimentary rock by compression. How is this related to the theory of plate tectonics and the location of geosynclines?

13. If sedimentary rocks in geosynclines at the margins of the continental plates begin the folding activity that usually starts orogenies, why do mountain chains generally have structures that include sizeable amounts of igneous and metamorphic rocks in their cores?

14. Because of the many different shapes layers of sedimentary rock assume when subjected to the severe compressive stresses associated with converging plates, we have a wide variety of names to express the relationships found within the compressed layers. In what direction does the axial line of the Sierra Nevada run? Of the Himalayas? Why?

15. Assuming that a series of ridges is produced by folding, can a particular slope be identified positively as either anticlinal or synclinal? Why or why not?

16. As a geographer examining a mountainous region of great relief and rough terrain, what clues would you seek to determine whether the forces that formed the mountains were related to orogenic activity or epeirogenic activity?

17. Normal faults are probably the most easily identified of the four major fault types. What are the distinguishing features of the normal fault, and how is its appearance different from the reverse fault? What causes this difference?

18. The major fault types can be related to particular plate boundaries. To which type of plate boundary would you expect each type of fault to correspond, and why?

19. Another major type of deformation of the earth's crust is the horst and graben landscape. How are the two features related, and with what type of faulting is each associated?

20. Unconformities and disconformities represent gaps in the continued development of a landscape or changes in that development. To what major geologic force might the presence of many disconformities in sedimentary sequences of rock be related?

WATER
IN THE
LITHOSPHERE

CHAPTER 17

Palm Springs, California; Idaho Falls, Idaho; Grand Rapids, Michigan; Salt Lake City, Utah—all have one thing in common with hundreds of other places in the United States. Their names express, in some way, water's effects on the surface of the land. Place names concerning water are common throughout the world. In England, the town of Winterbourne has taken as its name words that mean "a stream that flows in winter." The Maori name for South Island, New Zealand is *Te Wai Pounamu,* "the water of jade."

Water is without doubt the most widespread sculpturing force on the earth's surface. As such it will sooner or later affect all our lives. Yet water has many more-direct effects. For example, aspects of water flow must be taken into account whenever a farmer plows a field on sloping terrain or whenever an engineer builds a dam. Thousands of dollars are spent annually in legal fees for controversies over water rights. In short, the history of human settlement is in no small part the history of the search for, and use of, water.

This chapter focuses on the part of the hydrosphere found in the lithosphere, just as Chapter 4 focuses on the part of the hydrosphere found in the atmosphere. We will examine water at the surface of the earth, before it reaches rivers. We will look at the flow of water within rivers, and at floods. Then we will move our attention underground to discover the significance of groundwater. Finally we will consider the relationship between water in the lithosphere and the scope of human activity.

FIGURE 17-1
Raindrops intercepted by leaves.

WATER AT THE SURFACE

Our inquiry starts with a raindrop that arrives at the earth's surface. What happens to it depends on the nature and the state of the surface. In some cases raindrops never actually reach the ground. Drops fall on vegetation and evaporate before they can penetrate the soil, or they run off across the surface. As in Figure 17-1, *interception* occurs. The amount of water intercepted by vegetation depends on the vegetation's structure. In Australia, for example, eucalyptus trees intercept only 2% to 3% of the rain. But California hemlock and Douglas fir, with a different structure, intercept 40% of the rain.

If the rain water reaches the ground, it is either absorbed or repelled by the surface. Such surfaces as the concrete road or the leaf of a plant that do not permit water to pass through them are said to be *impermeable.* Other surfaces, such as newly plowed soil, absorb water that falls on them and are considered *permeable.* Concrete and similar surfaces are always impermeable, but soil may be permeable or impermeable, depending on preexisting conditions.

Water flow from the earth's surface through its pores and openings into the soil mass, called *infiltration,* is measured in millimeters per hour. The infiltration rate depends on several factors—the physical characteristics of the soil, how much moisture is already in the soil, the vegetation cover, the slope of the surface, and the nature of the rainfall. Usually water can infiltrate much faster into coarse, sandy soils than into clay soils. Soil structure and the closeness of the soil particles are also important. In the Emme Valley in Switzerland, forest soil absorbs 100 mm (4 in) of water in two minutes, but a pasture where cattle graze takes in the same amount of water in three hours. A soil that is already wet can absorb less water than a dry soil can because the soil surface becomes packed by the rain. The soil swells, closing small openings. Fine colloidal materials wash into the surface openings, and the air pores become filled with water and cannot accept any more. Infiltration rates are usually highest in vegetated areas because the vegetation prevents raindrops from compacting the soil. Furthermore, organic litter provides a home for burrowing animals, who turn over the soil. The type of vegetation also affects infiltration. The infiltration rate of a bluegrass meadow is markedly larger than that of a tilled cornfield. Figure 17-2, using

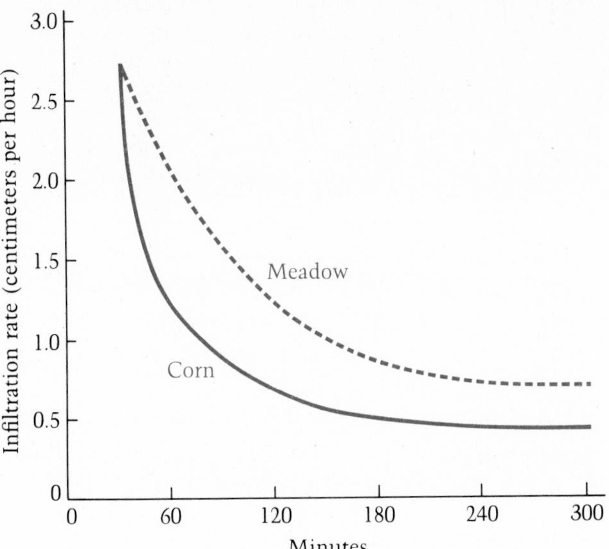

FIGURE 17-2

Infiltration rates for two different vegetated areas. Both show higher rates at the beginning than at the end of the rainfall, but the meadow has higher rates throughout.

this example, shows the typical shape of an infiltration curve. Infiltration rates are high at the start of a rainfall and then decrease with time. In addition, steep slopes drain quickly, before the water can be absorbed. The infiltration rate is therefore likely to be much higher on gentler slopes and flat surfaces.

The characteristics of the rainfall are also critical to the infiltration rate. Rainfall can vary in intensity, duration, and total amount. The *intensity* is the amount of water that falls in a given time. A rainfall intensity of 5 mm (0.2 in) per hour is quite heavy. If it kept up for a *duration* of 10 hours, it would produce a *total amount* of 50 mm (2 in) of rain.

If the rainstorm has a great enough intensity and duration, the soil eventually stops accepting water. It then starts acting as an impermeable surface. Water begins to accumulate into small puddles and pools. It collects in any hollow in a rough ground surface, detained behind millions of little natural dams. This situation is called *surface detention.*

When more rain causes the small detention hollows to overflow, surface *runoff,* or *overland flow,* begins. After a short period of detention, any rainfall that does not infiltrate the soil runs off it. Thus infiltration rates decrease, and runoff rates increase. Figure 17-3 shows this situation in a well-managed and a poorly managed pasture. These graphs also show the wisdom of sound land-management techniques. If the water is needed for plant growth, then the less that runs off the better. Because infiltration and runoff rates are inversely related, runoff depends on all the factors that affect infiltration. In addition, there are a few more factors that affect runoff.

Once surface runoff has started, the water continues to flow until it reaches a stream or river, or an area of permeable soil or rock. An area of land on a slope receives all the water that runs off the higher locations. The longer the total flow path, assuming that the runoff flows across uniform material, the greater the amount of runoff across the area. Therefore, the largest

FIGURE 17-3

Runoff and infiltration rates for two storms over well-managed and poorly-managed pasture.

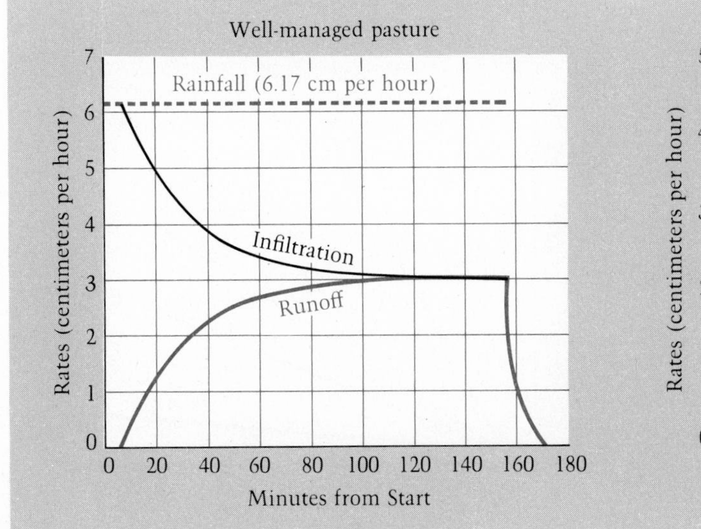

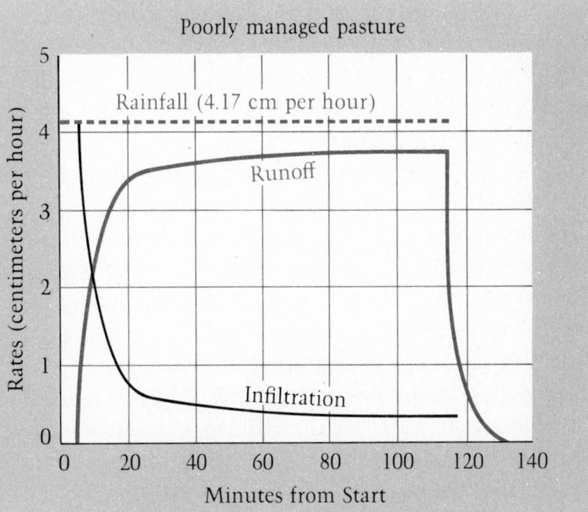

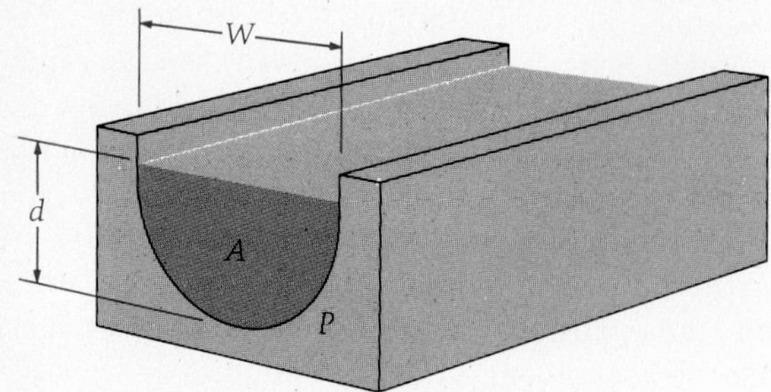

FIGURE 17-4
Common measurements pertaining to a river channel.

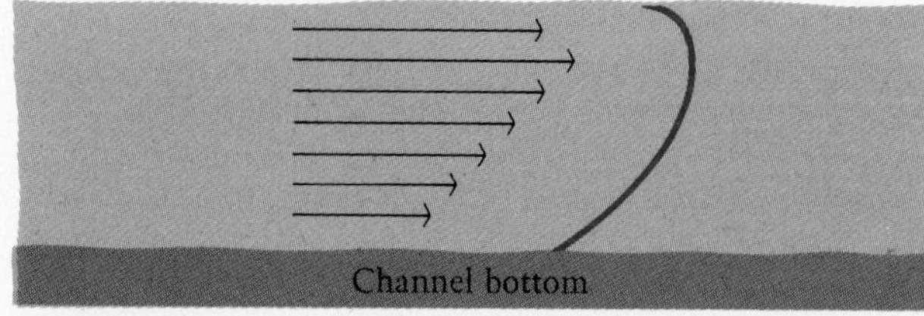

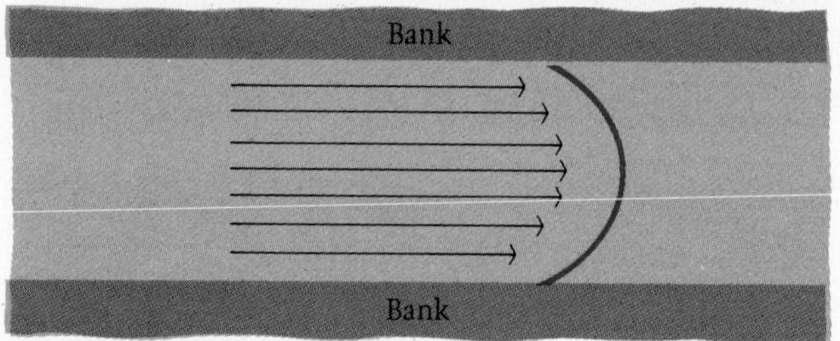

FIGURE 17-5
Velocity variation in a river.

runoff rates occur at the base of a slope, just before the runoff enters a river.

WATER FLOW IN RIVERS

Looking down at a flowing river holds the same kind of fascination as looking into the flames of a fire. The physical geographer's interest in river flow is less romantic but just as fascinated. A study of river flow helps him or her to make predictions about such items as pollution and floods. From the smallest flow in a tiny rivulet to the largest flow in the Mississippi River, certain general rules apply. All these rules have been found by observation. But so far, the theoretical reasons for only some of them have been discovered.

MEASURING A RIVER CHANNEL

A river channel is a trough in the earth's surface, usually occupied by running water. In Figure 17-4 the stream width *(W)* is the width of the water in the channel, and the depth *(D)* of the river is the distance from the water surface to the channel bottom. Scientists usually refer to the depth of a river at its deepest point. The *cross section* of the water *(A)* is the flat surface perpendicular to the river's flow. The edge where the water and channel meet is called the *wetted perimeter (P)*. The *stream gradient*, also called slope *(S)*, is the difference in elevation between two points on the bottom of a channel divided by the horizontal distance between them. These factors vary widely from river to river. A mountain stream might drop a meter within a meter, yet the Amazon River falls only about 6 m (20 ft) in its last 805 km (500 mi).

Using these measurements, *hydrologists*, scientists who work with water in the lithosphere, can derive two other important properties of the stream or river flow. If the cross-section is divided by the wetted perimeter, the result is a quantity called the *hydraulic radius (R)*. The *mean velocity (V)* of the water flow can be calculated if the hydraulic radius and the slope of the river are known. It is important to be able to calculate the mean velocity, because the actual velocity at any one point in the stream is rarely representative. As Figure 17-5 demonstrates, the actual velocity is fastest toward the center of the channel and just below the surface of the water.

If the stream velocity is multiplied by the cross-section of water flow, a parameter called *discharge* is obtained. The average discharge of a river and its extremes are the values a hydrologist most often needs

FIGURE 17-6
Changes in river width, depth, velocity, and sediment load with increasing discharge.

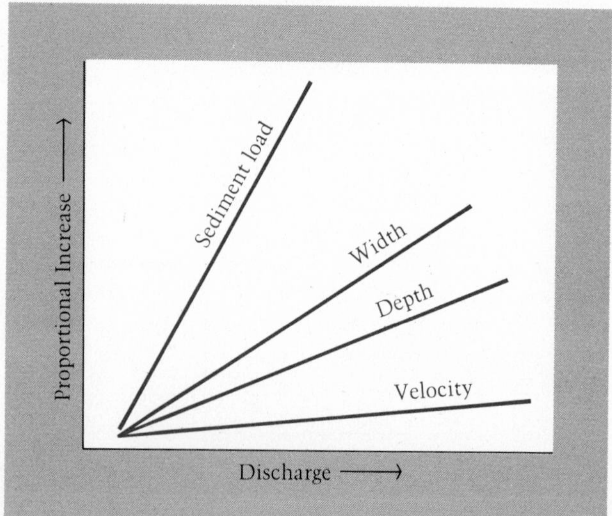

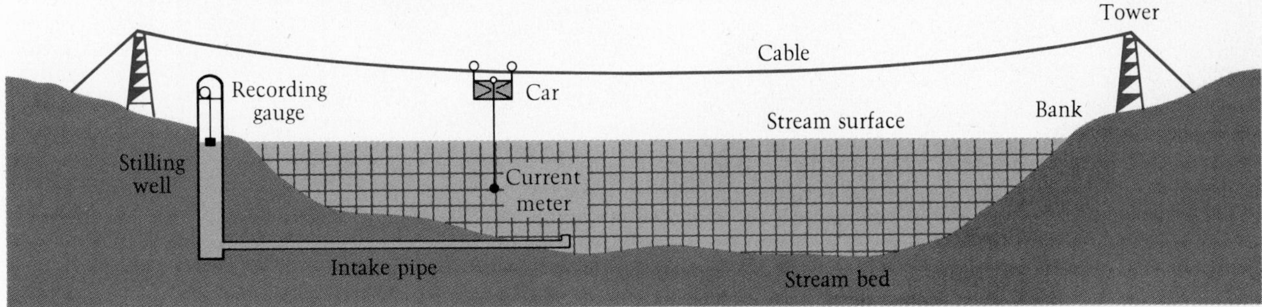

FIGURE 17-7
A stream-gauging installation.

to know. These are the properties that determine the size of both dams and floods.

RELATIONSHIPS IN A RIVER

Many scientists have spent a great deal of time examining the width, depth, and velocity of different rivers and different parts of the same rivers. Over a wide range of conditions, they found that width, depth, velocity, and the amount of material held in suspension in the river all increase when the discharge increases. As Figure 17-6 shows, width, depth, and velocity increase somewhat more slowly than the increase of suspended sediments (particles small enough to be kept in the water by its turbulent flow). These relationships hold true whether the discharge increases because of a storm or because the river becomes larger toward its mouth. Some of the relationships are surprising if you think about them. Who would have thought a sparkling mountain torrent would have a lower velocity than the placid lower sections of the Mississippi? Because mountain streams often flow in circular eddies, their backward movement is almost as great as their forward movement. These and similar characteristics make it hard to measure the actual velocity and discharge of a river.

MEASURING RIVER FLOW

In many cases the hydraulic radius of a river is not known or is only roughly approximated, so the mean velocity and discharge cannot be calculated accurately. The most accurate way of discovering their value is by direct measurement. This is achieved with an instrument called a *current meter* that measures the velocity of water flowing past it. The meter is lowered into the river at a large number of evenly spaced points on an imaginary grid, as suggested in Figure 17-7. The average velocity can be calculated from all the velocity measurements. The cross-section of the river can also be measured, and the discharge can then be computed. Measurements of this kind, called *stream gauging*, are usually performed from a bridge or cable car across the river.

Another frequent feature of a gauging site is a *stilling well*, a cylinder placed vertically in or near the river (see Figure 17-8). The water level inside the cylinder is held still, so the level of the surrounding river may be continuously checked against it. If enough current-meter measurements are made with the river at different levels, then a graph can be drawn comparing river discharge and river level. Such a graph, called a *rating curve*, appears in Figure 17-9. Once the rating curve has been established, it can be used to determine the amount of discharge directly from measurements of the river level. Rating curves are most reliable where there is a solid rock channel. If the channel bed is moved or eroded in time of flood, a new curve has to be laboriously established.

The discharge of a river has to be known before almost any engineering use of the river can be made. But discharge varies with time. A graph of river discharge against time is called a *hydrograph*, and hydrographs resulting from storms always take on the same characteristic shape. The hydrograph shown in Figure 17-10 for the St. Mary River at Stillwater, Nova Scotia, is typical. Before the storm, at day 1, the river discharge was composed of *base flow*, water derived from supplies underneath the ground. On the second day, a rainstorm yielding about 9.6 cm (3.8 in) of precipitation produced a great deal of surface runoff, reflected in the *rising limb* of the hydrograph. The greatest discharge in the river was not felt until the third day, when the peak, or *crest segment*, was experienced. From the fourth day on, discharge levels declined as indicated by the *recession limb*, until only the base flow remained.

Each storm in an area drained by a river creates a *flood wave* represented by the hydrograph. As you can see in Figure 17-10, the variation of discharge over time

FIGURE 17-8
A stilling well for continuous measurement of water level and discharge in a stream coming from the Peyto glacier in Canada.

looks like a wave. The magnitude of the wave varies with the intensity of the storm, its location in the drainage basin, and the preexisting soil moisture.

The hydrograph incorporates several inputs of water, including direct precipitation onto the river and surface runoff. In addition, water that reaches a shallow impermeable soil layer flows horizontally until it too joins the surface runoff. Such water flow is called *subsurface flow.* The final water input shown in the total discharge is groundwater.

FIGURE 17-9
Rating curve for the Churchill River above Granville Falls in Canada.

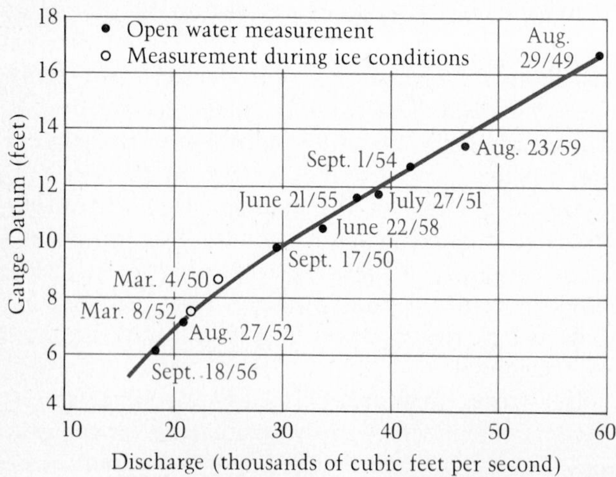

The rising limb and crest of the hydrograph are governed by the nature of the storm and the permeability of the river basin. If the basin tends to be impermeable, the hydrograph rises and falls relatively quickly. The stream that produces such a hydrograph is called *flashy.* The recession limb of the hydrograph shows the presence of subsurface runoff and the withdrawal of water from stream channels, stream banks, and underground reservoirs.

Hydrographs for individual storms describe the nature of both the storm and of the drainage basin. It is possible, with appropriate calculations, to derive a hydrograph that represents only the characteristics of the drainage basin. A hydrograph of this kind is called a *unit hydrograph.* These are valuable for engineers, who use them to determine the magnitude of possible floods from storms of various sizes, to estimate peak discharges, and sometimes to estimate the effect of snow melt.

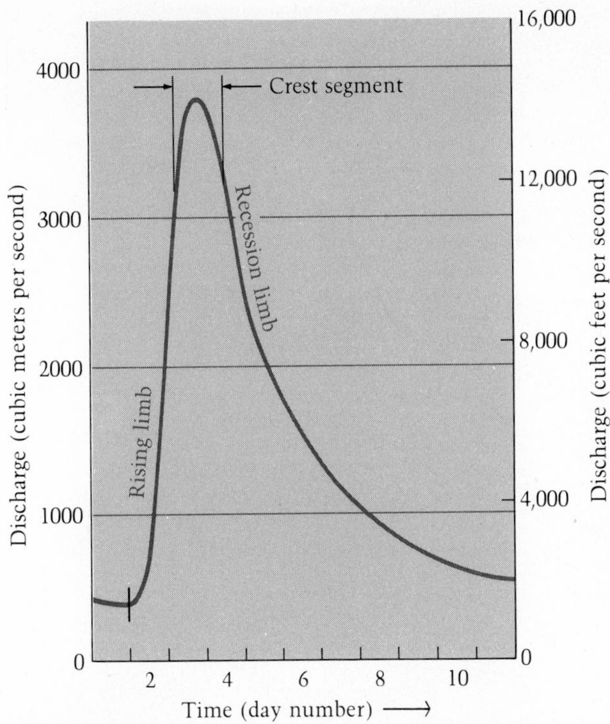

FIGURE 17-10

A hydrograph for the St. Mary River, Nova Scotia, Canada, resulting from an individual storm.

FLOODS

It is one thing for the hydrologist to be able to predict the size of floods; it is another for the landscape and people living on it to experience their effects. Chapter 19 examines the effects on the landscape, such as the great erosive power of flood waters. At this point let us investigate the impact of floods on humans.

About 12% of the United States population lives in areas of potential flooding. These people walk a thin line between the advantages and potential risks of living in a flood plain. The advantages include the fertile soils and flatness of the land that make building and development easy. The risks are obvious (see Figure 17-11). Recent flood losses in the United States have exceeded $1 billion per year. In 1976 over 60 lives were lost when the Big Thompson River flooded in Colorado. The decision that people make to live or not live on a flood plain is not strictly rational. The geographers Gilbert White and Ian Burton have shown that human adjustments to the known danger of flooding do not increase consistently as the risk becomes greater. People tend to make an optimistic rationalization for continuing to live in a potential flood area.

There are other decisions and trade-offs associated

FIGURE 17-11

a) View of Vanport, Oregon during the Columbia River flood of 1948.

FIGURE 17-11
b) Pepperwood, California on January 1, 1965. This small town on the Eel River in northern California was completely destroyed by flood.

with flood plain settlement and development. Floods are natural events, but we sometimes try unnatural methods to prevent their worst effects. Dams and high artificial banks or levees like those in Figure 17-12 can be built. But these artificial structures sometimes have some unwanted effects. Dams, although they help control flooding, can prevent the natural replacement of fertile alluvial soil. The sediment collects behind the dam instead. The Aswam Dam in Egypt holds back sediment that would otherwise be deposited in the agriculturally vital lower Nile area. Furthermore, artificial levees in flood plains can create a false sense of security. The levees are seldom designed to cope with the greatest possible flood, because this would be too expensive. But often the higher the levee, the more people who will live in the area and the greater the disaster when a major flood does occur.

Therefore, it is essential for us to have an understanding of flood plains, as well as other natural phenomena of the earth. Otherwise we cannot make rational decisions about where to live or the adjustments we should make in living there. One day you may have to vote on issues such as these. We will now look at the role of water within the lithosphere.

WATER BENEATH THE SURFACE

Despite the beauty and importance of rivers, they contain only 0.03% of all the fresh water in the world. Twice as much is stored as soil moisture, and ten times as much is held in lakes. By far the greatest proportion of the world's fresh water (75%) is locked in glaciers and ice sheets. But about a quarter of the total fresh water is available under certain conditions. This water is hidden beneath the ground, within the lithosphere, and is called groundwater.

A raindrop that falls to the earth may remain above ground as we have seen. But it may also sink into the soil. Two zones within the ground may hold this water. The top zone, usually unsaturated except at times of heavy rain, is called the *zone of aeration* or the *vadose zone*. Below this is the *zone of saturation*, sometimes called the *phreatic zone*. The raindrop infiltrates into the vadose zone first.

SOIL MOISTURE IN THE VADOSE ZONE

Once rain water has infiltrated into the soil, it is called *soil moisture*. Any further movement is by processes

FIGURE 17-12
Luling Levee was built to contain the Mississippi River in Louisiana. The white strip marks the high point of the levee.

other than infiltration. The downward movement of water through the pores and spaces in the soil under the influence of gravity, called *percolation*, is the most common method of water movement in a soil. In a contrary motion, described in Chapter 10, water moves upwards, like liquid in a straw, through capillary movement. Moisture can also move around in the soil through evaporation, movement of the vapor, and recondensation onto new surfaces.

The texture and structure of the soil determines the amount of water it can hold. *Field capacity* is the most water a soil can possibly contain without becoming waterlogged. Theoretically soil moisture can fall to zero. But it takes a large amount of drying to reach this limit, because a thin film of water clings tenaciously to most soil particles. This *hygroscopic* water is unavailable to plant roots.

A practical lower limit to soil moisture is termed the *wilting point*, and it varies for different soils and different crops. Below this point, a crop sustains permanent injury by drying out. The total *soil storage capacity* for agricultural purposes is the product of the average depth in centimeters to which roots grow and the water storage per centimeter for that soil type.

GROUNDWATER IN THE PHREATIC ZONE

If you dig a hole far enough into the earth, chances are that sooner or later you will come to a layer that is permanently saturated with water. This is the top of the phreatic zone, a surface called the *water table*. Instead of lying horizontally, the water table tends to follow the outline of the land surface as Figure 17-13 indicates. Where it intersects the surface, a spring, stream, river, or lake is exposed. Figure 17-13 shows the location of a stream in relation to the water table.

The material of the lithosphere can be classified according to its water-holding properties. Porous and permeable material that can be at least partially saturated is called an *aquifer*. Sandstone and limestone are good examples of aquifers. Other rocks, such as mudstone and shale, are usually quite impermeable and resist groundwater. These are called *aquicludes*. You might imagine aquifers and aquicludes in the earth as a mixture of spongelike material set amid plastic. An *unconfined aquifer* obtains its water from local infiltration, as if the sponge were at the surface. But a *confined aquifer* exists between aquicludes, like a sponge between two layers of plastic. It often obtains its water

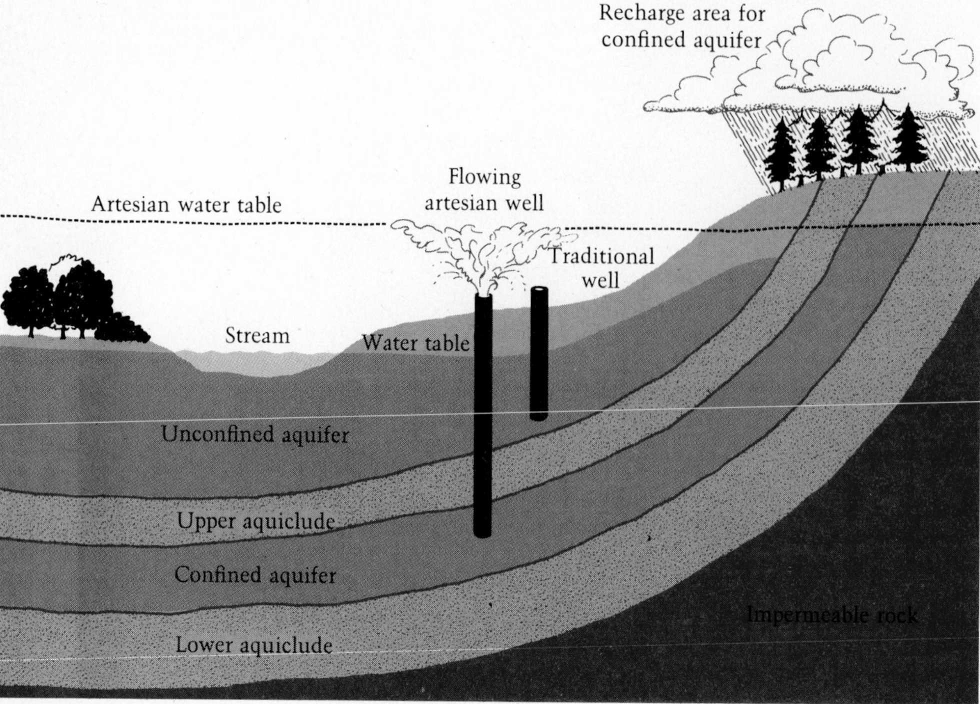

FIGURE 17-13

Aquifers, aquicludes, and their relationship to wells and the water table.

from a distant area where the rock layer of the aquifer eventually reaches the surface as in Figure 17-13.

WELLS AND SPRINGS

In many parts of the world, settlements are located near wells and springs. In this case geology directly influences human decisions. Wells can be dug wherever an aquifer lies below the surface. In most cases springs are formed when an aquifer and an aquiclude interact.

There are two kinds of wells—the traditional and the artesian. The traditional well is simply a hole in the ground that penetrates the water table. Water is then drawn or pumped to the surface. Usually traditional wells are sunk below the average level of the water table. The actual level of the water table may vary because of droughts and storms. So the deeper the well is sunk below the water table, the less chance there is of the well becoming dry when the water table falls.

The French region of Artois was blessed with a confined aquifer whose water supply is recharged from a remote location. Wells sunk into this aquifer produce water that flows under its own natural pressure to the surface. Figure 17-13 diagrams the mechanics. Artois has lent its name to this kind of well, and now *artesian wells* are common in many parts of the world. The pressure in artesian wells can sometimes be quite strong. One dug in Belle Plaine, Iowa, spouted water over 30 m (100 ft) into the air and had to be plugged with 15 wagonloads of rock before it could be brought under control.

Wells sometimes suffer from side effects that limit their use. In some cases the water is withdrawn faster than it can be replaced by water flow through the aquifer. When this happens, the local water table drops, forming a *cone of depression* like that in Figure 17-14. The amount of the drop in the local water table

FIGURE 17-14

The drawdown effect causing a cone of depression around a well.

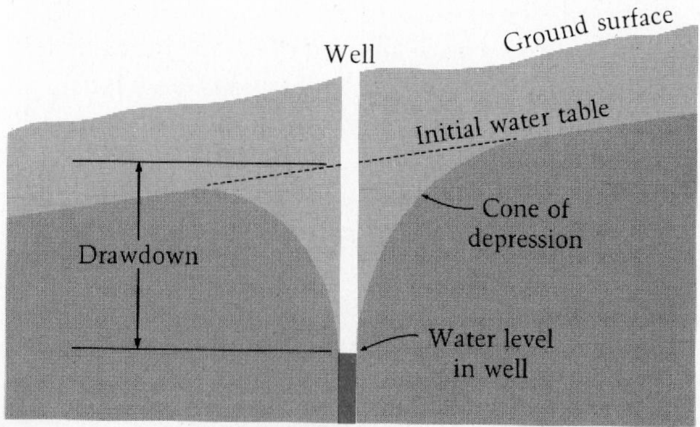

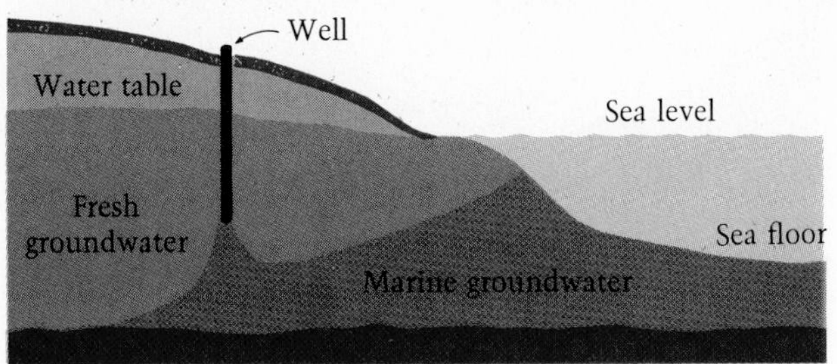

FIGURE 17-15
The contamination of a well by salt water.

is called the *drawdown*. When it drops, energy must be used to bring the water to the surface. In addition, salt water can intrude into a well near a coastline. Excessive pumping gradually moves denser saline sea water into the well as diagramed in Figure 17-15. Wells therefore must be used with care.

A surface stream of flowing water that emerges from the ground is called a *spring*. Springs can be formed in a number of ways. Most commonly an aquiclude stops the downward percolation of water, which is then forced to flow from a hillside, as indicated in Figure 17-16a. Sometimes the aquiclude leads to the formation of a separate water table, called a *perched water table*, at a higher elevation than the main water table. Sometimes water finds its way through joints in otherwise impermeable rocks, such as granite. Where it reaches the surface, springs form. Often faulting rearranges aquifers and aquicludes so that springs form, as shown in Figure 17-16b. And exceptionally high water tables after long periods of heavy rainfall can occasionally raise the water level high enough to cause temporary springs, as in Figure 17-16c.

An interesting case is the formation of *hot springs*. The water from these has an average temperature of above 10°C (50°F) and often comes from a fissure that brings hot magma from great depths, as in Figure 17-16d. Near Rotorua, New Zealand, hot springs are used for cooking and laundry, and steam from them is used for generating electricity. Much of Reykjavik, the capital of Iceland, is heated by water from hot springs. Hot springs containing large quantities of minerals dissolved from the surrounding rocks are sometimes called *mineral springs*, and the mineral water may be used for medicinal purposes. It is very pleasant to sit in the hot water of a mineral spring or bath, but this is one of the least important uses of water.

THE USES OF RAIN

On the average, 76.2 cm (30 in) of rain falls on each square centimeter of the United States every year. The United States covers about 9.5 million km^2 (3.6 million mi^2), so the annual rainfall amounts to a great deal of water: some 60 km^3 (14.4 mi^3). In addition, a much smaller quantity, 0.07 km^3 (0.02 mi^3), is mined from aquifers. Almost half of this water (48%) is used directly or indirectly in some human activity.

Let us call the annual amount of water reaching the surface of the earth in the United States 100 units, as in Figure 17-17. Eventually, 71 units of this evaporate back into the atmosphere, and 29 units go into rivers. More than half of the water that evaporates, 39 units, first serves a useful purpose. Most is used by farm crops and pastures. A significant amount is also used by forests, whose wood is later used, and by the vegetation eaten by such herbivores as cattle and sheep. This water is therefore indirectly associated with human activity. The remaining 32 units evaporate from "non-economic" vegetation, such as that found in deserts.

THE MINERAL SPRINGS AT BATH, ENGLAND

Hot mineral springs occur in many places in the world ranging from Rotorua, New Zealand to the Yellowstone National Park in the United States. Few springs of this kind have as long a history of use for medicinal reasons as those of Bath, England. Most fashionable in the eighteenth and nineteenth centuries, the resort built around the springs saw Lord Nelson, General Wolfe, Beau Nash, Queen Victoria, Charles Dickens, and Jane Austen. Later excavations, for extensions to the buildings around the baths filled by the spring, uncovered Roman remains.

Further excavations showed that the Romans had built on the same site a complex perhaps even more elaborate than that of the eighteenth century master architects. Visitors to the site today can easily imagine the Romans bathing to gain relief from rheumatism caused by the northern winters. There is also evidence that the site was used for the same purpose by the ancient Britons long before the Romans took up residence. Today, public baths still exist, supplied by the same hot springs, and they form part of one of the oldest national health services in the world.

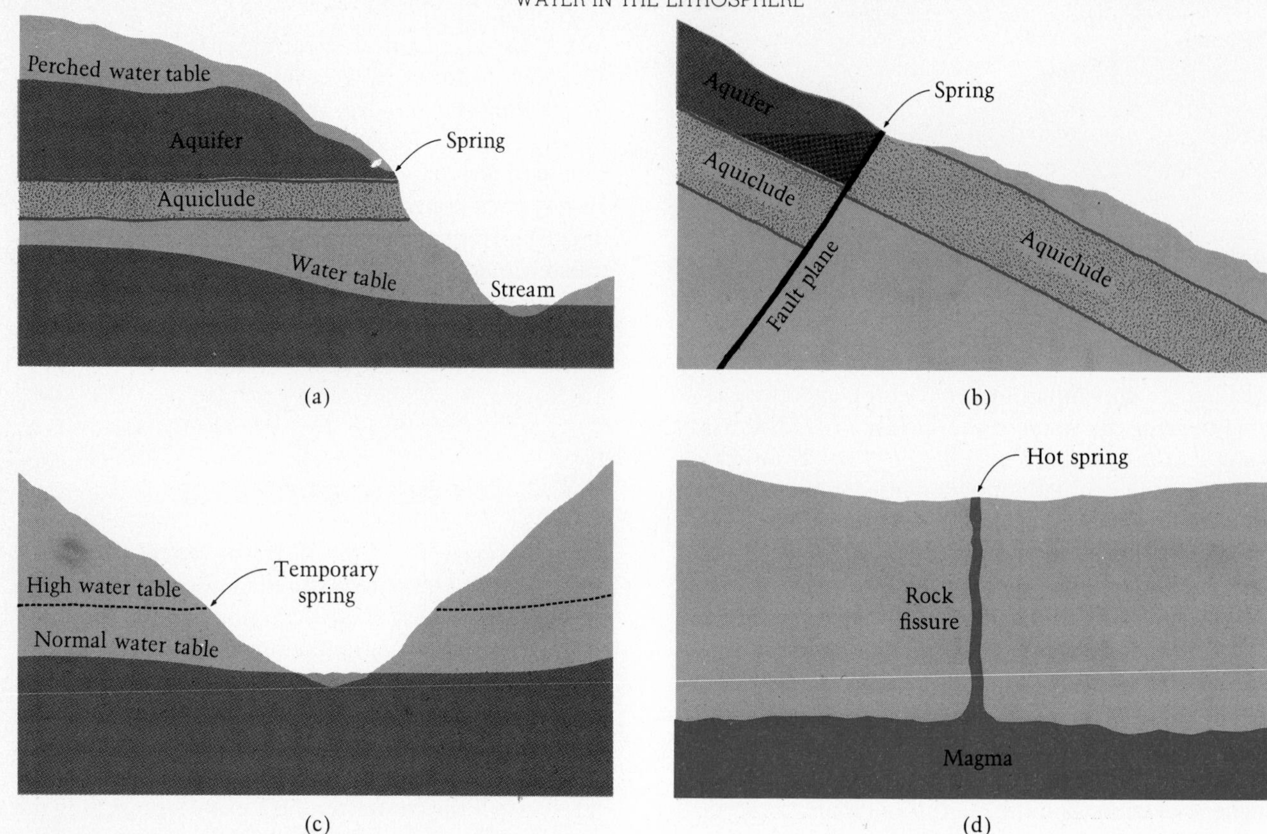

FIGURE 17-16
Conditions leading to the formation of springs.

Of the 29 units of water in rivers, 20 units are not withdrawn by humans. Industry withdraws more than half of the 9 units taken from rivers. Industrial water is used in many ways and at great rates. It takes 760 *l* (198 gal) to produce your Sunday newspaper and 1520 *l* (395 gal) to provide a tankful of gasoline. About three-quarters of the industrial water is used to generate electricity. Other major uses are metal processing and refining, food processing and canning, paper production, and petroleum refining. The amount of water used in industry has risen rapidly in recent years and now accounts for about 5 of the original 100 units of

THE DEATH OF LAKE ERIE

It is a sad fact that there are numerous examples of the failure of human beings to understand the mechanics of ecosystems. One of our more spectacular failures is the case of Lake Erie. A large amount of sewage has been placed in this lake in the present century from the heavily populated area around the shore. The sewage sinks to the bottom of the lake. Here the bacterial activity of the sewage is greatly increased and the respiration of the bacteria uses up oxygen from the water. Fish lay their eggs on the bottom, and the eggs require a minimum oxygen level to survive. The amount of dissolved oxygen available in the bottom water has fallen from 78% of saturation in 1928 to 25% in 1960. The effect on the commercially valuable blue pike has been disastrous. In 1956 the catch of the blue pike was 3.06 million kg (6.8 million lb) valued at $1.3 million. Seven years later the total catch was a mere 90 kg (200 lb) valued at $120. Many other fish such as trout, whitefish, herring, sauger, and walleye have almost vanished and the fishing industry has lost $3 million per year. Apart from the fishing industry, continuous dumping of sewage into the lake is now sometimes causing bacterial counts to go so high that beaches have to be closed. Will we save Lake Erie?

water reaching the surface of the United States.

Agriculture, through irrigation, is another major user, consuming 3 units, and municipal and rural use accounts for approximately another unit. These values may not mean much until you realize that it takes 15,200 *l* (3952 gal) of water to produce a pound of beef or 760,000 *l* (197,600 gal) to grow a ton of alfalfa. Meanwhile, the average city dweller uses 600 *l* (156 gal) of water in the home every day.

These figures may be surprising, but we must remember that they are averages. There was once a statistician who nearly drowned in a river whose average depth was only 10 cm (3.9 in). He could have explained that, although average values are useful, the distribution of the water in time and space is often a matter of life or death. In the United States, 95% of the water used for irrigation is withdrawn west of the Mississippi River—an obvious maldistribution in space. By far the greatest proportion of this water is used between April and September, when rainfall is almost absent—a clear maldistribution in time.

Such unequal distributions cause problems for human activities. The average daily consumption of water per person in the United States rose from 4940 *l* (1300 gal) in 1950 to 6840 *l* (1800 gal) in 1970. It has been estimated that by the year 2000, of all the major drainage basins in the contiguous United States, only the Northeast, the Southeast, and the Ohio Valley will have an assured water supply. Facts like these make it imperative for us to learn more about the role of water in the lithosphere.

SUMMARY

In this chapter we have examined the parts of the hydrosphere found on, and in, the lithosphere, that accompany the earth's landforms. First we looked at what happens to water when it reaches the earth's surface from the atmosphere. The water may be intercepted by vegetation, run off across the land surface, or permeate the soil.

A particular form of runoff is the water flow in rivers. We have discussed how this flow is measured and considered the human significance of high flow—floods.

Water lies beneath the surface in the zone of aeration and the zone of saturation. Groundwater supplies are important to us, and access to them by means of wells and springs is essential.

We do not use all the water that falls as rain, but the half we do use permeates all aspects of our lives. That is why we must begin thinking about the future of water in the lithosphere.

QUESTIONS

1. What main factors affect the permeability of a particular soil? Is it sufficient to know only the soil type and structure? Why or why not?

2. In Figure 17-2 we see the infiltration rates for fields of bluegrass and corn. What causes both curves to

FIGURE 17-17

Uses of water arriving at the land surface of the United States. The annual precipitation value of 55.9 km³ and the 0.07 km³ of water withdrawn annually from aquifers are taken as 100 units.

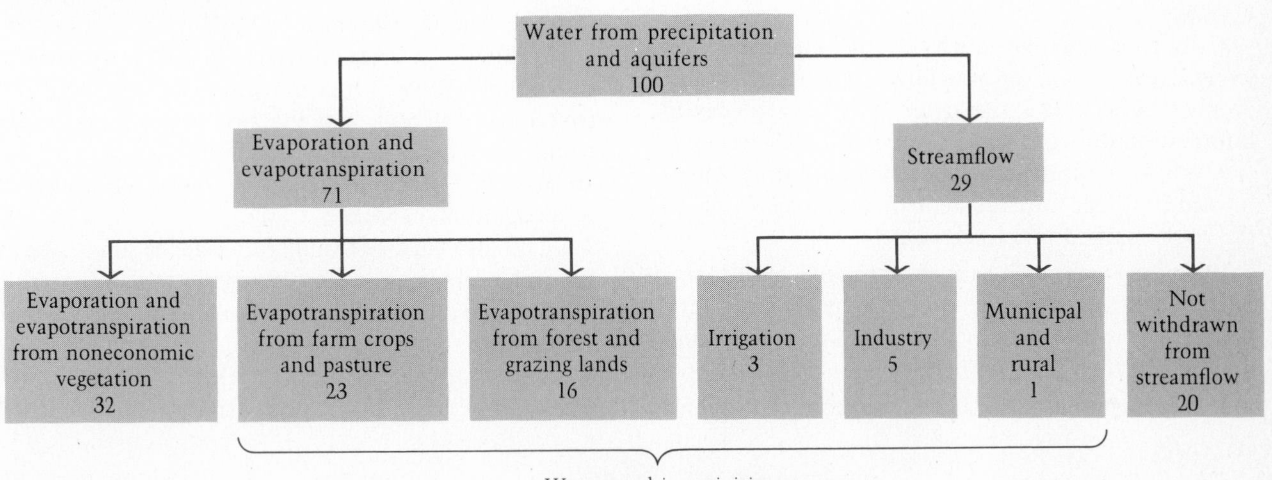

exhibit great changes in rate with increasing time of rainfall?

3. Two fields, identical in permeability, structure, and slope, each receive 5 cm of rain during a given week, even though the fields are geographically far apart. In one field, 1 cm of the water infiltrates the soil. In the other, 4 cm of the water infiltrates the soil. What can we infer about the nature of the rainfall in each location?

4. What is the relationship between infiltration rates and runoff rates for the same surfaces? Assuming that all other factors are held constant, how will an increase in compaction affect the two rates? For crops requiring a great deal of water in the soil around them, what would be sound farming practice given the relationship between compaction of the soil, infiltration, and runoff rates?

5. When we think of the weathering and erosion of areas of great relief, we assume that the highest areas are worn down first. Actually, in terms of total mass of material removed, the areas of lower (though not of *lowest*) elevation tend to be eroded most rapidly. How is this fact connected with the behavior of surface runoff?

6. According to Figure 17-5, where in a river would we expect to find the fastest flowing water? Why?

7. Streams having increases in discharge show increases in velocity, width, depth, and sediment load. Consider a single rainstorm of great intensity but short duration, occurring in an area with low infiltration rates and fairly pronounced relief. What would be its effect on stream velocity and sediment load? What does this imply for areas having small streams during most of the year, but occasional torrential rainfall? In Death Valley, for example, stream-carried debris extends far into the valley, with little sign of stream presence.

8. In stream gauging, why is the current meter lowered into the river at a large number of points, evenly spaced across the river? What two pieces of information do we need to compute the discharge?

9. Why is a stilling well used to measure stream level instead of direct measurement by means of a pole in the river, marked in a linear scale?

10. Stilling wells are relatively expensive monitoring devices, yet they are widely used. Why are they such popular devices among hydrologists, and what information can they provide, besides river level?

11. Hydrographs tell us a great deal about the behavior of streams under varying conditions, as well as the basin characteristics from which the stream receives its water. Assume that the St. Mary River has a basin of average permeability and slope. What can we conclude about a river basin that produced a hydrograph having a less steep rising limb, a higher base flow, and a less steep recession limb from the same storm that gave us the hydrograph in Figure 17-10?

12. A hydrograph with steeper limbs than those in the St. Mary River hydrograph is a bit harder to interpret than one with less steep limbs, because a variety of factors can influence increased runoff. What further information would be needed before the river basin could be characterized in the case of a hydrograph with steeper limbs than that in Figure 17-10? Could the hydrograph itself provide clues to this missing information, and if so, how?

13. Why is the recession limb of almost any hydrograph less steep than the rising limb? What components affecting the recession limb are not present or behave different in the rising limb?

14. Despite the potential for periodic disaster, people continue to live in flood plains, as they have in prehistoric times and throughout recorded history. What are the advantages of flood-plain settlement? Has the technology used to control flooding been totally successful?

15. Water can move through the soil in three ways, each related to different physical forces. What are the three processes of movement of soil moisture, and how do they work?

16. Assume that you are living on the side of a valley, 100 m above the valley floor. Your neighbor, who lives at the bottom of the valley, has just finished digging a well, and he reached water after drilling 85 m into the ground. Since you live 100 m above him, does this mean that you will have to drill 185 m to reach the water you need? Why or why not?

17. How is an artesian well different from a traditional well? What structural differences explain the functional differences of the two well forms? Where would you look for the infiltration source of the artesian well dug at Belle Plaine, Iowa, which had a water spout 30 m high?

18. Intensive use of a well often leads to a lowering of the water table in the vicinity of the well. When this happens, do all wells in the same area suffer? Why or why not?

19. How are the springs in Figure 17-16a, b, and d related to geologic formations within the earth's crust?

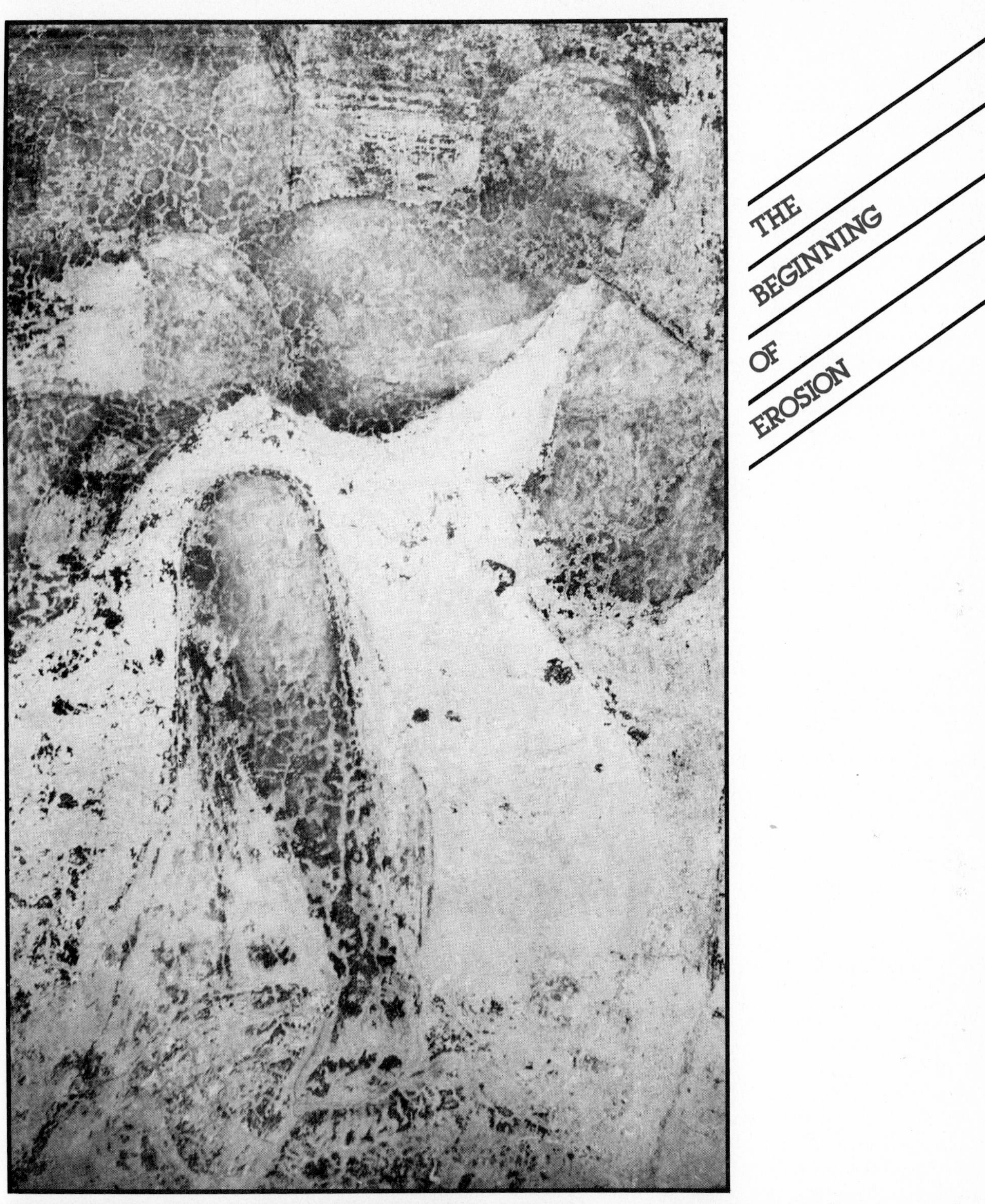

THE BEGINNING OF EROSION

CHAPTER 18

We turn now to the processes that work to wear down and destroy the many forms and shapes produced by geological forces at the surface of the crust. Every escarpment, every anticline, every horst is attacked by these processes even before it is fully formed. Although isostatic uplift keeps rejuvenating mountains that are being reduced, the forces of weathering and erosion eventually gain the upper hand. Soon nothing remains but the roots of ranges that once stood thousands of meters above sea level. Not only does nature have ways to disintegrate even the hardest, most resistant rocks; it even provides the most effective machinery for removal of the waste. Weathering, erosion, and mass movement may be less spectacular processes than those attended by earthquakes and volcanic eruptions, but they ultimately prevail.

In the breakdown of rocks, the most important agent is water. In one way or another, water is involved in the disintegration and decay of rocks almost everywhere on earth; practically all landforms have been sculptured by water at some stage of their development. In streams and rivers, in the form of ice, and in massive ocean waves, water attacks the land and wears it down. Even in deserts, water, not wind, is the most effective agent of change.

Three processes dominate the breakdown and removal of rocks: weathering, mass wasting, and erosion. As we note in detail later, *weathering* involves the decomposition and disintegration of rocks, with little or no movement of the particles produced when rocks break down. Weathering, for example, figures prominently in the formation of soil from rocks. When *mass movement* occurs, gravity is at work. Loose material such as boulders, blocks, and even soil may slide, roll, or flow to new positions. But the loosened material never gets far from where it started. The job of distant removal is done by *erosion*. In the strict meaning of the term, loosened material is carried away, ground into sediment, and eventually redeposited, sometimes thousands of kilometers from its source.

Sometimes the term *erosion*, used in the more general sense, includes not only the grinding up and transport of material but its weathering as well. The word *denudation* can also be used in this wide sense. In this chapter we will deal only with the beginning of erosion in its broad sense and hardly at all with erosion in its specific sense. We will look more closely at the weathering processes, some of which are introduced in Chapter 10. Then we will examine the curious features of limestone scenery that owe much of their origin to a particular form of weathering. Finally, we will discuss mass movement, the process that makes material ready for removal by rivers. But first we will turn to the initial breakdown of the rocks.

WEATHERING

The processes of weathering are continuous everywhere on the surface of the landmasses and even below. These are quiet, unspectacular processes, but their effect is considerable. Rocks exposed to humid air and to alternating daytime heat and nighttime cold inevitably break apart. Even below the surface, weathering continues as rocks decay into fine particles of soil.

There are several different kinds of weathering, with two major categories: *mechanical* weathering and *chemical* weathering. We may also distinguish *biological* weathering, which includes aspects of both mechanical and chemical weathering. This grouping should not mislead us into assuming that mechanical and chemical weathering are mutually exclusive and that if one occurs the other does not. Rather, they usually operate in some combination, with one prevailing as the most effective force. Sometimes both attack rocks at the same time, and then it is difficult to determine just which process has the greater role in the breakdown of the rocks.

MECHANICAL WEATHERING

Mechanical weathering, also called physical weathering, involves the destruction of rocks through the imposition of certain stresses. A prominent example is *frost action*. We are all aware of the power of ice to damage automobile radiators and split open water pipes. Similarly, the water contained by rocks—in cracks, joints, even pores—can freeze into crystals that shatter even the strongest igneous masses. The ice can produce about 1890 metric tons (2100 tons) of pressure

FIGURE 18-1
A talus cone on the west wall of Canyon Creek Valley in Glacier National Park, Montana.

THE DUST BOWL

A dust storm blows up early in the novel, *The Grapes of Wrath*. Author John Steinbeck describes how the wind in the Dust Bowl swept away the soil and the livelihood of a farming community in the 1930s:

"The air and the sky darkened and through them the sun shone redly, and there was a raw sting in the air. During the night the wind raced faster over the land, dug cunningly among the roots of the crops.

"The dawn came, but no day, and the wind cried and whimpered over the fallen corn.

"Men and women huddled in their houses, and they tied handkerchiefs over their noses when they went out, and wore goggles to protect their eyes.

"When the night came again it was black night, for the stars could not pierce the dust to get down. Now the dust was evenly mixed with the air, an emulsion of dust and air. Houses were shut tight, but the dust came in so thinly that it could not be seen in the air, and it settled like pollen on the chairs and tables and dishes. The people brushed it from their shoulders. Little lines of dust lay at the door sills."*

In the story, as in real life, the farmers could not grow enough crops to survive. The "Okies" packed up their families and headed to California.

How did this tragedy occur? After Congress passed the Homestead Act, families poured onto the Great Plains anxious to establish farms of their own. They plowed up the thick protective layer of vegetation to plant crops or put out cattle to graze on the dense grass cover.

The region usually receives sparse rainfall, marginal for agriculture, but for a few years heavier rain fell. The farmers moved quickly to cash in on the unexpected moisture. They plowed more fields to plant wheat or put more cattle out to graze. They gambled that the extra rain would continue and nourish their farms. But they lost their bet.

In the mid-1930s the weather turned dry. The wheat did not survive. The cattle devoured the grass and then went hungry. Farmers abandoned their plots and moved west. Absentee landlords ignored their land and hoped for better times. The land lay exposed without any plant cover to hold down the soil. When the wind began to blow, inches of topsoil were stripped off the earth and whisked away in dust storms. The natural drought conditions were aggravated by the extensive expansion of wheat farming and grazing.

By 1938 the Dust Bowl conditions began to subside. Heavier rainfall returned. The government encouraged farmers to use conservation techniques to protect the soil. Some land was retired from cultivation and planted with cover crops. Shelter belts of trees were planted to inhibit the wind. Less cattle pressured the grassland.

The Dust Bowl is part of our past, or is it? Some farmers have cut down their shelter belts to plant a few more rows of wheat. Ranchers add a few more head of cattle to their herd to increase their meager profits. Speculative farming on the Great Plains, overgrazing, and a few dry years could bring us to the brink of a dust bowl again. Only great care by farmers can prevent it from returning.

*Excerpts from John Steinbeck, *The Grapes of Wrath* (New York: Viking Press, 1939).

for every 0.1 m^2 (1 ft^2). Of course, frost action operates only where winter brings below-freezing temperatures. In high-altitude zones, where extreme cooling and warming alternate, the water that penetrates into the rocks' joint planes (see Chapter 14) freezes and thaws repeatedly during a single season, wedging apart large blocks and boulders and loosening them completely. Then, depending on the local relief, these pieces of rock may either remain more or less where they are, awaiting dislodgement by wind or precipitation, or they may roll or fall downslope and collect at the base of the mountains.

When the rock particles remain near their original location, they form a *rock sea*. Sometimes other words are used for this, such as *blockfield* or the German term *felsenmeer*. When the particles roll downslope, they create a *scree slope* or *talus cone* like that in Figure 18-1. If you have been in the Rocky Mountains or other mountainous areas in the western United States, you have probably seen such piles of loose boulders. They often lie at steep angles and seem ready to collapse.

Rocky soils can also reflect the action of frost. For a long time, physical geographers wondered what produced the remarkable hexagonlike structures of stone that mark the soils in arctic regions shown in Figure 18-2. These *stone nets* are created when ice forms on the underside of rocks in the soil, a process that tends to

FIGURE 18-2
This photo shows sorted stone circles in the Alaska Range, south central Alaska. These are caused by the action of frost.

wedge the rock upward and sideways. Eventually the rocks meet and form lines and patterns that look as though they were laid out by ancient civilizations for some ceremonial purpose.

In arid regions, too, mechanical weathering occurs. Here, the development and growth of salt crystals has an effect similar to that of ice. When water in the pores of such rocks as sandstone evaporates away, small salt crystals form. The growth of those tiny crystals pries the rocks apart and weakens their internal structure. As a result, caves and hollows form on the face of scarps, and the wind may remove the loosened grains and assist the process.

The mineral grains that rocks are made of have different rates of expansion and contraction in response to temperature changes, so the bonds between them may be loosened by continual temperature fluctuations. Although impossible to replicate in the laboratory, this kind of weathering may play a role in the mechanical disintegration of rocks. Over many thousands of years, daily temperature fluctuations may well have a weakening effect on exposed rocks. But other weathering processes enter the scene: As the contacts between the grains are loosened, moisture enters and promotes decay of the minerals.

CHEMICAL WEATHERING

The minerals that rocks are made of are subject to chemical alteration (see the statue, Figure 18-3). Some minerals, such as quartz, resist such alteration quite successfully, but others dissolve easily, as the calcium carbonate of limestone does. In any rock made up of a combination of minerals, the chemical breakdown of one set of mineral grains leads to the disintegration of the whole mass. In granite, for example, the quartz resists chemical decay much more effectively than the feldspar, which is chemically more reactive and yields to become clay. Often you can see a strongly pitted granite surface. In such a case the feldspar grains are likely to have been weathered to clay and blown or washed away. The quartz grains still stand up, but they may soon be loosened too. So even a rock as hard as granite cannot withstand weathering processes forever.

Three kinds of mineral alteration dominate in chemical weathering. When minerals are moistened, *hydrolysis* occurs, producing not only a chemical alteration but expansion in volume as well. This expansion can contribute to the breakdown of rocks. Hydrolysis, it should be noted, is not simply a matter of moistening; it is a true chemical alteration, and minerals change to other mineral compounds in the process. For example, feldspar hydrolysis gives a clay mineral, silica in solution, and a carbonate or bicarbonate of potassium, sodium, or calcium in solution. The new minerals tend to be softer and weaker than their predecessors. In granite boulders, hydrolysis combines with other processes to cause the outer shells to flake off in what looks like a small-scale version of exfoliation. This is *spheroidal weathering,* and it affects other igneous rocks besides granite.

When minerals in rocks react with the oxygen in the air, the chemical process is *oxidation.* We have plenty of evidence of this process, in the reddish color of soils in many parts of the world and in the reddish-brown hue of layers exposed in such places as the Grand Canyon. The products of oxidation are compounds of iron and aluminum, which account for the reddish colors seen in so many rocks and soils. In tropical areas, oxidation is the dominant chemical weathering process.

Various circumstances may convert water into a mild acid solution, thereby increasing its effectiveness as a weathering agent. With a small amount of carbon dioxide, for example, water forms carbonic acid, which in turn reacts with carbonate minerals such as limestone and dolomite (a harder relative of limestone, a carbonate of calcium and magnesium). (The actual chemical reaction appears in Chapter 14.) This form of chemical weathering, *carbonation*, is especially vigorous in humid areas, where limestone and dolomite formations are often deeply pitted and grooved and where the evidence of solution and decay are prominent. As we will soon see, this process even attacks limestone underground, contributing to the formation of caves and subterranean corridors. In arid areas, however, limestone and dolomite stand up much better, and although they may show some evidence of carbonation at the surface, they appear in general to be much more resistant strata.

Chemical weathering is the more effective agent of rock destruction in humid areas, because moisture promotes chemical processes. Physical weathering prevails to a greater extent in dry and cold zones. But water plays an important role in the dry as well as the moist areas. The growth of destructive salt crystals in porous rocks of arid areas, for example, takes place only after some moisture has entered the pores and then evaporated, triggering the crystals' growth. As we will find, the role of water in other rock-destroying processes is also paramount.

FIGURE 18-3

This statue is on the Cologne Cathedral in West Germany. The white patches on the statue are evidence of chemical weathering. The moist, chemical-containing atmosphere in Cologne is slowly turning the solid sandstone into powdery white gypsum.

BIOLOGICAL WEATHERING

We should remind ourselves at this point of the role of weathering in the formation of soils. It is through the breakdown of rocks and the accumulation of a layer of minerals that plants can grow—plants whose roots and other parts in turn contribute to the weathering processes. But it is likely that the role of plant roots in forcing open bedding planes and joints is rather overestimated. The roots follow paths of least resistance and adapt to every small irregularity in the rock. Roots certainly keep cracks open once they have been formed, but more importantly, areas of roots tend to collect decaying organic material that is involved in chemical weathering processes.

One of the most important aspects of biological weathering is the mixing of soil by burrowing animals and worms, mentioned in Chapter 10. Another interesting aspect is the action of the lichen, a combination of algae and fungi, that live on bare rock. Lichens draw minerals from the rock by ion exchange. The swelling and contraction of lichens as they get wet and dry may also cause small particles of the rock to fall off.

Of course, another agent of biological weathering that becomes more potent every year is human activity. In one sense quarrying can be regarded as weathering (see Figure 18-4). Another weathering effect humans have is in polluting the air with substances that vastly accelerate some chemical weathering processes, as demonstrated in Figure 18-5.

We turn now to an extreme form of weathering, the result of long-term carbonation and other processes. An area of limestone on the coast of Yugoslavia, called the Karst region, has lent its name to limestone landforms the world over.

KARST TOPOGRAPHY

Water erodes not only the surface but even the subsurface. Water below the ground dissolves rocks and carries them away in solution, creating subterranean tunnels and caverns. Of course, the circumstances for underground erosion must be there, because water does not dissolve most rocks. But limestone, a calcium

FIGURE 18-4
Quarrying the largest hole in the world—the Bingham Copper Mine, Utah.

FIGURE 18-5

Accelerated weathering during the twentieth century. The sandstone of a statue carved in Germany in 1702 remained relatively unharmed in the 206 years until 1908, when the photograph on the left was taken. In the next 61 years until 1969, the date of the right-hand photograph, chemical weathering has been extreme.

carbonate, is susceptible to solution by water (and more so by water that contains some carbon dioxide). Rain water and meltwater from snow and ice have a small quantity of carbon dioxide from the air, which makes such water a mild carbonic acid. Water acquires stronger doses of carbon dioxide as it percolates through the soil. This carbonic acid reacts with the calcium carbonate of limestone to form calcium bicarbonate. The calcium bicarbonate dissolves in and is carried away by the groundwater. Again, the actual chemical reaction is given in Chapter 14.

Imagine a series of sedimentary layers, with a porous sandstone on top and a limestone layer below. Water percolates through the porous sandstone, attacks the limestone, and carries away some of the limestone in solution, creating an underground system of drainage that is constantly enlarged as solution proceeds. Making use of joints and bedding planes in the limestone, the water forms a system of tunnels and caves. In parts of France, such caves form ideal places for storing champagne because of their cool, stable temperatures. The tunnels may extend for kilometers under the surface. The caves may consist of huge chambers at several levels, with iciclelike *stalactites* hanging from the roof and pillar-shaped *stalagmites* standing on the floor, as in Figure 18-6. These stalactites and stalagmites form when the water emerges from the rock and deposits some of the calcium bicarbonate it carries in solution.

When the surface shows evidence of underground erosion, it is called *karst* topography. Typically, the rain water does not flow through a well-integrated system of surface drainage but enters, and occasionally stands in, pits and basins, meanwhile seeping slowly downward. Some of these pits and basins form as the rocks dissolve. But others form when the underground tunnels are so extensive that the overlying rocks, left unsupported, collapse into *sinkholes* (see Figure 18-7). From the air, a karst landscape looks pockmarked, as though the area had been hit by numerous bombs. But in fact the area is being undermined. Every year, houses are damaged or lost as the ground they stand on

FIGURE 18-6
Stalactites and stalagmites in Carlsbad Caverns, New Mexico.

suddenly collapses. New sinkholes frequently form in areas of known underground erosion, including parts of Florida, southern Indiana, Kentucky, and sections of Missouri. The Mammoth Cave area of Kentucky is an example of karst topography and its associated features (Figure 18-8).

Although landscapes associated with underground erosion occur in many parts of the world, their total extent is not large. Certain karst landscapes are less common than even those landscapes that are attributable primarily to wind erosion and deposition. Erosion by running water at the surface remains by far the leading agent of landscape sculpturing. Before the water performs its work of transportation, weathering has usually taken place. After the weathering, the force of gravity is often helpful in moving the material toward a river.

MASS MOVEMENT

The movement of earth materials by the force of gravity is identified as *mass movement*. From this term it is clear that the materials involved are moved en masse, in bulk. A landslide is a case of mass movement. So is the imperceptible movement of soil, which tends to creep slowly downslope. You will sometimes see the term *mass wasting* employed to describe these various gravity-induced movements, but our interest focuses in part on the nature of the movement that displaces the materials: whether this movement is fast or slow, whether contained in a valley or unconfined, and so forth. So the term *mass movement* is probably preferable.

Mass movements are important in the breakdown of

rocks—for obvious reasons. If weathering manages to loosen a joint block from the face of a granite escarpment, it is up to the force of gravity to pull that block downslope and to expose a new target for weathering. If mass movements did not occur, the whole process of erosion would be slowed down.

CREEP MOVEMENTS

Mass movements produce characteristic marks and scars on the landscape. Sometimes this evidence is so slight that we have to assume our detective role again to prove that mass movement is indeed taking place. The slowest mass movement, *creep,* involves the slow, imperceptible motion of the soil layer downslope. We cannot observe this movement as it actually happens, but it is quite common. Soil creep reveals itself in trees, fence posts, and even gravestones that tilt slightly downslope as in Figure 18-9. The upper layer of the soil moves faster than layers below, and vertical objects are rotated in the process.

Soil creep (and rock creep, where upturned weathered rock strata are similarly affected) appears to result from the alternate freezing and thawing of soil particles or from alternate periods of moistness and drought. The particles are slightly raised or expanded during freezing and wetting and then settle slightly lower along the slope when they are warmed or dried. The total change during a single freeze–thaw or wetting–drying sequence is miniscule, but over periods of years the results are substantial. Creep is faster on less vegetated areas and steep slopes.

A special kind of soil creep is called *solifluction,* a term from Latin and French that literally means "soil flow." In solifluction the soil and rock debris are saturated with water and flow as one mass. The process is especially common in subpolar regions, where the ground below the thaw zone is impermeable because of

FIGURE 18-8
Mammoth Cave, Kentucky. Massive underground erosion is caused by constant moisture and slow seepage.

FIGURE 18-7
This large sinkhole developed in 1972 in central Alabama. US Geological Survey scientists say it might be the largest sinkhole formed in the United States. It is 128 m (425 ft) long, 105 m (350 ft) wide, and 45 m (150 ft) deep.

FIGURE 18-9
Commonplace evidence of the slow downslope creep of soil and weathered rock.

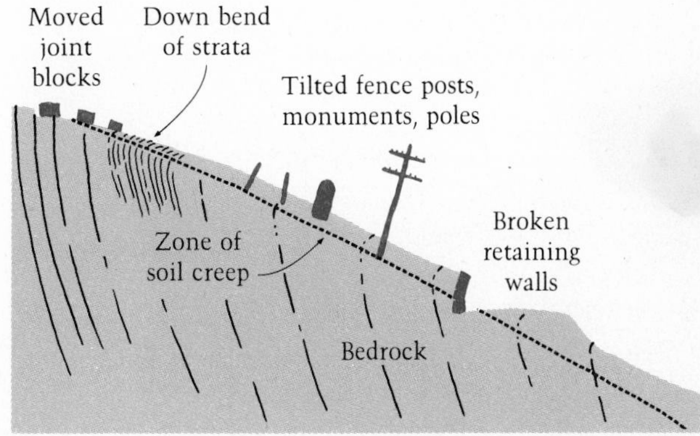

FIGURE 18-10
Solifluction lobes on an Alaskan hillside. Saturated soil does not lose its water because of the permafrost underneath.

permafrost. The results of solifluction on the landscape look as if a giant painter had painted it with a brush that was far too wet. Figure 18-10 depicts the effect.

FLOW MOVEMENTS

The impact of soil creep on the total landscape is obviously slight over a short period of time. The form of a hill changes barely at all even though soil creep may be occurring on it. Perhaps the valley between one hill and the next is slightly modified by some filling with soil creeping downslope from both sides. Although far less widespread, *flow* movements have a much stronger impact on the landscape. Once again, the critical ingredient (in addition to gravity, of course) is water.

In the case of *earth flows,* a section of soil or weak and weathered bedrock, lying at a rather steep angle on a hillside, becomes saturated by heavy rains and lubricated enough to flow. Figure 18-11 diagrams the result. Often this lubrication is helped by the presence of clay minerals, which are slippery and promote flow when they become wet. The earth flow forms a lobe-shaped mass that moves a limited distance downslope, perhaps leaving a small scarp at the top and pushing a tongue of debris into the valley below. Although earth flow can occur quite quickly, it is normally rather slow and not nearly so dramatic as, say, a landslide.

FIGURE 18-11
Typical earth flows on steep slopes.

An associated feature of hillside flow is *slumping.* A spectacular example of this motion blocked the highway at Oakland, California in December 1950 (see Figure 18-12). Again, saturation and lubrication cause rock masses to weaken and slump, as illustrated in Figure 18-13. Unlike the earth flow, slump movements involve major sections of weakened bedrock that stand in near-vertical scarps. As the figure shows, slumping includes a downslope motion and a backward rotation of the slump block. Slumping is normally a rather slow process—faster than creep but much less rapid than slides.

Still another form of flow is the *mudflow,* a stream of fluid, lubricated mud. The mudflow originates when a heavy rain comes to an area that has long been dry and where weathering has loosened a large quantity of relatively fine material. Very little of the rain water can percolate downward, and thus virtually all of it runs down the valleys, carrying the loose particles downslope. But so much material is brought with it that, instead of a stream, the channel soon contains a porridge-like mass of mud. When the water supply is ample, the mudflow is rather thin and moves rapidly. At other times there is an overload of debris (the mudflow soon accumulates rocks and boulders as well), and then it becomes stiff and moves rather slowly, eventually coming to a stop. But before this happens, the mudflows from several valleys may join in a larger valley, forming an enormous volume of material that is pushed along like advancing molten lava. This sort of mud flow causes great destruction when it reaches populated areas. Mudflows can move for kilometers, emerging from mountainous areas and extending far into adjacent flatlands.

SLIDE MOVEMENTS

Whenever rock and soil lie at a high angle on a steep slope, there is a possibility that a segment of such material will break away in a *landslide* or rockslide. The difference between landslides and the previously discussed mass movements is the speed with which the slide occurs. Landslides move much faster than flows and sometimes roar downslope with a thunderous

FIGURE 18-12
a) Slump blocking a major highway in Oakland, California, December 1950. b) A slump along a river bank in Stanley County, South Dakota.

sound and great force. Here water plays a lesser role, although it does help in the weathering and loosening that has gone before. A landslide is in effect a collapse, and no lubricants are necessary, although the weight of water saturation certainly does contribute. Many of southern California's feared landslides are triggered by heavy rains that overload slopes already near collapse.

Nature induces landslides in other ways as well. Earthquakes and earth tremors sometimes provide the vibrations necessary to set the downward slide in motion. A river cutting into the side of its valley can oversteepen the slope and initiate a collapse. Humans, too, play a role in starting landslides by cutting into mountain sides to construct roads and other works, thus weakening and eventually destroying the support needed to keep the slope intact. The Ministry of Construction in Japan made a study in 1976, locating 35,000 sites that were potentially dangerous because of

FIGURE 18-13
The rotational movement of a slump block and the resulting landform.

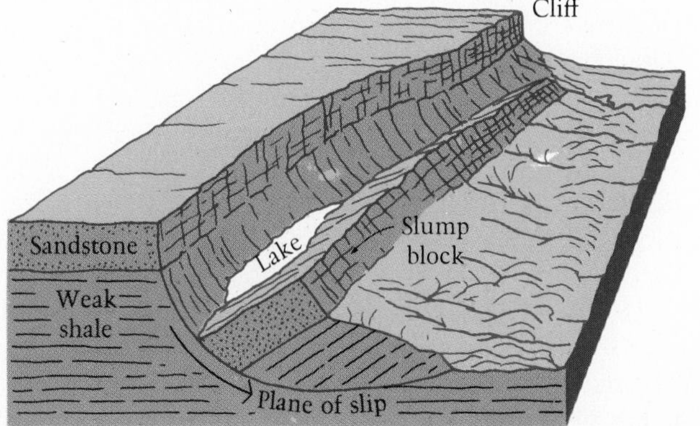

FIGURE 18-14
Landslide in Shimoda on the Izu peninsula, Japan. The landslide was caused by a torrential rain.

FIGURE 18-15

The landslide (avalanche) that destroyed the Peruvian cities of Yungay and Ranrahirca and killed 40,000 people. The avalanche covered the towns of Ranrahirca (on the right) and Yungay (on the left). The Rio Santa flows across the bottom.

land development by humans. Torrential rains in July 1976 on the Izu peninsula, near Tokyo, caused almost 500 landslides like the one in Figure 18-14. The landslides killed nine people in the popular holiday resort and left 9794 without homes.

The destructiveness of landslides makes them perhaps the best known of all mass movements. Tens of millions of tons of rock debris may break loose at once and in minutes bury whole towns. In the 1970 Huascaran landslide, triggered by an offshore earthquake, two Peruvian towns with a combined population of over 40,000 were buried under as much as 50 million m^3 (1765 million ft^3) of debris, snow, and ice (see Figure 18-15). This slide began as an avalanche of snow and ice, gathering rock, soil, and mud on its 15 km (9.3 mi) journey downslope. Other lesser catastrophes abound in the record of human occupation of mountainous zones.

Landslides and rockslides differ from creep and flow movements not only in terms of the speed of movement but also, frequently, in the nature of the material carried down. In a landslide, everything goes down—bedrock as well as overburden. Weathering loosens pieces of bedrock by attacking along joint planes and fracture zones. Any steep slope carries within it the parallel planes of weakness that may eventually fail and yield. As soon as the landslide has occurred, exposing a fresh scarp or slope, the cycle begins again. New joints and fractures are attacked, and slowly the slide threat returns.

FALL MOVEMENTS

The last of the mass movements we will discuss is the fastest of all. This is the free *fall*, or downslope rolling

LANDSLIDES

Areas of high relief, especially when they are subject to drenching rains, can be very risky as places of residence. Landslides frequently block California roads and highways, and not infrequently homes are swept down hillsides in southern California. People assist the process by scraping away the vegetation, loosening the earth, and not anchoring structures adequately.

East Asia's dense population clusters suffer casualties every summer as a result of landslides. In 1972 alone, more than 100 people were lost in mid-June in Hong Kong, another 80 in Japan when a train was buried by a landslide in July, and several hundred died in South Korea in August when heavy rains set off landslides in the region of Seoul. Although people know that they live in areas susceptible to landslides, the destruction usually comes suddenly and without warning.

of pieces of rock that have become loosened by weathering. Usually such rock fragments and boulders do not go far, ending at the base of the slope or cliff from which they broke away. They form a talus cone or talus slope, an accumulation of rock fragments, large and small, at the foot of the slope, as in Figure 18-16. Such an accumulation, also called a scree slope, lies at a surprisingly high angle—as high as 35° for large fragments. When an especially large boulder falls on the slope, the talus may slide some before attaining a new adjustment.

FIGURE 18-16

Talus cones at the foot of steeply sloped igneous rocks in northern Afghanistan.

THE IMPORTANCE OF MASS MOVEMENT

It is easy to underestimate the importance of mass movements in the processes of erosion. Often mass movement occurs almost imperceptibly. But it is taking place along the banks of every river, and a large part of the material swept out to sea by the world's rivers is brought to the water through mass movements. Figure 18-17 compares the volumes of rock eroded directly by a stream and the volume first moved by mass movement. Even slow, sluggish, meandering streams undercut their banks, causing the collapse of materials into their waters. Rapidly eroding streams in highland areas and streams in deep, canyonlike valleys cause much

FIGURE 18-17

A comparison of the amount of rock eroded directly by a stream (*A*) and that first moved to the stream by various kinds of mass movement (*B, B′*).

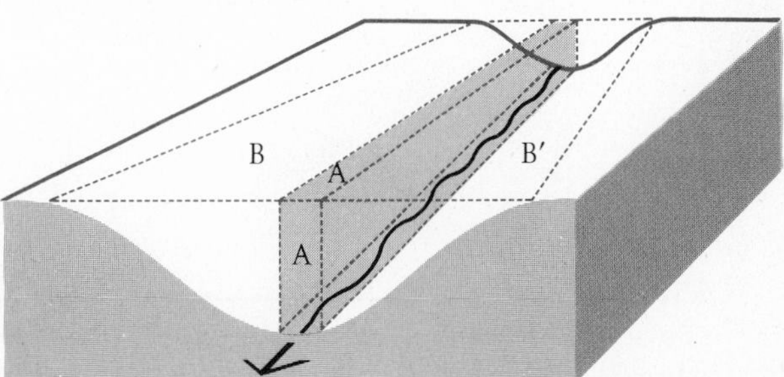

oversteepening and constant collapse along their valley walls.

We have dealt with four major categories of mass movement: creep, flow, slide, and fall. Within these categories, there is sometimes more than one type of movement. They may be summarized in terms of the rate of downslope movement and the amount of water involved. Figure 18-18 shows how the different types of mass movement relate to one another with respect to these two parameters.

Why should we spend so much time examining the sliding and flowing earth? A knowledge of mass movement can help decision making. It is quite costly to use public money to create highways in unstable areas. In the San Francisco Bay area alone, financial losses from landslides in 1968 and 1969 are estimated to have exceeded $25 million. In addition, many people build houses in locations chosen for their spectacular scenic views, with no heed to the stability of the land. Some of these people have rude awakenings, as Figure 18-19 shows.

Physical geographers are interested in mass movement for other reasons. In sculpturing the land surface, mass movements also perform the important job of exposing new bedrock to the forces of weathering. By the time the material sags and collapses in a flow or slide, weathering has become much less effective in attacking deeper rock layers. But the slide removes the weathered material and exposes new rock faces to the forces of weathering. Thus mass movements are vitally important in the total complex of erosional processes.

FIGURE 18-18

Common types of mass movement classified by rate of movement and water content.

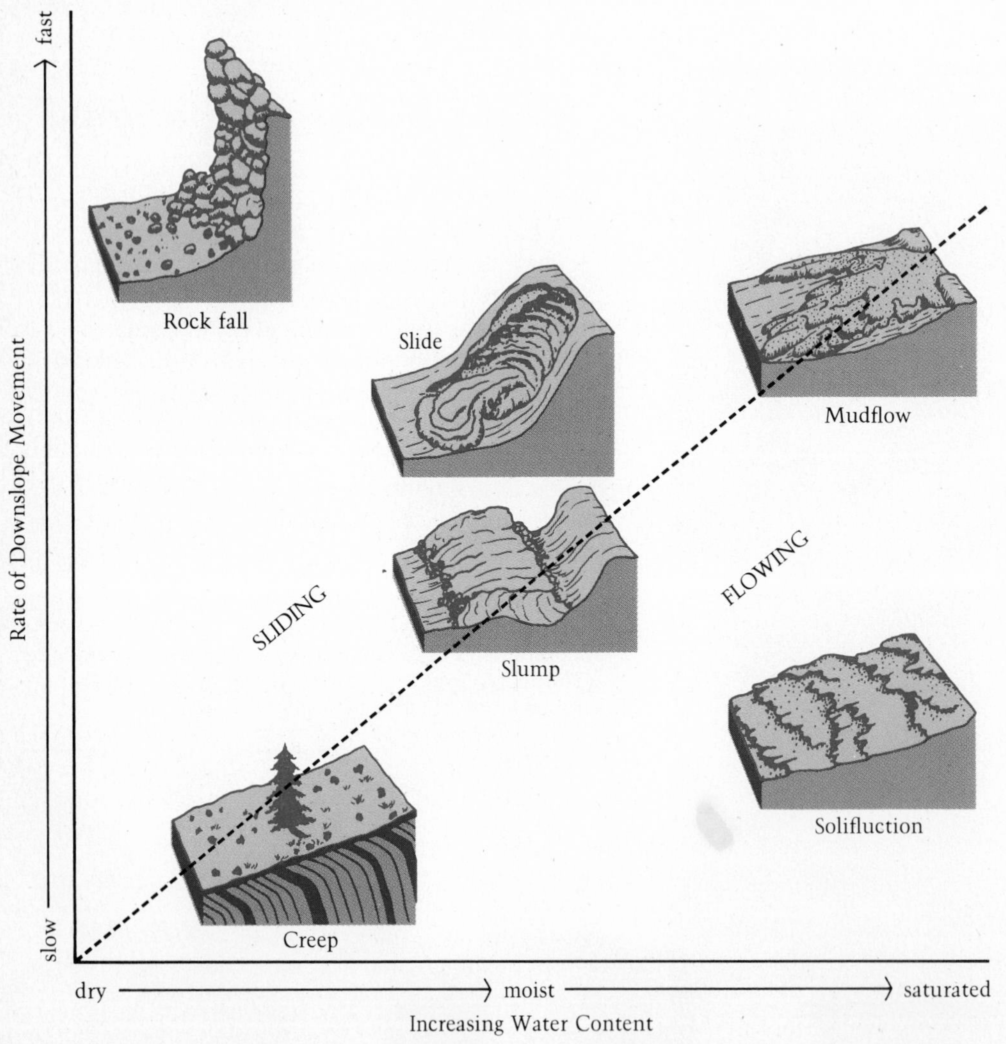

FIGURE 18-19
The destruction of a house by mass movement at Pacific Palisades, California.

SUMMARY

This chapter describes some of the first processes to occur in sculpturing the land surface after it has been uplifted. Weathering occurs the minute a new land surface is raised above the sea. There are two main kinds of weathering: mechanical and chemical. The major types of chemical weathering include hydrolysis, oxidation, and carbonation. Special landforms are derived from both mechanical and chemical weathering. Biological weathering includes aspects of both types. One of the unique sets of landforms arising from chemical weathering is that of karst topography.

After material has been weathered, and in some cases before, it is subject to the force of gravity, which results in various kinds of mass movement. Mass movement varies in the speed with which it takes place and in the amount of water in the material moved. There are many practical reasons for understanding mass movements, and physical geographers are interested in it because it represents, together with weathering, the beginning of erosion.

QUESTIONS

1. When we consider the removal of surface features by different processes, we most commonly think of erosion. How is this an oversimplification? What is the correct sequence of processes?

2. How does low temperature contribute to mechanical weathering processes? What is the difference, in weathering ability, between constant low temperatures and alternation of high and low temperatures? What is the reason for the difference?

3. High, alpine mountains tend to weather extremely rapidly, no matter where they are located. What accounts for this?

4. Stone nets are generally found in areas with arctic soils that have a deep layer of permafrost. Near Mount Shasta, in California, is a field of stone nets at an elevation of about 1400 m. Since there is no permafrost or arctic soil in this area, and since seasonal temperature variations are much less extreme than in arctic regions, how can we explain the presence of these stone nets? Explain how they form and when and why they formed here.

5. Mechanical weathering occurs in arid regions as well as in areas of extreme temperature fluctuation and ice formation. In what ways is this mechanical weathering similar to that associated with the effects of varying temperatures on water, and in what ways is it different?

6. Granite is a hard igneous rock, and it contains a large proportion of resistant quartz. Through what processes is granite finally decomposed into constituent minerals and altered compounds? How does the weathering of granite illustrate the interrelationship of many decompositional processes in the weathering of a complex mineral assemblage?

7. The location of various rock types often plays an important part in how they weather and at what rate. Limestone and dolomite in humid areas weather quite rapidly, while the same rocks are extremely resistant in arid regions. How would you explain this difference?

8. Is water an important weathering factor only in very humid areas? What roles does water play in weathering processes in arid regions?

9. For many years, geomorphologists believed that biological weathering proceeded mainly through the prying apart of massive rocks by roots. Is this a complete and accurate explanation? What other forms of biological weathering may also be important?

10. How have human activities influenced both mechanical and chemical weathering? What types of evidence do we have for each type of weathering and human effects on them?

11. What factors influence the development rate of limestone caves in areas such as southern Indiana, Kentucky, Puerto Rico?

12. What surface features are associated with underground erosion in limestone? How are they related to

the processes occurring below them?

13. Mass movement is vital in controlling the rate at which landscapes erode. What are the forces involved in mass movement, and how does this process control the rate of erosion in many areas?

14. Soil creep is the slowest form of mass movement. What evidence do we have that it actually occurs, and what mechanisms do we use to explain it?

15. How do flow movements differ from soil creep? Why do you think solifluction is considered a type of creep movement rather than a flow movement?

16. How are landslides and rockslides distinguished from the associated phenomena of flow movements and creep movements? Is the role of water the same in slide movements as in flow movements? How are slides generally more dangerous than the other two types of mass movement?

17. What is the significance of the scree slope (or the talus cone) in the process of mass movement? Does water play an important role in this type of movement? Why or why not?

18. Why do talus cones develop (instead of large areas of scattered rock fragments) during the free fall of rock material in alpine landscapes?

19. How does mass movement work in conjunction with stream erosion to denude landscapes? Which process is more important in terms of the amount of material actually moved? Could either process continue long without the other? Why or why not?

20. Ignoring isostasy for the moment, how can the information in Figure 18-18 on the rates of mass movements be used to explain the rapid downwearing of new mountain ranges compared with old ones?

THE STORY OF RIVERS

CHAPTER 19

Important as weathering and mass movements are in the lowering of the landmasses, they come nowhere near matching river erosion in their total impact on the continents. Rivers and streams are the dominant agents of erosion, the major forces that attack mountains and destroy plateaus. Chapter 21 shows how ice, ocean waves, and wind contribute to the sculpturing of the landmasses. But for the present, we will confine ourselves to the work of running water at the surface, in rivers and streams.

A river performs many jobs. The river dislodges particles of soil and bedrock. When streams and rivers cut down and erode their channels, *degradation* is said to occur. The river has to carry its load, and so *transportation* is an important factor. And of course, rivers deposit their sedimentary load in levees and deltas near and at the coast. There they function not as erosional agents but as builders, and the prevailing process is *aggradation*. We can hardly overestimate the significance of rivers: Virtually all the materials accumulated in those thousands of meters of sedimentary rocks on all the landmasses were, at one time or another, transported by rivers.

In performing all these tasks, the river behaves in special ways. It attempts to reach an ideal longitudinal profile from its source to its mouth. We will discuss its attempts to do this, its attempts to achieve what is called *grade*. If rivers could flow over homogeneous material, sometimes their perfect profile might be reached. But the rocks of the crust are neither homogeneous nor stable. Rock differences and tectonic and other movements affect the actual courses of rivers. The ways they do so, as we will see, are just one of the aspects of the fascinating story of rivers.

THE TASKS OF RIVERS

Rivers and streams do most of their eroding in their upper reaches, in the mountains and highlands where they often begin. There the waters rush downslope, and waterfalls and rapids signify the force and sometimes raging speed of the stream. Boulders are dislodged from valley walls, to be rolled and dragged along until, eventually, they are worn down to the fine sediments that make up deltas and silty river bottoms.

Rivers can be studied to help unravel the past of a region, because rivers have life cycles of their own and because rivers behave according to certain physical laws. Some rivers, tens of millions of years old, have witnessed all kinds of changes in their regional settings: changes of climate, of vegetation, of topography. They have contributed to these changes through their own erosional activities, but they have also had to contend with the internal forces of the earth itself, forces that altered slopes, threw up mountains, and fractured the surface. Many rivers survived such disruptions, perhaps relocating in another valley, forming new waterfalls, or adjusting in other ways to be discussed later. Other rivers undoubtedly disappeared under lava flows or in crumbling continental margins. But such great rivers as the Mississippi, Amazon, Niger, and Nile have prevailed for millions of years.

It would be impossible to trace the history of any major river back to its beginnings, but we can speculate how great drainage systems, such as those of the Mississippi and Amazon basins, were formed. As noted in Chapter 15, the Pangaea landmass had rivers, many of them oriented not toward the ocean but toward interior seas and lakes. Then when the supercontinent broke up, the outlines of the present drainage system began to take shape, becoming better organized and integrated as time went on. It is one of the characteristics of youthful rivers and streams that they are not yet adjusted to one another, not oriented to a common

FIGURE 19-1

A lone tree holding the soil below it, while all around the tree sheet erosion and gullying have washed away the valuable soil. Soil conservation measures have largely controlled extreme effects of this kind.

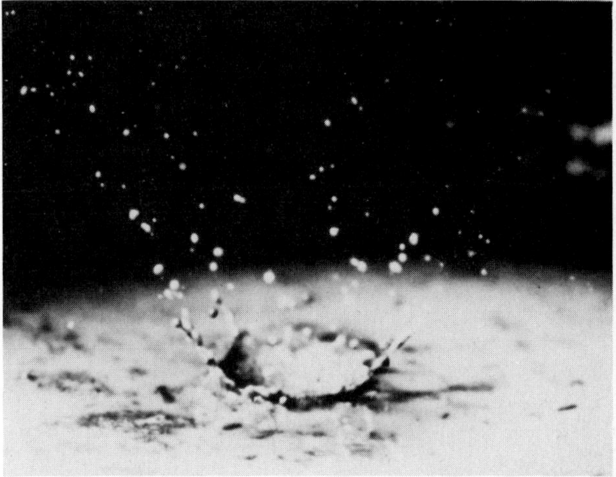

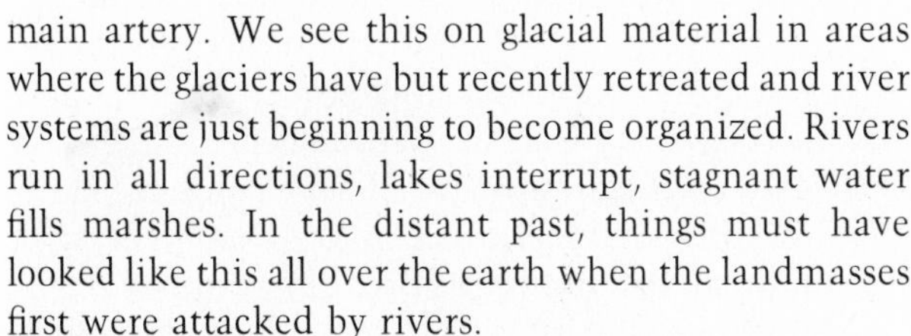

FIGURE 19-2

Splash erosion. The impact of a raindrop throws particles of soil into the air. If this occurs on a slope, the particles often fall downslope from their original position. The "crater" left after impact is about four times larger than the raindrop.

main artery. We see this on glacial material in areas where the glaciers have but recently retreated and river systems are just beginning to become organized. Rivers run in all directions, lakes interrupt, stagnant water fills marshes. In the distant past, things must have looked like this all over the earth when the landmasses first were attacked by rivers.

When rain falls on an area, some of it seeps downward through the soil and enters rock strata below as groundwater. Not all the rain water can always infiltrate the soil, however, because the capacity of the pore spaces to accept water is limited. So when rainfall intensity exceeds rates of infiltration, a certain amount of water begins to flow along the surface in a thin layer or in small rills. This produces *sheet erosion,* which carries a good deal of the fine particles of the soil surface away(Figure 19-1). The rills sometimes develop into larger rivulets and into gullies. Gullies can become so deep and permanent that they intersect a groundwater table and become permanent streams. Many a river can be traced to its beginning in a gully.

Even the rain can have erosional impact. Where the vegetation protects the soil, the drops are slowed and the rain's destruction is slowed. But on bare soil, the drops hit hard and often dislodge surface particles in the process (see Figure 19-2). On a slope, the particles come down slightly downslope from where they started. This *splash erosion* is effective in dry areas and in places where the natural vegetation has been removed. Imagine the aggregated effect of tens of millions of drops! An 8 hectare (20 acre) building site in Maryland lost 3420 metric tons (3800 tons) of sediment during the construction period. The 3 cm (1.2 in) of soil removed from the site would naturally take 1000 years to run off.

The runoff in sheets and rills and the seepage of

groundwater fill the channels. Then the water's journey to the river's mouth is confined, except in times of flood, to the channel. The valley's sides change character along its way. Where the river has its start, the valley walls are likely to be steep and high. Most frequently the cross-profile of the valley resembles a V, and the volume of water flowing in the bottom of the V varies with the amount of rain received in the area.

VALLEY MODIFICATION

The valleys of all rivers are constantly changing. Rivers oversteepen their banks, causing rock and soil to collapse into the water. When a river's course bends and curves, the erosive action of the water is especially strong on the outside of the curve, so the valley wall that the river bends against undergoes the greatest erosion. Near the river's source and even near its mouth, cross-profiles often show the asymmetry resulting from such differential erosion.

In the area where the river begins, where the valley shape tends most nearly to resemble a V, the process of valley *deepening* is usually most active. The narrowness of the valley, the volume of the water, and the large size of the rock particles being moved along all contribute to this deepening. In the meantime the valley is being lengthened, because the river is eating its way back into the countryside through a process called *headward erosion*.

Downstream, valley *widening* becomes much more important, and the valley opens up. *Interlocking spurs*, hills intruding into the valley like interlocking fingers that the river curved around in its earlier stage, are now rounded off. The river is calmer, more adjusted. Change is also less perceptible, although it occurs all the time. As Figure 19-3 indicates, the river begins to meander, the valley opens still wider, and rather than erode everywhere, the river actually begins to drop some of its load, filling the bottom of its valley with *alluvium*.We will return to this phase of deposition shortly.

FIGURE 19-3

Differences often found between the higher and lower parts of a river in its upstream area.

(a)

(b)

EROSION BY RIVERS

Stream erosion takes place in more than one way. We give the name *hydraulic action* to the force of the moving water that alone is enough to dislodge and drag away material from the valley bottom. These loosened materials, especially larger pieces, knock loose still other parts of the valley floor as they roll and bounce along in the water. In the process they are ground down to finer sediment. This is *abrasion*, an important contributor to the valley-deepening process in the upper reaches of river systems. Certain rocks that the river flows over are susceptible to solution. This chemical form of erosion is identified as *corrosion*, the least significant contributor to the overall erosion by rivers. Of course, we should not lose sight of the river's contribution to mass movement of the valley sides. Collapse of valley sides from the river's undercutting is a most significant process in the sculpturing of land surfaces.

TRANSPORTATION IN RIVERS

Rivers erode in different ways; they also transport their load in different ways. Anyone who has ever watched a mountain stream in action has noted that rock fragments of considerable size, even boulders, are carried along by the sheer force of the rushing water. Hikers know it is not safe to cross a mountain stream when they can hear boulders being moved along the stream bed. A river near its ocean mouth presents a very different picture. The water may be brown or gray with mud, but you can see no rocks or boulders. There the river carries only fine particles.

Where the river moves rock materials (and gravel and sand as well) by dragging them along the bottom, the process is one of *traction*. When the fragments are sufficiently reduced in size, they are carried along the channel floor for brief moments, as if bouncing: This is *saltation*. When the particles are reduced still further, part of the load is carried permanently along by the water itself, kept above the floor by currents and eddies. Now the load is in *suspension*. And of course rocks contain minerals that can be dissolved in water, so that *solution* is a part of the transportation work of rivers.

The dissolved load is often less than the suspended load. The Mississippi River, for example, carries 65% of its load in suspension and 29% in solution. Lakes in the St. Lawrence River system, dividing the United States and Canada, act as filters for suspended loads. In this system, the dissolved load accounts for 88% of the total load.

RIVER PROFILES

We can also learn much about the behavior of streams by studying their *profiles*—their downward curve from the interior highlands to the coast. A number of factors are involved here. First and perhaps most obvious is the lowest level to which a river can erode downward. A river that flows into the sea cannot erode more than a few meters below sea level. For all intents and purposes, sea level is its lowest limit, its *absolute base level*. A stream that happens to flow into a lake has that lake level as its base level, and one that emerges from mountains and dries up in an adjacent plain relates to the plain at its base level. The latter cases are referred to as *local base levels*. Figure 19-4 diagrams the concept of base level.

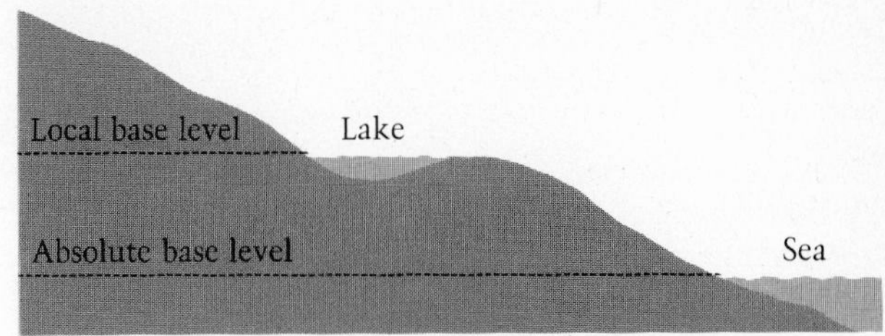

FIGURE 19-4
Longitudinal river profile illustrating local and absolute base levels.

The second factor is the river's volume of water. A river that begins as a small rill may end up as a kilometer-wide mass of water flowing into the ocean. As tributaries joint the main artery, more and more water is added. We already know that the channel and valley change, widening downstream, to accommodate all this water. We should also keep in mind that seasonal changes can cause great variation in the volume of water that passes a given point along the river's course. Rains in the interior cause the river to fill or even exceed its channel (melting of snow in the spring has the same effect); a prolonged drought may reduce volume to a fraction of what it is on the average. Many engineering projects along rivers try to ensure an even flow throughout the year.

Third among relevant conditions is the speed of the water. Obviously water runs faster through a steeply inclined valley than through a gently sloping one, but other factors affect its velocity as well. If the channel bottom is smooth, the water's speed is greater than when it is rough, because the roughness causes turbulence and eddies that tend to slow the water down. Volume also affects velocity: A larger volume, with its greater weight and its capacity to overcome obstacles, makes for faster flow than a small volume does.

The volume of water and the speed at which it moves down the valley determine the river's capacity to perform the job of removal or transportation (as we called it earlier). The critical factor is velocity. When, during a flood, the rate of flow of a river doubles, it can transport as much as 8 to 16 times the load it normally carries. This means, of course, that the erosional capacities of the river (especially hydraulic action and abrasion) are greatly increased during flood periods, and much of the change undergone by a valley occurs in the course of the river's flooding.

Another aspect of rivers that we will examine now is the concept of grade and river acting as a system. It may be that we are fascinated by rivers because they, like many of us, try to live their lives with the least possible work.

THE RIVER AS A SYSTEM

The river is one of the simplest and most easily understood examples of a system that we have on the earth's surface. It is an open system, with both matter and energy flowing through it as diagramed in Figure 19-5. The most obvious material flowing through the

THE NOAH FACTOR

River channels can usually control some increase of discharge above a normal level. If a river completely fills its channel, it enters what is called its *bankfull stage*. If it overtops the banks of the channel, it is said to be in *flood stage*. All rivers will flood at some time during their lifespans. As human beings we accept this fact in a strangely equivocal way. Scientists produce maps showing what elevations are likely to be flooded by a river by an average annual flood, as well as by a flood that might occur on the average once in 10, 25, 50, or 100 years. Dams and flood control structures are built to cope with a flood of a specified size, but seldom is the size the largest that could ever occur. It would be too costly, in most cases, to build a structure so large.

As a result of this and other quirks of human nature, flood control structures create a false sense of security in human beings. Furthermore, there is always a pressure to settle on the fertile alluvial soils of a river valley. The United States is now spending as much money annually on flood control structures as is lost in flood damage. But people continue to live in flood plains, accepting the risk.

system is water, entering as precipitation on any part of the drainage basin and leaving by evaporation and flow into the ocean.

The water carries with it energy. Through the action of the hydrologic cycle (see Chapter 4)—powered by energy from the sun—water vapor and, after condensation, liquid water are given potential energy. Any object possesses potential energy by virtue of being raised above the earth's surface and of work having been done against gravity. The motion of water in a river represents the transformation of potential energy into kinetic energy, the energy of movement. The water uses kinetic energy to carry its load and move itself. By the time the water has reached its base level, which is usually the sea, there is no more potential energy available. Thus no kinetic energy can be generated, and the river is unable to do further work. Entropy measures the inability to perform work. The entropy of the river system is at its maximum at the base level of the river.

Another kind of material, sediment, leaves the system. Some scientists would say that sediment (or potential sediment) enters the system when the land is tectonically uplifted. Others might say a new system is operating if the uplift occurs very quickly. However, slow uplift such as that from isostatic adjustment could reasonably be interpreted as representing a sediment input into the system.

The river system is a steady-state system, at least with regard to water and energy. A physical law states that energy and matter in a steady-state system move in a particular way. First, there is a tendency for the

FIGURE 19-5

A river as an open system.

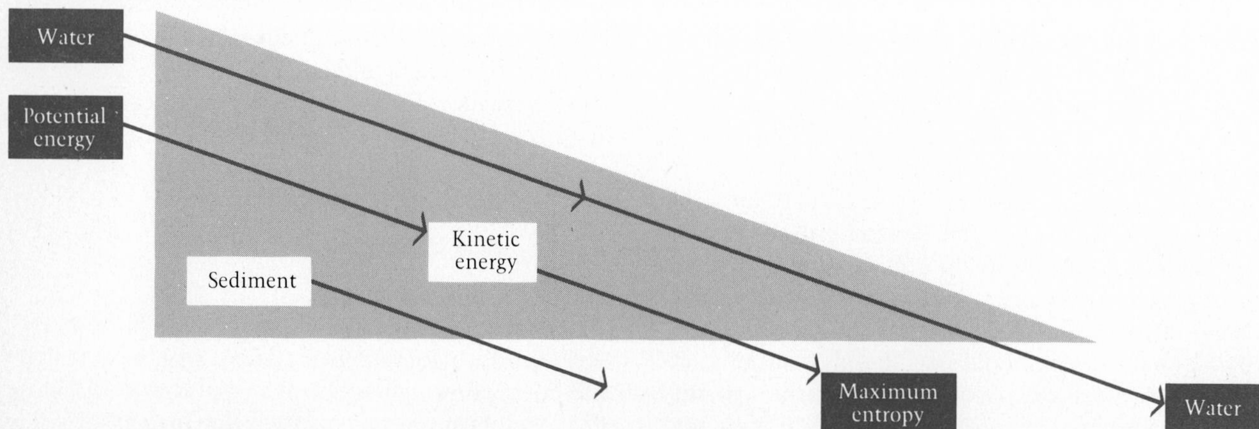

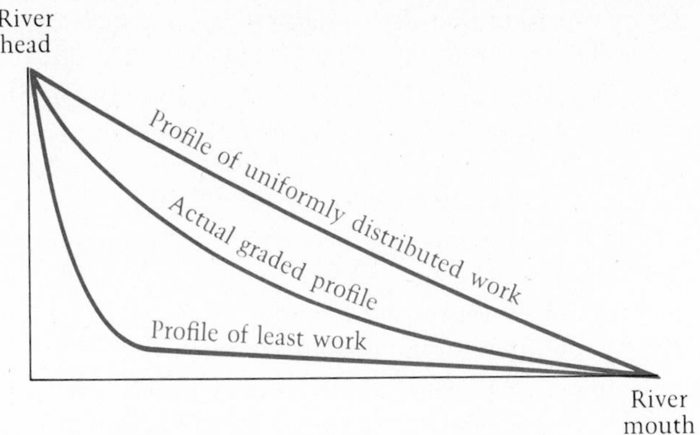

FIGURE 19-6
Formation of a graded profile as a compromise between the principles of uniformly distributed work and least work. This profile is reached if the material of the river channel is adjustable.

least work to be done. If we imagine a river where all the water starts at the top of one tributary, the least-work profile that the river would try to adopt would be a waterfall straight down to sea level. More practically, the least-work profile would be steep near the head and close to horizontal near the mouth of the river, as demonstrated in Figure 19-6. But a second tendency in a steady-state system is for work to be uniformly distributed. A river in which this happened would get wider downstream but have a nearly constant slope, as in the figure. The graded river profile is a compromise between the principles of least work and uniform distribution of work, but the graded profile is possible only when the channels are in a material that can be degraded and aggraded. The fact that this profile can be reached confirms that the graded river represents a steady-state system, but it also requires us to look more closely at the concept of grade.

A GRADED RIVER SYSTEM

Rivers attempt to reach a longitudinal profile that allows the load of the river to be transported without either degradation or aggradation taking place at any part of the profile. If this happens, the river is said to be *graded.* A graded river represents a balance among valley profile, water volume, water velocity, and transported load. As many as ten factors are involved in the tendency for a river to attain a graded state. Arthur Bloom of Cornell University has pointed out that these factors may be classed as either independent, semidependent, or dependent. Table 19-1 presents his scheme.

Independent factors are those that the river has little or no control over and to which it must simply adjust. The ultimate base level is a good example. The river must adjust to its ultimate base level because at this point it no longer has any potential or kinetic energy.

The semidependent factors are partly determined by the three independent factors but partly interact among themselves. The roughness of the river bed and the grain size of the sediment load, for example, interact together. The bed roughness is partly determined by the size of the sediment grains and partially determines the degree of mixing, known as *turbulence,* in the river. Sediment size generally decreases downstream because of abrasion. But only finer particles remain in suspension in the lower part of the river, where turbulence decreases. Thus sediment size and bed roughness interact via the turbulence factor.

Finally, there is only one variable that appears to

Table 19-1
Factors involved in the tendency of a river to achieve grade

INDEPENDENT	SEMIDEPENDENT	DEPENDENT
Discharge	Channel width	Slope
Sediment load	Channel depth	
Ultimate base level	Bed roughness	
	Grain size of sediment load	
	Velocity	
	Meander/braid tendency	

depend on all the others, and this is the slope of the river. The river has virtually complete control of this by either depositing or scouring away material as necessary. It was the propensity of a river to change its slope that first set investigators thinking about the concept of grade. Not until later was it completely realized that many other factors were involved in grade.

The modern definition of grade was established by J. Hoover Mackin in 1948 and has been slightly modified by Luna B. Leopold, Thomas Maddock, and Marie Morisawa. A consensus might be that a *graded river* is one in which, over a period of years, slope and channel are delicately adjusted to provide, with available discharge, just the velocity required for the transportation of the load supplied from the drainage basin. The graded stream is a steady-state system. Its diagnostic characteristic is that any change in any of its controlling factors causes a displacement of the equilibrium in a direction that tends to absorb the effect of the change. So if you tried to dam a graded stream with rocks, the stream would immediately begin to try to wear the rocks away.

Not all rivers are graded, and not all graded rivers assume a smooth longitudinal profile. A newly uplifted tectonic landscape would present a chaotic drainage system, with profiles far from graded. The semidependent variables are usually the first to adjust themselves to the process of moving material downstream. Furthermore, a river usually establishes grade in its lower reaches first, and then the graded condition is slowly extended toward the river head. Because of differences in rock resistance and other factors, it is quite possible for grade to exist in isolated segments of the complete river profile. The upper reaches of the Chialing Chiang River, a tributary of the Yangtze River of China, is an example. In other cases, individual reaches along a river may be graded but not form a smooth profile altogether. Brandywine Creek, in Delaware, manifests an irregular profile, and yet it has been shown to have achieved grade as Figure 19-7 demonstrates.

FIGURE 19-7

The longitudinal profile of Brandywine Creek, Delaware, a graded river without a smooth profile.

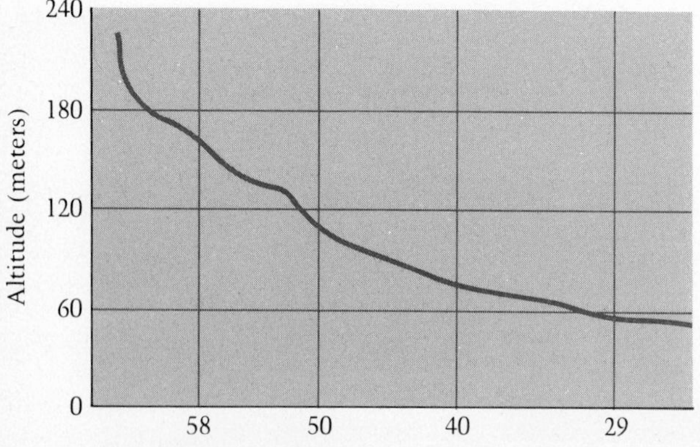

The most obvious criterion the casual observer can use to assess whether a river is graded or not is by judging the stability of a channel. If the channel has clearly remained in the same place for years and years, it is likely that the river is graded. But if there is evidence that the river occasionally changes its channel in some way, it is unlikely that grade has yet been achieved.

The establishment of grade has both practical and aesthetic consequences. Any human alteration of a river system, such as lining a channel with concrete or building a dam, must take into account the state of grade. Otherwise the dam or new channel might quickly become silted, and the expensive engineering work will be wasted. The stability of a graded river shows itself in a peaceful and pleasing landscape. Think, for example, of the calm generated in the human mind by looking at the lower course of a great river, such as that in Figure 19-8. The Greek philosopher Aristotle introduced the concept of the golden mean, a way of wisdom and safety between extremes. In the steady state of the graded river, nature has presented us with a fine example of the golden mean.

RIVER FORMS AND FEATURES

In its attempts to reach a graded profile, a river takes on and develops many forms and features. We will examine some of the more important of these—the meanders of rivers and their tendency to break into several channels, levees, and deltas.

MEANDERING RIVERS

To begin with, let us focus our attention on the middle and lower courses of rivers, from the point where evidence of a developing flood plain exists to the mouth, where a delta may exist. A *flood plain* constitutes a section of flat valley floor not covered by river water at all times but subject to inundation during periods of exceptionally high water. Such a flood plain is created by sideways, or lateral, erosion and is the product of valley widening. It normally begins to form in a bend of the river, where the river's actual course pushes against the outside wall of the valley, leaving a section of the inside unoccupied. This situation is

FIGURE 19-8
The Thames River upstream from London. The beauty of a graded river has inspired authors and artists for centuries.

illustrated at the top of Figure 19-3, where it is just starting. Then the river swings from side to side, widening its flood plain, which is covered by sand and gravel. As it does so, the valley walls against the outside of the river bends retreat through erosion and over-steepening, and the situation shown at the bottom develops. The river now *meanders* and its bends become rounded into smooth, regular curves like those in Figure 19-9. This pattern reflects the resistance of the local materials, the strength of the current, and sometimes the considerable age of the river, because it is usually achieved by prolonged erosion. This is why rivers in such a stage of development are sometimes called *mature*.

A flood plain thus consists of material deposited by the river during a flood stage, when it fills the valley cut into the bedrock below. This can be observed in the lower part of Figure 19-3. Because a river is almost always subject to fluctuations of volume, sections of this flood plain are alternately built up and carried away. During a flood, a river may overflow its meander course and completely cover its flood plain. When the flood abates, the flood plain is covered with new

FIGURE 19-9
A view of meanders in the flood plain of the Hay River, Alberta, Canada.

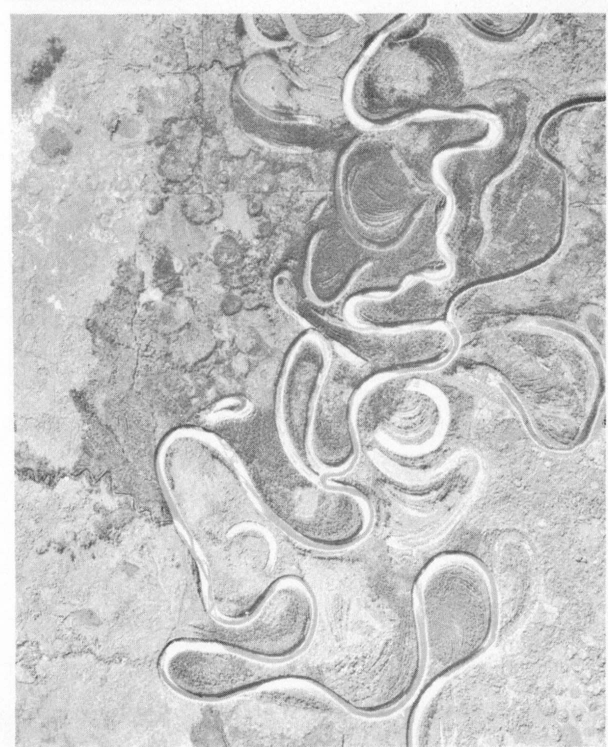

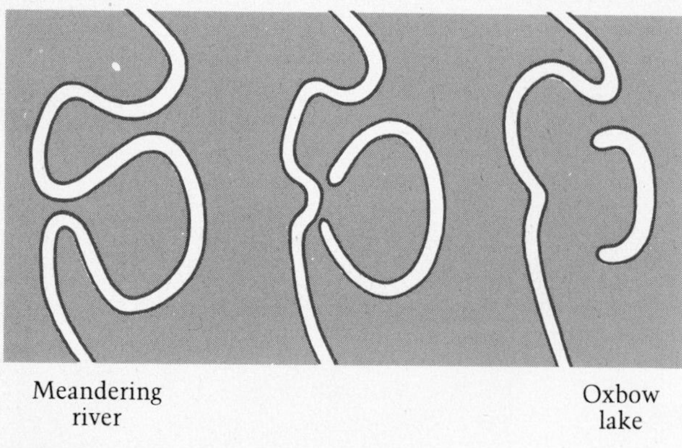

FIGURE 19-10

The development of an oxbow lake. Several oxbow lakes can be seen in Figure 19-9.

sediment, and the meander pattern may be disturbed. During the ensuing period of relative quiescence, the meander course is reestablished, perhaps with some changes. For example, two adjacent meanders may touch, cutting off one of the bends and creating an abandoned *oxbow lake* like those depicted in Figure 19-10.

BRAIDED RIVERS

With a flood plain established, a river has attained its aggradation phase, the stage where it begins to deposit as well as erode. Sometimes a river may be unable to maintain its channel within the flood plain, so overloaded does it become with sediments. Tributaries may bring an exceptional volume of gravel, sand, and silt. Meltwater from a glacier is usually laden with ground-up rock. When this happens or if, for some reason, the speed of the water flow is slowed down (by a dike across the valley, for example), then the river becomes disorganized. Its main channel subdivides into numerous distributaries that flow more or less parallel to one another across the accumulated sediment. Figure 19-11 shows this stream *braiding*, which sometimes also happens when a river reaches the delta at its mouth, the subdivision of the channel. But before the mouth, the braided stream usually collects itself again into a main channel. Examples of braided rivers can be found on the Great Plains. Early settlers, surprised by this channel form, described the rivers and their beds as "too thin to plow, too thick to drink."

NATURAL LEVEES

One of the most noteworthy characteristics of certain aggrading rivers is the formation of natural *levees*, strips of higher ground that develop next to the river's course as it winds its way across the flood plain. Such levees consist of sand, silt, and mud, and their origin is related to the flood periods of the river.

When a flood occurs and a river inundates its flood plain, much of the downstream flow still goes through the meander course that normally carries the water. Here water runs fastest, faster than it moves anywhere else on the flood plain. This means that the majority of sediments are also carried through this submerged river course. However, water does spread beyond the confines of the channel, and when it does, it mixes with the flood waters that stand on the flood plain. Quickly it loses its downstream velocity and drops much of its sedimentary load. People who live on flood plains sometimes have over a third of a meter (1 ft) of alluvium dropped in their yard during a single flood. The sediment is deposited immediately next to the meander channel, so that is where a ridge of sediments—a natural levee—develops. When the river returns to normal, confined once again to its meander course, and the flood plain has dried up, the natural

FLOOD!

The flood plain of a major river often is an area of rich, fertile soils, but it can also be a place of danger. Below the levees built up by the streams in various parts of the world (such as the Hwang River in China, the Ganges in India, and the Mississippi in the United States) live tens of millions of people, frequently at great risk. Sustained heavy rains can turn the quietly flowing river into a raging torrent that overflows its levees, breaks through them, and inundates the farmlands. All too frequently, there is no time for the local residents to escape.

Improved warning systems and better transport facilities have in recent years reduced the toll, but over the past century alone, many millions of people have died in flood plain inundations. The Johnston flood (Pennsylvania) in 1889 killed more than 2200 people. Among the more serious disasters of recent years is the September 1971 flood in Uttar Pradesh, India, when more than 300 persons died in floods in the vicinity of the capital, Lucknow.

FIGURE 19-11
A braided river, Peters Creek, in the Chugach Mountains near Anchorage, Alaska. The creek is fed by melt water from Eagle Glacier in the background.

levee remains, forming a belt of land higher than any other in the whole flood plain. Figure 19-12 diagrams this structure.

In an old river, whose wide flood plain has been inundated innumerable times, the levees may be built up so strongly that they contain a river whose surface actually lies above the level of the flood plain. The alluvial material of a flood plain is often excellent farmland, but a breach of the levee can bring a tremendous rush of floodwaters into homes and lands. Over the past century alone, millions of people have lost their lives in such calamitous breakthroughs in the flood plains of the great rivers of the world.

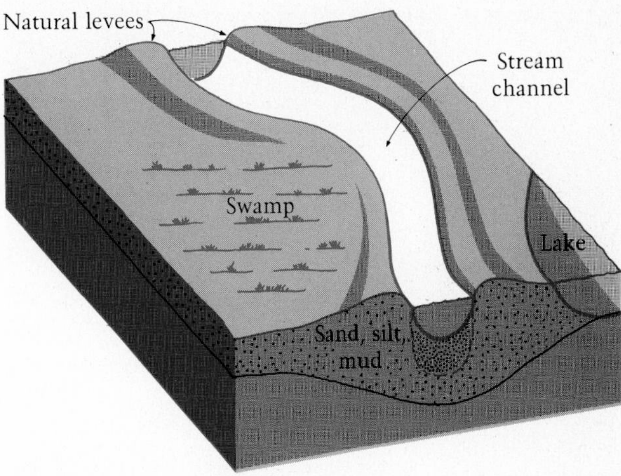

FIGURE 19-12
The relationship of the natural levee to the river and the flood plain.

DELTAS

When a sediment-laden river enters a lake or a sea, its velocity is first reduced, then brought to a virtual standstill, and the sediments are laid down. The result is a *delta*, a deposit surrounding and extending beyond the mouth of the river. Often such a fan had a vaguely triangular shape, as the famous Nile Delta does in Figure 19-13. The ancient Greeks used the fourth letter of their alphabet, shaped like a triangle, to denote this feature.

Many of the world's major rivers have deltas: the Nile, the Mississippi, the Amazon in South America, the Niger in Africa, and the Ganges in Asia. Some

FIGURE 19-13
The Nile delta, with a roughly triangular shape.

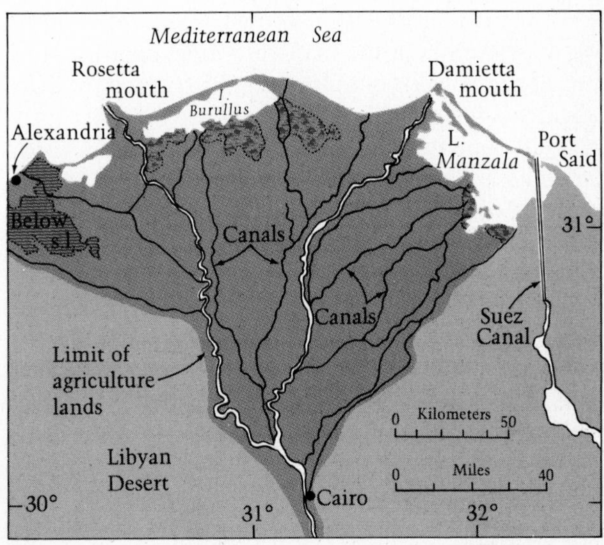

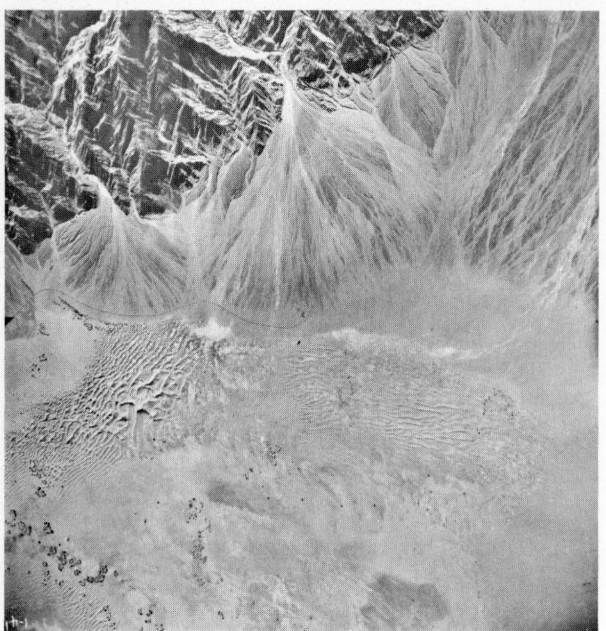

FIGURE 19-14
Several alluvial fans at the margin of Death Valley, California. Note the delta shape and the numerous stream channels that build up the fans.

rivers do not have deltas, however, at least not on the same scale. The Zäire (Congo) River is a conspicuous example. A combination of circumstances prevents the formation of a large delta. Rapid currents along the coast may disperse sediments before they can come to rest. Or no suitable shelf may exist where the river reaches the sea.

Rivers that generate deltas do so in several different shapes. But not all deltas are as nearly triangular as that of the Nile. The Mississippi River now extends more than 160 km (100 mi) beyond New Orleans, out into the Gulf of Mexico. It flows across delta lands it has built, just as it flows within its own levees upstream on the flood plain. On the map, the Mississippi lies on a long finger pointing out into the gulf. This type of delta is referred to as a *bird-foot delta*.

Rivers sometimes flow into dry plains and not into lakes or seas. Emerging from a mountain area following heavy rain storms, they are full of sediments ranging from coarse to fine, a load that proves too great when they reach the nearly flat surface. Again the river starts its aggradation: It deposits its coarser material first, then the medium-sized particles, and eventually the finer ones. The result looks a great deal like a dry-land delta, as Figure 19-14 shows, and it is called an *alluvial fan*. An important aspect of such alluvial fans is the groundwater they often contain. When streams reach the fan, part of the water begins to seep into the pile of loose sediment and becomes trapped lower down. Wells can retrieve it for human use, and the fan is a ready-made slope for distributing it to water-deficient areas below.

ACCIDENTS IN RIVER DEVELOPMENT

So far we have discussed the life cycle of rivers without consideration of one critical factor: the changes that occur in the crust that the river flows over. If the landmass being eroded by the stream stood completely still—without any tilting, warping, folding, or isostatic readjustment—then the sequence of events might be as we described it. But things are never that simple.

NEW LIFE TO OLD RIVERS

Consider, for instance, what would happen if the region where the river has its flood plain were suddenly tilted toward the ocean, increasing the slope—not enough to throw the river out of its meanders, but enough to increase its rate of flow. The river's capacity to erode downward, long a minor factor on the flood plain, would be quickly revived. Those shallow meanders would cut deeply into the flood-plain floor. The meanders may have formed on flood-plain sediments, but with its new power, the river etches them into the bedrock as well. Thus the meanders become *entrenched,* a sign that something has happened to give the erosive power of the river a new lease on life. Appropriately, we may speak of the *rejuvenation* of the river. Figure 19-15 portrays a rejuvenated river. Such rejuvenation may also occur when sea level falls.

A river may also increase in volume as a result, for example, of a change to a more humid climate in the region it drains. The river begins to cut down into alluvial deposits that fill its valley, changing its work from aggradation to degradation. Soon much of the alluvium has been removed, especially if the base level has been lowered as well, and only remnants of it remain along the valley sides. Those remnants are *alluvial terraces,* marked by a flat upper surface and a minor scarp facing the center of the river valley. Sometimes a river valley has a steplike succession of terraces that reflect a history of staged rejuvenation by uplift, climatic change, or some other process.

River terraces are common features of alluvial landscapes. Much of London is built on river terraces. A terrace may be underlain by either alluvium or bedrock. The lower parts of valleys near ice sheets or glaciers often have a large amount of debris that has been brought down by the ice. When the ice retreats, changes of base level may cause the valley streams to cut down and form terraces. Terraces on either side of a valley with their surfaces at the same altitude are called *paired terraces.* Sometimes individual, *unpaired terraces* develop when a meandering, down-cutting river removes all the alluvium on one side of the valley but leaves some on the other. Both paired and unpaired terraces are useful clues in determining the number of glacial advances and the sequence of events in an area. Where river terraces occur in a region that has been both glaciated and subjected to tectonic uplift, complex sets of terraces can occur.

There is still another way for a river to experience a kind of rejuvenation. This happens when the headwaters of a river, in their work of headward erosion, intersect the course of another stream and *capture* it. River capture adds the volume of that other stream to the existing water flow, leading to increased velocity, erosive capacity, and carrying ability.

ANTECEDENT RIVERS

It is possible, of course, that the forces of crustal deformation are so strong that a whole river system is entirely destroyed. River systems along coasts undergoing mountain building are not likely to survive unless the orogeny is unusually slow and mild. Still we see in the landscape evidence of rivers that have managed to maintain themselves throughout a period (or periods) of considerable geological change. When a river approaches a mountain range, cuts through it, and flows on the other side, it is possible that the river was there before the range was thrown up. The river might have been stopped or diverted if it had not had enough downcutting power to keep up with the rising land. When the mountain building finally stopped, the river was still in position, although interrupted by rapids and falls. Because this river predates (or *ante*dates) the

FIGURE 19-15

Rejuvenated landscape in Utah. Regional uplift has increased altitude and relief and has permitted the Colorado River to form entrenched meanders.

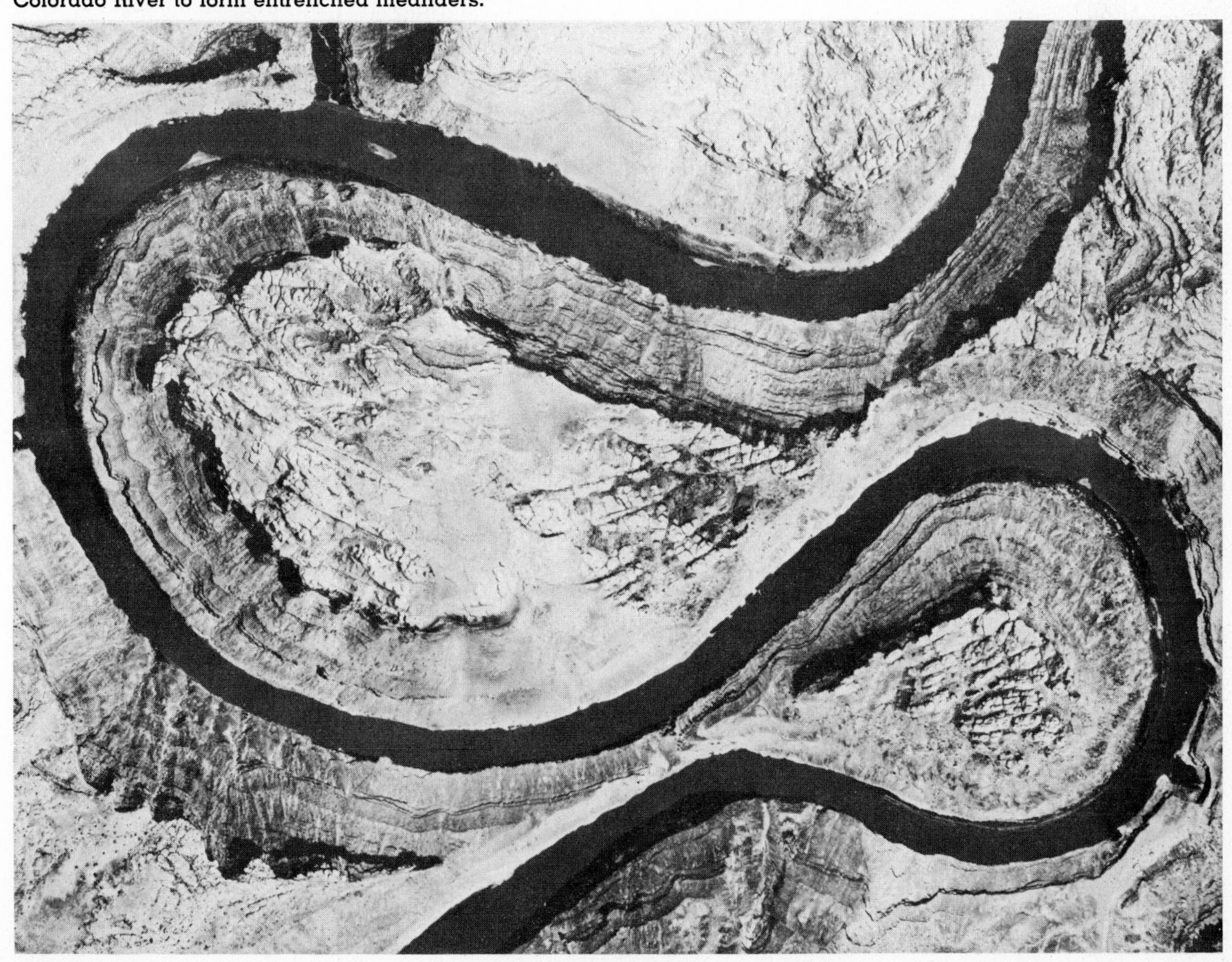

structure it flows across, it is identified as an *antecedent* stream.

It is not always easy to be sure that a river is indeed antecedent, because, as we will see, there are other ways for a river to cut through a mountain range. But we know that the sequence of events is possible. The Columbia River, where it cuts through the Cascade Mountains in the northwestern United States, may be antecedent to that range. In England, the gorge of the river Avon at Bristol is believed by some to be another example of antecedent drainage. The gorge, between Bristol and the sea, afforded good protection for the sailing ships that helped make the city a great port. But the modern larger ships could not navigate the gorge, so Bristol declined as a port.

SUPERIMPOSED RIVERS

Rivers can cut transversely, across topography, in other ways. What, for example, would be more reasonable than for rivers to flow in the valleys between the ridges of the Appalachians, northward to New England and southward to the coastal plain? But this is not the pattern at all: A whole system of rivers cuts through the Appalachians, draining the area from west to east. Although tributaries do conform to the relief, the direction of prevailing drainage in the Appalachians seems to have developed without any reference to the parallel ridge system.

Transverse drainage can develop in several ways. Imagine a series of parallel ridges, several hundred meters high, located on a plain—perhaps in the upper Midwest. In a glacial period, the ice sheets push hundreds of meters of ground-up rock across the ridges, filling the valleys and even covering the tops. The whole ridge topography is buried. Eventually the glaciers begin to melt, and a drainage pattern develops on the surface. This system does not correspond to the buried ridges at all, only to the slope of the existing surface. Rather soon this river system becomes organized, valleys develop, and erosion proceeds. At some point, those eroding rivers begin to expose the buried ridges, exhuming them. But the rivers are already firmly established, and they simply cut down into the ridges. Small gorges or passes develop in the emerging ridges, gorges that grow larger as the whole drainage system is *superimposed* on the formerly buried topography. By the time all the loose glacial material is removed and the ridges stand above the plain once again, they have been bequeathed with a river system they did not generate.

This imaginary sequence of events has many real-world parallels. It can happen when glacial debris covers an area, when lava flow buries the topography beneath it, or when sea level rises, the ocean invades, and sand and silt are deposited on the submerged relief. The case of parts of the Rocky Mountains is interesting. At one time, the Yellowstone, Bighorn, and Laramie rivers flowed over alluvium. The area was later uplifted, and the rivers were able to cut through both the alluvium and the large buried mountain ranges.

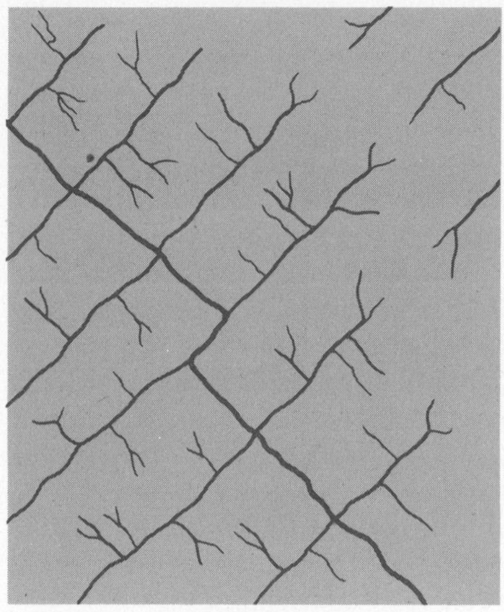

Trellis drainage

FIGURE 19-16

Trellis drainage pattern.

REGIONAL DRAINAGE PATTERNS

Although river systems can develop without structural controls and even escape those controls when they emerge, in numerous other instances geological structures do exert control over the regional drainage pattern. The drainage pattern is then not accidental. Sometimes we can deduce the form of the structure simply from looking at a map of the river pattern. A series of parallel ranges, for example, is likely to be drained by a set of rivers that conform to the topography in a *trellis* pattern, as in Figure 19-16. The main river is joined by a set of parallel tributaries, and they are in turn joined by other streams at marked right angles.

Much more common than the trellis pattern, however, is the treelike *dendritic* pattern shown in Figure 19-17. The dendritic pattern gives evidence that the drainage system has developed with little structural control, mainly on the same kind of rock (all soft sedimentary layers, for example). Tributaries join the major rivers at angles much less than the 90° seen in the trellis pattern. The whole structure indeed resembles the branches of a tree.

From a dome structure, or from a single large mountain such as a volcano, streams flow outward in all directions in a *radial* pattern. Structures such as basins, craters, and other depressions have the opposite effect and cause streams to converge in a *centripetal* pattern. Still other patterns of drainage develop under special circumstances, revealing structural controls such as joints, faults, eroded anticlines, and other features. The point is to be watchful for such evidence, in conjunction with other information, as we try to comprehend the shaping of the landscape and the role of rivers in that process.

As you may have gathered from the preceding pages, physical geographers often try to retrace the story of river development using the evidence still present in the landscape. But this is not the only concern of the geographer. Rivers have always held an interest for people—even beyond that of providing water for essential purposes. You may recall racing sticks down a stream as a child, to see which would arrive at a point first. The physical geographer who specializes in hydrology has important decisions to make about dam locations and sizes and the like. Even so he or she never quite loses the human fascination with the river.

FIGURE 19-17

Dendritic, radial, and centripetal drainage patterns.

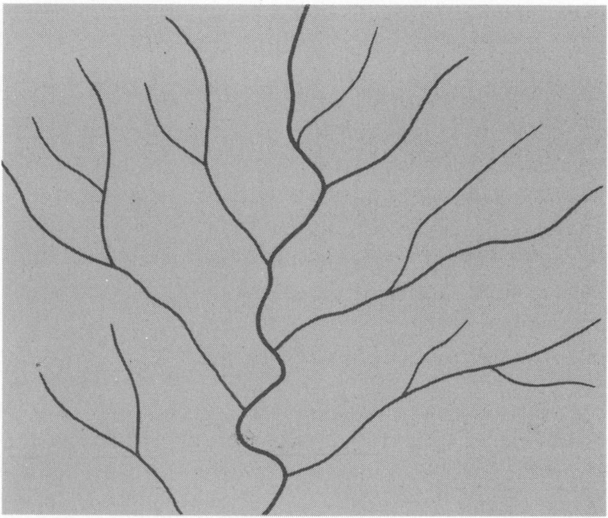

Dendritic drainage

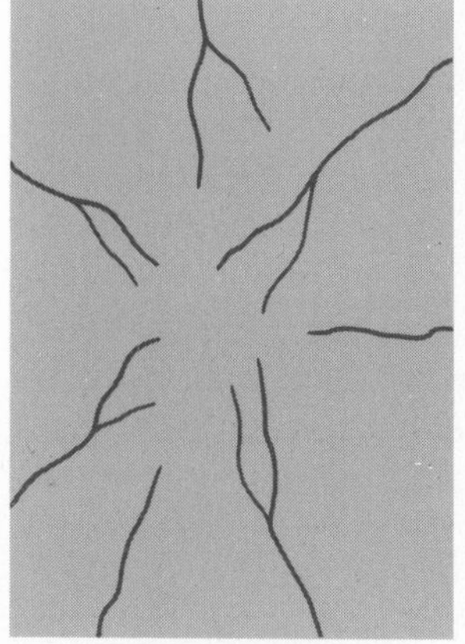

Radial drainage

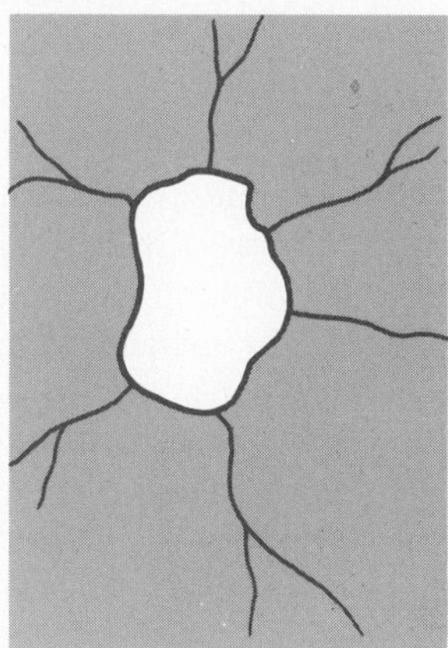

Centripetal drainage

SUMMARY

We have examined the complex and fascinating story of rivers. We saw how they begin, how their valleys are modified, how they erode, and how they transport.

All rivers try to cut down the earth materials to a base level that is sometimes a lake and more often the sea. The river system, through which matter and energy flow, tries to reach a state of grade in its longitudinal profile.

Many of the features of the river channel represent attempts of the river to reach the graded state. Rivers and their valleys have many forms and features, including meanders, braiding, levees, and deltas.

The story of rivers is not confined to their present processes. Many rivers have interesting histories. They can be rejuvenated, antecedent to, or superimposed on present-day landscapes. They also may show several different types of drainage patterns.

QUESTIONS

1. Rivers are said to have the greatest impact of any single force in shaping the surface features of landforms. What are the three basic processes involved in rivers' alteration of the landscape? How are these processes interrelated?

2. Why do most rivers and streams erode most rapidly in their upper reaches? How does the rate of erosion relate to valley shape?

3. Besides acting as a source of water for beginning streams, rain can be an erosional agent in its own right. What contributes to the most rapid removal of surface material by rain? What contributes to the greatest amount of runoff?

4. The different appearances of river valleys near the source of the river and near the mouth of the river are caused by several factors. Upstream, valleys have a characteristic V shape, while downstream they have a very broad U shape. What interrelated factors cause this difference in appearance?

5. In what ways do rivers erode, and what are the interrelationships among the three main types of stream erosion?

6. Rivers transport their load in four major ways—by traction, by saltation, in suspension, and in solution. Where in the general course of a river would we expect traction and saltation to be the principal means of transport? Where would suspension and solution be the main transport processes? What accounts for this difference in method of load movement?

7. The Mississippi River carries 65% of its load in suspension and 29% in solution, while the St. Lawrence River system, near its mouth, carries 88% of its load in solution. How does the morphology of these rivers explain this difference?

8. After studying the activity of rivers over several years, early geologists concluded that erosional processes took immense periods of time. They thus created geologic time scales for river valleys that were incorrect by three or four times the actual amount of time required for downwearing and widening processes to occur. What factors dealing with stream erosion and stream behavior did they fail to consider, and how might they have avoided making these incorrect assumptions about the average rate of erosion by rivers?

9. How does a river, as an open system, transform potential energy into kinetic energy? What factors are involved in this transformation, and how is this kinetic energy dispelled?

10. The graded profile of some rivers is a compromise between the two conflicting tendencies found in steady state systems. Identify these two tendencies. How is the graded profile a compromise between them? What final condition must be fulfilled before this compromise solution can take effect?

11. How do the semidependent variables of bed roughness, turbulence, and grain size relate to each other and to the independent variables in determining the slope of a river?

12. Is a stream with a graded profile, a steady state system, in a static state of balance? Why or why not?

13. Considering the factors that might cause a river to achieve a graded profile in either an upper or a lower reach, explain why most rivers achieve graded profiles in their lower reaches first, and then extend the graded condition slowly headward.

14. How does the state of grade of a channel affect the type of alteration engineers can make to the channel? Will the effects found in conjunction with dams, for instance, be greater or lesser with a graded stream than with a nongraded one? What factors affecting stream profile and achievement of a graded profile does the building of a dam change?

15. Mature rivers meander over the width of their flood plains, which can be considerable. How does meandering affect the slope of a river channel, and what effects might this have on the suspended load carried by mature streams in their lower reaches, where these wide, looping meanders form?

16. The development of natural levees, strips of higher ground that form next to a river's channel in a flood plain, is associated with changes in velocity of moving water carrying sediments in suspension. How is this process related to flood conditions in the formation of natural levees?

17. What similar processes affect both the formation of deltas and of alluvial fans? Why do alluvial fans tend to be more regular in shape than river deltas?

18. Rejuvenation of a river can occur either when the area through which the river flows is uplifted or when sea level falls, as might occur during an ice age. Incision is increased, sometimes dramatically, as in the Grand Canyon where uplift caused the Colorado River to cut deeply into the plateau. What factors related to both of these larger processes—the uplift of an area or a drop in sea level—give a rejuvenated river such new erosive power?

19. Structural control of drainage patterns is often not evident, or patterns may even appear surprising given the existing structure through which they flow. What are the two major types of atypical, unusual drainage patterns, and how are they different from each other?

20. In the majority of cases drainage patterns correspond to existing structural features. How can we figure out these structural features from a map that simply shows the drainage patterns of rivers? What type of structural feature is implied by radial drainage patterns? By centripetal patterns? By trellis patterns? What regular aspect of stream behavior, in most instances, allows us to see structural features from maps of their associated drainage patterns?

GEOMORPHOLOGY:
THE
EARTH
IN
PROFILE

CHAPTER 20

Other chapters in this book view the make-up of the landmasses and the various processes that operate to break them down: weathering, mass movements, river erosion and transportation. We know how weathering affects certain rocks, how mass movements alter particular slopes, how rivers adjust to specific base levels. But these chapters do not look at the whole landscape at once to try to explain how extensive landscapes develop. In this chapter we become physical geographers in the truest sense. We will take a regional view of landscape, seeking to discover how whole regions came to appear the way they do: the flatness of the Australian interior, the height and evenness of the Brazilian highlands, the low relief of the African plateau. In the United States, we will note that the numerous ridges of the Appalachians have summits of remarkably even height, that the Great Plains include several "plains" at different elevations, and that the even summits we find in the Appalachians prevail also in the Ouachita Mountains of Arkansas. How did these landscapes develop? This is the broad question of *geomorphology*, the study of landforms.

What we are looking for, of course, is a sort of life history of landscapes, a model sequence of events, or an alternative theory that would account for most of the variety we see. A number of scholars have attempted to develop such a model. In this chapter we will trace some of their efforts. We will find that highly intelligent people can look at the same landscape and arrive at different ideas of how it was formed. Which of the major different ideas are correct is still under debate. The answer to the debate lies in the surface of the earth. If you walk on the surface and look closely, perhaps you will be the one to find the answer.

In efforts to build models of landscape development, the critical element is the interpretation of the behavior of slopes. Landscapes are made of slopes, and even a surface that looks absolutely flat to the naked eye usually proves to be at least gently inclined.

ANATOMY OF SLOPES

Before we investigate the formation and modification of slopes, we should recognize the different types of slopes. We use such terms as *cliff, scarp, canyon,* and *ravine* (among others) to differentiate among them. The terms we use in physical geography may be a little less familiar, but they are easily understood. For example, a *free face* is a cliff or steep scarp. We see free faces on fault scarps, along coasts where the sea undercuts the land, and sometimes where glaciers have deepened valleys in hard rock. Exposed as the free face is to weathering, especially mechanical weathering, it often yields fragments of rock that fall to the base and accumulate there in what is called a talus slope or scree slope. The free-face portion of the slope, however, is the nearly vertical bare-rock segment above the scree slope, as shown in Figure 20-1.

Some slopes have a markedly concave, hollowed-out profile. The *concave slope* occurs frequently in dry or semiarid areas, where there is little soil and the slope is carved out of hard rock. In more humid areas, the accumulation of weathered material at the base of the hill may contribute to the concavity of the slope. In general, concave slopes in humid areas appear smoother in profile than those in arid areas, which often display a rather sharp break between the "foot" of the slope and the higher segment.

Other slopes have the opposite form; that is, they bulge. These *convex slopes* dominate many landscapes, especially those in more humid areas. Gently rolling plains are made up of such convex slopes, the convexity prevailing especially at the top of the hills.

As might be expected, the surface of some slopes is

FIGURE 20-1
Common slope elements.

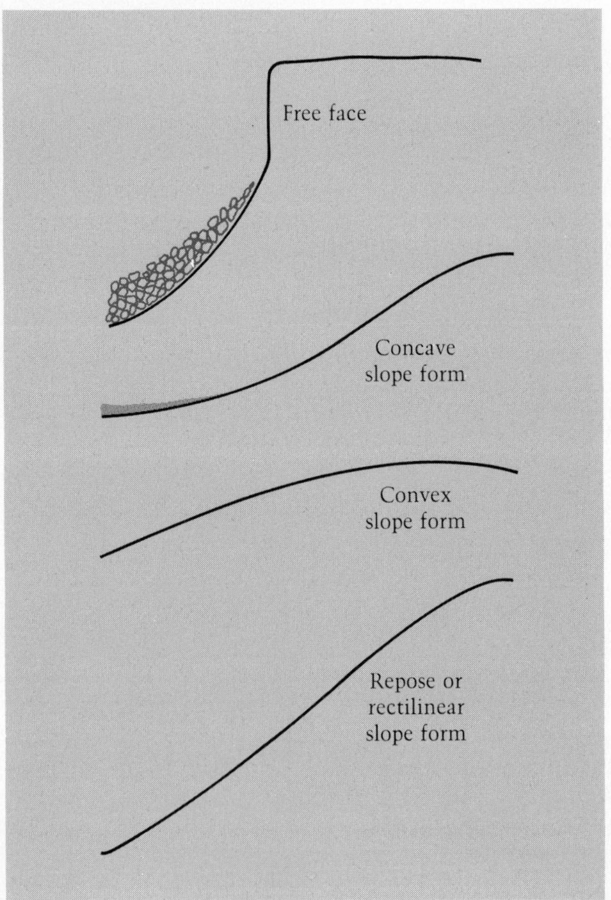

approximately straight, neither concave nor convex. Various names have been given to such slopes, including *rectilinear* (which simply means "straight line") and *repose* (because these slopes are often formed by weathered material lying at a certain angle). Although weathered material and material accumulated through mass movements often form such repose slopes, rectilinear slopes may develop in the same way concave and convex slopes form, by erosion or denudation.

Many slopes display more than one of these characteristics and are appropriately called *composite* slopes. Even without Figure 20-2, it is not difficult to imagine a hill with a convex top, a straight-line middle slope, and a concave slope at its base. A slope marked by a free face or scarp may also have a repose slope below the free face and a slightly concave foot, or *pediment,* still lower down.

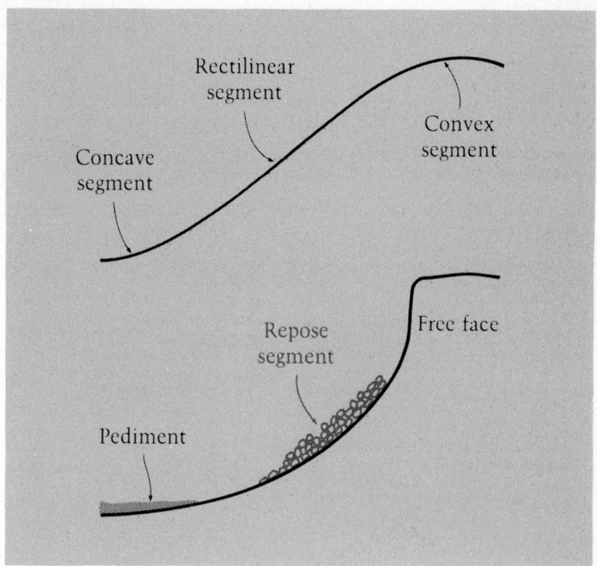

FIGURE 20-2
The segments of composite slope forms.

EVOLUTION OF SLOPE FORMS

Do the different slope forms reflect contrasting processes of formation? Will a concave slope remain concave until it is finally eroded away, and will a convex slope retain its form until the end? Or can a concave slope change to a convex one during erosion? Many physical geographers have tried to answer these questions, and still we are not absolutely sure. The problem is, of course, that no one lives long enough to witness the complete elimination of a single slope. This takes hundreds of thousands, even millions, of years. So it is necessary to observe different slopes in different places and deduce what changes have occurred.

The problem has many aspects: Soil creep and other mass movements affect slopes; the underlying geological structure plays a role; the degree of hardness, the resistance of the rocks have an influence; the climate may play a significant role. The initial slope angle is also of consequence: The slope we are studying may have begun as a steep fault scarp or as a comparatively slight incline resulting from an epeirogenic warp. Even this list is far from complete.

MASS MOVEMENT AND WATER FLOW

As long ago as 1908, American physical geographer Nevin M. Fenneman pointed out that, when rain water falls on a slope, there is little of it near the top, so erosion near a summit is slight. But downslope, there is more water. The water that runs from the summit is supplemented by additional rain on the upper slope, and erosional forces gain strength. Fenneman also pointed out that particles picked up by the running water contribute to the erosional process downslope, and with increasing effect. Thus, he concluded, a slope should become steeper—that is, convex—from the summit downward. Concavity at the bottom of the hill would result from the accumulation of particles carried downslope during the rain's "washing" of the slope or from the development of rills that would erode to a local base level at the foot of the hill. Such small hillside streams would develop miniature graded courses with concave cross-sections like that of the lower part of Figure 20-2.

Others, notably Glover K. Gilbert, pointed out that this idea of slope development did not account for soil creep and other forms of mass movements that occur on slopes. Some argued that soil creep alone might account for slope convexity. If a uniform thickness of soil is in motion, the farther downslope one goes, the greater the mass of soil that moves toward the base of the slope. And the greater the volume, the steeper must be the slope to maintain a steady flow of soil. This was the equivalent of general systems thinking as early as 1909. Again the conclusion is that slopes become steeper from the summit downward, taking a convex form.

Note that neither of these theories takes into account all the factors mentioned earlier: differences among underlying rocks, varying climates, and so forth. It took a long time for someone to suggest that both rain wash and soil creep might account for slope development, at least in humid areas.

The study of slope evolution got a major boost in

THE DAM ON THE TETON RIVER

On June 3, 1976, the massive earth and rockfill dam on the Teton River in southeast Idaho gave way. Eleven people were killed in the resulting flood, over 40,000 hectares (100,000 acres) of topsoil were washed away, and 13,000 head of cattle were drowned. The total amount of damage exceeded a billion dollars, but to many people, the worst thing about the disaster was the fact that it was probably predictable and preventable.

In the seven years of planning, testing, and litigation that preceded the start of construction in 1971, a wide variety of experts and other concerned citizens protested the building of the dam. Wildlife groups predicted the destruction of animal and fish preserves; seismologists pointed out that more than one major fault runs near the dam; politicians were concerned about escalating costs and a reduction in the amount of agricultural land as a result of construction.

But the most serious charges—and ultimately the ones that explain the collapse of the dam—involved the nature of the dam site itself. Most dams are designed to be as impervious as possible, permitting only minor leakage of water from the reservoir behind the dam. Shirley Pytlak, a former geologist with the Bureau of Reclamation (Department of Interior) which was building the dam, pointed out that the bedrock at the site was composed of a porous and easily cracked volcanic substance that absorbs water like a sponge. The Bureau countered that charge with the plan to fill the foundation with a cement mixture to plug the holes. This process, called grouting, while quite common to solve porosity problems in dams, is anything but a guarantee. The confidence the Bureau placed in the grouting of the Teton Dam proved unjustified. Water attacked the dam from the sides (along the canyon walls), percolated through its interior, and destroyed its structure. In the collapse, almost 50% of the earthfill used to construct the dam was washed away.

1942, when Alan Wood published an article suggesting that fully developed slopes consist of the four elements in Figure 20-3: the waxing slope, at the top; the free face; the debris or talus slope; and the pediment. Wood reported that each of these four elements undergoes semi-independent evolution in the course of a landscape's history. Not all four need be present in any given slope. According to Wood, a composite slope consisting of a convex upper segment and a concave lower segment, for example, would be an incompletely developed slope, lacking the free face and debris slope. A free face may have been present at some time but the rock below may not be strong enough to sustain a free face.

In a fully developed slope, the *waxing slope* is the convex crest or summit. As the name implies, this is a developing slope segment. As the free face encroaches on it, the waxing slope becomes increasingly convex. When there is no free face but only a two-segment composite slope, the waxing slope may constitute the rounded, convex hilltop.

The *free face,* as we noted earlier, is the scarp segment of the slope. Fragments of rock are pried from it by weathering, then fall down to the pile of talus or debris below. By implication, the free face is retreating, encroaching on the waxing slope while contributing to the lengthening of the debris slope below.

The *debris slope* is made of the rock fragments and boulders that have come from the free face, the angle of repose being related to the size of the pieces. This material is coarse, and so rain water passes between the fragments and partially erodes the rock floor the talus lies on. Weathering continues to attack the debris, reducing it in size and readying it for erosional removal.

The finer material temporarily lies on the *pediment,* the foot of the slope, the *waning slope* segment. Erosion on the pediment, as Fenneman stated decades earlier, produces a graded form similar to that of river valley profiles. Overall, the upper parts of slopes tend to

FIGURE 20-3

The four slope elements of which most slopes are made.

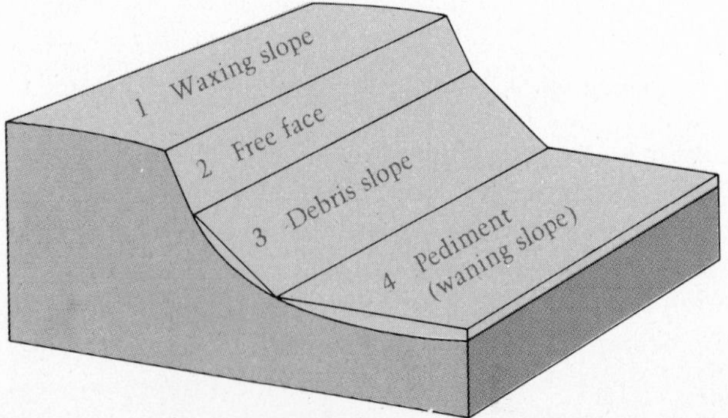

be dominated by mass movement, especially creep, and in the lower parts transportation by flowing water becomes more important.

THE RETREAT OF SLOPES

Wood identified the four elements of what he considered fully developed slopes, and he argued that, once formed (say by stream erosion in a valley or by the formation of a fault block), such slopes would retreat under erosion. Wood had a strong ally in this position. The famous German physical geographer Walther Penck had already suggested that slopes could retain their form while being driven back by weathering and erosion. Penck, in a book published in Germany in 1924 (it was not translated into English until 1953), had reasoned that the rate of retreat of a slope was related to its steepness.

A scarp, according to Penck, retreats much faster than a gentle slope for various reasons, among them the exposure of a scarp and the protective cushion of weathered material on a gentler slope. Thus if you look at the butte in Figure 20-4, you can visualize the steep sides of the hill retreating quite rapidly toward one another and the debris slope and pediment doing so slowly. Eventually the free-face segments meet, and the slope is replaced by the waxing and debris slopes. The pediment, too, gradually expands. Penck's idea of "slope replacement" is perhaps a better one than Wood's concept of fully and partially developed slopes.

Another scholar who supported Wood's analysis was South African Lester King, and he carried Wood's concept several steps further. King wrote that he saw a "uniformitarian nature" in slopes. That is, he envisioned Wood's four-element slope to be the world's basic landform, to be found in all regions of the globe and under all climates where running water is the dominant agent of erosion. Furthermore, the slope acted the same way in past geological periods as it does at present. Slopes, King said, are made of materials that respond more to laws of physics than to conditions of climate.

In support of his thesis, King pointed out something we can all recognize as true: Similar slope forms can be found in humid and in arid areas and under all kinds of climatic conditions. There are scarps in moist as well as dry areas, and you can find pediments in the Appalachians as well as the dry Southwest. But, as critics later emphasized, it is also possible to find different slope forms within the same climatic area. Thus there appears to be no alternative but to ascribe the contrasts to the factors King would dismiss: the number of streams in a given area, mass movements, rock type. The mode of slope change subscribed to by

FIGURE 20-4

A butte of horizontal red sandstones near Cambria, Wyoming.

Penck, Wood, King, and others—slope retreat—may not be as universal a process as they all assumed.

One alternative to slope retreat is slope *reduction*. Whereas Penck and King would have argued that the lower composite slope in Figure 20-2 would retreat under erosion, William Morris Davis, one of the most influential of all geomorphologists, suggested that such a slope would be progressively lowered. Weathering, mass movement, and erosion occur on all parts of the slope, reasoned Davis, and as a result it wears down, not back. The difference between these two views is diagramed in Figure 20-5. Davis agreed that, under certain circumstances, slope retreat could occur—for example, in dry areas where hills have thin soils, little vegetation, and often a hard caprock to sustain them, as mesas and buttes often do. But the "normal" sequence of events was that prevailing in humid areas, and there, Davis reasoned, the evidence was in favor of slope reduction.

Among the lessons to be learned from this controversy is that physical geographers can observe the same slopes and then draw completely different conclusions. Davis saw the Appalachians and concluded that they indicated a prevailing slope convexity and slope reduction. King saw the same Appalachians and found waxing slopes on the summits, poorly developed free faces, debris slopes, and pediments; and he thought that this proved slope retreat. Obviously measurement and mathematical analysis were in order so that the issue would involve less impressionism and more precision.

FIGURE 20-5

Contrasting ideas of slope formation. Penck and King advocated slope retreat, and Davis suggested general slope reduction.

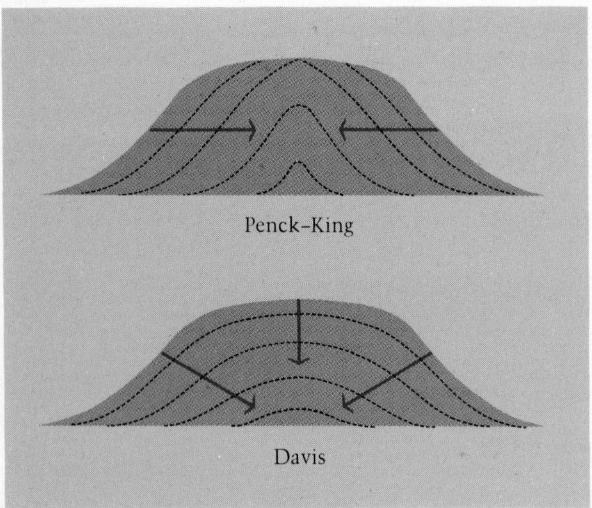

SLOPES IN BALANCE

This is where Arthur N. Strahler, the distinguished physical geographer now retired from Columbia University in New York, made significant contributions to the slope problem. Around 1950 he selected an area where a combination of conditions (rock type and structure, vegetation, soil, climate, overall relief) were uniform, and then he measured the maximum angle that slopes attained in various parts of the area. He found that this maximum angle was remarkably uniform throughout the study area. When Strahler calculated the average of all the maximum angles he had measured, he discovered that the great majority of all the slopes in the region had maximum inclines within a degree or two of that average. This situation, he reasoned, could not be a matter of accident.

Strahler concluded that this must be the angle that facilitates the most efficient removal of material loosened by weathering, mass movement, and rain wash. In other words, this is the angle of *equilibrium*, of balance. According to Strahler's equilibrium-slope hypothesis, the slope angle represents a condition of balance involving not only rock type and structure, but also climate, soil, vegetation, and other circumstances. If one of these conditions were to change (say an increase in annual rainfall), then we could expect the angle of equilibrium to change as well.

Here again is an application of the principle of dynamic equilibrium. The slope is regarded as an open system in Figure 20-6. Forces act on the materials of the slope: Gravity pulls down, and the underlying bedrock pushes up. Matter, in this case rock material, passes through the system from the top to the bottom. Energy also passes through the system. The upper rocks gain potential energy from previous tectonic uplift. Potential energy changes to kinetic energy when gravity pulls the rock materials downslope. As in any system, there are movements and changes of material and energy on a slope. Strahler suggests that the angle of the slope adjusts itself so that the system operates efficiently.

MODELS OF LANDSCAPE DEVELOPMENT

The same physical geographers who study slopes use their conclusions to develop large-scale models for the evolution of landscapes. In this context, the behavior of slopes is obviously critical. Take just one instance: the retreat as opposed to the reduction of slopes. If a certain region is uplifted and attacked by erosion, it follows that, under the *retreat* of slopes, the *original upland surface survives until retreating slopes meet.*

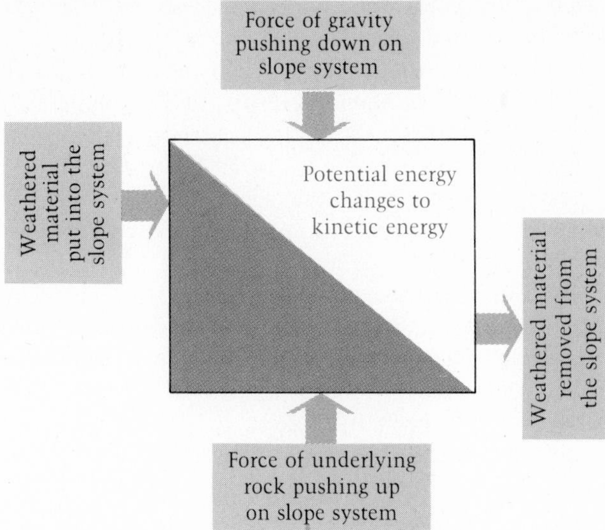

FIGURE 20-6
The slope system in dynamic equilibrium.

Similarly, the top of an outlier is reduced to mesa size and eventually to butte size, but the upper surface remains until near the very end. During the *reduction* of slopes, however, *original surfaces are destroyed relatively early in the cycle,* because the whole landscape is lowered almost as soon as erosion begins.

THE DAVIS MODEL

The first comprehensive model to gain wide acceptance in physical geography was that of Davis, who proposed it in a series of articles published in the 1890s and early in this century. By the time Davis produced his model, other physical geographers had already developed concepts of stream grading, base level, stream genetics (such as superimposition), and what we will call *planation,* the production of a surface of low relief as the final product of a long period of erosion. Davis integrated these ideas into what he called the *geological cycle.*

According to Davis, landscapes can be analyzed with reference to three variables: structure, process, and stage. We have already noted how *structure* can influence the appearance of landforms. Faults, rifts, domes, folds, and other structural features can be recognized with ease before erosion begins to destroy or modify them. By *process,* Davis meant the actual forces shaping landforms, including not only weathering, mass movements, and erosion by running water, but also glaciers, waves, and wind. The *stage* refers to the length of time the processes that fashion the landforms have prevailed. Davis, as we will see shortly, was much concerned with the ages of landscapes.

In order to illustrate his cycle of erosion, Davis made several assumptions. First, he postulated that a nearly flat surface, say a coastal plain, had been uplifted very rapidly in geologic terms, to perhaps several hundred meters above its previous level, and sloped gently toward the sea. Second, Davis assumed that there would be no further crustal movement to affect this surface, so that for the duration of the erosion cycle to follow, it would stand still. Third, Davis's model held base level to be fixed during the entire cycle. Now, Davis asked, what would be the sequence of events in the landscape developing on this elevated surface?

Youth Initially, drainage of the uplifted land surface is disoriented, but with ample water supply (Davis assumed, for his "normal" cycle, a humid-temperate environment), an integrated system of streams develops rather soon. Streams forming on the initial landmass are called *consequent* streams, and their direction of flow is a consequence of the slope of the land. The streams rapidly etch V-shaped valleys bounded by steep slopes. In the process, the streams cut into rocks of different hardness, and waterfalls and rapids develop. On the valley slopes, mass movement becomes active as undercutting occurs, and the rock fragments that fall and slide into the river provide ammunition for even more effective erosion.

The landscape's youthfulness reveals itself in the preservation of large sections of the original surface, still unaffected by the action of the streams. In later youth, the active headward erosion of the streams and the intensifying *stream texture,* the number of streams in a given area, combine to reduce the amount of original surface remaining (see Figure 20-7). The convex slopes Davis saw in the world's plains take shape quite early.

Maturity The rivers reach maturity first. After rapid downcutting in the youthful stage, they reach a condition where they are able to transport just the amount of load that enters them. Falls and rapids are worn down, the lengthwise profile is smoothed out, and the streams become graded. Relief is at a maximum, because small sections of the original surface still exist. The river's lateral erosion becomes more active, and the first evidence of flood-plain development appears.

The rivers and the whole landscape achieve a graded condition during maturity. Although lateral erosion does affect the valley sides of the rivers, weathering, soil creep, and rain wear down the *divides,* the higher ground between the rivers. These processes produce debris at a rate that is in balance with the capacity of the rivers to remove it, so the divide areas join in the

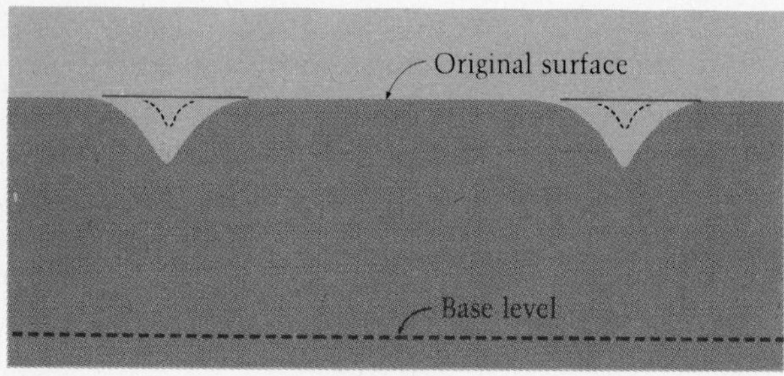

Early youth

Late youth

FIGURE 20-7

The removal of parts of the original surface in the youth stage of the Davis model of landscape development.

overall equilibrium believed to exist during maturity.

Later during maturity, as Figure 20-8 shows, the reduction of the divide areas eliminates the last remnant of the original surface. Cliffs, rock outcrops, and other free-face manifestations are rounded off, and a mantle of weathered debris and soil covers the countryside. The general appearance of the landscape is that of a system of deeply entrenched, graded streams separated by high, well-rounded divides. Now, however, the whole landscape is under erosional attack, and the divides are being lowered quite rapidly.

Old age During the last stage of the Davis model, the rivers reach their lowest possible gradients with the prevailing base level, and the divides are worn down until they are mere undulations. Rivers meander slowly over enormously wide flood plains. Everything is slowed down: the flow of the water in rivers, the erosion of remaining valley sides, the processes of downwearing on the lowered divides. This is by far the longest stage of the three in the Davis model. Youth ends quite quickly (by comparison); maturity lasts somewhat longer; but old age is longest of all.

Toward the end of old age, the landscape is nearly flat, with only a *monadnock* of exceptionally resistant rock still rising above the surface. Davis called this landscape a *peneplain* (from the Latin *paene,* "almost"), a nearly flat plain, like that in Figure 20-9. Some geographers write *peneplane* because Davis was describing a landscape that was almost a flat surface, and a flat surface in geometry is a plane. It is rumored, incidentally, that Davis described a peneplain as a surface that a horse could draw a carriage over at a trot in any direction.

Davis himself realized that some of the assumptions necessary for the completion of his proposed erosional cycle might rarely, perhaps never, exist. It is unlikely that an uplifted surface would remain unaffected by earth movements for the millions of years needed to transform it as the model suggests. Thus Davis, and others too, tried to modify the model by predicting what would happen if renewed uplift were to occur somewhere during the cycle. If rejuvenation were to occur during youth or maturity, there would be evidence in the river valleys—perhaps breaks in the cross-profiles of the valley sides or terraces on the valley floor. Uplift might also be recorded and imprinted on the divide slopes. Rejuvenation during old age, when peneplain conditions were being approached, would lead to the development of a second peneplain at a lower elevation.

Davis was a prolific author and a powerful teacher, and the impact of his model was enormous. Colleagues and students of Davis began to find peneplains and remnants of peneplains nearly everywhere. And indeed, although they have all now been reinterpreted,

FIGURE 20-8

The landscape profile during the stage of maturity in the Davis model.

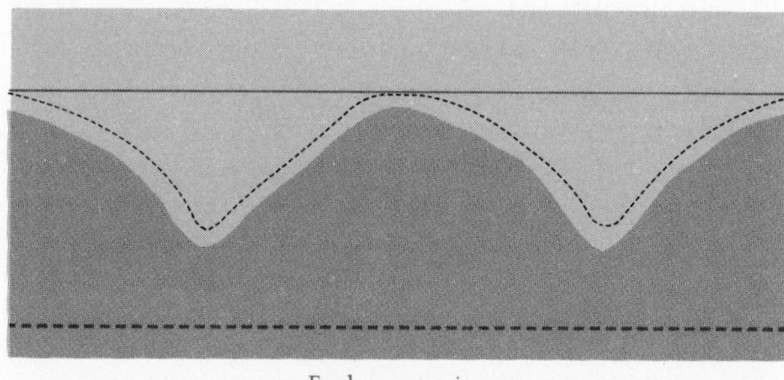

Early maturity

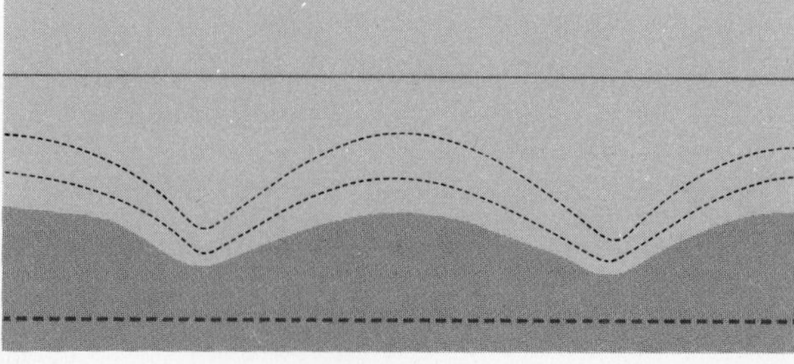

Late maturity

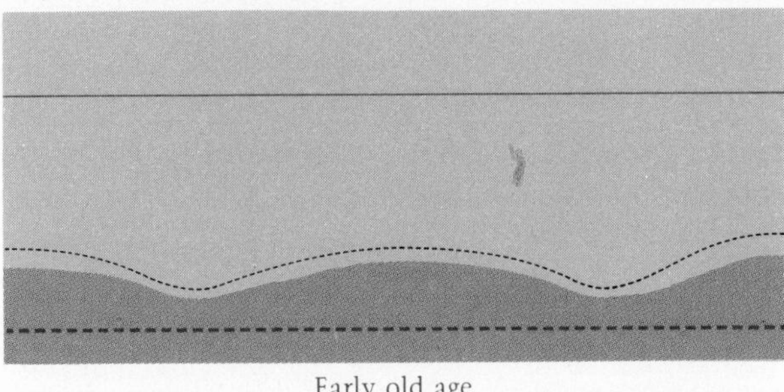

Early old age

Peneplain

FIGURE 20-9

The landscape profile during old age in the Davis model.

some of the examples they recorded seemed to prove Davis correct. In the Appalachians, for instance, the Davis cycle provided a single answer to two remarkable landscape features: the nearly accordant summit heights of the ridges and the apparently unrelated drainage pattern mentioned in Chapter 19.

THE PENCK-KING MODEL

Among the most prominent physical geographers who disagreed with aspects of the Davis model was the same Penck we met earlier in this chapter. Primarily, Penck held that Davis's assumptions relating to the stability of the crust were unrealistic, and he argued that any model of landscape development ought to incorporate the effects of uplift, tilting, and warping in the crust. Rivers eroding an area might produce landscapes giving evidence of one of three combinations of circumstances, as in Figure 20-10: *(a)* rivers eroding faster than the rate of uplift; *(b)* rivers eroding just about as fast as the rate of uplift; and *(c)* rivers eroding slower than the rate of uplift. For example, in case *(a)* rivers cut down more rapidly than the rate of uplift, so some erosive energy can be spent on lateral, sideways erosion. The result is concave slopes. In case *(b)*, straight slopes result, and in case *(c)*, rivers must spend all their energy in downcutting, so adjacent valley slopes are convex.

To view divides as only convex and as merely wearing down, as Davis had suggested, seemed to Penck an oversimplification. He was not convinced that slopes are subject to reduction. As we have seen, Penck believed that slopes are under retreat instead. As they retreat, they leave behind a gently sloping surface, at an angle related to prevailing base level, called the *pediment*. Penck and Davis engaged in active debate for years. They once met in the tectonically active landscape of California, and their conversation must have been priceless.

In the long run, King translated certain of Penck's ideas into a comprehensive model of landscape development. It is important here to recall what King said about the "uniformitarian" character of slopes. Davis had postulated a "normal" cycle of erosion for humid areas, and where this model would not work, in arid areas for instance, he proposed a quite different one.

FIGURE 20-10

Valley cross-sections in the Penck model of landscape development. In *(a)* rivers erode faster than the rate of land uplift; in *(b)* rivers erode just about as fast as the rate of uplift; and in (c) rivers erode slower than the rate of uplift.

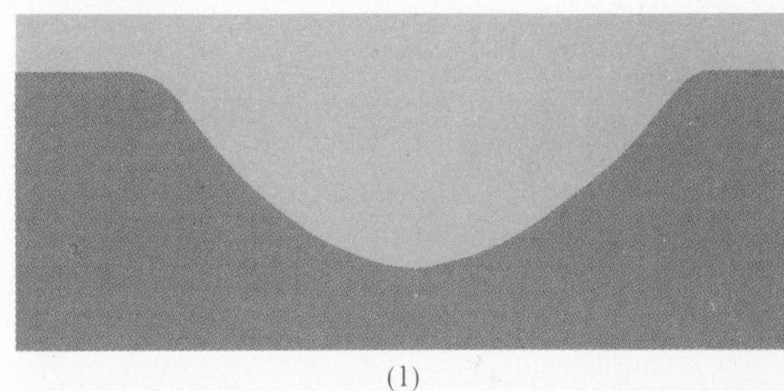

(1)

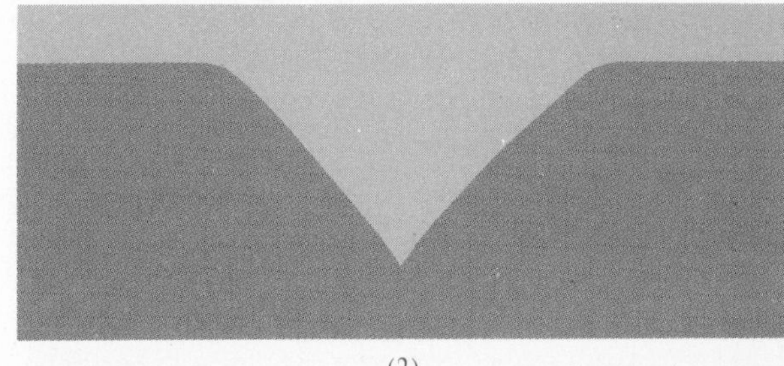

(2)

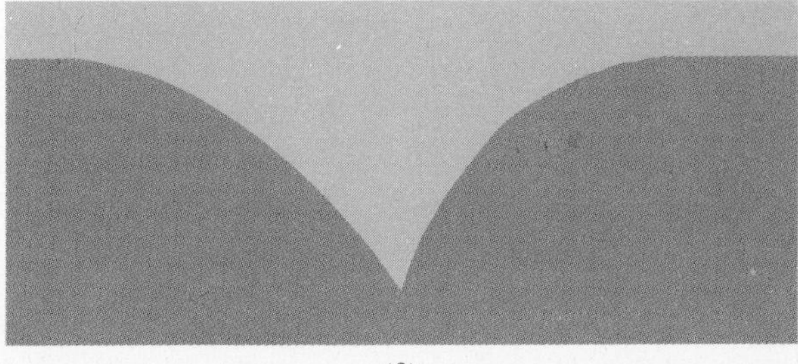

(3)

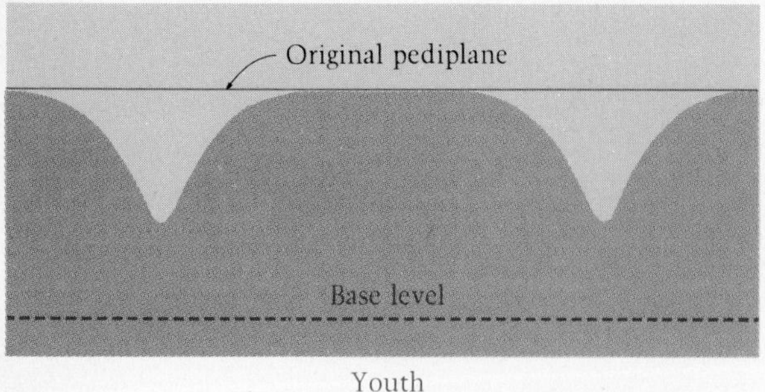

FIGURE 20-11

The landscape profile in the youthful stage of the Penck-King model of landscape development.

King argued that a comprehensive model for landscape development should really be comprehensive and account for all landscapes created by running water. In any case, King stated, the world map of climate argues against "normally" humid conditions. Perhaps dry climates are the norm, and the humid cycle should be considered the exception. The Penck-King model, outlined in a series of papers that appeared during the late 1940s and 1950s, is based on two main principles: slope retreat (not reduction) and the formation of pediments that form a *pediplane* when they coalesce.

Youth Although King did not apply Davis's stages to the model, it is useful to compare the two ideas in the time context Davis had introduced. Youth, of course, involves an uplifted pediplane (pedi*plain*). As in the Davis model, the streams commence their rapid downcutting, creating slopes that begin to retreat as soon as lateral erosion begins. Once created, the form of the slope remains more or less constant, like that in Figure 20-11. In late youth, small pediments begin to form in the river valleys. But substantial sections of the former pediplane still exist, unaffected by reduction and as yet untouched by slope retreat.

Maturity The pediments gradually broaden, and the new pediplane makes its appearance while large parts of the former, uplifted pediplane still survive. In some areas, retreating slopes intersect, and the older pediplane surface is wiped out as in Figure 20-12. Remnant hills, or *inselbergs,* make their appearance in such instances. These inselbergs are the same as the monadnocks Davis saw rising above the peneplain. On the African plateau, where King did much of his initial research, they do indeed look like small islands rising above the flat erosion surface, and hence the name *inselberg,* "island hill."

Even in this stage, substantial remnants of the older surface can still be observed and mapped. This is a major and critical difference between the Davis model and that of Penck and King. In the Davis view of the Appalachians, the surface that extended over the summits of the ranges lay some meters above the present tops of the ridges, because reduction has already taken place in the new cycle. But according to King, we can go to the Appalachians and actually stand on remnants of that older surface. The ridge summits are not just approximate accordances (the same height), they are actual pieces of the old surface, not yet destroyed by slope retreat. If the rock is hard enough, the pediplane surface can survive for millions, even tens of millions, of years. King believed he had mapped remnants of the 180-million-year-old Gondwana surface in Africa.

Old age Coalescing and interlocking pediments dominate the landscape, with an occasional inselberg rising above the area. The overall view suggests a slight concavity to the divide areas between streams, as in Figure 20-13, not the convexity Davis saw. The implication of the Davis model is slope reduction and the rapid elimination of old surfaces. The implication of the King model is slope retreat and the lengthy survival of old surfaces.

THE HACK MODEL

In the early 1950s, the central Appalachians, where Davis had formulated many of his ideas, were visited by geomorphologist John T. Hack. He was deliberately seeking a new approach to geomorphological problems and was making a conscious effort to abandon the cyclic theory as an explanation of landforms. He believed Davis to be wrong on a number of fundamental counts. For example, Davis was wrong to apply such concepts as lateral planation, the tendency for streams to laterally erode their banks, in all areas at all times. Davis was further in error in assuming that rivers have higher velocities in their upper courses than their

FIGURE 20-12

The landscape profile in the mature stage of the Penck-King model, with previous positions of retreating slopes also shown.

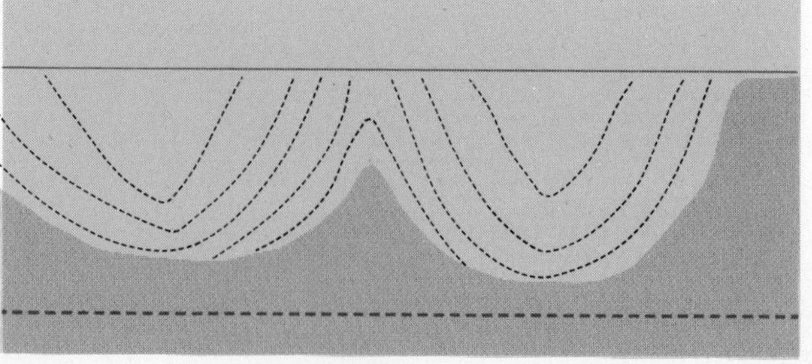

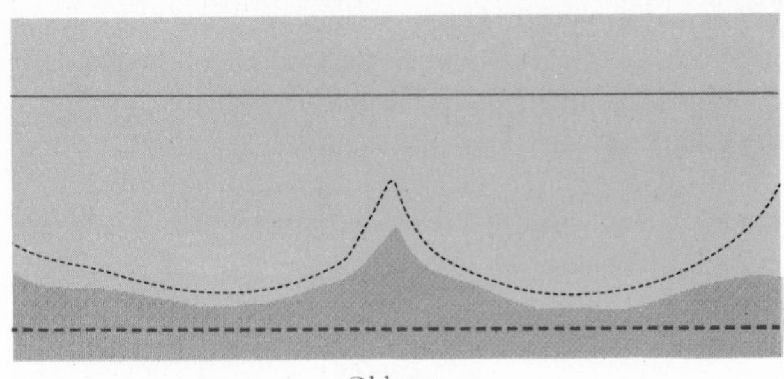

FIGURE 20-13

The landscape profile in old age in the Penck-King model.

washed down the face from its summit is exactly balanced by erosion at the foot.

It is the preservation of the balance of energy, says Hack, that accounts for different relief in different topographies. If an area were composed partly of hard quartzite and partly of soft shale, more energy would be required to break up and transport the quartzite at the same rate as the shale. Hack claims that the rates of removal must be the same to preserve the overall balance of energy. So greater energy is found in the quartzite area—in the form of high relief and steep slopes. Less energy is found in the shale area, as shown by its lower relief and shallower slopes.

lower courses, an idea that permeates his writings.

Hack reintroduced the principle of dynamic equilibrium. The idea had already been applied by Gilbert as long ago as 1877 and by Strahler just prior to Hack's work. As Hack described it in 1960,

> the concept of dynamic equilibrium requires a state of balance between opposing forces such that they operate at equal rates and their effects cancel each other to produce a steady state, in which energy is continually entering and leaving the system.

Examples of dynamic equilibrium in the landscape include an alluvial fan, where debris from the mountain behind it is deposited at the same rate that debris is removed by erosion, and a slope, where material

Erosionally graded topography Landscapes in which large areas are mutually adjusted and the variation of form is mainly the result of rocks that yield to weathering in different ways are called *erosionally graded landscapes.* In such landscapes, differences in the bedrock and the way it breaks up into different components on slopes and in streams are all-important.

Figure 20-14 shows an area in Virginia, on the northwestern side of the Shenandoah Valley. It contains three rivers—Dry River, Briery Branch, and North River—that rise in resistant rock and then travel over soft rocks. One stream, Mossy Creek, travels only over soft rocks.

There are many examples of equilibrium in the landscape of this area. First, in the resistant rock area, relief is high because there is high potential energy and

FIGURE 20-14

The Appalachians northwest of the Shenandoah Valley in Virginia.

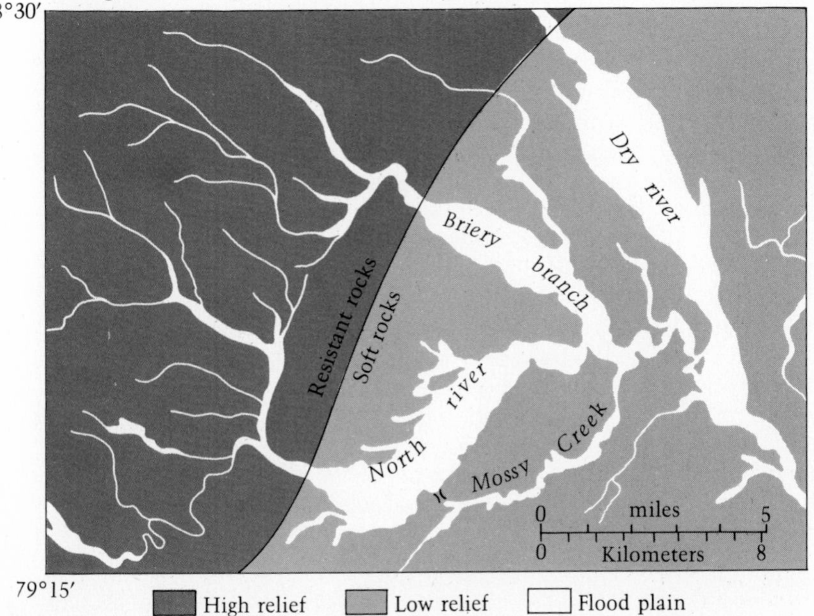

consequently high kinetic energy. Mechanical weathering processes are dominant, and there are many rock slides. In the soft rock area, chemical weathering is more important. Although the surface of the soft rock is being lowered at the same rate as the resistant rock, graded slopes are much gentler and the relief is lower in the soft-rock area.

Second, when the three streams that cross both hard and soft rocks enter the area of the soft rocks, they immediately adjust themselves. They develop broad flood plains and terraces composed of cobbles and sand from the resistant rock area.

Third, the stream that crosses only soft rock, Mossy Creek, has a flatter gradient than North River even though it is a smaller stream. It has adjusted to only the soft rock and not to both resistant and soft rocks.

Dynamic equilibrium and time The models of Davis and King are time-dependent; that is, everything depends on the stage in the cycle at which the landscape is observed and studied. Hack and others have argued that there may be times when landforms do not undergo changes of form. This is not to suggest that rainfall, weathering, mass movements, and other processes actually cease to act on the landscape. Rather, everything is changed at the same rate at the same time, so although a whole landscape may be lowered, all things relative—stream valley widths, slope angles, weathering rates—stay the same.

The only time when forms change is when there is a change of energy applied to the landscape system. This creates a problem in Hack's model, because the energy applied to the landscape does change in time. It becomes less as the land is gradually reduced in height and increases in times of tectonic uplift. New adjustments to these changes, so that dynamic equilibrium is reached, may produce different landforms. The question is still open.

MODELS AND HUMAN REALITY

You have every right to ask if there is any advantage in having such a long debate on the way landforms develop. But the answer, of course, is yes, from both the practical and the more abstract points of view. Let us look at a practical example first.

Manganese ore is a valuable mineral deposit. It is usually like iron but not magnetic. One place where it is found is under the quartzite gravels shed from highland areas of the Appalachians, such as those in Virginia. If the Davis cyclic theory of landscape evolution is correct, these ores were formed during a cycle of erosion that took place in Tertiary times. The implication is that they are not a renewable resource (unless we wait for another erosion cycle). But if the theory of the equilibrium, or erosionally graded, landscape is correct, ore deposits may be forming at the present time. In other words, the ore may be much more of a renewable resource than was previously thought, although a large amount of time is still involved in its renewal. You may now be able to see the significance of the argument among the scholars.

A NEW TOOL—NEW DISCOVERIES

Just as cartography was revolutionized by the development of aerial surveying, geomorphology is getting a dramatic boost from the data supplied by the twin earth satellites Landsat I and Landsat II, built and maintained by the United States. In addition to important data on agriculture, climate, pollution, shipping conditions, and population distribution, the powerful and sharp lenses in the satellites' cameras are revealing aspects of the earth's crust that literally would be undetectable by land-based instruments.

For example, *Newsweek* magazine quotes John Reinemund of the United States Geological Survey as reporting massive creases in the earth's surface, some of which run across one continent, probably under water, and across another continent. These areas might represent mineral rich "fractures" in the earth's surface that could be profitably mined without additional costly research.

The economic impact of the Landsat data is hard to calculate. For example, petroleum geologists can "read" color and density variations in the earth's crust from above, looking for the distinctive patterns that indicate oil-bearing rock below. In planning roads and other construction the Landsat cameras can be used to detect excessive moisture or other rock and soil characteristics that influence construction safety and efficiency, so that slides and collapses can be avoided.

The technology of the satellite cameras and their accuracy from almost 1000 km (600 mi) up is so refined that geologists now have at their disposal an extraordinary tool whose uses they have only begun to develop.

FIGURE 20-15
A contrast in slopes around Wast Water in the English Lake District.

Today Hack's model may seem to some to be rather revolutionary. Arthur Bloom has pointed out that

> only 150 years ago the concept of slow, orderly development of landscapes under the same conditions that operate today (Davis, Penck, King) was a dangerous challenge to the established religious and philosophical order.

This concept set the stage for the theory of biological evolution. From this, we may gather that there is a great value in such debate. If landscape does evolve, it is too slow to be witnessed. That is why there are so many opinionated writings on the subject. Again as Bloom says, "we still have the needs of the ancient philosophers who meditated on the dubious durability of the hills and thereby found strength for themselves."

The only way to decide which theory is correct is to go into the landscape, preferably many landscapes, and look, maybe measure, and think for yourself. Take the area of Wast Water, in the English Lake District, for example (see Figure 20-15). How is it that on the hill on the right of the lake, there is one type of slope on one side and a different type on the other? One thing physical geographers do know: Only 12,000 years ago the area was covered by ice. They may not be able to give a good answer concerning the slopes, but they do know what happens when ice scours the land, as we see in the next chapter.

SUMMARY

This chapter describes the search for a model that explains the variety of landscapes we see around us. The search begins with the question of slopes. Slopes have one or more of the following: a free face; a talus slope; a convex, concave, or rectilinear part.

The upper parts of slopes tend to be dominated by mass movement. In the lower parts, transportation by flowing water becomes more important. There is a debate as to whether slopes are generally reduced, retreat parallel to their original position, or are in a state of dynamic equilibrium.

When it comes to models of landscape development, the debate is even more pronounced. The Davis model suggests a cyclic development of landscape through stages of youth, maturity, and old age, with general slope reduction eventually producing a peneplain. The Penck-King model uses parallel retreat of slopes and relates the forms of the landscape to changes in the rate of uplift relative to the rate of erosion. Finally, the Hack model holds that all features of a landscape are in a state of dynamic equilibrium the majority of the time.

QUESTIONS

1. What are the four basic processes that cooperate to

break down landmasses? How are they all interrelated?

2. Why do slopes provide the key to the interpretation of landscapes? What factor of slope morphology tends to control processes that sculpt new landscapes?

3. Where do we find free-face portions of slopes in the United States? With what other features are they often associated, and why?

4. In what ways may a composite slope be related to convex, concave, and rectilinear slopes? Is more than one composite form possible? If so, describe them.

5. What are the major factors affecting slope form? Considering all that we know about the processes that cause slopes to take certain forms, why is there still so much uncertainty and debate among geomorphologists about the ways in which these forces operate?

6. What two early theories attempted to account for convex slopes, and on what processes did they depend? What were the limitations of both theories?

7. According to the model proposed by Wood, every slope implies four types of subslopes, each related to a specific process of formation. What are these elements, and how is each formed?

8. While Walther Penck's theories of slope retreat were similar in some respect to those of Wood, there was one major difference—the differential rate of retreat of various slopes. According to Penck's theory, what was the impact of slope steepness on rate of retreat, and what did this imply for various landforms and landscapes?

9. Lester King elaborated on Wood's analysis of slope elements. However, he ascribed slope development to physical laws rather than climatic influences. What evidence supported his theory? What evidence later contradicted it?

10. How was William Davis' theory of slope reduction different from the theories of slope retreat put forth by Penck and King? How did Davis explain the butte and mesa types of slopes used by Penck and King as the basis for their arguments on slope retreat?

11. Although most slopes are a composite of subslopes with individual angles, they usually attain a maximum slope angle within one or two degrees of one another if rock type, rock structure, vegetation, soil, climate, and overall relief are held constant. How did Arthur Strahler use this information to create a theory of slope formation based on systems theory? What are the principal components of slope formation, according to this theory, and how do they work together to maintain slope angles?

12. In Davis' model of landscape development, what three variables interact to influence the appearance of landforms? Which variable represented a departure from previous thought on landscape formation, and why?

13. According to the Davisian geological cycle theory, what accounts for the V-shaped valleys of a landscape in the youth stage? How are consequent streams, stream texture, and downcutting related in this stage of a landscape?

14. How is the concept of the peneplain central to Davis' theory of landscape formation? What evidence was taken as proof that Davis' theory was correct by colleagues searching for peneplains? According to the theory, especially the times required for each of the stages of landscape development, why aren't there more peneplains?

15. What aspects of Davis' model were considered unrealistic by Penck and King, and what evidence did they use to disprove Davis? Had Davis made any attempt to answer these criticisms of his model, and if so, how?

16. Penck and Davis disagreed on the roles of erosion and uplift in the formation of valleys of different shape. How did both men explain the V-shaped valley form and the U-shaped valley form? What are the basic controlling factors in each man's model?

17. King elaborated on the Penck model of landscape development, in which slope retreat is the principal mechanism. In what ways is his model different from that of Davis?

18. How were the respective models of Penck-King and Davis partially dependent on the areas in which these particular geomorphologists did their major field work—the Appalachians and the eastern United States for Davis and Africa's plateau and the mountains of South America for King and Penck, respectively?

19. How did Hack extend the concepts of dynamic equilibrium from Strahler's use in explaining slope formation to explaining landscape formation? What specific landforms are examples of dynamic equilibrium in the landscape?

20. Erosionally graded landscapes show mutual adjustment of large areas. This variety of form is principally the result of the characteristic rates of erosion of rocks that exhibit different weathering behavior. How is the change in gradient of Briery Branch in Figure 20-14 related to erosional grading of the different landscapes through which it passes?

ICE, WAVES AND WIND

CHAPTER 21

Many parts of the world were once covered by ice. In these areas, moving masses of ice shaped the landscape. By its sheer weight and incredible power, the ice scrapes and scours, gouges and deepens existing topography, leaving countless tons of debris when it finally melts away. Just as there are landforms obviously sculptured by glaciers, so there are features etched by the power of ocean waves. Along the coasts of continents, waves batter rocks and build sand barriers, destroying and building at the same time. There are also landforms distinctly associated with the action of wind, moving air with notorious strength in desert and steppe areas, as anyone who has ever seen a dust storm can verify.

In this chapter we will investigate the work of these erosional agents and the distinctive landscapes they create. In combination, their spatial expression comes nowhere near that of running water at the earth's surface, but the landscapes of ice, waves, and wind are widespread nevertheless.

LANDSCAPES SHAPED BY ICE

Not so long ago, the globe was engulfed by ice, a great deal more than it has today. Chapter 9 shows that the earth has gone through periodic temperature declines when the amount of ice in the high latitudes and mountainous regions increases enormously. It also notes that the ice will quite possibly return again.

ICE SHEETS AND MOUNTAIN GLACIERS

Many of us are familiar with the seasonal snowfalls that accompany the winter cold in northern areas of the United States and in Canada. With the coming of spring, we expect the snow to melt away, exposing the ground once again. But what would happen if the snow did not melt completely? The next winter accumulation would begin on the snow of the previous year and years before. In successive years the snow depth would increase. Soon the weight of the overlying snow would press the lowest layers into ice. When certain conditions of underlying slope and thickness existed, the mass of ice would begin to move.

During the ice ages, this is what happened on a huge scale in the high-latitude zones of North America, Eurasia, and Antarctica. *Ice sheets,* or *continental glaciers,* developed over the heart of northern Canada, over Greenland (this ice sheet still exists), over Scan-

FIGURE 21-1

The source areas of the major ice sheets in the Pleistocene Ice Age.

FIGURE 21-2
Mountain glaciers in Tongass National Forest, Alaska. This aerial view shows the north and south arms of North Sawyer Glacier joining to form the main glacier.

dinavia, and in several locales in northern Siberia. From these positions, the ice pushed southward. The glaciers grew ever larger as snow accumulations continued, as Figure 21-1 indicates. In North America the ice sheets reached the vicinity of the Ohio River, completely covering what is today Wisconsin, Michigan, Minnesota, and Iowa and partially covering several other states. At the same time, they scoured and gouged out the basins of the Great Lakes.

The process of ice formation can also occur in the snowy, cold summit areas of mountain ranges. The snow-thickening sequence takes place in the high reaches of a river valley, and the ice moves down the valley, assisted by the slope and the ready-made channel. These *mountain glaciers,* also called *valley glaciers* or *alpine glaciers,* are represented in Figure 21-2. The volume of ice in a mountain glacier is minimal compared to that in an ice sheet. A fully developed ice sheet during the Pleistocene Ice Age was as much as 3000 m (10,000 ft) thick, a cover of some 3 km (1.9 mi) of ice over a large region. But a mountain glacier is confined by the familiar dimensions of river valleys.

EROSION AND DEPOSITION BY ICE SHEETS

Ice, like running water, is a more efficient agent of erosion when it carries rock fragments. Pieces of rock embedded in the ice are dragged and pushed along the surface the ice moves over, wearing it down. We can see the evidence of this process preserved in especially hard and resistant rock surfaces, which may have scratches called *striations* and other gouge marks as a result of the movement of the ice. Ice sheets are especially powerful erosive agents where the surface they move over is uneven and where the rocks below are of varying resistance.

Eventually, when the ice has melted and the glaciers have gone, the hollows and trenches they have carved are filled with water. Northern Minnesota's lake country reveals the topography left by the base of an ice sheet. New York's Finger Lakes were once stream valleys that ran parallel to the direction of ice movement. The ice made lengthy troughs of them.

The ice sheet picks up rocks at its base, where it erodes. It drags the boulders away and grinds them

FIGURE 21-3
Kansan till in a bluff in Nebraska.

down. Eventually the ice sheet carries a whole layer of debris—some of it as large as boulders, other parts ground to a fine powder. All this material is deposited at the base of the ice sheet, often in a layer over 30 m (100 ft) thick, burying all or most of the underlying topography. The resulting accumulation is a mass of rubble, from the coarsest boulders to the finest silt, because ice does not have the capacity to sort its load as rivers can. This mass of material, called *till* is depicted in Figure 21-3. When it eventually becomes compacted and cemented, the rock layer is known as *tillite*. Tillite found in Africa, South America, and the other landmasses of Gondwana provided strong evidence for continental drift.

Another type of glacial deposit occurs when there is some running water. Ice sheets contain streams and meltwaters at their base, on the ice surface, and even within the ice itself. These streams pick up the rubble, carry it some distance, sort it considerably in the process, and then deposit it in layers. The coarse material lies with other coarse debris, and gravel, sand, silt, and clay are separated into layers or zones. This is referred to as *stratified drift*, and its layers differentiate this material from till.

When ice sheets invade an area, advance repeatedly, and finally wither away, they leave their load of till in the form of *moraines*. Extensive *ground moraines* cover large areas where till was deposited. Other moraines form ribbonlike belts of rubble marking the farthest advance of the ice sheet (the *terminal moraine)* or places where glacial retreat was temporarily halted *(recessional moraines)*. Sometimes moraines form the only well-marked topography in an area and so are the only places that can be used for ski resorts.

As you can imagine, the ice sheets and their meltwaters have left numerous landforms as evidence of their former presence—far too many to identify here. One of the most interesting, however, is the *drumlin*, a low oval-shaped, well-rounded hill made of till, as in Figure 21-4. Drumlins, whose origins are still under study, are fashioned from till that is especially rich in clay. A certain set of conditions for erosion and deposition at the base of the ice sheet seems to accompany their formation. Some drumlins may have been shaped when a new ice sheet overspread the ground moraine deposited by an earlier one, smoothing the debris and sculpturing it into low hills (their long axes are thought to parallel the direction of ice flow). In some areas, as in southern Wisconsin, extensive sections of the countryside were given a drumlin topography by the advancing ice sheets.

Another interesting feature resembles a snake, a long ribbonlike ridge extending across the countryside. This is an *esker*, which formed when water began to flow through a tunnel below the ice, bringing debris with it and eventually clogging its own passage. A tortuous, narrow belt of material is left standing above the local relief.

Numerous features beyond the edge of the former ice sheet remind us of its existence. The *outwash plain*

LONG ISLAND

One of the most valuable pieces of real estate in the world was created by the intersection of two portions of terminal moraine. If you look at a map of New York State, you will see an island stretching 220 km (118 mi) eastward from New York Bay. Long Island, which includes the New York City boroughs of Brooklyn and Queens in its western half and the counties of Nassau and Suffolk in its eastern half, is 4480 sq km (1723 sq mi) of glacial debris, left by the Labrador Ice Sheet.

The glacier actually left a wide-ranging span of material off the shore of New England, and circling all the way around Cape Cod, south of Connecticut and on across New Jersey and Pennsylvania. Long Island and several of the other islands off the northeast coast of the United States are simply the evidence of the moraine above sea level.

The two portions of terminal moraine that form Long Island are the Harbor Hill moraine and the Ronkonkoma moraine. In addition to forming the twin "spine" of the island, detectable as very gently rolling hills, the glacial origins of the island are observable in other ways.

For example, the light sand and gravelly particles pushed ahead by the glacier left a porous layer underneath the moraine surface. As a result, water readily trickles down through the island, leaving little lateral run-off; as a result, there are few streams on the island. It also results in a ready supply of fresh clean water.

had lakes (some still survive), deltas, terraces, small mounds at the edge of the ice called *kames,* and other landforms. Some of the landscape features deposited by an ice sheet are shown in Figure 21-5. Try to imagine the bleak landscape that must have prevailed in such areas as Wisconsin while the ice sheets were retreating.

FIGURE 21-4
A swarm of drumlins south of Lake Athabaska, Saskatchewan, Canada.

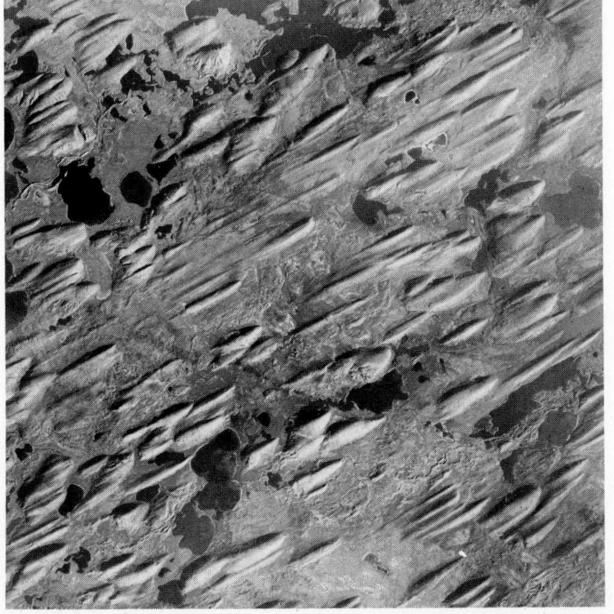

LANDFORMS OF MOUNTAIN GLACIERS

The continental glaciers covered enormous areas, but the mountain glaciers made by far the most spectacular landscapes. Mountain glaciers start the same way as ice sheets, that is, by the excess accumulation of snow and its transformation into permanent ice. This happens high up, near mountain summits, where mechanical weathering and streams did the erosive work before the ice age began. As more ice accumulates in the high valleys, the glacier begins to move, aided by the existing incline of the river valley. Thus whereas an ice sheet must accumulate an enormous thickness in order to begin to move, mountain glaciers can move with far less mass, because of the gradients that exist where they accumulate.

When glacial erosion takes over from river erosion in a mountainous area, the overall topography changes dramatically. As the glaciers grow and move down the valleys, they deepen, widen, and straighten them through their weight and erosive power. The source area, or *cirque,* grows larger as temperature-induced mechanical weathering breaks up the valley sides. Steep walls replace them, making the cirque look like an amphitheater, as Figure 21-6 shows. *Cirque* is in fact a French word meaning "circus." Also in French, the line where two surfaces intersect is called an *arête.* Glaciers, too, are capable of headward erosion, and where two or more cirques intersect, they create a sharp, jagged ridge called an *arête,* or a steep-sided peak called a *horn,* like the famous Matterhorn in Switzerland. So glaciers replace the roundness in the river-

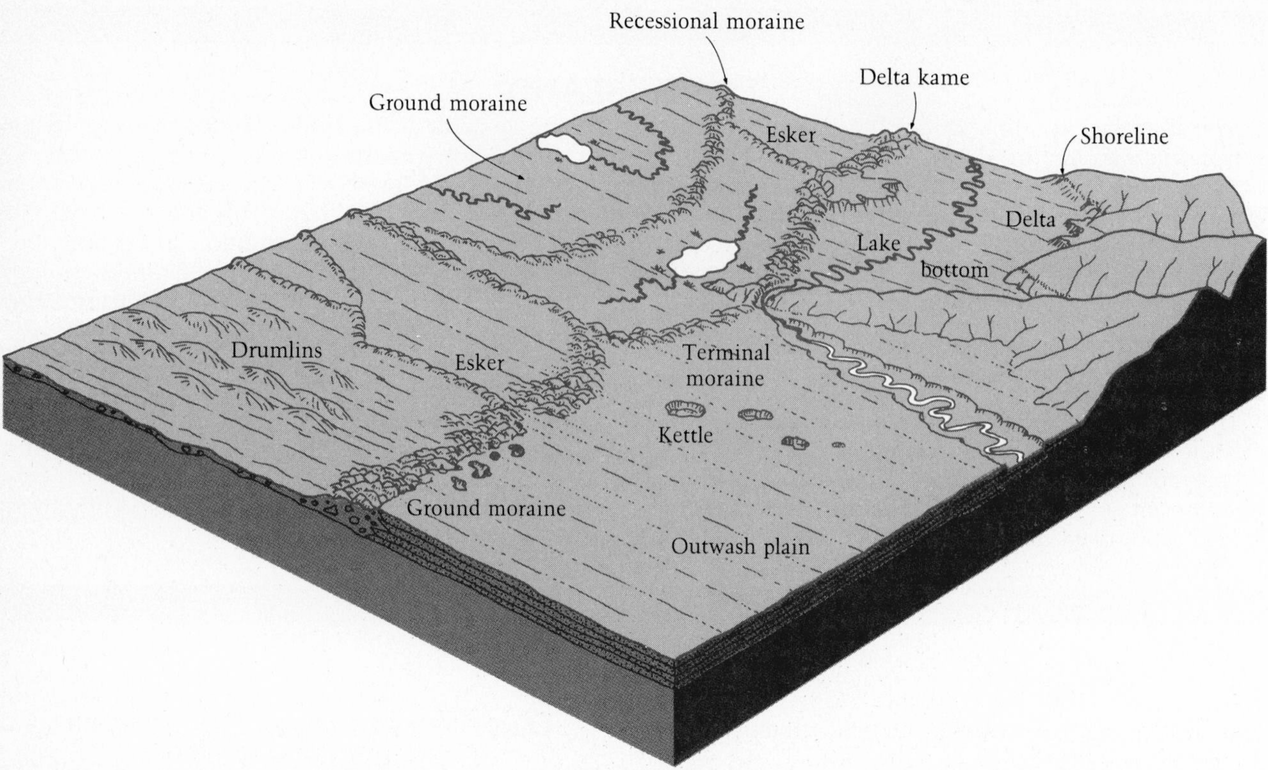

FIGURE 21-5

Depositional features resulting from the presence of an ice sheet.

FIGURE 21-6

Two cirques at the head of the Carbon Glacier bite deeply into Mount Rainier in Washington. Liberty Ridge (middle) is the sharp arête separating the two cirques.

sculptured landscape with angular roughness.

Glaciers, too, receive tributaries as rivers do, but the floors of the tributary valleys are not adjusted to those of the main glaciers. When the glaciers melt, there are waterfalls where those side glaciers entered the main valley. These *hanging valleys* on the side of the glacial trough are sure evidence that the glaciers existed and that they disrupted the drainage system.

The glacial valleys really do resemble troughs, because the glacier scoops them out and gives them a U shape that is quite unlike any river valley. Glacial valley sides are steep, sometimes nearly straight up, and the spurs that the river twisted and turned around are simply sliced off. The famous Half Dome in Yosemite National Park is an example of the valley glacier's capacity to wipe out obstacles. *Truncated spurs* can be seen on the valley sides in Figure 21-7. The valley floor also shows evidence of the glacier's irregular action. Moving ice responds to differences in elevation and often excavates a series of steps that, after the glacier has disappeared, may accommodate standing lakes.

In high latitudes and even midlatitudes—for example, in New Zealand—the glaciers continued to erode below sea level when they reached the sea. Thus

FIGURE 21-7
Hanging valley and waterfall in Yosemite Valley, California. The U-shaped glaciated valley has El Capitan on the left and Bridal Veil Falls on the right. Half Dome is in the center distance.

glaciers sometimes create gaping seaward valley mouths with characteristic steep sides. When the warming trend melted the glaciers, the sea penetrated those troughs. *Fiords* formed, often with quite breathtaking scenery (see Figure 21-8).

Mountain glaciers, like ice sheets, carry large amounts of rock debris. When glaciers erode, they attack obstacles by overriding them, creating a smooth, gradual incline. The sheltered side of the obstacle is "plucked" by the ice, as Figure 21-9 shows. Ice freezes onto the sheltered side of the rock, but when this ice moves with the total ice flow, small rocks are pulled from the main obstacle. Knoblike rock protuberances, smoothed on one side and rough and broken on the other, exist all across glaciated topography. Individual protuberances, looking like oversized lying sheep, are called *roche moutonnée* (again from the French).

When mountain glaciers contain rock debris, it is not distributed randomly throughout the ice; rather it is concentrated in zones that show up clearly as dark bands on the surface. The mountain glacier's *lateral moraine* consists of rock debris that has been acquired from the valley sides and is carried along near the glacier's edge. When two glaciers with lateral moraines meet, however, their inner lateral moraines lie in the middle of the joined valley, as a *medial moraine*. Add several other tributary glaciers, and a whole series of such medial moraines forms, as in Figure 21-2. When the glacier begins to melt, this material is left in a series of ridges in the valley. When the mountain glacier stops advancing, rock debris is left as a pile of rubble across the valley floor. This terminal moraine marks the farthest advance of the glacier. If the glacier's melting is interrupted, recessional moraines form where there was a stillstand.

Meltwater coming from the fading edge of the

FIGURE 21-8
A tourist liner sailing into the fiord of Milford Sound, New Zealand. The fiord was formed when the glacier that eroded the valley below sea level melted.

glacier carries some of the glacial debris downslope, forming landform features comparable to the outwash plains of the ice sheets. Deltalike river braiding prevails, with some stratification of the glacial rubble. Terraces, kames, and lake deposits exist in many glacial valleys. Eskers, too, wind across valley floors where the ice once stood. The overall landscape is unmistakably the result of forces other than running water alone. The cooperation of ice and water has produced landforms that are appropriately called *glaciofluvial* landforms.

WAVES AND SHORELINES

The landscapes of glaciers are certainly distinctive. But they are no more so than the landscapes of coastlines, whose cliffs, beaches, lagoons, and terraces have been created by the waves, currents, and tides of the oceans. Waves endlessly pound the edges of the land, sometimes driven by gale-force winds, eroding them so fast in places that the shoreline retreats by several meters per year—an incredibly high rate in the context of geological time. Elsewhere, the waves do not destroy but rather build, bringing gravel and sand and creating beaches and *offshore barriers*—in effect enlarging the landmass.

One of the determinants of the character of coastlines is the behavior of the crust of the earth. The edge

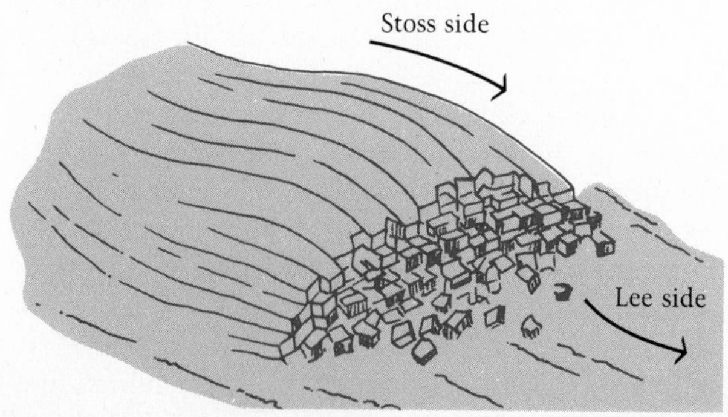

FIGURE 21-9
A glacially rounded and plucked rock knob, or *roche moutonnée*.

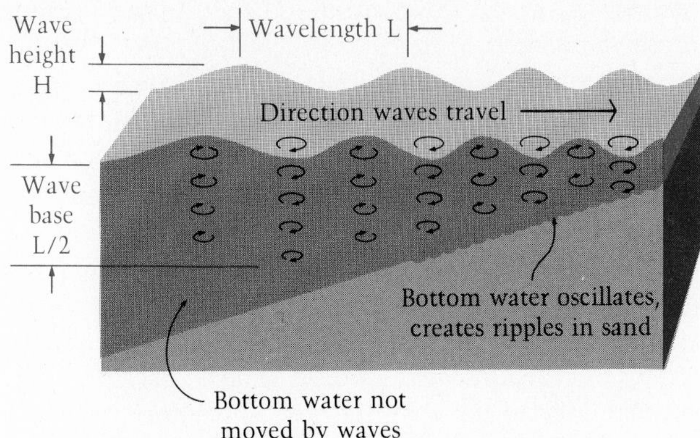

FIGURE 21-10

Waves on water surface passing from deep water (depth more than half the wavelength) to shallow water. Curved lines show orbits of water particles after passage of each wave. Orbits are circles in deep water and ellipses in shallow water.

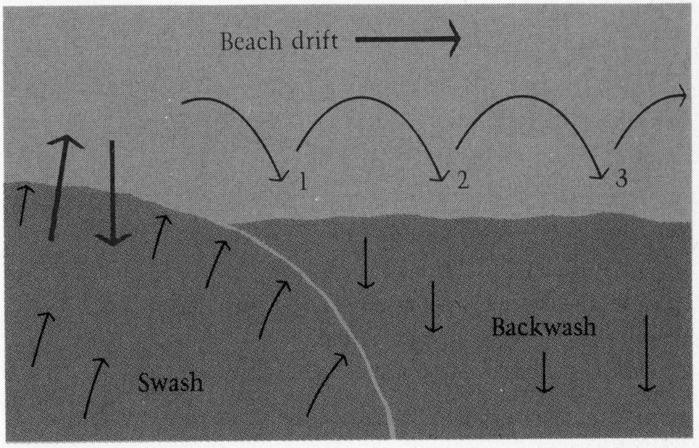

FIGURE 21-11

Swash and backwash on a beach and the resultant drift motion of beach sand.

of the landmass may be sinking or rising or temporarily stable. The erosive power of the waves along coasts depends greatly on the configuration of the adjacent sea bottom, which is the product of the recent geological history of the land margin. To appreciate this we need only compare the western coast of the United States, with its instability and active crustal deformation, to the edge of the comparatively quiet coastal plain of the southeastern United States. The western coast is cliff-studded, with peninsulas, islands, and deep bays; the southeastern coast is comparatively smooth, straight, and shallow offshore.

EROSIONAL PROCESSES

Although currents and tides also have an effect on the formation of coastal landscapes, most of the erosional work along shorelines is done by waves. Waves are generated by wind, which is produced by the sun's energy (see Chapter 5). Where winds are persistent and strong, wave erosion is greater than where winds are directionally variable and weak. Large waves are much more effective erosive agents than small ones. They develop when the wind speed is high; the wind blows for a long period in the same direction, and the stretch of water the wind blows over—the *fetch*—is large.

Wave actions When large waves from the open ocean approach the shallower water of a coastline, they tend to grow in height. As Figure 21-10 shows, the shape of the wave is altered by the water's decreasing depth. When it can no longer maintain its shape, the wave *breaks*. The top part of the wave still moves forward at speed, but the bottom part, where the water is returned, is slowed down by the sea bottom. The resulting *breakers*, in effect the collapse of the wave, end when a *swash* of water runs up the beach and then back to sea again as *backwash*, as diagramed in Figure 21-11. This process creates a certain slope on the beach—determined by the force and length of the swash and the amount and kind of beach drift—because waves are likely to reach the shore at a slight angle and the backwash to run straight down the slope.

Waves are likely to reach the shore at a slight angle for several reasons. One of them is that the wind generating the waves is unlikely to be exactly perpendicular to the shoreline. And when the *wave front*, the imaginary line formed by the wave, begins to interact with the bottom, it is bent or *refracted*. This refraction has three important effects. One is that the erosive energy of the waves is concentrated on the promontories. Second, wave refraction helps to create a *longshore* or *littoral drift*, as you can see in Figure 21-12.

FIGURE 21-12

Patterns of waves approaching an irregular shoreline.

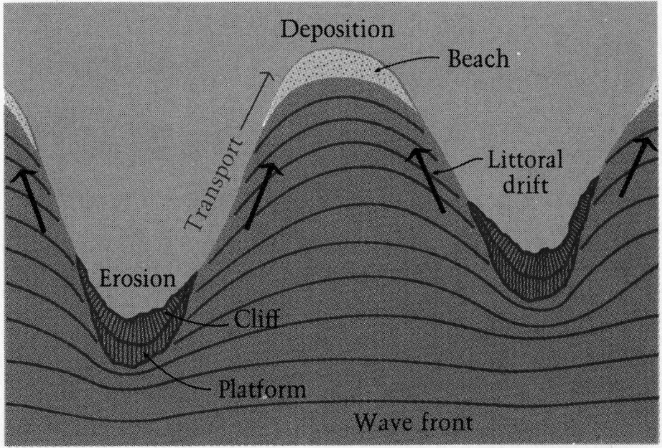

This works with the process of swash and backwash to transport sand or other beach material into bay heads. Third, the total effect of the first two processes is to straighten out the coastline by removing the promontories and filling in the bays.

Longshore currents can be caused in other ways—by large-scale ocean currents, for example. But however they form, they have important effects. In the past, humans have built many coastal structures to defend against them. *Groins,* for example, wall-like structures extending out from the shore, have been built to keep sand on some city beaches. But the real problem is often that the natural supply of sand to a beach has been stopped by dams and other structures inland. Groins sometimes lead to the erosion of other beaches downdrift.

Human response to beach problems is of two kinds. One view is exemplified by the US Corps of Engineers, who are charged to "defend the coast," almost at all costs. The other philosophy, ascribed to by the US National Park Service, is that no structures should be built on beaches and no public funds should be spent on restoring storm-damaged structures. A sound knowledge of the physical processes at work will help you to determine which of these viewpoints is correct in any particular case.

Factors in coastal erosion Water alone is a powerful erosive agent when it is hurled onto a beach or against a cliff. It was once estimated that the power of waves acting against Bikini Atoll was equivalent to over a quarter of the electric power generated by the Hoover Dam. But water erodes far more effectively when it carries rock fragments of various sizes. Like rivers and moving ice, ocean waves operate by abrasion. When a storm does a lot of damage along a coast, you can be sure that much of the damage was done by the boulders and pebbles thrown against the land by the waves.

Much depends, naturally, upon the resistance of the rocks along the coast. Highly resistant crystalline shores recede much more slowly than soft sedimentary layers do. The hard crystallines, in addition to their greater resistance, also yield projectiles far less easily than the soft sedimentaries. Imagine how quickly a coastline made of glacial till would erode, with its ready supply of cobbles and boulders.

Most significant of all factors, however, is the existing topography along the coast. Chapter 13 considers the continental shelves that extend below the water for distances from a few to several hundred kilometers. Where the continental plate extends well beyond the edge of the landmass and the continental shelf is wide, the gentle slope of the continental shelf may continue on the land itself, as in the southeastern United States. But where two plates are colliding, as along the western coasts of North and South America, coastal landscapes (and offshore seascapes as well) have much greater relief. Thus a rise or fall of sea level of, say, several dozen meters would have very different effects along the two coasts. A rise of sea level would inundate large parts of the eastern coastal plain, but still the waves would roll onto a gently sloping surface. Along the western coast, the waves would smash into the near-vertical cliffs of the mountainous shoreline.

Approaches to coastal erosion If you research further into the geography of shorelines, you will find that several physical geographers have tried to discern cycles of coastal erosion and to classify coastlines according to whether they display evidence of rising, sinking, or stability. These efforts may not be worthwhile, because coastlines are subject to almost endless change. Large quantities of water are periodically locked up by continental ice sheets, during ice ages. This causes a drop in sea level of several hundred meters, exposing formerly inundated areas and creating new coastlines. When the ice melts, the sea level rises. At the same time, the landmasses are also capable of vertical change because of isostatic adjustment (see Chapter 16). Nor are coasts immune to folding, faulting, and other kinds of crustal deformation.

Thus it may be more sensible to look at coastlines in two different ways. Coastlines where topography is rough, slopes are steep, and adjacent water is comparatively deep are called high-angle or high-relief coastlines. Where coastal topography is of low relief, coastal slopes are gentle, and the adjacent water tends to be shallow and to deepen gradually, the coastline may be considered low angle or low relief.

LANDSCAPES OF HIGH-RELIEF COASTS

When the ocean's waves can approach the land without the retarding effect of shallower water, their erosive power is at its peak. Thus when steep slopes descend below the water line and deep water lies next to the shore, erosion is rapid. The most characteristic feature in coastal landscapes where this occurs is the cliff. Steep-sided cliffs are especially likely to develop on exposed *headlands* or *promontories* extending into the sea. Often the cliff retreats in such a way that sections of it become separated and stand apart as *stacks,* as in Figure 21-13. The erosive action of the waves often produces caves, especially where the cliff is made of soft rocks below and harder rocks above.

FIGURE 21-13
Sea cliff and stack at the northeast end of Bell Island, Newfoundland, Canada.

The rapid retreat of cliffs occasionally produces a feature also found in glacial topography: hanging valleys. Streams that once led all the way to sea plunge off a cliff, because the slope they once flowed down has been cut back by the waves.

At the foot of a cliff, the waves and the agitated rock debris tend to create a smooth, seaward-sloping surface called a *wave-cut* or *abrasion platform*. If the area is uplifted (or sea level drops) after the erosional phase, such a platform, unmistakably wave-cut, forms a terrace that gives evidence of the coast's complex history.

The scenery of high-relief coasts is incomparable. Anyone who has seen the vistas along California's central and northern coast or that of Oregon can attest to the magnificence of this landscape. Where high-relief coasts have been glaciated, the steep walls created by the glaciers when they excavated their coastal troughs—the fiords—are maintained by the attacks of the waves. Even an atlas map, with its small scale, reveals the contrasts between such high-relief coastlines as those of Norway, southern Chile, and northwestern North America, and the low-relief coastlines of the southeastern United States and northeastern North America.

LANDSCAPES OF LOW-RELIEF COASTS

Where the land surface declines gently toward the coast and then continues its slight slope under the water, the associated landscapes are less spectacular but no less characteristic. Here the waves are less constantly destructive than they are along high-relief coasts. Except during storms, when they do most of their damage, the waves actually build, or aggrade.

The waves deposit belts of sand some distance from the original shoreline, and eventually these begin to emerge above sea level—in part because aggradation continues during high tides and even during all but the most destructive storms. These offshore barriers, or bars, stop the waves entirely and receive all the sediment being pushed landward. Figure 21-14 diagrams

CONTROLLING THE COASTLINE

Throughout the world, people who settle along low-lying ocean coastlines are constantly subject to the danger of flooding. The 890-km (430-mi) coastline of the Netherlands on the North Sea is one of the major exceptions to this rule; it is a marvel resulting from a thousand years of engineering.

The Dutch have two natural problems with which to contend. First, the low plain of the Netherlands ("low lands") is criss-crossed by three major rivers and countless streams. This area is constantly pounded by the tides of the North Sea. Second, most of the Dutch soil is of moderate to poor quality. Not a fraction of the dense population of that country could have been fed without somehow increasing both the quality and the amount of arable land.

Beginning around AD 900, the Dutch began building blocking structures across the numerous estuaries that mark the entrance of the sea into the land area. Called dikes, these structures effectively reduce the ocean-fronting coastline of the country by about two-thirds and make it easier to control. First with windmills and later with electrical pumps, the land areas behind the dikes were pumped out, creating polders, areas of usable land that are below sea level, but completely protected from the ravages of the North Sea tides. Today, over two-fifths of the Netherlands is actually below sea level.

The cost has been high. Over 11,000 engineers and technicians are required to maintain the system. Over 2000 pumping installations are located along the coast to reverse the sea's constant inroads into the Dutch land. It is estimated that almost 9% of the Dutch government's budget is devoted to coastal development and reinforcement. But for this crowded and land-poor country, it is an essential project.

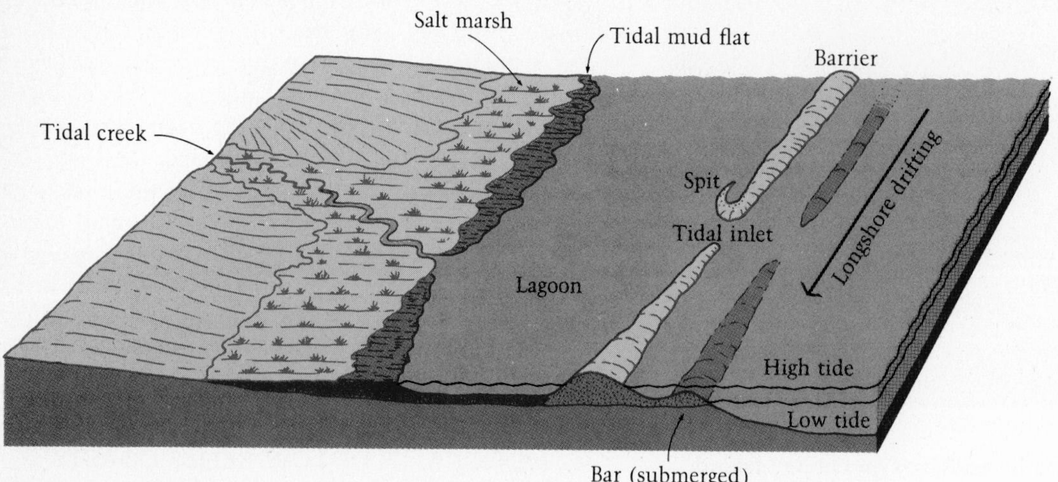

FIGURE 21-14

Some typical landforms of a low-relief coast.

the resulting landscape. Vegetation develops on the barrier or barrier island, and its appearance of permanence becomes stronger.

Between the offshore barrier and the original shoreline, a *lagoon* forms. If rivers reach the coast in this area, they no longer spill their sediments into the open sea but into the relatively confined lagoon. Thus the lagoon may slowly fill up, developing mud flats that eventually begin to emerge permanently above the tidewaters. What the waves have done, therefore, is extend the land rather than reduce it.

Earlier we learned about the coastal drift of loose particles, driven by the back-and-forth motion of swash and backwash and by currents. This may have a characteristic effect along irregular shallow coastlines. A combination of wave action and coastward drift produces a partial barrier, frequently curved, that is called a *spit*. The large spit at the end of Cape Cod, Massachusetts is a good example. Normally such a spit does not close off the bay completely, but this does occur occasionally. Then the spit becomes a bar, and the bay a lagoon.

In some ways these offshore barriers are like the levees in wide river valleys. Like levees, they can be breached during periods of high water. When a gale occurs during a time of exceptionally high tide, the ocean may invade the lagoon. The water pouring across the mud flats attacks even the coastline protected by the offshore barrier, which is seldom touched by wave action. When the tide recedes and the gale diminishes, the water rushes out of the overflowing lagoon, cutting across and through the barrier island. The outrush of the pent-up waters is not the only force that breaches an offshore bar; the storm itself can accomplish this, as Figure 21-15 shows. Once the

FIGURE 21-15

Storm-cut breach through Fire Island, Long Island, New York. Sediments have been carried from the sea into the lagoon to construct a tidal delta.

offshore barrier is broken, it takes time to repair. The tides can once again penetrate easily to the lagoon, and the back-and-forth motion of the tidewater keeps the breach open. Along the Gulf of Mexico, hurricanes that hit the shoreline often create gaping breaches in offshore barriers that have long protected the land, sometimes with disastrous results as the storm surge sweeps onto the mainland. In June 1957, the barrier along the western Louisiana coast was flooded by hurricane Audrey. The natural constructive forces, however, rebuilt it within two years.

Offshore barriers along shorelines of low relief may become densely settled. Miami Beach, for example, lies on an offshore barrier now reinforced by breakwaters, concrete sea walls, and even artificial beaches. Along much of the Florida coastline and along the coasts of other southeastern states, the towns and villages that have developed on offshore barriers exist in fear of storm erosion and hurricane attack. The string of lagoons between the offshore bars and the mainland's original coastline forms the famous Intracoastal Waterway, a nearly continuous barrier-protected route for smaller craft.

One other coastal feature that should be mentioned is the *estuary*. These are often formed by the submergence of a river valley. Chesapeake Bay in the eastern United States is an example, as is the Rio de la Plata between Argentina and Uruguay. Estuaries are often accompanied by salt marshes. These have been particularly vulnerable to human activity in recent years. It was estimated in 1970 that 23% of all such areas in the United States had been severely modified and that about 50% had been moderately altered. The importance of these areas as spawning grounds for fish and other wildlife has been all too frequently underestimated.

SOME OTHER COASTLINES

In addition to the two major types of shorelines, there are other kinds of coasts, but their total length is comparatively insignificant. It is not difficult to identify these. The coasts of deltas (Figure 21-16) are coasts of low relief, but they aggrade from the landward side. Where volcanic mountains stand on a coastline, their lava flows push into the sea, producing another kind of aggradation. High-relief coastlines tend to result from lava flows. Where a coastline is created by the drop of a block along a fault, the coast may be straight and steep. Madagascar's eastern coast is an example of this type of high-relief coastline. Note that each of these special cases can be placed in one of the major groups: Delta coasts (and coastlines of alluvial fans) tend to be low-relief coastlines, and volcanic and structural coastlines fall into the high-relief category.

One group of shorelines would defy such classifica-

FIGURE 21-16

An aerial view of the Mississippi Delta, looking south at the Gulf of Mexico. This is a low-relief coastline.

tion. *Coral reefs* are "coastlines" that do not build through the accumulation of sediments but through the build-up of the skeletons of organisms. Islands in tropical waters are often rimmed by reefs that, in favorable environmental circumstances, can attain widths of several hundred kilometers. Coral reefs may lie immediately offshore, as a continuation of the land, or they may lie away from the shoreline, about where an offshore bar would be located—and with the same effect; because barrier reefs, too, create lagoons. Australia's eastern coast is affected by reef development, and the eastern coast of the Yucatán Peninsula, Mexico has a well-developed barrier reef. Reef development can take place along coastlines of high relief as well as low relief.

LANDSCAPES OF WIND ACTION

The wind, moving air that drives the waves as they fashion the coastlines, also moves materials on the land. Stand on a beach on a windy day, and you will feel the sand scouring your legs. Drive through an arid area in the southwestern United States, and the dust that fills the air will scratch the finish on your car. By itself, the wind is not a powerful erosional agent. But when it is laden with sand and dust, the wind scours and sculptures. It would be difficult to mistake an area of dunes for anything but the product of the action of wind.

WIND EROSION

Wind, like other erosional agents, lifts, transports, and deposits. The capacity of the wind to move loose particles depends on its speed. The higher the velocity of the moving air, the greater the size of the particles that can be moved and the more effective the erosion. Even on a nearly still day, there is some dust in the air, small specks of clay too light to fall to earth as long as there are the slightest upward air currents. When wind eddies pick up and remove clay and silt-sized particles, the process is called *deflation*. As the wind speed increases, sand grains are moved along. Sand never travels as high as dust does, but it may be a meter off the ground if the wind has storm strength. Unlike the dust particles, which are in effect in suspension in the air, sand grains bounce along the ground, the process of saltation. Larger particles, such as gravel and pebbles, are occasionally rolled and pushed by gale-force winds, but they do not move more than a few meters through the action of wind alone. It takes running water to transport these larger fragments over long distances.

The erosional action of wind is a process of abrasion, the impact of the wind-driven particles against obstacles in the landscape. This process fashions such characteristic features as the self-explanatory *pedestal rocks* (shown in Figure 21-17) and *deflation hollows*, shallow depressions excavated by deflation. But it is not the wind alone that does the work. Weathering and erosion by running water are predominant processes even in climatically dry areas. A single rainstorm modifies the landscape more than months of wind action. Still, the wind removes particles loosened by weathering or left behind in the valley of a dried-up stream and thus contributes to the formation of typical arid-area topography.

Transportation by moving air can involve enormous quantities of material. In dry regions, strong winds sometimes generate *dust storms* like the one in Figure 21-18, huge masses of fine clay and silt particles that "saturate" the air to the point that sunlight is completely blocked out and darkness prevails. In moving across the countryside, dust storms transport tens of millions of tons of material at a time. Because this

FIGURE 21-17

Pedestal rock formed by sandblasting at base, Death Valley, California.

FIGURE 21-18
Leading edge of a dust storm in Union County, New Mexico.

material is so fine-textured and not all picked up at once from a single area, the effect on the landscape may be slight. But over the course of many thousands of years, the cumulative effect of these storms is substantial.

Dust penetrates air layers at very high altitudes, beyond the troposphere, and it travels thousands of kilometers. In 1973 meteorologists determined that high-altitude haze layers filtering the sunlight over the Caribbean Sea were made of dust from the Sahara, where intense dust storms and sand storms occur. Ships far off the African coast in the Atlantic Ocean are often covered with dust during the local wind called the *harmattan,* when air predominantly moves from the Sahara and western Africa.

By contrast, *sand storms* remain confined to the lowest few meters of air, because the sand particles are moved by saltation. The vast majority of the sand grains transported during a sand storm move in a layer just a few centimeters above the ground, and the wind's erosive power is greatest at this level. Sand storms can be blinding and viciously destructive. In desert areas, sand storms can invade oases or irrigated areas and destroy crops. Persistent sand storms accompanied the terrible drought in western Africa in the 1970s, hampering the relief effort.

DUNES

The idea that deserts are mainly made of drifting sand and shifting dunes is a common misconception. More often than not, desert areas are rocky—not sandy. But loose sand has accumulated in some large areas. Where this happens, the sand tends to collect in specific patterns, in mounds shaped and spaced certain ways. These mounds are called *dunes,* and dunes come in many different forms and dimensions. No landscape is more closely associated with wind action that a dune landscape is.

Some dunes constantly change their shape and even their location, as the wind adds sand on one side and blows it away on the other. Others are covered by vegetation that managed to take hold at some point in their history and fixed them permanently. Grass is often planted on coastal dunes to strengthen them as a line of defense against storm waves. But we are most interested in the active, live dunes, still built of loose sand that shifts and changes. Occasionally settlements are swamped and obliterated by such shifting sands.

A prominent dune shape is the crescent. The *barchan* looks like a quarter moon; the points of the barchan lie downwind and reveal the direction of the prevailing wind, as in Figure 21-19. The windward slope of the dune is gentle, and the leeward side is considerably steeper. As Figure 21-20 shows, the rim of the barchan is so sharp that it forms a thin line. A wind with any strength at all creates a smokelike cloud of sand off this crest.

Barchans that develop on a rocky surface stand alone or in small groups, in sharp contrast to the rest of the countryside. Barchans in areas with sufficient sand to cover the whole countryside develop other shapes. In such *sand seas,* as these places of complete sand inundation are called, *transverse dunes* develop frequently. These are sand ridges whose crests stand at right angles to the direction of the wind. The dunes shift slowly with the wind, and generally the windward side is consistently less steep than the lee side.

FIGURE 21-19
Barchan dunes. The arrow indicates wind direction.

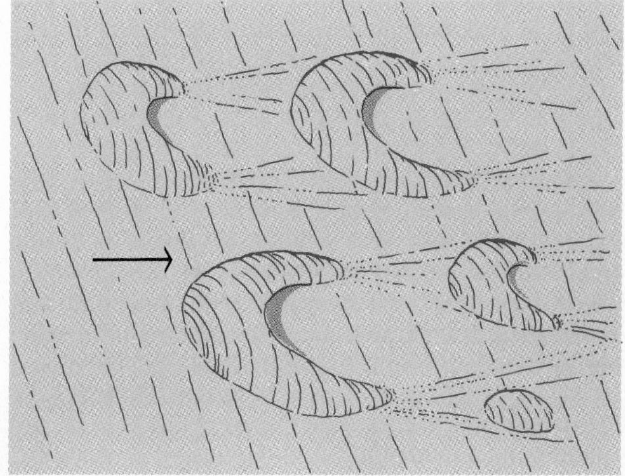

FIGURE 21-20
Barchan dunes formed on a terrace bordering the Columbia River, Oregon. They are formed by a wind of constant direction (from the left in this photo).

But transverse dunes are less regular and consistent than barchans; their form is more variable. It is, of course, possible to identify other dune shapes and to give them names. Often different kinds of dunes are found in association.

At times there is evidence that old dunes, formed thousands of years ago and long held in place by a vegetative cover, have become mobile again when this plant cover was destroyed. Others are in the process of becoming "fixed" as vegetation takes root in them. In some cases, ancient dunes become "fossilized." They tell us a great deal about past climatic conditions.

Dunes also develop along coastlines. The waves sweep the sand onto the beach, and the wind piles it in mounds above the high-tide line. Coastal dunes form a pile of protection against storm waves, so efforts are made to fix them by planting and fertilizing vegetation. When there is an ample supply of sand (for example, during a slight lowering of sea level), the coastal dunes may migrate inland, perhaps smothering stable vegetation there. When sea level rises, the coastal dunes are quickly attacked, broken up, and carried away.

LOESS

Another indication of the wind's work lies in extensive deposits of loess, shown in Figure 21-21. *Loess* is a wind-transported, wind-deposited, fine-grained silt that is widely distributed in several parts of the world. The loess deposits in the United States extend from western Indiana to eastern Colorado and from Minnesota and South Dakota to Mississippi. The Palouse region of Washington and Idaho also has loess deposits. The thickness of loess in these areas varies from a few centimeters to over 30 m (100 ft). When the yellowish loess is exposed, as in a road cut, no layering or stratification, typical of a water-deposited sediment, is visible. However, loess deposits have a vertical, jointlike cleavage. The loess layer breaks off along this cleavage when undercut, leaving the rest standing in vertical walls.

In China, loess deposits are over 90 m (300 ft) thick. The tendency of loess to stand in vertical sections is magnified by deep erosional valleys and the scars made by human movement. Roads long in use have been worn over 6 m (20 ft) below the surface. Loess is also very fertile, and in China this fertility has sustained a large population. The absence of building materials and the capacity of loess to stand up in vertical walls have induced the people of China to dig cavelike dwellings into the valley sides and roadsides.

The majority of loess deposits seem to have originated in the Pleistocene glaciation. China's loess may be derived from the Gobi Desert, however, Loess is made of ground-up quartz, feldspar, and other minerals. Apparently strong winds blew across the outwash plains of the retreating glaciers and ice sheets, throwing up gigantic dust storms (probably unlike anything witnessed today) and depositing the loess elsewhere. In North America, Europe, and other places where loess has been identified, there seems to be a strong spatial relationship between the loess areas and the glaciers. Of course, after the loess was deposited, water erosion attacked the loess and redistributed it somewhat. Thus the loess we see today is not solely the result of wind deposition. But it was the wind that picked up that enormous quantity of material and first removed it from the glacial margins to the regions of aggradation. Moving air can be a powerful erosional agent.

GLACIAL, COASTAL, AND WIND SYSTEMS

We have examined the landscapes formed by ice, waves, and wind by looking at the processes associated with the formation of landscapes. We do this not only to understand the spatial distribution of landforms but also, in many cases, to find rational answers to practical questions. We can understand the processes and the landscapes even better by applying the systems approach to aspects of glacial, coastal, and wind-affected landscapes.

We might have taken this approach in many areas in this chapter. For example, a glacier may be viewed as a system with inputs of ice in the form of snow and with an output of meltwater. The balance of input and output of water substance, the *mass balance* or budget of the glacier, determines the "health" of the glacier. If the mass budget over several years has a surplus, the glacier is likely to advance. During and slightly after (there is usually a lag) periods of deficit, the glacier may retreat.

Many problems associated with beaches can also be approached from a systems viewpoint. Figure 21-22 shows, in a simplified fashion, how a beach could be viewed as a system with gains and losses of material. A realization of the full implications of this simple approach may have saved thousands of dollars in coastal engineering projects.

One of the most fascinating examples of a system is one involving both wind and water transportation. On the southwestern side of the Sangre de Cristo mountains in Colorado lies an area of giant sand dunes, some 180 m (600 ft) high, now known as the Great Dunes National Monument. The predominant winds, from

FIGURE 21-21

Deposits of loess in Vicksburg, Mississippi. Note the vertical cleavage. This photo also shows how easily loess is eroded where water trickles over the side of the cliff.

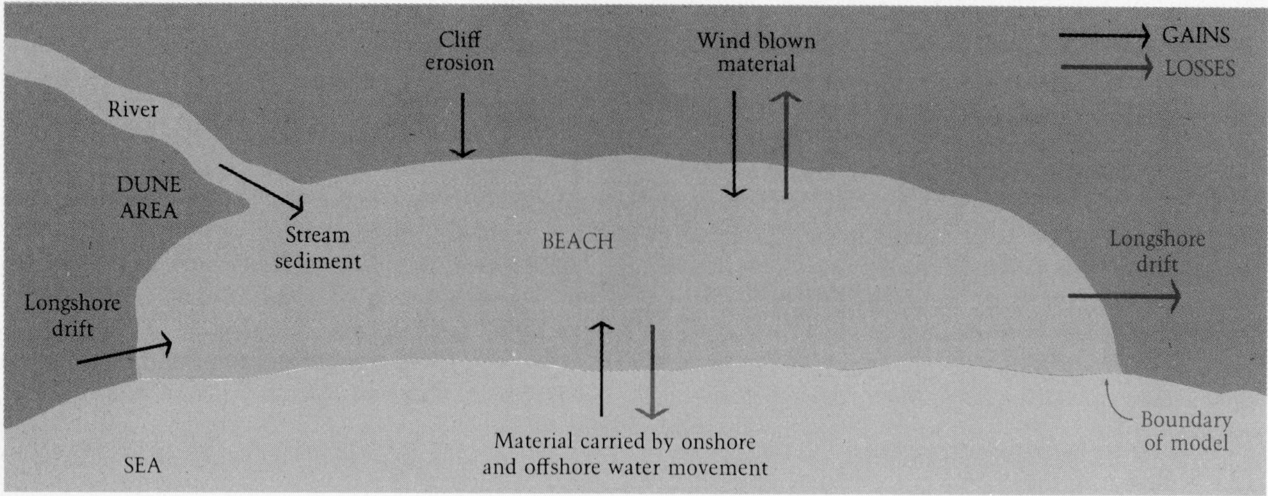

FIGURE 21-22

The beach system, showing gains and losses of materials.

the southwest, carry silt and sand toward the mountains, as diagramed in Figure 21-23. The winds drop the load at the foot of the mountains to form great transverse dunes. Sand from the dunes is carried eastward into the mountains through a pass acting as a wind funnel. Occasional rainfall in the mountains returns this material by stream action back to the valley floor, to begin the cycle again. One of the most intriguing aspects of this particular system is that, as far as the sand is concerned, it is a rare example of a closed system.

SUMMARY

We have examined the landscapes formed by the action of ice, waves, and wind. Moving ice sculptures the land both in mountain and lowland locations. The resulting landforms are not only erosional but depositional as well.

The same is true of the landforms resulting from the action of waves. A division between high-relief coasts and low-relief coasts helps us organize our description of coastal landforms. The features of high-relief coasts are usually erosional in origin, but the landforms of low-relief coasts may be either erosional or depositional.

The landscapes formed by wind action also possess erosional elements, usually formed when the wind is armed with abrasive material, and depositional landforms, such as dunes. Loess formation is one of the more important results of wind action because of the fertile soil it can produce.

Finally, we have seen examples of how aspects within the landscapes of ice, waves, and wind may be treated from a systems viewpoint. This approach gives

FIGURE 21-23

The closed sand system, Great Dunes National Monument, Colorado.

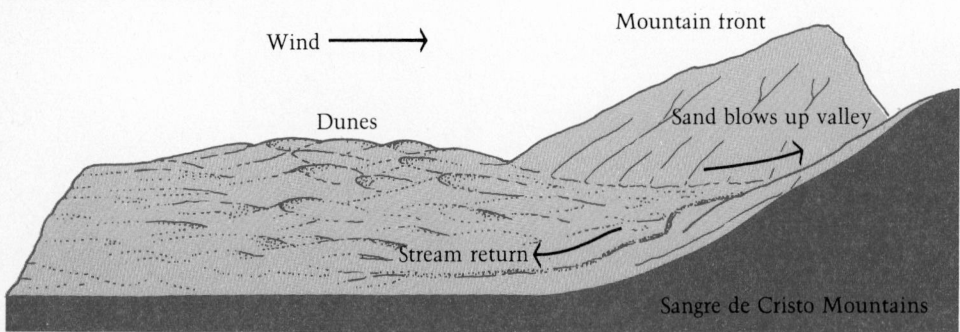

a greater understanding of the mechanisms involved in some of the most interesting and beautiful landscapes on the earth.

QUESTIONS

1. What conditions are necessary for ice sheet or alpine glacier formation? How did these conditions affect the initial positions of continental glacier formation during the last Pleistocene glaciation?

2. In Figure 21-1, two major ice sheets cover half of North America. The Laurentide Ice Sheet developed from a position near Hudson's Bay in the heart of Canada, spreading southward to cover much of the northern half of the United States. The other ice sheet, in the western half of the continent, seems to be related to another type of landscape. How did this western ice sheet probably form? What controlled its direction of spread? In what ways was its formation probably somewhat different from that of the Laurentide Ice Sheet?

3. What factors contribute to the high erosive ability of glaciers? What differences in erosive ability can we see in evidence from landscapes over which glaciers have passed?

4. Geomorphologists have been able to construct a fairly comprehensive picture of past glaciations, including relatively small-scale advances and retreats, and previous glaciations where later glaciations have occupied the same areas. What type of evidence have they used to determine the positions of glaciers at various times? How can they determine previous glaciations in the same area, if each glacier erodes the surface anew?

5. How is stratified drift differentiated from glacial till? What processes cause each of these two forms of glacial debris?

6. In many previously glaciated areas, roads are built on top of long, narrow ridges that stand above and snake through the surrounding countryside. These ridges, known as eskers, rarely extend past the terminal moraine. How are these ridges formed, and why do they halt rather abruptly when they reach the terminal moraine? (See Figure 21-5.)

7. Continental ice sheets and mountain glaciers create dramatically different landscapes. How are the landscapes created by each type of glacier different, and what causes this difference?

8. While some aspects of glacial activity parallel those of stream activity, differences exist. How do hanging valleys, features associated with alpine glaciers, illustrate one difference between stream erosion and glacial erosion?

9. Fiords are deep bays formed by glaciers that erode below present sea level. Where are these features generally located in terms of latitude, and why? What conditions allowed the glaciers to erode so far below present sea level?

10. Both medial moraines and eskers look somewhat similar in Figures 21-5 and 21-2, but in the field a geomorphologist would have little trouble in telling them apart. What clues might a geomorphologist look for to tell which type of landform was being observed? How are these differences in appearance related to differences in formation?

11. Why are many glacial landscapes actually called glaciofluvial? Why are there very few purely glacial landscapes?

12. Waves do most of the erosional work along shorelines, sculpting new shorelines continuously. On what does the erosive power of waves depend, and how are the factors interrelated?

13. What factors contribute to littoral drift? How does human activity interfere with the natural processes contributing to littoral drift, and what are the ultimate results?

14. How are the forms of the Atlantic and Pacific coastlines of the United States related to the geologic structure and tectonic activity of each area?

15. Why are high-relief coastlines, such as that of California or Oregon, maintained as high-relief landscapes for long periods of time, rather than being worn down by wave action to low-relief landscapes?

16. How is the basic action of the waves in coastal landscapes of low relief different from that found in coastal landscapes of high relief? What associated landforms are produced in low-relief coastal areas, and how are they related to this basic difference between wave action in high and low-relief areas?

17. Settlement patterns along the Atlantic and Pacific coasts of the United States differ markedly. This is not entirely explained by the fact that the Atlantic coast is closer to Europe, the home of many of the early colonizers. How can the early settlement patterns in the United States be related to the distinctive morphology of each coast?

18. Although transportation of material by moving air can involve enormous quantities of silt and sand, it is rarely, if ever, the dominant erosive agent in an area.

What factors limit the role wind erosion can play in denuding a landscape?

19. Wind, like other fluids we have studied, does not only erode landscapes—it can also deposit its transported load. What causes such features as dunes and loess deposits, and how are they related, each in a different way, to varying wind velocity?

20. What are the advantages of studying erosional and depositional processes and landscapes from a systems viewpoint? Are glaciers open or closed systems? How do we know? Is the same true of coastal systems?

PHYSIOGRAPHIC REGIONS: THE SPATIAL VARIATION OF LANDFORMS

CHAPTER 22

When you fly home from college or across the United States on vacation, you pass over a panorama of changing landscapes. Even from a height of 12 km (40,000 ft), the great canyons of the Colorado Plateau, the Rocky Mountains, the Great Plains, and the Appalachians leave clear impressions, and you would not mistake one for the other. The landscapes of the continents are not jumbled and disorganized. On the contrary, each area has its distinct and characteristic properties. In our everday language we use words to identify these properties: the *Rocky* Mountains, the Great *Plains*, the Hudson *Valley*. In other parts of the world, you might have used such examples as the Alps, the Amazon basin, or the Siberian plains. In so doing, you prove to be a good geographer, because physical geographers (like other geographers) use the regional concept to classify and categorize information. And the Rocky Mountains and Great Plains are just that: regions in the North American landscape.

Physical geographers are interested in the spatial arrangement of the phenomena on the earth's surface and the relationships among the phenomena. After an examination of the major processes that operate in the atmosphere in Chapters 3 through 6, Chapter 7 discusses the spatial distribution of climate. Similarly, following a discussion of some aspects of the biosphere, Chapter 12 considers the spatial distribution of soils and vegetation. The geological processes of the earth's crust—those that tend to build it and those that wear it down—also have a spatial component. This component is expressed in what are known as physiographic regions.

Physiographic regions are characterized by a certain uniformity or homogeneity of landscapes and landforms. When we think of the Rocky Mountains or the coastal plain of the southeastern United States, an image of a particular scenery comes to mind. It would be tedious to examine physiographic regions for the whole world, so we will concentrate on North America. It is fortunate that examples of almost every variety of landform found in the world can be found in this continent.

DEFINING A PHYSIOGRAPHIC REGION

Why do we attempt the regionalize our observations? Regionalization does for us what classification does in other fields. It permits us to simplify local details, to see the larger schemes and frameworks, to compare areas while keeping track of points of difference. A *physiographic region* (sometimes referred to as a physiographic *province)* is a part of the earth's surface within whose boundaries there is substantial uniformity and homogeneity of landscape.

But we should be mindful of the scale of our regionalizing. If we work only with *first-order* relief units of the earth, then our world map of regions would show only two sets: continental landmasses and ocean basins. At the *second order,* in North America we differentiate among such relief features as the Canadian shield, the Gulf-Atlantic coastal plain, the Appalachian highlands, the interior plains, the southern Caribbean mountains, and the western mountains—as Figure 22-1 does. But this regionalization still obscures several subdivisions we can recognize quite clearly. The western mountains, for example, include both the Rocky Mountains and the ranges of the Pacific Coast—as well as the Colorado Plateau and some other regions that are not mountainous at all. And in the vast region identified as the interior plains, we recognize not only the Great Plains and the region around the Great Lakes, but also the Ozarks and the Ouachitas, which are not plains by any means. So we proceed to *third-order* relief units to identify such regions as the northern, middle, and southern Rocky Mountains, the Columbia Plateau, the Piedmont, and the St. Lawrence River valley.

The level of regionalization, like the level of classi-

FIGURE 22-1

Major physical subdivisions of North America. These represent second-order relief features.

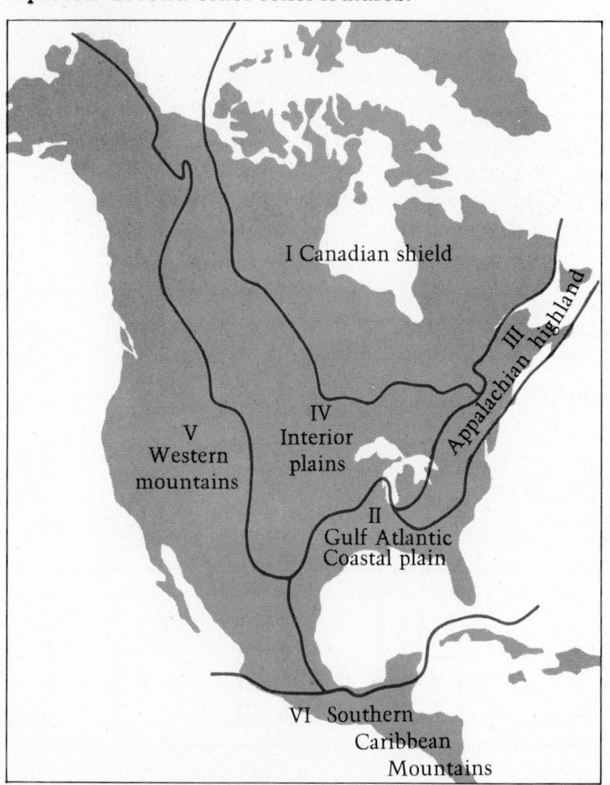

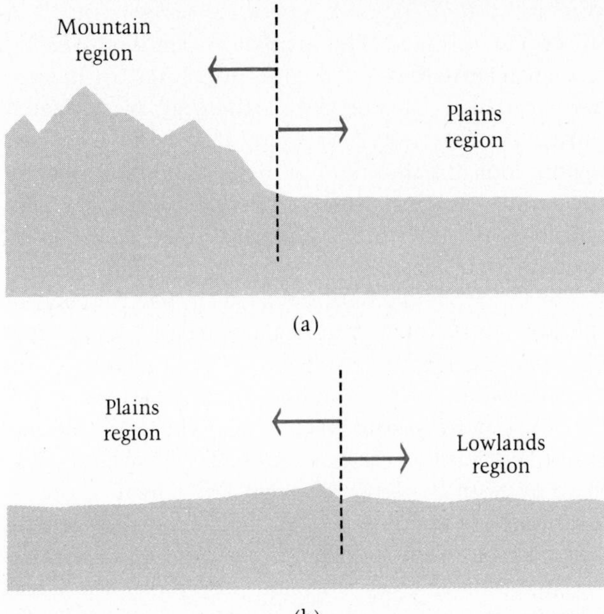

FIGURE 22-2
The topographic features that mark the physiographic boundaries of the Great Plains.

fication, should always be tied to our objectives. In this instance, we do not want to amplify every area of drumlins, every river valley, every group of hills. We want to get a picture of the continent's grand scheme of landscapes.

A question raised by our definition has to do with the criteria we employ in defining physiographic regions or provinces. The term *physiography* involves more than just the landscape and the landforms it is made of. It relates—check your geographical dictionary or any dictionary, for that matter—to all the natural features on the earth's surface, including not only the landforms but also the climate, soils, vegetation, hydrography, and whatever else may be relevant to changes in the overall natural landscape. For example, an area such as the interior plains in Figure 22-1 may be subdivided into several physiographic provinces because it extends through several climatic zones. The relief may remain generally the same, but the natural vegetation in one part of the area may be quite different from that of another section. This, of course, is a response to climatic transition, soil differences, and other regional contrasts.

This leads us to the crucial question: How are the boundaries of physiographic regions established? Occasionally there are no problems. Along much of the western edge of the Great Plains, the Rocky Mountains rise sharply, terminating the rather level surface of the Great Plains and marking the beginning of a quite different region in terms of relief, slopes, rock types, landforms, vegetation, and other aspects as well. But as Figure 22-2 indicates, the transition is far less clear and much less sharp in other instances. The eastern boundary of the Great Plains is a good case in point. Somewhere in the tier of states from North and South Dakota through Nebraska, Kansas, and Oklahoma, the Great Plains terminate and the interior lowlands (or central lowlands) begins. Exactly where this happens depends on the criteria we use to define the two adjacent regions and the method we use to draw the dividing line. Sometimes a persistent linear landform, such as an escarpment, may serve effectively, overshadowing by its prominence the other transitions in the landscape (see Figure 22-2). Where such a natural dividing line does not present itself, another solution must be found.

Nature rarely draws sharp dividing lines. Vegetation zones, climatic regions, and soil belts merge into one another in transition zones, and the lines we draw on maps to delineate them are the products of our calculations, not those of nature. In such a situation we can establish dividing lines for each individual criterion (soil change, vegetative change, climatic change, and so on), superimpose the maps, and draw an "average" line to delineate the physiographic province as a whole, as in Figure 22-3. Still another solution is to place a grid system over the area that the boundary is expected to lie in and to give numerical values to each of the criteria mapped. The physiographic region's boundary can thus be mapped by computer. But as always, everything ultimately depends on our choice of criteria and our ability to map them or attach values to them.

PHYSIOGRAPHIC REGIONS OF NORTH AMERICA

In the discussion that follows, we will view North America first at the highest level of regional generalization, to get a picture of the overall layout of the continent's landscapes. Then we will focus on several regions in the United States identified on the basis of third-order relief units.

THE INTERIOR PLAINS AND SHIELD

The map of the physiographic regions of the North American continent in Figure 22-4 shows the interior plains of the landmass divided into several regions: the Great Plains, the interior lowlands, the Canadian shield, and the Arctic coastal plain. The *Arctic coastal plain,* smallest of these regional units, fringes the

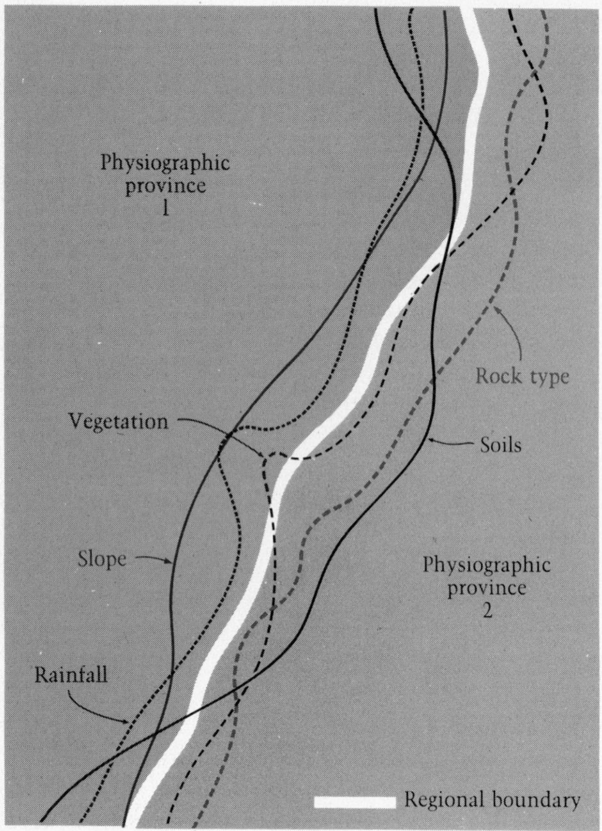

Regional boundary derived from several criteria

FIGURE 22-3

The derivation of a regional boundary from boundaries for several criteria.

southern side of Hudson Bay and parts of the Arctic Ocean. It is made of sediments covering the rocks that make up most of the heart of Canada, the Laurentian Shield.

Where these crystalline rocks are exposed is the *Canadian shield*, a province marked by low relief and unproductive soils but noted for its minerals, especially iron, copper, gold, and titanium. This is the region that spawned the great ice sheets of the Pleistocene Ice Age, which scoured the rocks and hollowed out the basins now filled with the water of Canada's many lakes. Low hills and a few isolated mountains diversify a scenery that is characteristically monotonous and slightly rolling. Where the meager soils are enough to sustain vegetation, stands of spruce and fir clothe the countryside. Here is much of the boreal forest described in Chapter 12.

The *interior lowlands*, from the edge of the Canadian Rockies to the margins of the Appalachian highlands, constitute another region of low relief, but it is underlain mainly by sedimentary rather than crystalline rocks. There are contrasts also in soils and vegetation (see Chapter 12) between this province and the Canadian shield. The interior lowlands, also called the central lowlands and the interior plains, are in large part sustained by the glacial till that accumulated during the ice sheets' advance. If you fly over this region, look for the moraine belts that stretch across the surface, bearing witness to periods of glacial standstill. It is not difficult to identify other glacial landforms as well.

The *Great Plains* occupy the southwestern sector of the low-lying North American heartland. The Great Plains are now known for their wheat and small-grain production and for livestock ranching. A look at the map of climatic regions (see Figure 7-4) tells us that this region is drier than the adjacent interior lowlands. The vegetation map confirms this. The Great Plains is essentially a region of short, rather sparse grasses, whereas the interior lowlands are covered by denser, longer grass and trees. The soil map shows the Great Plains to be a region of pedocals, especially mollisols, whereas the interior lowlands have alfisols as well as mollisols. Of course, these differences occur in transition zones, and the maps we are referring to are summaries based on artificial indexes. But the typical scenery of the Great Plains reflects a dryness that implies a number of significant contrasts to the interior lowlands, and there is ample justification for the identification of a discrete physiographic region here.

THE APPALACHIAN HIGHLANDS

The composite area made up of the four low-relief provinces of the North American interior is flanked on several sides by highlands and mountains. To the east lie the *Appalachian highlands*, which extend from northern Alabama and Georgia to New England and beyond to New Brunswick, Nova Scotia, and Newfoundland. As we will find later, this region contains several clear-cut subdivisions, but in general it is characterized by elongated ridges. Many of these ridges show a parallel or slightly zigzig pattern, rolling and sometimes rugged hills, some mountainous areas, and persistent valleys. Such a zigzag is shown in Figure 22-5.

Some of the contrasts within this region stem from the fact that, although the northern sector was glaciated during the Pleistocene Ice Age, the southern part was not. The rocks vary from folded sedimentary layers and maturely eroded horizontal strata to igneous intrusions. The familiar parallel, tree-clad ridges separated by populated valleys prevail, especially in the central zone of the region. Toward the west, plateau characteristics take over, and rugged, crystalline-supported topography dominates to the east. The Great Smoky Mountains, shown in Figure 22-6, part of this

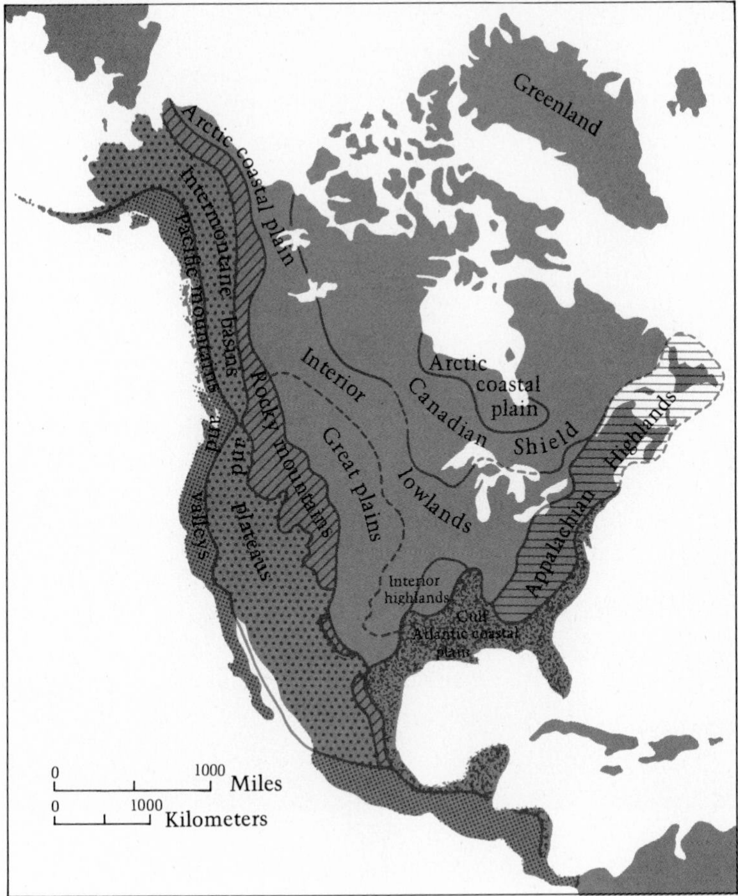

FIGURE 22-4
Physiographic regions of North America.

eastern zone of the Appalachian highlands, are sustained by ancient igneous and metamorphic rocks and are often referred to as the "older" Appalachians. On these old rocks, the Appalachian highlands reach their highest elevations, about 1800 m (6000 ft)—several hundred meters higher than the crestlines of the parallel ridges and valleys of the folded central zone.

To the south of the interior lowlands lie two smaller highlands, identified on Figure 22-4 as the *interior highlands* but usually separated into the Ozark Plateau in the north and the Ouachita Mountains in the south. The Ozark Plateau consists of slightly domed sedimentary layers, maturely eroded but still standing more than 450 m (1500 ft) above the average elevation of the interior lowlands to the north. This lowland is now occupied by the valley of the Mississippi River.

The Ouachitas are reminiscent of a small-scale Appalachian ridge-and-valley scene. Like much of the Appalachians, the area is suited to forestry, cattle raising, and the tourist resort. In fact, the Ouachitas are an

FIGURE 22-5
A zigzag pattern in plunging folds, south of Hollidaysburg, Pennsylvania.

FIGURE 22-6
The Great Smoky Mountains, part of the Appalachian highlands.

extension of the Appalachian highlands, and the intervening lowland somehow failed to undergo the same crustal readjustment that occurred in both the Appalachians to the east and the Ouachitas to the west. The Ouachitas generally are lower than the Appalachians, however, reaching no more than 780 m (2600 ft) in the central section and less than 300 m (1000 ft) near the region's boundaries.

HIGHLANDS OF THE WEST

Along its entire length, the interior low plain of North America is flanked to the west by high mountains, much higher and far more rugged than those to the east and south. The *Rocky Mountains* shown in Figure 22-7, extend from Alaska to New Mexico and beyond into Mexico as the Sierra Madre Oriental (Eastern Sierra Madre). The Rockies, formed much later than the Appalachians, rise as much as 2700 m (9000 ft) above their surroundings, to elevations that in places exceed 4200 m (14,000 ft).

None of the Appalachians' regularity exists in the Rocky Mountains, which consist of ranges extending in different directions for varying distances, valleys of diverse width and depth, and even large sediment-filled basins between the mountains. The underlying rocks include igneous batholiths (especially in the northern sector), volcanics, metamorphosed strata, and sedimentary layers. There is intense folding and large-scale block faulting, all reflected in the high-relief topography. The region carries all the evidence of Pleistocene mountain glaciation, with cirques, horns, arêtes, U-shaped valleys, and aggradational landforms. As a result it is one of the great tourist and ski resort areas of the world. There is still some permanent snow on the Rockies in the United States; in Alaska the glaciers still inhabit the valleys, as Figure 22-8 attests.

As Figure 22-4 shows, the Rocky Mountains physiographic province is actually a rather narrow but lengthy zone. To the west of it lies the region of *intermontane basins and plateaus.* This region, like the Rockies, begins in Alaska, crosses Canada and the United States, and extends into Mexico. It is a complex region that can be subdivided, as we will see later. The western boundary of the Rocky Mountain province is not as sharp as its eastern boundary, because the intermontane basins and plateaus lie at considerable elevations and do not provide the dramatic, wall-like effect that marks the Rockies from the lowlands of the interior.

The overall physiography of the intermontane region is quite different, however: There are expanses of low relief, underlain by only slightly disturbed sedimentary strata or by volcanic beds; isolated fault-block mountains; areas of internal drainage; spectacular canyons and escarpments; and in much of the region, landscapes of aridity, like that in Figure 22-9. Because it includes such diverse areas as the Colorado Plateau, the Mexican highlands, and the Columbia Plateau as well as the so-called basin-and-range region, the intermontane region could be considered not one physiographic province but several. Certainly the strongest bond among these areas are their "intermontane" location—their position between the mountains—with the Rockies on one side and the ranges of the Pacific coast on the other. But internally there is much basis for dividing the region into several discrete provinces.

The *Pacific mountains and valleys* form the westernmost physiographic region in North America, again extending all the way from Alaska to Central America. This region incorporates three zones that lie approximately parallel to the Pacific coastline. In the interior, away from the coast, lies a series of mountain ranges that includes the Cascade Range of Oregon and Washington and the Sierra Nevada of California. Adjacent to these are several large valleys, chiefly the central California valley and the Puget Sound—Willamette lowland. Both the valley and the lowland are famed agricultural areas, as shown in Figure 22-10. Separating these valleys from the Pacific Ocean are the Pacific

FIGURE 22-7

Rocky Mountains National Park, Colorado. The view is to the south from the high point on Sky Ridge Road. The elevation where the sightseers are standing is 3633 meters (12,110 feet).

FIGURE 22-8

South Sawyer Glacier, Alaska. Some high points of the mountains in the background are over 1800 meters (6000 feet).

FIGURE 22-9

An arid landscape in the southern part of the region of intermontane basins and plateaus, Death Valley, California.

FIGURE 22-10
Aerial view of orchards in the Willamette Valley, west of Portland, Oregon.

Coast ranges, such as the Olympic Mountains of Washington and the Klamath Mountains of northern California.

Except for the plains and low hills in the valleys, the topography in this province is rough, the relief high, and the rocks and geological structures extremely varied. Mountain building is in progress, earthquakes occur frequently, and movement can be observed along active faults. The coastline is marked by steep slopes and deep water, pounding waves, cliffs, and associated landforms of high-relief shores. Elevations in the interior mountains rival those of the US Rockies in places, reaching over 4200 m (14,000 ft) in the US sector of the region and exceeding the Canadian Rocky Mountains along the Alaskan coast, where some mountains are over 5400 m (18,000 ft) in height. This is the leading edge of the North American tectonic plate (refer to Chapter 15), and the landscapes of this coastal province reflect the forces at work here.

THE GULF-ATLANTIC COASTAL PLAIN

The major physiographic province remaining to be identified extends along much of the coastline of North America, from the vicinity of New York (Long Island is a part of the region) to the northern coast of Costa Rica in Central America. The *coastal plain* is bounded on the landward side by the Piedmont (the foothills of the Appalachians), the Ouachita Mountains, the Sierra Madre Oriental, and the Central American mountain ranges. It incorporates all of Florida and attains its greatest width in the area of the Mississippi River. The coastal plain extends far up the Mississippi's valley, and all of the river's delta forms part of the region (see Figure 22-11). In Texas, the boundary between the coastal plain province and the Great Plains region is in places rather indistinct, although a fault scarp separates the two provinces in southern Texas.

In the eastern United States, the boundary between the coastal plain province and the Appalachian region is marked by a series of falls and rapids on rivers leaving the Appalachian highlands and entering the coastal plain. These falls have developed along the line of contact between the soft sedimentary layers of the coastal plain and the harder, older rocks of the Piedmont. Appropriately, the boundary is called the *Fall Line*. In former times the places where rivers cross the Fall Line marked not only the head of ocean navigation but also the source of water power. Famous cities such as Philadelphia, Pennsylvania and Richmond, Virginia grew up on the Fall Line.

The seaward boundary of the coastal plain is, of course, the continental coastline. This coastline, however, has changed position in recent times as sea level has risen and fallen. A case can therefore be made that the coastal plain really continues under the ocean's

FIGURE 22-11
New Orleans, one of the many famous cities to have grown on the coastal plain province. The flat area of the plain can be seen stretching away in the distance.

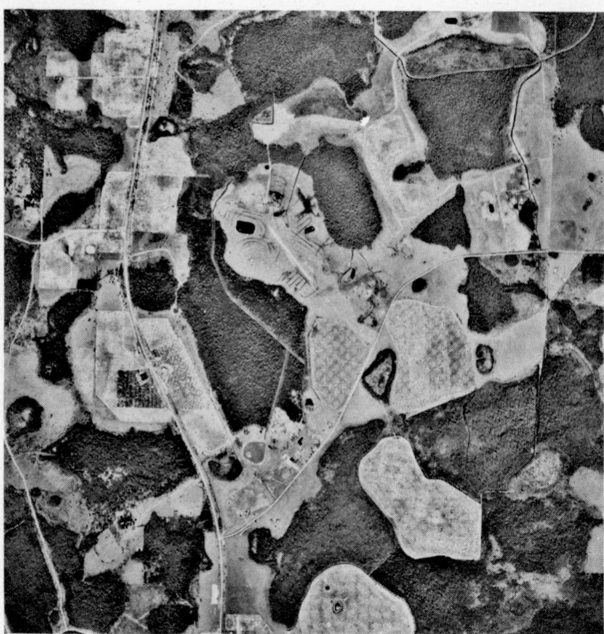

FIGURE 22-12
The Lake District of central Florida, southeast of Winter Haven, a leading orange and grapefruit area. Sinkholes form distinctive features of the karst topography.

water to the edge of the continental shelf, the continental slope. The coastline, as shown in Chapter 21, has the characteristics of a low-relief coast, with offshore bars, lagoons, beaches, and related landforms. The deepest indentations and strongest irregularities exist in the northern sector, where Chesapeake Bay is the largest recess. Otherwise the coastline is quite straight. Coral reefs appear off the coast in the warmer southern waters and are especially well developed off the Yucatán Peninsula. The coastline is rather changeable in the area where the Mississippi Delta is building outward.

The whole of the coastal plain lies below about 300 m (1000 ft) in elevation, and it is a region of low relief and little topographic variety. Sedimentary rocks sloping gently seaward underlie the province. In the northern sector, there is a cover of glacial material. Areas of low hills occur in the Carolinas and in central Florida, and west of the Mississippi River a series of low ridges lie more or less parallel to the coastline. The Mississippi River valley and delta form a distinct subregion in this province, as does the area of karst topography in Florida, shown in Figure 22-12, where numerous sinkholes mark the countryside.

The coastal plain lies substantially in near-tropical latitudes, and the soil map in Figure 12-10 reflects this: The characteristic reds and yellows of tropical soils occur here as well. The vegetation is quite varied. There are deciduous forests in the eastern and southern United States, pine forests in the southeast, grasslands in Texas and northern Mexico, and some savanna country in Yucatán and nearby Central America.

FOCUS ON THE UNITED STATES

In looking at the physiographic provinces of the entire North American continent, we have generalized considerably. For example, we viewed the Rocky Mountains from Alaska to Mexico as one region, and we combined the varied Appalachian physiography into one province. When we bring the matter down to the continental United States alone, however, we gain the opportunity to refine our scheme in several areas.

In the discussion that follows, you may find it useful

WHERE WILL THE GROUND COLLAPSE?

When you plan to buy a house, it is always a good idea to know something about the terrain on which it stands. Are landslides possible in the area? Can floodwaters reach the house? And—could the ground below collapse, swallowing the whole house up or, at least, damaging its foundations?

In the United States, areas of karst topography are widely dispersed, though not very large in area. Central Florida has limestone that is dissolving, as does an area on the Florida–Georgia border. Appalachian sediments include limestones, and Appalachian valleys exhibit collapse features intermittently from Pennsylvania to Alabama. In central Alabama, in 1972, a sinkhole formed that measured well over 125 m (400 ft) in diameter and nearly 45 m (150 ft) in depth (Figure 18-7).

Another area of karst topography extends from central Kentucky into southern Indiana, and central Missouri and eastern Kansas are susceptible as well. Lesser areas occur in Texas, New Mexico, Colorado, and Wyoming. However, most karst areas lie in the eastern half of the country.

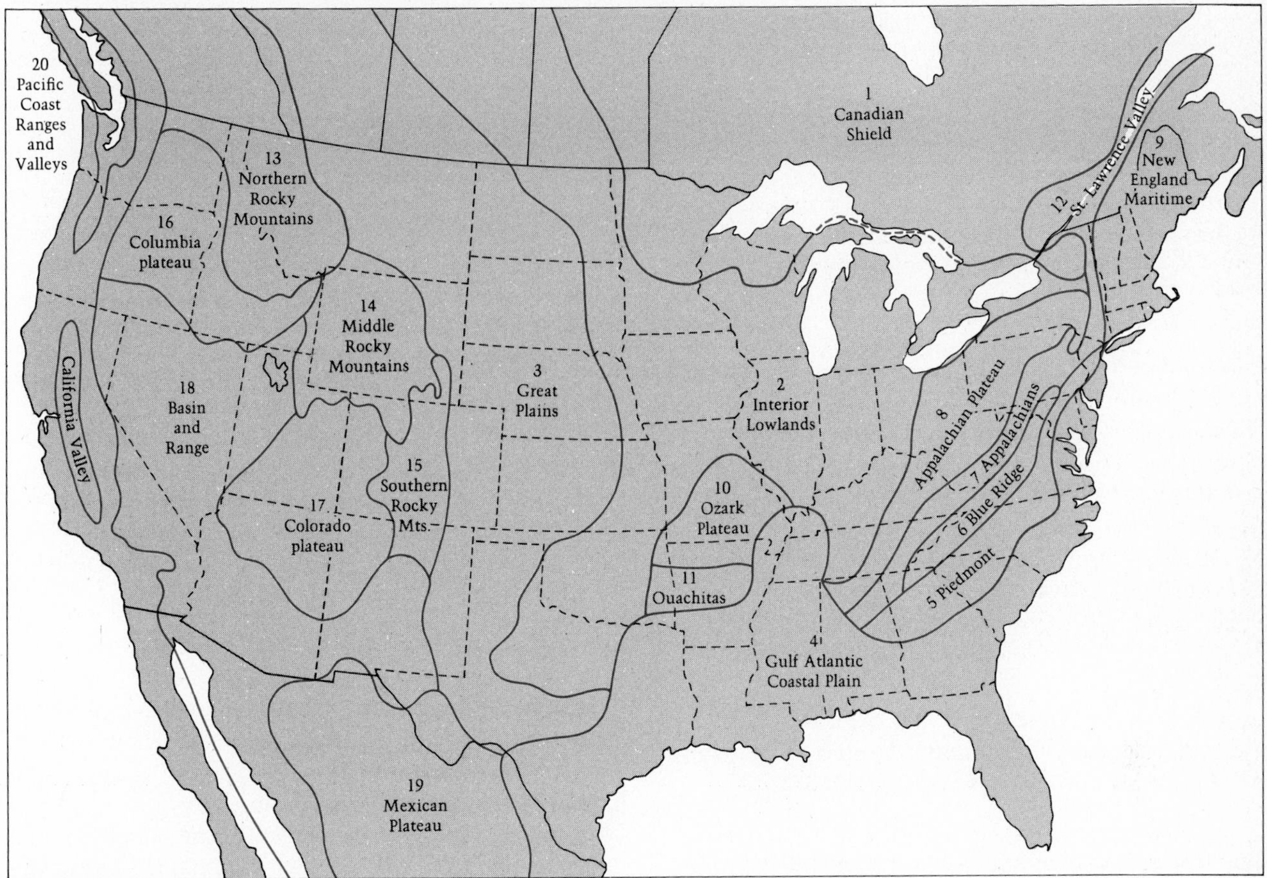

FIGURE 22-13
Generalized physiographic regions of the United States.

to compare Figures 22-13 and 22-4. Note, for example, that the United States map shows no fewer than five regions in what we earlier called the Appalachian highlands, that the Rocky Mountains are identified as not one but three provinces, and that the belt of plateaus and basins between the Rockies and the Pacific Coast ranges is also fragmented into several discrete provinces.

THE APPALACHIAN HIGHLANDS

The Appalachian highlands actually constitute a group of several physiographic provinces. Most familiar to us, perhaps, is the ridge-and-valley topography of the younger or "newer" Appalachians, a discrete region in Figure 22-13. But this topography, as the map shows, exists only in the central zone of the Appalachian region. In the east-central United States, there are four other different landscape regions, all part of the Appalachian province delimited in Figure 22-4. They are grouped together in Figure 22-4 because they have a common geologic history and because they lie, as highlands, between lowland regions.

Appalachian plateau There is much to differentiate the Appalachian highlands, as you can see in Figure 22-14. The Appalachian plateau, noteworthy for its coal deposits, has an irregular topography carved by erosion out of horizontal and slightly dipping strata. Parts of this province might be described as low mountains, because the rivers have cut valleys 450 m (1500 ft) deep and even deeper and because summit elevations reach 1200 m (4000 ft) in such areas as West Virginia. Despite forming a part of the Appalachian highlands region, the eastern (that is, the inside) boundary of the Appalachian plateau is quite well defined by major escarpments—the Allegheny Front in the north and the Cumberland Escarpment in the south. In fact, it is the western boundary, facing the interior lowlands, that is in places indistinct, because the Appalachian plateau loses elevation and prominence toward the west.

"Newer" Appalachians Compared to the Ap-

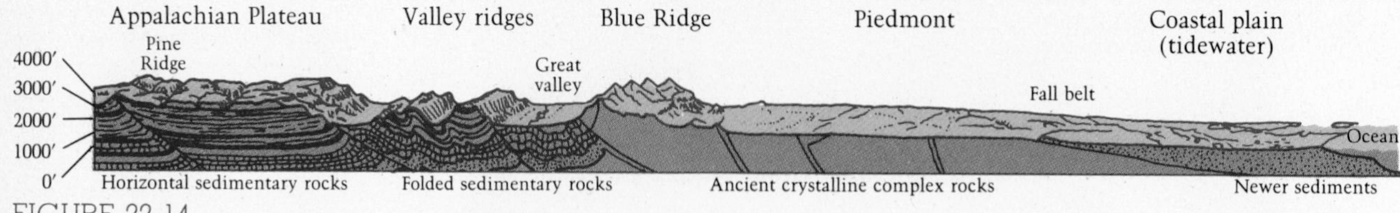

FIGURE 22-14

A cross-section of the Appalachian highlands at Virginia, showing the principal physiographic subdivisions.

palachian plateau's irregular topography, the "newer" Appalachians' ridge-and-valley topography forms a sharp enough contrast to delineate a discrete province. Parallel vegetation-clad ridges reach remarkably even summits between 900 and 1200 m (3000 and 4000 ft) above sea level. Note in Figure 22-13 that the "newer"-Appalachian province extends northward into the valley of the St. Lawrence River.

The eastern edge of the "newer" Appalachians is marked by a wide lowland called the *Great (Appalachian) Valley.* The Great Valley, with fertile limestone- and slate-derived soils, has been well tilled by the famous Pennsylvania Dutch farmers since before the American Revolution (see Figure 22-15).

Blue Ridge East of the Great Valley lies the Blue Ridge. This province is underlain by crystalline rocks, mostly highly metamorphosed and ancient. In this respect the Blue Ridge differs from the Appalachian plateau and the "newer" Appalachians; they are sustained by sedimentary rocks. The Blue Ridge is indeed the "older" Appalachians by virtue of the age of the rocks and the successive periods of mountain building. In the mountainous Blue Ridge, the Appalachian highlands reach their highest elevations. The Great Smoky Mountains typify this province

FIGURE 22-15

Contour farming, plowing around contour lines, in the part of the Great Valley farmed by Pennsylvania Dutch farmers.

Piedmont Toward the east, the mountainous topography of the Blue Ridge yields to the much lower relief of the Piedmont, which is still an upland compared to the coastal plain but not nearly so rugged as the "older" Appalachian ranges. In the Blue Ridge the Appalachians rise to over 1800 m (6000 ft), but in the Piedmont the maximum elevations are about 903 m (3100 ft) along the inner boundary, declining to the contact with the coastal plain. The underlying rocks of the Piedmont, however, are the same as those of the Blue Ridge: ancient, highly altered crystallines. Thus the difference is principally a matter of topography and relief. Were it not for the Fall Line, the Piedmont would merge almost imperceptibly into the coastal plain. The Piedmont is somewhat hillier, with several monadnocks rising above the general level of the countryside.

New England maritime province Figure 22-13 shows one other region within the Appalachian highlands—the New England maritime province. As the name of the province implies, this is a completely different environment. It is strongly affected by glaciation, dotted with lakes, penetrated by the waters of the Atlantic Ocean, and bounded by a high-relief coastline (Figure 22-16). Major topographic features are made, like the Blue Ridge, of ancient crystalline rocks. Elevations exceed 1200 m (4000 ft) in places, but the topography has none of the regularity and parallelism that marks the ridge-and-valley region in the eastern United States. In fact, the New England maritime province is extremely varied and in places quite rug-

THE CUMBERLAND GAP

The Appalachian region is naturally divided into three plateaus. The westernmost of the subdivisions is the Cumberland Plateau. The Cumberland Mountains, which are the highest and most rugged of the plateau's terrain, range in height from 600 to 1244 m (2000 to 4145 ft).

During the first hundred years of settlement by Europeans in the area that is now Virginia and the Carolinas, the Cumberland Mountains were an effective barrier against westward expansion. It was not until 1750, when Thomas Walker discovered a relatively low, clear passage through the mountains, that Kentucky was settled.

The Cumberland Gap, at an altitude of 495 m (1650 ft), is one of the most significant mountain passes in history. Through it passed Abraham Lincoln's grandparents and Daniel Boone's Wilderness Trail. It was so valuable in controlling the surrounding region that it was a strenuously contested site in the Civil War, changing hands between the armies of the North and the South three times.

Today, Cumberland Land Historical Park at the point where Virginia, Tennessee, and Kentucky meet, commemorates the pioneers who opened this major western route.

ged. Moraines left by the glaciers help to give this region its special identity. If you were to fly over it, you would also be impressed by the considerable extent of the forest cover that still prevails compared to that of the Appalachian highlands in the east-central United States.

FIGURE 22-16

A portion of the high-relief coastline of the New England Maritime Province. This is Acadia National Park in Maine.

THE ROCKY MOUNTAINS

In Figure 22-4, the Rocky Mountain physiographic province is represented as one region extending from Alaska to Mexico. But in Figure 22-13, no fewer than three regions are identified as Rocky Mountain provinces: northern Rocky Mountains, middle Rocky Mountains, and southern Rocky Mountains. On the basis of topography alone, there is enough contrast among different parts of the Rocky Mountains to justify the recognition of three discrete physiographic provinces. The southern Rocky Mountains display a compactness and north–south orientation that is quite different from the openness and directional diversity of the middle Rockies. The northern Rocky Mountain province does not display the dominance of linear ranges that we find in both the middle and southern Rockies.

Southern Rocky Mountains This province lies across the heart of Colorado, as Figure 22-17 shows. The mountains rise sharply out of the adjacent Great Plains along a range called the Front Range in Colorado but other names elsewhere. The Front Range and similar ranges in the southern Rockies are made of crystalline rocks, with upturned sedimentary layers along their sides. The crystalline rocks near the Cripple Creek area of the Front Range yielded half a billion dollars of gold from 1890 to 1962, when the last mine

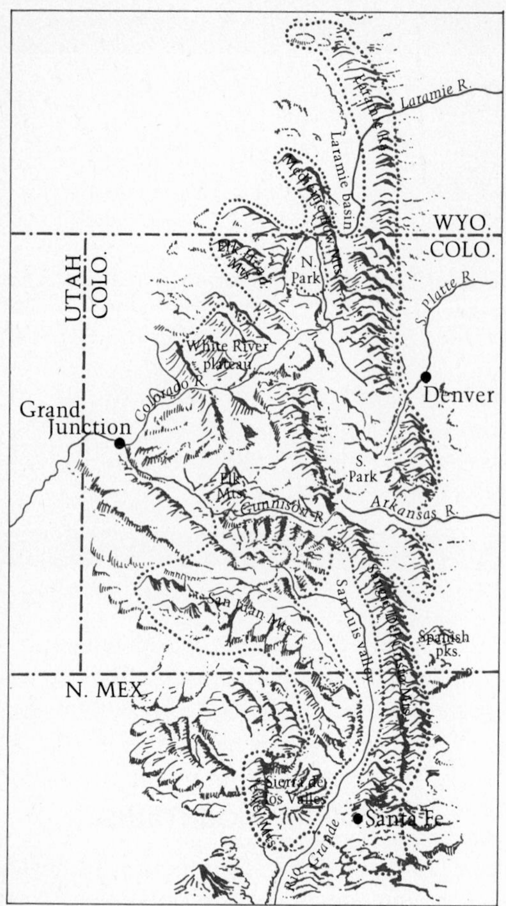

FIGURE 22-17
The topography of the southern Rockies.

closed. Elevations in many places exceed 4200 m (14,000 ft), and even the valleys between the ranges (the *parks)* lie at elevations of 2100 m (7000 ft) or more. Some large rivers start in this province, including the Rio Grande and the Colorado River, but it is the glacial period that gave this region its distinctive topography through scouring and deposition.

Middle Rocky Mountains The continuity of the Rocky Mountains northward is broken by the Wyoming Basin (also called the Green River basin), which on small-scale physiographic maps looks like a westward extension of the Great Plains. The basin signals a change in the character of the Rocky Mountains. The middle Rocky Mountain province consists of ranges lying in various directions, wide open valleys and basins, and a less-congested topography than in the southern Rockies, as Figure 22-18 indicates. The middle Rockies, the Uinta Mountains of Utah, lie east–west, but the next range, the Wasatch Range, runs north–south. In the northern sector of the province, the Bighorn Basin is almost encircled by a series of ranges. Most of the major ranges are anticlinal uplifts similar to those of the southern Rockies, but the Tetons are a fault block. Folded sedimentary strata play a role in the northern part of the province. Elsewhere, lava flows create some expanses of low relief, as in Yellowstone National Park, which is a part of this province.

Northern Rocky Mountains North of Yellowstone Park, the Rocky Mountain landscape changes considerably. Gone are the persistent, conspicuous, high-crested ranges that provide such spectacular scenery in the middle and southern Rockies, and lost also are the high-elevation parks, basins, and valleys that separate the ranges. The northern Rocky Mountains exhibit a confused topography, generally lower than that of the Rockies southward and sustained by an even more complex set of structures than exists in the south. The boundary of the region can be seen quite clearly in Figure 22-13. To the east lies the Great Plains province, to the south the Snake River plain forms a sharp topographic limit, and to the west the plain of the

FIGURE 22-18
The topography of the Middle Rockies.

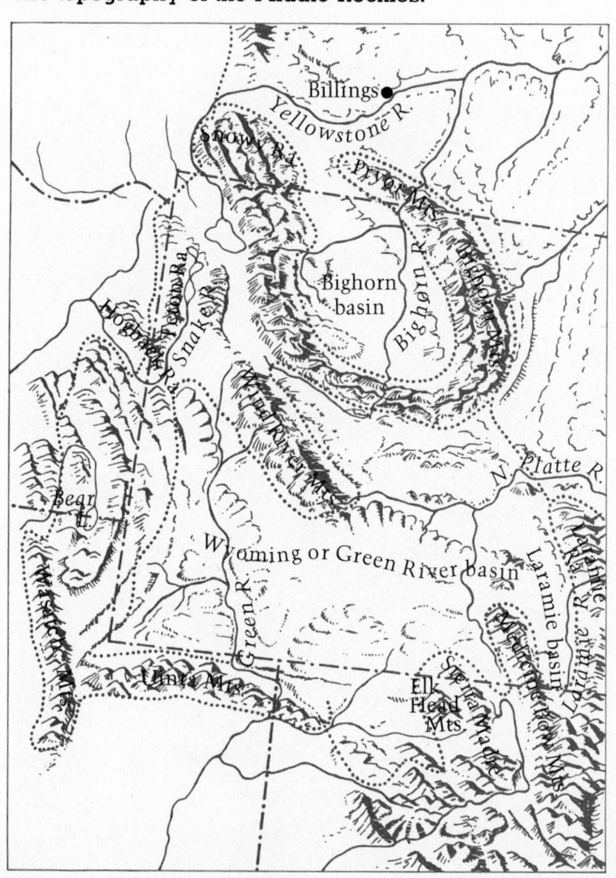

Columbia River borders the northern Rockies. The Snake River plain and the Columbia plain are actually lava plateaus that abut the mountains. Within this boundary, the northern Rocky Mountains are cut from large batholiths, folded sedimentary layers, and extrusives. Large-scale faulting and severe erosion, as well as glaciation, have combined to make this one of the country's most complex landscape regions.

THE INTERMONTANE BASINS AND PLATEAUS

The last physiographic province we will view in detail is that vast region lying between the Rocky Mountains to the east and the Pacific Coast ranges to the west, mapped in Figure 22-4 as the region of intermontane basins and plateaus. In the United States, this region subdivides quite clearly into three provinces: the Columbia Plateau, the Colorado Plateau, and the province called the basin-and-range. Figure 22-13 shows their arrangement.

Columbia Plateau This region lies wedged between the northern Rocky Mountains to the east and the Cascade Range of Washington and Oregon to the west. The plateau itself is one of the largest lava surfaces in the world. We call it a plateau only because it lies as high as 1800 m (6000 ft) above sea level. The plateau offers few obstructions to travel. The relief is rather low, and the countryside rolls, except where volcanic and other mountains rise above the flood of lava that engulfed them. The southern boundary of the province is determined by the limits of this basaltic lava field, and in many places it is not clear in the topography at all. There is not much doubt that this region differs basically from the mountainous provinces that flank it. The two major rivers that drain the plateau, the Columbia River and the Snake River, form deep canyons in the lava. The Snake River's canyon is one area well over a kilometer deep. Alluvial plains and loess accumulations date from the glacial period, when the rivers were blocked and when the winds piled dust in deep layers.

Colorado Plateau This region, by contrast, is underlain chiefly by flat-lying or nearly horizontal sedimentary strata, weathered in this arid environment into vivid colors (*colorado* is the Spanish word for "red") and carved by erosion into uniquely spectacular landscapes. The eastern boundary, of course, is the southern Rocky Mountains, a very different topography. On the western side, a lengthy fault scarp bounds the plateau and separates it from the basin-and-range region. To the south, the boundary is not marked by linear landscape features or by strong topographic contrasts. The southern boundary is traditionally taken as the divide between the plateau's major river, the Colorado, and the drainage area of the Rio Grande.

The Colorado Plateau is traversed by several sets of steep-walled canyons with multi-colored layers of sedimentary rocks. The Grand Canyon, shown in Figure 22-19, is of course the most famous. The dryness of this area keeps weathering processes to a minimum, preserving the valley walls and enhancing the coloration of the exposed strata. Such rivers as the Colorado, San Juan, and Little Colorado have cut down vigorously from the plateau surface, which lies as high as 3480 m (11,600 ft) above sea level. The flat upper surface is broken by fault scarps. Erosion has created numerous mesas, buttes, badlands, and much more. No brief description could adequately summarize the distinct character of the Colorado Plateau, which is unlike any other physiographic province in North America.

Basin-and-range province South of the Columbia Plateau and west of the Colorado Plateau lies the basin-and-range province. Actually, this is not just one basin but an entire region of basins, many of them internally drained and not connected to other depressions by permanent streams. Dryness is also the hallmark of this region.

A large number of linear mountain ranges rise above the basins. These mountain ranges are generally from 80 to 130 km (50 to 80 mi) long and from 8 to 24 km (5 to 15 mi) wide. They rise 600 to 1525 m (2000 to 5000 ft) above the basin floors, reaching heights of 2100 to 3000 m (7000 to 10,000 ft) above sea level. The mountains are mainly oriented north–south and have a steep side and a more gently sloping side, which suggests that they are structurally related. Faulting is the most likely explanation, because the basin-and-range province looks like a sea of fault blocks thrust upward along parallel axes.

Today, with aridity prevailing, the basins are mostly dry, streams are intermittent, and vegetation is sparse. Wind action plays a major role in sculpturing the landscape. This was not always so, because old shorelines of now-extinct lakes have been found along the margins of some of the basins in the province. Undoubtedly the most famous of all the existing bodies of water in the province is the Great Salt Lake in Utah, shown in Figure 22-20. This lake's volume has varied enormously in recorded times, and its high salt content tells a story of contraction and withering for all the province's bodies of water. Late in the nineteenth century, the Great Salt Lake's area was about 5000 km^2 (1900 mi^2), but early in the twentieth

FIGURE 22-19
a) The Grand Canyon, part of the Colorado Plateau.
b) A view of the Grand Canyon from 12,000 meters (40,000 feet).

century, during a drought, it almost disappeared. Today it is back to about 5000 km^2, but another serious drought could destroy this lake at any time.

THE END PRODUCT

These, then, are the major physiographic provinces of the United States. What we see as landscape results from all the forces and erosional systems we have considered in this book. Geologic forces produced the structures in the upper layer of the crust that exert so much control in the formation of landscape. Continuing crustal deformation from continental drift and plate tectonics continues to modify what exists. Solar radiation powers atmospheric circulation, and the hydrologic cycle rides on the movement of air. Water is brought to the landmasses, but once the precipita-

FIGURE 22-20

The salt crust, Great Salt Lake, Utah. Note the high shore lines on the distant mountains made by a former freshwater lake.

tion collects as running water, it comes under the force of gravity and various laws of physics as it carries out its erosional work. Finally, weathering processes and mass movements contribute to the sculpturing processes whose results we see summarized in a physiographic map.

SUMMARY

A physiographic region is a part of the earth's surface within whose boundaries there is substantial uniformity and homogeneity of landscape. It expresses the active processes of the lithosphere in spatial terms.

In this chapter we examined the physiographic regions of the North American continent at three distinct orders of generalization. We first looked briefly at a five-fold division. In particular, we examined the interior plains and shield, the Appalachian highlands, the highlands of the west, and the Gulf-Atlantic coastal plain.

Then we focused on some of the physiographic provinces of the United States. We found that the Appalachian highlands, the Rocky Mountains, and the intermontane basins and plateaus are composed of smaller physiographic subdivisions.

QUESTIONS

1. How does the concept of a physiographic region correspond to other kinds of scientific categorization? Why must scale always be carefully defined in dealing with physiographic regions?

2. What criteria are involved in defining physiographic regions or provinces? Are the structural features of an area alone sufficient criteria?

3. Why are the dividing lines between physiographic regions often difficult to define? How do geomorphologists overcome these difficulties?

4. How are the Arctic Coastal Plain and the Canadian Shield differentiated? What processes have contributed to the appearance of each?

5. What is the principal difference between the Interior Lowlands and the Great Plains? How does this affect the ways in which humans use each region?

6. Consider the Appalachian Highlands a single province. What are the features that characterize this region, and what were the processes that contributed to the broad differences between its northern and southern sectors?

7. What processes connect the Ouachita Mountains with the Appalachian Highlands, and how are the intervening Interior Lowlands somewhat unusual in this scheme? Structurally, to what is the Ozark Plateau probably connected?

8. What evidence remains of the processes that shaped the Rocky Mountain province, a long, narrow region stretching from the Sierra Madre Oriental in Mexico all the way to Alaska?

9. Why is the eastern boundary of the Rocky Mountain Province more visually spectacular than the western boundary, despite the presence of areas of low relief on both sides? What factors contribute to the different appearances of the landscapes in the Intermontane Basins and Plateaus and the Great Plains provinces?

10. The Pacific Mountains and Valley Province is one of the most active in the United States at the present time. What forces contribute to the present appearance of this province, and why?

11. The boundary between the Appalachian Highlands and the Atlantic Coastal Plain is distinct. What is this boundary, and why was it important in the early settlement patterns of North America?

12. Where is the seaward boundary of the coastal Plain province, and how is it related to changes in other

provinces? What features characterize the appearance of this province?

13. Using the United States as our framework and working with a more detailed scale, how are the Appalachian Highlands fragmented into several physiographic provinces? What distinguishes these provinces from one another?

14. Several criteria are used to separate the Appalachian Highlands into four provinces. What criteria are used for each region? Are they consistent throughout the categorization scheme?

15. What distinguishes the New England Maritime Province from the other Appalachian Highlands provinces? What processes cause this difference in appearance?

16. How are the three provinces in the Rocky Mountains distinguished? What is the principal criterion for separating this region into the Southern, Middle, and Northern Rocky Mountains?

17. Why is the Columbia Plateau a relatively easy physiographic region to define in terms of its boundaries? With what structural features is it associated?

18. How is the Colorado Plateau structurally distinct from the Columbia Plateau? How is the southern boundary of this region different from the other three boundaries?

19. What is the most commonly accepted explanation for the north–south orientation of the mountain ranges and basins of the Basin-and-Range province? What other factors characterize this province?

SUPPLEMENTARY READING FOR PART FOUR

Part Four discusses a considerable variety of topics, and for additional reading we must go to a number of different sources. For an overall view describing the continents and the ocean floors, mountains and plains, see *The Morphology of the Earth: A Study and Synthesis of World Scenery* by L.C. King, published by Hafner in 1967. Rocks and landscapes are discussed in *Rocks and Relief* by B.W. Sparks, published by St. Martins in 1972. A less formal book, beautifully illustrated, is *Volcano: Ordeal by Fire in Iceland's Westmann Islands.* The description of what happened near Iceland in the 1960s will suggest the incredible forces unleashed by major volcanic eruptions. This is an official *Iceland Review* book. On continental drift, start by looking at A. Wegener's *The Origin of Continents and Oceans* in a new translation by J. Biram, published by Methuen in 1966. You might also like to have a copy of A.L. du Toit's classic, *Our Wandering Continents,* published by Oliver and Boyd in 1937, and reprinted several times since then. Among the more recent books on continental drift, the most useful probably is a collection of readings from *Scientific American—Continents Adrift,* edited by J. Tuzo Wilson and published by W.H. Freeman in 1972. Crustal deformation is discussed by J. Verhoogen and others in *The Earth: An Introduction to Physical Geology,* published by Holt, Rinehart & Winston in 1970. Other useful books covering much of the material presented in this part are *Physical Geology* by J.E. Sanders and others, published by Harper's College Press in 1976, and *The Surface of the Earth* by A.L. Bloom, published by Prentice-Hall in 1969. The spatial variation of landforms is well documented in the *Atlas of World Physical Features* by R.E. Snead, a Wiley publication of 1972. Finally, on physiographic regions, we recommend W.W. Atwood's volume *The Physiographic Provinces of North America,* published in 1940 by Ginn & Co., and C.B. Hunt's *Natural Regions of the United States and Canada,* published in a most recent edition in 1974 by W.H. Freeman.

FROM THE GEOGRAPHER'S NOTEBOOK

MEASURING A RIVER

This field trip deals with part of the biosphere, and in particular the flow of water through and over the land. With the possible exception of the Southwest and parts of California, no place in the United States is far from a river. As Chapters 17 through 19 explain, running water represents one of the single most potent instruments of erosion. With the help of mass movement processes, rivers erode and transport material from the land surface to form the valleys that are so familiar a part of our landscape. Rivers themselves are partly fed by water that has seeped into the soil and has eventually found its way into the river channel. In this field trip we first look at the passage of water into soil—the process of infiltration—and especially the speed with which the process can occur. Then we examine a stream and try to understand something about its behavior by making some simple measurements and observations.

WATER THROUGH SOIL

Water enters a stream either by direct precipitation or runoff or by flow underneath the earth's surface. The amount of water that enters the soil itself is determined by the infiltration rates of the soil. In the first part of this field trip you should perform an experiment to determine approximate infiltration rates of the soil found in your locality.

First it is necessary to isolate the sample of soil to be tested. You can do this by using an old can. Select one with about a 10-cm (4-in) diameter. Cut both ends out of the can. Using a hammer or a mallet, knock the can into the ground where you want to test the soil so that about 5 cm (2 in) remain above the surface. Find a small container such as a chemistry beaker or a kitchen measuring jar that can be used for measuring small quantities of water, say about 50 ml (1.65 oz). You will also need a watch with a second hand. The experimental set up is diagrammed in Figure 1.

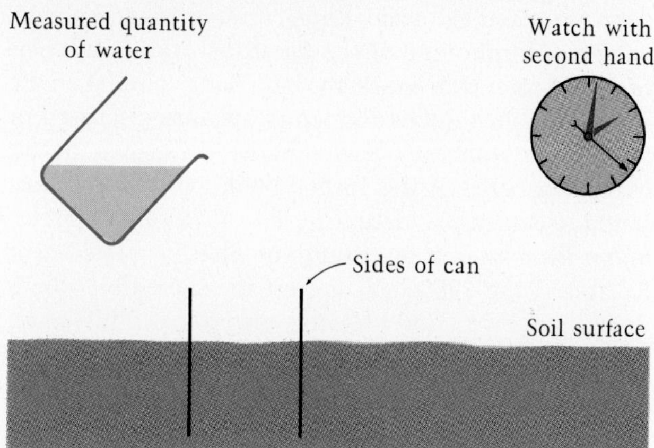

FIGURE 1

Experimental set up for measuring infiltration rates in the soil.

Pour one measure of water into the area surrounded by the can and note the time taken for it all to disappear into the soil. Repeat the procedure, noting each time how long it takes for the water to be absorbed by the soil. If the soil is dry to start with, the time for 50 ml of water to infiltrate should be quite short, but with further additions it should lengthen. When the time is more or less constant for three water additions, or it takes over five minutes for the water to disappear, stop the experiment.

Now calculate the depth of water added with each application. Do this by dividing the volume of the water added each time by the area defined by the can. The area is given by the square of the radius of the can multiplied by 3.14. For example, if you added 50 ml of water to a can of diameter 10 cm (radius 5 cm), then a depth of 0.64 cm [$50/(3.14 \times 5^2)$] would be added each time. In performing this calculation make sure that you work in the same units for both the volume of water and the radius of the can.

It is now possible to draw a graph of your results. Plot on the vertical axis the time in seconds that it took for each addition of water to disappear. Plot on the horizontal axis both the number of the water application (application number 1, 2, etc.) and the accumulated depth of water added. The resulting graph, which might look like Figure 2, will show how the infiltration rates change as more and more water is added. It is normal for the infiltration rate to be relatively high at the beginning and low at the end.

Now see if you can find a different soil type and repeat the experiment. Also repeat it on the same soil but remove the vegetation layer and see what difference you obtain in infiltration rates. Plot all the graphs to the same scales, and you will have a quantitative comparison of the infiltration rates under dif-

FROM THE GEOGRAPHER'S NOTEBOOK

ferent surface and soil conditions. This is an interesting way to begin to learn about the soils around you as well as part of the hydrologic cycle.

Water in the soil can flow beneath the surface and find its way to a river channel where it can add to the river flow. In the next part of this field trip you must try to measure the discharge of a river or stream and observe other factors concerning flowing water.

WHAT CAN WE MEASURE?

Because you have to measure the river or stream channel, it is important that you pick a stream shallow enough to walk in safely or a part crossed by a bridge from which you can probe the channel bottom. Start by measuring channel cross sections. Two cross sections can be computed—the channel used by the water at the time of the measurement and the channel used by water at times of flood. You may have to be very observant to complete the second observation. If it is not obvious where the top of the river bank is, look carefully for signs of occasional water presence such as lack of vegetation, flood debris lodged in stones or boulders, or a high tide mark on a beach.

Measure the present channel first. Stretch a string with marks at 50-cm (20-in) intervals across the surface of the water. Then every 50 cm, measure the depth of the water. You can use a stick for this and then measure the depth on the stick using a rule or tape measure. Complete the depth measurements all across the channel to the other side. If your stream is wider than about 10 m (11 yd) increase the horizontal interval of depth measurements.

The flood channel can be measured in a similar way but with a few additions. Stretch the string not on the water surface but from bank to bank or across the channel you assume to be inundated during a flood. Make any additional measurements you need over the dry parts of the channel, and for the area where water is present add the constant height of the distance between the water surface and the string to the channel measurements you have already taken. You now have all the measurements needed to calculate the channel area, in its present flow and in times of flood.

FIGURE 2
Graph of infiltration rates.

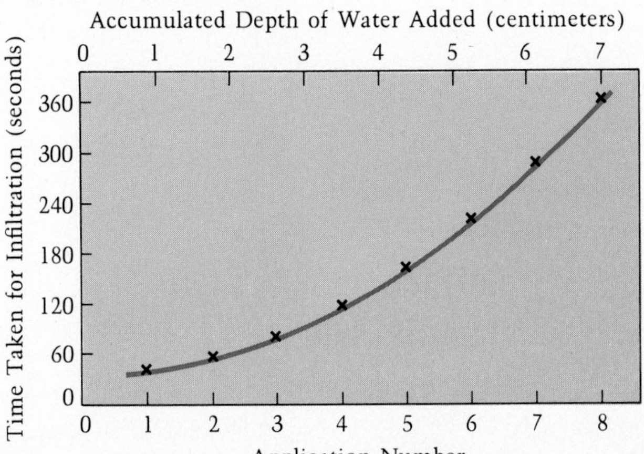

HOW FAST IS IT MOVING?

Next we want to measure the velocity of the stream. Use five floats. These can be of various materials—wood, polystyrene, plastic—experiment to see which you think works best—which seems to be carried at the same speed as the current of the river. Measure distances 5 m (5.45 yd) up and downstream from the point where you measured your cross section. Throw the floats into the stream at the point 5 m above the cross section, and measure the time it takes for each of them to pass the point 5 m below the cross section. Repeat the procedure five times to increase the chances of obtaining a representative sample. Then calculate the average time the floats take to cover the 10-m distance. Now calculate the average speed (in meters per second) indicated by the floats.

Make a sketch map of the part of the stream you are working with and mark on it the paths most often favored by the floats. This will help you assess how representative your computed average velocity is. Consider if the average float time is the time taken by most of the floats, or whether a few unrepresentative floats are affecting the computed average. If you judge the latter to be the case, omit these unrepresentative floats from your calculation.

As you are doing this you will see that it is quite difficult to assess the velocity of the river. The method you are using, although not very accurate, gives a far better estimate than you could get by casual inspection. We tend to overestimate the velocity if we guess.

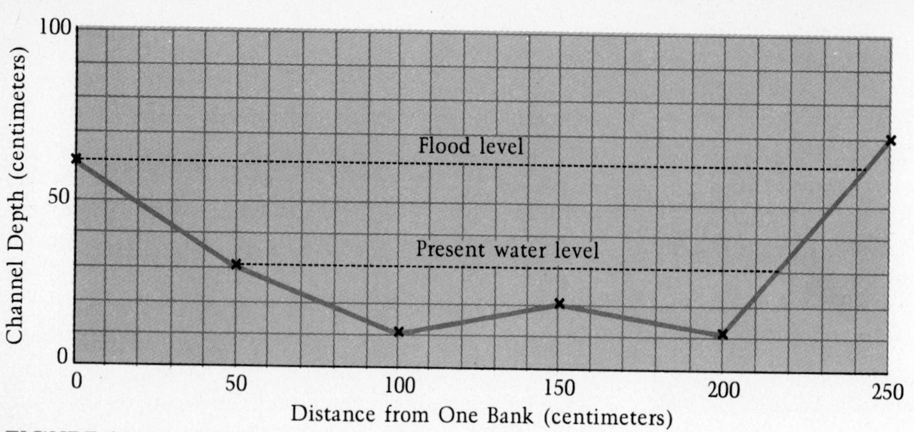

FIGURE 3

Plot of cross section of channels with actual water level and flood water level.

FIGURE 4

The effect of velocity on particle size in erosion, transportation, and deposition.

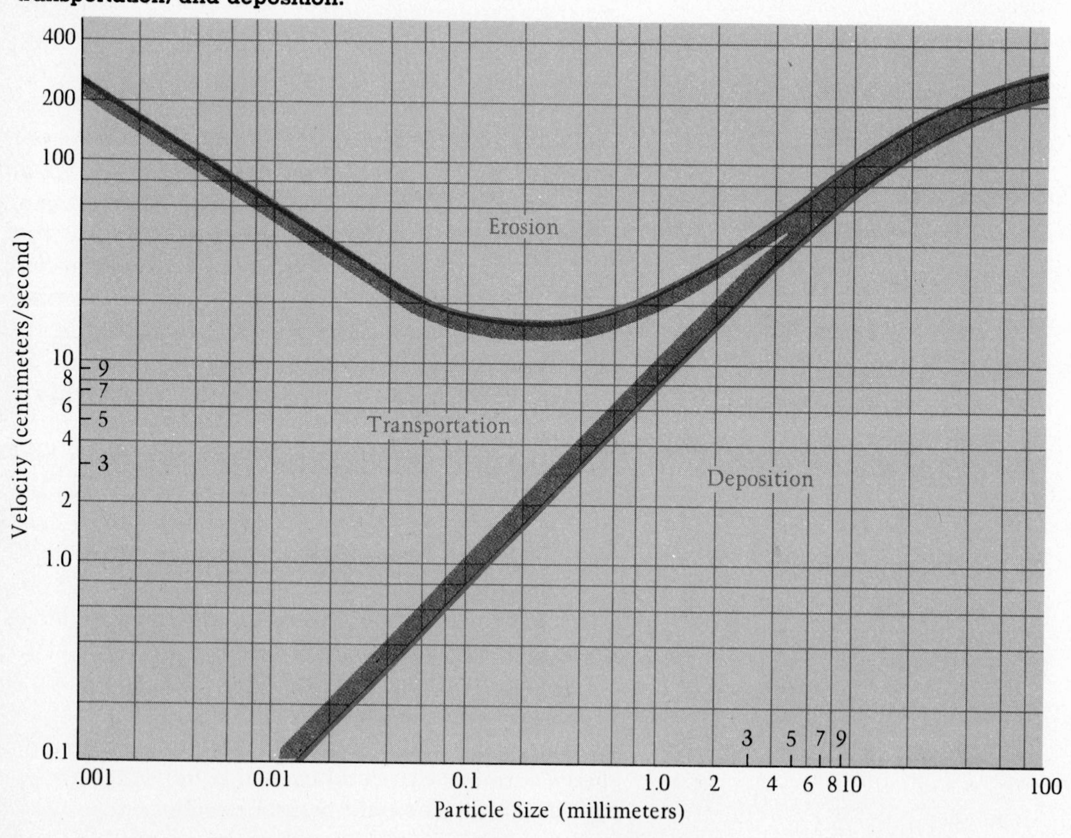

FROM THE GEOGRAPHER'S NOTEBOOK

Measuring flow on the Merced River below Stoneman Bridge in Yosemite Valley, California

Making surface water discharge measurements of the Columbia River for the Geological Survey at Grand Coulee, Washington.

The fastest moving currents tend to be at the center of the stream, but this is by no means always the case. You should also remember that we are making an error in assuming that the average velocity of the floats represents the average velocity of all of the water in the stream. However, by studying the stream this way you will begin to appreciate how the water flow operates and you will notice that the actual flow is quite complex.

When the velocity results have been obtained, the discharge of the stream can be computed. Do this back at home or at school. Take some graph paper and plot the channel cross section both with the present water level and with the flood level. Figure 3 shows how this might look. Use the same scale for the vertical dimension as for the horizontal. That is, if one small square on the graph paper represents 10 cm in the vertical dimension, let it equal 10 cm in the horizontal dimension as well. When you have plotted the profile of the channel bottom, count the number of all the small squares between the bottom and the water line and from this calculate the area of the channel by multiplying the number of squares by 100 square cm. Using the same procedure, you can compute the channel area for a flood flow. Notice how much greater this is.

You are now able to compute the discharge simply by multiplying the average velocity of the stream by the channel area. You may express the results in cubic meters per second. Again you will see that the flood discharge is very large compared with what you actually measured. Let us see what this means in terms of the erosive work of the river.

WHAT DO THE NUMBERS TELL US?

Figure 4 shows the relationships among erosion, transportation, and deposition within a river with respect to the river velocity and the size of particles in the channel. Note that the scales in this diagram are not linear. They are called logarithmic scales and the intervals increase in absolute size rapidly toward the top and toward the right of the diagram.

FROM THE GEOGRAPHER'S NOTEBOOK

Two curves are shown on the diagram. The upper curve shows the velocities required to move particles of various sizes. The velocities required decrease with particle diameter to sizes between about 0.1 and 1.0 mm (0.004 and 0.04 in) and then rather surprisingly rise again for particles less than 0.1 mm in diameter. The reason it takes very high velocities to move very small particles is that these particles tend to be very cohesive and also to form smooth channel floors where they exist.

The lower line on Figure 4 indicates whether a particle will continue to be transported or whether it will be deposited. Deposition will take place if the river velocity is below the line, but if the velocity is above the line particles that are already moving will continue to move.

Take your calculated velocity value and use these curves to theoretically determine the erosive ability of the stream you studied. For the velocity value you obtained, use the top curve to show the range (if any) of particle sizes that could be moved. Also find the largest size of particle that could be transported if it were already moving.

Having made these observations, hopefully you will find that your interest in the river will have increased enormously. Remember, the measurements are not very accurate. Consider ways in which your experimental technique could be improved. Precise determination of the mean velocity of the river requires the use of accurate velocity measuring instruments and a slightly complicated formula. Nevertheless, you will begin to see how rivers work as agents of landscape formation, and you will certainly see the fascination they hold for geographers.

EPILOGUE: THE EARTH IN PROFILE

CHAPTER 23

We have looked at the earth in profile and have examined its spheres. Now we can review what is particularly important to our field of inquiry—physical geography. Three items are outstanding. First are the distributions of the elements of the physical world and the relationships among these elements. Second is the two-way relationship between humans and their environment. Third is the overview of environmental problems provided by the study of physical geography. All these factors help develop our attitudes toward the environment we live in.

DISTRIBUTIONS AND RELATIONSHIPS

The most important maps in this book are those showing the distributions of climates, soils, vegetation, and physiographic regions. The climates, according to Köppen's classification, are shown in Figure 7-4. The distribution of soils, using the comprehensive classification, is shown in Figure 12-10, and the major vegetational biomes appear in Figure 12-11. The physiographic regions of North America are shown in Figure 22-4. If you return to these maps and consider them together, many relationships among them should be apparent.

One way to make the relationships even clearer is to construct a *transect.* This is an imaginary line across the earth's surface along which are included details taken from maps of physiography, soil, vegetation, and climate. Figure 23-1 shows the location of a transect from New Orleans, Louisiana to Nachvak, Newfoundland. Details of physical elements, together with the predominant economic activity, can be seen in Figure 23-2.

Many correspondences among the physical elements appear along the transect line. It is quite common to have a boundary between two of the elements

FIGURE 23-1
The location of a transect from New Orleans, Louisiana to Nachvak, Newfoundland.

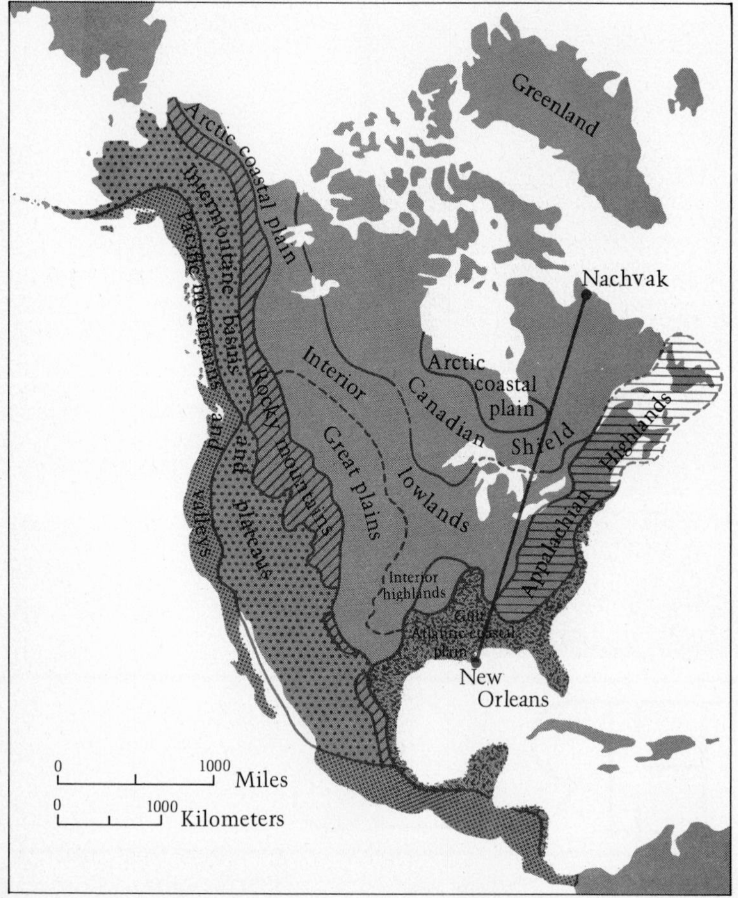

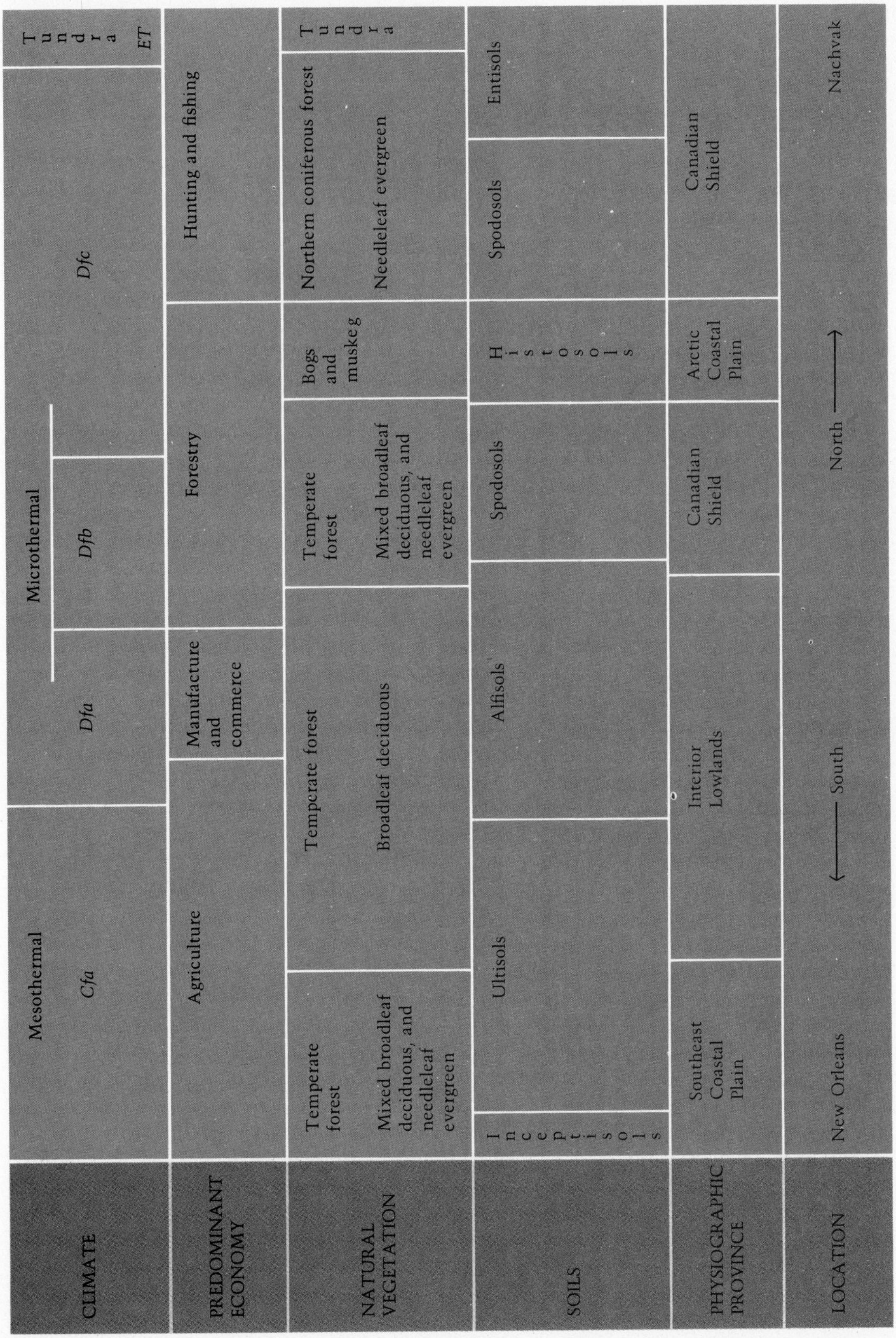

CLIMATE	Mesothermal *Cfa*	Microthermal *Dfa*	Microthermal *Dfb*	Microthermal *Dfc*	Tundra *ET*		
PREDOMINANT ECONOMY	Agriculture	Manufacture and commerce	Forestry	Hunting and fishing			
NATURAL VEGETATION	Temperate forest Mixed broadleaf deciduous, and needleleaf evergreen	Temperate forest Broadleaf deciduous	Temperate forest Mixed broadleaf deciduous, and needleleaf evergreen	Bogs and muskeg	Northern coniferous forest Needleleaf evergreen	Tundra	
SOILS	Inceptisols	Ultisols	Alfisols	Spodosols	Histosols	Spodosols	Entisols
PHYSIOGRAPHIC PROVINCE	Southeast Coastal Plain	Interior Lowlands	Canadian Shield	Arctic Coastal Plain	Canadian Shield		
LOCATION	New Orleans	⟵ South	North ⟶	Nachvak			

FIGURE 23-2

A transect from New Orleans, Louisiana to Nachvak, Newfoundland, showing correspondences among climate, economy, natural vegetation, soils, and the physiographic provinces.

at the same point of the transect. Examples are the vegetational change at the junction of the southeast coastal plain and the interior lowlands and the change in soil type at the *Cfa/Dfa* climatic boundary. Even more remarkable are the change in three of the factors at the southern boundary of the Canadian shield and the arctic coastal plain and the change in four of the factors on the northern side of the arctic coastal plain. If you draw a transect for yourself, say east–west across the United States, you will soon appreciate the fascination that geographers have for distributions and relationships.

The relationships among the atmosphere, biosphere, lithosphere, and hydrosphere are responsible for the changes in the distributions. Figure 23-2 shows the relationships on a large spatial scale, one the size of a continent. Chapters 7 and 10 show that such relationships can also occur on a smaller scale, such as along a hillside. One of the purposes of physical geography is to point out and to explain such distributions and relationships, regardless of their scale.

HUMANS AND THEIR ENVIRONMENT

If you look at the predominant economies listed in Figure 23-2, you can see that human activities are closely connected to the physical world. Throughout this book we have seen wide-ranging examples of the interaction between humans and their environment. We have seen how humans build shelters for protection from the weather and how these houses, when congregated in large towns, can affect the atmosphere. We have seen how natural soils, such as histosols, can be inhospitable to cultivation in their natural state and yet be quite productive when excess water is removed from their upper layers through drainage projects. Chapter 12 describes how natural tropical rainforest is relatively easy to walk through but how impenetrable undergrowth grows back following the destruction of the original forest by humans. Chapter 17 tells how human settlement has often been associated with natural water sources and how humans have manipulated the hydrologic cycle by building dams to provide a more even flow of water. The rocks of the lithosphere give humans many of their resources, but human pollution accelerates chemical weathering of some rock surfaces.

The human-environmental interaction, however, is not always clear. In many cases the interaction is filtered through human culture. For example, the human response to a long drought might vary from the performance of rain dances to the increase of budgets for cloud-seeding research or water redistribution projects. A drought almost always causes hardship to humans, but the degree of hardship depends on the culture and on how well organized the people are where the drought occurs. The culture factor acts through time as well as space. The area in Figure 23-2 where the predominant economy is now agriculture once had a hunting and fishing economy.

In many cases the environment directly influences human behavior, and human behavior directly influences the environment. Environmental effect on human behavior is clearly seen in the Kalahari Desert. Gikwe Bushmen, who live in the more barren parts of this environment, have learned to locate individual plants in the dry season. In contrast, the Kung Bushmen, who live nearby in a better-endowed environment, need only learn the location of groups of plants. The thoroughness of the Gikwe's search for food is greater than that of the Kung because of the comparative poorness of their environment.

You might be surprised to learn that humans have been asking for a long time the question so prominent today: How are we affecting our environment? The Greek philosopher Plato decried deforestation, which he believed had led to the deterioration of the Mediterranean landscape. Since then, the detrimental effects of cutting down trees have often been a point of discussion. But deforestation is just one way that humans alter their environment. We have seen many other examples in the preceding pages. These examples have shown that environmental alteration may be either detrimental or beneficial. The important thing to remember is that our alterations are more likely to be beneficial if we first study the natural processes of the earth.

THE OVERVIEW OF ENVIRONMENTAL PROBLEMS

Physical geography is not the only subject to deal with distributions, nor is it the only subject to consider human-environmental interaction. But it is the only subject to focus on these matters in a comprehensive way. The physical geographer has to be aware of all parts of the physical world because of the interactions among these parts. He or she does not have to be an expert simultaneously in climatology, pedology, biology, geology, and geomorphology but must know enough of their basic principles to be able to talk to the people who are experts. In this way, he or she is in a unique position to contribute to the solution of environmental problems.

A recent report on the role of environmental impact assessments pointed out that "scientists from different disciplines do not yet know quite how to work with each other or with sociologists or economists." Geographers, both physical and cultural, have been aware of this problem for decades and have had practice in dealing with it. Their solution is to specialize in one area, say climatology, while making a deliberate attempt to follow the major advances in the fields of their colleagues. The approach works well. Although most professional research in geography is done within a specialized field, the majority of geographers are able to describe and explain their observations in terms of the regional subsystem discussed in Chapter 1. They are also able to give constructive advice to their fellow workers at seminars and meetings. This leads to many supportive situations, all stemming from the initial interdisciplinary training of the physical geographer.

Let us look at an example of the physical geographer's overview in action. Chapter 21 describes the construction of groins along beaches to prevent the disappearance of sand from resort areas. Vast amounts of money have been spent in attempts to hold sand on such beaches. But much of the money is wasted, because the constructors do not recognize that the beach is an open system and is usually in dynamic equilibrium, as Figure 21-22 shows. In many cases the sand disappears because of urban development back from the shore. This land was formerly the source of sand input to the beach, but the construction of houses and roads inhibits natural erosion. An overview that takes account of all the inputs to and outputs from the beach might be much more effective in controlling erosion than such short-term solutions as the construction of groins. The physical geographer's advantage lies in understanding that everything affects everything else. This brings us to the main point of this book.

A NEW WAY OF SEEING THINGS

One of the greatest benefits of studying physical geography is that it helps change our attitudes toward our environment. This may happen in small or in large ways.

Let us take the small ways first. Even after this brief introduction to physical geography, you may notice many more items that once caused you no thought at all. You may be more aware of the climate where you live. You might find yourself wondering what kind of soil exists in your yard and how it is related to the plants that grow there. When you pass an outcrop of rock in a road cut, you may look at the rock with new interest. You may have a greater appreciation of landscape. What was to you a boring area of plains may now be a regional subsystem with complex interactions of rocks, soil, vegetation, fauna, and climate. Spectacular mountain scenery has even more meaning if we can imagine past glaciers gouging away at the rocks from the valleys.

Physical geography also changes attitudes toward the environment in large ways. This comes about through new or increased appreciation of the basic themes of physical geography:

1. The special distributions existing in the physical world—distributions of climate, soils, plants and animals, and physiography
2. The relationships among these distributions and among the physical elements themselves

These are the factors that lead to a better understanding of the two-way relationship between humans and their environment, and these are the concepts that help develop a broad approach to environmental matters.

In the last 2500 years, the word *nature* has become more and more limited in its meaning. To the Greeks before Socrates, the equivalent word meant "all" or "everything." In the Middle Ages, *nature* came to refer to all the regions of change beneath the moon. Today when we speak of nature, we tend to think of the countryside or a national park. We do not conceive of nature existing in the middle of New York City. This is a pity, because the "natural" world—the physical world—does exist in New York City and everywhere else. The town is subject to rainstorms just as the countryside is. If the present ice sheets were to melt, New York and many other cities would be submerged.

And so we return to the point where we started this book: the continual presence of the physical world. The physical world is an ever-present backdrop to all our daily activities, wherever we live. If we are to live successfully *on* the earth, we must live successfully *with* it. To do this, we must continue to search for its order, to learn more about it. We must continue to examine the earth in profile.

GLOSSARY

Abrasion the grinding action of fragments on other fragments and on rocks, wearing their surfaces down and creating ever finer sediments.

Absolute vorticity the sum of relative and planetary vorticity, which remains constant in any spinning system.

Abyssal plains the flat, plainlike areas of the ocean floor, covered by millions of years of accumulation of sand, silt, and mud, and lying at depths up to 6000 m (20,000 ft) below sea level.

Accordant intrusion an intrusion that does not significantly disturb the rock into which it intrudes.

Actual evapotranspiration the amount of water that can be lost to the atmosphere from a land surface with particular (sometimes limiting) soil moisture.

Adiabatic occurring without gain or loss of heat to the surrounding environment.

Adiabatic lapse rates the constant lapse rates of vertically moving air in the troposphere. When air is completely dry, it cools at the dry adiabatic lapse rate (DALR) of 1°C/ 100 m. When air is moist or saturated, it cools at a rate slower than the DALR, the saturated adiabatic lapse rate (SALR).

Advection the horizontal movement of material in the atmosphere.

Aggradation the building up of the surface over which the river flows through deposition of sediment.

Air mass an extensive portion of air in the lower troposphere with relatively uniform qualities of temperature and moisture.

Air-pollution potential a measure of the possible amount of air pollution an area can receive, estimated by calculating the vertical range of vigorously mixed pollutants and the average wind speed through the mixing layer.

Albedo the ratio of radiation that is reflected by a surface to the incoming radiation that reaches the surface.

Allogenic succession plant succession where vegetation change is brought about by some outside environmental factor, such as disease.

Alluvial fan an aggradational deposit surrounding and extending beyond a point where a river descends from highlands to a relatively flat plain, suddenly losing velocity.

Alluvial terraces forms on the sides of river valleys, marked by flat upper surfaces and minor scarps facing the river, and caused by a period of downcutting following aggradation by a river.

Alluvium soil particles that have washed downslope and have come to rest in a valley.

Altostratus high, layered clouds that form below 6 km.

Amensalism the inhibition of one species by another.

Anabatic wind air blowing up a slope as a result of warming.

Analog model a model that resembles reality, but involves a level of simplification or abstraction.

Angular momentum a force on a particle on a rotating body that is directly proportional to the rate of spin and the distance of the particle from the axis of rotation.

Annuals plants completing their entire life cycle in a single growing season.

Antarctic Circle latitude 66½°S.

Antecedent river a river exhibiting transverse drainage across a structural feature that would normally impede its flow formed because the river predates the structural feature.

Anticline a fold in which the slopes of the two limbs face away from each other, tilting downward and away from the axis of the fold.

Anticyclone an atmospheric high pressure area of a generally circular shape. The isobars around an anticyclone are circular, with lower values toward the outside of the anticyclone.

Anvil top the drawn-out top of towering cumulonimbus clouds, caused by increased horizontal winds at the fringes of the air column, and resembling a blacksmith's anvil.

Aquicludes rocks that are impermeable and resist groundwater flow.

Aquifer material that is porous and permeable and can be at least partially saturated.

Arctic Circle latitude 66½°N.

Arête a jagged ridge formed by the intersection of two or more cirques.

Artesian well a well sunk into a confined aquifer, producing water that flows under its own natural pressure to the surface.

Aspect the direction a mountain slope faces.

Asthenosphere the portion of the mantle below the tectonic plates that is plastic in nature.

Atmosphere the layer of air that begins a few meters in the soil or at a water surface and extends approximately 60,000 km above the earth.

Aurora borealis and aurora australis vivid shows of light occurring in the northern and southern hemispheres, respectively, caused by the penetration of ionized particles through the thermosphere.

Autotroph an organism that manufactures its own organic materials from inorganic chemicals.

Axial plane the plane that bisects a fold longitudinally, at a 90-degree angle to the surface in which the fold occurs.

Axis the imaginary line drawn through the earth from the North Pole to the South Pole, around which the earth rotates. The axis of a fold is an imaginary line that divides the fold into two equal halves, lengthwise.

Barchan a dune that looks like a quarter moon, with the points of the crescent lying downwind.

Barometer an instrument used to measure atmospheric pressure, usually according to the weight of a column of liquid of known density that can be supported by the atmosphere at any given altitude.

Base exchange the process of the release of the bases to plants from colloids.

Base flow river water derived from supplies existing underneath the ground.

Base level the lower limit below which a river cannot erode. A river's ultimate base level is sea level, but local base levels may be lakes or plains.

Base a mineral or soil containing ions of elements such as calcium, magnesium, hydrogen, aluminum, and potassium.

Batholith an igneous rock mass covering enormous area, formed well beneath the earth's surface and exposed by erosion.

Biomass the mass of plant matter, which can be expressed at a number of different scales, such as the biomass of a single plant, of a field, or of the earth. Also known as phytomass.

Biome a major grouping of ecosystems, defined on the basis of unique overall appearance and climatic characteristics and occurring over a wide geographical area.

Biosphere the part of the earth containing all living organisms.

Blocky structure soil structure in which irregularly shaped aggregates of earth tend to have straight edges.

Braided river a subdivided river channel to accommodate high sediment load.

Breakers waves in which the top part of the wave moves toward shore unimpeded, while the bottom part of the wave is slowed by frictional contact with the sea or lake bottom.

Breccia a metamorphic rock formed by cementation of fractured angular rock bits from fault or volcanic activity.

Calcification the process of formation of soils having a characteristic dark brown color in the A horizon with a whitish B horizon; found in the grasslands, steppes, and semidry regions of the world.

Caliche the dense stony layers of white calcium carbonate left in the soil during calcification.

Calorie the amount of heat energy required to raise the temperature of 1 g of water 1°C.

Capillary action the rise of water upward through soil air gaps owing to the tension between water molecules.

Carbonation the reaction of weak carbonic acid with minerals.

Catena a series of soil types in homogeneous parent material arranged in a characteristic pattern.

Celsius scale the temperature scale on which water freezes at 0°C and boils at 100°C. Also known as the Centigrade scale.

Centripetal force the force on a rotating particle or body directed toward the axis of rotation.

Centripetal pattern a drainage pattern in which a series of rivers extend inward to a central point, usually a basin or depression.

Chaparral a type of vegetation found in the temperate forest biome, typically consisting of hard leaf scrub of low-growing woody plants, usually evergreen with thick waxy leaves well adapted to long dry summers. Also called maquis and mattoral, in the Mediterranean region and in Chile, respectively.

Chernozem rich, black soil.

Chlorophyll a green pigment found in plant surfaces that insures that the correct wavelength of light is absorbed for the process of photosynthesis, and that enables this process to take place.

Cirque the amphitheaterlike source area of an alpine glacier.

Cirrostratus high, layered clouds that form above 6 km.

Cirrus high, thin, wispy clouds formed above 6 km.

Clastic sediments sediments made up of particles of other rocks.

Climate the average values of weather elements such as temperature, precipitation, and radiation over a thirty-year period, and the important departures from these average values.

Climatic normal the average value for one of many weather parameters during the period 1931 to 1960 for one of many worldwide locations, published in tables by the World Meteorological Organization.

Climatic state the average, together with the variability and other statistics, of the complete set of atmospheric, hydrospheric, and cryospheric variables over a specified period of time in a specified domain of the earth-atmosphere system.

Climax community the final sere of any particular succession, in which the vegetation and its ecosystem are in complete dynamic equilibrium with the other environmental factors.

Cold anticyclone a high-pressure area formed in the mid and high latitudes owing to cold subsiding polar air. The Siberian high is an example of a cold anticyclone.

Cold front the transition zone between two masses of air, in which the colder is advancing, moving under and replacing the warmer.

Colloids substances within the soil in a state of fine subdivision, with particles less than 0.0001 mm in diameter, formed of minute mineral or organic materials.

Colluvium soil particles that have washed downslope and come to rest on the slope.

Comfort-index classification a climatic classification system based on the subjective feelings of comfort in large groups of individuals at certain combinations of temperature and humidity.

Composite slope a slope formed of more than one characteristic, as in a hill having a convex top, a straight side, and a concave bottom.

Composite volcanic cone a steep volcanic cone formed by alternating layers of thick magma and pyroclasts.

Condensation the process by which a substance changes from the gaseous to the liquid state.

Condensation nuclei small particles usually present in the air around which liquid droplets can form when water vapor condenses. Condensation nuclei are usually present in the atmosphere in the form of dust and salt particles.

Conduction the transport of heat energy from one molecule to the next.

Cone of depression the cone-shaped drop in the water table around a well.

Confined aquifer an aquifer that exists between two aquicludes, obtaining its water from an area a great distance away where the rock strata of the aquifer reach the surface.

Conformal having the property of preserving shape around a point.

Conglomerate the coarsest grained of all sedimentary rocks, composed of pebbles and sometimes even larger fragments of rock which tend to be well rounded.

Consequent streams streams forming on an initial landmass, whose direction of flow is a consequence of the slope of the land.

Conservation of absolute vorticity the principle that the total amount of spin in a system remains constant. Given a spinning column of air on the earth, the sum of the spin of the column and the spin of the planet, divided by the depth of the air column, is constant.

Conservation of angular momentum the principle that in any rotating system, the total angular momentum remains constant.

Contact metamorphism metamorphic change induced by local contact with extremely hot magma or lava.

Continental glacier immense sheets of ice, hundreds or thousands of meters thick, that covered the continental landmasses in high and midlatitudes during periods of general cooling on the earth, especially during the Pleistocene.

Continental plate a tectonic plate that includes all or part of a continental landmass.

Continental rise the point where the ocean floor begins to gently slope upward toward the continental slope.

Continental shelf the plain that extends from the conjunction of the ocean water and the continental landmass beneath the ocean.

Continental slope the actual margin of the continental landmasses; a submarine scarp that drops sharply from the continental shelf.

Convection movement of a gas or fluid in a vertical direction, in contrast to advection.

Convective heat flow the flow of sensible heat to or from the body through the mechanism of moving air of a temperature above or below body temperature.

Convergent plate boundary a plate boundary that occurs when two plates move toward each other and collide, one plate usually plunging beneath the other.

Coral reef an aggradational reef formed from the skeletal remains of marine organisms.

Core the spherical center portion of the earth, separated into the inner and outer core, with a radius of about 2400 km (1500 mi).

Coriolis force the apparent deflecting force acting on a body in circular motion.

Corrosion the chemical solution of rocks by running water.

Counter radiation longwave radiation emitted by the atmosphere and clouds.

Creep the slow, imperceptible downslope motion of the topmost soil layer under the influence of gravity.

Crest segment the part of a hydrograph with maximum values for runoff-produced discharge.

Cross-stratified sedimentary rocks that exhibit nonhorizontal bedding owing to the direction of currents when they were deposited.

Crust the surface layer of the earth, composed of the lightest earth rocks, and between 15 and 40 km thick.

Cryosphere the collective name for the ice system of the earth.

Cumulonimbus cumulus clouds that produce rain.

Cumulus towering puffy clouds that form in unstable air.

Cup podsols local podsols that form through the localized leaching action of water running down tree trunks.

Cyclic autogenic succession plant succession in which one kind of vegetation is replaced by another, which is in turn replaced by the first, with other series possibly intermixed.

Cyclone an atmospheric low-pressure area. The isobars around a cyclone are circular in shape, with higher values toward the outside of the cyclone.

Cylindrical projection a projection formed when an imaginary cylinder encompasses the globe and having rectangular shape, with parallels and meridians intersecting at right angles.

D, E, and F layers belts of ions found in the mesosphere and thermosphere that reflect radio waves sent from the earth's surface.

Daylight saving time time moved forward one hour from standard time within a time zone to conserve daylight hours for human activities.

Deciduous trees and plants that seasonally lose their leaves.

Deflation the removal of clay and silt sized particles by wind.

Deflation hollows shallow depressions excavated by deflation.

Degradation the downcutting and erosional activity carried out by streams.

Degree the angle formed by radii from the center of a circle to the ends of an arc that is 1/360 of the circle's circumference. Degrees are further divided into 60 minutes of 60 seconds each.

Delta an aggradational deposit surrounding and extending beyond the mouth of a river, where it empties into a larger body of water.

Dendritic pattern a drainage pattern developed on a surface with very little structural control, exhibiting a treelike pattern.

Density the number of molecules of a given substance per unit volume of that substance, usually expressed in weight per unit volume: grams per cubic centimeter, for example.

Desert an arid region supporting only sparse vegetation, usually exhibiting extremes of heat and cold owing to lack of moisture as a moderating influence.

Dew point temperature the temperature at which the air becomes saturated and condensation occurs.

Diffuse radiation rays that reach the earth's surface only after scattering in the atmosphere.

Dike an igneous intrusion that penetrates rock layers vertically by inserting itself along jointing planes.

Direct capture the incorporation of smaller precipitation droplets into larger ones in the coalescence process of droplet formation.

Direct radiation rays that travel directly to the earth's surface.

Discharge a parameter found by multiplying a stream's velocity by the cross section of water flow.

Disconformity an interruption in the sequence of accumulation of the geologic history in an area not accompanied by angular deformation of the previously produced layers.

Discordant intrusion an intrusion that disturbs the rocks into which it intrudes.

Divergent plate boundary a boundary where plates move away from each other and basaltic magmas emerge to fill the gap formed by this divergence.

Dormant volcano a volcano that has erupted in recorded times but shows no sign of recent activity.

Drawdown the amount of drop in the local water table around a well.

Drumlin an oval shaped, low, well rounded hill made of glacial till and shaped by the movement of a glacier over the till.

Dunes mounds of sand in characteristic shapes formed by wind.

Dust dome the characteristic shape taken by the large quantities of dust and gaseous pollutants in a city's atmosphere.

Dynamic equilibrium the state of a system when it is neither growing nor getting smaller, but remaining in complete, balanced operation.

Dyne the force that produces an acceleration of 1 cm per second on an object 1 g in weight.

Easterly waves wavelike phenomena in the flow of the easterly trade winds.

Ecological efficiency the percentage measure of the ability of organisms in a food chain to convert the chemical food energy from one trophic level to the next.

Ecological niche the way a group of organisms makes its living in nature, or the space in the environment in which an organism operates most efficiently.

Ecosystem an open energy system in which plants and animals are linked to their environment through a series of feedback loops.

Edaphic factors factors affecting plant growth having to do with the soil conditions.

Eemian Period the warm interglacial period that occurred about 125,000 years ago and lasted for about 10,000 years.

Effective temperature the temperature of still air saturated with water vapor at which a certain number of people experience a subjectively equivalent sensation of comfort.

Ekman spiral the pattern of change in wind direction and velocity with height, seen from above. The Ekman spiral operates in both the atmosphere and the oceans.

Entrenched meanders meanders that have been downcut in place in bedrock.

Entropy a measure of the amount of energy in a system or process that is unavailable for work.

Environmental determinism the view that the environment determines aspects of human life such as physical attributes, culture, and institutions.

Environmental lapse rate the lapse rate that actually exists at any particular time or place.

Epeirogeny the process of mountain building through regional uplift without accompanying tectonic activity.

Epicenter the place at the surface of the crust directly above the focus of an earthquake.

Equation of time a mathematical equation that corrects an actual solar day to a mean solar day.

Equator the parallel around the middle of the earth defined as 0° latitude.

Erosion (1) the distant removal of weathered rock material. (2) the suite of processes by which the earth is worn down.

Erosion surface the surface produced by a period of downwearing or erosion.

Erosionally graded landscape a landscape in which large areas are mutually adjusted, and the variation of form is mainly the result of differential erosion of rocks that yield to weathering in different ways.

Esker a long ribbonlike ridge extending across the countryside, formed by the clogging of a river course within the glacier, the debris from which remains after the glacier melts.

Estuary a feature formed by the submergence of a river valley by rising sea level.

Evaporation the process by which a substance changes from the liquid to the gaseous state.

Evapotranspiration the combined moisture loss from direct evaporation of water from the earth's surface and transpiration of plants.

Evergreen trees and plants that retain their leaves.

Exfoliation a joint pattern resembling a series of concentric shells, causing outer layers to peel away progressively, leaving rounded or dome-shaped forms.

Exponential notation a method of expressing a large number as a simple number times a power of ten, rather than using many zeros before or after the decimal place.

Extinct volcano a volcano with no recorded instances of activity that shows evidence of uninterrupted long-term erosion, unbroken by further accumulations of lava.

Extrusive rock rock formed when magma escapes to the surface through fissures or vents before cooling into igneous rock.

Fahrenheit scale a temperature scale on which water freezes at 32°F and boils at 212°F.

Fall movement the largely unimpeded downward movement of rock material from nearly vertical slopes.

Fault a fracture of the earth's crust along which earthquakes tend to occur.

Fault plane the surface of a fault along which the blocks of rock move.

Feedback a situation in which a change in one part of a system causes a change in another part of the system. Negative feedback mechanisms operate to keep a system in its original condition. Positive feedback mechanisms make an increasingly greater change away from the system's original state.

Ferrel cell the atmospheric cell hypothesized by Rossby in his model of global atmospheric circulation that was thought to transfer heat poleward from, and be driven by friction with, the Hadley cell.

Fetch the uninterrupted distance traveled by a wind or an ocean wave.

Field capacity the upper limit of water a soil can possibly contain without becoming waterlogged.

Fiords deep glacial valleys that extend below sea level in coastal areas.

Flashy the name given to a stream having a hydrograph that rises and falls relatively quickly owing to impermeability of basin material, and the nature of the precipitation in the catchment.

Flood plain a section of flat valley floor created by lateral erosion and aggradation that is subject to inundation during periods of exceptionally high water.

Floristic classification the classification system of Linnaeus, based on the structure of plants, and other features.

Flow movement the downslope movement of material under the combined influence of water and gravity.

Focus the point of origin of an earthquake.

Föhn wall cloud the wall-like cloud of descending air that is found with the föhn wind.

Föhn wind the warm, dry wind experienced on the leeward side of mountains owing to warming of air with loss of altitude.

Foliation the property of banding in metamorphic rocks caused by the lining up of various crystals within the rock as a response to greatly increased pressures.

Food chain the stages that energy in the form of food goes through within an ecosystem.

Foredeeps the trenches found associated with the abyssal plains of the ocean floor, which can be as deep as 9000 m (30,000 ft) below sea level.

Free face a cliff or steep scarp.

Freezing nuclei the small particles of matter necessary for the formation of ice particles in the atmosphere.

Frictional force a force common to two bodies in contact that resists the motion of one relative to the other.

Front the zone of contact between warm and cold air masses that face each other horizontally.

Gauging measuring stream velocity directly with a current meter at regular intervals of the cross section, or by the use of other methods.

Geomorphology the study of landforms.

Geostrophic wind an upper-atmosphere wind produced by the balancing of the pressure gradient force on air by the Coriolis force on the same air. Geostrophic winds blow parallel to isobars in relatively linear paths.

Geosyncline a large subsiding trough of deposition in the earth's crust that slowly fills with sediments from adjacent higher land areas.

Glaciofluvial landforms landforms created by a combination of glacial and fluvial erosion and deposition.

Gneiss a metamorphic rock of essentially the same composition as granite, from which it has been transformed.

Gnomonic projection a zenithal projection on which all straight lines are arcs of great circles.

Graben a rift valley formed between two parallel normal faults when a large section of the earth's crust is depressed as a unit.

Graded stream a stream in which slope and channel characteristics are adjusted over a period of years to provide just the velocity required for the transportation of the load supplied from the drainage basin, with available discharge.

Gradient wind a wind above the surface found in an area of curved air flow such as around a cyclone and an anticyclone and produced by the balancing of pressure gradient and centripetal forces by the Coriolis force.

Granular structure soil structure in which individual soil grains are separate.

Gravity the attractive force between two bodies that can be determined by dividing the size of the product of the mass of the two bodies by the square of the distance between them.

Great circle the circle formed when a sphere is cut in half by a plane.

Greenhouse effect the blanket effect of the atmosphere by which surface temperatures on the earth are increased through absorption of outgoing radiation and its reradiation as counter radiation.

Greenwich Mean Time the base time for the earth's 24 time zones, defined as the time at 0° longitude, the meridian of the Royal Observatory at Greenwich, England.

Groins wall-like structures extending out from the shore, built to keep sand from being removed from designated beaches.

Ground heat flow the heat conducted into and out of the earth's surface.

Gyres ocean currents resembling cyclones and anticyclones in behavior. Gyres are usually large enough to fill entire oceans.

Habitat the environment normally occupied by an organism within its geographic range.

Hadley cell a vertical current of air in tropical latitudes in which warm air rises and cool air descends in a circular path and moves warm air poleward.

Hanging valley a valley formed by the intersection of a tributary glacier with a main glacier, when the valley floors are not adjusted to each other.

Headward erosion the lengthening process of a river valley in which the river erodes back into the countryside by undermining banks and causing collapse.

Heat a form of energy often manifested by the vibrations of the molecules of a substance.

Heat capacity the heat required to raise the temperature of 1g of a substance 1°C.

Heat island intensity the maximum difference in temperature between matched urban and rural environments.

Herbivore an animal that lives on plants, or more generally the first consumer stage of a food chain.

Heterosphere the layer of the earth's atmosphere above the homosphere, characterized by a chemical composition that changes with height.

Historical climatic data information on past climates contained in indirect references such as crop reports and subjective interpretations of climate.

Holocene Period the warm interglacial interval during the last 10,000 years approximately.

Homolographic projection an equal-area projection developed by K. B. Mollweide on which parallels are straight lines and all meridians except the central one are halves of ellipses.

Homolosine projection an equal-area projection developed by Paul Goode combining features of the sinusoidal and homolographic projections.

Homosphere the layer of the earth's atmosphere extending from the surface to 80 or 100 km, characterized by a uniform composition.

Horn a peak formed by the intersection of three or more cirques.

Horst a block mountain, formed between two parallel reverse faults when a large section of the earth's crust is raised as a unit owing to compressional forces.

Hot springs water from fissures which contain hot magmatic material existing at great depths. The water is usually at an average temperature above 10°C (50°F).

Humus the decomposed and partially decomposed organic matter in the soil that forms a dark layer at the top of the soil.

Hurricane the name given to the tropical cyclones formed in the southern North Atlantic Ocean.

Hydraulic action the force of moving water in a stream sufficient to dislodge and drag away material forming the valley or channel bottom.

Hydraulic radius a stream's cross section divided by its wetted perimeter.

Hydrograph a graph of river discharge against time.

Hydrographic map map used for marine-related purposes.

Hydrologic cycle the system in which water circulates between the atmosphere, lithosphere, and hydrosphere through the mechanisms of precipitation, evapotranspiration, and flow.

Hydrolysis the chemical combination of water molecules with minerals, causing expansion of their volume.

Hydrosphere the water at the surface of the earth and surrounding the earth, including the water in the oceans and in the atmosphere.

Hygrophytes plants adapted to areas having an excess of water, such as swamps, marshes, and bogs.

Hygroscopic water a thin film of water that clings to soil particles and is unavailable to plant roots.

Ice the solid form of water in which molecules are linked together in a regular latticelike fashion.

Iconic model a model that is a smaller representation of reality, containing most of the detail found in the original.

Igneous rocks rocks formed when molten magma cools and crystallizes.

Impermeable surfaces that do not permit water to pass through them.

Index cycle the change in the upper westerlies from low amplitude to high amplitude Rossby waves and back. The entire cycle takes from three to eight weeks.

Inert gases gases that do not combine with other gases or compounds.

Infiltration water flow or movement from the soil surface through the pores and openings into the soil mass.

Infrared radiation longwave radiation emitted by

the sun and the earth, which is heat radiation, longer than about 0.7 microns.

Inselberg remnant hills, similar to monadnocks, except formed by the intersection of retreating slopes.

Intensity of precipitation the amount of water that falls in a given time.

Interception the situation in which raindrops fall on vegetation and evaporate before reaching the soil surface.

Interfaces the boundaries of systems and subsystems where the transfer or exchange of energy and matter takes place.

International date line longitude 180° E or W, the meridian at which travelers must either repeat or skip a day in order to maintain the 24 one-hour increments of Greenwich Mean Time over the surface of the entire globe. The line diverges from the 180th meridian to miss land areas.

Intertropical convergence zone (ITCZ) the upward limb of the Hadley cell found in the tropics where surface northeasterly and southeasterly winds converge.

Intrusive rock rock formed when magma is forced upward into and through certain layers of the upper crust, crystallizing beneath the surface.

Inversions positive lapse rates; situations in which air temperature rises with height.

Ionization the process by which molecules are reduced to individual electrically charged particles.

Ions individual electrically charged particles formed by the reduction of molecules.

Isobars imaginary lines of equal pressure used for mapping pressure surfaces.

Isostasy a condition of equilibrium between floating landmasses and the mantle beneath them, maintained in spite of the forces continually tending to change the landmasses.

Isotopes related forms of an element having the same number of protons and electrons but different numbers of neutrons.

Jet streams fast-flowing currents of air in both the Northern and Southern Hemispheres. Major ones are located above the subtropical high-pressure zone and above the polar front in each hemisphere.

Jointing the pattern of fractures within a rock mass.

Kame a ridge or mound of glacial debris or stratified drift at the edge of a glacier.

Karst topography the topography of surface features associated with underground chemical erosion of limestone.

Katabatic wind air moving down a slope because it has cooled. Also called cold air drainage.

Kelvin scale a temperature scale that is based on the temperature that a gas would have when it hypothetically no longer has molecular motion, volume, or pressure. Kelvin degrees are the same size as Celsius degrees, and water freezes at 273°K and boils at 373°K. This scale is also known as the absolute scale.

Kinetic energy the energy of movement.

Köppen system of climate classification one of the first and most widespread climate classification systems, based on the relationship between plant types and climate types. The Köppen system is divided into categories based on temperature and precipitation.

Laccolith a lenticular dome-shaped feature formed when an igneous intrusion of thick, viscous magma forces overlying strata upward.

Lagoon the body of water that exists between an offshore barrier and the original shoreline.

Landforms individual products of the earth's internal geologic forces as well as of forces of weathering, breakdown, and erosion.

Landscape the total scenery of a portion of the earth's surface including numerous landforms.

Landslide the sudden fall of rock and soil from steep slopes.

Lapse rate the rate of a drop in temperature, expressed in degrees Celsius per unit height.

Latent heat of fusion the heat energy involved in the transformation of a solid to a liquid, or vice versa. It takes 80 cal/g at 0°C to melt ice into water and a similar amount of heat is given off when water freezes into ice.

Latent heat of vaporization the heat energy involved in the transformation of a liquid into a gas, or vice versa. It takes 597 cal/g at 0°C to change water into water vapor and a similar amount of heat is given off when vapor condenses into water droplets.

Laterization the process of soil formation in warm, moist climates, in which there is complete decay of the organic layer.

Latitude the angular distance north or south of the equator, measured on a meridian in degrees.

Lava magma extruded onto the earth's surface from a fissure, vent, or other conduit.

Lava dome an irregular mass formed by stiff lava forcing preexisting rock upward.

Leaching the process in which percolating water dissolves and washes away many of the soil's mineral substances.

Lenticular clouds clouds resembling flying saucers, formed by the vertical wave motion of low-amplitude waves at high altitudes.

Limbs the two sides of a fold.

Limestone a sedimentary rock that can be either clastic or nonclastic, containing calcium carbonate as its main component, formed in marine environments through the activity of marine organisms.

Linear autogenic succession the plant succession that occurs when the plants themselves initiate changes in the environment that cause consequent vegetational changes.

Lithification the process of rock formation, especially the cementation process in sedimentary rock formation.

Lithosphere the crust of the earth, the rock layers at and beneath the earth's surface.

Little Ice Age a period of cooler temperatures between AD 1430 and 1850 during which glaciers in most parts of the world were advancing.

Littoral drift the movement of material parallel to a shoreline. Also known as longshore drift.

Loam soil that contains sand, silt, and clay.

Loess wind-transported and wind-deposited, fine-grained silt, usually existing in widespread deposits.

Longitude the angular distance east or west of the prime meridian, expressed in degrees and measured on a parallel.

Longwave radiation radiation from the sun and the earth with wavelengths longer than 0.7 microns, usually described as heat radiation.

Magma the liquid, molten mass from which igneous rocks are formed, existing in extremely hot form in the subsurface of the earth.

Magnetosphere the outermost layer of the atmosphere, so named because the earth's magnetic field is often more important than its gravitational field at this height.

Mantle the portion of the earth's interior between the crust and the core approximately 3000 km (1800 mi) thick.

Map projection an orderly arrangement of meridians and parallels produced by a systematic method that can be used for drawing a map of the spherical earth on a flat surface.

Maritime effect the moderating effect of marine air on coastal climates that keeps coastal areas from becoming extremely warm or cold.

Mass balance the balance of input and output of material in a landform or glacier when viewed in terms of systems.

Mass movement the movement of lithospheric surface material under the influence of gravity.

Mean free path of a molecule the average distance molecules move before they collide with one another.

Mean solar day the average rate at which the earth rotates relative to the sun.

Mean velocity of a river the average speed of flow of a river's water.

Meanders the smooth, rounded curves of a river.

Megatherms plants adapted to conditions of high heat.

Melting the process by which a substance changes from a solid to a liquid state through the addition of heat energy.

Mercator projection a mathematically derived conformal projection on which straight lines are true compass bearings, and where meridians and parallels join at right angles.

Meridional circulation the north–south movement of air along the meridians, as in the Hadley cell.

Meridians lines on the earth's surface running north–south, converging toward the poles where they meet.

Mesopause the boundary between the mesosphere and the thermosphere.

Mesophytes plants adapted to areas of intermediate moisture availability

Mesosphere the atmospheric layer above the stratosphere, in which temperatures decrease with height.

Mesotherms plants adapted for intermediate temperatures.

Metabolic heat heat produced by the human body in changing the chemical energy in food to heat energy.

Metamorphic rocks rocks made from preexisting rocks through modification by increases in temperature and pressure.

Microtherms plants adapted to withstand low temperatures.

Millibar the unit of pressure used in atmospheric studies, equal to 1000 dynes per square centimeter.

Mineral a naturally occurring chemical element or compound with a crystalline structure.

Mineral springs springs containing dissolved minerals.

Minimum tillage an agricultural technique practiced in some areas of the United States that uses chemicals for weed control and a minimum of plowing to maintain the organic content and fertility of alfisols.

Mixing ratio the ratio of the mass of water vapor present to the total mass of the dry air containing the vapor.

Monadnock a hill of exceptionally resistant rock rising above a peneplain.

Monsoon a large-scale wind that reverses with the seasons, especially the one found in the Indian Ocean and in southern Asia.

Monsoon rainforest a rainforest that occurs along tropical coasts influenced by monsoon rainfall, often backed by highlands, with evergreen and grassland vegetation predominating.

Moraine a ridge or mound of glacial debris deposited by a retreating glacier. Moraines are classified as terminal moraines, which mark the furthest advance of the glacier; recessional moraines, marking places where glacial retreat was temporarily halted; lateral moraines, composed of material carried along the edges of mountain glaciers; and medial moraines, formed by the intersection of two lateral moraines when glaciers meet and join.

Mountain glaciers huge rivers of ice that form in mountainous regions as a result of cooling temperatures and increased snowfall coupled with a lower rate of melting of accumulated snow. Also known as valley or alpine glaciers because of their location.

Mudflow a stream of fluid, lubricated mud flowing down stream valleys in semiarid and dry regions.

Muskeg a type of northern bog vegetation consisting of sphagnum mosses, sedges, and sometimes stunted spruce and tamarack trees.

Mutations variations in reproduction in which the message of heredity contained in the genes is imperfectly passed on and from which new species may originate.

Mutualism the living together of two or more different species because one or both are absolutely necessary to the survival of the other.

Net productivity the increase of living tissue in a given time period.

Net radiation incoming flows of radiation minus outgoing flows.

Nitrogen-fixing bacteria bacteria found in the roots of some plants that are able to transform atmospheric nitrogen into compounds usable by plants and animals.

Noctilucent clouds very high, wispy night clouds found in the mesosphere and thought to be sunlight reflected from meteoric dust particles coated with ice crystals.

Nonclastic sediments sediments formed from chemical deposits and other nonparticulate processes.

Nonvariant gases gases in the atmosphere that are present in the same percentages at all times.

Normal fault a fault that results from tensional forces within the earth's crust in which the fault plane is inclined in the direction of the block that has moved downward.

Oblate ellipsoid a spherical shape that has an

equatorial diameter somewhat greater than the polar diameter.

Oceanic plate a tectonic plate that consists only of a segment of crust beneath an ocean.

Occluded front a front formed when a cold front overtakes a warm front and forces it above the surface. It can occasionally be formed when a cold front rises above a warm front.

Offshore bars depositional features usually formed of sand that protrude above sea level and stop the movement of waves toward coasts of low relief.

Open wave the second stage of development of a midlatitude cyclone in which cyclonic motion develops.

Order of magnitude estimate of size expressed as a power of ten.

Orogenic belt site of a present or former mountain range.

Orogeny the process of active mountain building.

Orographic rainfall rainfall caused by the cooling and condensing of moisture-laden air traveling over mountains.

Orthographic projection a zenithal projection on which scale decreases outward from a central point giving a visual effect of a globe in three dimensions.

Outliers pieces of upthrown block that remain standing after much of the escarpment has retreated past them.

Outwash plain the plain formed by the removal of material carried in the glacier by meltwater, exhibiting both erosional and depositional features.

Overland flow another name for runoff.

Overthrust fault a low-angle variant of the reverse fault in which the fault plane is nearly horizontal and one limb lies over the other.

Overthrusting a condition in which a fold is broken by faulting and one limb overlays the other.

Oxbow lake a lake formed when two adjacent meanders touch and one of the bends in the channel, shaped like a bow, is cut off.

Oxic horizon the soil horizon in tropical soils from which weathering has removed or altered a large part of the silica previously combined with iron and aluminum, leaving mainly oxides of these two metals in this layer.

Oxidation the chemical name for the combination of oxygen with other materials to create new products.

Oxisols soils characteristically formed under laterization, having oxides of aluminum and iron present in large quantities.

Ozonosphere a layer of the atmosphere between 15 and 50 km above the earth's surface that contains a concentration of ozone.

Parallels imaginary lines on the earth's surface running east–west, parallel to the equator, and meeting the meridians at right angles.

Parent material the rocks of the earth and the deposits formed from them from which soil forms.

Pedalfers soils formed under podsolization, having large quantities of aluminum and iron.

Pedestal rocks erosional forms carved in rocks through the impact of wind-driven particles against obstacles in the landscape.

Pediment a slightly concave foot, or lower slope, associated with composite slopes.

Pediplane a surface formed by the coalescence of numerous pediments after a long period of erosion has led to parallel slope retreat.

Pedocals soils characteristically formed under calcification, having large quantities of calcium.

Pedology the study of soils, done by pedologists.

Peneplain a nearly flat landscape formed by extensive erosion over long periods of time.

Perched water table a water table formed when an aquiclude cuts off flow of groundwater to the main water table at lower elevation.

Percolation the downward movement of water through the pores and spaces in the soil under the influence of gravity.

Perennials plants that take more than a single season or year to complete their life cycle.

Perihelion the point of the earth's elliptical orbit where it is closest to the sun, usually occurring around January 3.

Permafrost a permanently frozen layer of the subsoil in arctic regions up to about 300 m in depth. It may also occur in mountains.

Permeable surfaces that absorb water.

Perspiration the passing of water through the skin to a point where it can be evaporated and used to cool the body temperature.

Photosynthesis the plant process in which carbon dioxide and water are converted to carbohydrates and oxygen through the addition of solar energy.

Physiographic region a part of the earth's surface within whose boundaries there is substantial uniformity and homogeneity of landscape.

Phytoplankton microscopic green autotrophic plants at the beginning of the food chain.

Plain an area of low relief.

Planation the production of a surface of low relief as the final product of a long period of erosion.

Plane of the ecliptic the plane in space defined by the earth's orbit around the sun.

Planetary vorticity the spin of the earth, which is defined as zero at the equator and a maximum at the poles.

Plankton the billions of small organisms that float in the oceans.

Plant nutrients the life-giving nutrients that circulate through the soil and the plant life connected with them, including nitrogen compounds, phosphates, and potassium.

Plant succession the process of one type of vegetation replacing another.

Plateau a plain found at higher elevations.

Plates the large, rigid sections of the earth's crust and the upper mantle that behave as single units, slowly moving over the earth's surface.

Platy structure soil structure in which soil has a flakelike appearance, with soil particles arranged in overlapping horizontal planes.

Pleistocene Ice Age the complete set of glacial and interglacial periods that has occurred within the Pleistocene geological epoch of the last 2.5 million years.

Plume the elongated form taken by an urban dust dome in the presence of winds greater than 13 km per hour. Also used to describe individual emissions of smoke or other pollutants.

Podsolization the process of soil formation in cooler climates with inhibited bacterial action, giving soils a characteristic ashy appearance.

Polar easterlies winds in the high latitudes caused by the sinking of cold polar air and flowing in a westward direction under the influence of the Coriolis force.

Polar front jet stream the meandering jet stream that forms in the midlatitudes in association with the polar front.

Polar high-pressure zone a surface high-pressure zone caused by the subsidence of cool polar air. The polar high-pressure zone is not present in the upper atmosphere.

Polar night westerlies winter stratospheric winds that blow from the west.

Potential energy the energy an object has by virtue of its position relative to another object.

Potential evapotranspiration the maximum amount of water that can be lost to the atmosphere from a land surface without water availability limitations.

Precession the wobble of the earth's axis, much like that in a spinning top, having a periodicity of 21,000 years.

Precipitation the collective name for the forms in which water is returned from the atmosphere to the earth surface, including rain, snow, hail and sleet.

Predation a situation in which one species preys upon another.

Pressure the force exerted by moving molecules upon a surface divided by the area of the surface.

Pressure gradient force the force that acts on air to make it move from areas of relatively high pressure toward areas of relatively low pressure.

Primary pollutants gaseous or solid pollutants produced by industrial and domestic sources and the internal combustion engine.

Primary succession linear autogenic successions that form on previously unvegetated ground.

Primary waves the directly propagated waves of an earthquake radiating out in all directions from the focus. Known as *P* waves.

Prime meridian the meridian that passes through the Royal Observatory at Greenwich, England, and defined as 0° longitude.

Prismatic structure soil structure in which soil particles are arranged in columns.

Proxy climatic data indirect evidence of climate in past times, usually found in the biological or geological record of an area.

Psychrometer an instrument used to measure relative humidity, specific humidity, and mixing ratio. It consists of a wet-bulb and a dry-bulb thermometer.

Puddled soil soil in which the structure has been destroyed.

Pyroclastics volcanic debris formed by the explosive extrusion of viscous magmatic material containing pent-up gases.

Radial pattern a drainage pattern in which a series of rivers flow outward from a central point, owing to domelike structure.

Radiational index of dryness the net radiation received at a place divided by the heat required to evaporate the precipitation at the same place.

Range of optimum the area in which a species can survive and maintain a large healthy population.

Rating curve a graph of stream discharge plotted against level, which allows discharge to be calculated directly from stream level as long as bed configurations do not change.

Recession limb the limb of a hydrograph signaling the decline of runoff-produced discharge.

Rectilinear slope a slope with a straight-line profile. Also known as a repose slope because it is often formed by weathered material lying at a certain angle.

Recumbent folds that are heavily overturned.

Regional metamorphism metamorphic change induced by widespread folding or structure changes in a large area.

Rejuvenation a new increase in erosive power in a river caused by a relative fall in base level.

Relative humidity the proportion of water vapor present in a sample of air relative to the maximum the air could hold at the same temperature.

Relative vorticity the spin of an object on a larger spinning object such as a planet.

Relief the vertical difference between the highest and lowest elevations in an area.

Remote sensing the technique of viewing the earth from remote observation stations such as airplanes and satellites.

Residual soil soil that is formed directly from parent material of the underlying rock.

Respiration the process of oxidation within cells or organic substances that is accompanied by the release of energy.

Reverse fault a fault that results from compressional forces within the earth's crust, in which the fault plane is inclined in the direction of the upthrown block.

Rhumb lines lines of true compass bearing.

Rift valley the continental trough or trench formed when land sinks between parallel faults in strips.

Rising limb the limb of a hydrograph signaling the advent of runoff.

Roche moutonnée individual protuberances, smoothed on one side and rough on the other, created by a glacier as it passes over and around the obstacle.

Rock sea the area of rock fragments formed when weathered rock pieces remain near their original location. Also known as a block field, or a felsenmeer.

Rossby waves meanders in the upper altitude westerlies form low-pressure troughs and high-pressure ridges. Also known as planetary long waves.

Runoff water that moves across a soil or ground surface; removed from land surfaces in the form of rivers and streams.

Sahel Arabic for shore, the name given to the southern boundary of the Sahara Desert.

Saltation the rolling or bouncing motion of smaller rock particles along the channel bottom of a stream.

Sand sea the complete inundation of a landscape by sand.

Saturated the state of a gas when it contains more vapor molecules than it can hold in given temperature, pressure, and volume conditions.

Savanna a grassland region with scattered trees, usually in the subtropical zones.

Scavenging the washout effect in the atmosphere achieved by precipitation in which the atmosphere rids itself of pollutants.

Schist a metamorphic rock formed from sedimentary rocks that have been altered beyond recognition.

Seamounts isolated or clustered submarine mountains, often of volcanic origin.

Secondary pollutants pollutants produced in the air by the interaction of two or more primary pollutants, or by reaction with normal atmospheric constituents.

Secondary succession plant succession that begins on ground previously vegetated.

Secondary waves the transverse waves propagated by an earthquake having their principal component of motion perpendicular to the direction of propagation. Known as *S* waves.

Sedimentary rocks rocks formed from the deposition and accumulation of matter from organic compounds, other rocks, and chemicals, which undergo compaction and cementation to a hard rock form.

Seismogram the record of an earthquake at a seismograph.

Sensible heat the heat energy that can be sensed, as distinguished from latent heat energy. Specifically the product of the temperature and the specific heat of a gas at constant pressure.

Seral stages the series of stages through which the vegetation in an ecosystem passes in reaching the climax community. Also know as seres.

Shale the finest grained clastic sedimentary rock, formed of compacted mud.

Sheet erosion the erosion that occurs when rainfall intensity exceeds the rate of infiltration on a slope, and fine particles of the soil are carried away by thin layers or rills of water.

Shield the oldest parts of the continents around which landmasses were built up.

Shield volcano a broad, gently sloping volcano formed by the accumulation of liquid lavas that can travel great distances before cooling.

Shortwave radiation radiation coming from the sun, including the ultraviolet, visible, and part of the infrared portions of the spectrum.

Sial a shorthand name for granitic rocks containing mainly aluminum-rich silicates.

Sidereal day, and sidereal time the measurement of the rotation of the earth relative to distant stars, equal to 23 hours 56 minutes of the mean solar day. Sidereal time is based on the sidereal day.

Sill an igneous intrusion that inserts itself between horizontal beds of preexisting rock.

Sima a shorthand name for granitic rocks containing mainly magnesium-rich silicates.

Sinkholes surface forms created by the collapse of unsupported underground chambers.

Slickensides mirrorlike surfaces of fault scarps formed by the sandpaper-like action of fault movement.

Slumping flow movement in which a backward rotating motion accompanies the downward movement.

Small circle the circle formed when a sphere is cut by a plane into two unequal parts.

Soil a mixture of fragmented and weathered grains of minerals and rocks with variable proportions of air and water. The mixture usually has a fairly distinct layering, and its development is influenced by climate and living organisms.

Soil horizon each of the layers found in a soil profile.

Soil profile the collection of distinct layers or horizons found with depth in the soil of a particular area.

Soil storage the product of the rooting depth (average depth to which roots grow) and the water storage per centimeter from that soil type.

Solar constant the average energy received at the earth's outer atmosphere, measured as 1.95 calories per square centimeter per minute.

Solution load the material of a river in chemically dissolved form.

Species a group of organisms with the potential to interbreed and produce fertile offspring.

Specific humidity the ratio of the weight of water vapor in the air to the total weight of the air plus the vapor.

Spectrum an array of light waves in accordance with the magnitudes of their wavelengths.

Spheroidal weathering a weathering process in some rocks in which hydrolysis combines with other processes to cause the outer shells to flake off as if small-scale exfoliation were occurring.

Spit a curved partial barrier formed by wave action and coast wise drift.

Splash erosion the erosion of soil particles on slopes owing to the impact of raindrops and the effective downward movement on the particles by gravitational force.

Spring a surface stream of flowing water that emerges from the ground where the water table intersects the ground surface.

Squall-line storm thunderstorms with exaggerated circulatory systems, that scoop hailstones back into the main cloud formation over and over.

Stable air air that returns to its original position after having been given a force and in the absence of strong, sustained mechanical movement.

Stacks parts of steep-sided cliffs along coasts of high relief that have become separated from the main land body and stand apart in the ocean.

Stalactites iciclelike forms of rock hanging from the roofs of caves.

Stalagmites pillarlike forms of rock standing on the floors of caves.

Stationary front the first stage in the development of a midlatitude cyclone in which a mass of cold air and a mass of warm air lie side by side with neither encroaching on the other.

Steppe an extensive grassland plain without trees, in high latitudes.

Stereographic projection a zenithal projection on which scale increases outward from the map center, and which is conformal.

Stock a small batholith, approximately 100 square km that forms as an offshoot of a major batholith.

Stomata small holes in leaf surfaces through which continuous moisture is provided to dissolve carbon dioxide in the process of photosynthesis.

Stone nets hexagon-shaped structures of stone that mark the soils in arctic regions.

Stratified drift material transported by glaciers and later sorted and deposited by the action of running water, either within the glacier or as it melts.

Stratopause the boundary between the stratosphere and the thermosphere.

Stratosphere the atmospheric layer above the troposphere that has a positive lapse rate.

Stratospheric biannual wind regime a wind reversal that occurs in the stratosphere every 12 to 13 months.

Stratus layered cloud sheets that form below 3 km.

Stream gradient the slope of a stream over a particular distance, or the difference in elevation between two points on the bottom of a stream channel divided by the horizontal distance between them.

Stream texture the number of streams in a given area.

Striations scratches or gouges caused by the abrasive movement of rock-carrying glaciers over resistant rock surfaces.

Strike-slip fault a fault that results from horizontal movement in the earth's crust as two blocks move past one another.

Subduction zone the area where material from the crust is reincorporated into the mantle at a convergent plate boundary.

Sublimation the process of changing a substance from a solid to a gaseous state, or a gaseous to a solid state, without apparent liquefaction. The heat involved in this process is the sum of the latent heats of fusion and vaporization.

Subsurface flow water that infiltrates the soil and flows horizontally upon reaching an impermeable layer.

Subtropical high-pressure zones the downward limbs of the Hadley cell that produce clear weather around 30° N and S owing to the piling up of poleward moving air as it is deflected westward and angular momentum is conserved.

Succulents perennial plants having fleshy, water-storing leaves or stems, usually found in desert areas.

Summer and winter solstices the points of the earth's orbit around the sun when the sun appears highest and lowest in the sky, creating the longest and the shortest days of the year, respectively.

Superimposed river a river exhibiting transverse drainage across a structural feature that would normally impede its flow because the feature was at some point buried beneath alluvium, lava, or sediments of another kind, upon which the river developed, cutting through the buried feature.

Surface detention the collection of water in small puddles and pools on an impermeable soil surface.

Suspension the movement of sediments above the channel bottom by river water.

Swash the water of a broken wave that runs up the beach and back to sea.

Symbolic model a model that represents reality by abstract, verbal, or mathematical expressions.

Syncline a fold in which the slopes of the two limbs face each other, tilting toward the axis of the fold.

System a set or collection of interrelated events that take place within defined boundaries. General systems theory is a body of laws and an approach relating to systems. In a closed system, there is no exchange of energy or matter across the system's boundaries. Energy and matter can be exchanged across the boundaries of an open system. Any system can have one or more subsystems, which are a part of a larger system, and which act within and are related to the larger system.

Taiga the coniferous evergreen forests covering large areas of subarctic lands in Asia and North America.

Talus cone the form created by rock particles that weather and fall downslope, concentrating in one area.

Temperature the index used to measure the kinetic energy of molecules.

Thematic maps maps that serve a specialized purpose or show particular information.

Thermometer an instrument used to measure temperature, usually according to the known properties of expansion of a liquid placed in contact with the substance to be measured.

Thermosphere the fourth atmospheric layer from the earth's surface, above the mesosphere, in which temperatures increase with height.

Till the unsorted mass of material deposited by a melting glacier.

Tillite a rock layer formed from compacted and cemented glacial till.

Topographic map map showing the relationships of surface features.

Tornado a small, vicious storm, consisting of a small vortex of air with extremely low central surface pressure and extremely high wind speeds.

Traction load the rock materials dragged along the river bottom.

Trade winds the predominating northeasterly and southeasterly winds found in the tropics.

Transcurrent boundary the boundary that occurs when two plates are in contact along a fault and tend to move sideways relative to each other.

Transpiration the passage of water through leaf pores in plants to the atmosphere during photosynthesis.

Transportation the movement of eroded sediments and fragments of rock by a stream.

Transported soil soil in which the parent material has been brought from some other location.

Transverse dunes sand ridges whose crests are at right angles to the direction of the wind.

Trellis pattern a drainage pattern in which a series of nearly parallel rivers drain an area of parallel ridges, joined by other streams at markedly right angles.

Triangulation a method of fixing the location of points based on properties of triangles. Given a base line of known length, the length of the other two sides of the triangle can be derived if the angle from one end of the base line to the third vertex of the triangle can be accurately measured.

Trophic level each of the stages along the food chain in which food energy is passed through the ecosystem.

Tropic of Cancer latitude 23½° N, the northern limit of vertical rays from the sun on the earth.

Tropic of Capricorn latitude 23½° S, the southern limit of vertical rays from the sun on the earth.

Tropical cyclone a moving low-pressure system drawing heat energy from the warm ocean and having wind speeds greater than 32 m per second and a central surface pressure less than 900 mb.

Tropical rainforest a rainforest in the tropical regions showing the greatest effects of heat and moisture, with tall evergreen vegetation completely covering the land surface.

Tropopause the boundary between the troposphere and the stratosphere.

Troposphere the bottom layer of the atmosphere where temperature decreases with height.

Truncated spurs spurs of hillsides that have been cut off by glacial or fluvial activity.

Tsunamis huge, fast-moving waves propagated by submarine earthquakes, which can cause great damage to coastal areas.

Tundra a vast, nearly level marshy plain in the arctic regions of Europe, Asia, and North America, with sparse vegetation consisting of mosses, lichens, and stunted trees.

Typhoon the name given to the tropical cyclones formed in the China Sea.

Ultraviolet radiation short wavelength solar radiation that can have harmful effects on living organisms.

Unconfined aquifer an aquifer that obtains its water from local infiltration.

Unconformity a gap in the geologic history of an area as found in the rock record, owing to a hiatus in deposition or other ongoing activity, followed by deformation of the surface, with further deposition or geologic activity continuing later.

Unit hydrograph a hydrograph solely representative of the characteristics of a drainage basin.

Urban heat island the form taken by an isotherm representation of the heat distribution within a city that appears as an area of high land or an island of higher temperatures on a plain of more homogeneous temperatures.

Van Allen radiation belts concentrations of ions occurring high in the thermosphere, approximately 4000 and 20,000 km above the earth's surface.

Vapor pressure the pressure exerted by the molecules of water vapor in air.

Vapor pressure gradient the difference in vapor pressure between two locations that facilitates the movement of water molecules from locations of high pressure to those of low pressure.

Variant gases gases in the atmosphere that are present in different quantities at different times.

Varves layers of alternating sediments caused by seasonal variations in deposition.

Vernal and autumnal equinoxes the points in the earth's orbit around the sun when the sun's rays fall vertically over the equator and the sun appears to set and rise due west and east. This produces days and nights of equal length.

Waning slope the pediment section in a fully developed slope.

Warm anticyclone a high-pressure area formed in the subtropical regions by subsidence, usually in the Hadley cell. The Bermuda and Hawaiian highs are examples of warm anticyclones.

Warm front a transition zone between a mass of warm air and the cold air it is replacing.

Water table the top of the phreatic zone.

Wave front the imaginary line perpendicular to the direction a wave is traveling.

Wave-cut platform the smooth, seaward-sloping surface at the foot of a coastal cliff created by waves and agitated rock debris. Also known as an abrasion platform.

Waxing slope the convex crest or summit in a fully developed slope.

Weathering the decomposition and disintegration of rocks involving little or no movement of the particles produced when the rocks break down. Weathering can be mechanical or chemical.

Wetted perimeter the outline of the edge where the stream channel and its water meet.

Willy-willy the name given to the tropical cyclones formed near Australia.

Wilting point the soil moisture below which a crop will sustain permanent injury by drying out.

Wind-chill index the subjective amount of cold experienced by a percentage of people at certain temperature and wind speed combinations.

Xerophytes plants adapted to dry areas.

Younger Dryas event a sharp return to glacial conditions between 10,800 and 10,100 years ago.

Zenithal projections a class of projections centered around a point where an imaginary surface touches the globe and having a wheel-like symmetry. Straight lines on zenithal projections are arcs of great circles. Also known as azimuthal projections.

Zonal circulation the movement of air in a basic east-west direction often in the form of planetary long waves.

Zone of aeration the top zone of the ground, usually unsaturated except at times of heavy rain. Also known as vadose zone.

Zone of eluviation the area in the soil from which the mineral substances have been leached.

Zone of illuviation the layers in the soil where the leached minerals from the zone of eluviation accumulate.

Zone of increasing physiological stress the area in which a species can survive, though not without hardship, and in small numbers.

Zone of intolerance the area in which a species is unable to survive, except for short, intermittent periods.

Zone of saturation the lower zone of the ground that is generally saturated with water, except during periods of drought or long spells without rain. Also known as the phreatic zone.

Part and Chapter Opening Photos

Part One: Inland sea viewed from Mt. Washu, Japan. Chapter 1: Tundra in Canada. Chapter 2: Sun and solar flare. Part Two: Switzerland. Chapter 3: The eye of Typhoon Ida at 15,000 m (50,000 ft). Chapter 4: Falls on the Ouitetchouan River, Quebec. Chapter 5: The Rocky Mountains. Chapter 6: Hurricane Ava, June 7, 1973. Chapter 7: Winter scene in Japan. Chapter 8: San Francisco street scene. Chapter 9: Aerial view of the Weisshorn in the Valais Alps, Switzerland. Part Three: Yosemite National Park. Chapter 10: Alaskan soil. Chapter 11: Olympic Rainforest. Chapter 12: Farmhouse in Japan. Part Four: Athabaska Glacier, Alberta, in summer. Chapter 13: Steam issuing from caverns recently melted in glacier in Sherman Crater, Mt. Baker volcano, Washington. Chapter 14: Vertical view of Giant's Causeway. Chapter 15: Surtsey. Chapter 16: Switzerland. Chapter 17: Hot springs at Beppu, Kyushu. Chapter 18: Weathering of sixteenth-century fresco by Titian, in Venice. Chapter 19: Giant network of braided streams from melting glacier on the southeast coast of Iceland,. Chapter 20: A farm near Somerset West, South Africa, with Cape Mountains in the background. Chapter 21: White Sands National Monument. Chapter 22: Great Smoky Mountain National Park. Epilogue: The Chatanika River northwest of Fairbanks.

CREDITS

Photographs

Part One Pages 2–3: Japan National Tourist Organization.
Chapter 1 Page 5: National Air Photo Library. Page 7: Port of Long Beach. Page 8: NASA. Page 10, top: Robert C. Frampton. Page 10, bottom: Wide World Photos. Page 11: The Frank Lloyd Wright Memorial Foundation.
Chapter 2 Page 17: Haleakala Observatory. Page 18: Royal Geographical Society, London. Page 19, left: The University Museum, University of Pennsylvania, Philadelphia. Page 19, right: The British Museum. Page 30: Aerofilms, Ltd. Page 40: David Greenland.

Part Two Pages 42–43: Swiss National Tourist Office.
Chapter 3 Page 45: U.S. Air Force. Page 47, left: Wide World Photos. Page 47, right: The Nitragin Company, Inc., Milwaukee. Page 48: Wide World Photos. Page 54: The Mansell Collection. Page 57: NASA. Page 60: Air Force Geophysics Laboratory. Page 61: NOAA.
Chapter 4 Page 63: Canadian Government Travel Bureau. Page 67: Sun Valley News Bureau. Page 68: NOAA. Page 69: NOAA.
Chapter 5 Page 83: David Greenland. Page 89: NOAA.
Chapter 6 Page 101: NOAA. Page 103: FAO. Page 107: NASA. Page 108: NOAA. Page 115: NASA. Page 117: NOAA. Page 120: Professor R. S. Scorer, Imperial College, London.
Chapter 7 Page 123: Japan National Tourist Organization. Page 132, top: United Nations/IDA/Pickerrell. Page 132, bottom: U.S. Forest Service. Page 134: FAO. Page 138: British Information Services. Page 142: O. J. Ferrians, Jr., USGS. Page 145: U.S. Navy.
Chapter 8 Page 149: San Francisco Convention and Visitor's Bureau. Page 154, left: Fujihira, Monkmeyer Press Photo Service. Page 154, right: Courtesy Canadian Consulate. Page 155, left: Pro Pix, Monkmeyer Press Photo Service. Page 155, right: Ewing Galloway. Page 156: Canadian Government Travel Bureau. Page 161: David Greenland.
Chapter 9 Page 165: Swiss National Tourist Organization. Page 168: E. C. Hardin, USGS. Page 169, top left: Swiss National Tourist Organization. Page 169, top right: Department of Hydrology and Glaciology, Federal Institute of Technology, Zurich. Page 169, bottom left: Department of Hydrology and Glaciology, Federal Institute of Technology, Zurich. Page 169, bottom right: Department of Hydrology and Glaciology, Federal Institute of Technology, Zurich. Page 170, top left: Russell R. Reynolds, USDA. Page 170, bottom right: SATOUR. Page 175: Royal Greenwich Observatory. Page 179: Leon Davico, UNICEF.

Part Three Pages 188–189: National Park Service.
Chapter 10 Page 191: Steve McCutcheon, Alaska Pictorial Service. Page 192: B. C. McLean, USDA. Page 193: Thomas G. Meier, Soil Conservation Service, USDA. Page 196: J. Olsen, FAO. Page 198: Roy Simonson, Soil Conservation Service, USDA. Page 199: David Greenland.
Chapter 11 Page 209: U.S. Forest Service. Page 211: T. E. Weier, from *Botany: An Introduction to Plant Biology* by T. E. Weier, C. R. Stocking, and M. G. Barbour (New York: John Wiley & Sons, 1974). Page 218: H. B. D. Kettlewell, Oxford University. Page 219: P. Fregosi, FAO. Page 222: C. H. Muller, from "Aerial View of Aromatic Shrubs," by J. Ward, *Science*, Cover, vol. 143, January 31, 1964, © 1964 by the American Association for the Advancement of Science.
Chapter 12 Page 227: Japan National Tourist Organization. Page 228: WHO. Page 231: Jack Boucher, National Park Service. Page 232: C. Colton, Soil Conservation Service, USDA. Page 233: C. Colton, Soil Conservation Service, USDA. Page 234: C. Colton, Soil Conservation Service, USDA. Page 239, top: U.S. Forest Service. Page 239, bottom: P. Fregosi, FAO. Page 240: Rex King, U.S. Forest Service. Page 241: Jack Dermid. Page 242: Douglass F. Roy, U.S. Forest Service. Page 243: Canadian Government Travel Bureau.

Part Four Pages 252–253: Canadian Pacific Railway.
Chapter 13 Page 255: D. M. Miller, USGS. Page 260: Courtesy Vincent Renard, Centre Océanologique de Bretagne. Page 262: Smithsonian Institution. Page 267: Hawaiian Volcano Observatory, Hawaii National Park.
Chapter 14 Page 271: The British Tourist Authority. Page 273: American Museum of Natural History. Page 274: Northern Ireland Tourist Board. Page 276: F. C. Calkins, USGS. Page 277, top: National Park Service. Page 277, bottom: J. R. Stacy, USGS. Page 278: G. W. Stose, USGS. Page 279: The Mansell Collection. Page 280: National Park Service. Page 281: Japan National Tourist Organization. Page 282, top: P. Jay Fleisher, © 1972 Harper & Row. Page 282, bottom: G. K. Gilbert, USGS. Page 283: J. R. Stacy, USGS. Page 284: Geological Survey of Canada.
Chapter 15 Page 289: Wide World Photos. Page 294: Icelandic Photo and Press Service. Page 295, top: Swiss National Tourist Office. Page 295, bottom: NASA. Page 299: SATOUR. Page 301: Harm de Blij.
Chapter 16 Page 307: Swissair. Page 311: David Greenland. Page 312: Bureau of Reclamation. Page 314: Swissair. Page 315: American Museum of Natural History. Page 318: R. E. Wallace, USGS. Page 319: J. R. Balsley, USGS. Page 320: D. J. Miller, USGS.
Chapter 17 Page 323: Japan National Tourist Organization. Page 324: William Grimaud. Page 328: David Greenland. Page 329: Dale L. Swartz, Soil Conservation Service, USDA. Page 330: U.S. Army Corps of Engineers, San Francisco District. Page 331: U.S. Army Corps of Engineers, New Orleans District.
Chapter 18 Page 337: Dominique Roger, UNESCO. Page 338: H. E. Malde, USGS. Page 340: O. J. Ferrians, Jr., USGS. Page 341: Wide World Photos. Page 342: Kennecott Copper Corporation. Page 343: Der Landeskonservator von Westfalen-Lippe, Munster. Page 344: New Mexico Department of Development. Page 345, top: George Grant, National Park Service. Page 345, bottom: USGS. Page 346: P. S. Smith,

USGS. Page 347, top left: William S. Young, San Francisco Chronicle. Page 347, top right: D. R. Crandell, USGS. Page 347, bottom: Wide World Photos. Page 348: NASA. Page 349: United Nations. Page 351: E. F. Patterson, USGS.
Chapter 19 Page 353: Icelandic Survey Department. Page 354: U.S. Forest Service. Page 355: USDA. Page 361, top: J. Allan Cash. Page 361, bottom: National Air Photo Library. Page 363: Steve McCutcheon, Alaska Pictorial Service. Page 364: Teledyne Geotronics. Page 365: USGS.
Chapter 20 Page 369: SATOUR. Page 373: Aerofilms, Ltd. Page 381: N. H. Darton, USGS.
Chapter 21 Page 383: National Park Service. Page 385: Leland Prater, U.S. Forest Service. Page 386: R. D. Miller, USGS. Page 387: Geological Survey of Canada. Page 388: U. S. Fiske, USGS. Page 389: F. E. Matthes, USGS. Page 390: National Publicity Studios, courtesy of New Zealand Consulate General. Page 393: Geological Survey of Canada. Page 394: U.S. Air Force. Page 395: Wide World Photos. Page 396: American Airlines. Page 397: Soil Conservation Service, USDA. Page 398: G. K. Gilbert, USGS. Page 399: E. W. Shaw, USGS.
Chapter 22 Page 403: National Park Service. Page 407: John S. Shelton. Page 408: Jack E. Boucher, National Park Service. Page 409: Bluford Muir, U.S. Forest Service. Page 410, top: Leland J. Prater, U.S. Forest Service. Page 410, bottom: American Airlines. Page 411, top: Aerial Photography Department, USDA. Page 411, bottom: U.S. Corps of Engineers, New Orleans District. Page 412: Aerial Photography Department, USDA. Page 414: Aerial Photography Department, USDA. Page 415: National Park Service. Page 418, top: National Park Service. Page 418, bottom: U.S. Air Force. Page 419: American Airlines. Page 425, left: Water Supply Papers, USGS. Page 425, right: Water Resources Division, USGS.

Epilogue Page 427: Steve McCutcheon, Alaska Pictorial Service.

Line Art

Chapter 2 Figure 2–12(b): Copyright by The University of Chicago Department of Geography. Page 30: From Yi-Fu Tuan, *Topophilia: A Study of Environmental Perception, Attitudes and Values,* © 1974, pp. 35–36. Adapted by permission of Prentice-Hall, Inc., Englewood Cliffs, New Jersey.
Chapter 3 Figure 3–2: Data after G. M. Plass, "Carbon Dioxide and Climate," *Scientific American*, vol. 201, pp. 41–47.
Chapter 4 Figures 4–2, 4–3: Data after R. J. More, "Hydrological Models and Geography," in *Models in Geography*, R. J. Chorley and R. Hoggett, ed. (London: Methuen, 1967), pp. 145–185. Figure 4–7: After W. D. Sellers, *Physical Climatology* (Chicago: University of Chicago Press, 1965), p. 88. Figures 4–8, 4–9, 4–12: After J. G. Lockwood, *World Climatology: An Environmental Approach* (New York: St. Martin's Press, 1974), p. 62, fig. 2–12; p. 62, fig. 2–16; p. 43. Figures 4–13, 4–14, 4–15: After M. I. Budyko et al., "The Heat Balance of the Earth," *Soviet Geography*, vol. 3, no. 5, 1962, pp. 3–15.
Chapter 5 Figure 5–11: After Y. Mintz, "The Observed Zonal Circulation of the Atmosphere," *Bulletin of the American Meteorological Society*, vol. 35, no. 5, 1954, p. 209. Figure 5–12: After H. R. Byers, *General Meteorology*, 4th ed. (New York: McGraw-Hill, 1974), p. 267. Figure 5–13: After D. E. Greenland, "Heat Balance Climatology," in *Dynamic Relationships in Physical Geography*, W. B. Johnston, ed. (Canterbury: University of Canterbury, Department of Extension Studies and Geography, 1967), p. 28. Figure 5–15: After R. G. Barry and R. J. Chorley, *Atmosphere, Weather and Climate* (New York: Holt, Rinehart and Winston, 1970), p. 117. Figure 5–16: After H. Riehl, "On the Role of the Tropics in the General Circulation of the Atmosphere," *Weather*, vol. 24, 1969, p. 288.
Chapter 6 Figure 6–8: After F. Defant and H. Taba, "The Threefold Structure of the Atmosphere and the Characteristics of the Tropopause," *Tellus*, vol. 9, 1957, pp. 259–274. Figures 6–12, 6–14: After A. N. Strahler, *Introduction to Physical Geography*, 1st ed. (New York: John Wiley, 1965), p. 89, fig. 6–2; p. 95, fig. 6–11. Figure 6–16: After H. R. Byers and R. R. Braham, *The Thunderstorm* (U.S. Weather Bureau, 1949). Figure 6–17: After F. H. Ludlum, "The Hailstorm," *Weather*, vol. 16, 1961, pp. 152–162.
Chapter 7 Figures 7–2, 7–5: After W. Terjung, "World Patterns of the Distribution of the Monthly Comfort Index," *International Journal of Biometeorology*, vol. 12, no. 2, 1968, p. 120; 146–147. Figures 7–4, 7–10: After A. N. Strahler, *Introduction to Physical Geography*, 2nd ed. (New York: John Wiley, 1970).
Chapter 8 Figures 8–2, 8–3: Adapted after L. P. Herrington, "Biophysical Adaptations of Man Under Climatic Stress," Meteorological Monographs (American Meteorological Society), vol. 2, no. 8, 1954, pp. 30–42. Figure 8–7: After S. A. Changnon, Jr., "Recent Studies of Urban Effects on Precipitation in the United States," in *Urban Climates* (Geneva: World Meteorological Organization, no. 254, 1970), pp. 325–341.
Chapter 9 Figures 9–5, 9–6, 9–11: After National Academy of Sciences, *Understanding Climatic Change*, U.S. Committee for the Global Atmospheric Research Program, NAS, 1975, p. 148; p. 130; p. 22. Figure 9–12: After A. H. Bunting, M. D. Dennett, J. Elston, and J. R. Milford, "Rainfall Trends in the West African Sahel," *Quarterly Journal of the Royal Meteorological Society*, vol. 102, 1976, pp. 59–64.
Chapter 10 Figure 10–2: After D. Steila, *The Geography of Soils* (Englewood Cliffs, NJ: Prentice-Hall, 1976), p. 17. Figure 10–13: After R. E. Gabler, R. Sager, S. Brazier, and J. Pourciau, *Introduction to Physical Geography* (San Francisco: Rinehart Press, 1975), p. 322. Figure 10–17: Simplified from D. H. Yallon, Transactions of the International Congress of Soil Science, vol. 16, 1960, pp. 119–123.
Chapter 11 Figure 11–3: After H. Lieth and E. Box, "Evapotranspiration and Primary Productivity," C. W. Thornthwaite memorial model, C. W. Thornthwaite's Laboratory of Climatology, Publications in Climatology, vol. 25, no. 3, 1972, p. 42, fig. 14. Figure 11–4: After A. V. Drozdov, "The Productivity of Zonal Terrestrial Plant Communities and the Moisture and Heat Parameters of an Area," *Soviet Geography*, vol. 12, no. 1, 1971, pp. 54–60. Figures 11–5, 11–6: After H. T. Odum, "Trophic Structure and Productivity of Silver Springs, Florida," *Ecological Monographs*, vol. 27, 1957, pp. 55–112. Figure 11–7: After W. A. Williams, "Range Improvement as Related to Net Productivity, Energy Flow, and Foliage Configuration," *Journal of Range Management*, vol. 19, 1966, pp. 29–34. Figure 11–10: Data after G. M. Woodwell, "The Ecological Effects of Radiation," *Scientific*

American, vol. 208, 1963, pp. 40–49. Figure 11–18: After J. T. Curtis, in W. L. Thomas, Jr., ed., *Man's Role in Changing the Face of the Earth* (Chicago: University of Chicago Press, 1956), p. 726, fig. 147. Copyright © 1956 by The University of Chicago.
Chapter 12 Figure 12–8: After D. Steila, "The Comprehensive Soil Classification," *The Professional Geographer*, vol. 26, no. 2, 1974, p. 200. Figure 12–9: From the National Cooperative Soil Survey Classification of 1967, USGS, Dept. of the Interior. Figure 12–10: From the Soil Geography Unit, Soil Conservation Service, USDA, May 1972. Figure 12–19(b): After A. Z. Gregoryev, "The Heat and Moisture Regime and Geographic Zonality," *Soviet Geography*, vol. 2, no. 5, 1961, pp. 3–12. Figure 12–19(c): After I. P. Gerasimov, "The Moisture and Heat Factors of Soil Formation," *Soviet Geography*, vol. 2, no. 5, 1961.
Chapter 13 Figure 13–7: After *Investigating the Earth*, Earth Science Curriculum Project (Boston: Houghton Mifflin, 1972), p. 425. Figure 13–12: After L. D. Leet, *Earth Waves* (Cambridge: Harvard University Press, 1950). Figure 13–13: After B. Gutenberg, *Internal Constitution of the Earth* (New York: Dover, 1951).
Chapter 15 Figure 15–9: After A. Wegener, *The Origin of Continents and Oceans* (New York: Dutton, 1924).
Chapter 17 Figure 17–3: After J. Hidore, *Physical Geography: Earth Systems* (Glenview, IL: Scott, Foresman, 1974), p. 136. Figures 17–9, 17–10: After J. P. Bruce and R. H. Clark, *Introduction to Hydrometeorology* (Elmsford, NY: Pergamon Press, 1966), p. 90; p. 179. Figure 17–17: After L. Laporte, *Encounter with the Earth* (San Francisco: Canfield Press, 1975), p. 252.
Chapter 19 Figure 19–7: After M. Morisawa, *Streams: Their Dynamics and Morphology* (New York: McGraw-Hill, 1968), p. 127.
Chapter 20 Figure 20–14: After J. Hack, "Interpretation of Erosional Topography in Humid Temperate Regions," *American Journal of Sciences*, vol. 258, 1960, p. 92.
Chapter 21 Figure 21–5: After A. N. Strahler, *Introduction to Physical Geography*, 1st ed. (New York: John Wiley, 1965), p. 328. Figure 21–23: After J. F. Kolars and J. D. Nystuen, *Physical Geography: Environment and Man* (New York: McGraw-Hill, 1975), p. 55.
Chapter 22 Figure 22–14: After W. W. Atwood, *The Physiographic Provinces of North America* (Boston: Ginn, 1940).
From the Geographer's Notebook, Part Four Figure 4: After D. Hjulstrom, "Studies of the Morphological Activity of Rivers as Illustrated by the River Fyris," University of Uppsala Geographical Institute *Bulletin*, vol. 25, 1935, pp. 221–557.

Tables

Chapter 3 Table 3–1: After M. Neiburger, J. G. Edwiger, and W. D. Bonner, "*Understanding Our Atmospheric Environment* (San Francisco: W. H. Freeman, 1973), p. 25, table 2.1.
Chapter 4 Table 4–1: After W. D. Sellers, *Physical Climatology* (Chicago: University of Chicago Press, 1965), pp. 32, 47.
Chapter 5 Table 5–1: After M. I. Budyko, "The Heat Balance of the Earth's Surface," translated by N. A. Stepanova (Washington, DC: U.S. Dept. of Commerce, Office of Technical Services, 1958).
Chapter 7 Table 7–1: After W. Terjung, "World Patterns of the Distribution of the Monthly Comfort Index," *International Journal of Biometeorology*, vol. 12, no. 2, 1968, p. 121.
Chapter 8 Table 8–1: After J. H. Westbrook, "A Method of Predicting the Frequency Distribution of Windchill," Technical Report EP-143 (Natic, MA: Quartermaster Research and Engineering Center, 1961), p. 26. Table 8–2: After H. E. Landsberg, "Climates and Urban Planning," in *Urban Climates* (Geneva: World Meteorological Organization, no. 254, 1970), pp. 364–374.

INDEX